ELASTIC LIDAR

ELASTIC LIDAR

Theory, Practice, and Analysis Methods

VLADIMIR A. KOVALEV
WILLIAM E. EICHINGER

WILEY-
INTERSCIENCE

A JOHN WILEY & SONS, INC., PUBLICATION

Published by John Wiley & Sons, Inc., Hoboken, New Jersey.
Published simultaneously in Canada.

For general information on our other products and services please contact our Customer Care Department within the U.S. at 877-762-2974, outside the U.S. at 317-572-3993 or fax 317-572-4002.

Wiley also publishes its books in a variety of electronic formats. Some content that appears in print, however, may not be available in electronic format.

Library of Congress Cataloging-in-Publication Data is available.

ISBN 0-471-20171-5

MIX
Paper from responsible sources
FSC® C013604
www.fsc.org

CONTENTS

PREFACE

It has been 20 years since the last comprehensive book on the subject of lidars was written by Raymond Measures. In that time, technology has come a long way, enabling many new capabilities, so much so that cataloging all of the advances would occupy several volumes. We have limited ourselves, generally, to elastic lidars and their function and capabilities. Elastic lidars are, by far, the most common type of lidar in the world today, and this will continue to be true for the foreseeable future. Elastic lidars are increasingly used by researchers in fields other than lidar, most notably by atmospheric scientists. As the technology moves from being the point of the research to providing data for other types of researchers to use, it becomes important to have a handbook that explains the topic simply, yet thoroughly. Our goal is to provide elastic lidar users with simple explanations of lidar technology, how it works, data inversion techniques, and how to extract information from the data the lidars provide. It is our hope that the explanations are clear enough for users in fields other than physics to understand the device and be capable of using the data productively. Yet we hope that experienced lidar researchers will find the book to be a useful handbook and a source of ideas.

Over the 40 years since the invention of the laser, optical and electronic technology has made great advances, enabling the practical use of lidar in many fields. Lidar has indeed proven itself to be a useful tool for work in the atmosphere. However, despite the time and effort invested and the advances that have been made, it has never reached its full potential. There are two basic reasons for this situation. First, lidars are expensive and complex instruments that require trained personnel to operate and maintain them. The second reason is related to the inversion and analysis of lidar data. Historically, most lidars have been research instruments for which the focus has been on the development of the instrument as opposed to the use of the instrument. In recent years, the technology used in lidars has become cheaper, more common, and less complex. This has reduced the cost of such systems, particularly elastic lidars, and enabled their use by researchers in fields other than lidar instrument development.

The problem of the analysis of lidar data is related to problems of lidar signal interpretation. Despite the wide variety of the lidar systems developed for periodical and routine atmospheric measurements, no widely accepted method of lidar data inversion or analysis has been developed or adopted. A researcher interested in the practical application of lidars soon learns the following: (1) no standard analysis method exists that can be used even for the simplest lidar measurements; (2) in the technical literature, only scattered practical recommendations can be found concerning the derivation of useful information from lidar measurements; (3) lidar data processing is, generally, considered an art rather than a routine procedure; and (4) the quality of the inverted lidar data depends dramatically on the experience and skill of the researcher.

We assert that the widespread adoption of lidars for routine measurements is unlikely until the lidar community can develop and adopt inversion methods that can be used by non-lidar researchers and, preferably, in an automated fashion. It is difficult for non-lidar researchers to orient themselves in the vast literature of lidar techniques and methods that have been published over the last 20–25 years. Experienced lidar specialists know quite well that the published lidar studies can be divided into two unequal groups. The first group, the smaller of the two groups, includes some useful and practical methods. In the other group, the studies are the result of good intentions but are often poorly grounded. These ideas either have not been used or have failed during attempts to apply them. In this book, we have tried to assist the reader by separating out the most useful information that can be most effectively applied. We attempt to give readers an understanding of practical data processing methodologies for elastic lidar signals and an honest explanation of what lidar can do and what it cannot do with the methods currently available. The recommendations in the book are based on the experience of the authors, so that the viewpoints presented here may be arguable. In such cases, we have attempted to at least state the alternative point of view so that reader can draw his or her own conclusions. We welcome discussion.

The book is intended for the users of lidars, particularly those that are not lidar instrument researchers. It should also serve well as a useful reference book for remote sensing researchers. An attempt was made to make the book self-contained as much as possible. Inasmuch as lidars are used to measure constituents of the earth's atmosphere, we begin the book in Chapter 1 by covering the processes that are being measured. The light that lidars measure is scattered from molecules and particulates in the atmosphere. These processes are discussed in Chapter 2. Lidars use this light to measure optical properties of particulates or molecules in the air or the properties of the air (temperature or optical transmission, for example). Chapter 3 introduces the reader to lidar hardware and measurement techniques, describes existing lidar types, and explains the basic lidar equation, relating lidar return signals to the atmospheric characteristics along the lidar line of sight. In Chapter 4, the reader is briefly introduced to the electronics used in lidars. Chapter 5 deals with the basic analytical solutions of the lidar equation for single- and two-component atmospheres. The most important sources of measurement errors for different solutions are analyzed in Chapter 6. Chapter 7 deals with the fundamental problem that makes the inversion of elastic lidar data difficult. This is the uncertainty of the relationship between the total scattering and backscattering for atmospheric particulates. In Chapter 8, methods are considered for one-directional lidar profiling in clear and moderately turbid atmospheres. In addition, problems associated with lidar measurement in "spotted" atmospheres are included. Chapter 9 examines the basic methods of multiangle measurements of the extinction coefficients in clear atmospheres. The differential absorption lidar (DIAL) processing technique is analyzed in detail in Chapter 10. In Chapter 11, hardware solutions to the inversion problem are presented. A detailed review of data analysis methods is given in Chapters 12 and 13. Despite an enormous amount of literature on the subject, we have attempted to be inclusive. There will certainly be methods that have been overlooked.

We wish to acknowledge the assistance of the lowa Institute for Hydraulic Research for making this book possible. We are also deeply indebted to the work that Bill Grant has done over the years in maintaining an extensive lidar bibliography and to the many people who have reviewed portions of this book.

Vladimir A. Kovalev
William E. Eichinger

DEFINITIONS

$\beta_{\pi,m}$ Molecular angular scattering coefficient in the direction $\theta = 180°$, relative to the direction of the emitted light (m^{-1} steradian^{-1})

$\beta_{\pi,p}$ Particulate angular scattering coefficient in the direction $\theta = 180°$ relative to the direction of the emitted light (m^{-1} steradian^{-1})

$\beta_{\pi,R}$ Raman angular scattering coefficient in the direction $\theta = 180°$ relative to the direction of the emitted light

$\beta_\pi = \beta_{\pi,p} + \beta_{\pi,m}$ Total of the molecular and particulate angular scattering coefficients in the direction $\theta = 180°$

β_m Molecular scattering coefficient (m^{-1}, km^{-1})

β_p Particulate scattering coefficient (m^{-1}, km^{-1})

β Total (molecular and particulate) scattering coefficient, $\beta = \beta_m + \beta_p$

$\Delta\sigma = \sigma_{on} - \sigma_{off}$ Differential absorption cross section of the measured gas

$\kappa_{A,m}$ Molecular absorption coefficient

$\kappa_{A,p}$ Particulate absorption coefficient

κ_A Total (molecular and particulate) absorption coefficient, $\kappa_A = \kappa_{A,m} + \kappa_{A,p}$

κ_m Total (scattering + absorption) molecular extinction coefficient, $\kappa_m = \beta_m + \kappa_{A,m}$

κ_p Total (scattering + absorption) particulate extinction coefficient, $\kappa_p = \beta_p + \kappa_{A,p}$

κ_t Total (molecular and particulate) extinction coefficient, $\kappa_t = \kappa_p + \kappa_m$

λ Wavelength of the radiant flux

λ_l Wavelength of the laser emission

λ_{off} Wavelength of the off-line DIAL signal

λ_{on} Wavelength of the on-line DIAL signal

λ_R Wavelength of the Raman shifted signal

Π_m Molecular backscatter-to-extinction ratio, $\Pi_m = \beta_{\pi,m}/(\beta_m + \kappa_{A,m})$ (steradian^{-1})

Π_p Particulate backscatter-to-extinction ratio, $\Pi_p = \beta_{\pi,p}/(\beta_p + \kappa_{A,p})$ (steradian^{-1})

$\sigma_{\theta,p}$ Particulate angular scattering cross section

σ_{N_2} Nitrogen Raman cross section (m^2)

$\sigma_{S,p}$ Particle scattering cross section

$\sigma_{S,m}$ Molecular scattering cross section

$\sigma_{t,p}$ Particulate total (extinction) cross section (m^2)

$\sigma_{t,m}$ Molecular total cross-section (m^2)

$\tau(r_1,r_2)$ Optical depth of the range from r_1 to r_2 in the atmosphere

h Height

n_m Molecular density (number/m^3)

$P(r, \lambda)$ Power of the lidar signal at wavelength λ created by the radiant flux backscattered from range r from lidar with no range correction

$P_{\pi,p}$ Particulate backscatter phase function, $P_{\pi,p} = \beta_{\pi,p}/\beta_p$ (steradian^{-1})

$P_{\pi,m}$ Molecular backscatter phase function, $P_{\pi,p} = \beta_{\pi,m}/\beta_m = 3/8\Pi$ (steradian^{-1})

r_0 Minimum lidar measurement range

r_{max} Maximum lidar measurement range

$Z(r) = P(r)\, r^2\, Y(r)$ Lidar signal transformed for the inversion

$Z_r(r)$ Range-corrected lidar return

$T(r_1, r_2)$ One-way atmospheric transmittance of layer (r_1, r_2)

T_0 One-way atmospheric transmittance from the lidar ($r = 0$) to the system minimum range r_0 as determined by incomplete overlap

$T_{max} = T(r_0, r_{max})$ One-way atmospheric transmittance for the maximum lidar range, from r_0 to r_{max}

u Angstrom coefficient

$Y(r)$ Lidar signal transformation function

1

ATMOSPHERIC PROPERTIES

It is our intention to provide in this chapter some basic information on the atmosphere that may be useful as a quick reference for lidar users and suggestions for references for further information. Many of the topics covered here have books dedicated to them. A wide variety of texts are available on the composition and structure, physics, and chemistry of the atmosphere that should be used for detailed study.

1.1. ATMOSPHERIC STRUCTURE

1.1.1. Atmospheric Layers

The atmosphere is a relatively thin gaseous layer surrounding the earth; 99% of the mass of the atmosphere is contained in the lowest 30 km. Table 1.1 is a list of the major gases that comprise the atmosphere and their average concentration in parts per million (ppm) and in micrograms per cubic meter. Because of the enormous mass of the atmosphere (5×10^{18} kg), which includes a large amount of water vapor, and its latent heat of evaporation, the amount of energy stored in the atmosphere is large. The mixing and transport of this energy across the earth are in part responsible for the relatively uniform temperatures across the earth's surface.

There are five main layers within the atmosphere (see Fig. 1.1). They are,

Elastic Lidar: Theory, Practice, and Analysis Methods, by Vladimir A. Kovalev and William E. Eichinger.
ISBN 0-471-20171-5

TABLE 1.1. Gaseous Composition of Unpolluted Wet Air

	Concentration, ppm	Concentration, μg/m³
Nitrogen	756,500	8.67×10^8
Oxygen	202,900	2.65×10^8
Water	31,200	2.30×10^7
Argon	9,000	1.47×10^7
Carbon dioxide	305	5.49×10^5
Neon	17.4	1.44×10^4
Helium	5.0	8.25×10^2
Methane	1.16	7.63×10^2
Krypton	0.97	3.32×10^3
Nitrous oxide	0.49	8.73×10^2
Hydrogen	0.49	4.00×10^1
Xenon	0.08	4.17×10^2
Organic vapors	0.02	—

Bouble et al. (1994).

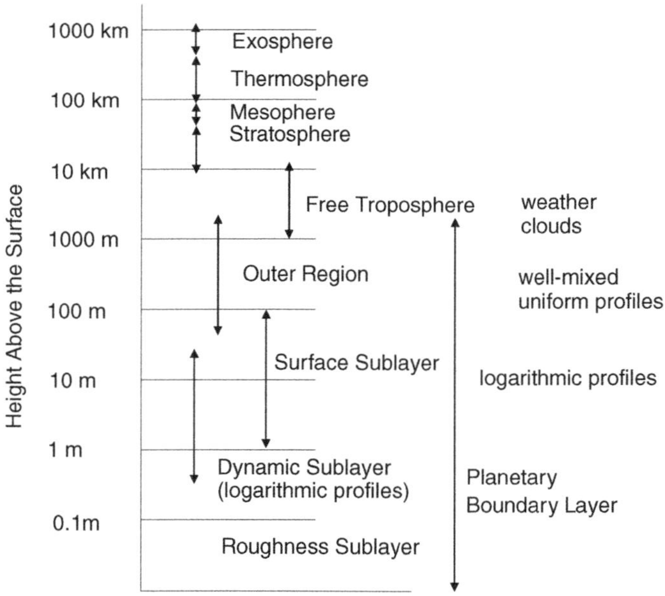

Fig. 1.1. The various layers in the atmosphere of importance to lidar researchers.

from top to bottom, the exosphere, the thermosphere, the mesosphere, the stratosphere, and the troposphere. Within the troposphere, the planetary boundary layer (PBL) is an important sublayer. The PBL is that part of the atmosphere which is directly affected by interaction with the surface.

Exosphere. The exosphere is that part of the atmosphere farthest from the surface, where molecules from the atmosphere can overcome the pull of gravity and escape into outer space. The molecules of the atmosphere diffuse slowly into the void of space. The lower limit of the exosphere is usually taken as 500 km, but there is no definable boundary to mark the end of the thermosphere below and the beginning of the exosphere. Also, there is no definite top to the exosphere: Even at heights of 800 km, the atmosphere is still measurable. However, the molecular concentrations here are very small and are considered negligible.

Thermosphere. The thermosphere is a relatively warm layer above the mesosphere and just below the exosphere. In this layer, there is a significant temperature inversion. The few atoms that are present in the thermosphere (primarily oxygen) absorb ultraviolet (UV) energy from the sun, causing the layer to warm. Although the temperatures in this layer can exceed 500 K, little total energy is stored in this layer. Unlike the boundaries between other layers of the atmosphere, there is no well-defined boundary between the thermosphere and the exosphere (i.e., there is no boundary known as the thermopause). In the thermosphere and exosphere, molecular diffusion is the dominant mixing mechanism. Because the rate of diffusion is a function of molecular weight, separation of the molecular species occurs in these layers. In the layers below, turbulent mixing dominates so that the various molecular species are well mixed.

Mesosphere. The mesosphere is the middle layer in the atmosphere (hence, mesosphere). The temperature in the mesosphere decreases with altitude. At the top of the mesosphere, air temperature reaches its coldest value, approaching –90 degrees Celsius (–130 degrees Fahrenheit). The air is extremely thin at this level, with 99.9 percent of the atmosphere's mass lying below the mesosphere. However, the proportion of nitrogen and oxygen at these levels is about the same as that at sea level. Because of the tenuousness of the atmosphere at this altitude, there is little absorption of solar radiation, which accounts for the low temperature. In the upper parts of the mesosphere, particulates may be present because of the passage of comets or micrometeors. Lidar measurements made by Kent et al. (1971) and Poultney (1972) seem to indicate that particulates in the mesosphere may also be associated with the passage of the earth through the tail of comets. They also show that the particulates at this level are rapidly mixed down to about 40 km. Because of the inaccessibility of the upper layers of the atmosphere for in situ measurements, lidar remote sensing is one of the few effective methods for the examination of processes in these regions.

In the region between 75 and 110 km, there exists a layer containing high concentrations of sodium, potassium, and iron (~3000 atoms/cm^3 of Na maximum and ~300 atoms/cm^3 of K maximum centered at 90 km and ~11,000 atoms/cm^3 of Fe centered about 86 km). The two sources of these alkali atoms are meteor showers and the vertical transport of salt near the two poles when stratospheric circulation patterns break down (Megie et al., 1978). A large number of lidar studies of these layers have been done with fluorescence lidars (589.9 nm for Na and 769.9 nm for K). A surprising amount of information can be obtained from the observation of the trace amounts of these ions including information on the chemistry of the upper atmosphere (see for example, Plane et al., 1999). Temperature profiles can be obtained by measurement of the Doppler broadening of the returning fluorescence signal (Papen et al., 1995; von Zahn and Hoeffner, 1996; Chen et al., 1996). Profiles of concentrations have been used to study mixing in this region of the atmosphere (Namboothiri et al., 1996; Clemesha et al., 1996; Hecht et al., 1997; Fritts et al., 1997). Illumination of the sodium layer has also been used in adaptive imaging systems to correct for atmospheric distortion (Jeys, 1992; Max et al., 1997).

The mesosphere is bounded above by the mesopause and below by the stratopause. The average height of the mesopause is about 85 km (53 miles). At this altitude, the atmosphere again becomes isothermal. This occurs around the 0.005 mb (0.0005 kPa) pressure level. Below the mesosphere is the stratosphere.

Stratosphere. The stratosphere is the layer between the troposphere and the mesosphere, characterized as a stable, stratified layer (hence, stratosphere) with a large temperature inversion throughout its depth. The stratosphere acts as a lid, preventing large storms and other weather from extending above the tropopause. The stratosphere also contains the ozone layer that has been the subject of great discussion in recent years. Ozone is the triatomic form of oxygen that strongly absorbs UV light and prevents it from reaching the earth's surface at levels dangerous to life. Molecular oxygen dissociates when it absorbs UV light with wavelengths shorter than 250 nm, ultimately forming ozone. The maximum concentration of ozone occurs at about 25 km (15 miles) above the surface, near the middle of the stratosphere. The absorption of UV light in this layer warms the atmosphere. This creates a temperature inversion in the layer so that a temperature maximum occurs at the top of the layer, the stratopause. The stratosphere cools primarily through infrared emission from trace gases. Throughout the bulk of the stratosphere and the mesosphere, elastic lidar returns are almost entirely due to molecular scattering. This enables the use of the lidar returns to determine the temperature profiles at these altitudes (see Section 12.3.1). In the lower parts of the stratosphere, particulates may be present because of aircraft exhaust, rocket launches, or volcanic debris from very large events (such as the Mount St. Helens or Mount Pinatubo events). Particulates from these sources are seldom found at altitudes greater than 17–18 km.

The stratosphere is bounded above by the stratopause, where the atmosphere again becomes isothermal. The average height of the stratopause is about 50 km, or 31 miles. This is about the 1-mb (0.1 kPa) pressure level. The layer below the stratosphere is the troposphere.

Troposphere. The troposphere is the lowest major layer of the atmosphere. This is the layer where nearly all weather takes place. Most thunderstorms do not penetrate the top of the troposphere (about 10 km). In the troposphere, pressure and density rapidly decrease with height, and temperature generally decreases with height at a constant rate. The change of temperature with height is known as the lapse rate. The average lapse rate of the atmosphere is approximately 6.5°C/km. Near the surface, the actual lapse rate may change dramatically from hour to hour on clear days and nights. A distinguishing characteristic of the troposphere is that it is well mixed, thus the name troposphere, derived from the Greek *tropein*, which means to turn or change. Air molecules can travel to the top of the troposphere (about 10 km up) and back down again in a just a few days. This mixing encourages changing weather. Rain acts to clean the troposphere, removing particulates and many types of chemical compounds. Rainfall is the primary reason for particulate and water-soluble chemical lifetimes on the order of a week to 10 days.

The troposphere is bounded above by the tropopause, a boundary marked as the point at which the temperature stops decreasing with altitude and becomes constant with altitude. The tropopause has an average height of about 10 km (it is higher in equatorial regions and lower in polar regions). This height corresponds to about 7 miles, which is approximately equivalent to the 200-mb (20.0 kPa) pressure level. An important sublayer is the PBL, in which most human activity occurs.

Boundary Layer. This sublayer of the troposphere is the source of nearly all the energy, water vapor, and trace chemical species that are transported higher up into the atmosphere. Human activity directly affects this layer, and much of the atmospheric chemistry also occurs in this layer. It is the most intensely studied part of the atmosphere. The PBL is the lowest 1–2 km of the atmosphere that is directly affected by interactions at the earth's surface, particularly by the deposition of solar energy. Stull (1992) defines the atmospheric boundary layer as "the part of the troposphere that is directly influenced by the presence of the earth's surface, and responds to surface forcings with a time scale of about an hour or less." Because of turbulent motion near the surface and convection, emissions at the surface are mixed throughout the depth of the PBL on timescales of an hour.

Figure 1.2 and the figures to follow are lidar vertical scans that show the lidar backscatter in a vertical slice of the atmosphere. The darkest areas indicate the highest amount of scattering from particulates, and light areas indicate areas with low scattering. Figure 1.2 illustrates a typical daytime evolution of the atmospheric boundary layer in high-pressure conditions over land. Solar heating at the surface causes thermal plumes to rise, transporting moisture, heat, and particulates higher into the boundary layer. The plumes rise and expand adiabatically until a thermodynamic equilibrium is reached at the top of the PBL. The moisture transported by the thermal plumes may form convective clouds at the top of the PBL that will extend higher into the tropos-

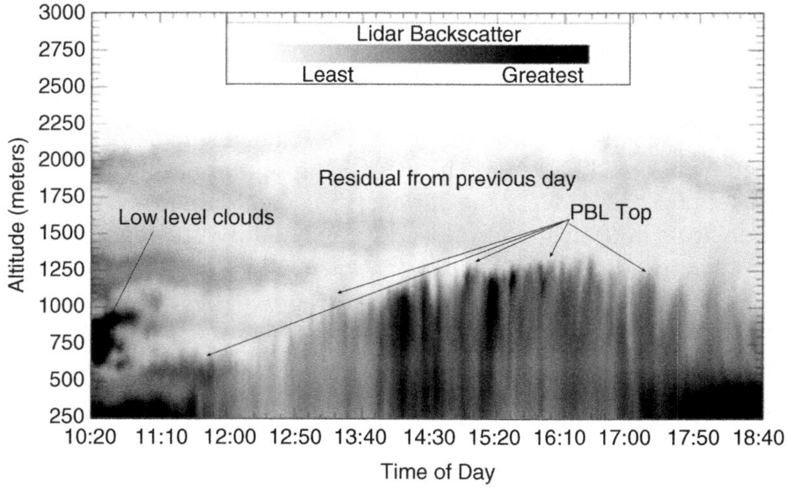

Fig. 1.2. A time-height lidar plot showing the evolution of a typical daytime planetary boundary layer in high-pressure conditions over land. After a cloudy morning, the top of the boundary layer rises. The rough top edge of the PBL is caused by thermal plumes.

phere. The top of the PBL is characterized by a sharp increase in temperature and a sudden drop in the concentration of water vapor and particulates as well as most trace chemical species. As the air in the PBL warms during the morning, the height at which thermal equilibrium occurs increases. Thus the depth of the PBL increases from dawn to several hours after noon, after which the height stays approximately constant until sundown. Figure 1.3 is an example of a lidar scan showing convective thermal plumes rising in a convective boundary layer (CBL).

The lowest part of the PBL is called the surface layer, which comprises the lowest hundred meters or so of the atmosphere. In windy conditions, the surface layer is characterized by a strong wind shear caused by the mechanical generation of turbulence at the surface. The gradients of atmospheric properties (wind speed, temperature, trace gas concentrations) are the greatest in the surface layer. The turbulent exchange of momentum, energy, and trace gases throughout the depth of the boundary layer are controlled by the rate of exchange in the surface layer.

Convective air motions generate turbulent mixing inside the PBL above the surface layer. This tends to create a well-mixed layer between the surface layer at the bottom and the entrainment zone at the top. In this well-mixed layer, the potential temperature and humidity (as well as trace constituents) are nearly constant with height. When the buoyant generation of turbulence dominates the mixed layer, the PBL may be referred to as a convective boundary layer. The part of the troposphere between the highest thermal plume tops and deepest parts of the sinking free air is called the entrainment zone. In this

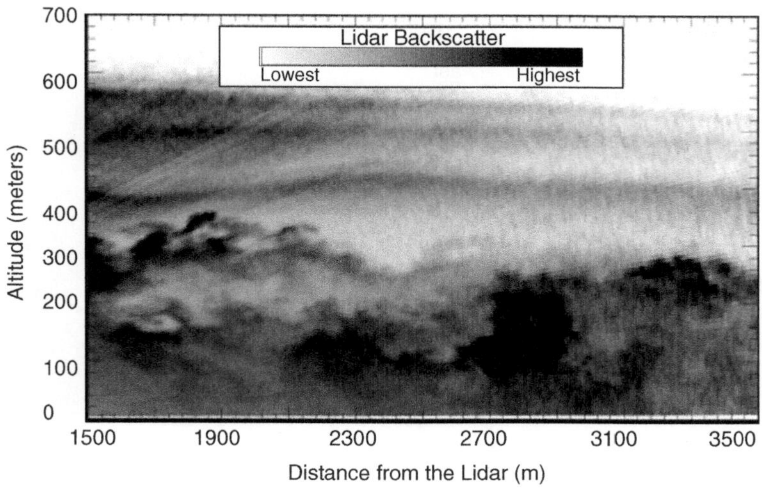

Fig. 1.3. A vertical (RHI) lidar scan showing convective plumes rising in a convective boundary layer. Structures containing high concentrations of particulates are shown as darker areas. Cleaner air penetrating from the free atmosphere above is lighter. Undulations in the CBL top are clearly visible.

region, drier air from the free atmosphere above penetrates down into the PBL, replacing rising air parcels.

1.1.2. Convective and Stable Boundary Layers

Convective Boundary Layers. A fair-weather convective boundary layer is characterized by rising thermal plumes (often containing high concentrations of particulates and water vapor) and sinking flows of cooler, cleaner air. Convective boundary layers occur during daylight hours when the sun warms the surface, which in turn warms the air, producing strong vertical gradients of temperature. Convective plumes transport emissions from the surface higher into the atmosphere. Thus as convection begins in the morning, the concentrations of particulates and contaminants decrease. Conversely, when evening falls, concentrations rise as the mixing effects of convection diminish. These effects can be seen in the time-height indicator in Fig. 1.2. The vertical motion of the thermal plumes causes them to overshoot the thermal inversion. As a plume rises above the level of the thermal inversion, the area surrounding the plume is depressed as cleaner air from above is entrained into the boundary layer below. This leads to an irregular surface at the top of the boundary layer that can be observed in the vertical scans (also known as range-height indicator or RHI scans) in Figs. 1.3 and 1.4. This interface stretches from the top of the thermal plumes to the lowest altitude where air entrained from above can be found. The top of a convective boundary layer is thus more of a region

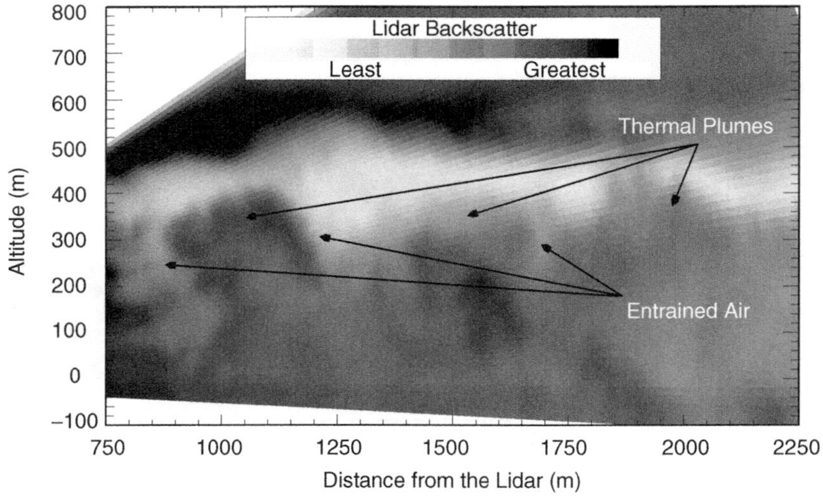

Fig. 1.4. A vertical (RHI) lidar scan showing convective plumes rising in a convective boundary layer.

of space than a well-defined location. Lidars are particularly well suited to map the structure of the PBL because of their fine spatial and temporal resolution. As the plumes rise higher into the atmosphere, they cool adiabatically. This leads to an increase in the relative humidity, which, in turn, causes hygroscopic particulates to absorb water and grow. Accordingly, there may be a larger scattering cross section in the region near the top of the boundary layer and an enhanced lidar return. Thus thermal plumes often appear to have larger particulate concentrations near the top of the boundary layer. The free air above the boundary layer is nearly always drier and has a smaller particulate concentration. Potential temperature and specific humidity profiles found in a typical CBL are shown in Fig. 1.5. Normally, the CBL top is indicated by a sudden potential temperature increase or specific humidity drop with height.

It is increasingly clear that events that occur in the entrainment zone affect the processes at or near the surface. This, coupled with the fact that computer modeling of the entrainment zone is difficult, has led to intensive experimental studies of the entrainment zone. When making measurements of the irregular boundary layer top with traditional point-measurement techniques (such as tethersondes or balloons), the measurements may be made in an upwelling plume or downwelling air parcel. The vertical distance between the highest plume tops and lowest parts of the downwelling free air may exceed the boundary layer mean depth. Nelson et al. (1989) measured entrainment zone thicknesses that range from 0.2 to 1.3 times the CBL average height. Thus there may be cases in which single point measurements of the CBL depth may vary more than 100 percent between individual measure-

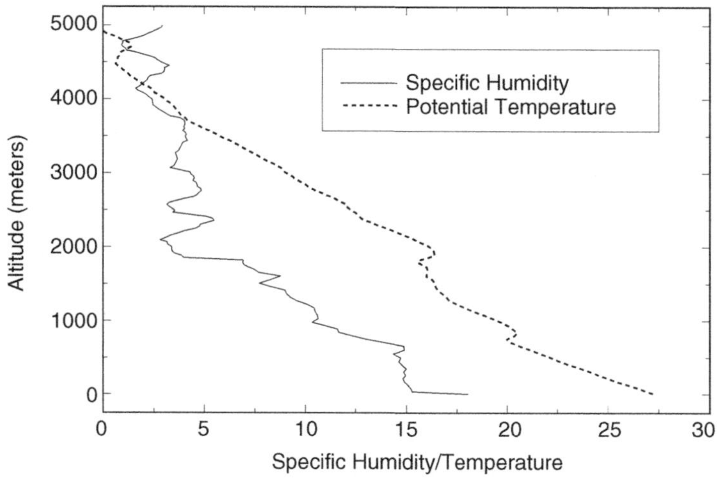

Fig. 1.5. A plot of the temperature and humidity profile in the lower half of the troposphere. A temperature inversion can be seen at about 800 m. Below the inversion the water vapor concentration is approximately constant (well mixed), and above the inversion, the water vapor concentration falls rapidly.

ments. Therefore, to obtain representative CBL depth estimates, relatively long averaging times must be used. Again, scanning lidars are ideal tools for the study of entrainment and the dynamics of PBL height. Section 12.4 discusses these measurement techniques in depth.

Because clouds scatter light well, they are seen as distinct dark formations in the lidar vertical scan. This allows one to precisely determine the cloud base altitude with a lidar pointed vertically. However, cloud top altitudes can be determined only for clouds that are optically thin, because it is impossible to determine whether the observed sharp decrease in signal is due to the end of the cloud or due to the strong extinction of the lidar signal within the dense cloud. However, a scanning lidar can often exploit openings in the cloud layer and other clues to determine the elevation of the cloud tops.

Stable Boundary Layers. The boundary layer from sunset to sunrise is called the nocturnal boundary layer. It is often characterized by a stable layer that forms when the solar heating ends and the surface cools faster than the air above through radiative cooling. In the evening, the temperature does not decrease with height, but rather increases. Such a situation is known as a temperature inversion. Persistent temperature inversion conditions, which represent a stable layer, often lead to air pollution episodes because pollutants, emitted at the surface, do not mix higher in the atmosphere. Farther above, the remnants of the daytime CBL form what is known as a residual layer.

Stable boundary layers occur when the surface is cooler than the air, which often occurs at night or when dry air flows over a wet surface. A stable bound-

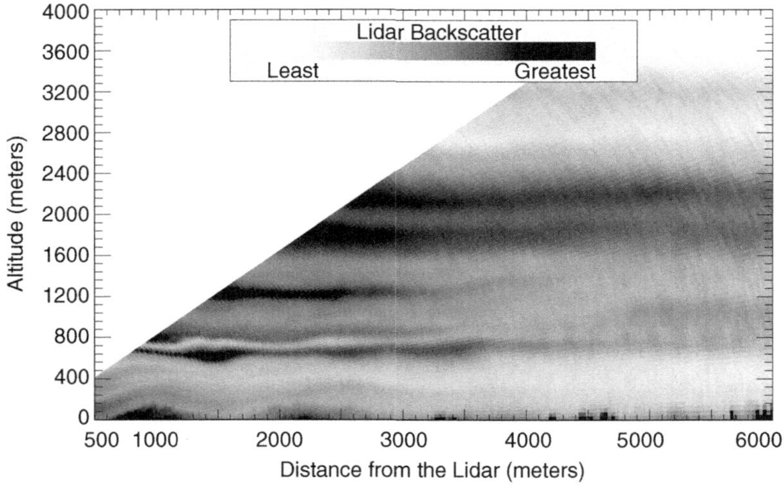

Fig. 1.6. A vertical (RHI) lidar scan showing the layering often found during stable atmospheric conditions. The wavelike features in the lower left are caused by the flow over a large hill behind the lidar.

ary layer exists when the potential temperature increases with height, so that a parcel of air that is displaced vertically from its original position tends to return to its original location. In such conditions, mixing of the air and turbulence are strongly damped and pollutants emitted at the surface tend to remain concentrated in a layer only a few tens of meters thick near the surface. Stable boundary layers are easily identified in lidar scans by the horizontal stratification that is nearly always present (Fig. 1.6). The bands are associated with layers that will have different wind speeds (and, possibly, directions), temperatures, and particulate/pollutant concentrations.

There has been a great deal of work and a number of field experiments in recent years that developed the present state of understanding of the physics of stable boundary layers and offered a significant research opportunity for lidars (for example, Derbyshire, 1995; McNider et al., 1995; Mahrt et al., 1997; Mahrt, 1999; Werne and Fritts, 1999; Werne and Fritts, 2001; Saiki et al., 2000). A stable boundary layer is characterized by long periods of inactivity punctuated by intermittent turbulent bursts that may last from tens of seconds to minutes, during which nearly all of the turbulent transport occurs (Mahrt et al., 1998). These intermittent events do not lead to statistically steady-state turbulence, a basic requirement of all existing theories. As a result, the underlying turbulent transfer mechanisms are not well understood and there is no adequate theoretical treatment of stable boundary layers. In stable atmospheres, turbulent quantities, like surface fluxes, are not adequately described by Monin–Obukhov similarity theory, which is the major tool applied to the study of convective boundary layers (Derbyshire, 1995). The vertical size of the turbulent eddies in a stable boundary layer is strongly damped, and

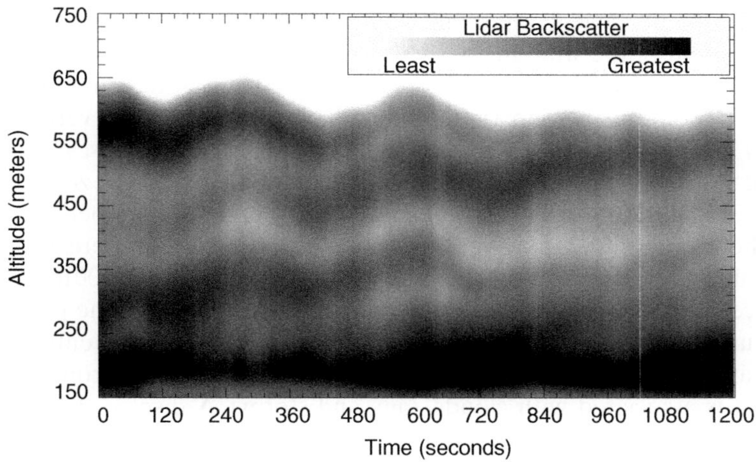

Fig. 1.7. A time-height lidar plot showing a series of gravity waves. Note that the passage of the waves distorts the layers throughout the depth of the boundary layer. (Courtesy of H. Eichinger)

turbulence above the surface is only minimally influenced by events at the surface. Thus turbulent scaling laws do not depend on the height above the surface as they do for convective conditions. This is known as z-less stratification (Wyngaard, 1973, 1994).

It is believed that the intermittence, found in stable boundary layers, is associated with larger-scale events, such as gravity waves (Fig. 1.7), overturning Kelvin–Helmholtz (KH) waves, shear instabilities, or terrain-generated phenomena. Much of the vertical transport that occurs near the surface is then related to events that occur at higher levels. These events are difficult to model or incorporate into simple analytical models. To compound the problem, internal gravity waves and shear instabilities may propagate over long distances. (Einaudi and Finnigan, 1981; Finnigan and Einaudi, 1981; Finnigan et al., 1984). As a result, a turbulent event at the surface may occur because of an event that occurred tens of kilometers away and a kilometer or more higher up in the atmosphere.

Under clear skies and very stable atmospheric conditions, the dispersion of materials released near the ground is greatly suppressed. This has a wide range of practical implications, including urban air pollution episodes, the long-range transport of objectionable odors from farms and factories, and pesticide vapor transport. Thus stable atmospheric conditions are a topic of intensive study.

1.1.3. Boundary Layer Theory

In the boundary layer, the mean wind velocity components are given differently by various communities. Boundary layer meteorologists commonly use

$\bar{u}$, $\bar{v}$, and $\bar{w}$ to indicate wind direction, where the bar indicates time averaging. The compontent of the wind in the direction of the mean wind (which is also taken as the x-direction) is denoted as u, the component in the direction perpendicular to the mean wind (y-direction) is v, and that in the vertical (z-direction) is w. Meteorologists and modelers working on larger scales often divide the wind into a meridional (east-west) component, u, and a zonal component, v. Temperature is usually taken to be the potential temperature, θ_p. This is the temperature that would result if a parcel of air were brought adiabatically from some altitude to a standard pressure level of 1000 mb. Near the surface, the difference between the actual temperature and the potential temperature is small, but at higher altitudes, comparisons of potential temperature are important to stability and the onset of convection. Tropospheric convection is associated with clouds, rain, and storms. A displaced parcel of air with a potential temperature greater than that of the surrounding air will tend to rise. Conversely, it will tend to fall if the potential temperature is lower than that of the surrounding air. The potential temperature is defined to be

$$\theta_p = T\left(\frac{P_0}{P}\right)^{\alpha}$$

where P_0 is 100.0 kPa, and P is the pressure at the altitude to which the parcel is displaced. The exponent α is $R_d(1 - 0.23q)/C_p$, here R_d is the gas constant for dry air, R_d = 287.04 J/kg-K, R_v is the gas constant for water vapor, R_v = 461.51 J/kg-K. C_p is the specific heat of air at constant pressure (1005 J/kg-K). The density of dry air is given by $N_{dry} = \dfrac{P - e_w}{R_d T}$, and the water vapor density is given by $N_{water} = \dfrac{0.622 e_w}{R_d T}$ (here 0.622 is the ratio of the molecular weights of water and dry air, i. e., 18.016/28.966). The factor e_w is the vapor pressure of water, an often-used measure of water vapor concentration. The saturation vapor pressure, e_w^* is the pressure at which water vapor is in equilibrium with liquid water at a given temperature. The latter is given by the formula (Alduchov and Eskridge, 1996)

$$e_w^* = 6.1094 e^{\left(\frac{17.625\text{T}}{243.04+\text{T}}\right)} \tag{1.1}$$

Water vapor concentration is normally given as q, the specific humidity. This is the mass of water vapor per unit mass of moist air

$$q = \frac{0.622 e_w}{\text{P} - 0.378 e_w}$$

The specific humidity q is similar to the mixing ratio, the mass of water vapor per unit mass of dry air. The relative humidity, Rh, is the ratio of the actual mixing ratio and the mixing ratio of saturated air at the same temperature. Rh

is not a good measure of water concentration because it depends on both the water concentration and the local temperature.

The addition of water to air decreases its density. The density of moist air is given by

$$\rho_{\text{air}} = \frac{P}{R_d T}\left(1 - \frac{0.378 e_w}{P}\right) \tag{1.2}$$

Because of the change in density with water content, water vapor plays a role in atmospheric stability and convection. It should be noted that air behaves as an ideal gas, provided the term in parenthesis in Eq. (1.2) is included. Treating air as an ideal gas may also be accomplished through the use of a virtual temperature, T_v, defined as $T_v = T(1 + 0.61q)$ so that $P = \rho R_d T_v$. The virtual temperature is the temperature that dry air must have so as to have the same density as moist air with a given pressure, temperature, and water vapor content. Virtual potential temperature θ_v is defined as $\theta_v = (1 + 0.61q)\theta_p$. It is common to consider the virtual potential temperature as a criterion for atmospheric stability when water vapor concentration varies significantly with height.

Vertical transport of nonreactive scalars in the lowest part of the atmosphere is caused by turbulence and decreasing gradients of concentration of the scalars in the vertical direction. Turbulent fluxes are represented as the covariance of the vertical wind speed and the concentration of the scalar of interest. With Reynold's decomposition (Stull, 1988), where the value of any quantity may be divided into mean and fluctuating parts, the wind speed, for example, can be written as $u = (\bar{u} + u')$ where the bar indicates a time average. Advected quantities are then determined by *advected water vapor* = $\overline{uq}$, for example, and that portion of the water transported by turbulence in the mean wind direction as *turbulent water vapor transport* = $\overline{u'q'}$. The surface stress in a turbulent atmosphere is $\tau = \sqrt{-\overline{u'w'}}$. The vertical energy fluxes are the sensible heat flux, $H = \rho C_p \overline{w'\theta'}$ and the surface latent heat flux, $E = \rho\lambda_e \overline{w'q'}$ where C_p is the specific heat of air at constant pressure and λ_e is the latent heat of vaporization of water (2.44×10^6 J/kg at 25°C). The surface friction velocity, u^*, is defined to be $u^* = (\overline{u'w'}^2 + \overline{v'w'}^2)^{1/4}$. The friction velocity is an important scaling variable that occurs often in boundary layer theory. For example, the vertical transport of a nonreactive scalar is proportional to u^*. The Monin–Obukhov similarity method (MOM) (Brutsaert, 1982; Stull, 1988; Sorbjan, 1989) is the major tool used to describe average quantities near the earth's surface. The average horizontal wind speed and the average concentration of any nonreactive scalar quantity in the vertical direction can be described using Monin–Obukhov similarity. With this theory, the relationships between the properties at the surface and those at some height h can be determined. Within the inner region of the boundary layer, the relations for wind, temperature, and water vapor concentration are as follows

$$u(h)=\frac{u^*}{k}\left[\ln\left(\frac{h}{h_{\text{om}}}\right)+\psi_{\text{m}}\left(\frac{h}{L_{\text{mo}}}\right)\right]$$

$$T_{\text{s}}-T(h)=\frac{H}{C_{\text{p}}ku^*\rho}\left[\ln\left(\frac{h}{h_{\text{oT}}}\right)+\psi_{\text{T}}\left(\frac{h}{L_{\text{mo}}}\right)\right]$$

$$q_{\text{s}}-q(h)=\frac{E}{\lambda_{\text{e}}ku^*\rho}\left[\ln\left(\frac{h}{h_{\text{ov}}}\right)+\psi_{\text{v}}\left(\frac{h}{L_{\text{mo}}}\right)\right] \tag{1.3}$$

where the Monin–Obukhov length L_{mo} is defined as

$$L_{\text{mo}}=-\frac{\rho\left(u^*\right)^3}{kg\left[\dfrac{H}{Tc_{\text{p}}}+0.61\text{E}\right]} \tag{1.4}$$

$h_{0\text{m}}$ is the roughness length for momentum, $h_{0\text{v}}$ and $h_{0\text{T}}$ are the roughness lengths for water vapor and temperature, q_{s} and T_{s} are the specific humidity and temperature at the surface, $q(h)$ is the specific humidity at height h, H is the sensible heat flux, E is the latent heat flux, ρ is the density of the air, λ_{e} is the latent heat of evaporation for water, and u^* is the friction velocity (Brutsaert, 1982); k is the von Karman constant, taken as 0.40, and g is the acceleration due to gravity; ψ_{m},ψ_{v}, and ψ_{T} are the Monin–Obukhov stability correction functions for wind, water vapor, and temperature, respectively. They are calculated as

$$\psi_{\text{m}}\left(\frac{h}{L_{\text{mo}}}\right)=2\ln\left[\frac{(1+x)}{2}\right]+\ln\left[\frac{(1+x^2)}{2}\right]-2\arctan(x)+\frac{\pi}{2} \qquad L_{\text{mo}}<0$$

$$\psi_{\text{m}}\left(\frac{h}{L_{\text{mo}}}\right)=5\frac{h}{L_{\text{mo}}} \qquad L_{\text{mo}}>0$$

$$\psi_{\text{v}}\left(\frac{h}{L_{\text{mo}}}\right)=\psi_{T}\left(\frac{h}{L_{\text{mo}}}\right)=2\ln\left[\frac{(1+x^2)}{2}\right] \qquad L_{\text{mo}}<0$$

$$\psi_{\text{v}}\left(\frac{h}{L_{\text{mo}}}\right)=\psi_{\text{T}}\left(\frac{h}{L_{\text{mo}}}\right)=5\frac{h}{L_{\text{mo}}} \qquad L_{\text{mo}}>0 \tag{1.5}$$

where

$$x=\left(1-16\frac{h}{L_{\text{mo}}}\right)^{1/4} \tag{1.6}$$

The roughness lengths are free parameters to be calculated based on the local conditions. Heat and momentum fluxes are often determined from measurements of temperature, humidity, and wind speed at two or more heights. These relations are valid in the inner region of the boundary layer, where the atmosphere reacts directly to the surface. This region is limited to an area between the roughness sublayer (the region directly above the roughness elements) and

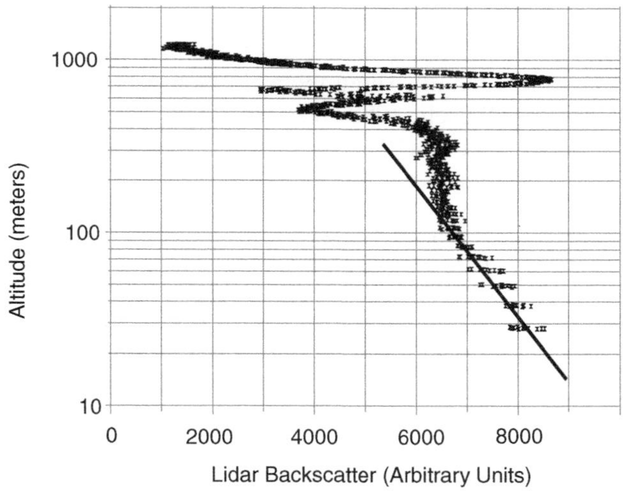

Fig. 1.8. A plot of the elastic backscatter signal as a function of height derived from the two-dimensional data shown in Fig. 3.6. The lidar data covers a spatial range interval of 100 meters in the horizontal direction. The data, on average, converge to the logarithmic curve in the lowest 100 m. From 100 m to 400 m, the atmosphere is considered to be "well mixed." Between 400 m and 500 m there is a sharp drop in the signal that is indicative of the top of the boundary layer. Above this is a large signal from a cloud layer.

below 5–30 m above the surface (where the passive scalars are semilogarithmic with height). The vertical range of this layer is highly dependent on the local conditions. The top of this region can be readily identified by a departure from the logarithmic profile near the surface. Figure 1.8 is an example of an elastic backscatter profile with a logarithmic fit in the lowest few meters above the surface. Suggestions have been made that the atmosphere is also logarithmic to higher levels and may integrate fluxes over large areas (Brutsaert, 1998). Similar expressions can be written for any nonreactive atmospheric scalar or contaminant.

Monin–Obukhov similarity is normally used in the lowest 50–100 m in the boundary layer but can be extended higher up into the boundary layer. There are various methods by which this can be accomplished involving several combinations of similarity variables (Brutsaert, 1982; Stull, 1988; Sorbjan, 1989). Each method has limitations and limited ranges of applicability and should be used with caution.

Monin–Obukhov similarity can also be used to describe the average values of statistical quantities near the surface. For example, the standard deviation of a quantity, x, u^*, and the surface emission rate of x, $(\overline{w'x'})$ are related as

$$\frac{\sigma_x u^*}{\overline{w'x'}} = f_x\left(\frac{h}{L_{mo}}\right) \tag{1.7}$$

where σ_x is the standard deviation of x, and f_x is a universal function (to be empirically determined) of h/L_{mo}, where h is the height above ground and L_{mo} is the Monin–Obukhov length. The universal functions have several formulations that are similar (Wesely, 1988; Weaver, 1990). For unstable conditions, when $L_{mo} < 0$, DeBruin et al. (1993) suggest the following universal function for the variance of nonreactive scalar quantities

$$f_x\left(\frac{h}{L_{mo}}\right) = 2.9\left(1 - 28.4\frac{h}{L_{mo}}\right)^{-1/3} \tag{1.8}$$

Another quantity that scanning lidars can measure is the structure function for the measured scalar quantity. A structure function is constructed by taking the difference between the quantity x at two locations to some power. This quantity is related to the distance between the two points, the dissipation rate of turbulent kinetic energy, ε, and the dissipation rate of x, ε_x, as:

$$\overline{[x(r_1) - x(r_2)]^n} = \text{constant } \varepsilon^{-n/6}\varepsilon_x^{n/2} r_{12}^{n/3} = C_{xx}^n r_{12}^{n/3}, \tag{1.9}$$

where r_1 and r_2 are the locations of the two measurements, r_{12} is the distance between r_1 and r_2, C_{xx} is the structure function parameter, and n is the order of the structure function. Structure function parameters may also be expressed in terms of universal functions, the height above ground h, u^*, and the surface emission rate of x, $(\overline{w'x'})$. For the second-order structure function

$$\frac{C_{xx}^2 h^{2/3}\left(u^*\right)^2}{\overline{w'x'}^2} = f_{xx}\left(\frac{h}{L_{mo}}\right) \tag{1.10}$$

For unstable conditions, $L_{mo} < 0$, DeBruin et al. (1993) suggest the following universal function for nondimensional structure functions of nonreactive scalar quantities

$$f_x\left(\frac{h}{L_{mo}}\right) = 4.9\left(1 - 9\frac{h}{L_{mo}}\right)^{-2/3} \tag{1.11}$$

The relations for various structure functions and variances can be combined in many different ways to obtain surface emission rates, dissipation rates, and other parameters of interest to modelers and scientists. Although these techniques have been used by radars (for example, Gossard et al., 1982; Pollard et el., 2000) and sodars (for example, Melas, 1993) to explore the upper reaches of the boundary layer, they have not been exploited by lidar researchers. We believe that this is an area of great opportunity for lidar applications.

Buoyancy plays a large role in determining the stability of the atmosphere at altitudes above about 100 m. If we assume a dry nonreactive atmosphere that is completely transparent to radiation, with no water droplets in hydrostatic equilibrium, then buoyancy forces balance gravitational forces and it can be shown that

$$\frac{dT}{dh} = -\frac{g}{C_p} = -\Gamma_d, \tag{1.12}$$

where g is the acceleration due to gravity, C_p is the specific heat at constant pressure (1005 J/kg-K), and Γ_d is the dry adiabatic lapse rate, about 9.8 K/km. The temperature gradient dT/dh determines the stability of the real atmosphere; if $-dT/dh < \Gamma_d$ the atmosphere is stable and conversely, if $-dT/dh > \Gamma_d$ the atmosphere is unstable. As previously noted, the average lapse rate in the atmosphere, $-dT/dh$ is about 6.5 K/km. A more complete analysis includes the effects of water vapor and the heat that is released as it condenses. Such an analysis will show that

$$\Gamma_s = \Gamma_d\left[1+\frac{\lambda_e e_w M_{wv}}{PRT}\right]\left[1+\frac{\lambda_e e_w M_{wv}}{PRT}\frac{0.622 L_{mo}}{C_p T}\right] \tag{1.13}$$

where λ_e is the latent heat of evaporation, e_w is the vapor pressure of water, M_{wv} is the molecular weight of water, R is the gas constant, and Γ_s is the wet adiabatic lapse rate. It can be seen from Eq. (1.13) that $\Gamma_s \geq \Gamma_d$ for all conditions. Γ_s determines the stability of saturated air in the same way that Γ_d determines the stability of dry (or unsaturated) air.

1.2. ATMOSPHERIC PROPERTIES

When modeling the expected lidar return for a given situation, it is necessary to be able to describe the conditions that will be encountered. To accomplish this, the temperature and density of the atmosphere and the particulate size distributions and concentrations must be known or estimated. We present here several "standard" sources for this type of information. It should be recognized that these formulations represent "average" conditions (which are useful to know when making analyses of lidar return simulations in different atmospheric conditions) and that the actual conditions at any point may be quite different.

1.2.1. Vertical Profiles of Temperature, Pressure and Number Density

The number density of nitrogen molecules, $N(h)$, at height h can be found in the *U.S. Standard Atmosphere* (1976). The temperature $T(h)$, in degrees Kelvin and pressure $P(h)$, in pascals, as a function of the altitude h, in meters, for the first 11 km of the atmosphere can be determined from the expressions below:

$$T(h) = 288.15 - 0.006545 * h$$
$$P(h) = 1.013 \times 10^5 * \left[\frac{288.15}{T(h)}\right]^{\left(\frac{0.034164}{0.006545}\right)} \tag{1.14}$$

The temperature and pressure from 11 to 20 km in the atmosphere can be determined from:

$$T(h) = 216.65$$
$$P(h) = 2.269 \times 10^4 * e^{\left(\frac{-0.034164(h-11000)}{216.65}\right)} \tag{1.15}$$

The temperature and pressure from 20 to 32 km in the atmosphere can be determined from:

$$T(h) = 216.65 + 0.0010 * (h - 20{,}000)$$
$$P(h) = 5528.0 * \left[\frac{216.65}{T(h)}\right]^{\left(\frac{0.034164}{0.0010}\right)} \tag{1.16}$$

The temperature and pressure from 32 to 47 km in the atmosphere can be determined from:

$$T(h) = 228.65 + 0.0028 * (h - 32{,}000)$$
$$P(h) = 888.8 * \left[\frac{228.65}{T(h)}\right]^{\left(\frac{0.034164}{0.0028}\right)} \tag{1.17}$$

$P(h)$ and $T(h)$ having been determined, the number density of molecules can be found from:

$$N(h) = \left(\frac{28.964\,\text{kg/kmol}}{8314\,\text{J/kmol} - \text{K}}\right)\frac{P(h)}{T(h)} = 0.003484 * \frac{P(h)}{T(h)}\,\text{kg/m}^3 \tag{1.18}$$

1.2.2. Tropospheric and Stratospheric Aerosols

In addition to anthropogenic sources of particulates, there are three other major sources of aerosols and particulates in the troposphere. These sources include large-scale surface sources, volumetric sources, and large-scale point sources. Large-scale surface sources include dust blown from the surface, salts from large water bodies, and biological sources such as pollens, bacteria, and fungi. Volumetric sources are primarily due to gas to particle conversion (GPC), in which trace gases react with existing particulates or undergo homogeneous nucleation (condensation) to form aerosols. The evaporation of cloud droplets is also a major source of particulates. Point sources include large events such as volcanoes and forest fires. Each of these sources has a major body of literature describing source strengths, growth rates, and distributions. Particulates will absorb water under conditions of high relative humidity and absorb chemically reactive molecules (SO_2, SO_3, H_2SO_4, HNO_3, NH_3). The size and chemical composition of the particulates and, thus, their optical properties may change in time. This makes it difficult to characterize even average conditions. The effects of humidity on optical and chemical properties have led to increased interest in simultaneous measurements of particulates and water vapor concentration (see, for example, Ansmann et al., 1991; Kwon et al., 1997). The number distribution of particulates also varies because of the rather short lifetimes in the troposphere. Rainfall and the coagulation of small particulates are the main removal processes. In the lower troposphere, the maximum lifetime is about 8 days. In the upper troposphere, the lifetime can be as long as 3 weeks.

The largest sources of tropospheric particulates are generally at the surface. The particulate concentrations are 3–10 times greater in the boundary layer than they are in the free troposphere (however, marine particulate concentrations have been measured that increase with altitude). Lidar measured backscatter and attenuation coefficients change by similar amounts. The sharp drop in these parameters at altitudes of 1–3 km is often used as a measure of the height of the PBL. There is evidence for a background mode for tropospheric particulates at altitudes ranging from 1.5 to 11 km from CO_2 lidar studies (Rothermel et al., 1989). At these altitudes there appears to be a constant background mixing ratio with convective incursions from below and downward mixing from the stratosphere. These inversions can increase the mixing ratio by an order of magnitude or more.

Stratospheric aerosols differ substantially from tropospheric aerosols. There exists a naturally occurring background of stratospheric aerosols that consist of droplets of 60 to 80 percent sulfuric acid in water. Sulfuric acid forms from the dissociation of carbonyl sulfide (OCS) by ultraviolet radiation from the sun. Carbonyl sulfide is chemically inert and water insoluble, has a long lifetime in the troposphere, and gradually diffuses upward into the stratosphere, where it dissociates. None of the other common sulfur-containing chemical compounds has a lifetime long enough to have an appreciable concentration in the stratosphere, and thus they do not contribute to the formation of these droplets. In addition to the droplets, volcanoes (and in the past, nuclear detonations) may loft large quantities of particulates above the tropopause. Because there are no removal mechanisms (like rain) for particulates in the stratosphere, and very little mixing occurs between the troposphere and stratosphere, particles in the stratosphere have lifetimes of a few years. Because of the long lifetime of the massive quantities of particulates that may be lofted by large volcanic events, these particulates play a role in climate by increasing the earth's albedo. Size distributions of droplets and volcanic particulates as well as their concentration with altitude and optical properties can be found in Jager and Hofmann (1991).

TABLE 1.2. Atmospheric particulate characteristics

Atmospheric Scattering Particulate Type	Range of Particulate Radii, μm	Concentration, cm^{-3}
Molecules	10^{-4}	10^{19}
Aitken nucleus	10^{-3}–10^{-2}	10^{4}–10^{2}
Mist particulate	10^{-2}–1	10^{3}–10
Fog particulate	1–10	100–10
Cloudy particulate	1–10	300–10
Rain droplet	10^{-2}–10^{4}	10^{-2}–10^{-5}

McCarney (1979).

1.2.3. Particulate Sizes and Distributions

As shown in Table 1.2, particulates in the atmosphere have a large range of geometric sizes: from 10^{-4} μm (for molecules) to 10^{4} μm and even higher (for rain droplets). Natural particulate sources include smoke from fires, wind blown dust, sea spray, volcanoes, and residual from chemical reactions. Most manmade particulates are the result of combustion of one kind or another. Particulate concentrations vary dramatically depending on location, time of day, and time of year but generally decrease with height in the atmosphere. Because many particulates are hygroscopic, the size and distribution of these particles are strongly dependent on relative humidity.

A number of analytical formulations are in common use to describe the size distribution of particulates in the atmosphere. These include a power law or Junge distribution, the modified gamma distribution, and the log normal distribution (Junge, 1960 and 1963; Deirmendjian, 1963, 1964, and 1969). For continuous model distributions, the number of particles with a radius ρ between ρ and $(\rho + d\rho)$ within a unit volume is written in the form

$$dN = n(\rho)d\rho \tag{1.19}$$

where $n(\rho)$ is the size distribution function with the dimension of L^{-4}. Integrating Eq. (1.21), the total number of the particles per unit volume (the number density) is determined as

$$N = \int_{0}^{\infty} n(\rho)d\rho \tag{1.20}$$

In practical calculations, a limited size range is often used, so the integration is made between the finite limits from ρ_1 to ρ_2:

$$N = \int_{\rho 1}^{\rho 2} n(\rho)d\rho \tag{1.21}$$

where ρ_1 and ρ_2 are the lower and upper particulate radius ranges based on the existing atmospheric conditions (see Table 1.2).

Among the simplest of the size distribution functions that have been used to describe atmospheric particulates is the power law, known as the Junge distribution, originally written as (Junge, 1960 and 1963; McCartney, 1977),

$$dN/d\log\rho = c\rho^{-v} \tag{1.22}$$

where c and v are constants. The other form of presentation of the distribution can be written as (Pruppacher and Klett, 1980)

$$n_N(\log D_p) = \frac{C_s}{(D_p)^{\alpha}} \tag{1.23}$$

where C_s and α are fitting constants and D_p is the particulate diameter. For most applications, α has a value near 3. Although this distribution may fit measured number distributions well, in a qualitative sense, it performs poorly when used to create a volume distribution (particulate volume per unit volume of air), which is

$$n_v(\log D_p) = \frac{\pi C_s}{6} D_p^{3-\alpha} \tag{1.24}$$

Both of these functions are straight lines on a log-log graph. They fail to capture the bimodal (two humped) character of many, especially urban, distributions. These bimodal distributions have a second particulate mode that ranges in size from about 2 to 5 μm and contains a significant fraction of the total particulate volume. Because the number of particles in the second mode is not large, the deviation from the power law number distribution is, generally, not large, and they appear to adequately describe the data. However, when used as a volume distribution, they do not include the large particulate volume contained in the second peak and thus fail to correctly determine the particulate volume and total mass. These distributions are often used because they are mathematically simple and can be used in theoretical models requiring a nontranscendental number distribution. However, because environmental regulations often specify particulate concentration limits in terms of mass per unit volume of air, the failure to correctly reproduce the volume distribution is a serious limitation.

To account for the possibility of multiple particulate modes, particulate size distributions are often described as the sum of n log-normal distributions as (Hobbs, 1993)

$$n_N(\log D_p) = \sum_{i=1}^{n} \frac{N_i}{(2\pi)^{1/2}\log\sigma_i} \exp\left[-\frac{\left(\log D_p - \log \overline{D_{pi}}\right)^2}{2\log^2\sigma_i}\right] \tag{1.25}$$

TABLE 1.3. Model Particulate Distributions—Three Log Normal Modes

Type	Mode I			Mode II			Mode III		
	N, cm^{-3}	D_p, μm	$\log\sigma$	N, cm^{-3}	D_p, μm	$\log\sigma$	N, cm^{-3}	D_p, μm	$\log\sigma$
Urban	9.93×10^4	0.013	0.245	1.11×10^3	0.014	0.666	3.64×10^4	0.05	0.337
Marine	133	0.008	0.657	66.6	0.266	0.210	3.1	0.58	0.396
Rural	6650	0.015	0.225	147	0.054	0.557	1990	0.084	0.266
Remote continental	3200	0.02	0.161	2900	0.116	0.217	0.3	1.8	0.380
Free troposphere	129	0.007	0.645	59.7	0.250	0.253	63.5	0.52	0.425
Polar	21.7	0.138	0.245	0.186	0.75	0.300	3×10^{-4}	8.6	0.291
Desert	726	0.002	0.247	114	0.038	0.770	0.178	21.6	0.438

Jaenicke (1993).

where N_i is the number concentration, $\overline{D_{pi}}$ is the mean diameter, and σ_i is the standard deviation of the ith log normal mode. Table 1.3 lists typical values for the relative concentrations, mean size, and standard deviation of the modes for a number of the major particulate types.

In many studies, the distribution used was proposed by Deirmendjian (1963, 1964, and 1969) in the form

$$n(\rho) = a\rho^g \exp(-b\rho^\gamma) \tag{1.26}$$

where a, b, α, and γ are positive constants. The distribution was named by a modified gamma distribution as it reduces to conventional gamma distribution when $\gamma = 1$. The modified gamma distribution of Deirmendjian is often used to describe the droplet size distribution of fogs and clouds. This function is given by

$$n(\rho) = N\frac{6^6}{5!}\frac{1}{\rho_m}\left[\frac{\rho}{\rho_m}\right]^6 e^{-6\rho/\rho_m} \tag{1.27}$$

where ρ_m is the mean droplet size (mean radius) and N is the total number of droplets per unit volume. This distribution with $\rho_m = 4\,\mu m$ fits fair weather cumulus cloud droplets quite well. In general, a linear combination of two distributions is required to fit measured cloud sizes (Liou, 1992). For example, stratocumulus droplet size distributions are often bimodal (Miles et al., 2000). This situation can be modeled as the sum of two or more gamma distributions or as the sum of multiple log-normal distributions. Miles et al. (2000) have accumulated a collection of more than 50 measured cloud droplet distributions.

1.2.4. Atmospheric Data Sets

In this section we present a number of data sets or programs that are often used to represent "standard" conditions in the atmosphere. The *U.S. Standard Atmosphere* (1976) is a source for average conditions in the atmosphere, and the rest are sources for optical parameters in the atmosphere. A number of radiative transfer models exist that can calculate radiative fluxes and radiances. The four codes that are used most often for atmospheric transmission are HITRAN (high resolution transmittance), MODTRAN (moderate-resolution transmittance), LOWTRAN (low-resolution transmittance), and FASCODE (fast atmospheric signature code). LOWTRAN, MODTRAN, and FASCODE are owned by the U.S. Air Force. Copies may be purchased on the internet at http://www-vsbm.plh.af.mil/. At least one vendor (http://www.ontar.com) is licenced to sell versions of these codes.

HITRAN is a database containing a compilation of the spectroscopic parameters of each line for 36 different molecules found in the atmosphere originally developed by the Air Force Geophysics Laboratory approximately 30 years ago. A number of vendors offer computer programs that use the HITRAN data set to calculate the atmospheric transmission for a given wavelength. As might be expected, the usefulness of the programs varies considerably and depends on the features incorporated into them. Perhaps the best place for information on HITRAN is the website at http://www.HITRAN.com.

LOWTRAN is a computer program that is intended to provide transmission and radiance values for an arbitrary path through the atmosphere for some set of atmospheric conditions (Kneizys et al., 1988). These conditions could include various types of fog or clouds, dust or other particulate obscurants, and chemical species and could incorporate the temperature and water vapor content along the path. In practical use, sondes are often used to provide information on temperature and humidity instead of a model atmosphere. Several types of aerosol models are included in the program. MODTRAN was developed to provide the same type of information albeit with a higher ($2\,cm^{-1}$) spectral resolution than LOWTRAN can provide (Berk et al., 1989). The molecular absorption properties used by both programs use the HITRAN database.

The Air Force Philips Laboratory has developed a sophisticated, high-resolution transmission model, FASCODE (Smith et al., 1978). The model uses the HITRAN database and a local radiosonde profile to calculate the radiance and transmission of the atmosphere with high spectral resolution. The radiosonde provides information on temperature and water vapor content with altitude. The model incorporates various types of particulate conditions as well as cloud and fog conditions.

For many modeling applications, information on the meteorology of the atmosphere with altitude is required. A number of standard atmospheres exist, but the most commonly used one is the *U.S. Standard Atmosphere*. The most current version of the *U.S. Standard Atmosphere* was adopted in 1976 by the

United States Committee on Extension to the Standard Atmosphere (COESA). The work is essentially a single profile representing an idealized, steady-state atmosphere with average solar activity. In the profile, a wide range of parameters are given at each altitude. These parameters include temperature, pressure, density, the acceleration due to gravity, the pressure scale height, the number density, the mean particle velocity, the mean collision frequency, mean free path, mean molecular weight, speed of sound, dynamic viscosity, kinematic viscosity, thermal conductivity, and geopotential height. The altitude resolution of the profile varies from 0.05 km near the surface up to as much as 5 km at high altitudes. The work can be obtained in book form from the National Geophysical Data Center (NGDC) or the U.S. Government Printing Office in Washington, D.C. Fortran codes that will generate the values can be obtained from many sites on the Internet including Public Domain Aeronautical Software.

For many lidar applications, detailed transmission data such as that provided by HITRAN or MODTRAN are not required. Information on the average particulate concentration and scattering/absorption properties may be found in several different compilations. These include Elterman (1968), McClatchey et al. (1972), and Shettle and Fenn (1979). Atmospheric constituent profiles can be found in Anderson et al. (1986). Penndorf (1957) has a compilation of the optical properties for air as a function of wavelength.

2

LIGHT PROPAGATION IN THE ATMOSPHERE

Transport, scattering, and extinction of electromagnetic waves in the atmosphere are complex issues. Depending on the particular application, transport calculations may become quite involved. In this chapter, the basic principles of the scattering and the absorption of light by molecules and particulates are outlined. The topics discussed here should be sufficient for most lidar applications. For further information, there are many fine texts on the subject (Van der Hulst, 1957; Deirmendjian, 1969; McCartney, 1977; Bohren and Huffman, 1983; Barber and Hill, 1990) that should be consulted for detailed analyses.

2.1. LIGHT EXTINCTION AND TRANSMITTANCE

A number of quantities are in common use to quantify or characterize the amount of energy in a beam of light.

Radiant flux: The radiant flux, F, is the rate at which radiant energy passes a certain location per unit time (J/s, W).

Spectral radiant flux: The spectral radiant flux, F_λ, is the flux in a narrow spectral width around λ per unit spectral width (W/nm or W/μm).

Radiant flux density: The radiant flux density is the amount of radiant flux intercepted by a unit area (W/m^2). If the flux is incident to the surface,

Elastic Lidar: Theory, Practice, and Analysis Methods, by Vladimir A. Kovalev and William E. Eichinger.
ISBN 0-471-20171-5

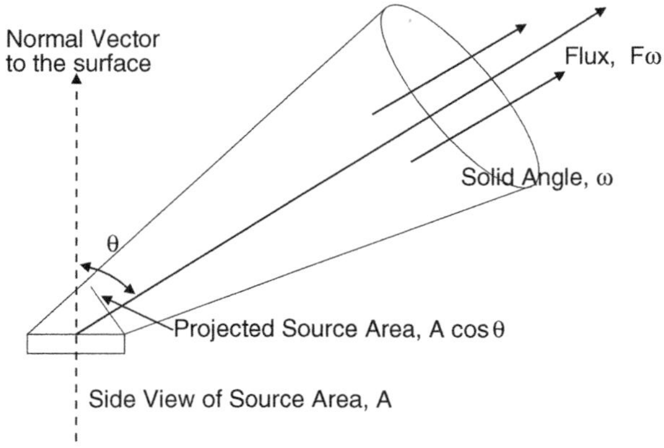

Fig. 2.1. The concept of radiance.

it is called irradiance. If the flux is being emitted by the surface it is called emittance or exitance.

Solid angle: The solid angle ω, subtended by an area on a spherical surface is equal to the area divided by the square of the radius of the sphere (steradians).

Radiance: The radiance is the radiant flux per unit solid angle leaving an extended source in a given direction per unit projected area in the direction (W/steradian-m^2) (Fig. 2.1). If the radiance does not change with the direction of emission, the source is called Lambertian.

The theory of scattering and absorption of electromagnetic radiation in the atmosphere is well developed (Van de Hulst, 1957; Junge, 1963; Deirmendjian, 1969; McCartney, 1977; Bohren and Huffman, 1983; Barber and Hill, 1990, etc.). Thus only an outline of this topic is considered here. In this chapter, the analytical relationships between atmospheric scattering parameters and the corresponding light scattering intensity are primarily discussed. Details of the scattering process depend significantly on the wavelength and the width of the spectral interval (band) of the light. When a light source emitting over a wide range of wavelengths is used, more complicated methods must be applied to obtain estimates of the resulting light scattering intensity (see, for example, Goody and Yung, 1989; Liou, 1992; or Stephens, 1994). These methods generally involve complex numerical calculations (MODTRAN, for example) rather than analytical formulas. This dramatically complicates the analysis of the relationships between the various scattering parameters and the intensity of the scattering light. This difficulty is not encountered when a narrow band light source, such as a laser, is used.

Although exceptions exist, most lidars use a laser source with a narrow wavelength band (as narrow as 10^{-7} nm). Because of this, lidars are considered to be monochromatic sources of light so that simple formulations for the scat-

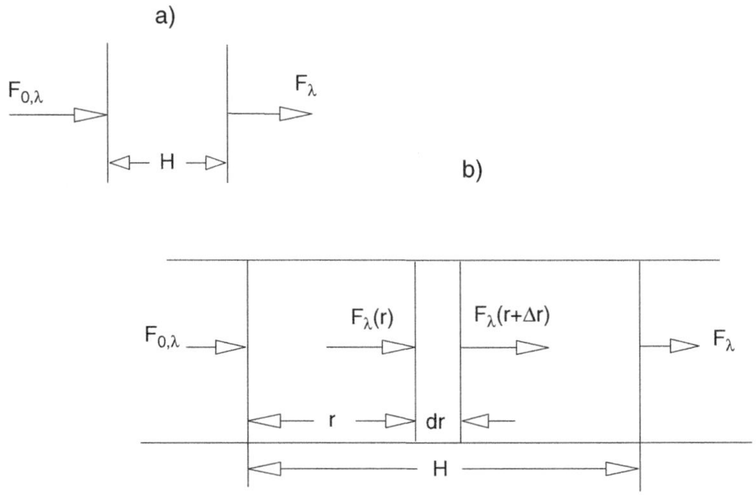

Fig. 2.2. The propagation of light through a turbid layer.

tering characteristics can be applied. There are circumstances when the finite bandwidth of the laser emitter must be considered [for example, in some differential-absorption lidars (DIAL) or high-spectral-resolution lidars], but they are the exception. For nearly all applications, considering the laser to be monochromatic is a simple, yet effective approach for lidar data processing. This approximation is assumed in the discussion to follow. These single wavelength theories must be used with care over wider ranges of wavelengths.

When light scattering occurs, a portion of the incoming light beam is dissipated in all directions with an intensity that varies with the angle between the incoming light and the scattered light. The intensity of the scattering in a given angle depends on physical characteristics of the scatterers within the scattering volume. Similarly, the intensity of light absorption depends on presence of the atmospheric absorbers, such as carbonaceous particulates, water vapor, or ozone, along the path of the emitted light. Unlike scattering, the light absorption process results in a change in the internal energy of the gaseous or particulate absorbers.

Figure 2.2 illustrates how light interacts with a scattering and/or absorbing atmospheric medium. A narrow parallel light beam travels through a turbid layer with geometric thickness H (Fig. 2.2 (a)). Because the intensity of both scattering and absorption depends on the light wavelength, the quantities in the formulas below are functions of the wavelength of the radiant flux, λ. The radiant flux of the beam is $F_{0,\lambda}$ as it enters the layer H. After the light has passed through the layer, it decreases to the value F_λ, such that $F_\lambda < F_{0,\lambda}$. The ratio of these values, $F_\lambda/F_{0,\lambda}$, defines the optical transparency T of the layer H. The transparency describes the fraction of the original radiant (or luminous) flux that passed through the layer. Thus, the ratio

$$T(H) = \frac{F_\lambda}{F_{0,\lambda}} \tag{2.1}$$

is defined to be the transmittance of the layer H. The transmittance is a measure of turbidity of a layer that may range in value from 0 to 1. The transmittance of a layer is equal to 0 if no portion of the light passes through the layer H. Transmittance $T(H) = 1$ for a medium in which no scattering or absorption occurs. The particular value of the transmittance depends on the depth of the layer H and its turbidity, which, in turn, depend on the number and the size of the scattering and absorption centers within the layer.

To establish the relationship for the transmittance of a heterogeneous medium, a differential element dr located within the layer H is defined at a range r from the left edge (Fig. 2.2 (b)). A monochromatic beam of collimated light of wavelength λ with a radiant flux $F_\lambda(r)$ enters dr at the left edge of the element. Defining $\kappa_{t,\lambda}(r)$ to be the probability per unit path length that a photon will be removed from the beam (i.e., either scattered or absorbed), then the reduction in the radiant flux in the differential element is $dF_\lambda(r)$ and is equal to

$$dF_\lambda(r) = -\kappa_{t,\lambda}(r)F_\lambda(r)dr \tag{2.2}$$

After dividing both the parts of Eq. (2.2) by $F_\lambda(r)$ and integrating both sides of the equation in the limits from 0 to H, one obtains Beer's law (often referred to as the Beer–Lambert-Bouger's law), which describes the total extinction of the collimated light beam in a turbid heterogeneous medium:

$$F_\lambda = F_{0,\lambda} e^{-\int_0^H \kappa_{t,\lambda}(r)dr} \tag{2.3}$$

The transmittance of a layer of thickness H can be written as

$$T(H) = e^{-\int_0^H \kappa_t(r)dr} \tag{2.4}$$

where the subscript λ is omitted for simplicity and with the understanding that this applies to narrow spectral widths. In the above formulas, $\kappa_t(r)$ is the extinction coefficient of the scattering or absorbing medium. In the general case, the removal of light energy from a beam in a turbid atmosphere may take place because of the following factors: (1) scattering and absorption of the light energy by the aerosol particles, such as water droplets, mist spray, or airborne dust; (2) scattering of the light energy by molecules of atmospheric gases, such as nitrogen or oxygen; and (3) absorption of the light energy by molecules of atmospheric gases, such as ozone or water vapor. For most lidar applications, the contributions of such processes as fluorescence or inelastic (Raman) scattering are small, so that the extinction coefficient is basically the sum of two major contributions, the elastic scattering coefficient β and the absorption coefficient κ_A:

$$\kappa_t(r) = \beta(r) + \kappa_A(r) \tag{2.5}$$

The light extinction of the collimated light beam after passing through a turbid layer of depth H depends on the integral in the exponent of Eq. (2.4):

$$\tau = \int_0^H \kappa_t(r)\mathrm{d}r \tag{2.6}$$

which is defined to be the optical depth of layer (0, H).

For a collimated light beam, the optical depth of the layer, rather than its physical depth, H, determines the amount of light removed from the beam as it passes through the layer.

Taking into account the theorem of mean, one can reduce Eq. (2.6) into the form

$$\tau = \bar{\kappa}_t H \tag{2.7}$$

where $\bar{\kappa}_t$ is the mean extinction coefficient of the layer H, determined as

$$\bar{\kappa}_t = \frac{1}{H}\int_0^H \kappa_t(r)\mathrm{d}r \tag{2.8}$$

In a homogeneous atmosphere $\kappa_t(r) = \kappa_t = const$; thus for any range r, Eq. (2.7) reduces to

$$\tau(r) = \kappa_t r \tag{2.9}$$

Note that if the range r is equal to unity, the extinction coefficient κ_t is numerically equal to the optical depth τ [Eq. (2.9)]. The extinction coefficient shows how much light energy is lost per unit path length (commonly a distance of 1 m or 1 km) because of light scattering and/or light absorption. With $\kappa_t = const.$, the formula for total transmittance [Eq. (2.4)] reduces to

$$T(r) = e^{-\kappa_t r} \tag{2.10}$$

Equation (2.3) is the attenuation formula for a parallel light beam. However, any real light source emits or reemits a divergent light beam. This observation is valid both for the propagation of a collimated laser light beam and for light scattering by particles and molecules. Collimating the light beam with any optical system may reduce the beam divergence. Therefore, when determining the total attenuation of the light, the additional attenuation of the light energy due to the divergence of the light beam should be considered. In other words, when a real divergent light beam passes the turbid layer, an attenuation of the light energy occurs because of both the extinction by the atmospheric particles and molecules and the divergence of the light beam. Thus the true transport equation for light is more complicated than that given in Eq. (2.3). Fortunately, in such situations, a useful approximation known as the point source of light may generally be used. Any real finite-size light source can be considered as a "point source" of light if the distance between the source and the photoreceiver is much larger than the geometric size of the light source. For such a point source of light, the amount of light captured by a remote light detector is inversely proportional to square of the range from the source location to the detector and directly proportional to the total transmittance over the range. The light entering the receiver from a distant point source of the light obeys Allard's law:

$$\mathrm{E}(r) = \frac{IT}{r^2} = \frac{I}{r^2} e^{-\int_0^r \kappa_t(r')dr'} \tag{2.11}$$

where $E(r)$ is the irradiance (or light illuminance) at range r from the point light source, and I is the radiant (or luminous) intensity of the light energy source.

2.2. TOTAL AND DIRECTIONAL ELASTIC SCATTERING OF THE LIGHT BEAM

When a narrow light beam passes through a volume filled by gas molecules or particulates, light scattering occurs. Scattering theory states that the scattering is caused by the difference between the refractive indexes of the molecular and particulate scatterers and the refractive indexes of the ambient medium (see Section 2.3). During the scattering process, the illuminated particulate reemits some fraction of the incident light energy in all the directions. Thus, in the scattering process, the particulate or molecule acts as a point source of the reemitted light energy.

Accordingly, some portion of the light beam is dissipated in all directions. The intensity of the angular scattering depends on the angle between the scattering direction and that of the original light beam and on the physical characteristics of the scatterers within the scattering volume. For any particular set of scatterers, the scattered light is uniquely correlated with the scattering angle. Let us consider basic formulas for the intensity of a directional scatter-

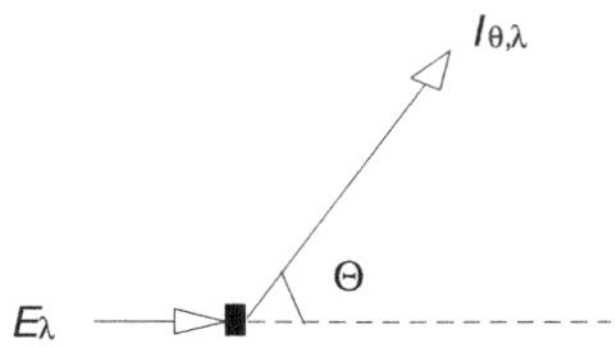

Fig. 2.3. Directional scattering of the light beam.

ing when a narrow light beam of wavelength λ propagates over a differential volume. The radiant spectral intensity of light with wavelength λ, scattered per unit volume in the direction of θ relative to the direction of the incident light (Fig. 2.3) is proportional to spectral irradiance E_λ and a directional scattering coefficient for scattering angle θ:

$$I_{\theta,\lambda} = \beta_{\theta,\lambda} E_\lambda \tag{2.12}$$

The directional scattering coefficient $\beta_{\theta,\lambda}$ determines the intensity of light scattering in the direction θ. In the above formula, the coefficient is normalized over the unit of the length and on the unit solid angle; thus its dimension is ($cm^{-1}\,sr^{-1}$) or ($m^{-1}\,sr^{-1}$) for the unit volume 1 cm^3 or 1 m^3, respectively. In general case, the scattered light may have a number of sources. First, it may include molecular and particulate elastic scattering constituents, which have the same wavelength λ as the incident light. Second, under specific conditions, resonance scattering may occur with no change in wavelength. Third, the scattered light may have additional spectral constituents, such as a Raman or fluorescence constituent, in which wavelengths are shifted relative to that of the incident light λ (Measures, 1984). In this section, only the first elastic scattering constituent is considered. Let us consider a purely scattering atmosphere, assuming that no light absorption takes place so that the light extinction occurs only because of scattering. The total radiant flux scattered per unit volume over all solid angles can be derived as the integral of Eq. (2.12). Omitting the index λ for simplicity, one can write the equation for the total flux as

$$F_{(4\pi)} = \int_0^{4\pi} I_\theta d\omega = \beta E, \tag{2.13}$$

where

$$\beta = \int_0^{4\pi} \beta_\theta d\omega \tag{2.14}$$

is the total volume scattering coefficient.

The angular dependence of the scattered light on the angle θ is defined by the phase function P'_θ. The phase function is formally defined as the ratio of the energy scattered per unit solid angle in the direction θ to the mean energy per unit solid angle scattered over all directions (Van der Hulst, 1957; McCartney, 1977). The latter is equal $\beta/4\pi$ so that the phase function for the not polarized light is defined as

$$P'_\theta = \frac{\beta_\theta}{\beta/4\pi} = \frac{4\pi\beta_\theta}{\int_0^{4\pi} \beta_\theta d\omega} \tag{2.15}$$

It follows from the above equation that P'_θ obeys the constraint

$$\int_0^{4\pi} P'_\theta d\omega = 4\pi \tag{2.16}$$

The angular distribution of scattered light for atmospheric particulates and molecules as a function of their relative size is discussed later. Scattering that occurs from molecules and small-size particulates has approximately the same distribution and scatters light equally in the forward and backward hemispheres. As the particulate radii become larger, they scatter more total energy and a larger fraction of the total in the forward direction as compared to small particulates. Several examples of the angular distribution are shown in the next section.

In the practice of remote sensing, the phase function P_θ is often normalized to 1, so that

$$\int_0^{4\pi} P_\theta d\omega = 1 \tag{2.17}$$

Such a normalization defines the phase function, P_θ, as the ratio of the angular scattering in direction θ to the total scattering:

$$P_\theta = \frac{\beta_\theta}{\beta} \tag{2.18}$$

2.3. LIGHT SCATTERING BY MOLECULES AND PARTICULATES: INELASTIC SCATTERING

A principal feature of the particulate scattering process is that the scattering characteristics are different for different types, sizes, shapes, and compositions of atmospheric particles. What is more, the intensity and the angular shape of the scattering phase function are also dependent on the wavelength of the light.

2.3.1. Index of Refraction

The index of refraction, m, is an important parameter for any scattering or absorbing media. The index of refraction is a complex number in which the real part is the ratio of the phase velocity of electromagnetic field propagation within the medium of interest to that for free space. The imaginary part is related to the ability of the scattering medium to absorb electromagnetic energy. The real part of the index for air can be found from (Edlen, 1953, 1966):

$$10^8(m_s - 1) = 8342.13 + \frac{2406030}{130 - \nu^2} + \frac{15997}{38.9 - \nu^2} \tag{2.19}$$

where m_s is the real part of the refractive index for standard air at temperature T_s = 15°C, pressure P_s = 101.325 kPa, and $\nu = 1/\lambda$, where λ is the wavelength of the illuminating light in micrometers. The effect of temperature and pressure on the refractive index is described by Penndorf (1957):

$$(m - 1) = (m_s - 1)\left(\frac{1 + 0.00367T_s}{1 + 0.00367T}\right)\frac{P}{P_s} \tag{2.20}$$

where m is the real part of the refractive index at temperature T and pressure P. According to Penndorf (1957), water vapor changes the refractive index of air only slightly. For a change of water vapor concentration on the order of that found in the atmosphere, $(m - 1)$ changes less than 0.05 percent.

The variations of the refraction index with wavelength are described in a study by Shettle and Fenn (1979). For the visible and near-infrared portions of the spectrum, the real component of the refractive index varies from 1.35 to 1.6, whereas the imaginary component varies approximately from 0 to 0.1. In clean or rural atmospheres, where the particulates are primarily mineral dust, absorption at the common laser wavelengths is not significant, and the imaginary part is often ignored. However, relatively extreme values may occur in urban particulates having a soot or carbon component for which the corresponding values of the real and imaginary refraction indices at 694 nm are 1.75 and 0.43, respectively. Gillespie and Lindberg (1992a, 1992b), Lindberg and Gillespie (1977), Lindberg and Laude (1974), and Lindberg (1975) have also published a number of papers on the imaginary component of various boundary layer particulates.

2.3.2. Light Scattering by Molecules (Rayleigh Scattering)

If we ignore depolarization effects and the adjustments for temperature and pressure, the molecular angular scattering coefficient at wavelength λ in the direction θ relative to the direction of the incident light can be shown to be

$$\beta_{\theta,m} = \frac{\pi^2(m^2-1)^2 N}{2N_s^2\lambda^4}(1+\cos^2\theta) \tag{2.21}$$

where m is the real part of the index of refraction, N is the number of molecules per unit volume (number density) at the existing pressure and temperature, and N_s is the number density of molecules at standard conditions ($N_s = 2.547 \times 10^{19}\,\text{cm}^{-3}$ at $T_s = 288.15\,\text{K}$ and $P_s = 101.325\,\text{kPa}$). The form of the Rayleigh phase function as $(1 + \cos^2\theta)$ assumes isotropic air molecules.

The amplitude of the scattered light is symmetric about direction of travel of the light beam. For the case of symmetry about one axis, a differential solid angle can be written as

$$d\omega = 2\pi \sin\theta \, d\theta \tag{2.22}$$

where $d\theta$ is a differential plane angle. Integrating over all possible angles, one can obtain the molecular volume scattering coefficient as

$$\beta_m = \int_{\phi=0}^{2\pi}\int_{\theta=0}^{\pi} \beta_{\theta,m} \sin\theta \, d\theta \, d\phi \tag{2.23}$$

and after substituting Eq. (2.21) into Eq. (2.23), the following expression for the molecular volume scattering coefficient can be obtained:

$$\beta_m = \frac{8\pi^3(m^2-1)^2 N}{3N_s^2\lambda^4} \tag{2.24}$$

The intensity of molecular scattering is sensitive to the wavelength of the incident light: the scattering is proportional to λ^{-4}. Therefore, the atmospheric molecular scattering is negligible in the infrared region of the spectrum and dominates scattering in the ultraviolet region. For example, with other conditions being equal, light scattering at wavelength 0.25 µm (the ultraviolet region) differs from that at wavelength 1 µm (the infrared region) by a factor of 256!

The values of m and N in Eq. (2.24) must be adjusted for temperature. Failure to adjust for temperature may lead to errors on the order of 10 percent. With the adjustment for the pressure P and temperature T, the total molecular scattering coefficient at wavelength λ can be shown to be (Penndorf, 1957; Van de Hulst, 1957; McCartney, 1977; Bohren and Huffman, 1983)

$$\beta_m = \frac{8\pi^3(m^2-1)^2 N}{3N_s^2\lambda^4}\left(\frac{6+3\gamma}{6-7\gamma}\right)\left(\frac{P}{P_s}\right)\left(\frac{T_s}{T}\right) \tag{2.25}$$

where γ is the depolarization factor. Published tables over the years (Penndorf, 1957; Elterman, 1968; Hoyt, 1977) have used a number of different values of the depolarization factor, which largely accounts for the differences between them. A discussion of the topic can be found in Young (1980, 1981a, 1981b). The current recommended value is $\gamma = 0.0279$, which includes effects from Raman scattering.

As follows from Eqs. (2.21) and (2.24), the molecular phase function $P_{\theta,m}$, normalized to 1, is

$$P_{\theta,m} = \frac{\beta_{\theta,m}}{\beta_m} = \frac{3}{16\pi}(1+\cos^2\theta) \tag{2.26}$$

From this, it follows that the molecular phase function is symmetric, that is, it has the same value of $3/8\pi$ for backscattered light ($\theta = 180°$) and for the light scattered in forward direction ($\theta = 0°$).

For the atmosphere at sea level, where $N \approx 2.55 \times 10^{19}$ molecules-cm^{-3}, the volume backscattering coefficient at the wavelength λ is given by

$$\beta_m = 1.39\left[\frac{550}{\lambda(\text{nm})}\right]^4 \times 10^{-8}\,\text{cm}^{-1}\text{sr}^{-1}$$

In scattering theory, the concept of a *cross section* is also widely used. For molecular scattering, the cross section defines the amount of scattering due to a single molecule. The molecular cross section σ_m is the ratio

$$\sigma_m = \frac{\beta_m}{N} \tag{2.27}$$

where N is the molecular density. The molecular cross section σ_m specifies the fraction of the incoming energy that is scattered by one molecule in all directions when the molecule is illuminated. The dimensions of the molecular scattering coefficient β_m is inverse range (L^{-1}); the molecular density N has dimension L^{-3}, accordingly, the dimension of the cross section σ_m is L^2. As follows from Eqs. (2.27) and (2.24), the molecular cross section may be presented in the form

$$\sigma_m = \frac{8\pi^3(m^2-1)^2}{3N_s^2\lambda^4} \tag{2.28}$$

The basic characteristics for the molecular scattering may be summarized as follows:

(1) The total and angular molecular scattering intensity is proportional to λ^{-4}. Therefore, atmospheric gases scatter much more light in the ultraviolet region than in the infrared portion of the spectrum. Accordingly,

a clear atmosphere, filled with only gas molecules, is much more transparent for infrared than for ultraviolet light.

(2) The molecular phase function is symmetric. Thus the amount of forward scattering is equal to that in the backward direction.

The type of scattering described in this section, commonly known as Rayleigh scattering, is inherent not only to molecules but also to particulates, for which the radius is small relative to the wavelength of incident light.

2.3.3. Light Scattering by Particulates (Mie Scattering)

As the characteristic sizes of the particulates approach the size of the wavelength of the incident light, the nature of the scattering changes dramatically. For this case, one may visualize the scattering as an interaction between waves that wrap themselves around and through the particle, constructively interfering in some cases, destructively interfering in others. This scattering process is often called Mie scattering after the first to provide a quantitative theoretical explanation (Mie, 1908). In the scattering diagrams to follow, for situations in which the circumference of the particle is a multiple of the wavelength, that is, where the waves constructively interfere as they wrap around the particle, the cross sections are large. For those cases in which the circumference is a multiple of a wavelength and a half, destructive interference occurs and the magnitude of the cross section is a minimum. Although the preceding sentences are true for ideal conducting spheres, real particles are generally not ideal and are not conductors. Because the wave travels through the particle as well as around it, the peaks in the angular scattering are often offset from exact multiples of the wavelength, depending on the magnitude of the index of refraction of the scattering material. For situations in which the size of the particles is much greater than the wavelength, the laws of geometric optics govern.

The laws that govern particulate scattering are quite complex, beyond what is covered here, and they exist only for a limited number of particle shapes. However, there are a number of computer programs that will calculate the cross sections quite easily. The formulas in general use are usually approximations to complex functions, which make it possible to calculate the desired parameters. Thus convergence is an issue, and such programs should be used with care (Bohren and Huffman, 1983). Recognizing that particulates in the atmosphere are always found with some size and composition distribution that is seldom known, one begins to understand the magnitude of the problem of inverting lidar data to obtain information on the size and number of particles present.

The intensity of light scattering by particulates depends upon the particulate characteristics, specifically, the geometric size and shape of the scattering particle, the refractive index of the particle, the wavelength of the incident light, and on the particulate number density. In this section, it is assumed that the scatterers are spherical. This excludes from consideration many common types of particles such as ice crystals or dry dust particles. Formulations do exist for some particulate shapes such as rods and hexagons (for example, Mulnonen et al., 1989; Barber and Hill, 1990; Wang and Van de Hulst, 1995; and Mishchenko et al., 1997), but their use in practical situations is often a challenge. It is also assumed that the incident light is spectrally narrow, similar to the light of a conventional laser. Finally, it is assumed that multiple scattering is negligible and can be ignored.

2.3.4. Monodisperse Scattering Approximation

At first, the simplest case is considered, when the scattering volume under consideration is assumed to be filled uniformly by particles of the same size and composition. These particulates each have the same index of refraction and, thus, scattering properties. Similar to molecular scattering, the total particulate scattering coefficient can be written in the form

$$\beta_p = N_p \sigma_p \tag{2.29}$$

where N_p is the particulate number density and σ_p is the single particle cross section. In particulate scattering theory, two additional dimensionless parameters are defined. The first is the scattering efficiency, Q_{sc}, which is defined as the ratio of particulate scattering cross section σ_p to the geometric cross-sectional area of the scattering particle, i.e.,

$$Q_{sc} = \frac{\sigma_p}{\pi \rho^2} \tag{2.30}$$

where ρ is the particle radius. The second dimensionless parameter is the size parameter ϕ, defined as

$$\phi = \frac{2\pi\rho}{\lambda} \tag{2.31}$$

where λ is the wavelength of the incident light. As follows from Eqs. (2.29) and (2.30), the total particulate scattering coefficient can be written as

$$\beta_p = N_p \pi \rho^2 Q_{sc} \tag{2.32}$$

In Fig. 2.4, the dependence of the factor Q_{sc} on size parameter ϕ for four different indexes of refraction, $m = 1.10$, $m = 1.33$, $m = 1.50$, and $m = 1.90$, is shown. The third curve with $m = 1.5$ is typical for a particulate on which little moisture is condensed. The second curve with $m = 1.33$ applies to conditions in which condensation nuclei accumulate large quantities of water, for example, for

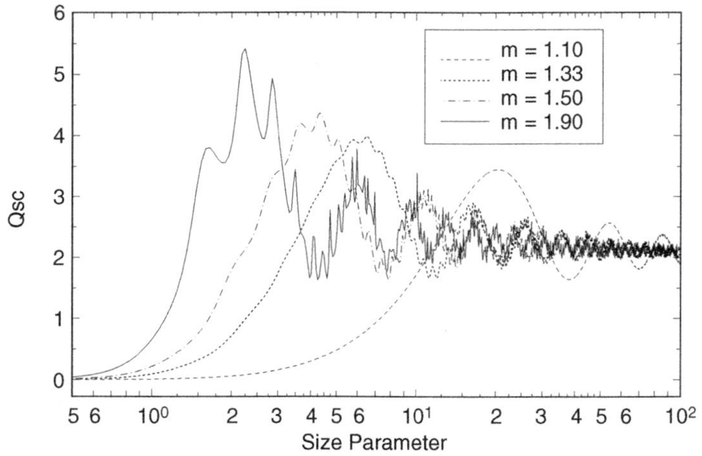

Fig. 2.4. The dependence of particulate scattering factor Q_{sc} on the size parameter ϕ for different indexes of refraction without absorption.

droplets in a fog or cloud. If the size parameter ϕ is small ($\phi < 0.5$), the particulate scattering efficiency is also small. As the parameter ϕ increases, the scattering efficiency factor increases, reaching maximum values of $Q_{sc} = 4.4$ (for $m = 1.50$) and $Q_{sc} = 4$ (for $m = 1.33$). Then it decreases and oscillates about an asymptotic value of $Q_{sc} = 2$. In the range where $\phi > 40$–50, the efficiency factor Q_{sc} varies only slightly from 2. This type of scattering is inherent to the scattering found in a heavy fog or in a cloud. For these values of the size parameter, the scattering does not depend on the wavelength of incident light. Carlton (1980) suggested a method of using this property to determine cloud properties. Note that Q_{sc} converges to the value of 2 rather than 1. From the definition of the efficiency factor, it follows that the particulate interacts with the incident light over an area twice as large as its physical cross section. A detailed analysis of this effect, which is explained by the laws of refraction, is beyond the scope of this book but may be found in most college-level physics texts.

Thus particulate scattering can be separated into three specific types depending on size parameter ϕ. The first type, where $\phi << 1$, characterizes scattering by small particles, such as those in a clear atmosphere. This type of scattering is somewhat similar to molecular or Rayleigh scattering. The region where $\phi > 40$–50 characterizes scattering by large particles, such as those found in heavy fogs and clouds. The intermediate type, with ϕ between 1 and 25, characterizes scattering by the sizes of particles that are commonly found in the lower parts of the atmosphere.

For sizes $\phi < 0.2$ (i.e., when $\rho < 0.03\lambda$), the molecular and particulate scattering theories yield approximately the same result. According to particulate scattering theory, the cross section of small isotropic particulates converges to an asymptotic relation in which the scattering intensity from small particulates is also proportional to λ^{-4}. Accordingly, small particulates scatter more light in the ultraviolet region than in the infrared range of the spectrum. Just as with molecules, scattering from small particulates is symmetric in the forward and backward hemispheres.

$$\sigma_p = \frac{128\pi^5\rho^6}{3\lambda^4}\left(\frac{m^2-1}{m^2+2}\right)^2$$

As defined on page 32, the angular distribution of scattering, commonly called the phase function, is the amplitude of the scattered light as a function of the scattering angle. This function, which is important in the study of most diffuse scatterers, most notably clouds, is a function of the size parameter ϕ. For small values of the scattering parameter, the angular distribution is symmetric, similar to that for molecular scattering (Fig. 2.5). As the size parameter increases, the fraction of the light scattered in the forward direction increases. For large particles, the scattering at a given angle may change dramatically for relatively small changes in the size of the particle. Figure 2.6 shows details of the angular distribution of scattering and the local peaks, at which scattering is enhanced. However, when scattering occurs from an ensemble of different size particulates in a real finite volume, these peaks are significantly smoothed.

The basic characteristics for particulate scattering in the regions where $\phi > 1$ can be summarized as:

- The amount of scattering in the forward direction is much greater than scattering in the backward direction. As the size parameter ϕ increases, scattering in the forward direction increases.
- The angular dependence of particulate scattering is more complicated than for molecular scattering. As ϕ increases, additional directional lobes of radiation appear.
- Scattering by large particles is relatively insensitive to wavelength compared with molecular or small particulate scattering.

It is often useful to know a simple approximation of the wavelength dependence of atmospheric particulate scattering. The Ångstrom coefficient, u, is a parameter that describes this approximated dependence. This coefficient is defined by the relation

$$\beta_p = \frac{const}{\lambda^u} \tag{2.33}$$

For a real atmosphere, u ranges from $u = 4$ (for purely molecular scattering) to $u = 0$ (for scattering in fogs and clouds). Because u is obtained by an

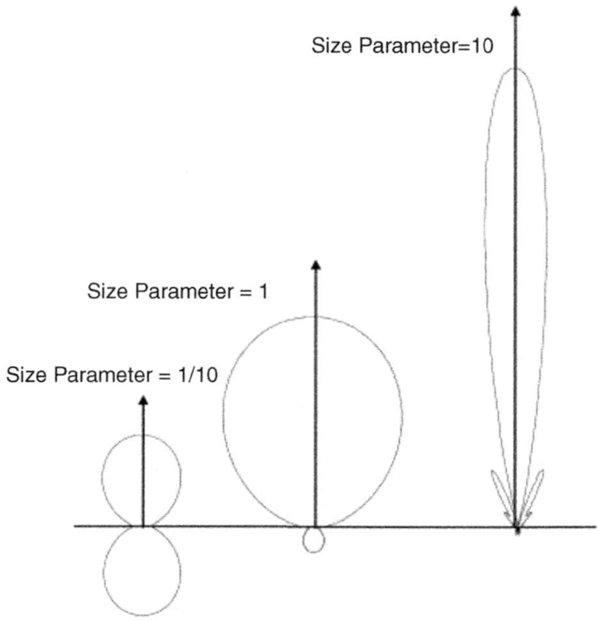

Fig. 2.5. The angular distribution of scattered light intensity for the particles of different sizes for three different size parameters. As the scattering parameter ϕ increases, the scattering in the forward direction also increases in magnitude. The amount of backscattering also increases dramatically, the size of the rightmost distribution has been reduced by a factor of 10,000 to show the shapes of all three parameters.

empirical fit to experimental data rather than derived from scattering theory, the use of a specific value of u is limited to a restricted spectral range or certain atmospheric conditions.

2.3.5. Polydisperse Scattering Systems

The assumption of uniformity in particulate size and composition made above is generally not practical for the real atmosphere. This approximation, however, provides a theoretical basis for the case of the more practical polydispersion scattering. Actually, any extended volume in the atmosphere contains particulates that differ in composition and geometric size. As shown in Table 1.2, the radius of particulates in a clear atmosphere can range from 10^{-4} to 10^{-2} μm, in mist from 0.01 to 1 μm, etc. Therefore, scattering within the real atmospheres always involves a distribution of particulates of different compositions and sizes. No unique particulate distribution exists that is inherent to the atmosphere. To determine the particulate size distribution, it is necessary to make in situ measurements of the total number of scattering particulates with instruments designed for the task. The total number of par-

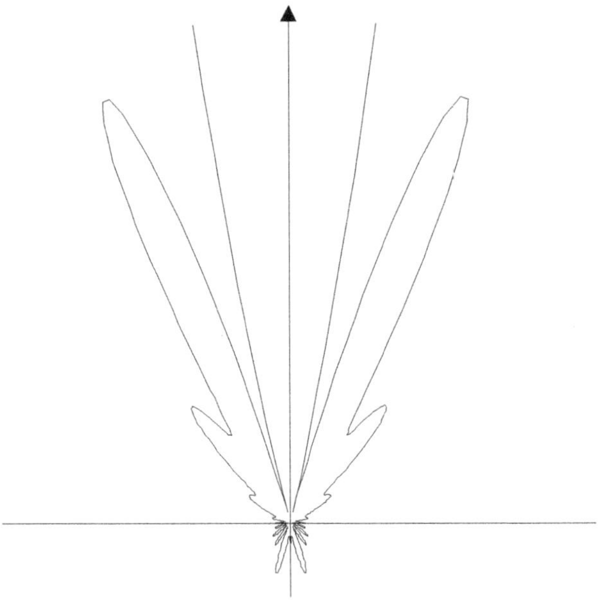

Fig. 2.6. This figure is an enlargement of the angular distribution of scattered light intensity for the particles with a size parameter of 10. The angular distribution of scattered light is complex for particles large with respect to the wavelength of light.

ticles in a unit volume of air may generally be determined as the sum of all scatterers in the volume:

$$N = \sum_{i=1}^{k} N(\rho_i) \tag{2.34}$$

here $N(\rho_i)$ is the number of particulates with radius ρ_i. The total scattering coefficient can be determined as the sum of the appropriate constituents:

$$\beta_p = \sum_{i=1}^{k} N(\rho_i)\pi\rho_i^2 Q_{sc,i} \tag{2.35}$$

In general, the scatterers may have different shapes, but our analysis here is restricted to spherical scatterers. In the general situation, this will not be the case except for water droplets or water-covered particulates (which occur in high relative humidity). Knowing the particulate size distribution, one can determine the attenuation or scattering coefficients through the application of Eq. (2.35). Although any appropriate distribution can be used to approximate a real distribution, a modified gamma distribution or a variant (Junge, 1963;

Deirmendjian, 1969) is often used because of the relative mathematical simplicity. The integral form of Eq. (2.35) for the total scattering coefficient in a polydispersive atmosphere is

$$\beta_p = \int_{\rho_1}^{\rho_2} \pi\rho^2 Q_{sc} \ell_{sc} n(\rho)\, d\rho \tag{2.36}$$

where some sensible radius range from ρ_1 to ρ_2 is used to establish the lower and upper integration limits. In the same manner as for molecular scattering, the relative angular distribution of scattered light from particulates can be described by the particulate phase function $P_{\theta,p}$. Such a phase function, normalized to 1, is defined in the same manner as in Eq. (2.18), i.e.,

$$P_{\theta,p} = \frac{\beta_{\theta,p}}{\beta_p} \tag{2.37}$$

Knowledge of the numerical value and spatial behavior of this parameter in the backscatter direction ($\theta = 180°$) is very important for lidar data processing. In lidar measurements, it is common practice to assume that backscattering is related to the total scattering or extinction. The most commonly used assumption is a linear relationship between the extinction coefficient and the backscatter coefficient (Chapter 5). Such a relationship is not supported by any theoretical analysis based on the Mie theory unless the size distribution and composition of the particulates are constant. On the contrary, the backscatter coefficient, when calculated by Mie theory, is a strongly varying function of the size parameter and indices of refraction. However, in lidar measurements, this variation is reduced considerably where polydispersion of different-size particles is involved (Derr, 1980; Pinnick et al., 1983; Dubinsky et al., 1985). In other words, in real atmospheres, some smoothing of the backscatter-to-extinction ratio occurs. For example, for typical cloud size distributions, the extinction coefficient is a linear function of the backscatter coefficient within an error of ~20%. This dependence is independent of droplet size (Pinnick et al., 1983). The validity of a linear approximation for the relationship between extinction and backscatter coefficients was also shown by calculating these parameters for a wide range of droplet size distribution and in laboratory measurements with a He-Ne laser and polydisperse clouds generated in scattering chambers. Similar results were obtained by Dubinsky et al. (1985). However, further comprehensive investigations revealed that the linear relationship between particulate extinction and backscatter coefficients may take place only in relatively homogeneous media with no significant spatial change of particulate scatterers. This question is considered further in Chapter 7.

The most important characteristics of light scattering by the atmospheric particulates may be simply summarized. All of the basic characteristics of the total and angular scattering depend on the ratio of the particulate radius to the wavelength of incident light rather than on the geometric size of the scattering particle. In other words, the same scattering particulate has a different angular shape and a different intensity of angular and total scattering when illuminated by light of different wavelengths. On the other hand, particulates with different geometric radii ρ_1 and ρ_2 may have identical scattering characteristics if they are illuminated by light beams with the appropriate wavelengths λ_1 and λ_2. As follows from the above analysis, the latter observation is valid if $\rho_1/\lambda_1 = \rho_2/\lambda_2$. Therefore, when particulate scattering characteristics are investigated, any analysis requires that the wavelength of the incident light be taken into consideration. If the size of the scattering particulate is small compared with the wavelength of the incident light, that is, the particulate radius $\rho \le 0.03\lambda$, the scattering is termed Rayleigh scattering. Note that the spectral range that is mostly used in atmospheric lidar measurements includes the near-ultraviolet, visible, and near-infrared range, that is, it extends approximately from 0.248 to 2.1 μm. In this range, Rayleigh scattering occurs for both air molecules and small particles, such as Aitken nuclei. For larger particles with radii $\rho > 0.03\lambda$, light scattering is described by particulate scattering theory. Knowledge of the value and spatial behavior of this parameter in the backscatter direction ($\theta = 180°$) is important for lidar data processing. It is common practice to assume that the backscatter cross section is proportional to the total scattering or extinction. Such a relationship is not obvious from a general theoretical analysis based on Mie theory unless the particulate size distribution remains constant over the examined area and time.

All expressions above are only valid for single scattering, that is, if the effects of multiple scattering are negligible. Single scattering takes place if each photon arriving at the receiver has been scattered only once. For practical application, the approximation of single scattering means that the amount of scattered light of the second, third, etc. order that reaches the receiver is negligibly small in comparison to the single (first order) scattered light.

The influence of multiple scattering depends significantly on the optical characteristics of the atmospheric layer being examined by a remote sensing instrument, on the optical depth of the layer, and on homogeneity of the particulates along the measurement range. The multiple scattering intensity also depends on the diameter and divergence of the light beam, on the wavelength of the emitted light, on the range from the light source to the scattered volume, and on the field of view of the photodetector optics. The rigid formulas to determine the intensity of multiply scattered light are quite complicated and, what is worse, are practical, at best, only for a homogeneous medium.

2.3.6. Inelastic Scattering

Although the dominant mode of molecular scattering in the atmosphere is elastic scattering, commonly called Rayleigh scattering, it is also possible for the incident photons to interact inelastically with the molecules. Raman scattering occurs when the scattered photons are shifted in frequency by an amount that is unique to each molecular species. The Raman scattering cross section depends on the polarizability of the molecules. For polarizable molecules, the incident photon can excite vibrational modes in the molecules, meaning that the molecule is raised to a higher energy state in which its vibrational amplitude is increased. The scattered photons that result when the molecule deexcites have less energy by the amount of the vibrational transition energies. This allows the identification of scattered light from specific molecules in the atmosphere. Two commonly used shifts are $3652\,cm^{-1}$ for water vapor and $2331\,cm^{-1}$ for nitrogen molecules.

The Raman scattering process can be understood in a completely classical sense. The explanation begins with the concept of a dipole moment. When two particles with opposite charges are separated by a distance r, the electric dipole moment p, is given by $p = er$, where e is the magnitude of the charges. As an example, heteronuclear diatomic molecules (such as NO or HCI) must have a permanent electric dipole moment because one atom will always be more electronegative than the other, causing the electron cloud surrounding the molecule to be asymmetric, leading to an effective separation of charge. In contrast, homonuclear diatomic molecules will not have a permanent dipole moment because both nuclei attract the negative elections equally, leading to a symmetric charge distribution.

It is easy to see that a heteronuclear diatomic molecule in an excited state will oscillate at a particular frequency. When this happens, the molecular dipole moment will also oscillate about its equilibrium value as the two atoms move back and forth. This oscillating dipole will absorb energy from an external oscillating electric field if the field also oscillates at precisely the same frequency. The energy of a typical vibrational transition is on the order of a tenth of an electron volt, which means that light in the thermal infrared region of the spectrum will cause vibrational transitions.

However, when an external oscillating electric field with a magnitude of $E = E_0 \sin(2\pi\nu_{ext}t)$, (where E_0 is the amplitude of the wave and ν_{ext} is the frequency of the applied field) is applied to any molecule, a dipole moment p is induced in the molecule. This occurs because the nuclei tend to move in the direction of the applied field and the electrons tend to move in the direction opposite the applied field. The induced dipole will be proportional to the field strength by $p = \alpha E$, where the proportionality constant, α, is called the polarizability of the molecule. All atoms and molecules have a nonzero polarizability even if they have no permanent dipole moment.

For most molecules of interest, the polarizability of a molecule can be assumed to vary linearly with the separation distance, r, between the nuclei as

$$\alpha = \mathrm{a}_0\left(\frac{d\alpha}{dr}\right)\delta r \tag{2.38}$$

where δr is the distance between the nuclei, which for a molecule that is oscillating harmonically is $\delta r = r_0 \sin(2\pi\nu_v t)$, r_0 is the maximum amplitude of the oscillation, and ν_v is the frequency at which the molecule is oscillating before the application of the external electric field. In the presence of an externally applied oscillating electric field, the induced dipole moment p for a linearly polarizable molecule becomes

$$p = \alpha_0 E_0 \sin(2\pi\nu_{ext}) + E_0 r_0 \left(\frac{d\alpha}{dr}\right) \sin(2\pi\nu_{ext} t) \sin(2\pi\nu_v t) \tag{2.39}$$

which can be rewritten as

$$p = \alpha_0 E_0 \sin(2\pi\nu_{ext}) + \left(\frac{E_0}{2}\right) r_0 \left(\frac{d\alpha}{dr}\right) \cos[2\pi(\nu_{ext} - \nu_v)t] + \left(\frac{E_0}{2}\right) r_0 \left(\frac{d\alpha}{dr}\right) \cos[2\pi(\nu_{ext} + \nu_v)t] \tag{2.40}$$

The first term in Eq. (2.40) represents elastic (Rayleigh) scattering, which occurs at the excitation frequency ν_{ext}. The second and third terms represent Raman scattering at the Stokes frequency of $\nu_{ext} - \nu_v$ and the anti-Stokes frequency of $\nu_{ext} + \nu_v$. Thus on each side of the laser frequency there may be emission lines that result from inelastic scattering of photons because of molecular vibrations in the scattering material.

If the internuclear axis of the molecule is oriented at an angle ϕ to the electric field, the result of Eq. (2.40) must be multiplied by $\cos\phi$. Similarly, when the molecule is rotating with respect to the applied field, the dipole moment calculated in Eq. (2.40) must be multiplied by the same $\cos\phi$. Because the molecule is rotating, the angle ϕ changes as $\phi = 2\pi\nu_\phi t$. Multiplying Eq. (2.40) by $\cos(2\pi\nu_\phi t)$ leads to terms with frequencies of ν_{ext}, $\nu_{ext} \pm \nu_v$, $\nu_{ext} \pm \nu_\phi$, $\nu_{ext} + \nu_v \pm \nu_\phi$, and $\nu_{ext} - \nu_v \pm \nu_\phi$. Because there multiple vibrational and rotational states may be populated at any given time, a spectrum of frequencies will occur. The result is shown in Fig. 2.7. The vibrationally shifted lines are successively less intense, generally by an order of magnitude of more. At normal temperatures found on the surface of the earth, there is not sufficient collisional energy to excite molecules to vibrational states above the ground level. Thus anti-Stokes vibrationally shifted lines are seldom observed. Similarly, vibrationally shifted states beyond the first order are sufficiently weak so that they are seldom (if ever) used in lidar work.

2.4. LIGHT ABSORPTION BY MOLECULES AND PARTICULATES

Depending on the wavelength of the incident light, atmospheric particulates and molecules can also act as light-absorbing species. Water vapor, carbon dioxide, ozone, and oxygen are the main atmospheric gases that absorb light energy in the ultraviolet, visual, and infrared regions of spectra. In addition,

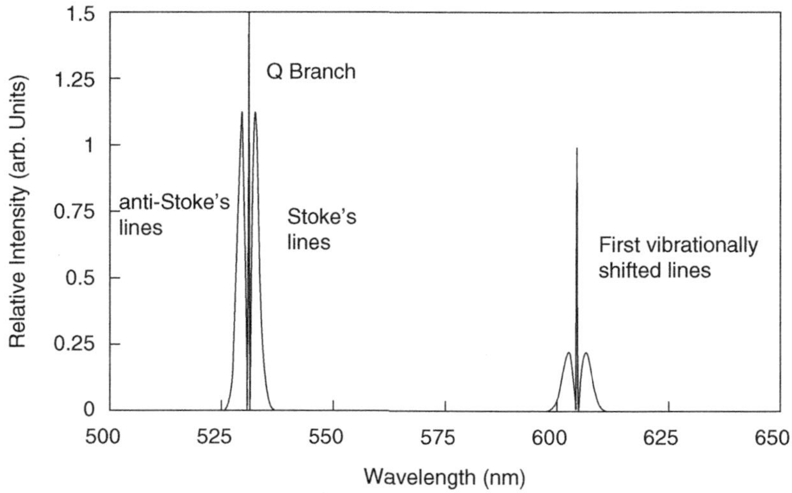

Fig. 2.7. A diagram showing the Raman scattering lines from the 532 laser line. The lines shown centered on 532 nm are purely rotational lines. The lines centered on 609 nm are the same lines but shifted by the energy of the first vibrational state.

trace contaminants such as carbon monoxide, methane, and the oxides of nitrogen are found in the atmosphere that absorb strongly in discrete portions of the spectrum. A major type of lidar, a differential absorption lidar or DIAL, uses these concepts to determine the concentration of various absorbing gases. In this section, we outline the main aspects of atmospheric absorption characteristics, which may be useful for the reader of Chapter 10, in which the determination of the absorbing gas concentration with the differential absorption lidar is discussed.

As shown in the previous section, absorbing particles are characterized by a complex index of refraction m, which is comprised of real and imaginary quantities. The real part is commonly referred to as the index of refraction (the ratio of the speed of light in a vacuum to the speed of light inside the medium), and the imaginary part is related to the absorption properties of the medium. These parameters depend on the particulate type and the wavelength of the incident light. In the troposphere, different types of absorbing particulates are found, such as water and water-soluble particulates, and insoluble particulates, for example, minerals and soot (carbonous).

Figure 2.8 shows effect of the variations in the imaginary part of the index of refraction (which is related to attenuation) on the scattering parameter, Q_{sc}. The graph is given for an index of refraction of 1.33 (i.e., water droplets) and for various values of the complex part of the index. The complex part of the index (the part responsible for absorption or attenuation) can have a large impact on the Q_{sc} factor. Note that the magnitudes of Q_{sc} in Fig. 2.8 are much different than those of Fig. 2.4.

With Mie scattering theory, an expression can be written for the absorption coefficient in a unit volume filled by absorbing species. For the species of

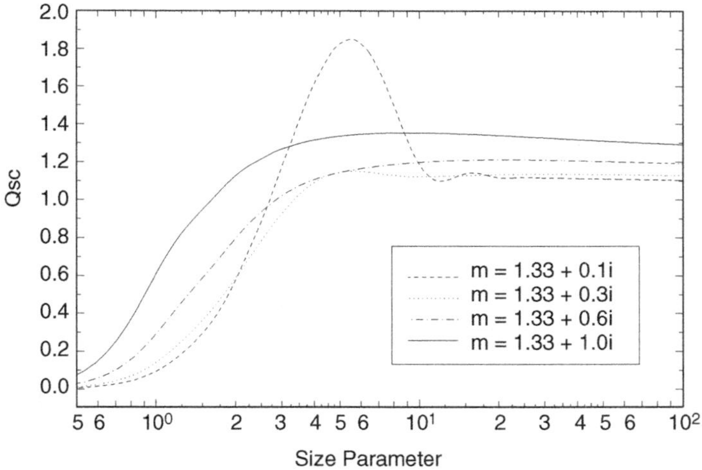

Fig. 2.8. The dependence of particulate scattering factor Q_{sc} for an index of 1.33 (typical of liquid water) with varying values of absorption.

the same size and type, the formula is similar to that for the scattering coefficient [Eq. (2.32)]

$$\kappa_A = N\pi\rho^2 Q_{abs} \tag{2.41}$$

where κ_A is the absorption coefficient, Q_{abs} is the absorption efficiency factor, and N is the number of absorbing particles per unit volume. The absorption efficiency factor is related to the absorption cross section in the same way as the scattering efficiency factor, i.e.,

$$Q_{abs} = \frac{\sigma_A}{\pi\rho^2} \tag{2.42}$$

where σ_A is the absorption cross section of the absorbing particle. The absorption coefficient can be written in terms of the absorption cross section as

$$\kappa_A = \sigma_A N \tag{2.43}$$

The absorption coefficient for a collection of particles of different sizes and types with a radius range from ρ_1 to ρ_2 can be found as

$$\kappa_{A,p} = \int_{\rho_1}^{\rho_2} \pi\rho^2 Q_{abs}(\rho, m) n_A(\rho, m) d\rho \tag{2.44}$$

where $n_A(\rho, m)$ is the number density of the absorbing particles as a function of radius and complex index of refraction, and $Q_{abs}(m)$ is the absorption efficiency factor for the complex index of refraction m.

For the wavelengths normally used by elastic lidars, molecular absorption generally occurs in groups or bands of discrete absorption lines. Most of the common laser wavelengths are not coincident with molecular absorption lines, so that molecular resonance absorption is not an issue. There are exceptions, however. For example, the Ho:YAG laser at 2.1 μm must be tuned to avoid the many water vapor lines found in the region over which it may lase.

There are three main mechanisms by which an electromagnetic wave can be absorbed by a molecule. In order of decreasing energy the mechanisms are electronic transitions, vibrational transitions, and rotational transitions. There are three properties that characterize absorption/emission lines. These are the absorption strength of the line, S, the central position of the line (the most probable wavelength to be absorbed), ν_o, and the shape/width of the line. The central position of an absorption/emission line is a function of the quantum mechanical states of the particular molecule in question. Thus it does not vary for situations that are commonly found in the atmosphere. The strength of the line is the total absorption of the line, or the integral of the line shape. The integral under the shape is constant, regardless of how the line may change shape and width as a function of temperature. The strength of a given line is related to the population density of the beginning and ending states involved in the transition. The population density of a given state is, in turn, related to the temperature of the molecule. Although temperature effects may be a problem for particular applications, comparisons between the strengths of various lines in an absorption band have been used to determine temperature.

The shape and width of absorption and emission lines are functions of several things. First of all, there is a natural lifetime to the excited quantum mechanical state. This lifetime may vary from state to state and from molecule to molecule. By the Heisenberg uncertainty principle, there is a fundamental relationship between the ability to accurately determine both the lifetime and the energy of a given state simultaneously. The product of the uncertainties in time and energy must be greater than $h/2\pi$, which leads to the following conclusion:

$$\Delta t_{\text{lifetime}} \Delta E \approx \frac{h}{2\pi} \quad \Rightarrow \quad \Delta \nu = \frac{1}{2\pi \Delta t_{\text{lifetime}}} \approx \frac{\Delta E}{h} \tag{2.45}$$

In addition to the natural widening of the line because of the finite lifetimes of the states, the lines are also widened by the effects of the Doppler shift of the frequency due to the velocity of the molecules. The Maxwell–Boltzmann distribution function governs the distribution of molecular velocities for a given temperature. The probability that a molecule in a gas at temperature T has a given velocity V in a particular direction is proportional to

$$\exp[-MV^2/2kT] \tag{2.46}$$

where k is the Boltzmann constant, 8.617×10^{-5} eV/degree and M is the mass of the molecule. The shift caused by the motion of an emitter with velocity, V and emissions with frequency, ν_0, is known as the Doppler shift, the magnitude of which is given by

$$\Delta\nu = \pm\frac{V}{c}\nu_0 \tag{2.47}$$

Combining the last two expressions, one can show that the extinction at a given wavelength is related to the peak extinction, κ_{D_0} by

$$\kappa_D(\nu) = \kappa_{D_0}\exp\left[-\frac{Mc^2}{2kT}\left(\frac{\nu-\nu_0}{\nu_0}\right)^2\right] \tag{2.48}$$

which is a Gaussian-shaped distribution with a half-width of

$$\Delta\nu_D = \nu_0\xi\sqrt{\frac{T}{M}} \tag{2.49}$$

where the mass of the molecule M is in gram-atoms and the temperature T is in Kelvin; the quantity ν_0 denotes the centerline frequency, and ξ is a constant (3.58×10^{-7} degree$^{-1/2}$). The shape of the width due to Doppler broadening is Gaussian and is proportional to the square root of temperature and inversely proportional to the square root of the mass of the molecule.

The third mechanism that acts to broaden the spectral absorption lines is collisional or pressure broadening. This type of broadening dominates for most wavelengths and pressures in the lower atmosphere. In this mechanism, it is assumed that the vibrational or rotational state is interrupted by a collision with another molecule. The frequencies of the oscillation before and after the collision are assumed to have no relationship to each other. This acts to greatly reduce the lifetimes of the excited states, and thus increase the width of the lines. Because the amount of shortening is related to the time between collisions, the width will be related to the pressure, P, and temperature of the gas, T. The line shape due to collisional broadening is given by the formula (Bohren and Huffman, 1983; Measures, 1984)

$$\kappa_c(\nu) = \kappa_{c_0}\frac{P}{T}\nu^2\frac{\Delta\nu_c}{(\nu-\nu_0)^2+(\Delta\nu_c)^2} \tag{2.50}$$

where the half-width due to molecular collisions, $\Delta\nu_c$, is also a function of temperature and pressure and is given by

$$\Delta\nu_c = \Delta\nu_{c_0} \frac{P}{P_0}\left(\frac{T_0}{T}\right)^n \tag{2.51}$$

where P_0 and T_0 are the reference pressures and temperatures for collisions $\Delta\nu_{c_0}$. The shape of the absorption lines for collisional broadening is Lorentzian.

For most short-wave radars and visible light, collisional broadening dominates over Doppler broadening. The ratio of the line widths is given approximately as

$$\frac{\Delta\nu_{\text{Doppler}}}{\Delta\nu_{\text{collisional}}} \approx 10^{-12}\frac{\nu_0}{P} \tag{2.52}$$

where ν_0 is in hertz, and P is in millibars. For the region in which the line widths are approximately equal, the total line width is given by $\Delta\nu \approx (\nu_{\text{Doppler}}{}^2 + \nu_{\text{collisonal}}{}^2)^2$. The shape in this region is known as the Voight line shape.

In Section 2.1, the assumption was made that Beer's law of exponential attenuation is valid for both scattering and absorption. For remote sensing measurements, where the concentration of absorbing gases of interest is generally small, such a condition is reasonable and practical. In this case, the dependence of light extinction on the absorption coefficient can be written in the same exponential form as for scattering

$$\frac{F_\nu}{F_{0,\nu}} = e^{-\kappa_A(\nu)r} = e^{-N\sigma_A(\nu)r} \tag{2.53}$$

where N is the number density of absorbing molecules and, for simplicity, the dependence is written for a homogeneous absorption medium. Equation (2.53) is valid under the condition that the absorption cross section $\sigma_A(\nu)$ depends neither on the concentration of the absorbing molecules nor on the intensity of the incident light. The first condition means that every molecule absorbs light energy independently from other molecules. This holds when the concentration of the absorbing molecules is small. An increase in the molecular concentration increases the partial pressure and enhances intermolecular interactions. The increased pressure in the scattering volume can change the molecular cross section, causing a bias in the attenuation calculated by Beer's law. On the other hand, the actual light absorption is less than that determined by Eq. (2.53) if the power density of the incident light becomes larger than approximately $10^7\,\text{Wm}^{-2}$.

Changes in atmospheric pressure can also influence the behavior of the absorption. Atmospheric pressure is caused mainly by nitrogen and oxygen gases. Pressure varies insignificantly for the same altitudes. The partial pressure of all the other gases in the atmosphere is small. Because the total and partial pressure and temperature are correlated with altitude, gas absorption cross sections are different at different altitudes. This effect is quite significant, for example, for the measurement of water vapor concentration. When making the measurement within a gas-absorbing line, one should keep in mind that the parameters of the gas-absorbing line depend on the temperature and total and partial gas pressure and that the lidar-measured extinction is a convolution of the laser line width and the absorption line parameters. Apart from that, in the same spectral interval, a large number of spectral lines generally exist, and their profiles have wide overlapping "wings." To achieve acceptable accuracy in the measurement of the absorption of a particular gas, one must carefully select the best lidar wavelength to use. In practice, this requirement often meets large difficulties.

Measurement of the concentration of gaseous absorbers with the differential absorption lidar (DIAL) is currently the most promising technique for environmental studies. The method works by using the measurement of the absorption coefficient at two adjacent wavelengths for which the absorption cross sections of the gas of interest are significantly different (see Chapter 10).

3

FUNDAMENTALS OF THE LIDAR TECHNIQUE

3.1. INTRODUCTION TO THE LIDAR TECHNIQUE

Lidar is an acronym for light detection and ranging. Lidar systems are laser-based systems that operate on principles similar to that of radar (radio detection and ranging) or sonar (sound navigation and ranging). In the case of lidar, a light pulse is emitted into the atmosphere. Light from the beam is scattered in all directions from molecules and particulates in the atmosphere. A portion of the light is scattered back toward the lidar system. This light is collected by a telescope and focused upon a photodetector that measures the amount of back scattered light as a function of distance from the lidar. This book considers primarily the light that is elastically scattered by the atmosphere, that is, the light that returns at the same wavelength as the emitted light (Raman scattering is discussed in Section 11.1).

Figure 3.1 is a schematic representation of the major components of a lidar system. A lidar consists of the following basic functional blocks: (1) a laser source of short, intense light pulses, (2) a photoreceiver, which collects the backscattered light and converts it into an electrical signal, and (3) a computer/recording system, which digitizes the electrical signal as a function of time (or, equivalently, as a function of the range from the light source) as well as controlling the other basic functions of the system.

Lidars have proven to be useful tools for atmospheric research. In appropriate circumstances, lidars can provide profiles of the volume backscatter

Elastic Lidar: Theory, Practice, and Analysis Methods, by Vladimir A. Kovalev and William E. Eichinger.
ISBN 0-471-20171-5

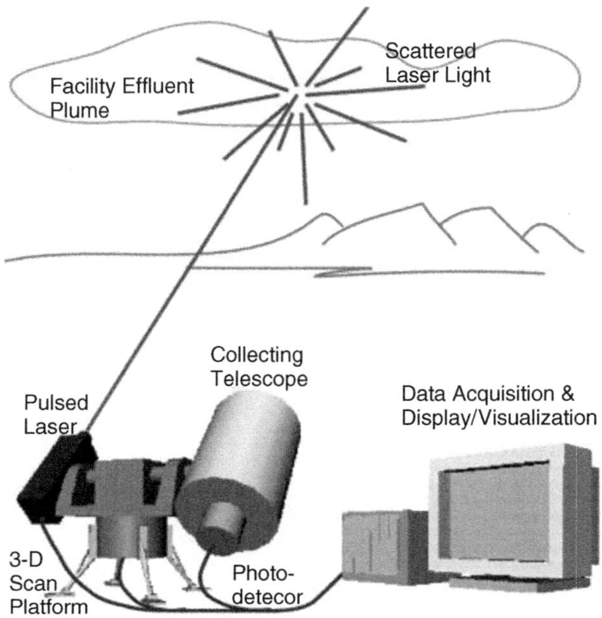

Fig. 3.1. A conceptual drawing of the major parts of a laser radar or lidar system.

coefficient, the volume extinction coefficient, the total extinction integral, and the depolarization ratio that can be interpreted to provide the physical state of the cloud particles or the degree of multiple scattering of radiation in clouds. The altitude of the cloud base, and often the cloud top, can also be measured. Elastic backscatter lidars have been shown to be effective tools for monitoring and mapping the sources, the transport, and the dilution of aerosol plumes over local regions in urban areas, for studies of contrails, boundary layer dynamics, etc. (McElroy and Smith, 1986; Balin and Rasenkov, 1993; Cooper and Eichinger, 1994; Erbrink, 1994). Because of the importance of the impact of clouds on global climate, many studies have been made of the radiative and microphysical properties of clouds as well as their distribution horizontally and vertically. Lidars have played an important role in this effort and have been operated at many different sites throughout the world.

Understanding the physiochemical processes that occur in the atmospheric boundary layer is a necessary requirement for prediction and mitigation of air pollution events. This in turn, requires understanding of the dynamic processes involved. Determination of the relevant parameters, such as the average boundary layer height, wind speeds, and the entrainment rate, is critical to this effort. A description of the boundary layer structure from conventional soundings made twice a day is not sufficient to obtain a thorough understanding of these processes, especially in urban regions. Elastic lidars that can trace the

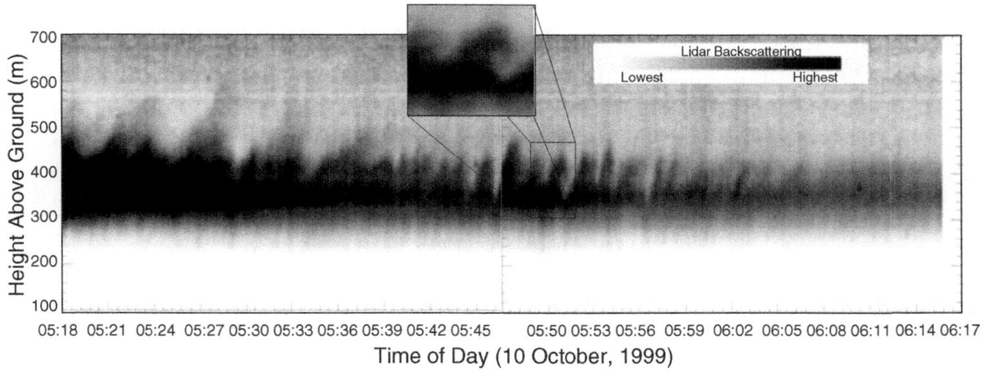

Fig. 3.2. An example of Kelvin–Helmholtz waves detected by a vertically staring lidar during the CASES99 experiment over a period of about an hour. The waves are generated in a thin particulate layer that has a layer of air directly above it which is moving faster than the layer below. This causes waves (similar to water waves) in the denser air mass containing the particulates. The vertical scale has been exaggerated so that the waves might be clearly seen. The inset has one of the waves in approximately equal scale horizontally and vertically. These types of waves are believed to be a cause of intense turbulent bursts in the nighttime boundary layer.

movement of particulates are valuable instruments to support these types of measurements. The varying particulate content of atmospheric structures allows their differentiation so that a wide variety of measurements are possible.

Perhaps the greatest contribution of lidars has been in the visualization of atmospheric processes. In particular, the lidar team at the University of Wisconsin, Madison has made great strides toward making visualization of time-resolved, three-dimensional processes a reality (see, for example, the website at http://lidar.ssec.wisc.edu/). Even lidars that do nothing but stare in the vertical direction can provide time histories of the evolution of processes throughout the depth of the atmospheric boundary layer (the lowest 1–2 km). Figure 3.2 is an example of Kelvin–Helmholtz waves taken over a period of an hour at an altitude of about 400 m. Depending on the wavelength of the laser used, the type of scanning used, and the optical processing done at the back of the telescope, many different types of information can be collected concerning the properties of the atmosphere and the processes that occur as a function of spatial location.

Lidar light pulses are well collimated, so that generally, the beam cross section is less than 1 m in diameter at a distance of 1 km from the lidar. Because of extremely short pulses of the emitted light, the natural spatial resolution offered by lidar systems is many times better than that offered by other atmospheric sensors, for example, radars and sodars. Exceptionally high spatial resolution is a common characteristic of elastic lidars. Because the cross sections for elastic scattering are quite large in comparison to those for other types of scattering, the amount of returning light is comparatively large for an elastic lidar. The result is that elastic lidars can be quite compact and that the time required to scan a volume of space is relatively short. The result is a class of tools that can examine a large volume of space with fine spatial resolution in short periods of time. The possibility exists then of mapping and capturing atmospheric processes as they develop.

The laser light is practically monochromatic. This enables one to use narrow-band optical filters to eliminate interference or unwanted light from other sources, most notably the sun. Such filtering allows significant improvement in the signal-to-noise ratio and, thus, an increase in the lidar measurement range. The maximum useful range of lidar depends on many things but is generally between 1 and 100 km, although most elastic lidar have maximum ranges of less than 10 km.

3.2. LIDAR EQUATION AND ITS CONSTITUENTS

3.2.1. The Single-Scattering Lidar Equation

A schematic of a typical monostatic lidar, one in which the laser and telescope are located in the same place, is presented in Fig. 3.1. A short-pulse laser is used as a transmitter to send a light beam through the atmosphere. The emitted light pulse with intensity F propagates through the atmosphere, where it is attenuated as it travels. At each range element, some fraction of the light that reaches that point is scattered by particulates and molecules in the atmosphere. The scattered light is emitted in all directions relative to the direction of the incident light, with some probability distribution, as described in Section 2.3. Only a small portion of this scattered light, namely, the backscattered light F_{bsc}, reaches the lidar photoreceiver through the light collection optics. The telescope collects the backscattered light and focuses the light on the photodetector, which converts the light to an electrical signal. The analog output signal from the detector is then digitized by the analog-to-digital converter and processed by the computer. The lidar may also contain a scanning assembly of some type that points the laser beam and telescope field of view in a series of desired directions.

In Chapter 2, the backscatter coefficient was defined to be the fraction of the light per unit solid angle scattered at an angle of 180° with respect to the direction of the emitted beam. Light scattering by particulates and molecules in the atmosphere may be divided into two general types: elastic scattering, which has the same wavelength as the emitted laser light, and inelastic scattering, where the wavelength of the reemitted light is shifted compared with emitted light. A typical example of an inelastic scattering process is Raman scattering, in which the wavelength of the scattered light is shifted by a fixed amount. For both types of a scattering, the shape of the backscattered signal in time is correlated to the molecular and particulate concentrations and the extinction profile along the path of the transmitted laser beam.

For a monostatic lidar, the backscattered signal on the photodetector, the total radiant flux F_{bsc}, is the sum of different constituents, namely

$$F_{bsc} = F_{elas,sing} + F_{elas,mult} + \sum F_{inelas} \tag{3.1}$$

where $F_{elas,sing}$ is the elastic, singly backscattered radiant flux, $F_{elas,mult}$ is the elastic multiply scattered radiant flux, and ΣF_{inelas} is the sum of the reemitted radiant fluxes at wavelengths shifted with respect to the wavelength of the emitted light. Note that each of the scattering components is that portion of the scattered light which is emitted in the 180° direction. The intensity of the inelastic component of the backscattered light F_{bsc} is significantly lower (usually several orders of magnitude) than the intensity of the elastically scattered light and can be easily removed from the signal by optical filtering. Some lidar systems derive useful information from the inelastic components of the returning light. Measurement of the frequency-shifted Raman constituents is generally used for atmospheric studies in the upper troposphere and the stratosphere. This topic is examined in Chapter 11. The development that follows here ignores the inelastic component, assuming that it will be eliminated by the appropriate use of filters.

For relatively clear atmospheres, the amount of singly scattered light, $F_{elas,sing}$, is far larger than the multiply scattered component, $F_{elas,mult}$. Only when the atmosphere is highly turbid, the multiple-scattered component becomes important. On the other hand, there is an additional component to the signal not shown in Eq. (3.1) that exists during daylight hours, specifically, the solar background. This component, F_{bgr}, results in a constant shift in the overall flux intensity that may be large in relation to the amplitude of the backscattered light. The signal noise originated by the solar background, F_{bgr}, may be significant. For most daylight situations, the noise will eventually overwhelm the lidar signal at distant ranges and is one of the principal system limitations. The total flux on the photodetector is the sum of these two components:

$$F_{tot} = F_{bsc} + F_{bgr} \tag{3.2}$$

Although some lidar systems derive useful information from the inelastic components of the returning light, generally, the singly backscattered signal, $F_{elas,sing}$, is considered to be the carrier of useful information. All of the other contributions to the signal, including the multiply scattered constituents and the random fluctuations in the background, are considered to be components that distorts the useful information. When lidar measurement data are processed, the backscattered signal is separated from the constant background and then processed as a function of time, which is correlated to the distance

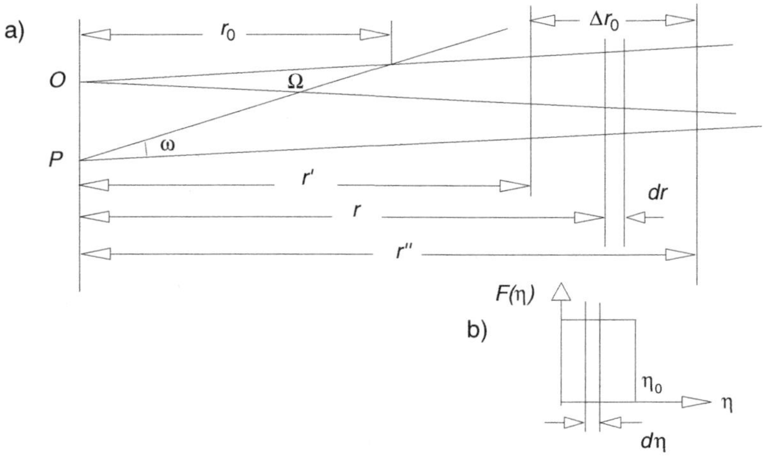

Fig. 3.3. A diagram of the geometry of the processes relevant to the analysis of the light returning from the laser pulse in a lidar.

from the lidar by the velocity of light. Unfortunately, there are no effective ways to suppress either the daylight background noise or the multiple scattering contribution. All of the methods to reduce these effects, such as reducing the field of view of the telescope, the use of narrow-spectral-band filters, the use of lidar wavelengths shifted beyond the most intense parts of the solar spectrum, and increasing laser power, only provide a moderate improvement in suppressing the background contribution to the signal (Section 3.4.2).

In Fig. 3.3 (a), a diagram of the processes along the lidar line of sight is shown. The laser, which emits a short light pulse with a full angle divergence of Ω, is located at the point O, and the photodetector with a field of view subtending the solid angle ω is located alongside of the laser, at point P. The light pulse from the laser has a width in time, η_0 [Fig. 3.3 (b)], which is equivalent to a width in space, Δr_0. In other words, the scattering volume that creates the instantaneous backscattered signal on the photodetector is located in the range from r' to r''. The laser thus illuminates a slightly divergent conical volume of space that is Ωr^2 in cross section, where r is the distance from the laser to the illuminated volume. In practice, the illuminated volume is often considered to be cylindrical and r as the mean distance to the scattering volume, that is, $r = 0.5\,(r' + r'')$. As this illuminated volume propagates through the atmosphere, it scatters light in all directions. Light scattered in the 180° direction is captured by the telescope and transformed to an electric signal by a photodetector. The light intensity at any moment t depends both on the scattering coefficient within the illuminated volume and on transmittance over the distance from the lidar to the scattering volume. Assuming that $t = 0$ when the leading edge of the laser pulse is emitted from the laser, let us consider the input signal on the photodetector at any moment in which $t >> \eta_0$. The scattering volume that creates the backscattered signal on the photodetector at moment t is located in the range from r' to r''. The relationship between the time and the scattering-volume-location range is as follows,

$$2r'' = ct \tag{3.3}$$

and

$$2r' = c(t - \eta_0) \tag{3.4}$$

where c is the speed of light. The light pulse passes along the path from lidar to scattering volume twice, from the laser to the corresponding edge of the scattering volume and then back to the photodetector. Therefore, the factor 2 appears in the left side of both Eq. (3.3) and Eq. (3.4). As follows from Eqs. (3.3) and (3.4), the geometric length of the region from r' to r'', from which the backscattered light reaches the photoreceiver, is related to the emitted pulse duration η_0 as

$$\Delta r_0 = r'' - r' = \frac{c\eta_0}{2} \tag{3.5}$$

Generally speaking, the lidar equation is a conventional angular scattering equation, as described in Chapter 2, for a scattering angle $\theta = 180°$. The instantaneous power in the emitted pulse at moment $d\eta$ is $F(\eta) = dW/d\eta$, where W is radiant energy in the laser beam and the time $d\eta$ corresponds to the scattering volume in dr at distance r from the lidar [Fig. 3.3 (b)]. The radiant flux at the photodetector, created by the molecular and particulate elastic scattering within volume of depth, dr, is determined by

$$dF_{\mathrm{elas,sing}} = C_1 F(\eta) \frac{\beta_{\pi,\mathrm{p}}(r) + \beta_{\pi,\mathrm{m}}(r)}{r^2} \exp\left\{-2\int_0^r [\kappa_\mathrm{p}(x) + \kappa_\mathrm{m}(x)]dx\right\} dr \tag{3.6}$$

where $\beta_{\pi,\mathrm{p}}$ and $\beta_{\pi,\mathrm{m}}$ are the particulate and molecular angular scattering coefficients in the direction $\theta = 180°$ relative to the direction of the emitted light; κ_p and κ_m are the particulate and molecular extinction coefficients. $F(\eta)$ is the radiant flux emitted by the laser. C_1 is a system constant, containing all system constants that depend on the transmitter and receiver optics collection aperture, on the diameter of the emitted light beam, and on the diameter of the receiver optics. The exponential term in the equation is defined to be the two-way transmittance of the distance from lidar to the scattering volume

$$[T(0,r)]^2 = e^{-2\int_0^r \kappa_t(x)dx} \tag{3.7}$$

here κ_t is the total (particulate and molecular) extinction coefficient.

Because the emitted pulse duration is always a small finite value, the backscattered input light at the photoreceiver at any time t is related to the properties of a relatively small volume of the atmosphere between r' and $r'' = r' + \Delta r_0$. Therefore, the total radiant flux at the photodetector at time t is created by the scattering inside the entire volume of the length Δr_0

$$F_{\text{elas,sing}} = C_1 \int_{r'}^{r'+\Delta r_0} \left\{ F(\eta) \frac{\beta_{\pi,p}(r) + \beta_{\pi,m}(r)}{r^2} \exp\left[-2\int_0^r \kappa_t(x)dx\right] \right\} dr \tag{3.8}$$

The length of the emitted pulse in time, normally on the order of 10 ns, depends on the type of laser used and varies in the range from a few nanoseconds to microseconds. The use of a long-pulse laser, which emits light pulses of long duration (on the order of microseconds), complicates lidar data processing and reduces the spatial resolution of the lidar so that the minimum size that can be resolved by the system is much larger. Attempts to resolve distances smaller than the effective pulse length of the lidar are discussed in Section 3.4.4.

Assuming that the laser emits short light pulses of rectangular form (i.e., that $F(\eta) = F = const.$), and that the attenuation and backscattering coefficients are invariant over Δr_0, an approximate form of Eq. (3.8) may be obtained for times much longer than the pulse length of the laser. This equation, generally referred to as the *lidar equation*, is written in the form

$$F(r) = C_1 F \frac{c\eta_0}{2} \frac{\beta_{\pi,p}(r) + \beta_{\pi,m}(r)}{r^2} \exp\left[-2\int_0^r \kappa_t(x)dx\right] \tag{3.9}$$

The subscript that indicates that the equation is valid for singly and elastically scattered light is omitted for simplicity.

Note that the approximate form of the lidar equation in Eq. (3.9) assumes that the pulse spatial range Δr_0 is so short that the term in the rectangular brackets of Eq. (3.8) can be considered to be constant. This can only be valid under the following conditions:

(1) All of the atmospheric parameters related to backscattering must be constant within the spatial range of the pulse, $\Delta r_0 = c\eta_0/2$. This requirement, equivalent to assuming that the number density and composition of the particulates in the scattering volume are constant, must be true at every range r within the lidar operating range. In practice this requirement may be reduced to the requirement of the absence of sharp changes in the particulate properties over the range Δr_0.

(2) The equation is applied to a distant range r, in which $r >> \Delta r_0$ so that the difference between the square of both ranges, i.e., between r^2 and $(r + \Delta r_0)^2$, is inconsequential, and

(3) The optical depth of the range Δr_0 is small within the lidar operating range, i.e.,

$$\int_{r'}^{r'+\Delta r} \kappa_t(x)\mathrm{d}x \leq 0.005 \tag{3.10}$$

This requirement is caused by the presence of the second integral in the exponent of Eq. (3.8). The transformation of Eq. (3.8) into Eq. (3.9) is only valid when the integral in the exponent of Eq. (3.8) can be assumed to be constant in the range of integration from r' to $r' + \Delta r$. If this requirement is neglected in conditions of strong attenuation, the convolution error may exceed 5%.

(4) In the lidar operating range, the field of view (FOV) of the photodetector optics must be larger than the laser beam divergence so that the lidar "sees" the entire illuminated volume. This means that the atmospheric volume being examined must be at a range greater than r_0, where r_0 is the range at which the collimated laser beam has completely entered the FOV of the telescope [Fig. 3.3 (a)]. The range up to r_0 is often defined as the lidar incomplete-overlap zone (Measures, 1984). Section 3.4.1 discusses the lidar overlap problem.

The instantaneous power $P(r)$ of the analog signal at the lidar photodetector output created by the singly scattered, elastic radiant flux $F(r)$ at range $r > r_0$ can be obtained by transforming Eq. (3.9) into the form

$$P(r) = \gamma_{an} F(r) = C_0 \frac{\beta_\pi(r)}{r^2} \exp\left[-2\int_0^r \kappa_t(x)\mathrm{d}x\right] \tag{3.11}$$

where γ_{an} is the conversion factor between the radiant flux $F(r)$ at the photodetector and the power $P(r)$ of the output electrical signal; $\beta_\pi(r)$ is the total (i.e., molecular and particulate) backscattering coefficient, and $\kappa_t(r)$ is the total extinction coefficient. The factor C_0 is the lidar system constant, which can be written as

$$C_0 = C_1 F_0 \frac{c\eta_0}{2} \gamma_{an}$$

One of the implications of this expression is a rule of thumb that lidar capability should be compared on the basis of the product of the laser energy per pulse, and the area of the receiving optics, sometimes called the power-aperture product. In other words, the energy per pulse of the laser can be reduced by a factor of four if the telescope diameter is doubled. A corollary to this rule of thumb is that the maximum range of the lidar varies approximately as the square root of the power aperture product. In practice, the range resolution of a lidar is also influenced by properties of the digitizer and other electronics used in the system.

On a fundamental level, the best range resolution that can be achieved by a lidar is a function of the length of the laser pulse and the time between digitizer measurements. Because the lidar pulse has some physical size, about 3 m for a typical q-switched laser pulse of 10 ns, the signal that is received by the lidar at any instant is an average over the spatial length of the pulse. This 3-m-long pulse will travel some distance between measurements made by the digitizer. For a given time between digitizer measurements, η_d, the distance the pulse travels is $c\eta_d/2$. The total distance that has been illuminated between digitizer measurements is thus $c(\eta_0 + \eta_d/2)$, where η_0 is the time length of the laser pulse. Historically (with the exception of CO_2 lasers with pulse lengths longer than 200 ns), the detector digitization rates and electronics bandwidth have been the limiting factors in range resolution. In an effort to improve the signal-to-noise ratio, the bandwidth of the electronics is often reduced or limited by a low-pass filter. The range resolution is also limited by the electronics bandwidth. For a perfect noiseless system, the digitization rate should be twice the detector electronics bandwidth. However, real systems with noise require sampling rates several times faster than this to reliably detect a signal. It follows that the real range resolution is limited to perhaps five times the distance determined by the digitization rate, $c\eta_d/2$. The effect of limited bandwidth on range resolution is complex and beyond the scope of this text. To our knowledge, it has not been dealt with in any detail in the literature. It is probably fair to say that most lidar systems in use today using analog digitization are limited by the bandwidth of the detectors and electronics. Spatial averaging that is used to reduce noise also limits the range resolution in ways that are dependent on the details of the smoothing technique used. A good discussion of basic filtering techniques and the creation of filters is given by Kaiser and Reed (1977).

A number of difficulties must be overcome to obtain useful quantitative data from lidar returns. As follows from Eq. (3.11), the measured power $P(r)$ at each range r depends on several atmospheric and lidar system parameters. These parameters include the following: (1) the sum of the molecular and particulate backscattering coefficients at the range r, (2) the two-way transmittance or the mean extinction coefficient in the range from $r = 0$ to r, and (3) the lidar constant C_0. Thus, in the above general form, the lidar equation includes more than one unknown for each range element. Therefore, it is considered to be mathematically ill posed and thus indeterminate. Such an equation cannot be solved without either a priori assumptions about atmospheric properties along the lidar line of sight or the use of independent measurements of the unknown atmospheric parameters. Unfortunately, the use of independent measurement data for the lidar signal inversion is rather challenging, so that the use of a priori assumptions is the most common method.

It is of some interest to consider attempts to use lidar remote sensing along with the use of appropriate additional information. The study made by Frejafon et al. (1998) is a good example of what can be accompished. In the study, a 1-month lidar measurement of urban aerosols was combined with a size distribution analysis of the particulates using scanning electron microscopy and X-ray microanalysis. Such a combination made it possible to perform simultaneous retrieval of the size distribution, composition, and spatial and temporal dynamics of aerosol concentration. The procedure of extracting information on atmospheric characteristics with the lidar was as follows. First, urban aerosols were sampled with standard filter technique. To check the spatial variability of the size distribution, 30 volunteers carried special transportable pumps in places of interest and took sampling. The sizes of the particulates were determined with scanning electron microscopy and counting. In addition, the atomic composition of each type of particles was found by X-ray microanalysis. These data were used to compute the backscattering and extinction coefficients, leaving as the only unknown parameter the particulate concentration along the lidar line of sight. Mie theory was used to determine backscattering and extinction coefficients for the smooth silica particles. The lidar data were inverted with the backscattering and extinction coefficients computed from the actual size distribution.

Even under these conditions, several additional assumptions were required to invert the lidar data. First, they assumed that the particulate size distribution is homogeneous over the measurement field. This hypothesis is, generally, much more appropriate for horizontal than for slant and vertical directions. To overcome this problem, it would be more appropriate to sample particles at several altitudes. Unfortunately, this is unrealistic in practice. Second, it was assumed that the water droplets can be neglected because of the low relative humidity during the experiment. Thus the described method can be applied only in dry atmospheres. The third approximation was in the application of spherical Mie theory to unknown particle shapes, which may be nonspherical, especially in dry atmospheres. The authors of this study believe that this disparity introduces no significant errors.

Two optical parameters can potentially be extracted from elastic lidar data, the backscatter and extinction coefficients. As follows from the lidar equation, the elastic lidar signal is primarily a function of the combined molecular and particulate backscatter cross section with a relatively small contribution from the extinction coefficient. This is especially true for clear and moderately turbid atmospheres. Consider the effect of a 10 percent change in both parameters over the distance of one range bin. A 10 percent change in the backscatter coefficient changes the signal by 10 percent. A 10 percent change in the extinction coefficient over a typical range bin of 5 m changes the magnitude of the signal by a factor that is not measurable. Unfortunately, as pointed out by Spinhirne et al. (1980), the backscatter cross section is not a fundamental parameter that can be directly used in atmospheric transfer studies. Although it is intuitive that backscatter is in some way related to the extinction coefficient, determining the extinction coefficient from the backscattered quantities is always fraught with difficulty. Despite this, some studies (Waggoner et al., 1972; Grams et al., 1974; Spinhirne et al., 1980) have used backscatter measurements to infer an aerosol absorption factor.

Generally, the extinction coefficient profile is the parameter of primary interest to the researcher. The extinction cross section is a fundamental parameter often used in radiative transfer models of the atmosphere. Basic aerosol characteristics such as number density or mass concentration are also more directly correlated to the extinction than the backscatter. The basic problem of extracting the extinction coefficient from the lidar signal is related to significant spatial variation in the particulate composition and size distribution, particularly in the lower troposphere. Therefore, a range-dependent backscatter coefficient should be used to extract accurate scattering characteristics of atmospheric particulates from the lidar equation. This greatly complicates the solution of the lidar equation. A potential way to overcome this difficulty might be to make independent measurements of backscattering along the line of sight of the elastic lidar. This can be achieved by the use of a combined Raman-elastic backscatter lidar method, proposed by Mitchenkov and Solodukhin in 1990. In spite of difficulties associated with small scattering cross-sections of inelastic scattering as compared to that of elastic scattering, such systems are now widely implemented in practice (Ansmann, et al., 1992 and 1992a; Müller et al., 2000; Mattis et al., 2002; Behrendt et al., 2002).

To extract the extinction coefficient values along the lidar line of sight, the calibration factor C_0, relating the return signal power $P(r)$ to the scattering, must also be known. The absolute calibration of the lidar system is quite complicated. What is more, it determines only one constant factor in the lidar equation, whereas in practice, an additional factor appears in the lidar equation. As mentioned above, a part of the lidar operating range exists, located close to the lidar, in which the collimated laser beam has not completely entered the FOV of the receiving telescope (Fig. 3.3). That part of the lidar signal that can be used for accurate data processing is limited to distances beyond this area, that is, in the zone of the complete lidar overlap, $r \geq r_0$. Setting the minimum range of the complete lidar overlap, r_0, as the minimum measurement range of the lidar is most practical. Therefore, the conventional form of the lidar equation, used for elastic lidar data processing, includes the transmission term over the range $(0, r_0)$ separately. With the corresponding change of the lower limit of the integral in Eq. (3.11), the equation is now written as

$$P(r) = C_0 T_0^2 \frac{\beta_\pi(r)}{r^2} \exp\left[-2\int_{r_0}^{r} \kappa_t(x)\mathrm{d}x\right] \tag{3.12}$$

where r_0 is the minimum range for the complete lidar overlap and T_0 is the total atmospheric transmittance of the zone of incomplete overlap, that is

$$T_0 = e^{-\int_0^{r_0} \kappa_t(x)\mathrm{d}x} \tag{3.13}$$

Thus transmittance of the overlap range from $r = 0$ to r_0 is also an unknown parameter, which must be somehow estimated to find the exponent term in Eq. (3.12). It is shown in Chapter 5 that to extract the extinction coefficient from the lidar return the product $C_0T_0^2$ must be determined as a boundary value rather than these two constituents separately.

Even the simplified lidar equation given in Eq. (3.12) requires special methodologies and fairly complicated algorithms to extract the extinction coefficients or related parameters from the recorded signal. The principal difficulty in obtaining reliable measurements is related to both the spatial variability of atmospheric properties and the indeterminate nature of the lidar equation.

3.2.2. The Multiple-Scattering Lidar Equation

In many applications, lidar data processing may be accomplished with acceptable accuracy by using the single-scattering approximation given in Eq. (3.12). However, in optically dense media, such as fogs and clouds, the effects of multiple scattering can significantly influence measurements, so that the single-scattering approximation leads to severe errors in the quantities derived from lidar signals. Unfortunately, this is one of the significant, not-well-solved problems in the field of radiation transport. A large collection of literature exists on the subject. The problem is considered here only to outline the issue and methods of mitigating its effects.

The origin of the effects of multiple scattering is easily understood as an effect of turbid media (Fig. 3.4). Various optical parameters influence the intensity of multiply scattered light. First, the intensity of multiple-scattered light depends on the properties of the scattering medium itself, such as the size and distribution of the scattering particles, and on the optical depth of the atmosphere between the scattering volume and the lidar. As the particles become larger, more light is scattered in all directions, but especially in the forward direction. In the development of the lidar equation in Section 3.2.1, we assumed that this light that was scattered in the forward direction was small enough and can be ignored. However, in a turbid medium, the amount of the forward-scattered light becomes a significant compared with the amount of light directly emitted by the laser and thus cannot be ignored. This additional light increases backscattering in comparison to that caused only by single scattering of the light from the laser beam. If the effect of multiply scattered light is ignored, the increased light return, for example, from inside the cloud makes the calculated extinction coefficient of the scattering medium be less than it actually is.

The intensity of multiply scattered light depends significantly on the lidar measurement geometry. The amount of multiply scattered light increases dramatically with increasing laser beam divergence, the receivers field of view, and the distance between the lidar and scattering volume. For example, if the lidar system is situated at a long distance from the cloud, as would be the case for a space-based lidar system, the amount of multiple scattering could be extremely high, even for a small penetration range in the cloud (Starkov et al., 1995). Thus the measurement of the single-scattering component from clouds often can be quite complicated or even impossible.

The multiple-scattering contribution to the return signal has been estimated in many comprehensive theoretical studies, for example, in studies by Liou and Schotland (1971), Samokhvalov (1979), Eloranta and Shipley (1982),

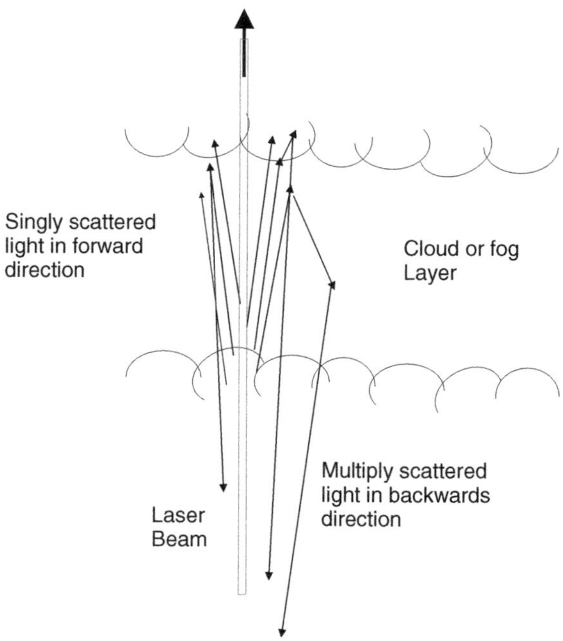

Fig. 3.4. A diagram showing the origins of multiple scattering. In an optically dense medium, both the fraction and absolute amount of light that is scattered in the forward direction become large. Some fraction of this forward-scattered light is scattered again, partly back toward the lidar. The intensity of this backscattered light may become a significant fraction of the total intensity of backscattered light collected by the lidar.

Bissonnette and Hutt (1995), Bissonnette (1996), and Krekov and Krekova (1998). These studies show that the various scattering order constituents are different for different optical depths into the scattering medium. When the optical depth τ of the scattering medium is less than about 0.8, single scattering generally prevails. This is true under the condition that a typical (somewhat optimal) lidar optical geometry is used. At an optical depth of ~0.8–1, the reflected signal consists primarily of first-order scattering with only a small contribution from second-order scattering. When the optical depth is equal or slightly higher than 1, the multiple-scattering contribution to the total return signal becomes comparable with that from single scattering. For the larger optical depths the amount of multiple scattering increases, and it becomes the dominant factor at optical depths of 2 and higher. Generally, these estimates are the same for both fog and cloud measurements, when no significant scattering gradients occur, but are highly dependent on the field of view of the lidar system.

Because of the high optical density of clouds, these became the first media in which the effects of multiple scattering in the lidar returns were investigated, beginning in the early 1970s. Two basic effects caused by multiple scattering may be used for the analysis of this phenomenon. The first effect is the change in the relative weight of the multiple-scattering component with the change of the receiver's field of view. This effect is caused by the spread of the forward-propagating beam of light because of multiple scattering. Accordingly, a segmented receiver that can detect the amount of backscattered light as a function of the angular field of view of the telescope can be used to detect the presence of and relative intensity due to multiple scattering. The second opportunity to investigate multiple scattering arises from lidar light depolarization in the cloud. Depolarization of the linearly polarized light from the laser occurs when the scattering of the second and higher orders takes place. Both of these effects have been thoroughly investigated by lidar researchers. Allen and Platt (1977) investigated the effects of multiple scattering with a center-blocked field stop, whereas Pal and Carswell (1978) demonstrated the presence of a multiple-scattering component in the lidar signal by detection of a cross-polarized component in the returning light. Both of these effects were also demonstrated in the study by Sassen and Petrilla (1986). In 1990s, special lidars were built to make experimental investigations of multiple scattering effects. Bissonnette and Hutt (1990), Hutt et al. (1994), Eloranta (1988), and Bissonnette et al. (2002) reported on the backscatter lidar measurement made at different receiver fields of view simultaneously. The authors concluded that not only is multiple scattering measurable but it can yield additional data on aerosol properties. By observing multiple scattering, the authors attempted to measure the extinction and the particle sizes. In Germany, Werner et al. (1992) investigated these multiple-scattering effects with a coaxial lidar.

Unfortunately, despite the huge amount of potentially valuable information contained in the multiple-scattering component, such measurements are difficult to interpret accurately. A large number of studies have been published concerning the extraction of information on multiple scattering from lidar signals. The simplest method to obtain this kind of information was based on the use of analytical models of doubly scattered lidar returns. Such an approach assumes the truncation of the multiple-scattering constituents to the second scattering order (see, for example, Eloranta, 1972; Kaul and Samokhvalov, 1975; Samokhvalov, 1979). After these initial efforts, during the 1980s much more sophisticated methods were developed. Detailed discussion and analysis of these methods is beyond the scope of this text. Here only an outline of the general methods is given to provide the reader some knowledge of the basic principles and models used in multiple-scattering studies.

Generally, the lidar multiple-scattering models that currently exist have two different applications. First, they may be used to estimate likely errors in lidar measurements caused by the single-scattering approximation used in data processing. A working knowledge of the amount of multiple scattering is very helpful when estimating the accuracy of the parameter of interest determined with the single-scattering approximation. For this use, even approximate multiple-scattering estimates are often acceptable. For example, it is a common practice to introduce a multiplicative correction factor into the transmission term of the lidar equation when investigating the properties of thin clouds or other inhomogeneous layering (Platt, 1979; Sassen et al., 1992; Young, 1995). This is done to reduce the extinction term in the lidar equation toward its true value (see Chapter 8). Different models can also be applied to lidar measurements of multiple scattering to infer information about the characteristics of the scattering media. Here the requirements for the models are much more rigorous. Moreover, model comparisons generally reveal that even small differences in the models or in the initial assumptions can yield significant differences in the estimates of the scattering parameters. In 1995, the international cooperation group, MUSCLE (multiple-scattering lidar experiments), organized an annual workshop, where such a comparison was made for seven different models of calculations (Bissonnette et al., 1995). The approaches included Monte Carlo simulations using different variance-reduction methods (Bruscaglioni et al., 1995; Starkov et al., 1995; Winker and Poole, 1995) and some analytical models based on radiative transfer or the Mie theory (Flesia and Schwendimann, 1995; Zege et al., 1995). In particular, Bissonnette et al. (1995) used the so-called radiative-transfer model in a paraxial-diffusion approximation. Flesia and Schwendimann (1995) applied extended Mie theory. In their approach, the spherical wave scattered by the first particle was considered as the field influencing the second one, and this procedure was repeated at all scattering orders. Starkov et al. (1995) used the Monte Carlo technique, which allowed a comparison of the transport-theoretical approach with a stochastic model, and Zege et al. (1995) presented a simplified semianalytical solution to the radiative-transfer equations. To compare the methods, all participants were to calculate the lidar returns for the same specified 300-m-thick cloud with some established particle size distribution, using the same assumed lidar instrument geometry. The comparison revealed that Monte Carlo calculations generally compared well with each other. Moreover, the study confirmed that some analytical models, such as that used by Zege et al. (1995), produced results in close agreement with Monte Carlo calculations. However, as summarized later in a study by Nicolas et al. (1997), a restricted number of inversion methods exist that can handle the problem of calculating multiple scattering with good accuracy and efficiency. These methods are invaluable when making different theoretical simulations and numerical experiments. On the other hand, these methods are, generally, complex and not enough reliable for the inverse problem to directly retrieve cloud properties from measured lidar data.

One should note the existence of inversion methods based on the so-called "phenomenological representation of the scattering processes" published in a study by Bissonnette and Hutt (1995) and later by Bissonnette (1996). A simplified formulation of a multiple-scattering equation was proposed that is explicitly dependent on the range-dependent extinction coefficient and on an effective diameter, d_{eff} of the scattering particles. It is assumed that the aerosols are large compared with the wavelength of the laser light, so that the size parameter $\pi d_{eff}/\lambda$ (see Chapter 2) is large enough for diffraction effects to make up half of the extinction contribution. The second assumption is that the multiply scatteied photons within a small field of view originate mainly from the forward diffraction peak and from backscattering near 180°. The remaining wide-angle scattering is assumed to be small enough that it can be ignored. However, for the near-forward direction, all of the contributing scatterings are taken into consideration, except those at the angles close to 180°. A variant of such a method was tested in two field experiments, in which the cloud microphysical parameters were independently measured with in situ sensors (Bissonnette and Hutt, 1995).

The first way used to overcome the complexity of the estimates for multiple scattering was to "correct" in some way the single-component lidar equation. The purpose of such a correction was to expand the application of the single-scattering lidar equation for the measurements in which the multiple scattering cannot be ignored. Platt (1973, 1979) proposed a simple extension of the single-scattering equation for cirrus cloud measurements. After making combined measurements of the clouds by lidar and infrared radiometer, he established that the presence of the multiple scattering produces a systematic shift in the measurement data obtained with the single-scattering lidar equation. As mentioned above, multiple scattering is additive. It causes more of the scattered light to return to the receiver optics aperture than for a single-scattering atmosphere. This effectively reduces the calculated optical depth at large distances if single-scattering Eq. (3.12) is used. Although this is mostly inherent in measurements of thick clouds, this effect also influences measurement accuracy in thin clouds. To avoid the necessity of using complicated formulas to determine the amount of multiple scattering, Platt proposed to include an additional factor when calculating optical depth of clouds examined by lidar. His approach was as follows. If the actual optical depth of the layer between cloud base h_b and height h is $\tau(h_b, h)$, and the effective optical depth obtained from the lidar return with the single-scattering approximation is $\tau_{eff}(h_b, h)$, then a multiple-scattering factor may be defined as

$$\eta(h_b, h) = \frac{\tau_{eff}(h_b, h)}{\tau(h_b, h)} \tag{3.14}$$

where the factor $\eta(h_b, h)$ has a value less than unity. After that, in all of the lidar equation transformations, one can replace the term $\tau_{eff}(h_b, h)$ with the product $[\eta(h_b, h)\tau(h_b, h)]$. This is in some ways a questionable procedure, but it may produce meaningful information. For example, the procedure is reasonable when one investigates a particular problem other than multiple scattering, but the optical medium under investigation is sufficiently turbid so that the multiple-scattering contribution cannot be ignored (Del Guasta, 1993; Young, 1995). Obviously, this factor may vary as the light pulse penetrates into the cloud, and the optical depth $\tau(h_b, h)$ increases. However, only the assumption that $\eta(h_b, h) = \eta = const.$ is practical in application. The parameter η for cirrus was estimated first by Platt (1973) to be $\eta = 0.41 \pm 0.15$. This value is related to the backscatter-to-extinction ratio, and therefore, the latter also must be in some way estimated (Platt, 1979; Sassen et al., 1989; Sassen and Cho, 1992).

The study of cirrus clouds with lidar technique dates back to the development of the first practical lidar systems. The reason for this was that cirrus clouds significantly contribute to the earth's radiation balance. However, there is no general agreement concerning the influence of the cirrus clouds on the climate. As shown, for example, in studies by Cox (1971) and by Liou (1986), clouds can produce either a warming or a cooling effect, depending on their microphysical and optical properties. The very first lidar studies of the cirrus clouds revealed the significant contribution of the multiple-scattering component in the lidar returns. This effect, which significantly complicates the interpretation of lidar signals, causes researchers to pay serious attention to the general problem of multiple scattering.

The seeming simplicity of the use of a variant of the single-scattering equation for the multiple-scattering medium makes it attractive to use such an approach for lidar data processing. The difficulty is that the required correction factor, has no simple, direct relationship with the properties of the cloud. The errors in the correction factor may cause large uncertainties in the resulting inversion of the lidar data. To have some physical basis on which to develop such a variant, some approximations must be made to extend the single-scattering equation to situations in which multiple scattering may be important. The assumptions that are generally made concern the relative amounts of forward and backward scattering. Alternately, some typical phase function shape in the forward and backward directions is assumed for the particulate scatterers. In Platt's (1973) modification, the single-scattering lidar equation is applied with the assumption that the phase function is, approximately, constant about the angle π. The assumption of a smooth phase function in the backward direction and a sharp peak in the forward direction is the most common approach (for example, Zuev et al., 1976; Zege et al., 1995; Bissonnette, 1996; Nicolas et al., 1997). When considering the problem of strongly peaked forward scattering in cirrus clouds, most researchers base the estimate of the parameter η as dependent on the forward phase function of the cloud. Some authors apply the single-scattering approximation in the intermediate regime between single and diffuse scattering. In this approximation, it is assumed that the total scattering consists of single large-angle scattering in the backward direction, which is followed by multiple small-angle forward scattering. Such an approximation may be valid for visible and near-infrared lidar measurements in clouds. Because of the presence of large particles in the clouds with a size parameter much greater than 1, the effective phase function has a strong peak in the forward direction. Following the study by Zege et al. (1995), the authors of the study by Nicolas et al. (1997) derived a multiple-scattering lidar equation in the limit of a uniform backscattering phase function. This makes it possible to obtain a formal derivation of η for the regime in which the field-of-view dependence of the multiple scattering reaches a plateau. The parameter η is established as a characteristic of the forward peak of the phase function, and it is taken as independent of the field of view and range.

Formally, for optical depths greater than approximately 1, the multiple-scattering equation may be reduced to the single-scattering equation by using the so-called "effective" parameters. In the most general form, the multiple-scattering equation for remote cloud measurement can be written with such effective parameters as (Nicolas et al., 1997)

$$P(r) = C_o \frac{\beta_{\pi,\mathrm{eff}}(r)}{(r_b + r)^2} T^2(0, r_b + r) T_p^2(0, r_b) \exp[-2\tau_{p,\mathrm{eff}}(r)] \tag{3.15}$$

where r_b is the range to the cloud base and r is the penetration depth in the cloud. $T^2(0, r_b + r)$ is the transmission over the path from the lidar to the range $(r_b + r)$ that accounts for the total (molecular and particular) absorption and molecular scattering, that is,

$$T(0, r_b + r) = \exp\left[-\int_0^{r_b+r} [\kappa_A(r') + \beta_m(r')] dr'\right] \tag{3.16}$$

Two path transmission terms remaining in Eq. (3.15), $T_p(0, r_b)$, and $\exp[-2\tau_{p,\mathrm{eff}}(r)]$, define the particulate scattering constituents. $T_p(0, r_b)$ is the path transmission over the range from $r = 0$ to r_b, which accounts for the particular scattering up to the cloud base, that is,

$$T_p(0, r_b) = \exp\left[-\int_0^{r_b} \beta_p(r')dr'\right] \tag{3.17}$$

and $\tau_{p,eff}(r)$ is the effective scattering optical depth within the cloud, that is, over the range from r_b to $(r_b + r)$, which is the product of two terms

$$\tau_{p,eff}(r) = \eta \int_{r_b}^{r_b+r} \beta_p(r')dr' \tag{3.18}$$

where $\beta_p(r)$ is the particulate scattering within the cloud. The effective backscattering coefficient $\beta_{\pi,eff}(r)$ in Eq. (3.15), introduced in the study by Nicolas et al. (1997), is related to the field of view of the lidar. Clearly, the practical value of such a parameter depends on how variable the phase function is over the range and what its shape is near the π direction. There is a question as to whether it can be used, for example, for the investigation of high-altitude clouds, where the presence of ice crystals is quite likely. Here the shape of the backscattering phase function is strongly related to the details of the ice crystal shape, and no estimate of $\beta_{\pi,eff}(r)$ is reliable (Van de Hulst, 1957; Make, 1993).

In studies by Bissonnette and Roy (2000) and Bissonnette et al. (2002), another transformation of the single-scattering equation is proposed. Unlike the correction factor, η introduced by Platt (1973) into the exponent of the transmission term of the lidar equation. Here a multiple-scattering correction factor, $M(r, \theta)$, related to the multiple-to-single scattering ratio, is introduced as an additional factor for the backscattering term. As shown in studies by Kovalev (2003a) and Kovalev et al. (2003), such a transformation allows one to obtain a simple analytical solution to invert the lidar signal that contains multiple scattering components. In these studies, two variants of a brink solution are proposed for the inversion of signals from dense smokes. Under appropriate conditions, the brink solution does not require an a priori selection of the smoke-particulate phase function in the optically dense smokes under investigation. However the solution requires either the knowledge of the profile of the multiple-to-single scattering ratio (e.g., determined experimentally with a multiangle lidar), or the use of an analytical dependence between the smoke optical depth and the ratio. In the latter case, an iterative technique is used.

The use of additional information on the scattering properties of the atmosphere may be helpful in the evaluation of multiple scattering. High-spectral-resolution and Raman lidars, which allow measurements of the cross section profiles (see Chapter 11), can provide such useful information. The opportunities offered by these instruments to improve our understanding of multiple scattering are discussed in the study by Eloranta (1998). The author proposed a model for the calculation of multiple scattering based on the scattering cross section and phase function specified as a function of range. Such an approach has a great deal of merit. Nevertheless, the problem of multiple-scattering evaluation remains a quite difficult problem, and there is no suggestion that it will soon be solved. To help to the reader to form an idea of how complicated the problem is, even when the additional information is available, one can give the list of the assumptions used by Eloranta (1998) for the applied model. The model assumes (1) a Gaussian dependence of the phase function on the scattering angle in the forward peak, (2) a backscatter phase function that is isotropic near the π direction, (3) a Gaussian distribution of the laser beam within the divergence angle, (4) multiply scattered photons at the receiver have encountered only one large-angle scattering event, (5) the extra path length caused by the small-angle deflections is negligible, and therefore the multiple- and single-scattered returns are not shifted in time, and (6) the receiving optics angle is small so that the transverse section of the receiver field of view is much less than the photon free path in the cloud. Apart from that, the question also remains of how instrumental inaccuracies influence the signal inversion accuracy when the inverted signal is strongly attenuated. As shown by Wandinger (1998), the information obtained by Raman instrumental systems may also be distorted by multiple scattering. The model calculations of Wandinger (1998) revealed that the different shape of the molecular and particulate phase functions causes different influence on multiple scattering in the molecular and particulate backscatter signals. The intensity of multiple scattering is generally larger in the molecular backscatter returns than in the particulate backscatter return. The estimates of multiple scattering in water and ice clouds revealed that in Raman measurements the largest errors may occur at the cloud base. This error may be as large as ~50%. It was established also that extinction and backscattering measurements have different error behavior. The estimates made for the ground lidar system showed that the extinction coefficient measurement error decreases with increasing penetration depth, whereas the error in the backscatter coefficient increases.

To summarize the previous discussion, many optical situations occur in which the contributions of multiple scattering cannot be ignored. Unfortunately, there are no simple, reliable models available for lidar data processing when multiple and single scattering become comparable in magnitude. Comparisons between the different models for processing such lidar data have shown that the problem is far from being solved, even although the models may often show good agreement. The comparisons also revealed that large systematic disagreements may occur between the models themselves. The basic reason is that higher-order scattering depends unpredictably on a large number of local and path-integrated particulate parameters and on the geometry of the lidar system. Obviously, it is very difficult, or perhaps even impossible, to reproduce all aspects of the multiple-scattering problem with uniform accuracy. Multiple scattering is a difficult problem, one for which, at the present time, there is no clear way to determine which model and solution are the best (Bissonnette et al., 1995).

3.3. ELASTIC LIDAR HARDWARE

3.3.1. Typical Lidar Hardware

We consider first the most typical type of elastic lidar system used for atmospheric studies. In particular, we will follow the light from the emission in the laser through collection and digitization. The miniature lidar system of the University of Iowa (Fig. 3.5) will be used as an example of one approach to engineering a lidar system. More sophisticated systems exist and offer certain advantages in accuracy or range, but this is achieved at the cost of size, portability, and price.

The light source used is a Nd:YAG laser operating at a wavelength of 1.064 μm. A doubling crystal in the laser allows the option of using 0.532 μm as the lidar operating wavelength. The pulse is 10 ns long with a beam divergence of approximately 3 mrad. The laser pulse energy is a maximum of 125 mJ with a repetition rate of 50 Hz. Because the length of the laser pulse is

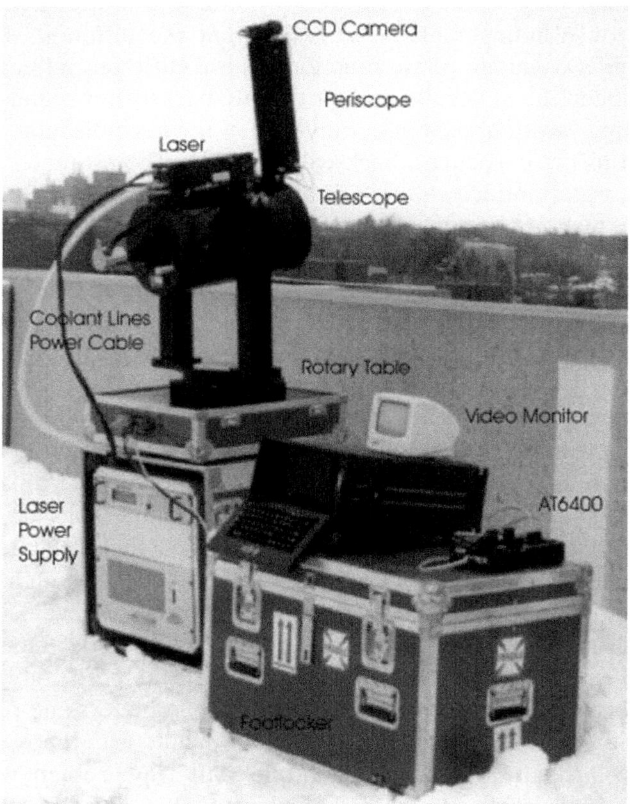

Fig. 3.5. The lidar set up in a typical data collection mode. The major components are labeled.

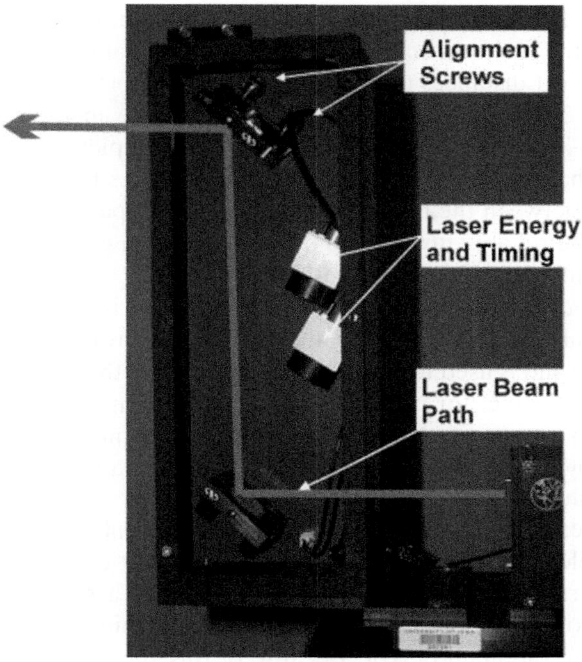

Fig. 3.6. Photograph of the periscope showing the mirrors and detectors inside. This is normally covered for eye safety reasons and to keep dust away from the mirrors.

one of the parameters that sets the minimum range resolution for a lidar, q-switched lasers with pulse lengths of 5–20 ns are normally used. (CO_2 lasers are one notable exception, having pulse lengths on the order of 250 ns for the main part of the pulse).

Light from the laser enters the periscope (Fig. 3.6), where it is reflected twice before exiting the periscope. The laser beam is emitted parallel to the axis of the receiving telescope at a distance of 41 cm from the center of the telescope. The periscope serves two functions. The first is to make the process of aligning the axes of the laser beam and telescope field of view simpler. The upper mirror shown in the figure is used for the alignment. The second function is related to reducing the dynamic range of the lidar receiver. Because the intensity of the light captured by the telescope is inversely proportional to the square of the distance r from the lidar [Eq. (3.12)], the difference in the intensity of the light between short and far distances is large and increases dramatically at very short distances (see Fig. 3.8a). Large variations in the magnitude of the intensity of the returning light in the same signal may become a design issue in that they require that the light detector, signal amplifier, and digitizer have large dynamic ranges. To minimize the problem, one can increase the distance at which the telescope images the entire laser beam, that is, increase the distance to complete overlap [in Fig. 3.3(a), this distance is marked as r_0]. Because both the telescope and laser have narrow divergences (typically on the order of milliradians), the laser beam is not "seen" by the telescope at short distances (see, for example, the short-range portions of the signal in Fig. 3.8). The application of the periscope in the miniature lidar system makes it possible to obtain distances of incomplete overlap from 50 to 400 m. Only that portion of the lidar signal that comes from the area of complete overlap between the field of view of the telescope and the laser beam ($r > 400$ m) can be reliably inverted to obtain extinction coefficient profiles (see Section 3.4.1 for more details of the overlap issue).

Two small detectors are mounted inside the periscope. These detectors detect the small amount of light scattered by the mirrors. One detector has a 1.064-μm filter and is used to measure the intensity of the outgoing laser pulse. This is used to correct for pulse-to-pulse variations in the laser energy when the lidar data are processed. The second detector has no filter and simply produces a fast signal of large amplitude that is used as a timing marker to start the digitization process.

The receiver telescope is a 25-cm, f/10, commercial Cassegrain telescope. Cassegrain telescopes are often used because they can be constructed to provide moderate *f*-numbers in a compact design. A Cassegrain telescope uses a second mirror to reflect the light focused by the main mirror back to a hole in the center of the main mirror. Because of this, the length of the telescope is half that of a comparable Newtonian telescope. The light is focused to the rear of the telescope, where it passes through a 3-nm-wide interference filter and two lenses that focus the light onto a 3-mm, IR-enhanced silicon avalanche photodiode (APD) (Fig. 3.7). An iris located just before the APD serves as a stop to limit the field of view of the telescope. Opening the iris allows light from near ranges to reach the detector. Closing the iris limits the telescope field of view (important in turbid conditions or clouds) and makes the location of complete overlap farther out, limiting the magnitude of the near field signal. This will allow the use of more gain in the electronics or more laser power so that a longer maximum range may be achieved. The characteristics of avalanche photodiodes allow a relatively noise-free gain of up to 10 inside the diode itself. Basic parameters of the transmitter and receiver of the miniature lidar system of the University of Iowa are given in Table 3.1.

A high-bandwidth (60 MHz) amplifier is located inside the detector housing. The signal is amplified and fed to a 100-MHz, 12-bit digitizer on an IBM PC-compatible data bus. A portable computer is used to control the system and to take the data. The computer controls the system by using high-speed data transfer to various cards mounted on the PC bus. For example, the azimuth and elevation motors are controlled through a card on the PC bus. The use of the PC bus confers a rapid scanning capability to the system. Similarly, a general-purpose data collection and control card is used to measure the laser pulse energy. This same multipurpose card is used to both set and measure the high voltage applied to the APD. The digitizers on the PC data bus are set up for data collection by the host computer and start data collec-

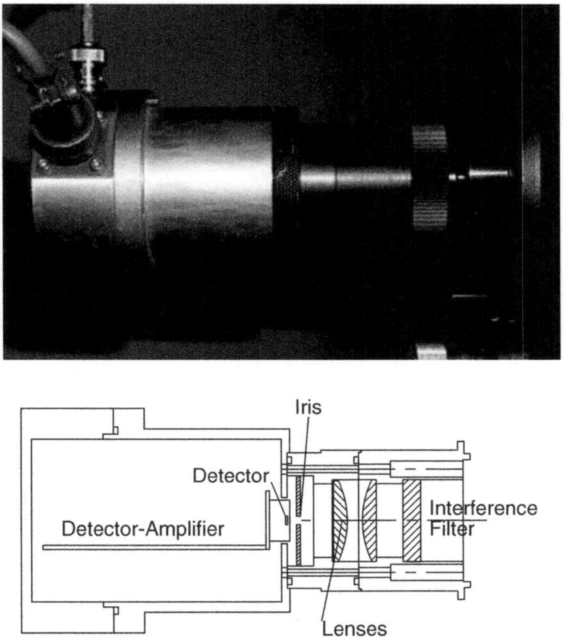

Fig. 3.7. An example of a detector amplifier housing containing focusing optics and an interference filter. This assembly is bolted to the back of the telescope. A 3-nm-wide interference filter is used to eliminate background light. The iris serves to limit the field of view of the telescope.

TABLE 3.1. Operating Characteristics of the Miniature Lidar System of the University of Iowa

University of Iowa Scanning Miniature Lidar (SMiLi)			
Transmitter		Receiver	
Wavelength	1064 or 532 nm	Type	Schmidt–Cassegrain
Pulse length	~10 ns	Diameter	0.254 m
Pulse repetition rate	50 Hz	Focal length	2.5 m
Pulse energy	125 mJ maximum	Filter bandwidth	3.0 nm
Beam divergence	~3 mrad	Field of view	1.0–4.0 mrad adj.
		Range resolution	1.5, 2.5, 5.0, 7.5 m

tion on receipt of the start pulse from the detector mounted inside the periscope. When the digitization of the pulse has been completed, a bit is set in one of the computer memory locations occupied by the digitizer. The computer scans this memory location and transfers the data from the digitizer to the faster computer memory when this bit is set and then resets the system for the next laser pulse. The return signals are digitized and analyzed by a computer to create a detailed, real-time image of the data in the scanned region.

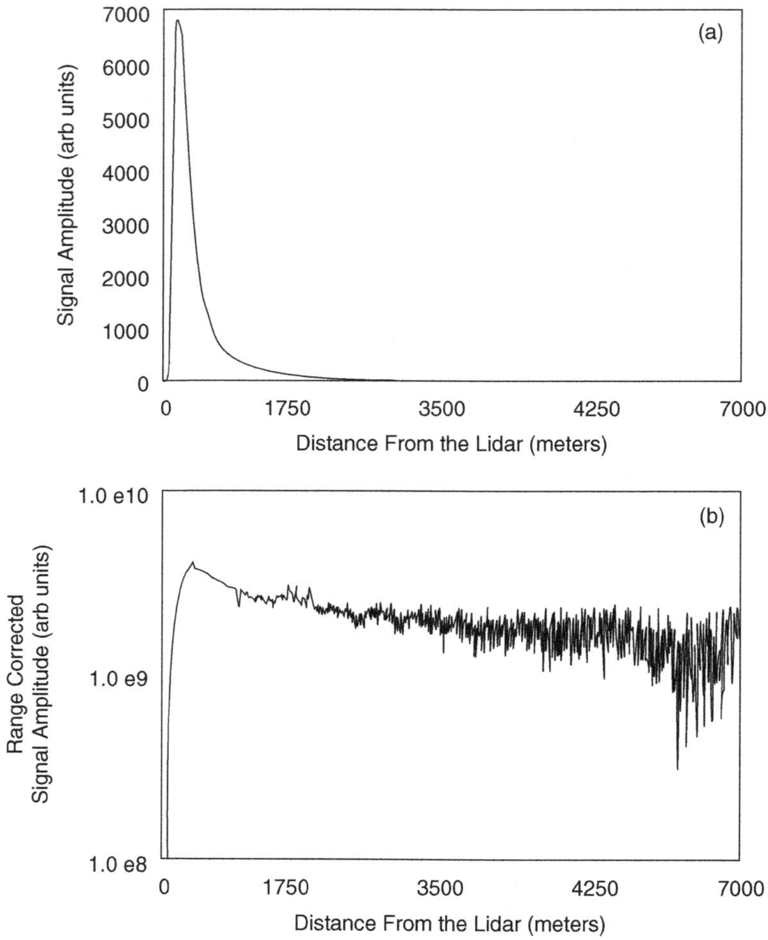

Fig. 3.8. The top part of the figure is a typical lidar backscatter signal from a line of sight parallel to the surface of the earth. The bottom part of the figure is the same signal corrected for range attenuation and shown in a logarithmic *y*-axis.

The lidar used as an example is intended to be disassembled and boxed so that it may be shipped and easily transported. The small size and weight also enable the lidar to be erected in locations that best suit the particular project. However, versatility has a price. The small size limits the maximum useful range to about 6–8 km.

A typical lidar backscatter signal along a single line of sight is shown in Fig. 3.8(a). At long ranges, the signal falls off as $1/r^2$, as implied by Eq. (3.12). At short ranges, the telescope does not "see" the laser beam. As the beam travels away from the lidar, more and more of the laser beam is "seen" by the telescope until, near the peak of the signal, the entire beam is inside the telescope field of view. Correcting for the decrease in signal with range, one obtains the range-corrected lidar signal, shown in Fig. 3.8(b). This lidar signal is often plotted in a semilogarithmic form to emphasize the attenuation of the signal with range. If the amount of atmospheric attenuation is small, the amplitude of the range-corrected signal is roughly proportional to the aerosol density. Although not strictly true, this approximation is useful in interpreting the lidar scans. Note that the signal immediately following the signal peak decreases more or less linearly with range. This is the source of the slope method of determining the average atmospheric extinction. The variations in the signal are due to variations in the backscatter coefficient along the path and signal noise.

Pulse averaging is often used to increase the useful range of the system. Because the size of the backscattered signal rapidly decreases with range, while the noise level remains approximately constant over the length of the pulse, the signal-to-noise ratio also decreases dramatically with range. This effect is aggravated by the signal range correction [Fig. 3.8(b)]. Averaging a limited number of pulses increases the signal-to-noise ratio and can significantly increase the useful range of a system. A series of pulses are summed to make a single scan along a given line of sight. A number of scans are used to build up a two-dimensional map of the range-corrected lidar return.

A wide range of scanning products can be made with lidars possessing that capability. By changing the elevation angle while holding the azimuth constant, a range height indicator (RHI) scan is produced showing the changes in the range-corrected lidar return in a vertical slice of the atmosphere (see Fig. 3.9 for an example). Conversely, holding the elevation constant while changing the azimuth angle produces a plan project indicator (PPI) scan showing the relative concentration changes over a wide area. Figure 3.10 is an example of such a horizontal slice of the atmosphere. Three-dimensional scanning can also be accomplished by changing the azimuth and elevation angles in a faster pattern.

The lidar system shown here is able to turn rapidly through 210° horizontally and 100° vertically by using motors incorporated into the telescope mount and arms. Because the operator of the lidar is normally sited behind the lidar during use, the range of azimuths through which it can scan is deliberately limited for safety reasons. Normally, the lidar programming controls the positioning of the telescope and synchronizes it with the data collection.

The lidar is entirely contained in five carrying cases. The first case contains the laser power supply and chiller and serves as the base for the second case. The second case contains the bulk of the lidar including the scanner motor power supplies and controllers as well as the power supply for the detector. The telescope is easily removed from the arms, and the arms are similarly removed from the rotary stage. The third case is a carrying case for the telescope and is used only for transportation. The portable computer, periscope, telescope arms, and all of the other required equipment are shipped in a footlocker-sized case that is used in the field as a table.

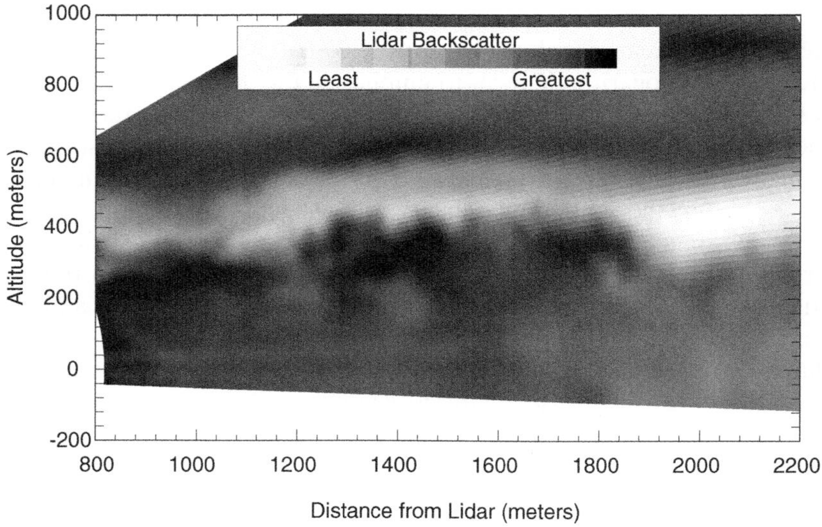

Fig. 3.9. An example of a RHI or vertical scan showing the relative particulate density in a vertical slice of the atmosphere over Barcelona, Spain. Black indicates relatively high concentrations, and light grays are lowest. The range resolution of this image is approximately 7.5 m.

Fig. 3.10. An example of a PPI or horizontal scan showing the relative particulate density in a horizontal slice of the atmosphere over Barcelona, Spain. Black indicates relatively high concentrations, and light grays are lowest. The range resolution of this image is approximately 7.5 m. The dark lines generally follow the lines and intersection of two major highways.

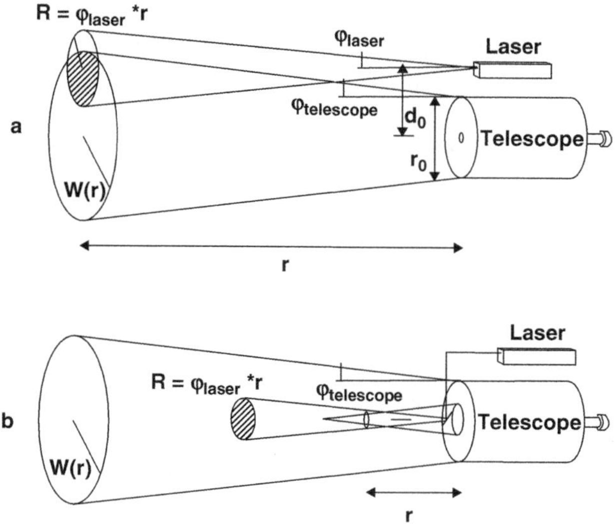

Fig. 3.11. A diagram showing the two types of overlap that may occur in lidar systems. (a): the type of overlap that occurs when the laser beam is emitted parallel to and outside the field of view of the telescope. (b): the type of overlap that occurs when the laser beam is emitted parallel to and inside the field of view of the telescope. In this case, the beam originates at the center of the central obscuration of the telescope.

3.4. PRACTICAL LIDAR ISSUES

In this section, some of the issues that afflict real lidar systems are discussed. Real systems have limitations that may not be obvious in a theoretical development. These systems have issues that affect their performance and often require trade-offs in the design of the systems. Although most of the lidars commonly used are monostatic (the telescope and laser are collocated) and short pulsed, this is by no means the only type that can be constructed.

3.4.1. Determination of the Overlap Function

There are two basic situations, shown in Fig. 3.11. The first is when the laser and telescope are biaxial and the axes of the two systems are parallel, but offset by some distance, d_0. This orientation is used in staring lidar systems and in scanning systems when the telescope moves. The second situation occurs when the laser beam exits the system in the center of the central obscuration of the telescope. The laser beam and telescope field of view are coaxial in this case. The central obscuration of the telescope shields the telescope from the large near-field return. This orientation is often used when a large mirror is used at the open end of the telescope to direct the field of view of the system and the laser beam.

Although the existence of the overlap function is a hindrance (information can be reliably obtained only from the region in which the overlap function is 1), it can serve a valuable function. Because the magnitude of the signal is dependent on $1/r^2$, the signal increases dramatically as the distance of complete overlap is reduced. For example, reducing the overlap distance from 200 m to 50 m increases the magnitude of the signal at the overlap by a factor of 16 and reduces the effective maximum range by a factor of about 4. Thus it may be desirable to increase the offset between the beam and the telescope (in the lidar of Section 3.3, a periscope is used to accomplish this). The overlap distance may also be adjusted by controlling the field of view of the telescope or the divergence of the laser beam. The field of view of the telescope may be adjusted through the use of an iris at the point of infinite focus at the back of the telescope. Kuse et al. (1998) should be consulted for a detailed explanation of the effect of stops on the lidar signal.

The existence of a region of incomplete overlap creates problems in processing remotely sensed data from lidars. This is especially true for transparency measurements in sloping directions made by ground-based lidars. The problem generally arises with respect to practical methods to extract atmospheric parameters in the lowest atmospheric layers, close to the ground surface (see Chapter 9). In principle, the data obtained in the incomplete overlap zone of the lidar can be processed if the overlap function $q(r)$ is determined. Nevertheless, researchers generally avoid processing lidar data obtained in the incomplete overlap zone. The reasons for this are as follows. First, to obtain acceptable measurement accuracy in this zone, the overlap function $q(r)$ must be precisely known. However, no accurate, practical methods exist to determine $q(r)$, so it can be found only experimentally. Second, any minor adjustment or the realignment of the optical system may cause a significant change in the shape of the overlap function. Therefore, after all such procedures, a new overlap function must be determined. Third, the intensity of scattered light in the zone, close to the lidar, is high. It should also be mentioned that the lidar signals measured close to the lidar may be corrupted because of near-field optical distortions. Also, some measurement errors may be aggravated in the near field of the lidar, for example, by an inaccurate determination of the lidar shot start time (a fast or slow trigger). Despite this, determination of the length of the incomplete overlap zone should be considered to be a necessary procedure before the lidar is used for measurements. First, the optical system must be properly aligned, and the researcher needs to know the minimum operating range r_0 of the lidar. This allows the development of relevant procedures and methods for measuring specific atmospheric parameters. Second, the determination of the shape of the overlap function in a clear atmosphere makes it possible to examine whether latent instrumental defects exist that were not detected during laboratory tests. Before measurements are made, the researcher must have certainty that, over the whole operating range, complete overlap occurs. This is quite important because the conventional lidar equation assumes that the function $q(r)$ is constant over the range. Finally, the knowledge of the function $q(r)$ for $r \leq r_0$ makes it possible to invert the signals from the nearest areas, where $q(r)$ is close but less than unity. In other words, in case of a rigid requirement for a short overlap distance, the minimum operating range of the lidar can be reduced and established at the range where $q(r) \approx 0.7$–0.8 rather than 1. All of these arguments show the value of a knowledge of $q(r)$. However, as pointed by Sassen and Dodd (1982), no practical method exists to determine the lidar overlap function except experimentally. The spatial geometry of the lidar system cannot be accurately determined until the system is used in the open atmosphere. The reason is that the function $q(r)$ depends both on the lidar optical system parameters and on the energy distribution over the cross section of the light beam cone. The distribution may be different at different distances from the lidar. Note also that before the overlap function is determined, the zero-line offset should be estimated and the corresponding signal corrections, if necessary, made. It is convenient to do all of these tests together when the appropriate atmospheric conditions occur.

Using an idealized approximation, one can derive analytical functions that describe the overlap function. These functions tend to be quite complex and generally consider only geometric effects (in particular, they either ignore or use oversimplified expressions for the energy distribution in the laser beam and exclude near-field telescope effects). As an example, consider the instrument geometry of Fig. 3.11(a), in which the laser beam is emitted parallel to and offset from the line of sight of the telescope. For this case, and assuming that the energy in the lidar beam is constant over its radius, the overlap function can be written as (Measures, 1984)

$$
\begin{aligned}
q(\zeta) = {} & \frac{1}{\pi}\cos^{-1}\left[\frac{S^2(\zeta)+Y(\zeta)X^2(\zeta)-X^2(\zeta)}{2S(\zeta)X(\zeta)\sqrt{Y(\zeta)}}\right] \\
& + \frac{1}{\pi Y(\zeta)}\left(\cos^{-1}\left[\frac{S^2(\zeta)+X^2(\zeta)-Y(\zeta)X^2(\zeta)}{2S(\zeta)X(\zeta)}\right]\right. \\
& \left. + \frac{S(\zeta)}{X(\zeta)}\sin\left\{\cos^{-1}\left[\frac{S^2(\zeta)+X^2(\zeta)-Y(\zeta)X^2(\zeta)}{2S(\zeta)X(\zeta)}\right]\right\}\right)
\end{aligned}
\tag{3.19}
$$

where

$$
\zeta = \frac{r}{r_0} \qquad S(\zeta) = \frac{d_0}{r_0} - \zeta\delta
$$

$$
Y(\zeta) = \frac{(1+\zeta^2\phi_{\text{laser}}^2\, r_0/w_0)}{(1+\zeta\phi_{\text{telescope}})^2\left(\dfrac{r_0}{W_0}\right)^2} \qquad X(\zeta) = 1 + \zeta\phi_{\text{telescope}}
$$

here r is the distance from the lidar to the point of interest, r_0 is the radius of the telescope, W_0 is the initial radius of the laser beam, ϕ_{laser} is the half-angle divergence of the laser beam, $\phi_{\text{telescope}}$ is the half-angle divergence of the telescope field of view, δ is the angle between the line of sight of the telescope and the laser beam, and d_0 is the distance between the center of the telescope and the center of the laser beam at the lidar.

In practice, analytical formulations of this type are not very useful. The behavior of real overlap function is very sensitive to small changes in the angle between the laser and telescope, δ, an angle that is seldom known precisely. The situation becomes even more complex for the more realistic assumption of a Gaussian distribution of energy in the laser beam. Sassen and Dodd (1982) discuss these effects as well as the effects of small misalignments. These formulations also assume that the telescope acts as a simple lens. A more detailed analysis of the telescope response can be performed that eliminates some of the limitations of the simple form of Eq. (3.19) (Measures, 1984; Velotta et al., 1998). The addition of more realistic assumptions makes the expressions even more complex but does not eliminate the problem that they are extremely sensitive to parameters that are not known to the accuracy required to make them useful.

The determination of an overlap correction to restore the signal for the nearest zone of the lidar has been the subject of a great deal of effort. The efforts have included both analytical methods (Halldorsson and Langerboic, 1978; Sassen and Dodd, 1982; Velotta et al., 1998; Harms et al., 1978; Harms, 1979) and experimental methods (Sasano et al., 1979; Tomine et al., 1989; Dho et al., 1997). The use of an analytical method requires the use of assumptions such as those made in the paragraph above. They also implicitly assume the presence of symmetry in the problem, an absence of aberrations in the optics, and a well-defined nature of the distribution of energy in the laser beam as it propagates through the atmosphere. The overlap function is extremely sensitive to all of these assumptions and parameters and to the accuracy of the angles involved. Attempts to measure laser beam divergence, the telescope field of view, and the angle between the telescope and laser to calculate the overlap function, $q(r)$, are not usually successful. Because of the mathematical complexity of the expressions, attempting to fit these functions to the data is difficult and requires complicated fitting algorithms. The bottom line is that these analytical expressions are not generally useful to determine a correction that may be applied to real lidar data.

In 1979, Sasano et al. proposed a practical procedure to determine $q(r)$ based on measurements in a clear, homogeneous atmosphere. Three approximations were used to derive the overlap function. First, the unknown atmospheric transmission term in the lidar equation was taken as unity. Second, the assumption was used that no spatial changes in the backscatter term exist that distort the profile. Third, it was implicitly assumed that no zero-line offset remained in the lidar signal after the background subtraction. Under these three conditions, the behavior of the function $q(r)$ may be determined from the logarithm of the range-corrected signal, $P(r)r^2$, at all ranges, including these close to the lidar. The approximate range of the incomplete overlap zone, r_0. may be determined as the range in which the logarithm of $P(r)r^2$ reaches a

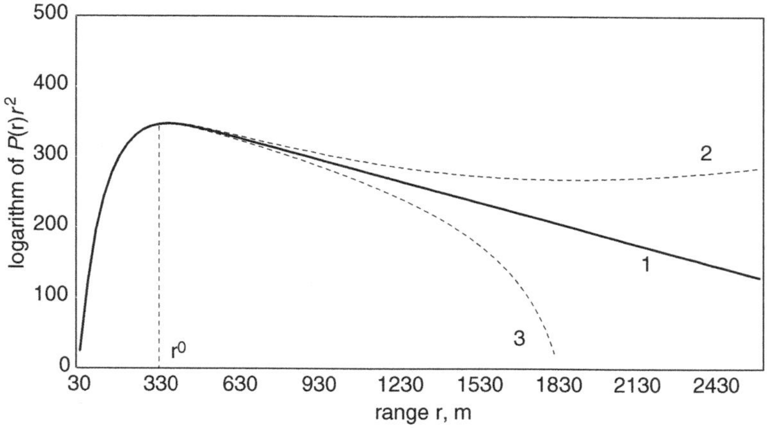

Fig. 3.12. Logarithms of the simulated range-corrected signal calculated for a relatively clear atmosphere with an extinction coefficient of $0.5\,km^{-1}$ (*curve 1*). *Curves 2* and *3* represent the same signal but corrupted by the presence of a positive and a negative zero-line shift, respectively.

maximum value, after which the curve transitions to an inclined straight line. In Fig. 3.12, the logarithm of $P(r)r^2$ is shown as *curve 1*, and the range r_0 is, approximately 350 m.

A similar method to determine $q(r)$, which can be used even in moderately turbid atmospheres, was proposed in studies by Ignatenko (1985a) and Tomino et al. (1989). Here the basic assumption is that a turbid atmosphere can be treated as statistically homogeneous if a large enough set of lidar signals is averaged. In other words, the average of a large number of signals can be treated as a single signal measured in a homogeneous medium. This assumption can be applied when local nonstationary inhomogeneities in the single lidar returns are randomly distributed. The extinction coefficient in such an artificially homogeneous atmosphere can be determined by the slope method over the range, where the data forms a straight line (see Section 5.1). This area is considered to be that where $q(r) = const.$ Then the lidar signal $P(r_q)$ is determined at some distance r_q, far enough to meet the condition $q(r_q) = 1$. The overlap function is determined as (Tomino et al., 1989)

$$\ln q(r) = 2\bar{\kappa}_t(r - r_q) + \ln\overline{(P(r)r^2)} - \ln\overline{(P(r_q)r_q^2)} \qquad (3.20)$$

where the averaged quantities are overlined. It should be noted however that the above procedure of the determination of $q(r)$ in a moderately turbid atmosphere cannot be recommended for the lidar that is assumed be used for measurements in clear atmospheres. For example, if a lidar is designed for the measurements in clear atmospheres, where the extinction coefficient may vary, from 0.01 km^{-1} to 0.2 km^{-1}, the investigation of the shape of $q(r)$ over the lidar operative range should be performed in the atmosphere with k_t close to the minimal value, 0.01 km^{-1}.

In the method used by Sasano et al. (1979) and by Tomino et al. (1989), the principal deficiency lies in the assumption that no systematic offset ΔP exists in the measured signals. Meanwhile, because of the possible background offset in the averaged signals, the shape of the logarithm of $q(r)$, determined by Eq. (3.20), may be distorted, similar to that shown in Fig. 3.12 (Curves 2 and 3). To avoid such distortion, the systematic residual shift remainder must be removed. A method for the determination of $q(r)$ with the separation of the residual shift was proposed by Ignatenko (1985a). A variant of this technique using a polynomial fit to the data instead of a linear fit was used by Dho et al. (1997). It should be recognized that in the incomplete overlap zone, the function $q(r)$ is useful mostly for semiqualitative restoration of the lidar data. Any values obtained as the result of an inversion are tainted by the assumptions built into the model by which the overlap function is obtained. For example, in the methods described, it is assumed that the average attenuation in the overlap region is the same as the average attenuation in the region used to fit the function.

The techniques described above are useful when the intended measurement range of the lidar is restricted to several kilometers. More difficult problems appear when adjusting the optical system of a stratospheric lidar, operating at altitudes from 50 to 100 km. Such systems generally operate in the vertical direction, so the alignment of the optical system can be made only in a cloud-free atmosphere. The principles of the optical adjustment of such a system are described by McDermid et al. (1995). The authors describe the methods used for a biaxial lidar system with a separation of 3.5 m between the laser and receiving telescope. The lidar system was developed for the measurements of stratospheric aerosols, ozone concentration, and temperature. During routine adjustments, the atmospheric backscattered signals at the wavelengths 308 and 353 nm were observed in the altitude range between 35 and 40 km. The position of the laser beam was changed so as to sweep through the field of view of the telescope in orthogonal directions, and the backscattered signal intensity was determined as a function of angular position. To adjust the beam to the center of the telescope field of view, the angle position corresponding to the centroid of the resulting curve was used. The signal was determined at 20 different angular positions. This operation required approximately 3.5 min. The authors of the study assumed that no signal biases occurred because of atmospheric variability when no clouds were present within the line of sight of the lidar. To monitor the changes that occur during routine experiments, both signals were monitored and plotted as a function of time. This made it possible to monitor the general situation during the experiment. For example, a simultaneous decrease in the signals in both channels was considered to be evidence of the presence of clouds whereas a change in only one channel showed alignment shifts.

3.4.2. Optical Filtering

There are many ways in which optical filtering can be accomplished, only a few of which are commonly found in lidars. The amount of scattered light collected by the telescope is normally small, so that the receiving optics must have a high transmission at the laser wavelength. Most elastic lidars operate during the day, so that a narrow transmission band is required along with strong rejection of light outside the transmission band. These requirements limit the practical filters to interference filters and spectrometers. Although there are a limited number of lidars using etalons as filters in high-spectral-resolution systems (Chapter 11), nearly all lidars use interference filters because of convenience and cost. Spectrographic filters are occasionally used because they offer the advantages of wavelength flexibility, high transmission at the wavelength of interest, and very strong rejection of light at other wavelengths.

Interference filters are relatively inexpensive wavelength selectors that transmit light of a predetermined wavelength while rejecting or blocking other wavelengths. The filters are ideal for lidar applications where the wavelengths are fixed and known and high transmission is important. They consist of two or more layers of dielectric material separated by a number of coatings with well-defined thickness. The filters work through the constructive and destructive interference of light between the layers in a manner similar to an etalon (Born and Wolf, 1999). The properties of a filter depend on the number of layers, the reflectivity of each layer, and the thickness of the coatings. The transmission band of a typical filter used in a lidar is Gaussian-shaped with a width of 0.5–3 nm. As the number of layers increases, the width of the transmission interval increases. When the number of layers reaches 13–16, the width can be as large as 200 nm in the visible portion of the spectrum. These types of filters can also be used to block light. A complete filter will consist of a substrate with the coatings bonded to other filters and colored glass used to block light outside the desired transmission band.

Blocking refers to the degree to which radiation outside the filter passband is reflected or absorbed. Blocking is an important specification for lidar use that generally includes the wavelength range over which it applies. Insufficient blocking will result in increased amounts of background light (leading to detector saturation and higher noise levels), whereas too much blocking will decrease the transmission of the filter at the wavelength of interest. Filters are usually specified by the location of the centerline wavelength, the width of the transmission band, and the amount of blocking desired. The width of the transmission band is most often measured as the width of the spectral interval measured at the half-power points (50% of the peak transmittance). It is often referred to as the full-width half-maximum (FWHM) or the half-power bandwidth (HPBW). Blocking is normally specified as the fraction of the total background light that is transmitted through the filter.

An interference filter requires illumination with collimated light perpendicular to the surface of the filter. The filter will function with either side facing the source; however, the side with the mirrorlike reflective coating should be facing the incoming light. This minimizes thermal effects that could result from the absorption of light by the colored glass or blockers on the other side. The central wavelength of an interference filter will shift to a shorter wavelength if the illuminating light is not perpendicular to the filter. Deviations on the order of 3° or less result in negligible wavelength shifts. However, at large angles, the wavelength shift is significant, the maximum transmission decreases, and the shape of the passband may change. The amount of shift with angle is determined as

$$\frac{\lambda_\theta}{\lambda_{\text{normal}}} = \left(\frac{n^2 - \sin^2\theta}{n^2}\right)^{\frac{1}{2}}$$

where λ_{normal} is the centerline wavelength at normal incidence, λ_θ is the wavelength at an angle θ from the normal, and n is the index of refraction of the filter material. Changing the angle of incidence can be used to "tune" an interference filter to a desired wavelength within a limited wavelength range. The central wavelength of an interference filter may also shift with increasing or decreasing temperatures. This effect is caused by the expansion or contraction of the spacer layers and by changes in their refractive indices. The changes are small over normal operating ranges (about 0.01 nm/°C). When noncollimated light falls on the filter, the results are similar to those at angle and depend on the details of the cone angle of the incoming light.

Spectrometers are occasionally used as filters in lidar systems. These are used because they offer the advantages of wavelength flexibility (they can be "tuned") and can service several wavelengths at a time. In general, spectrometers have a high transmission at the wavelengths of interest, relatively narrow transmission bands, and very strong rejection of light at other wavelengths. These instruments, however, are far more expensive than interference filters and require servicing and calibration to work properly. Figure 3.13 is a conceptual diagram of a simple spectrometer used as a filter. Light collected by the telescope falls on a slit. The light passing through the slit is collimated and directed to a diffraction grating. A lens at the proper angle captures the first-order diffraction peak and focuses the light on a detector. The spectrometer is tuned to different wavelengths by changing the angle between the lens and the incoming light. Multiple detectors mounted at the appropriate angles can detect multiple wavelengths simultaneously. More sophisticated systems use concave gratings that focus the light as well as diffract it. They may also include multiple gratings to increase the amount of light rejection at other wavelengths.

3.4.3. Optical Alignment and Scanning

There are two basic ways in which the lidar beam can be made parallel to the field of view of the telescope. The laser beam can be made collinear with

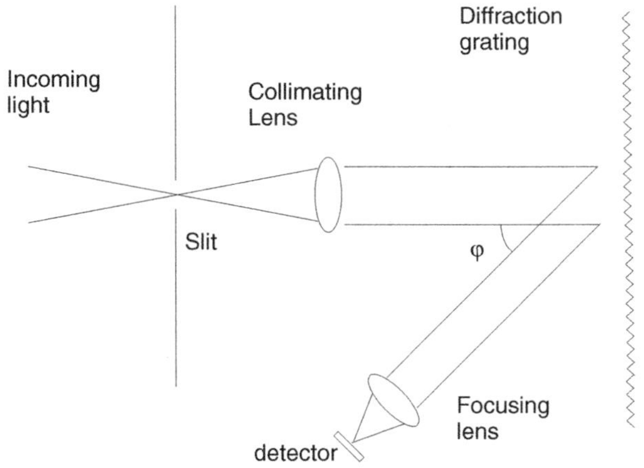

Fig. 3.13. A diagram of a simple spectrograph used as a filter. This type of filter offers tunability, high rejection of ambient light, and high spectral resolution.

the telescope in ways similar to the periscope used in the lidar in Section 3.3. The beam is made parallel to the telescope by using mirrors located outside the barrel of the telescope. The use of mirrors in a periscope fashion makes the problem of alignment simpler. If multiple lasers are used, they may be located at any convenient location and high-power mirrors may be used to direct the beam. Mirrors capable of withstanding the high power levels in the laser beam are not often found for widely separated laser wavelengths that are not harmonics. Thus damage to the mirrors is an issue for systems that have multiple wavelengths reflecting from a single mirror. Multiple mirrors specific to certain wavelengths can be used to align the beam and telescope.

The alternative is to locate the alignment mirror on the secondary of the telescope. The laser beam is then directed across the front of the telescope and then out parallel to the center of the telescope field of view. The secondary obscures the beam in the near field of the telescope so that there is a near-field overlap function. Because the beam must pass across the front of the telescope, there is often an initial intense pulse of scattered light seen by the detector when the laser is fired. This may be a problem for detectors because of the intensity of this pulse. The pulse can be considerably reduced by enclosing the laser beam across the front of the telescope, but this may reduce ihe effective area of the telescope.

The last method of alignment is to use the telescope as both the sending and the receiving optic. This method is most commonly used in systems where the amount of backscattered light is so small that photon counting methods must be used. In these systems, the solar background light must be considerably reduced. This is accomplished by reducing the telescope (and thus the laser) divergence to the smallest values possible. The major issue with using the telescope as the sending optic is the possibility of just a small fraction of the emitted light being scattered into the detector. Some method must be used to block this light to prevent the overloading of the detector and the nonlinear behavior (or afterpulse effects) that are associated with a fast but intense light pulse. Mechanical shutters or rotating disks with apertures have been used but are useful only for very long-range systems in which information from parts of the atmosphere close to the lidar are not needed. For a boundary layer depth on the order of a kilometer, a mechanical system must go from a fully closed to a fully open position on the time scale of 5 μs to detect even the top of the boundary layer. Although this is not impossible, response times this fast are extremely difficult for mechanical systems. If the desired information is at stratospheric altitudes, even longer shutter times may be desirable to reduce the effects of the larger, near-field signal.

Another solution to the shutter problem is to use an electro-optic shutter. If a polarizing beamsplitter is placed in front of the detector, light of only one linear polarization will be allowed to pass. This beamsplitter can be used to direct the light from the laser into the telescope. The laser is linearly polarized in the direction orthogonal to the detector pass polarizer. The problem with this method is that the only backscattered light that will be detected is that which has changed its polarization; the primary lidar signal maintains the original polarization. A Faraday rotator is placed between the polarizing beamsplitter and the telescope to change the polarization of the incoming scattered light by 90°. Because these electro-optic crystals can have response times on the order of 10 ns, none of the backscattered light need be lost because of the system response time. By activating the Faraday rotator in some alternate pattern with the laser pulses, the signals from the two orthogonal polarizations may be detected. This method, or variants of the method, are used in micropulse lidars (Section 3.5.2).

The choice of method used for alignment is often determined by the method that is to be used for scanning. If the system is not intended to scan, the collinear method is the simplest method to use and the least fraught with difficulty. If the scanning system moves both the telescope and laser as with the UI lidar system (Section 3.3), a collinear system is again the simplest method. If moving both the telescope and laser, care must be taken to rotate the system about the center of gravity. There are two reasons for this. The first is mechanical. Rotation about the center of gravity reduces the amount of torque required for the motion (so the motors are smaller), and it puts less strain, and thus wear, on the gears used to drive the system. The second reason is that when scanning, short, abrupt motions are often used and rotation about the center of gravity will reduce the amount of jitter produced at an abrupt stop. As a rule, only small telescopes and lasers are scanned in this way. Although larger systems have moved both telescope and laser head, they tend to be slow and cumbersome.

The most common form of scanning system is the "elevation over azimuth" scanning system shown in Fig. 3.14. These scanners can be purchased commercially and, although expensive, can be interfaced to a master lidar com-

Fig. 3.14. An example of an "elevation over azimuth" scanning system. The telescope is located under the center of the scanner, pointing vertically. A mirror in the center of the scanner directs the beam to the left and allows scanning in horizontal directions. A mirror behind the scanner exit on the left allows scanning in vertical directions.

puter and can scan rapidly over all angles in azimuth or elevation. Two mirrors are used in this type of scanner. One mirror is centered above the telescope aperture and is at a 45° angle to the telescope line of sight. This mirror rotates about an axis that is the same as the telescope line of sight. Thus this mirror allows the telescope to view any azimuthal angle parallel to the ground. A short distance from the first mirror, a second is placed at a 45° angle to and along the line of sight of the telescope. This mirror rotates on a horizontal axis that is perpendicular to the line of sight of the telescope. This mirror allows scanning in any vertical angle. An alternative scanning method is to use a single mirror located above the telescope field of view as shown in Fig. 3.15. This mirror is made to rotate about the axis that is telescope field of view and also about an axis perpendicular to the ground and in the plane of the mirror. This type of scanner can view any azimuthal angle but is limited to a maximum elevation angle that is determined by the relative sizes of the scanning mirror and telescope diameter. Note that the minimum size for the scanning mirror is to have the width to be the telescope diameter and the length to be 1.4 × telescope diameter. The longer the mirror, the greater the possible elevation angle. No similar limitation exists for the elevation over azimuth scanning method.

When the scanning mirrors are dirty or dusty, as often happens in field conditions, or have defects, they may reflect a great deal of light back into the telescope, producing a short, intense flash on the detector. This short but intense flash of light may cause detector nonlinearities. This flash can be minimized by controlling the amount of light scattered by the mirrors. Because

Fig. 3.15. An example of a single mirror scanner. The entire mirror assembly rotates to allow scanning in horizontal directions. The mirror rotates to allow scanning in vertical directions. The maximum vertical angle is limited by the size of the scanning mirror.

the scanning mirrors used with these scanners are large, they are seldom coated to handle high-power laser beams. Thus the beams must be expanded to lower the energy density to avoid damage to the scanning mirrors. Scanning systems like these generally place the alignment mirror in the center of the telescope, on the secondary mirror. This alignment method is the most likely to produce an alignment in which the laser beam and telescope field of view are parallel. A collinear method could be used, but it is not uncommon to have a small angle between the laser beam and the telescope field of view. Each mirror reflection will double the size of this angle. The result is that the alignment could change depending on the mirror directions.

Another scanning method moves the telescope. The Coude method places the telescope in a mount that rotates in azimuth and is located above the elevation axis (Fig. 3.16). Two high-power laser mirrors located on the axes of rotation direct the beam to be collinear with the telescope field of view. The laser beam is directed vertically on the horizontal axis of rotation. The first mirror is placed at the intersection of the two axes of rotation and reflects the laser beam from the horizontal axis of rotation to the elevation axis. A second mirror is placed at a 45° angle to direct the beam parallel to the telescope. This method is difficult to align, particularly in field situations, but allows the use of high-power laser mirrors. The laser beams must be directed exactly on the axes of rotation. Any deviation will cause misalignment as the system

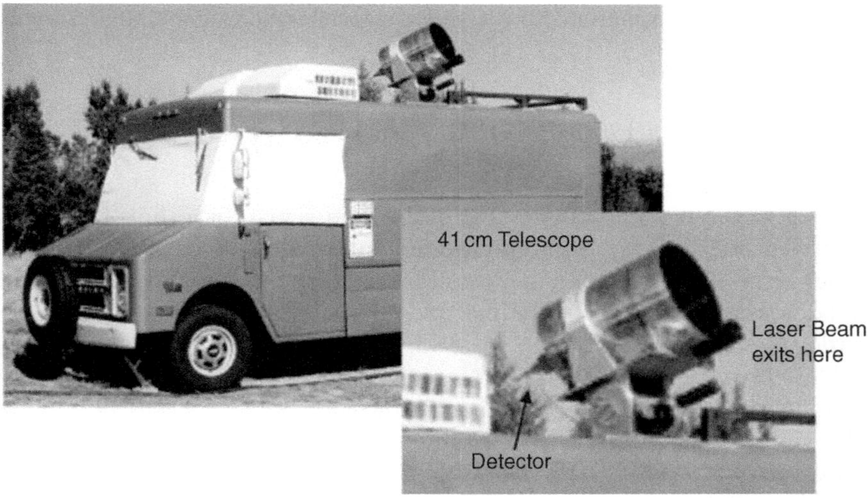

Fig. 3.16. An example of a scanning system using Coude optics. The beam enters the scanner from below and exits from the tube on the right side.

scans. For situations in which a moderately large telescope is desired and the high-energy laser beams cannot be expanded enough to avoid damage to scanning mirrors, the Coude method is a solution. These kinds of scanners can be constructed to scan rapidly and accurately.

3.4.4. The Range Resolution of a Lidar

The spatial averaging that is used to reduce noise also limits the range resolution in ways that are dependent on the details of the smoothing technique used. A good discussion of basic filtering techniques and the creation of filters is given by Kaiser and Reed (1977). We note also that the averaging of multiple laser pulses is a temporal average that limits spatial resolution as the structures move and evolve in space. The limits on resolution due to temporal averaging have also not been discussed in the literature to any great degree but are strongly dependent on the timescales involved and the wind speed at the point in question.

As detectors and electronics become faster (digitization rates of 10 GHz are currently available), and particularly for lasers that have very long pulse lengths, it is the size of the laser pulse that limits range resolution. For this case, methods have been devised to measure structures smaller than the physical length of the laser pulse. These methods assume that the light collected by the telescope is a convolution of the light from an infinitesimally short laser pulse and a normalized shape function, $T_L(t)$, representing the intensity of the laser pulse in time. Lidar inversion methods when applied to signals from long pulses may result in considerable error (Baker, 1983; Kavaya and Menzies,

1985). To develop a method to retrieve the proper lidar signal, the convolution is written as

$$P_c(r) = \int_0^\infty T_L(t')P(t-t')\mathrm{d}t' \quad \text{where} \quad 1 = \int_0^\infty T_L(t')\mathrm{d}t' \tag{3.21}$$

and P_c is the convoluted pulse and P is the lidar signal for a short laser pulse as derived in Eq. (3.12). Some inversion method must be used to obtain the proper form of the lidar signal. Several investigators have published methods for addressing the problem (Zhao and Hardesty, 1988; Zhao et al. 1988; Gurdev et al. 1993; Dreischuh et al. 1995; Park et al. 1997b). Of these, Gurdev et al. (1993) gave the most complete description of the available methods. In all of the inversion methods, a detailed knowledge of the intensity of the laser pulse with time is required. Dreischuh et al. (1995) have an excellent discussion of the uncertainty in the inverted signal due to inaccuracy in the shape of the laser pulse.

The simplest and most straightforward method to deconvolute the long pulse signal is to put the signal into a matrix format. This is a natural method considering the digital nature of the available data. Considering $T_L(t)$ to be constant between the measurement intervals, Eq. (3.21) can be written as (Park et al. 1997)

$$\begin{bmatrix} P_c(t_1) \\ P_c(t_2) \\ P_c(t_3) \\ \cdot \\ \cdot \\ \cdot \\ P_c(t_n) \\ P_c(t_{n+1}) \\ \cdot \\ \cdot \\ \cdot \end{bmatrix} = \begin{bmatrix} T_L(t_1) & 0 & 0 & 0 & 0 & 0 & \cdots \\ T_L(t_2) & T_L(t_1) & 0 & 0 & 0 & 0 & \cdots \\ T_L(t_3) & T_L(t_2) & T_L(t_1) & 0 & 0 & 0 & \cdots \\ \cdot & \cdot & \cdot & \cdot & \cdot & \cdot & \cdots \\ \cdot & \cdot & \cdot & \cdot & \cdot & \cdot & \cdots \\ \cdot & \cdot & \cdot & \cdot & \cdot & \cdot & \cdots \\ T_L(t_m) & T_L(t_{m-1}) & T_L(t_{m-2}) & \cdots & T_L(t_1) & 0 & \cdots \\ 0 & T_L(t_m) & T_L(t_{m-1}) & T_L(t_{m-2}) & \cdots & T_L(t_1) & \cdots \\ \cdot & \cdot & \cdot & \cdot & \cdot & \cdot & \cdots \\ \cdot & \cdot & \cdot & \cdot & \cdot & \cdot & \cdots \\ \cdot & \cdot & \cdot & \cdot & \cdot & \cdot & \cdots \end{bmatrix} \begin{bmatrix} P(t_1) \\ P(t_2) \\ P(t_3) \\ \cdot \\ \cdot \\ \cdot \\ P(t_n) \\ P(t_{n+1}) \\ \cdot \\ \cdot \\ \cdot \end{bmatrix} \tag{3.22}$$

where t_1, t_2, ... t_n, etc. are the number of times since some reference point in the lidar signal. The laser pulse is m number of digitizer samples in length. This matrix formulation can be simply solved by using a recurrence relationship or using banded matrix inversion methods for the general case. However, the formulation in Eq. (3.22) is not the only one that can be created. Because any reference point must be at some distance from the lidar, the assumption made implicitly by Eq. (3.21) is that the data at the first point are due only to scattering from a small portion of the beam. Depending on the assumptions that are made about the conditions at the beginning and ending of the examined area, the construction of the matrix may be different, but is banded in every case. These assumptions do not much affect the data far from the ends but do affect data near the ends. A consequence is that the inversions are not unique. Other inversion methods, for example, a Fourier transform convolution, must also make assumptions concerning the conditions on the ends, which lead to similar issues. A variation on this approach to enhanced resolution was accomplished by Bas et al. (1997), who offset the synchronization of the laser and digitizer from pulse to pulse by a small amount. The technique allows resolution at scales smaller than that allowed by the digitizer rate by subdividing the time between digitizer measurements. For example, to increase the resolution by a factor of four, the digitizer is synchronized to the laser pulse for the first pulse. For the second pulse, the digitizer start is delayed by one-quarter of the time between measurements. For the third pulse, the digitizer start is delayed by one-half of the time between measurements, and the fourth is delayed by three-quarters of that time. With the fifth pulse the sequence begins anew. The data from each laser pulse are slightly different from the others, enabling a set of matrix equations to be written and solved.

A deconvolution of this type should be done only after considering the bandwidth of electronics used in the lidar system. Deconvolution of data taken with a digitization rate of a gigahertz is not meaningful if the bandwidth of the detector-amplifier is limited to 50 MHz, for example. Information at frequencies much above 50 MHz is strongly attenuated by the electronics and simply is not present at the input to the digitizer. No amount of postprocessing can recover this signal. Maintaining the bandwidth of the entire electronics system at gigahertz-class bandwidths is quite difficult. Noise increases approximately as the square of the bandwidth, and the potential for reflections and feedback increases dramatically as the bandwidth increases.

3.5. EYE SAFETY ISSUES AND HARDWARE

In the United States, the accepted document that regulates laser eye safety issues is the American National Standard for the Safe Use of Lasers, ANSI Z136.1, dated 1993, by the American National Standard Institute. This document can be obtained from the Laser Institute of America (Suite 125, 12424 Research Parkway, Orlando, FL 32826). If a lidar is operating in the outdoors, permission should also be obtained from the Federal Aviation Administration (FAA). The appropriate FAA field office should be contacted before field experiments and written permission should be obtained. This section outlines the exposure limits for the safe use of lasers and several methods for attaining eye-safe conditions. Eye safety issues are a major obstacle to the practical use of elastic lidars. Should lidars ever be permanently installed for some practical application(s) (for example, for wind shear measurements at airports),

TABLE 3.2. Maximum Permissible Exposure (MPE)

Wavelength (μm)	Exposure Duration, t (s)	Maximum Permissible Exposure (J/cm²)	Notes
0.180–0.302	10^{-9}–3×10^4	3×10^{-3}	or $0.56\, t^{1/4}$, whichever is lower
0.303	10^{-9}–3×10^4	4×10^{-3}	or $0.56\, t^{1/4}$, whichever is lower
0.304	10^{-9}–3×10^4	6×10^{-3}	or $0.56\, t^{1/4}$, whichever is lower
0.305	10^{-9}–3×10^4	10^{-2}	or $0.56\, t^{1/4}$, whichever is lower
0.306	10^{-9}–3×10^4	1.6×10^{-3}	or $0.56\, t^{1/4}$, whichever is lower
0.307	10^{-9}–3×10^4	2.5×10^{-2}	or $0.56\, t^{1/4}$, whichever is lower
0.308	10^{-9}–3×10^4	4×10^{-2}	or $0.56\, t^{1/4}$, whichever is lower
0.309	10^{-9}–3×10^4	6.3×10^{-2}	or $0.56\, t^{1/4}$, whichever is lower
0.310	10^{-9}–3×10^4	0.1	or $0.56\, t^{1/4}$, whichever is lower
0.311	10^{-9}–3×10^4	0.16	or $0.56\, t^{1/4}$, whichever is lower
0.312	10^{-9}–3×10^4	0.25	or $0.56\, t^{1/4}$, whichever is lower
0.313	10^{-9}–3×10^4	0.40	or $0.56\, t^{1/4}$, whichever is lower
0.314	10^{-9}–3×10^4	0.63	or $0.56\, t^{1/4}$, whichever is lower
0.315–0.400	10^{-9}–10	$0.56\, t^{1/4}$	or $0.56\, t^{1/4}$, whichever is lower
0.400–0.700	10^{-9}–1.8×10^{-5}	5×10^{-7}	
0.700–1.050	10^{-9}–1.8×10^{-5}	$5 \times 10^{-7} * 10^{2(\lambda-0.700)}$	
1.050–1.400	10^{-9}–5.0×10^{-5}	$5 * C_c \times 10^{-6}$	$C_c = 1.0$ $\lambda = 1.050$–1.150 $C_c = 10^{18(\lambda-1.15)}$ $\lambda = 1.150$–1.200 $C_c = 8.0$ $\lambda = 1.200$–1.400
1.400–1.500	10^{-9}–10^{-3}	0.1	
1.500–1.800	10^{-9}–10	1.0	
1.800–2.600	10^{-9}–10^{-3}	0.1	
2.600–10^3	10^{-9}–10^{-7}	10^{-2}	

Extracted from Table 5, ANSI Z136.1.

they will have to operate in an automated and unattended mode and thus will have to be eye-safe.

For the most part, elastic lidars use short (~10 ns)-pulse lasers with the primary danger being ocular exposure to the direct laser beam at some distance. Table 3.2 lists the maximum permitted exposure (MPE) limits for various laser wavelengths and pulse durations.

For repeated laser pulses, such as those used with most lidars, an additional correction must be applied. The MPE per pulse is limited to the single-pulse MPE, given in Table 3.2, multiplied by a correction factor, C_p. This correction factor, C_p is equal to the number of laser pulses, n in some time period, t_{max}, raised to the one-quarter power, as $C_p = n^{-1/4}$. The time period, t_{max}, is the time over which one may be exposed. For visible light or conditions in which intentional staring into the beam is not expected, this time is taken to be 0.25 s. For situations in which it might be expected that someone would deliberately stare into the beam, a time period of 10 s is used. For a scanning lidar where the beam is moving, the time required for the beam to pass a spot would also be a reasonable time to use. For a 50-Hz laser, using the 0.25-s time interval, the correction factor reduces the MPE by a factor of 2. More detailed discussions can be found in ANSI standard Z136.

For some lidar systems, other dangers can exist. For example, lidars working in the ultraviolet region of the spectrum produce a great deal of scattered ultraviolet light in and around the lidar. The scattered light can lead to a situation in which there is a low background level of ultraviolet light in and around the lidar that is hazardous to both the skin and the surface of the eye. Similarly, nonvisible lasers may produce unintended reflections that can be many times the danger level. It should also be noted that lasers are sources of safety issues other than eye safety. The high-voltage currents used to pump many systems can be lethal if the power supplies are opened or mishandled. Other lasers contain solvents such as ethyl alcohol that are flammable or dyes that are carcinogenic. The handling of compressed gasses presents a problem in addition to the danger from toxic gasses or the potential danger from the displacement of oxygen in work areas.

3.5.1. Lidar-Radar Combination

Several approaches have been attempted to confront the eye safety issue with technology. One solution is to use a radar beam coaxially mounted with the lidar beam (Thayer et al., 1997; Alvarez et al., 1998). During the lidar measurement, the radar works in the alert mode. If an aircraft approaching the laser beam is detected by the radar, then the laser may be interrupted as the aircraft passes through the danger area. Such a system can be made completely automatic. The radar must examine regions on all sides of the laser beam that are large enough to provide sufficient time for detection of the aircraft and interruption of the laser. For rapid scanning systems this can be a problem in that the alignment of the two systems must be maintained as the lidar scans the sky.

A novel solution to this problem was accomplished by Kent and Hansen (1999), who mounted a radar coaxially with the lidar and used the lidar scanning mirrors to direct both the laser and the radar beams. A dichroic mirror made from fine copper wire and threaded rod was used to reflect the radar beam while passing light in both directions (Fig. 3.17). The aluminum front surface mirrors used in the scanner are capable of reflecting both the radar and visible/IR light with efficiencies on the order of 85–90 percent. With a radar beam divergence of 14°, the system was capable of providing 4–8 seconds of warning and automatic shutdown of the laser. The scattering of microwave radiation from exposed metal surfaces inside the lidar is a potential safety issue for the operators of the system. Lightweight microwave absorbers are available that can be used to cover exposed metal surfaces to reduce the risk of exposure.

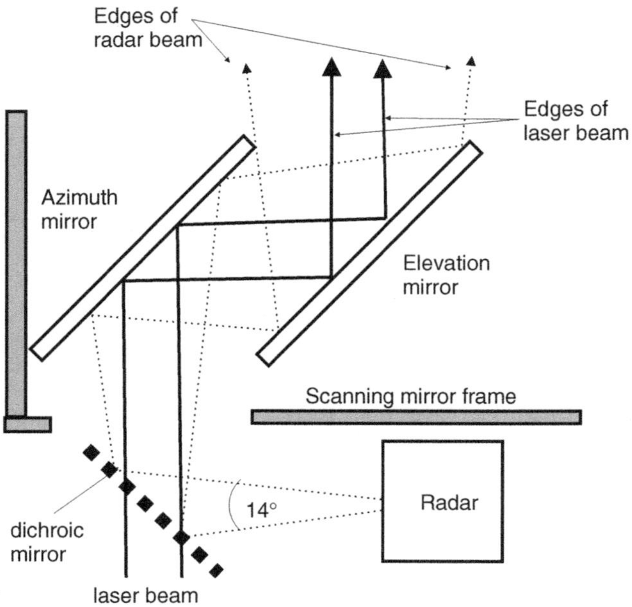

Fig. 3.17. An example of a radar beam inserted into the scanner and parallel to the lidar beam. Because the divergence of the radar beam is much larger than that of the lidar, it provides early warning of the approach of an aircraft (Kent and Hansen, 1999).

3.5.2. Micropulse Lidar

The requirements for eye safety for short-pulse lasers primarily limit the amount of laser energy per area. The idea behind the micropulse lidar is to both expand the area of the laser beam and reduce the energy per pulse to achieve an eye-safe irradiance. Expanding the cross-sectional area of the beam also allows one to reduce the beam divergence, which turns out to be a critical requirement in such a system. As a rule, reducing the energy of the laser pulse to eye-safe limits reduces the amount of the backscattered signal at the lidar receiver to the point that photon counting is required to achieve reasonable ranges. To limit the amount of scattered light from the sun entering the receiver, the telescope must have a narrow field of view. Because the amount of scattered sunlight allowed into the system is proportional to the square of the telescope angular field of view, reducing the field of view will result in significant reductions in background light. However, reducing the field of view increases the problems associated with incomplete overlap of the telescope field of view and the laser beam (discussed in Section 3.4.1). It can also make a system exceptionally difficult to align, particularly for photon counting systems.

Perhaps the most successful of the micropulse lidars (MPL) is the system originally developed at NASA-Goddard Space Flight Center (GSFC) as a

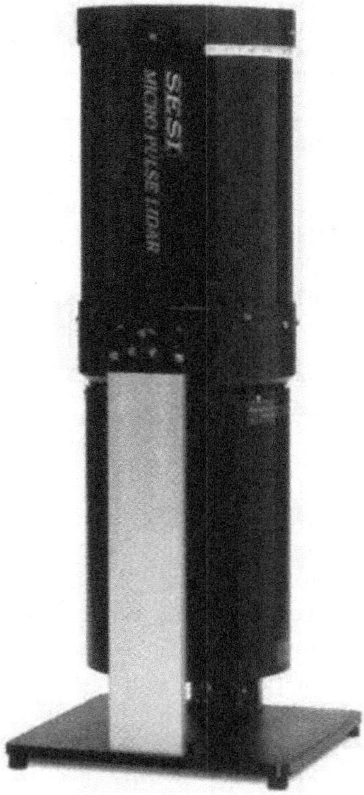

Fig. 3.18. A photograph of the micropulse lidar system. The telescope in this system both transmits the laser pulse and acts as a receiver. The system is compact, rugged, and eye safe, enabling unattended operation.

result of research on efficient lidars for space-borne applications by Spinherne (1993, 1995, 1996), which is now commercially available. This instrument is shown in Fig. 3.18. It has been deployed at a number of long-term measurement sites, particularly at the Atmospheric Radiation Measurement (ARM) program sites in north-central Oklahoma, Papua New Guinea, Manus Island, and the North Slope, Alaska. The instrument was also used during Aerosol99 cruise (Voss et al., 2001) and during the Indian Ocean Experiment (INDOEX) (Sicard et al., 2002).

The basic characteristics of the micropulse lidar are given in Table 3.3. The current design is capable of as little as 30-m vertical resolution. The micropulse lidar is fully eye-safe at all ranges. Eye-safe operation is achieved by transmitting low-power (10 μJ) pulses in an expanded beam (0.2-m diameter). To reduce the scattered solar input, an extremely narrow receiver field of view (100 μrad) is required. Because of the small amount of scattered light, photon counting is used to achieve a relatively accurate signal at medium and long

TABLE 3.3. Operating Characteristics of the Micropulse Lidar System

Micropulse Lidar (MPL)			
Transmitter		Receiver	
Wavelength	523 nm, Nd:YLF	Type	Schmidt-Cassegrain
Pulse length	10 ns	Diameter	0.2 m
Pulse repetition rate	2500 Hz	Focal length	2.0 m
Pulse energy	~10 μJ	Filter bandwidth	3.0 nm
Beam divergence	~50 μrad	Field of view	~100 μrad
		Range resolution	30–300 m
		Detector bandwidth	12 MHz
		Averaging time	~60 s

ranges. A high pulse repetition frequency (2.5 kHz) is used to build up photon counting statistics in a relatively short period of time. Corrections are required to account for afterpulse effects and detector deadtime.

Another variation of a low-power, eye-safe lidar system, the depolarization and backscatter-unattended lidar (DABUL) was developed by the NOAA Environmental Technology Laboratory (Grund and Sandberg, 1996; Alvarez II et al., 1998; Eberhard et al., 1998). In this system, a Nd:YLF laser beam at 523 nm is expanded by using the receiver optics as the transmitter to reduce the energy density to achieve eye safety. The large beam diameter (0.35 m) and low pulse energy (≤40 μJ) make the system eye-safe at all ranges including at the output aperture. To suppress the daytime background light, a narrow field of view of receiver is used in combination with a narrow spectral bandpass filter. The receiver comprises two receiving channels, separated by a beam-splitter, with different fields of view that are in full overlap by 4 km. The two channels have different fields of view, wide (640 μrad) and narrow (100 μrad), to provide signals over different range intervals. For most applications, the data from the narrow channel are used. For this, approximately 90% of the backscattered light is detected. The wide channel allows for a near field signal while the narrow channel provides increased dynamic range in situations with strong backscatter, for example, from dense clouds. Photomultipliers are used in photon-counting mode as the detectors. The DABUL system is able to scan from zenith down to 15° below the horizon. This makes it possible to obtain data close to the horizon, which are often quite useful as reference data. In the operating (unattended) mode, the lidar periodically scans to the horizon, once every 30 minutes, recording the horizontal profile. The horizontal backscatter measurements, made in homogeneous conditions, can be used to determine and monitor the overlap function. In Table 3.4, the basic characteristics of the DABUL system are presented.

TABLE 3.4. Operating Characteristics of DABUL

Depolarization and Backscatter-Unattended Lidar (DABUL)			
Transmitter		Receiver	
Wavelength	523 nm	Telescope diameter	0.35 m
Pulse energy	0–40 μJ	Spectral bandpass	0.3 nm
Pulse repetition rate	2000 Hz	Field of view	100 and 640 μrad
Beam diameter	0.3 m	Detectors	PMT, s (APD)
Beam divergence	<20 μrad	Detection	Photon counting
Spectral width	0.2 nm	Averaging time	~1–60 s
		Range resolution	30 m

Grund and Sandberg (1996); Eberhard et al. (1998).

3.5.3. Lidars Using Eye-Safe Laser Wavelengths

In principle, the best way to achieve eye-safety at short distances from the laser transmitter would be the use of a laser wavelength that the eye does not effectively focus. It could be achieved by the use of wavelengths shorter than 400 nm or longer than 1400 nm, where the maximum permissible exposure is much higher than within this range (ANSI Z136.1). However, the most lidars work within the range from, approximately 350 nm to 1064 nm. Wavelengths shorter than 350 nm are generally used in differential absorption (DIAL) measurements of ozone concetrations in the atmosphere (Chapter 10). The wavelength range 300–500 nm also is not often used for particulate measurements. The scattering at these wavelengths is primarily molecular, so the lidar signals contain less useful data on particulate concentrations. This leaves eye-safe wavelengths longer than 1400 nm.

There are issues with these wavelengths that limit the effectiveness of such lidar systems. The first issue is related to the availability of good detectors at these wavelengths. Until very recently photomultipliers have not been available for the wavelengths longer than 1 μm and have had very low quantum efficiencies at 1 μm. Solid-state detectors (generally InGaAs) at these longer wavelengths are generally small (on the order of 200 μm in diameter) and have detectivities, D^*, that are a factor of approximately 10 smaller than similar silicon detectors in visible and near-infrared wavelengths. Furthermore, the number density of particulates falls exponentially with diameter so that backscatter coefficients at wavelengths longer than 1400 nm may be an order of magnitude smaller than backscatter coefficients at visible wavelengths. Because Rayleigh scattering is proportional to $1/\lambda^4$, the amount of light from molecular scattering is also considerably reduced. The farther into the infrared, the more detectors are subject to thermal noise (or require cooling) and suffer from decreasing bandwidth. Lastly, water vapor and CO_2 absorption bands are common in this spectral region and can strongly attenuate the laser beam.

Ho:YAG/Er:YAG Lasers. The Nd:YAG laser uses a yttrium-aluminum garnet crystal doped with neodymium as the lasing material to produce light at 1.064 µm. Doping the garnet with other rare earth materials results in lasing at different wavelengths. Holmium (1.5 µm)- and erbium (2.1 µm)-doped garnet crystals have been suggested for eye-safe lasers because they operate in a region of the spectrum in which the eye does not focus light well. However, both of these materials have thermal properties that limit the rate at which they can be pulsed. Sugimoto et al. (1990) demonstrated a 2.0875-µm lidar system in a laboratory setting. With a pulse energy of 20 mJ per pulse into a 30-cm telescope, they achieved a signal-to-noise ratio of 1 at about 800 m. The system had a pulse repetition frequency of 2 Hz. Some of these materials show an excessive absorption of the laser beam that has been addressed, at least for thulium-doped lasers, by altering the host garnet, $Y_3Al_5O_{12}$ (YAG) crystal, so that the lasing occurs in particular windows. Kmetec et al. (1994) used varying amounts of Lu in place of yttrium in a Tm:YAG (Tm:$Y_3Al_5O_{12}$) to produce a laser rod operating in the spectral region near 2 µm. Because this spectral region also contains strong water vapor absorption lines, the laser must be "tuned" so that it lases at a wavelength between the water vapor lines. Using a mixture of Lu and Y, they managed to get quite close to a relatively clear window at 2022.2 nm. Because these crystals have similar absorption spectra and operating properties, they are a one-for-one replacement for existing Tm:YAG rods.

Methane Shifting of Nd:YAG. The 1980 version of ANSI standard Z136.1 for laser eye safety contained a single exception at 1540 nm for which an energy density of 1 J/cm^2 was allowed. This generated a great deal of effort to obtain this particular wavelength. One method of achieving this was through Raman shifting 1064-nm light from a Nd:YAG laser to 1540 nm by the use of methane gas. Energy conversion efficiencies up to 30 percent can be achieved through the use of methane under high pressure. Raman shifting has been used to generate additional wavelengths for particulate size determination or for ozone differential absorption lidars (see, for example, Chu et al. 1991; Hanser and McDermid, 1990; Grant et al., 1991).

Patterson et al. (1989) and later Chu et al. (1990) were among the first to demonstrate a working eye-safe system using the methane shifting technique. The level of Raman light production is a function of the molecular number density and energy density of the light, so high pressures and focused high-power lasers are required. Patterson et al. were able to achieve a 16 percent energy conversion efficiency from 1-µm light to 1.5-µm light. This was done with a 75-cm gas cell filled with methane gas at high pressure and illuminated by a 1.2 J/pulse Nd:YAG laser. The laser light was focused at the center of the cell and then recollimated as the light exited the cell. The divergence of the light from the Raman cell was measured as 2 mrad. With a 40-cm Newtonian telescope coupled to a 0.300-mm-diameter lnGaAs PIN diode and amplifier, the system was shown to be able of detecting particulates at distances of 6 km and with averaging 1000 laser pulses, thin cirrus at distances of 11 km.

The use of methane cells has several severe limitations. Because the efficiency of the cell increases with the energy density in the pump beam, high-energy laser pulses are often focused inside the cell. This leads to heating of the cell and dissociation of the methane gas, producing carbon soot. Heating of the gas leads to defocusing and low beam quality. The carbon soot tends to coat optical elements, producing damage to the elements. High-energy density of the laser also tends to damage optical elements. Mixing the gas in the cell can reduce the effects of heating and dissociation but is not a solution. Low pulse repetition rates can reduce the heating in the cell but affect the ability of the lidar system to take data at with even moderate temporal resolution.

Carnuth and Tricki (1994) achieved a maximum of 140 mJ per pulse of eye-safe light by Raman shifting with deuterium. A 1.0-J, 10-Hz, line-narrowed Nd:YAG laser was used with a 1.7-m-long Raman cell to generate 1560-nm light with an average energy of 120 mJ per pulse. The 1.5-km range was achieved with this light by using a 38-cm telescope.

4

DETECTORS, DIGITIZERS, ELECTRONICS

This chapter examines the electronic devices that are used to convert an optical signal to a series of digital numbers. In the early days of lidar, photographs of oscilloscope screens were made of the signals from photomultiplier tubes and data were derived from measurements made off of the photographs (see, for example, Cooney et al., 1969; Collis, 1970). Today, high-speed digitizers capable of measuring transient voltage signals at rates in excess of 2 GHz are commercially available. However, despite a great deal of progress with semiconductor detectors and amplifiers, photomultipliers remain an attractive option for many applications, particularly in the ultraviolet and near-ultraviolet portion of the spectrum. In many ways, the electronics that detect the light signal and then amplify and digitize it are still the limiting factors for system performance. The detector efficiency and noise level, coupled with the dynamic range of the digitizer, are nearly always the factors that limit the maximum range of lidar systems and set the precision limits for measurements.

4.1. DETECTORS

The purpose of a detector is to convert electromagnetic energy into an electrical signal. Detectors fall into two broad classes: photon detectors and thermal detectors. Photon detectors use the interaction of a quantum of light

Elastic Lidar: Theory, Practice, and Analysis Methods, by Vladimir A. Kovalev and William E. Eichinger.
ISBN 0-471-20171-5

energy with electrons in the detector material to generate free electrons that are collected to form a measurable current pulse that is proportional to the intensity of the incoming light pulse. To produce a signal, the quantum of light must have sufficient energy to free an electron from the molecule or lattice in which it resides. Thus the wavelength response of photon detectors shows a long-wavelength cutoff. When the wavelength is longer than a cutoff wavelength (which is material dependent), the amount of energy in the photon is insufficient to liberate an electron and the response of the detector drops to zero. Thermal detectors respond to the amount of energy deposited in the detector by the light, resulting in a temperature change in the material. The response of these detectors involves some temperature-dependent effect, often a change in the electrical resistance. Because thermal detectors respond to the amount of energy deposited by the photons, their response is independent of wavelength.

A number of different semiconductor materials are in common use as optical detectors. These include silicon in the visible, near ultraviolet, and near infrared, germanium and indium gallium arsenide in the near infrared, and indium antimonide, indium arsenide, mercury cadmium telluride, and germanium doped with copper or gold in the long-wavelength infrared. The most frequently encountered type of photodiode is silicon. Silicon photodiodes are widely used as the detector elements in optical systems in the spectral range of 400–1100 nm, covering the visible and part of the near-infrared regions.

Detectors used in the ultraviolet, visible, and infrared respond to the amount of energy in the optical signal, which is proportional to the square of the electric field. Thus they are often referred to as "square-law" detectors because of this property. In contrast, microwave detectors measure the electric field intensity directly.

4.1.1. General Types of Detectors

Detectors may be divided into several broad types. Photoconductive and photovoltaic detectors are commonly used in circuits in which there is a load resistance in series with the detector. The output is read as a change in the voltage drop across the resistor. Photoemissive detectors generally have internal gain and are essentially current sources.

Photoconductive. The electrical conductivity of a photoconductive detector material changes as a function of the intensity of the incident light. Photoconductive detectors are semiconductor materials that are characterized by an energy gap that separates the electron valence band from the conduction band. A semiconductor normally has no or few electrons in the conduction band, so that the material has few free elections and conducts electricity poorly. When an electron in the valence band absorbs a photon having an energy greater than the energy gap, it can move from the valence band into the conduction band. This increases the number of free electrons and increases

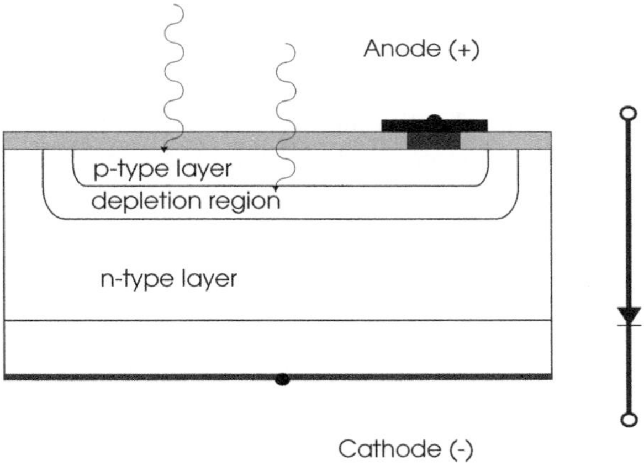

Fig. 4.1. A cross section of a typical silicon photodiode.

the conductivity of the semiconductor. Moving the electron into the conduction band leaves an excess positive charge, or hole, in the valence band, which can also contribute to conductivity. The conductivity of a photoconductor increases (resistance decreases) as the number of absorbed photons increases. These devices are normally operated with an external electrical bias voltage and a load resistor in series (Section 4.2). When the device is connected in a biased electric circuit, the current through the material is proportional to the intensity of the light absorbed by the material.

Photovoltaic. These detectors contain a p-n semiconductor junction and are often called photodiodes. The operation of photodiodes relies on the presence of a p-n junction in a semiconductor. When the junction is not illuminated, an internal electric field is present in the junction region because there is a change in the energy level of the conduction and valence bands in the two materials. This gives the diode a low forward resistance (anode positive) and a high reverse resistance (anode negative). A cross section of a typical silicon photodiode is shown in Fig. 4.1. N-type silicon is the starting material and forms most of the bulk of the device. The usual p-type layer for a silicon photodiode is formed on the front surface of the device by the diffusion of boron to a depth of approximately 1 μm. This forms a layer between the p-type layer and the n-type silicon known as a p-n junction. The electric field across the p-n junction causes the free electrons to move out of the region, depleting it of electrical charges and leading to the name "depletion region." The depth of the depletion region may be increased by the application of a reverse-bias voltage across the junction. When the depletion region reaches the back of the diode, the photodiode is said to be "fully depleted." The depletion region is important to photodiode performance because most of the sensitivity to radiation originates there. By varying and controlling the thickness of the various layers and the doping concentrations, the spectral and frequency response can be controlled. Small metal contacts are applied to the front and back surfaces of the device to form the electrical connections. The back contact is the cathode; the front contact is the anode. The active area is generally coated with a material such as silicon nitride, silicon monoxide, or silicon dioxide for protection, which may also serve as an antireflection (AR) coating. The thickness and type of this coating may be optimized for particular wavelengths of light.

When the junction is illuminated, photons pass through the p-type layer, are absorbed in the depletion region, and, if the photon energy is large enough, produce hole-electron pairs. The electric field in the junction separates the pairs and moves the electrons into the n-type region and the holes into the p-type region. This leads to a change in voltage that may be measured externally. This process is the origin of the photovoltaic effect used in solar cells, which may be used to generate energy. The photovoltaic effect is the generation of voltage when light strikes a semiconductor p-n junction. In the photovoltaic and zero-bias modes, the generated voltage is in the diode forward direction. Thus the polarity of the generated voltage is opposite to that required for the biased mode.

A p-n junction detector with a bias voltage is known as a photodiode. For lidar purposes, one generally applies a reverse-bias voltage to the junction. The reverse direction is the direction of low current flow, that is, a positive voltage is applied to the n-type material. The current that passes through an external load resistor increases with increasing light level. In practice, the voltage drop appearing across the resistor is the measured parameter. A reverse-biased photodiode has a linear response as long as the photodiode is not saturated and the bias voltage is higher than the product of the load resistance and the current. A reverse-biased photodiode has higher responsivity, faster response time, and greater linearity than a photodiode operated in the forward-biased mode. A drawback is the presence of a small dark current. In a forward-biased mode, the dark current may be eliminated. This makes photovoltaic devices desirable for low-level measurements in which the dark current would interfere. However, the responsivity and speed decrease in the forward-biased mode and the response becomes nonlinear for large values of the load resistance.

The capacitance of the diode, and thus the frequency response of a p-n junction, depends on the thickness of the depletion region. Increasing the bias voltage increases the depth of this region and lowers capacitance until a fully depleted condition is achieved. Junction capacitance is also a function of the resistivity of silicon used and the size of the active area.

Photoemissive. These detectors use the photoelectric effect, in which incident photons free electrons from the surface of a detector material. Operational devices have these materials on the inside of a glass vacuum tube where the freed electrons are collected with high-voltage electric fields. These devices include vacuum photodiodes, bipolar photomultiplier tubes, and photomultiplier tubes.

4.1.2. Specific Detector Devices

PIN Diodes. The PIN photodiode was developed to increase the frequency response of photodiodes. This device has a layer of intrinsic material between the thin layer of p-type semiconductor and the thick layer of n-type semiconductor that normally constitute a photodiode. A sufficiently large reverse bias voltage is applied so that the free carriers are swept out of the depletion region, spreads to occupy the entire volume of intrinsic material. This region has a high and nearly constant electric field. Light that is absorbed in the intrinsic region produces free electron-hole pairs, provided that the photon energy is high enough. These carriers are swept rapidly across the region and collected in the heavily doped regions. The carriers that are generated in the intrinsic region experience the highest electric field, are swept out the most rapidly, and provide the fastest response. The pin photodiode has a large intrinsic region designed to absorb light and minimize the contributions of the slower p- and n-type material. The frequency response of p-n junctions with an intrinsic region can be very high, on the order of 10^{10} Hz.

Photoconductor. Photoconductive detectors are most widely used in the infrared spectrum, at wavelengths where photoemissive detectors are not available and the wavelengths are much longer than the cutoffs of the best photodiodes (silicon and germanium). Because semiconductors will operate only over a relatively narrow wavelength range, many different materials are used as infrared photoconductive detectors. Typical values of spectral detectivity as a function of wavelength for some common devices operating in the infrared are shown in Fig. 4.2. The exact value of detectivity for a specific photoconductor depends on the operating temperature and on the field of view of the detector. Most infrared photoconductive detectors operate at cryogenic temperatures (<100 K), which may involve some inconvenience in practical applications.

In its most simple form, a photoconductive detector is a crystal of semiconductor material that has low conductance in the dark and an increased value of conductance when it is illuminated. In a series circuit with a battery and a load resistor, the detector element has a lower resistance, passing more current when exposed to light. The amount of light falling on the detector is proportional to the magnitude of the current, and thus the voltage drop across the load resistor. It is also possible to use photodiodes in a photoconductive mode.

Charge-Coupled Device (CCD). A more sophisticated photodetector most often used as part of a large array of detectors, the CCD is a small capacitor composed of metal, oxide, and semiconductor (MOS) layers, capable of both

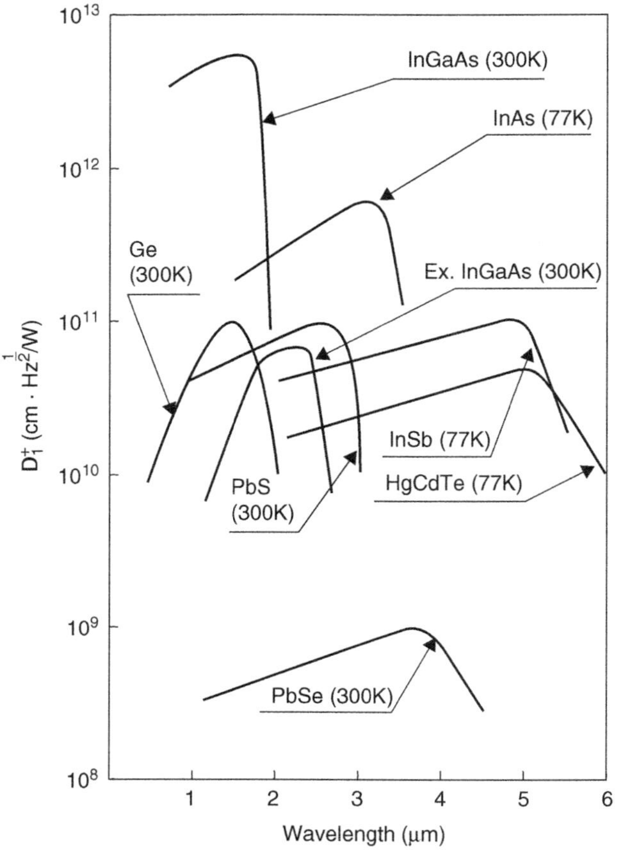

Fig. 4.2. Typical values of spectral detectivity for some common devices operating in the infrared.

photodetection and storage of charge. When a positive voltage is applied to the metal layer (called the gate), electron-hole pairs created in the semiconductor by the absorption of a photon are separated by an electric field and the electrons become trapped in the region under the gate. This trapped charge represents a small portion of an image known as a pixel. The complete image can be recreated by reading out a sequence of pixels from an array of CCDs. These arrays are used to capture images in video and digital cameras.

Avalanche Photodiode (APD). An avalanche photodiode is a p-n junction photodetector that is operated at a high reverse-bias voltage so that charges are rapidly swept from the depletion region. The applied voltage is close to the breakdown voltage of the material. Avalanche photodiodes are designed to have uniform junction regions so that they are able to handle the high electric fields generated in the depletion region. Gain occurs as electrons and holes accelerate inside the depletion region and cause ionizations (releasing more electrons or holes) as they collide with electrons in the material. A large current may be produced when light strikes the diode. The larger the applied voltage, the greater the number of ionizations achieved and the larger the amplification.

The most widely used material for avalanche photodiodes is silicon, but they have been fabricated from other materials, most notably germanium. An avalanche photodiode has a diffuse p-n junction, with surface contouring to permit the application of a high reverse-bias voltage without breakdown. The large internal electric field leads to multiplication of the number of charge carriers through ionizing collisions. The signal is increased, by a factor of 10–50 typically, but can be as much as 2500 times that of a nonavalanche device. High multiplication values can be achieved, but the process is generally noisy. Avalanche photodiodes cost more than conventional photodiodes, and they require temperature-compensation circuits to maintain the optimum bias, but they represent an attractive choice when high performance is required.

Phototransistors. are also used to amplify light signals. Their construction is similar to conventional transistors except that one of the transistor's junctions is exposed to light. In bipolar phototransistors, it is the base-emitter junction that is exposed to radiation; in field-effect phototransistors it is the gate junction.

Photomultiplier Tubes. A photomultiplier tube is an electron tube composed of a photocathode coated with a photosensitive material. Light falling upon the cathode causes the release of electrons into the tube through the photoelectric effect. These electrons are attracted to and accelerated toward the positively charged first dynode. The dynodes are arranged so that electrons from each dynode are directed toward the next dynode in the series. Electrons emitted from each dynode are accelerated by the applied voltage toward the next dynode, where their impact causes the emission of numerous secondary electrons. These electrons are accelerated to generate even more electrons in the next dynode. Finally, electrons from the last dynode are accelerated to the anode and produce a current pulse in the load resistor (representing an external circuit). Figure 4.3 shows a cross-sectional diagram of a typical photomultiplier tube structure. These tubes have a transparent end window coated on the inside with a photocathode material (a material with a low work function). With a good design, emitted photoelectrons can produce between one and eight secondary electrons at each dynode impact. The resulting flow of electrons is proportional to the intensity of the light falling on the photocathode. A photomultiplier tube is capable of detecting extremely low intensity levels of light and even individual photons.

The current gain of a photomultiplier is defined as the ratio of anode current to cathode current. Typical values of gain range from 100,000 to 10,000,000. Thus 100,000 or more electrons reach the anode for each photon

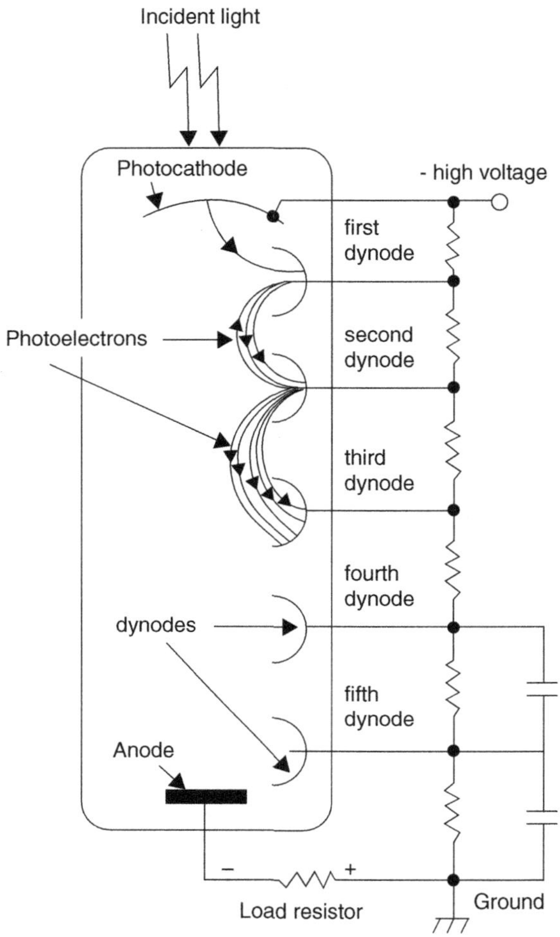

Fig. 4.3. A conceptual diagram of a photomultiplier with five dynodes. Electrons released from the photocathode are accelerated toward the next dynode, releasing additional electrons with each impact.

striking the cathode. This high-gain process means that photomultiplier tubes offer the highest available responsivity in the ultraviolet, visible, and near-infrared portions of the spectrum. Photomultiplier tubes come in two common types, end-on tubes, where the photocathode is on the end of the cylindrical tube, and side-on tubes, where the photocathode is on the side of the tube. In general, end-on tubes have higher gain, a faster time response, and more uniform response across the photocathode, whereas side-on tubes have higher quantum efficiency.

The spectral response curves (the amount of current per watt of light on the detector) for photomultipliers are governed by the materials used in the

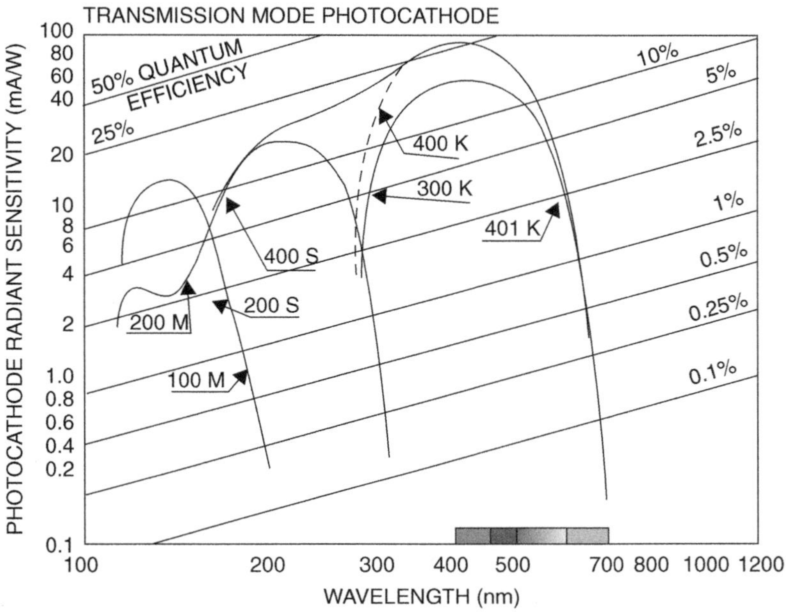

Fig. 4.4. A plot of the spectral response of several types of photomultipliers. Numbers indicate types of photocathode materials; 100M, CsI, 200M, 200S, CsTe, 300K, SbCs, 400K, alkali, 400S, multialkali. Courtesy of Hamamatsu.

cathode (Fig. 4.4). These materials have low work functions, that is, incident light with longer wavelengths may cause the surfaces to emit an electron. The cathodes are often mixtures containing alkali metals, such as sodium, cadmium, cesium, tellurium, and potassium. The usefulness of these devices extends from the ultraviolet to the near infrared. For wavelengths longer than 1.2 μm, few photoemissive materials are available. The short-wavelength end of the response curve is determined by the material used in the window in the tube. Common window materials include MgF_2 (50% transmission at 120 nm), synthetic quartz (50% transmission at 160 nm), UV glass (50% transmission at 210 nm), and borosilicate glass (50% transmission at 300 nm). With a wide range of materials available, one selects a device with a window and photocathode material that maximizes the response in the desired portion of the spectrum.

The circuitry used in photomultiplier tubes requires high voltages, in the kilovolt range. Because the gain of photomultiplier tubes is a strong function of the applied voltage, a small change in power supply voltage may result in a large change in the gain. Thus one must use a well-regulated, stable power supply for photomultiplier applications that is capable of supplying the maximum current required. The base in which the photomultiplier is mounted also contains a voltage-divider circuit, as illustrated in Fig. 4.3 for a five-stage photomultiplier. Voltages on the order of 100–300 V are required to accelerate electrons between the dynodes, so that the total tube voltage ranges from 500 to 3000 V, depending on the number of dynodes used. A string of resistors of equal value is connected in parallel with the dynodes. The relative values between the resistors determine the voltage that is applied from one dynode to the next. This arrangement is called a voltage-divider network. This arrangement is normally used with photomultipliers, instead of applying separate voltage sources to each dynode. The response of the photomultiplier at high counting rates may become nonlinear as the impedance of the tube changes (Zhong et al., 1989). Capacitors are often added across the last few dynodes to maintain the desired voltage when high current and high gain are needed. The capacitors help to maintain the desired voltage drop across the last dynodes. The total current amplification obtained in the tube is given by:

$$\text{amplification} = \left[C\left(\frac{V}{n+1}\right)^{\alpha}\right]^{n} \tag{4.1}$$

where C is a constant, n is the number of dynodes in the tube, V is the voltage applied across each dynode, and α is a coefficient determined by the dynode material and the geometry of the dynode chain (~0.75). Thus the amount of gain in the tube is governed by the number of dynodes and the applied voltage.

A small amount of current (known as the dark current) flows even when the face of the tube is not illuminated. This current flows because the materials used as the photocathode have low work functions and will emit thermal electrons at room temperature. The magnitude of the dark current is a function of the photcathode material, the temperature of the tube, and the applied voltage. Most manufacturers sell thermoelectric coolers for applications where a low dark current is desired.

Photomultipliers may be susceptible to magnetic fields. The dynode chains are designed and shaped to create electric fields that guide the electrons along preferred pathways to maximize the gain. The presence of external magnetic fields deflects the electrons from the preferred trajectories and lowers the overall gain. The more compact the photomultiplier, the less sensitive it is to magnetic fields. Most photomultiplier tube bases are equipped with shields made of materials with large magnetic permeability. These shields should be connected to the electrical ground.

The signal-to-noise ratio of an analog signal level in a photomultiplier is given by (Inaba and Kobayasi, 1972)

$$\text{SNR} = \frac{n(p\tau)^{1/2}}{[n + 2(n_{\text{b}} + n_{\text{d}})]^{1/2}} \tag{4.2}$$

where SNR is the signal to noise ratio, n is the number of photoelectrons emitted per unit time, p is the number of summed signal pulses, τ is the sampling time interval, n_b is the number of photoelectrons due to background light, and n_d is the number of effective photoelectrons due to dark current in the photomultiplier. Lidar signals from several laser pulses are often added (or averaged) to obtain a greater signal-to-noise ratio.

For lidar purposes, photomultipliers offer the largest amount of gain with the smallest amount of noise. However, they are susceptible to overloading, usually from background sunlight (Keen, 1965; Lush, 1965; Fenster et al., 1973; Hunt and Poultney, 1975; Hartman, 1978; Pitz, 1979). In the region from 300 to 1000 nm, a 3-nm filter allows enough sunlight through to saturate the photomultiplier unless steps are taken to limit the field of view of the telescope. Because of this, most systems using photomultipliers operate only at night. If the voltage between the cathode and the first dynode is turned off between the individual laser pulses, the electrons emitted from the cathode will not travel to the first dynode, effectively turning the tube off. This procedure is known as gating the photomultiplier. It has been used by some to overcome the problem of saturation, and several methods of high speed switching have been developed (Barrick, 1986; Lee et al., 1990). Some manufacturers sell gated bases, requiring only a transistor-transistor logic (TTL) pulse to turn the tube on. Gating helps reduce the effects of saturation but will not solve the saturation problem unless it is used as part of a larger effort to reduce the spectral width of the filter and the field of view of the telescope.

Although many consider photomultipliers to be an old, dead technology, they generally offer the highest degree of amplification with the lowest noise, and work continues to improve their capabilities. New photocathode materials will greatly increase photomultiplier capabilities. GaAsP, GaAs, and blue-enhanced GaAs may increase the quantum efficiency of photocathodes by as much as a factor of 2. Quantum efficiencies over 50% may be possible in the visible portion of the spectrum with GaAsP photocathodes. GaN is a promising material as a high-efficiency solar blind photocathode. Similar improvements are occurring in the infrared. Photomultipliers are currently available that are sensitive out to 1700 nm.

In addition to changes in photodiode materials, changes in materials and design are also improving photomultiplier performance. Metal channel dynodes have made it possible to construct extremely small photomultipliers. Multiple-element detectors are becoming increasingly available, offering increasing opportunities for low-light-level imaging. Improvements have also been made to reduce noise in the detectors. The use of low-potassium glass (which eliminates radioactive ^{40}K), new electro-optics designs, minimizing feedback and cooling the photocathode have resulted in significant noise reductions. The ability to cool the photocathode will become increasingly important as new materials increase the sensitivity at longer wavelengths. Photomultipliers are increasingly being packaged as a complete assembly. These packages require only a single, low-voltage power supply to operate them. Photon counting modules are available that provide the photomultiplier, high-voltage power supply, and discriminator all in a smaller box. These devices require only a low-voltage power supply and output a standard TTL pulse used by photon counters.

Calorimeter. A calorimeter is not really intended for use as a lidar detector but is often used as a calibration device for laser energy. Calorimetric measurements yield a simple determination of the total energy in a laser pulse but usually do not respond rapidly enough to follow the pulse shape. Calorimeters designed for laser measurements usually use a blackbody absorber with a low thermal mass and with temperature-measuring devices in contact with the absorber to measure the temperature rise. With knowledge of the thermal mass, measurement of the temperature change allows determination of the energy in the laser pulse. The temperature-measuring devices include thermocouples, bolometers, and thermistors. Bolometers and thermistors respond to the change in electrical resistivity that occurs as temperature rises. Bolometers use metallic elements; thermistors use semiconductor elements.

4.1.3. Detector Performance

The performance of optical detectors is described by several figures of merit that are used to describe the ability of a detector to respond to a small signal in the presence of noise. Detectors are rated in terms of their responsivity, $R(\lambda)$ at a given wavelength λ, by their noise, by their linearity, and by their temporal characteristics. The responsivity is defined as the ratio of the output current of the detector, in amperes, to the incoming light flux in watts. $R(\lambda)$ ranges from 0.4 to 0.85 A/W for Si PIN diodes and from 8 to 100 A/W for avalanche photodiodes. The responsivity is a characteristic that is usually specified by a manufacturer and is dependent on the wavelength of light used. Responsivity gives no information about the noise characteristics of the detector.

Also common is the quantum efficiency, η, defined as the average number of photoelectrons generated for each incident photon; η is related to the responsivity as

$$\eta(\lambda) = \frac{1.2399R(\lambda)}{\lambda} \tag{4.3}$$

It should be noted that for sensors with the ability to amplify internally, such as avalanche photodiodes, the quantum efficiency is quoted only for the primary photosensor and does not include the internal gain. Thus quantum efficiencies are numbers less than 1.

The response of a given detector material is a strong function of wavelength. Thus the desired range of wavelengths of the radiation to be detected is an important design parameter. On the long-wavelength end of the spectrum, there is a rapid drop in the detector response because the photons at these wavelengths lack the energy to free an electron. Silicon, for example,

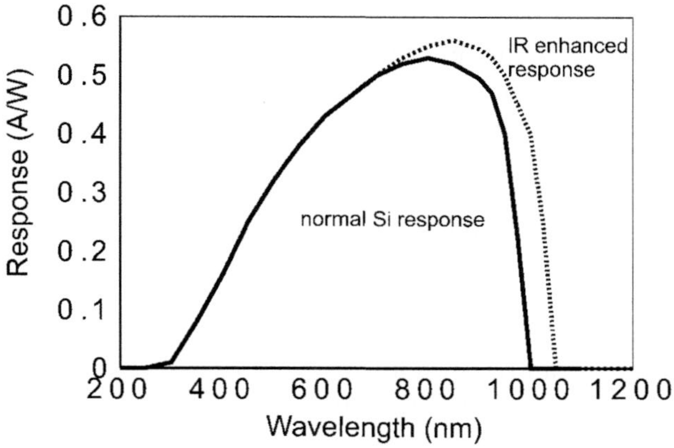

Fig. 4.5. The spectral responsivity of a typical commercial silicon photodiode (solid line) and the IR-enhanced version of the same diode (dashed line).

becomes transparent to radiation longer than 1100 nm wavelength and is thus not suitable for use at wavelengths appreciably longer than this. Detectors also exhibit a gradual decrease in response as the wavelength becomes shorter as well. This is due to the decreasing ability of short-wavelength photons to penetrate into the material. Protective surface coatings also affect the spectral response of the detector. Many photodiodes have antireflection coatings that can enhance the response at the desired wavelength but may reduce efficiency at other wavelengths that are preferentially reflected. The window on the case holding the photodiode may also modify the spectral response. A standard glass window absorbs wavelengths shorter than 300 nm. Special filter windows are also available to make it possible to adjust the spectral response to suit the application. The spectral responsivity of a typical commercial silicon photodiode is shown in Fig. 4.5. The responsivity reaches a peak value around 0.55 A/W near 900 nm, decreasing at longer and shorter wavelengths. Other materials provide somewhat extended coverage in the infrared or ultraviolet regions. Silicon photodiodes are useful for the detection of signals at many of the most common laser wavelengths, including argon ion (418–514 nm), copper ion (510–578 nm), He-Ne (632 nm), ruby (694 nm), Ti:sapphire (600–950 nm), and Nd:YAG (355, 532, and 1064 nm). As a practical matter, silicon photodiodes have become the detector of choice for many laser applications. They represent well-developed technology and are widely available.

Another important characteristic of detectors is their linearity. Photodetectors are characterized by a response that is linear with incident light intensity over a broad range, perhaps several orders of magnitude. If the output of the detector is plotted versus the input power, there should be no change in the slope of the curve. Then noise will determine the lowest level of incident light that is detectable. The upper limit of the input/output linearity is determined by the maximum current that the detector can handle without becoming saturated. Saturation is a condition in which there is no further increase in detector response as the input light is increased. Linearity may be quantified in terms of the maximum percentage deviation from a straight line over a range of input light levels. For large current pulses, amplifier circuits may also recover in a manner that oscillates about true voltage for some period after the pulse. The oscillations may be short or long with respect to the original voltage pulse and depend on the circuit characteristics. These oscillations can often be seen in the response of the lidar detector to the light pulse from low-level clouds.

When the incident light level is low, the range over which a linear response may be maintained can be as much as nine orders of magnitude, depending on the type of photodiode and the operating circuit. The lower limit of this linearity is determined by the noise equivalent power (NEP), (the lowest amount of light signal for which the signal-to-noise ratio is 1), whereas the upper limit depends on the load resistance, reverse-bias voltage, and saturation voltage of the amplifier. A manufacturer often specifies a maximum allowable continuous light level. Light levels in excess of this maximum may cause saturation, hysteresis effects, or irreversible damage to the detector. If the light occurs in the form of a very short pulse, it may be possible to exceed the continuous rating by some factor (perhaps as much as 10 times) without damage or noticeable changes in linearity.

An AC-coupled receiver has a capacitor in series with the load resistor so that it has no response at DC. These receivers may be useful when a small signal must be detected in the presence of a large cw component (such as in measuring a lidar return in a large solar background). An AC-coupled detector will be insensitive to the large cw component which, in a DC-coupled detector, would saturate the receiver's internal amplifier. Typically, a low-frequency cut-off is specified for a detector-amplifier system, below which there is little response.

4.1.4. Noise

The detection of any electromagnetic signal of interest must be performed in the presence of noise sources, which interfere with the detection process. The limit to the ability to detect weak signals is determined by the amount of noise in the system. Noise is defined as any undesired signal that masks the signal that is to be detected. Sources of noise can be external or internal. External noise involves those disturbances that appear in the detection system because of actions outside the system. Examples of external noise could be pickup of hum induced by 60-Hz electrical power lines or static caused by electrical storms. Internal noise includes all noise generated within the detector-amplifier system.

Noise cannot be described in the same manner as usual electric currents or voltages. Current or voltage is normally described as a function of time, a sine-wave (alternating current) voltage, for example. The noise output of an electrical circuit as a function of time is completely random. The output at any time cannot be accurately predicted. Thus there will be no regularity in the waveform (a flat power spectrum is indicative of white noise). Because of the random nature of the noise, the voltage of interest fluctuates about some average value V_{ave}. Because the average value of the noise over some period of time is zero, the time average of the squares of the deviations around V_{ave} is used to quantify the magnitude of the noise. The average must be made over a period of time much longer than the period of the fluctuations.

A photodetector-amplifier combination consists of three parts: the detector, an operational amplifier, and a feedback resistor (see Section 4.2). This model will have three contributions to noise: detector noise, amplifier noise, and thermal noise. One commonly used measure of system noise is the noise equivalent power (NEP). NEP is defined to be the minimum incident power needed to generate a photocurrent I equal to the total noise of the system at a specified frequency f within a specified frequency bandwidth Δf.

$$\mathrm{NEP}_{\mathrm{total}} = \frac{I_{\mathrm{noise\,(total)}}}{R(\lambda)} \tag{4.4}$$

where $R(\lambda)$ is the detector responsivity at wavelength λ. Related to the NEP is the detectivity, D, which is the inverse of the NEP. However, the specific detectivity, D^*, is most often quoted. In most infrared detectors, the NEP is proportional to the square root of the sensitive area A and bandwidth Δf. D^* then allows comparisons between detectors of different areas and bandwidths. D^* is defined as:

$$D^* = \frac{\sqrt{A\Delta f}}{\mathrm{NEP}} \tag{4.5}$$

A high value of D^* means that the detector is suitable for detecting weak signals in the presence of noise.

For detectors with no gain the NEP is not very useful, and when specified for these types of devices it should only be used to compare similar detectors. The amplifier or instrument that follows the detector will almost always produce additional noise exceeding that produced by the detector with no illumination. Attention should always be paid to obtain a low-noise amplifier in order to improve the overall sensitivity.

A photodiode can be operated in either a photovoltaic mode or a biased mode. In the photovoltaic mode, no bias voltage is applied. In this mode, detectors have as much as a factor of 2–5 less noise but the frequency response is significantly degraded. The noise spectrum versus frequency is nearly flat from DC to the cutoff frequency of the photodiode. Lidar detectors are operated in a biased mode to achieve the highest possible frequency response. The applied voltage causes the photoelectrons generated by the incoming photons to be rapidly swept from the region in which they are generated. However, this causes the noise to be greater because the bias voltage causes a leakage or dark current resulting in shot noise. The dark current is that current which flows in the detector in the absence of any signal or background light. The detector shot noise is generated by random fluctuations in the total current. The shot noise is given by

$$I_{\text{noise (shot)}} = \sqrt{2q(I_{\text{dark}} + I_{\text{background}} + I_{\text{photocurrent}})\Delta f} \tag{4.6}$$

where $q = 1.6\ 10^{-19}$ C is the charge of the electron, I_{dark} is the dark current (amperes), $I_{\text{background}}$ is the background current, $I_{\text{photocurrent}}$ is the signal photocurrent (amperes), and Δf is the bandwidth (Hertz). It is implicitly assumed that the individual currents are statistically independent so that the noise contributions can be added in this way. The shot noise may be minimized by keeping any DC component to the current small, especially the background light levels and the dark current, and by keeping the bandwidth of the amplification system as small as possible.

The term "shot noise" is derived from fluctuations in the stream of electrons in a vacuum tube. These variations create noise because of the random fluctuations in the arrival of electrons at the anode at any moment. It originally was likened to the noise of a hail of shot striking a target; hence the name shot noise. In semiconductors, the major source of noise is random variations in the rate at which charge carriers are generated and recombine. This noise, called generation recombination is the semiconductor counterpart of shot noise.

For avalanche photodiodes that have internal amplification, noise can be viewed as a statistical process creating electron-hole pairs. If the ionization rates for electrons and holes are the same, then the root-mean-square noise current at high frequencies is given by (McIntyre, 1966)

$$I_{\text{APDnoise}} = M\sqrt{2qM(I_{\text{dark}} + I_{\text{background}} + I_{\text{photocurrent}})\Delta f} \tag{4.7}$$

where M is the multiplication factor achieved in the diode and the currents, I_{dark}, $I_{\text{background}}$, and $I_{\text{photocurrent}}$ are the currents before amplification. The noise is increased by a factor of $M^{1/2}$ above noise-free amplification.

When connected to a circuit, particularly an amplifier, several other sources of noise should also be considered. The detector thermal (also known as the Johnson) noise is a function of the feedback resistance of the detector-amplifier combination and the temperature of the resistor. Thermal noise is a type of noise generated by thermal fluctuations in conducting materials. It results from the random motion of electrons in a conductor. The electrons are in constant motion, colliding with each other and with the atoms of the material. Each motion of an electron between collisions represents a tiny current.

The sum of all these currents taken over a long period of time is zero, but their random fluctuations over short intervals constitute Johnson noise

$$I_{\text{johnson}} = \sqrt{\frac{4kT\Delta f}{R_{\text{feedback}}}} \tag{4.8}$$

where $k = 1.38 \times 10^{-23}$ J/K is the Boltzmann constant, T is the absolute temperature, and R_{feedback} is the resistance of the feedback resistor. This expression suggests methods to reduce the magnitude of the thermal noise. Reducing the value of the load resistance will decrease the noise level, although this is done at the cost of reducing the available signal. Reduction of the bandwidth of the amplification to the minimum necessary level will also lower the noise level. Because temperature plays a role in this type of noise generation, cooling the detector-amplifier can significantly reduce the overall noise. Cooling will not help a detector-amplifier combination in which noise is dominated by the amplifier noise. If long-term stability is required, as for example in a calibrated lidar system, thermal stabilization may be required to eliminate variations in the detector-amplifier output with changes in outside temperature.

The last contribution to noise is the amplifier noise. Amplifier noise is a function of frequency as

$$I_{\text{amp noise}} = \sqrt{<I_{\text{amp}}>^2 + <V_{\text{amp}}\, 2\pi f C_{\text{T}}>^2} \tag{4.9}$$

where I_{amp} is the amplifier input leakage current, V_{amp} is the amplifier input noise voltage, and C_{T} is the total input capacitance as seen by the amplifier. I_{amp} and V_{amp} are characteristics of the amplifier and are normally specified by the manufacturer.

The total noise of the detector-amplifier system can be estimated by

$$I_{\text{totalnoise}} = \sqrt{<I_{\text{ampnoise}}>^2 + <I_{\text{noise(shot)}}>^2 + <I_{\text{johnson}}>^2} \tag{4.10}$$

The term $1/f$ noise ("one over f") is used to describe a number of types of noise that may be present when the modulation frequency is low. This type of noise is also called excess noise because it is larger than the shot noise at frequencies below a few hundred hertz. In photodiode detector-amplifier systems, it is sometimes called boxcar noise, because it may suddenly appear and then disappear in small boxes of noise observed over a period of time. The mechanisms that result in $1/f$ noise are poorly understood, and there is no simple mathematical expression that may be used to predict or quantify the amount of $1/f$ noise. The noise power is inversely proportional to the frequency, which results in the name for this type of noise. To reduce $1/f$ noise, a photodetector should be operated at a reasonably high frequency; 1000 Hz is often taken as a minimum. This value is high enough to reduce the contribution of $1/f$ noise to a negligibly small amount.

Even if all the sources of noise discussed here could be eliminated, there would still be some noise present in the output of a photodetector because of the random arrival rate of backscattered photons and from the sky background. This contribution to the noise is called photon noise, and it is a noise source external to the detector. It imposes a fundamental limit to the detectivity of a photodetector. The noise associated with the fluctuations in the arrival rate of photons in the signal is not something that can be reduced. The contribution of fluctuations in the arrival of photons from the background, a contribution that is called background noise, can be reduced. In lidar systems, the background noise increases with square of the field of view of the telescope-detector system and with the brightness of the sky. In general, it is recommended that the field of view of the telescope-detector system be reduced so as to match or slightly exceed the divergence of the laser beam. The field of view must not be reduced below the laser beam divergence. Should the application require that the field of view be further reduced, the laser beam can be expanded with a corresponding reduction in the divergence. The use of an extremely narrow field of view and expanded laser beam is the method used by the micropulse lidar (Chapter 3) to reduce the amount of background light. A consequence of the use of a narrow field of view is that the lidar system becomes increasingly difficult to align. The effects of background light can be reduced by inserting an optical filter between the collection optics and the light detector. The amount of light hitting the detector must be dramatically reduced to produce a sizable reduction in the induced noise. This requires the use of narrow-band interference filters, which are selected to match the wavelength of the laser (or the desired return wavelength) to reduce the amount of background light while passing the maximum amount of the desired light signal. Even with a reduced field of view, it is not uncommon to overload the detector when the lidar signal becomes stronger than expected, such as when encountering low-level clouds. Figure 4.6 is an example showing a ringing detector response above a dense layer of low-level clouds. The amplified signal from the clouds is about 10^4 times larger than the air just below the clouds. This is larger than the dynamic range of the amplifier and produces a decaying sinusoidal response, often referred to as "ringing."

4.1.5. Time Response

Most detectors are rated in terms of their rise time or their response time. Both are a measure of the amount of time required for the detector to respond to an instantaneous change in the input light level. Because photodetectors often are used for detection of fast pulses, the time required for the detector to respond to changes in the light levels is an important consideration. The response time is the time it takes the detector current to rise to a value equal to 63.2% of the steady-state value in response to an instantaneous change in the input light level. The recovery time is the time photocurrent takes to fall to 36.8% of the steady-state value when the light level is lowered instanta-

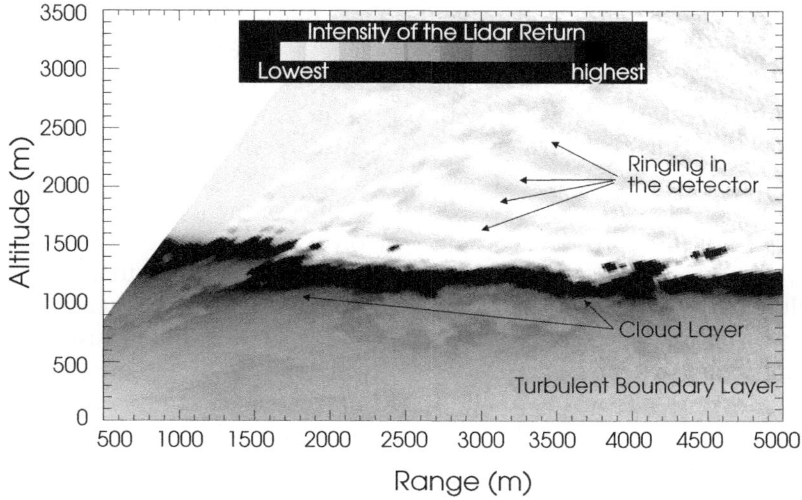

Fig. 4.6. A lidar return (r^2 corrected) from a convective boundary layer in New Jersey. The darkest returns indicate the largest lidar returns. Note the periodic nature of the returns above the cloud layer. This is an example of the nonlinear response of a detector-amplifier combination to a signal larger than the dynamic range of the combination.

neously. The rise time t_r of a diode is the time difference between the points at which the detector has reached 10% of its peak output and the point at which it has reached 90% of its peak output when it is exposed to a short pulse of light. The fall time is defined as the time between the 90% point and the 10% point on the trailing edge of the pulse. This is also known as the decay time. We note that the time required for a signal to respond to a decrease in the light level may be different from the time required to respond to an increase in the light level. Another measure of time response is the 3-dB frequency specification. If the light input to a diode is modulated sinusoidally and the frequency increased, then the point at which the output signal power falls to ½ of a low-frequency reference is the 3 dB point. An optical 3-dB specification is equivalent to an electrical 6-dB frequency and therefore is larger than the electrical 3-dB frequency, f_{3db}. The rise time is related to the 3-dB frequency by the approximation

$$t_r = 0.35/f_{3db} \tag{4.11}$$

For photodiodes, the response time is determined by the amount of time required to generate and collect the photoelectrons as well as the inherent capacitance and resistance associated with the device. To obtain the fastest response times, the resistivity of the silicon and an operating voltage must be chosen to create a depletion layer of sufficient size so that the majority of the charge carriers are generated inside the layer. Because the depth of the depletion region increases rapidly as the wavelength increases, the charge collection time increases as the wavelength increases. Thus rise times can be as much as 10 times shorter at a wavelength of 900 nm compared to 1064 nm for the same device. Thus the wavelength at which the response time is specified is also important.

Response times are also affected by the value of the load resistance that is used. The selection of a load resistance involves a trade-off between the speed of the detector response and high sensitivity. It is not possible to achieve both simultaneously. Fast response requires a small load resistance (generally 50 Ω or less), whereas high sensitivity requires a high value of load resistance. It is also important to keep any capacitance associated with the circuitry or display device as low as possible to keep the RC time constant [1/(system resistance * system capacitance)] low. Rise times are also limited by electrical cables and by the capabilities of the recording device.

The best response is obtained through the use of fully depleted detectors (using a bias voltage) and with a small load resistance. Increasing the bias voltage increases the carrier velocity inside the depletion region and decreases the response time. Because the diode has a capacitance related to the size of the detector, the response may be limited to the RC time constant of the load resistance and the diode capacitance. As the active area A of the detector increases, the capacitance rises as

$$C_{\text{detector}} \propto \frac{A}{\sqrt{(V_{\text{bias}} + 0.5)\rho}} \tag{4.12}$$

where V_{bias} is the detector bias voltage and ρ is the resistivity of the detector.

Because of the bandwidth dependence on detector area, the tendency is to use the smallest detector size possible. However, small detectors require high-quality optics to focus the light, may limit the lidar system field of view, and may have problems with near-field versus far-field focusing if the optical system is not fast. The alignment of the laser-telescope system with a narrow field of view is sometimes difficult. In general, the use of a higher bias voltage will also increase the bandwidth but will also increase the dark current, I_{dark}, and thus increase the noise. However, in PIN diodes, the normal bias voltage fully depletes the detector, so increasing the bias voltage further is ineffective.

Manufacturers often quote nominal values for the rise times of their detectors. These should be interpreted as minimum values, which may be achieved only with careful circuit design and avoidance of excess capacitance and resistance. It should also be noted that there is a fast component and a slow component to the charge collection time. In some devices the slow component may be significant or even dominate and be a limiting factor for high-speed applications.

4.2. ELECTRIC CIRCUITS FOR OPTICAL DETECTORS

The design of electric circuits is a dynamic field in which new capabilities are constantly being developed. There are also a number of difficulties associated with the design and construction of high-bandwidth circuits that limit the ability of novices in the field to construct detector-amplifier circuits that are useful for lidar systems. The discussion below is intended to discuss such devices only in basic terms.

There are three basic design components to a photodiode-amplifier circuit that must be considered; the photodiode, the amplifier, and the R-C amplifier feedback network. A photodiode is primarily selected because of its response characteristics to incoming light. However, the intrinsic capacitance and resistance of the photodiode may also have an effect on the noise level, stability, and linearity of the circuit and must also be considered. An operational amplifier should have a low-input bias current so as to preserve the linearity of the diode. Again, the characteristics of the amplifier can affect the stability and fidelity of the response. The R-C feedback network is used to establish the gain of the circuit and sets one of the fundamental bandwidth limits. The network may also influence the stability and noise performance of the circuit.

Fundamentally, a photodiode functions as a current generator in which the magnitude of the current generated is proportional to the amount of light incident on the device. The equivalent electrical circuit for a photodiode is shown in Fig. 4.7.

The junction capacitance, C_d, is the result of the width of the depletion region between the p-type and n-type material in the photodiode. A deeper depletion region will increase the size of the junction capacitance. However, the deeper depletion regions found with PIN photodiodes have a greater frequency response. The junction capacitance of a silicon photodiode may range from approximately 20 pF to several thousand picofarads. The junction capacitance affects the photodiode stability, bandwidth and noise. The parasitic

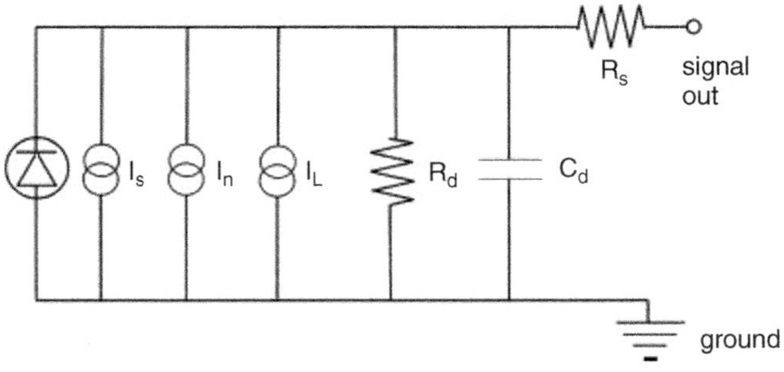

Fig. 4.7. An equivalent circuit model of a nonideal photodiode showing the signal current source, I_s, leakage current I_L, noise current I_n, junction capacitance C_d, series resistance R_s, and shunt resistance R_d.

resistance, R_d, is also called the shunt resistance. The shunt resistance is the resistance of the detector element in parallel with the load resistor in the circuit. This resistance is measured with the photodiode at zero bias. At room temperature, this resistance normally exceeds a hundred megohms. The shunt resistor, R_d, is the dominant source of noise inside the photodiode and is modeled as a current source, I_n. The noise generated by the shunt resistor is known as Johnson noise and is due to the thermal generation of carriers. The magnitude of this noise in terms of volts is (RCA 1974):

$$V_{\text{noise}} = \sqrt{4kTR_{\text{feedback}}\Delta f} \tag{4.13}$$

where k is the Boltzmann's constant, 1.38 10^{-23} J/K, T is temperature in Kelvin, and Δf is the bandwidth in Hertz.

The parasitic diode resistance, R_s, is known as the series resistance of the diode. This resistance typically ranges from 10 to 1000 Ohms. Because of the small value of this resistor, it only has an affect on the frequency response of the circuit at frequencies well above the operating bandwidth. Another source of error is due to the leakage of current across the photodiode, I_L. If the offset voltage of the amplifier is zero volts, the error due to the leakage current may be small.

When operated in its most basic form, without a bias voltage, the device acts in a photovoltaic mode. Figure 4.8 is an example of such a circuit. It produces a voltage proportional to the incident light intensity. In the circuit shown, an increase in light intensity increases the amount of current and thus the voltage drop across the load resistor, yielding a signal that may easily be monitored. This circuit is a low-noise circuit because it has almost no leakage current, so that shot noise is greatly reduced. An unbiased diode is used for maximum light sensitivity and linearity and is best suited for precision applications.

Because there is no amplification in this circuit, the value of the load resistor should be large in order to produce a large voltage drop. It is normal to

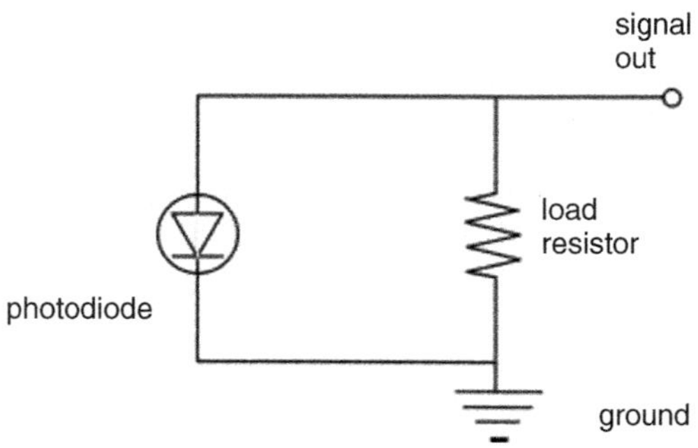

Fig. 4.8. The simplest form of an unbiased diode circuit. This type of circuit has the largest signal-to-noise ratio of the various types of circuits.

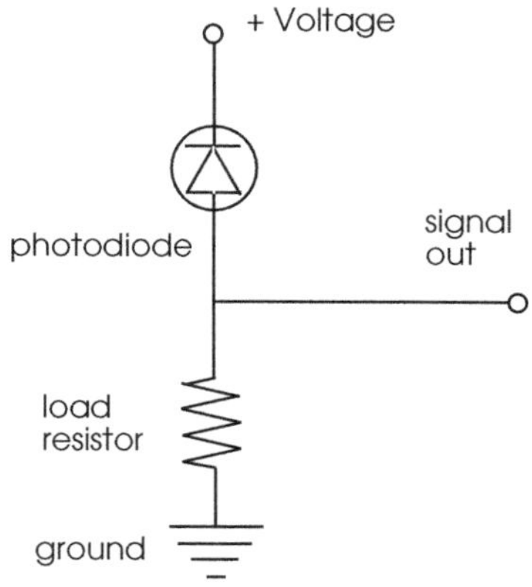

Fig. 4.9. The simplest form of a biased diode circuit. This type of circuit may be used in a trigger used to detect the firing of the laser and trigger the data collection process.

have the value of the load resistor much larger than the value of the shunt resistance of the detector. The value of the shunt resistance is specified by the manufacturer and for silicon photodiodes may be a few megohms to a few hundred megohms. However, the characteristics of the depletion region change as free carriers are deposited in the depletion region. The value of the detector shunt resistance drops exponentially as the light intensity increases. The output voltage then increases as the logarithm of the light intensity for intense light levels. Thus the response of this circuit may be nonlinear in nature and the magnitude of the signal depends on the shunt resistance of the detector. The value of the shunt resistance may be different from different production batches of detectors. This type of circuit has the highest signal to noise ratio. The bandwidth of the circuit is determined by the load resistance and the junction capacitance as bandwidth = $1/(2\pi R_L C)$.

To overcome these disadvantages, a photovoltaic photodiode is often used in a biased circuit such as shown in Fig. 4.9 or with an operational amplifier as in Fig. 4.10. Biasing the circuit enables high-speed operation; however, this comes at the cost of an increased diode leakage current (I_L) and linearity errors. In the case of Fig. 4.10, the photocurrent is fed to the virtual ground of an operational amplifier. In this case, the load resistance has a value much less than the shunt resistance of the photodiode. This provides amplification to counter the decreased voltage drop resulting from the low value of the load resistor. The use of a transimpedance amplifier in this circuit does not bias the photodiode with a voltage as the current starts to flow from the photodiode. One lead of the photodiode is tied to ground, and the other lead is kept at virtual ground by connection to the minus input of the transimpedance amplifier. This causes the bias across the photodiode to be nearly zero. This minimizes the dark current and shot noise and increases the linearity and detectivity of the detector. Because the input impedance of the inverting input

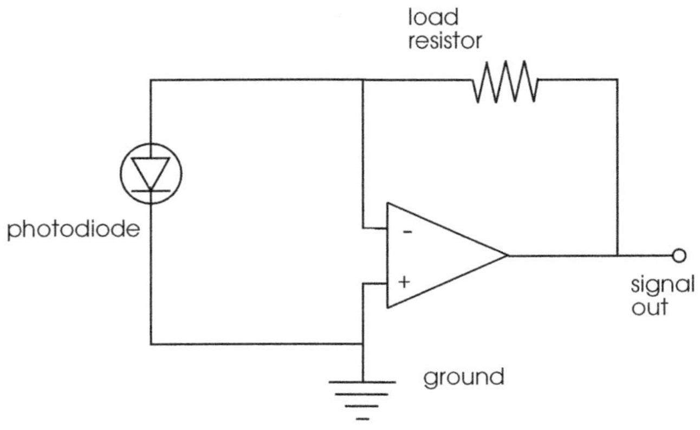

Fig. 4.10. Zero bias circuit with amplification.

of the CMOS amplifier is extremely high, the current generated by the photodiode flows through the feedback resistor $R_{feedback}$. The voltage at the inverting input of the amplifier tracks the voltage at the noninverting input of the amplifier. Thus the current output will change in accordance with the voltage drop across the resistor $R_{feedback}$. Effectively, the transimpedance amplifier causes the photocurrent to flow through the feedback resistor, which creates a voltage, $V = IR$, at the output of the amplifier.

This type of amplifier produces an inverted pulse; an increased level of light produces a voltage that is larger in the negative direction. In the photovoltaic mode, the light sensitivity and linearity are maximized and are best suited for precision applications. The key parasitic elements that influence circuit performance are the parasitic capacitance, C_D, and $R_{feedback}$, which affect the frequency stability and noise performance of the photodetector circuit.

An exceptionally fast time response is required for lidar applications. To achieve this, the detector circuitry uses a bias voltage and a feedback resistor in series with the detector, also known as a photoconductive mode. Figure 4.11 is an example of the simplest such circuit. The incident light changes the conductance of the detector and causes the current flowing in the circuit to change. The output signal is the voltage drop across the load resistor. The use of a feedback resistor is necessary to obtain an output signal. If the value of the load resistor were zero, all of the bias voltage would appear across the detector and there would be no distinguishable signal voltage. This type of circuit is capable of very high-frequency response. It is possible to obtain rise times on the order of a nanosecond. The biggest disadvantage of this circuit is that the leakage current is relatively large so that the shot noise may be significant. The basic power supply for a photodetector consists of a bias voltage applied to the detector and a load resistor in series with it. Figure 4.11 is an example of a negatively biased photodiode-amplifier circuit. This type of circuit produces a positive voltage signal for an increase in the light level.

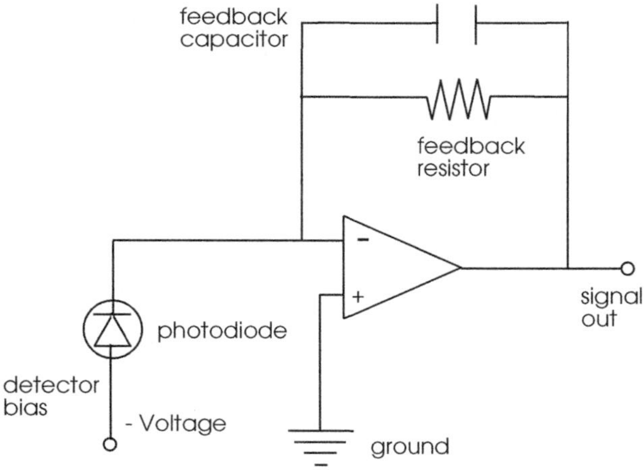

Fig. 4.11. A reverse-bias circuit with amplification.

In the photoconductive mode, the shunt resistance is nearly constant. Thus it is possible to use large values of load resistance, to obtain large signal values, and still maintain a linear output. The magnitude of the available signal increases as the value of the load resistor increases. However, this increase in available signal must be balanced against a possible increase in Johnson noise and a possible decrease in the frequency response because of the increased RC time constant of the circuit. The width of the depletion region is reduced when a voltage is applied across the photodiode. This reduces the parasitic capacitance (C_D) of the device. The reduced capacitance enables high-speed operation; however, the linearity, offset, and diode leakage current (I_L) characteristics may be adversely affected. A circuit designer must trade off each these effects against each other to obtain the best result for a particular application.

A low-input current operational amplifier with a field effect transistor (FET) at the input is the most often used in high-speed photodiode circuits to convert the diode current to a voltage to be measured. The bandwidth of these circuits is given by

$$\text{bandwidth} = \frac{1}{2\pi R_{\text{feedback}} C_{\text{feedback}}} \tag{4.14}$$

where R_{feedback} and C_{feedback} are the resistance and capacitance of the feedback elements shown in Fig. 4.11. It is often necessary to follow the amplifier with a low-pass filter to reduce the amplitude of noise at frequencies above the maximum signal frequencies. The use of a single-pole, low-pass filter can improved the signal to noise by several decibels. To improve the signal-to-noise ratio of the detector-amplifier system, one can use a lower-noise amplifier, reduce the size of the feedback resistor (effectively reducing the amplitude of the output voltage proportionally), adjust the capacitance characteristics of the system (effectively changing the bandwidth of the system), or reduce the bandwidth of the system with a filter. Another technique for lower noise is to change to an amplifier with a lower bandwidth. Adjustment of the capacitance of the system may mean the selection of a diode with a smaller parasitic capacitance C_D or an increased input capacitance of the operational amplifier, C_{DIFF}. A photodiode is selected primarily because of its light response characteristics. Each of the options to reduce noise comes at a price, either in gain or bandwidth.

It is reasonable to ask how much noise is too much noise in a photodiode-amplifier circuit. One point of reference is the capability of the digitizer used to measure the signal. For example, using a 12-bit digitizer with a 0- to 2-V input range, the least significant bit measures about 0.5 mV. Reducing the noise level below the least significant bit (or quantization level) is wasted effort because it cannot be measured.

4.3. A-D CONVERTERS / DIGITIZERS

4.3.1. Digitizing the Detector Signal

For a lidar to be useful, the signal from the detector must be measured, that is, converted to numbers that can be analyzed further. To accomplish this conversion, transient digitizers and, occasionally, digital oscilloscopes are used. These instruments sample voltage signals with a fast analog-to-digital converter (ADC). At evenly spaced intervals (determined by a clock), the ADC measures the voltage at the input and then stores the measured value in high-speed memory. The shorter the interval between measurements, the faster the digitizing rate and the higher the signal frequency that can be resolved. Once the digitizer is "armed," the ADC digitizes the signal continuously and feeds the samples into the memory with circular addressing. When the last memory location is filled, the system will start again at the lowest memory location, overwriting any data stored there. When a trigger is generated, the digitization continues until the memory is filled with a user-selected number of posttrigger samples. At that point the ADC stops digitizing. With some digitizers, it is possible to obtain data before the trigger event. In lidars, this is useful because these data are a good measure of the background light signal, that is, the value to which the signal should decay at long range. The time required to decay to this value as well as any undershooting can be used to evaluate problems in the detector-amplifier combination. In a well-functioning system, the pretrigger values can be used in background subtraction routines.

A trigger is required to start the digitization process. The trigger provides a timing mark indicating that the laser beam has left the lidar. Many lidars use a detector near the exit of the laser to provide this signal. Most digitizers fire when the leading edge of the trigger signal rises above some (usually programmable) level. The trigger must be a fast rising signal and well behaved in the sense that it does not ring or have other abnormalities that could cause false triggering of the digitizer.

The ADC in a digitizer is capable of measuring over some fixed voltage range, dividing that range into a number of equally spaced intervals. An N-bit digitizer has $2^N - 1$ intervals. Thus an 8-bit digitizer has 255 intervals. The width of each interval is the digitizer voltage range divided by the total number of intervals. The width of the interval represents the minimum voltage difference that can be resolved. An ideal digitizer has uniform spacing between each of the intervals. The greater the resolution of the ADC, the greater the sensitivity to small voltage changes. Many digitizers have a programmable amplifier in front of the ADC to better match the size of the signal to the voltage range of the ADC. Matching the size of the signal to the full ADC range is important in lidar systems where the dynamic range of the signal is large.

Most digitizers also have a programmable DC offset. The offset is used by the digitizer to shift the signal into the ADC desired voltage range. The offset that is selected contributes to the true baseline value of the signal. For lidar purposes, the DC level of the background light signal should be adjusted so that the background signal is a few intervals above zero. In this way, portions of the raw signal from the detector are not truncated by the digitizer. If the lowest parts of the signal were truncated, the lidar signal would be biased. A nonzero offset is also of value in determining whether the amplifier has problems with the zero level.

The sampling rate sets an upper limit on the frequencies that may be measured. To avoid aliasing (which distorts the captured waveforms) the sample rate must be at least twice as fast as the highest frequencies present in the signal (the Nyquist criterion) (Oppenheim and Schafer, 1989). Given an ideal, noiseless digitizer and a bandwidth-limited signal, the Nyquist criterion sets a sufficient sampling rate. The Nyquist criterion states that at least two samples must be taken for each cycle of the highest input frequency. In other words, the highest frequency that can be measured is one-half the sample rate. However, real systems have noise and distortion and require additional samples to adequately resolve the signal. If the signal is reconstituted by straight-line interpolation between data points, 10 or more samples per cycle are required. For a lidar, the sampling rate sets one limit on the range resolution of the lidar system.

The bandwidth of the front end amplifier also sets an upper limit to the maximum frequency that can be measured. Attenuation of the signal occurs at all frequencies, not just past the cutoff (−3 dB) frequency. Thus bandwidth is an important specification for digitizers. A digitizer's input amplifier and filters determine the bandwidth. A common practice is to have the bandwidth of the input amplifier be one-half the sampling rate of the digitizer.

One issue that may be of importance to lidar applications is the speed with which a digitized signal can be transferred to the control computer. Although some digitizers can automatically average successive signals, most can only digitize one laser pulse at a time. Thus the data in the digitizer memory must be transferred to the control computer between each laser pulse so that summing can be done by the control computer. As the laser pulse rate nears 100 Hz, data transfer rates may approach a megabyte per second, which may tax the ability of the particular method used to transfer data between digitizer and computer memory. Digitizers that share the same memory address space as the control computer are generally faster in transferring data. Digitizers that reside in an external configuration generally require a card in the computer to transfer data, although some use a GPIB or RS-232 interface. In this case, data transfer may be considerably slower. A computer may also reside on the bus in a CAMAC (computer automated measurement and control; IEEE Standard 583), VME, or VXI (VME extensions for instrumentation; IEEE Standard 1155) data collection system. These systems are essentially a high-speed computer bus in which a wide variety of cards can be inserted to accomplish a wide variety of tasks. Again, because the digitizer and computer share the same memory address space, data transfer rates are high.

4.3.2. Digitizer Errors

All digitizers contain sources of error that limit the accuracy of a measurement. Accuracy consists of three parts: resolution, precision, and repeatability. Resolution is a measure of the uncertainty associated with the smallest voltage difference capable of being measured. Precision is a measure of the difference between the measured voltage and the actual voltage. Repeatability is a measure of how often the same measurement occurs for the same input voltage. The types of errors that may occur include DC errors, differential nonlinearity, phase distortion, noise, aperture jitter, and amplitude changes with frequency.

DC errors occur when the digitizer fails to measure static or slow-moving signals accurately. The input amplifier, and not the ADC, determines the DC accuracy. Digitizers typically will have a DC accuracy on the order of 1–2 percent. Signals of all frequencies are attenuated. In a good amplifier, the attenuation of each frequency will be the same until the high-frequency cut-off is reached. The high-frequency cut-off is actually a gradual decrease in the transmitted signal with frequency. The 3-dB point is generally taken to be the cut-off.

Differential nonlinearity is a measure of the uniformity in the spacing between adjacent measurement intervals in a digitizer. The differential nonlinearity is defined as the worst-case variation, expressed as a percentage, from this nominal interval width. If voltage interval is 2 mV and the worst-case bin is 3 mV, then the differential nonlinearity is 50%. Differential nonlinearity typically causes significant errors only for small signals because the error is usually only one digitizer interval.

Phase distortion is the result of different phase shifts of the input signal for different frequencies. Pulses of complex shapes are composed of a spectrum of frequencies. The shape of the pulse can be maintained during the measurement process only if the relative phase of all the components at all of the frequencies remains the same at the digitizer output. Phase distortion results in erroneous overshoots and slower rise times on edges.

Amplitude noise is random or uncorrelated to the input signal. The amplifier associated with the digitizer inserts noise into the digitizing process. Noise can mask subtle input signal variations on transient events. For repetitive signals when the results from several laser pulses will be averaged, noise can be reduced by averaging several digitized waveforms.

Aperture jitter or uncertainty is the result of sampling time noise, or "jitter on the clock." The amplitude noise induced by clock jitter equals the time error multiplied by the slope of the input signal. The error in the measured amplitude increases for fast signal transitions, such as pulse edges or high-frequency sine waves. Aperture uncertainty also affects timing measurements such as rise time, fall time, and pulse width. Aperture uncertainty has little effect on low-frequency signals. Most digitizers have a continuous clock, so that on receipt of the trigger pulse, the digitization process will begin on the next rising edge of the clock signal. Thus there will be an average error of one-half the clock interval in the timing, even for perfect systems.

A figure of merit called effective bits is often used to compare the accuracy of two digitizers. It is a measure of dynamic performance. The number of effective bits estimator includes errors from harmonic distortion, differential nonlinearity, aperture uncertainty, and amplitude noise. The effective bits measurement compares the digitizer under test to an ideal digitizer of identical range and resolution. The use of effective bits as a measure of performance has many limitations. Effective bits measurements change with input frequency and amplitude. Because the effects of harmonic distortion, aperture uncertainty, and slewing are larger at higher signal frequencies, the number of effective bits decreases with frequency. To represent overall performance under a wide variety of conditions, the number of effective bits must be plotted for as a function of frequencies. Perhaps most significantly, the number of effective bits does not measure worst-case scenarios, nor does it indicate which source of error is responsible for the distortion. A detailed discussion of effective bits and digitizer errors can be found in the application note by Girard (1995).

4.3.3. Digitizer Use

The input signal should be matched to the digitizer characteristics. At least two major adjustments to the signal must be considered, the amplitude of the signal, and the dc offset of the signal. The digitizer will have an input range over which it is designed to operate. For example, the DA60 digitizer made by Signatec has a –2 to +2 V input range, a total of 4 V. The signal then should be amplified so that the signal spans a range that is slightly less than 4 V from the highest peak to the lowest part of the signal. In the case of the DA60, this can be done by programming the digitizer for the desired amount of amplification. In other cases, external amplifiers may have to be used. Matching the signal amplitude to the digitizer input makes maximum use of the dynamic range of the digitizer. For lidar purposes, this translates into greater range and greater sensitivity.

Having matched the amplitude of the signal to the digitizer input, the offset must also be adjusted. Lidar signals are either entirely positive or entirely negative in nature depending on the type of amplifier or photomultiplier circuit used. So for the case of the DA60, which desires an input from –2 to +2 V, a positive lidar signal (from 0 to 4 V) must be added to a constant dc offset of –2 V so that the signal input to the digitzer exactly matches the desired input range. The digitizer will truncate any signal that is above or below its input range. Because the digitizer can only measure voltages between –2 and +2 V, the offset value must be adjusted to put the raw input into this range. Examination of the digitized lidar signal without any processing or background subtraction will allow an operator to make the necessary adjustments to the signal. Figure 4.12 is an example of such a signal. The offset should also be set so that a 0-V signal has a value that is not the maximum or minimum of the digitizer. For example, in Fig. 4.12, 0 (the value of the lidar signal at long range) is set for a digitizer value of about 250. Because of variations in the background brightness of the sky, this may not have a constant value from shot to shot or between directions into the sky. There are several reasons for the selection of a nonzero baseline. One of the things that must be done in processing the signal is to remove the constant background signal. If the offset is set so that 0 V is a digitizer zero value, noise on the signal with values below 0 will be truncated. This will cause the signal at long ranges to be biased to a small positive value. At long ranges, this becomes significant because of the r^2 range correction and will affect any inversion method attempted. Several common detector problems such as a baseline shift, ringing, or feedback could show up at long ranges as a negative signal. Detection and correction of these problems requires that the entire signal be digitized.

By these criteria, the signal shown in Fig. 4.12 is not well matched to the digitizer. The signal is above the maximum level digitized for the ranges between 100 and 400 m and is truncated to 4095, the maximum level of a 12-bit digitizer. No meaningful data are available for these ranges. However, if the intent is to acquire high-resolution data at long ranges, this could be done by sacrificing data at short ranges. Amplifying the signal even more than was done in Fig. 4.12 would result in higher digitizer values (more resolution) at long ranges, at the cost of increasing the size of the region at short ranges with no data.

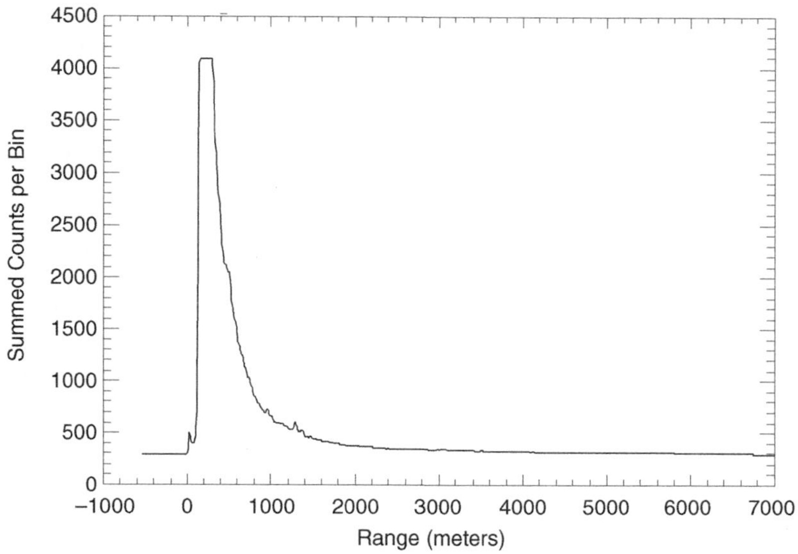

Fig. 4.12. Raw lidar data signal without background subtraction. Digitizer bin numbers on left correspond to 0–4095 for a 12-bit digitizer and span the −2 to +2 V input voltage range. The digitizer variables should be set to obtain the greatest dynamic range from the signal while keeping the signal significantly above zero in the far field (where the signal flattens out). Note that this signal is too large in the near field, i.e., the top of the signal is cut off at 4095 counts.

4.4. GENERAL

4.4.1. Impedance Matching

Coaxial cables are used to connect the photomultiplier tube base to the digitizer. Impedance matching of these cables is important. Cables with a characteristic impedance (usually 50 Ω) matching the impedance of the digitizer must be used. If the cables and termination are not matched, part of the energy in the pulse from the photomultiplier may be reflected back and forth along the cable. This produces what is commonly known as "ringing." Distortion of the original waveform may also occur. One method of addressing the problem is to add a resistor at the digitizer end of the cable. Although this may eliminate the ringing, it will reduce the size of the signal (Knoll, 1979).

4.4.2. Energy Monitoring Hardware

A significant improvement in two-dimensional lidar data sets can be obtained if the amplitude of the data is corrected for the shot-to-shot variations in the laser pulse energy. This can be done by monitoring and recording the energy of the laser pulse as it exits the system and then using that information to correct the digitized data (Fiorani et al., 1997; Durieux and Fiorani, 1998). Often this is done with a simple detector mounted so as to catch the off-angle reflection from a mirror used to direct the laser beam. Because the amount of light available for sampling is usually large and the detector can be positioned to catch the maximum amount of light, amplification is normally not necessary. A simple, unamplified, biased photodiode detector can be used to maximize the speed and linearity of the output pulse. The output pulse is input to a sample and hold circuit that follows the amplitude of the signal to its maximum value and then maintains that value long after the signal has decayed away. The output of the sample and hold circuit is held at the peak value of the pulse for as long as milliseconds so that it may be sampled by an analog-to-digital converter. Measurements of laser pulse energies on the order of 1–2 percent are relatively easily accomplished. Reagan et al. (1976) describe the construction of a detector with a sample and hold circuit. Today, high-quality detectors and sample and hold circuits are commercially available for a few hundred dollars.

4.4.3. Photon Counting

There are two ways in which the signal from a lidar can be recorded: current mode and photon counting mode. Current mode operation uses direct, high-speed digitization of the signal from the photodetector. The use of a current mode maximizes the near-field spatial resolution for lidars and is particularly useful for boundary layer observations. However, direct digitization of the signal is only good for a few-kilometer range because the signal decreases as the square of the range. Photon counting is required to obtain long-range soundings high into the troposphere or stratosphere. The returning photons are counted over time periods that are long in comparison to the digitizing rates used for current mode operation. Counting photons requires summing the results from a large number of laser pulses to obtain statistical significance in the measurements. Thus long range is exchanged for greatly decreased range and time resolution.

Counting photons is usually done only for wavelengths shorter than about 1 μm. The technology to photon count at significantly longer wavelengths (at least to about 1.6 μm) has been demonstrated (see, for example, Levine and Berthea, 1984; Lacaita et al., 1996; Owens et al., 1994; Rarity et al., 2000), albeit with significant difficulties. Because thermal or dark currents generally become larger as the wavelengths lengthen, it is possible to saturate the detector with only the dark current. Cooling is necessary to reduce the dark current, but reductions beyond a certain point may result in an increased number of afterpulses (Rarity et al., 2000). Photomultipliers and avalanche photodiodes are currently the only devices capable of detecting single photons and generating a signal fast enough and large enough to use conventional discrimination and counting equipment.

Detectors/Devices. To detect single photons, the one electron freed in the detector by the absorption of a photon must be amplified to the point that it may be unambiguously detected and counted. To achieve a millivolt level, signal into a 50 Ohm load requires an amplification on the order of 10^8. This can be done by using a photomultiplier tube with 10 or more stages or through the use of an avalanche photodiode (APD) in what is known at the Geiger mode of operation.

APDs can be used to detect single photons in the Geiger mode, in which the diode is operated above its breakdown voltage. At this voltage, the absorption of a single photon will initiate an avalanche breakdown inside the detector, producing a current that allows the detection of single photons. To maintain a high detection probability, the threshold level for obtaining a Geiger mode avalanche must be set to a low value. This can only be done if the dark current is very low. This requires that the device be cooled. If the threshold is set too low, thermal noise in the front-end amplifier and load may increase the apparent background and noise floor. Because the dark count rate is strongly dependent on temperature, cooling the detector from room temperature to about −25°C with a Peltier thermoelectric cooler can reduce the dark count by a factor of 50. The dark count rate is proportional to exp (−0.55 eV/kT) so that a moderate amount of cooling can make a significant difference. Because breakdown of the diode over an extended period can damage the diode, quenching the avalanche effect is also an issue that must be addressed. Several methods of active and passive quenching have been attempted (Brown et al., 1986, 1987; Cova, 1982).

The APDs used in the Geiger mode must be specially selected because they are sensitive to defects in the crystal, which cause dark counts and afterpulsing. Dark counts are caused by thermal generation in the depletion layer. Because of the high field strength in the APDs, this effect is often enhanced. The electrons released by thermal generation will be accelerated and generate an avalanche that imitates an incident photon. Afterpulsing is caused when one of the charge carriers, released by the avalanche breakdown, is captured by a trapping center in the depletion layer of the diode. If this carrier is released by the trap, it will initiate an avalanche breakdown as it accelerates across the depletion region. Afterpulsing and residual signals are also observed in photomultipliers because of different effects inside the tube (Coates, 1973a, 1973b; Riley and Wright, 1977; Yamashita et al., 1982).

There is a maximum voltage that may be applied to a photodiode in the reverse direction. The application of a voltage greater than this voltage may cause breakdown and/or severe degradation in the performance of the device. This voltage is a function of the material, size, and design of the material and thus must be specified by the manufacturer.

Photomultipliers are simpler to use for photon counting. In some portions of the spectrum (for example, the ultraviolet), photomultipliers are the only photon counting method currently available. Their inherently high gain and fast response makes photomultipliers ideal for photon counting. However, the quantum efficiency of photomultipliers, especially at longer wavelengths, is significantly less than for photodiodes. At 1064-nm (Nd: YAG laser) wavelengths, for example, a silicon photodiode may have a quantum efficiency over 10 percent whereas a photomultiplier with an S1 photocathode may have an efficiency on the order of a tenth of a percent.

Dead Time Corrections. In any detector system, there is a certain amount of time that is required to discriminate and process an event. If a second event occurs during this time, it will not be counted. The minimum amount of time that must separate two events such that both are counted is referred to as the "dead time." Because of the random nature of the arrival times of photons, there is always some dead time with some events that will not be counted. A dead time correction is required to account for those photons that arrive during the time required for the scalar to record a previous photon (generally about 9 ns). When recording the first photon, the scalar is effectively "dead" or incapable of recording the second photon. In lidar applications, the number of uncounted photons is significant at short ranges from the lidar and decreases in importance with range. There are two basic models for the behavior of counting systems. The one to be used depends on the details of the electronics used in a particular application. The models are somewhat idealized and are described in detail by Knoll (1979).

In a nonparalyzable detection system, a fixed mount of dead time follows a given photon and any photon that arrives during that time is ignored and does not increase the amount of overall dead time. Thus two photons that are separated in time by more than the dead time will both be counted (Fig. 4.13). If N_m is taken to be the system measured count rate, N_a is the actual count rate, and τ is the dead time, then the total fraction of the time that is dead is $N_\mathrm{m}\tau$, so that the rate at which event are lost is $N_a N_m \tau$. The corrected count rate is determined by

$$N_a = \frac{N_m}{(1 - N_m T)} \tag{4.15}$$

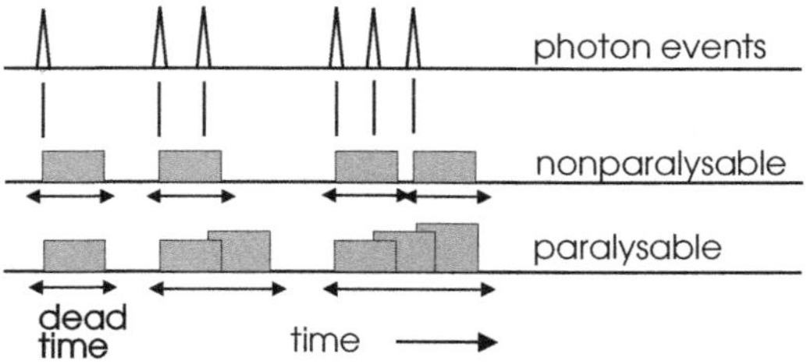

Fig. 4.13. Plot showing the difference between a paralyzable and a nonparalyzable detector. Note that the nonparalyzable detector registers four counts whereas the paralyzable detector registers only three.

In a paralyzable detection system, a fixed mount of dead time follows each photon and any photon that arrives during the dead time of another extends the dead time of the first by its own dead time (Fig. 4.13). The measured count rate for this type of electronic system is given by

$$N_m = N_a e^{-N_a T} \tag{4.16}$$

This expression is not invertible to determine the actual count rate, and for a given measured count rate there exist two values of the actual count rate that will produce the measured rate for a given dead time. Which value is correct must be determined from the context of the data. Methods to determine the paralyzability of electronics systems are covered in detail by Knoll (1979). A more detailed discussion of the dead time effect and the necessary corrections can be found in Funck (1986) and Donovan et al. (1993).

Photon Counting Electronics. A pulse from the absorption of a photon having been generated, the signal is fed to a discriminator or single channel analyzer (SCA). The bulk of the pulses from noise or afterpulsing are lower in amplitude than those from actual photon events (Helstrom, 1984). These pulses can be rejected by setting a minimum amplitude level for a pulse to be counted. A discriminator counts only those pulses with an amplitude above some adjustable level and outputs a TTL level pulse for counting. Careful adjustment of the discriminator level is required to pass the largest fraction of the true events while rejecting the largest fraction of the spurious or noise events. Some discriminators also have an adjustable upper limit as well as a lower limit so that pulses that are too large (such as two photons arriving nearly similtaneously) are also rejected.

These pulses are counted with a scalar. The scalar counts the number of TTL pulses that occur between successive clock pulses (essentially square waves of fixed frequency). At the beginning of each clock pulse, the number of counted pulses is saved to memory, the counter is zeroed, and counting is restarted. These devices are remarkably flexible and able to respond to clock pulses of arbitrary frequency up to some maximum rate. The time between successive clock pulses sets the range resolution of the system. This is usually on the order of 250–500 ns (37.5- to 75-m resolution). Because the pulses from single photons are generally on the order of 4–12 ns long, counting times shorter than 250 ns are not long enough to count a significant number of events. Faster photomultipliers and counting hardware can be obtained at significantly higher cost. Clocks are generally programmable, being capable of generating square waves with frequencies that are integer fractions of a fundamental frequency determined by an oscillator in the device. Depending on the hardware, either the clock or the scalar can be programmed for the number of range elements (or clock pulses) that will be counted for each laser pulse. Most scalars will sum the counts for successive laser pulses so that this need not be done by the control computer. The scalar-clock combination is started with a trigger pulse similar to that used to start a digitizer. It should be remembered that the clocks are free running. This causes a timing ambiguity that is, on average, half the time between clock pulses. In other words, the clock runs at a steady rate that is continuous. When a start pulse is received, the beginning of the next clock cycle will start the counting process. Because a start pulse could be received at any time during a clock cycle, counting could start as long as a full cycle after the start pulse. This effect further degrades the range resolution of photon counting lidar systems. A more complete discussion of the type of electronics used in photon counting systems can be found in Knoll (1979).

Although most photon counting equipment uses TTL logic for counting, there are several others that are in common use. Several of the most common are ECL (emitter coupled logic), NIM (nuclear instrument module), CAMAC (computer automated measurement and control; IEEE Standard 583), and TTL (transistor-transistor logic). ECL levels are a low or Boolean false at –1.75 V and a high or true state at –0.9 V with respect to ground. The NIM standard is actually a current specification that, with a 50-Ω load, equates to a Boolean false at 0 V and a Boolean true at –0.8 V. The CAMAC logic levels are a Boolean true equal to 0 V and a Boolean false equal to 2 V. TTL levels are a Boolean false (TTL low) equal to 0 V and a Boolean true (TTL high) equal to 5 V.

4.4.4. Variable Amplification

A significant problem with lidars is the extremely large dynamic range of the signals because of the r^{-2} fall-off (Chapter 3). This causes difficulties in maintaining linearity of the response both in the design of amplifiers and in the digitization of the signals. A number of efforts have been made to compress the lidar signal in order to reduce the dynamic range. The gain of a photomultiplier or avalanche photodiode can be varied through changes in the bias voltage (Allen and Evans, 1972). To obtain accurate quantitative information, one must have extremely accurate information on the shape of the voltage pulse used to bias the detector and of the response of the detector to that pulse. On a practical level, it is difficult to generate precise voltage waveforms, particularly at the high voltages required for the operation of a photomultiplier. The response of the detector is highly dependent on the characteristics of the individual device and may change as the detector ages. Logarithmic amplifiers are another method that has been used and are available from several electronic or lidar companies. When the digitized signal from a logarithmic amplifier is inverted to obtain the original signal, small errors in analog-to-digital conversion will be exaggerated. Furthermore, over large dynamic ranges, the fidelity of the logarithmic amplification is questionable. Thus the compression-expansion process may be significantly nonlinear. The use of a gain-switching amplifier has also been demonstrated by Spinhirne and Reagan (1976). A gain-switching amplifier avoids issues of linearity by applying different values of fixed gain to the signal that keep the amplitude of the signal within a given range. The demonstration by Spinhirne and Reagan achieved 3 percent linearity with a bandwidth of 2.5 MHz. Although not an electronic method of signal compression, the geometric form factor of the lidar has been suggested (Harms et al., 1978) as a means of reducing the dynamic range of the lidar signal. This concept uses the optical design of the lidar to reduce the size of the signal in the near field. We are not aware that any lidar has been constructed with this concept. However, Zhao et al. (1992) used multiple laser beams emitted at various distances from the telescope and parallel to its line of sight. This effectively reduces the dynamic range but introduces other issues such as alignment and interpretation of the data.

5

ANALYTICAL SOLUTIONS OF THE LIDAR EQUATION

As mentioned in Section 3.2.1, the atmospheric extinction coefficient $\kappa_t(r)$ rather than the backscatter coefficient $\beta_\pi(r)$ is the fundamental parameter that is generally extracted from an elastic lidar signal. Unfortunately, the lidar equation contains more than one unknown value and is thus undetermined. To overcome this problem and to be able to extract the extinction coefficient from the signal $P(r)$, the lidar equation constant must be estimated. In addition, the relationship between backscatter and total extinction must in some way be established or assumed. The problem of determining the relationship is considered in Chapter 7. In this chapter, we present methods for the inversion of lidar signals to obtain profiles of the extinction coefficient.

The simplest inversion technique, based on an absolute calibration of the lidar system, can use only the lidar system constant C_0, whereas the other factors in the lidar equation solution remain unknown, for example, the two-way atmospheric transmittance over the incomplete overlap zone (see Eq. (3.12)). Therefore, this technique is generally used in conjunction with other methods rather than separately. All self-sufficient elastic lidar signal inversion methods developed to date require the use of one or more a priori assumptions that are chosen according to the particular optical situation. The differences between the various retrieval methods lie in the ways of determining boundary conditions and in the selection of a priori assumptions concerning other missing information. There are three basic inversion methods, commonly

Elastic Lidar: Theory, Practice, and Analysis Methods, by Vladimir A. Kovalev and William E. Eichinger.
ISBN 0-471-20171-5

in practice, to find the unknown extinction coefficient. These methods are as follows:

1. The slope method. This method is useful for homogeneous atmospheres. In many cases, atmospheric horizontal homogeneity is a reasonable assumption. What is more, this assumption can be checked easily by an analysis of the lidar signal shape. With the slope method, a mean value of the extinction coefficient over the examined range in a homogeneous atmosphere is obtained.
2. The boundary point solution. This variant requires knowledge of or an a priori estimate of the extinction coefficient at some point within the measurement range and can be used in both homogeneous and inhomogeneous atmospheres.
3. The optical depth solution. Here the total optical depth or transmittance over the lidar measurement range should be known or assumed. This inversion technique can be used in both homogeneous and inhomogeneous atmospheres.

More complicated data processing methods are used for lidar multiangle measurements in the atmosphere. These methods, which are applied to a number of lidar signals measured under different elevation angles, are considered in Chapter 9. This chapter presents practical lidar inversion techniques that may be used to determine particulate-extinction-coefficient profiles in any desired direction. In Section 5.1, the slope method of retrieving information from lidar signals measured in a homogeneous atmosphere is examined. The method determines a mean value of the extinction coefficient over the range. There are some potential applications of this method, such as visibility measurements at airports or along highways, where the mean extinction coefficient (or atmospheric transmittance) is the desired information (see Chapter 12.1). In the other sections of this chapter, lidar equation solutions based on some assumed (or estimated) boundary conditions for the lidar equation are examined. These methods make it possible to extract local values of the extinction coefficient for any specified range and, accordingly, obtain profiles of the extinction coefficient as a function of range or altitude.

5.1. SIMPLE LIDAR-EQUATION SOLUTION FOR A HOMOGENEOUS ATMOSPHERE: SLOPE METHOD

It was shown in Chapter 3 that an area exists close to the lidar where the overlap of the collimated laser light beam with the receiving optics field of view is incomplete. In this area, signal intensity is less than that defined by Eq. (3.12). The lidar equation, which takes this effect into consideration, can be written as

$$P(r) = C_0 q(r) \frac{\beta_\pi(r)}{r^2} \exp\left[-2\int_0^r \kappa_t(r')dr'\right] \tag{5.1}$$

Eq. (5.1) is similar to Eq. (3.12) but includes the overlap function $q(r)$. In the areas of the complete overlap, the maximum value of $q(r)$ is, generally, normalized to unity. In the areas close to the lidar, where the laser beam and the field of view of the receiving optics do not intersect, no signal is obtained, so that here the factor $q(r) = 0$. Thus, with the increase of r, the function $q(r)$ in Eq. (5.1) ranges from zero to unity. The latter value is valid for the ranges $r > r_0$, where the laser beam is completely within the field of view of the receiving optics (Fig. 3.3). In Fig. 5.1, a typical form of the overlap function is shown as a function of range; here r_0 can be taken as approximately 550–600 m.

The knowledge of the shape of $q(r)$ over the incomplete overlap zone allows one to exclude the unknown term T_0^2 in Eq. (5.1). However, in practice, the data obtained within the region of incomplete overlap where $q(r) < 1$ are generally excluded from data processing (see Section 3.4.1). This is because of the difficulties associated with accurate correcting the measured signal for the overlap. Therefore, the range r_0 is considered to be the minimum range at which useful lidar data may be obtained. For the ranges $r \geq r_0$, the factor $q(r)$ is normalized to unity and therefore can be omitted from consideration (this assumes that the lidar optical system is properly adjusted, so that the laser beam remains within the receiver's field of view at all distances larger than r_0). By restricting the measurement range in the near field, difficulties associated with determining the shape of $q(r)$ may be avoided. On the other hand, no useful information can then be obtained from the lidar signal for this nearest zone, from $r = 0$ to r_0. Because of this, the equation used for lidar data processing, generally differs from Eq. (5.1) by the presence of an additional transmittance term T_0^2, whereas the term $q(r)$ is omitted

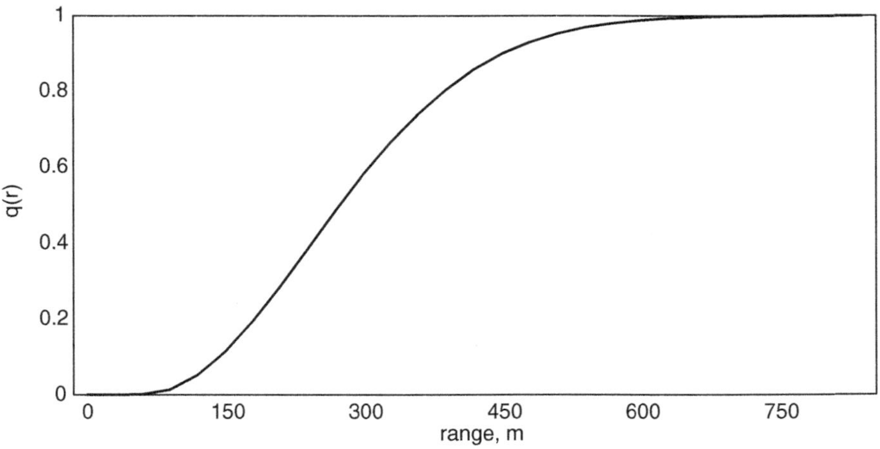

Fig. 5.1. Typical dependence of the overlap function $q(r)$ on the range.

$$P(r) = C_0 T_0^2 \frac{\beta_\pi(r)}{r^2} \exp\left[-2\int_{r_0}^{r} \kappa_t(r')dr'\right] \tag{5.2}$$

Here T_0^2 is an unknown, two-way atmospheric transmission over the incomplete overlap zone, from the lidar to r_0.

A simple mathematical solution for Eq. (5.2) is achievable for the unknown extinction coefficient κ_t if the examined atmosphere is or may be considered to be homogeneous. For a valid homogeneous atmosphere solution, the following two conditions must be met:

$$\kappa_t(r) = \kappa_t = const. \tag{5.3}$$

and

$$\beta_\pi(r) = \beta_\pi = const. \tag{5.4}$$

With Eqs. (5.3) and (5.4), the lidar equation for a homogeneous atmosphere then reduces to

$$P(r) = C_0 T_0^2 \frac{\beta_\pi}{r^2} e^{-2\kappa_t(r-r_0)} \tag{5.5}$$

The term $1/r^2$ in the lidar equation causes the measured signal $P(r)$ to diminish sharply with range because of the decreasing solid angle subtended by the receiving telescope with range (Fig. 3.8a). To compensate for this effect, the lidar signal $P(r)$ is commonly transformed into a range-corrected signal before lidar signal inversion is begun. This is accomplished by multiplying the original signal $P(r)$ by the square of the range, r^2. After multiplying by r^2, the range-corrected signal, denoted further as $Z_r(r)$, can be written as

$$Z_r(r) = P(r)r^2 = C_0 \beta_\pi e^{-2\kappa_t r} \tag{5.6}$$

Taking the logarithm of the transformed signal in Eq. (5.6), and denoting it as $\Phi(r) = \ln Z_r(r)$, one can rewrite the above equation as

$$\Phi(r) = \ln(C_0 \beta_\pi) - 2\kappa_t r \tag{5.7}$$

As follows from the homogeneity assumptions given in Eqs. (5.3) and (5.4), the product $C_0\beta_\pi$ and the extinction coefficient κ_t in Eq. (5.7) can be considered to be constants. Under such conditions, the dependence of $\Phi(r)$ on r can be rewritten as a linear equation

$$\Phi(r) = A - 2\kappa_t r \tag{5.8}$$

here $A = \ln(C_0\beta_\pi)$. The linear dependence of $\Phi(r)$ on range, r, is a key factor when seeking the simplest solution to the lidar equation (Collis, 1966). It allows determination of the attenuation coefficient κ_t in a least-squares sense. The use of optimal curve-fitting routines is the most effective manner to determine the average attenuation coefficient. What is more, the estimate of the standard deviation of the linear fit for $\Phi(r)$ can be used to estimate the degree to which the assumption of atmospheric homogeneity is valid. These features have great practical application when the lidar system is initially set up and tested in the atmosphere before actual experimental use. Note also that, formally, both constants from the linear fit can be found from Eq. (5.7), the extinction coefficient κ_t, and the backscatter term β_π. To find the latter, the constant C_0 must, in some way, be determined.

The lidar equation solutions can be expressed in the terms of either variable $Z_r(r) = P(r)r^2$ or its logarithm, $\Phi(r) = \ln[P(r)r^2]$. The latter form, which stems from the slope method and the direct Bernoulli solution (Klett, 1981; Browell et al., 1985), can be inconvenient for practical application. For example, when the logarithmic form is used, the ratio of the signal $Z_r(r)$ at r to that at the reference range, r_b, which is often used in the lidar equation solution, results in an awkward form

$$\frac{Z(r)}{Z(r_b)} = \exp[\Phi(r) - \Phi(r_b)] = \exp[\ln Z_r(r) - \ln Z_r(r_b)] \tag{5.9}$$

The other disadvantage of the logarithmic form was pointed out by Young (1995). In practice, before lidar data processing, a signal offset P_{bgr}, originating from a background light signal, F_{bgr}, is always subtracted; thus the range-corrected signal is determined as $Z_r(r) = [P_\Sigma(r) - P_{bgr}]r^2$. The use of the logarithmic form may create problems in areas of the lidar measurement range that are corrupted by noise (Kunz and de Leeuw, 1993). For example, in the regions above thin clouds, low signal-to-noise ratios and systematic errors can result in condition in which $P_\Sigma(r) < P_{bgr}$, and, accordingly, can produce local negative values of $Z_r(r)$. Rejecting such ranges from analysis is not acceptable because it may bias the results of the inversion. On the other hand, heavy smoothing of the signal to remove the negative values of $Z_r(r)$ is also not always acceptable. It degrades the range resolution of the lidar in regions where the signal is strong. The lidar measurements have revealed that the use of nonlogarithmic variables in the lidar equation are preferable, and these will be used in the further analysis.

An analytical solution of Eq. (5.7) for the unknown extinction coefficient κ_t can be obtained by taking the derivative of the logarithm of $Z_r(r)$

$$\kappa_t = -\frac{1}{2}\frac{d}{dr}[\ln Z_r(r)] \tag{5.10}$$

The practical application of Eq. (5.10) to determine the extinction coefficient requires the use of discrete numerical differentiation. As shown in Section 4.3, a continuous analog lidar signal is transformed into digital form at discrete intervals, Δt, which correspond to a spatial range resolution, $\Delta r_d = c\Delta t/2$. Accordingly, Eq. (5.10) must be applied to finite spatial intervals, $\Delta r = m\Delta r_d$, where m is an integer. For the finite range from r to $r + \Delta r$, Eq. (5.10) may be reduced to a form of numerical differentiation

$$\kappa_t(\Delta r) = \frac{-1}{2\Delta r}[\ln Z_r(r+\Delta r) - \ln Z_r(r)] \tag{5.11}$$

The main problem that arises in practice is that the solution obtained by numerical differentiation with small range increments Δr is extremely sensitive to signal noise and to the presence of local heterogeneity. Because of the presence of the factor $1/(2\Delta r)$ in Eq. (5.11), small uncertainties or systematic shifts in the quantities $Z_r(r)$ and $Z_r(r + \Delta r)$ may cause large errors in the extinction coefficient κ_t. This effect, which is considered in detail in Chapter 6, makes the use of the slope method impractical for short range intervals Δr.

On the other hand, the application of the slope method is limited by the degree of atmospheric heterogeneity. Actually, no absolutely homogeneous atmosphere exists in which the conditions given by Eqs. (5.3) and (5.4) are strictly valid. Even in horizontal directions, the conditions of homogeneity may be taken to be only approximate. Generally, this assumption may be valid when the lidar light beam is directed parallel to flat and uniform horizontal areas of the earth's surface, where no atmospheric disturbances occur and where no local sources of plumes exist.

The approximation of a homogeneous atmosphere may be useful in horizontal direction measurements and in lidar atmospheric tests. However, before the lidar-equation solution in Eq. (5.11) is applied, one should establish whether the optical conditions of the measurement are appropriate for the slope method. In other words, one must estimate the degree of atmospheric homogeneity and determine whether it is possible to achieve an acceptable measurement accuracy with this method. This is why the practical application of the slope method requires a definition of the concept of a homogeneous atmosphere. The general notion of the term homogeneity means "the quality or state of being uniform throughout in structure." In a strict sense, the atmosphere is never uniform. Particulates in the atmosphere never have uniform spatial distribution, and at least small-scale particulate heterogeneity is always present. However, the concept of atmospheric homogeneity over the distance examined by the lidar only assumes that the spatial scale of random heterogeneous structures is small. More precisely, the atmosphere can be considered as horizontally homogeneous if the horizontal sizes of the randomly distributed local heterogeneities are much less than the selected range Δr in Eq. (5.11).

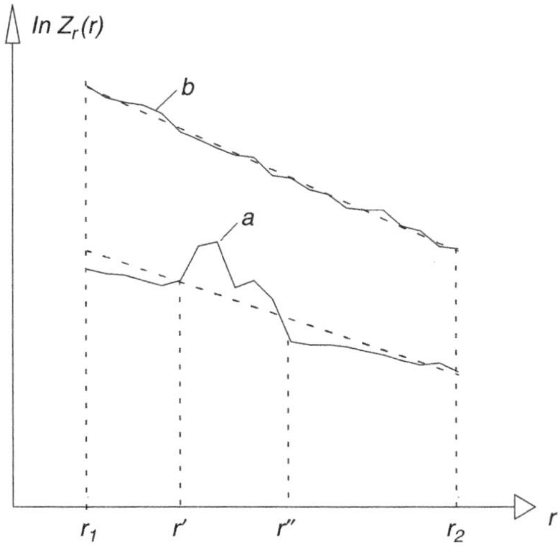

Fig. 5.2. Dependence of the logarithm of the square-corrected lidar signal on the range for inhomogeneous (a) and homogeneous (b) atmospheres.

The notion of a homogeneous atmosphere, as applied to a lidar measurement, differs from the general concept of homogeneity of the scattering medium. In particular, in the slope method, the assumption of the homogeneous atmosphere means only that the local heterogeneities do not significantly influence the mean linear fit over selected Δr, so that the slope method solution (Eq. 5.11) provides an acceptably accurate measurement result.

To understand, in a practical sense, when the slope method is applicable, let us consider typical examples of the logarithm of $Z_r(r)$ as a function of measurement range, shown in Fig. 5.2 (solid *curves a* and *b*). It can be seen that both *curves a* and *b* are not absolutely linear. For *case a*, the atmosphere cannot be considered as homogeneous, because a heterogeneous layer is clearly seen in the range from r' to r''. For *case b*, the optical situation is not so obvious, as no significant heterogeneous layer can be visualized. Here only local deviations of the function $[\ln Z_r(r)]$ from the linear approximation (dotted line) exist, which may be caused by either small-scale atmospheric heterogeneity or signal noise. The principal question that should be answered is whether the atmosphere for *b* can be considered as homogeneous over the range from r_1 to r_2, and accordingly, whether the slope method is applicable for this signal. Obviously, when using the slope method for the range interval $\Delta r = r_2 - r_1$, the difference between $\kappa_t(\Delta r)$ obtained with the slope method and its actual value, $\kappa_t(r_1, r_2)$, must be acceptable. In other words, some basis must be established to ensure that $\kappa_t(\Delta r)$ calculated with Eq. (5.11) does not differ significantly from the actual mean value

$$\kappa_t(r_1, r_2) = \frac{1}{r_2 - r_1} \int_{r_1}^{r_2} \kappa_t(r')dr'$$

so that the measurement error of $\kappa_t(\Delta r)$ calculated with the slope method is acceptable. There is thus a need to establish some criteria to evaluate the degree to which the assumption of homogeneity is valid. When the least-squares technique is used, the standard deviation obtained from the linear fit of the logarithm of $Z_r(r)$ may be considered as a criterion of the degree of atmospheric homogeneity. Although this technique is repeatable, the irregularities may skew the estimate of $\kappa_t(r_1, r_2)$ significantly without large changes in the standard deviation. Therefore to extract reliable information with the slope method, lidar data must be examined in light of all of the other available information on the conditions during which the data were collected. Particularly, the following questions should be addressed: (i) Was the measurement made in a horizontal or an inclined direction? (ii) What is the optical depth of the total range (r_1, r_2) estimated by the slope method? Is this value reasonable considering the measurement conditions? (iii) How large is the difference between the length of the measured distance (r_1, r_2) and the prevailing visibility? (iv) Were additional lidar measurements made in the same or shifted azimuthal directions? How do these data compare?

Such an analysis can be used for *case b* in Fig. 5.2. Generally, the atmosphere may be considered to be sufficiently homogeneous under the condition that the length of the linear range (r_1, r_2) is extended enough so that for moderately turbid atmospheres, the estimated optical depth of the measured interval is not less than $\tau(r_1, r_2) \approx 1$. In relatively clear atmospheres, with a visual range of more than 10–15 km, the use of the slope method is reasonable if the length of the interval over which the logarithm of $Z_r(r)$ is linear is at least 2–5 km. These conclusions are based on 2 years of simultaneous lidar and transmissometer measurements. These measurements were made at the experimental site of the Main Geophysical Observatory in Voeikovo (U.S.S.R.); a short outline of this investigation was published in a study by Baldenkov et al. (1988). These estimates are close to the result of the theoretical study by Kunz and de Leeuw (1993), who investigated the influence of random noise in the slope method. This theoretical analysis was made for a typical lidar system with the total range of 10 km. The authors' conclusion is that the extinction coefficient cannot be determined accurately when $\kappa_t < 0.1\,\text{km}^{-1}$. This is close to the conclusion above that one cannot accurately determine κ_t with the slope method if the total optical depth is less than ~1.

It should be stressed that these estimates cannot be considered to be universal; they are only estimates for a particular measurement site. Nevertheless, with this determination as a first rough criterion, it can be used to determine whether the slope method is applicable to *curve b* in Fig. 5.2. Assume, for example, that the measurement was made in a horizontal direction and that the mean extinction coefficient, obtained with Eq. (5.11), is $\kappa_t(\Delta r) \approx 1\,\text{km}^{-1}$. In this case, one can conclude that the slope method solution can be used for the *curve b* if the range (r_1, r_2) is not less than ~1 km, so that the optical depth $\tau(r_1, r_2) \approx 1$. Note also that the reliability of the slope-method data may be significantly increased if a number of signals measured in different azimuthal directions are used in the analysis. If the optical depth of the range under investigation is small, the application of the homogeneity approximation becomes questionable. Therefore, analyzing *curve a* in Fig. 5.2, obtained under the same conditions, one can conclude that the atmosphere cannot be considered to be homogeneous for the short range intervals (r_1, r') and (r', r_2). For these ranges, the slope method is not recommended to determine the mean values of κ_t. This is because the range intervals (r_1, r') and (r', r_2) are not enough extended to provide accurate data, at least for the optical conditions under consideration. One should always keep in mind that over short range intervals, the linear dependence of the logarithm of $Z_r(r)$ on r cannot be considered to be a reliable criterion of the degree of local atmospheric homogeneity.

An important specific of the slope method must be discussed. It was stated above that the dependence of the logarithm of the range-corrected signal on the range is linear if the extinction and backscatter coefficients are invariant within the measurement range. However, the inverse assertion may not be correct. In other words, the linear dependence of $\ln Z_r(r)$ on range r is necessary but not sufficient mathematical evidence of atmospheric homogeneity. Nevertheless, on a practical level, the linearity of the logarithm of $Z_r(r)$ can be used as an estimate of atmospheric homogeneity, at least in horizontal directions. One can show the validity of the above statement by using a "proof by contradiction." Suppose that the linear dependence of the logarithm of $Z_r(r)$ on r in Fig. 5.2, shown as *Curve b*, is obtained in a heterogeneous atmosphere over an extended range. For example, let us assume that the range (r_1, r_2), where κ_t and β_π are not constant, is 1 km or more. For this case, Eq. (5.2) can be rewritten as

$$Z_r(r) = C_0 T_1^2 \beta_\pi(r) e^{-2\int_{r1}^{r2} \kappa_t(r')dr'} \tag{5.12}$$

where T_1^2 is the two-way atmospheric transmission over the range $(0, r_1)$. As follows from Eq. (5.12), the following formula is then valid for the logarithmic curve

$$\ln Z_r(r) = \ln(C_0 T_1^2) + \ln \beta_\pi(r) - 2\int_{r_1}^{r_2} \kappa_t(r')dr' = A_1 - A_2 r$$

where A_1 and A_2 are constants of the linear fit. It follows from the above equation that for such a specific heterogeneous atmosphere, the following condition is required over the extended range (r_1, r_2)

$$\ln \beta_\pi(r) - 2\int_{r_1}^{r_2} \kappa_t(r')dr' = const. - A_2 r \tag{5.13}$$

that is, the algebraic sum of two range-dependent values must be linear over a distance of 1 km! Obviously, such an optical situation is unrealistic, so the existence of a linear logarithmic signal over extended horizontal ranges is normally indicative of homogeneous conditions.

The dependence of the logarithm of $Z_r(r)$ on range r is linear for atmospheres for which both κ_t and β_π are constant. The converse statement may be practical for extended atmospheric ranges, but it may be not valid for short ranges. For example, the linear relationship between ln $Z_r(r)$ and r does not provide a guarantee of atmospheric homogeneity over short distances as the lengths $[r_1, r']$ or $[r'', r_2]$ in Fig. 5.2 (*Curve a*). The linearity criterion cannot, generally, be used also for lidar measurements in directions not parallel to the ground surface.

Nevertheless, the slope method of lidar signal analysis is a basic method used for lidar system tests and as a diagnostic (see Section 3.4.1). Note that this method may be used successfully in both turbid and clear homogeneous atmospheres.

Compared with the other methods, the slope method often is the best method for the extraction of the mean particulate-extinction coefficient in homogeneous atmospheres. This statement is especially true for moderately turbid atmospheres, in which the particulate constituent is small, so that the attenuation due to particulates and molecules has the same order of magnitude. Unlike many other methods, in the slope method, it is not necessary to select a priori a numerical value of the particulate backscatter-to-extinction ratio to separate the aerosol contribution to extinction. However, the application of the slope method for routine atmospheric measurements is limited by the necessity of specifying formal criteria for the atmospheric homogeneity. A related problem, which is essential to obtain good estimates of the extinction coefficient, is the reliable selection of the homogeneous zones within the lidar measurement range that can be used in the analysis. Note also that the application of the slope method in clear atmospheres requires extremely accurate determination of the background component in order to minimize the signal offset remaining after the background component subtraction. A precise adjustment of the lidar optics is another requirement. This is necessary to avoid systematic distortions of the overlap function $q(r)$ over the range where the slope of the logarithm of $P(r)r^2$ is determined.

5.2. BASIC TRANSFORMATION OF THE ELASTIC LIDAR EQUATION

The slope method described above can only be used to determine the mean extinction coefficient over an extended measurement range in a homogeneous atmosphere. The determination of the extinction-coefficient profile or its value at a local point in an inhomogeneous atmosphere is significantly more difficult. To obtain local values of the extinction coefficient in homogeneous or heterogeneous atmospheres, more complicated retrieval methods are used. Generally, the measurement errors also become larger when local extinction coefficients are extracted.

To retrieve local values of the extinction or backscatter coefficient from lidar returns, the range-corrected lidar signal must be transformed by one of several methods. Different variants of the lidar signal inversion, published in numerous lidar studies, are in fact, similar and may be obtained with different forms for the lidar signal transformation. In this book, the general transformation that is used is based on the study by Weinman (1988). The application of the same type of transformation of the lidar signal throughout the book is done to provide continuity and enable discussion of the basics of elastic lidar data analysis. For the range of complete overlap, where $q(r) = 1$, the most general form of the elastic lidar equation is written as

$$P(r) = C_0 T_0^2 \frac{\beta_{\pi,p}(r) + \beta_{\pi,m}(r)}{r^2} \exp\left\{-2\int_{r_0}^{r} [\kappa_p(r') + \kappa_m(r')] dr'\right\} \tag{5.14}$$

where $\beta_{\pi,p}(r)$ and $\beta_{\pi,m}(r)$ are the particulate and molecular backscatter coefficients and $\kappa_p(r)$ and $\kappa_m(r)$ are the particulate and molecular extinction coefficients, respectively. Thus, in two-component (particulate and molecular) atmospheres, the lidar equation contains four unknown variables, $\beta_{\pi,p}(r)$, $\beta_{\pi,m}(r)$, $\kappa_p(r)$, and $\kappa_m(r)$. Obviously, to find any one of these variables, the other variables must be defined or relationships between the variables must be established. There is no problem in determining the relationship between the molecular extinction and backscattering, at least when no molecular absorption takes place (Section 2.3.2). For the particulate scatterers, the relationship between the backscattering term $\beta_{\pi,p}(r)$ and the extinction term $\kappa_p(r)$ depends on the nature, size, and other parameters of the particulate scatterers (Section 2.3.5). In real atmospheres, both quantities, $\beta_{\pi,p}(r)$ and $\kappa_p(r)$, may vary over an extremely wide range. Meanwhile, the particulate backscatter-to-extinction ratio has a much smaller range of values than the backscattering or the extinction. The most typical values for the backscatter-to-extinction ratio vary, approximately, by a factor of 5–10 (see Chapter 7). This is why it is reasonable to apply a numerical or analytical relationship between the values $\beta_{\pi,p}(r)$ and $\kappa_p(r)$ to invert the data from the lidar signal. The opportunity to replace the backscatter term $\beta_{\pi,p}(r)$ in the lidar equation by a slowly varying backscatter-to-extinction ratio significantly simplifies the lidar signal inversion. This replacement is widely used in elastic lidar measurements, both for particulate and molecular constituents. To accomplish this, the relationship between the extinction and backscatter coefficients must first be defined. For a pure scattering atmosphere, the particulate and molecular phase functions $P_{\theta,p}$ and $P_{\theta,m}$ given in Chapter 2 [Eqs. (2.26) and (2.37)] can be used. For backscattered light, the scattering angle $\theta = \pi$, so that the particulate and molecular phase functions are defined as

$$P_{\pi,p}(r) = \frac{\beta_{\pi,p}(r)}{\beta_p(r)} \tag{5.15}$$

and

$$P_{\pi,m} = \frac{\beta_{\pi,m}(r)}{\beta_m(r)} \tag{5.16}$$

Note that both functions, $P_{\pi,p}$ and $P_{\pi,m}$, are normalized to 1. Thus the molecular 180° phase function is $P_{\pi,m} = 3/8\pi$ [Chapter 2, Eq. (2.26)].

In processing lidar data, a more general form of these functions is generally used. Here the backscatter-to-extinction ratio is introduced, which can be used in both scattering and absorbing atmospheres. For an atmosphere in which both components exist, the particulate and molecular backscatter-to-extinction ratios should be written as

$$\Pi_p(r) = \frac{\beta_{\pi,p}(r)}{\kappa_p(r)} = \frac{\beta_{\pi,p}(r)}{\beta_p(r) + \kappa_{A,p}(r)} \tag{5.17}$$

and

$$\Pi_m(r) = \frac{\beta_{\pi,m}(r)}{\kappa_m(r)} = \frac{\beta_{\pi,m}(r)}{\beta_m(r) + \kappa_{A,m}(r)} \tag{5.18}$$

where $\kappa_{A,p}(r)$ and $\kappa_{A,m}(r)$ are the particulate and molecular absorption coefficients, respectively. In some studies, to relate extinction and backscatter, a so-called *S*-function is used that is the reciprocal of the backscatter-to-extinction ratio above. However, in the text of this book, the parameters defined in Eqs. (5.17) and (5.18) are used. The basic reasons for the use of these rather than the *S*-functions in this book are as follows. First, the particulate and molecular backscatter-to-extinction ratios in the lidar equation are physically motivated, as they show the fractions of the total particulate and molecular energy that are returned back, to the receiver's telescope. Accordingly, the use of these will make it easier for readers to understand physical processes underlying the lidar measurements and the structure of the lidar equation. Second, the functions $\Pi_{\pi,m}(r)$ and $\Pi_{\pi,p}(r)$ are more convenient when performing some lidar-signal transformations or error analyses. Third, they are directly proportional to the phase functions $P_{\pi,m}(r)$ and $P_{\pi,p}(r)$, introduced and used many tens years in classic scattering theories and studies. The relationship between the backscatter-to-extinction ratio and the phase function is

$$P_{\pi}(r) = \Pi(r)\left[1 + \frac{\kappa_A(r)}{\beta(r)}\right] \tag{5.19}$$

As follows from Eq. (5.19), in a purely scattering molecular atmosphere, $\kappa_{A,m} = 0$; thus, $\Pi_m = P_{\pi,m}$; similarly, $\Pi_p = P_{\pi,p}$ in a purely scattering particulate atmosphere, where $\kappa_{A,p} = 0$. With Eqs. (5.17) and (5.18), the lidar equation can be rewritten in the form

$$P(r) = C_0 T_0^2 \frac{\Pi_p(r)\kappa_p(r) + \Pi_m(r)\kappa_m(r)}{r^2} \exp\left\{-2\int_{r_0}^{r} [\kappa_p(r') + \kappa_m(r')]dr'\right\} \tag{5.20}$$

The particulate extinction term in the integrand of the exponential term is generally the main subject of the researcher's interest. The profile of $\kappa_p(r)$ rather than its integrated value generally must be determined. To determine the integrand in Eq. (5.20), the Bernoulli solution (Wylie and Barret, 1982) may be used. The unknown $\kappa_p(r)$ in the equation can also be found through transformation of the original lidar signal into a specific form (Weinman, 1988; Kovalev and Moosmüller, 1994). In this book the latter variant is used because of the simplicity of the interpretation of the mathematical operations with the functions involved. The initial lidar signal given in Eq. (5.20) must be transformed into the function $Z(x)$ with the following structure

$$Z(x) = Cy(x)\exp\left[-2\int y(x)dx\right] \tag{5.21}$$

where C is an arbitrary constant and $y(x)$ is a new variable of the lidar equation obtained after the transformation. Note that this equation contains only one independent variable, $y(x)$. This variable must be uniquely related to the unknown parameters in the initial lidar equation [Eq. (5.20)], so that these parameters can be later extracted from $y(x)$. The solution of Eq. (5.21) for $y(x)$ can be obtained by implementing an intermediate variable, $\zeta = \int y(x')dx'$, so that $d\zeta = y(x)dx$. With this intermediate variable, Eq. (5.21) can be transformed into the form

$$Z(x) = C\exp^{-2\zeta}\frac{d\zeta}{dx} \tag{5.22}$$

After integrating functions in both sides of Eq. (5.22), the relationship between the integrals of $Z(x)$ and $y(x)$ can be obtained in the form

$$\int Z(x)\mathrm{d}x = \frac{-C}{2}\exp\left[-2\int y(x)\mathrm{d}x\right] \tag{5.23}$$

With Eq. (5.23), the general solution for Eq. (5.21) is obtained in the form

$$y(x) = \frac{Z(x)}{C - 2\int Z(x)\mathrm{d}x} \tag{5.24}$$

The first step that must be accomplished in data processing is to transform the initial Eq. (5.20) into the form of Eq. (5.21). There are several different ways to effect such a transformation. The simplest way is the transformation of the exponential term in Eq. (5.20). Before such transformation, the range correction of the initial lidar signal is made, so that Eq. (5.20) can be rewritten into the form

$$Z_r(r) = P(r)r^2 = C_0 T_0^2 \Pi_\mathrm{p}(r)[\kappa_\mathrm{p}(r) + a(r)\kappa_\mathrm{m}(r)]\exp\left\{-2\int_{r_0}^{r}[\kappa_\mathrm{p}(r') + \kappa_\mathrm{m}(r')]\mathrm{d}r'\right\} \tag{5.25}$$

where $a(r)$ is the ratio

$$a(r) = \frac{\Pi_\mathrm{m}(r)}{\Pi_\mathrm{p}(r)} \tag{5.26}$$

To transform Eq. (5.25) into the form given in Eq. (5.21), the range-corrected lidar signal in Eq. (5.25) should be multiplied by some correction function, which transforms the exponential term. The correction function can be determined as

$$Y(r) = C_Y \frac{1}{\Pi_\mathrm{p}(r)}\exp\left\{-2\int_{r_0}^{r}\kappa_\mathrm{m}(r')[a(r') - 1]\mathrm{d}r'\right\} \tag{5.27}$$

where C_Y is an arbitrary scaling factor. The reciprocal of $\Pi_\mathrm{p}(r)$ is also included in the correction function as an additional factor. This makes it possible to remove factor $\Pi_\mathrm{p}(r)$ from Eq. (5.25) after the transformation is made. Note that to calculate $Y(r)$, the molecular extinction coefficient profile and the molecular and particulate backscatter-to-extinction ratios over the examined path must be known.

After the range-corrected lidar signal in Eq. (5.25) is multiplied by $Y(r)$, a new function $Z(r)$ is found, which has a structure similar to Eq. (5.21)

$$Z(r) = Z_r(r)Y(r) = C[\kappa_\mathrm{p}(r) + a(r)\kappa_\mathrm{m}(r)]\exp\left\{-2\int_{r_0}^{r}[\kappa_\mathrm{p}(r') + a(r')\kappa_\mathrm{m}(r')]\mathrm{d}r'\right\} \tag{5.28}$$

where the constant C is the product of an arbitrarily selected scale factor C_Y, the lidar constant C_0, and the unknown two-way transmittance T_0^2 over the range from $r = 0$ to r_0

$$C = C_Y C_0 T_0^2 \tag{5.29}$$

The lidar signal can be multiplied by any constant C_Y when the transformation of $P(r)$ into $Z(r)$ is made. This transformation makes it possible to define new variable to be the synthetic extinction coefficient, κ_W, as

$$\kappa_w(r) = \kappa_p(r) + a(r)\kappa_m(r) \tag{5.30}$$

The transformation results in the replacement of four variables, $\kappa_p(r)$, $\kappa_m(r)$, $\Pi_p(r)$, and $\Pi_m(r)$, in the original Eq. (5.20) by a new variable, which also has the dimension of an inverse length, $[L^{-1}]$, namely, the same as that for the extinction coefficient.

The variable κ_W of the transformed lidar equation is a weighted sum of the molecular and particulate components; the particulate extinction constituent κ_p is taken with a weight of 1, and the molecular constituent κ_m is taken with the weighting factor $a(r)$. With the new variable, κ_W, Eq. (5.28) becomes similar to Eq. (5.21)

$$Z(r) = C\kappa_w(r)\exp\left[-2\int_{r_0}^{r} \kappa_w(r')dr'\right] \tag{5.31}$$

The transformation of $Z_r(r)$ into $Z(r)$ changes the slope of the range-corrected signal, $Z_r(r)$, over the operating range. The change in slope is related to $a(r)$, so that smaller values of the particulate backscatter-to-extinction ratio Π_p cause larger changes in the original profile $Z_r(r)$ and its logarithm (Fig. 5.3). The relationship between the integrals of $Z(r)$ and κ_W is similar to that in Eq. (5.23); thus integrating $Z(r)$ in the limits from r_0 to r gives the formula

$$\int_{r_0}^{r} Z(r')dr' = \frac{C}{2}\left\{1 - \exp\left[-2\int_{r_0}^{r} \kappa_W(r')dr'\right]\right\} \tag{5.32}$$

Accordingly, the general solution for the new variable is similar to that in Eq. (5.24)

$$\kappa_W(r) = \frac{Z(r)}{C - 2\int_{r_0}^{r} Z(r')dr'} \tag{5.33}$$

Thus processing lidar data involves the following steps. First, the transformation function $Y(r)$ is calculated with Eq. (5.27). Note that before this can

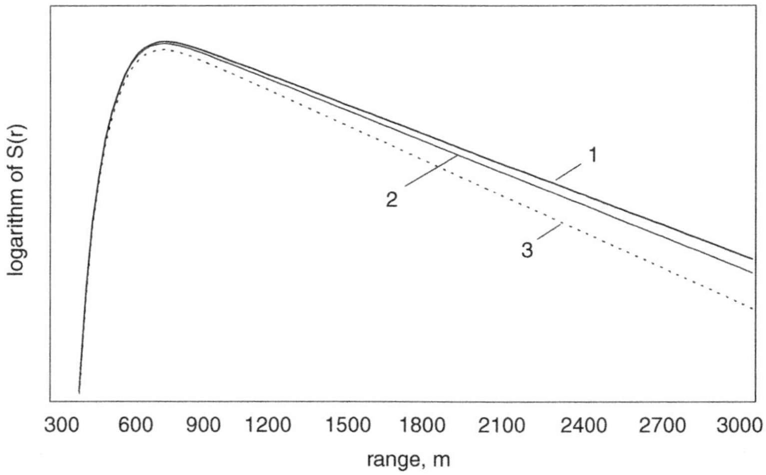

Fig. 5.3. Logarithm of the range-corrected signal $Z_r(r) = P(r)r^2$ (*curve 1*) calculated with the lidar system overlap function shown in Fig. 5.1 and the logarithms of this function after its transformation (*curves 2* and *3*). The corresponding functions $Z(r) = Z_r(r)Y(r)$ are calculated with the transformation functions $Y(r)$ using constant values of $\Pi_p = 0.05\,\text{sr}^{-1}$ (*curve 2*) and $\Pi_p = 0.02\,\text{sr}^{-1}$ (*curve 3*).

be done, the backscatter-to-extinction ratio $\Pi_p(r)$ must be somehow estimated (or taken a priori) to obtain $a(r)$. The profile of the molecular attenuation coefficient, $\kappa_m(r)$, must also be determined. In practice, the molecular profile is obtained either from balloon measurements or from a standard atmosphere tabulation. Second, the original lidar signal is range corrected and transformed into the function $Z(r)$ by multiplying the range-corrected signal by $Y(r)$. Then the weighted extinction coefficient $\kappa_W(r)$ is found with Eq. (5.33). The solution requires that the constant C in Eq. (5.33) be determined. Methods to determine this constant are given in the next sections. After the weighted function $\kappa_W(r)$ is found, the particulate extinction coefficient can be extracted by the simple formula

$$\kappa_p(r) = \kappa_W(r) - a(r)\kappa_m(r) \tag{5.34}$$

in which the same values of $\kappa_m(r)$ and $a(r)$ must be used as when calculating $Y(r)$.

Some comments must be made regarding the constant C in the lidar equation solution in Eq. (5.33). First, the constant C and the lidar system constant C_0 are not the same [see Eq. (5.29)]. Second, the constant C is uniquely related to the integral of $Z(r)$. The exponential term in Eq. (5.32) vanishes to zero when the range r tends to infinity. Accordingly, as $r \Rightarrow \infty$, the right side of Eq. (5.32) reduces to $C/2$, so that the constant C is related to the integral of $S(r)$ as

$$C = 2\int_{r_0}^{\infty} Z(r')dr' \tag{5.35}$$

Note that the constant C is actually constant only for a fixed lower limit of the integration, r_0. As follows from Eq. (5.29), its value depends on the transmission term T_0^2. When the near end of the examined path is moved away from the lidar, the corresponding transmission term in Eq. (5.29), and accordingly, the constant C, is reduced. The most general theoretical solution of the lidar equation for any range r may be obtained by substituting Eq. (5.35) into Eq. (5.33). This general form of the solution for $\kappa_W(r)$ is

$$\kappa_W(r) = \frac{Z(r)}{2\int_r^{\infty} Z(r')dr'} \tag{5.36}$$

The solution given in Eq. (5.36) was derived by Kaul (1977). Some aspects of this solution were considered later by Zuev et al. (1978a). Kaul's solution was derived for a single-component turbid atmosphere, but it is easily adapted for clear, two-component atmospheres (Kovalev and Moosmüller, 1994).

The lidar signal transformation considered in this section is the most practical, but it is not unique. There are other ways to transform the lidar signal, which can be used in specific cases. For example, an alternate way of transforming the exponential term in Eq. (5.25) exists, where the transformation function is determined with the particulate extinction-coefficient profile rather than with the molecular profile. In this case, the transformation function is found as

$$Y'(r) = C'_Y \frac{1}{\Pi_m(r)} \exp\left\{-2\int_{r_0}^{r} \kappa_p(r')\left[\frac{1}{a(r')} - 1\right]dr'\right\} \tag{5.37}$$

Note that the transformation function $Y'(r)$ can be calculated only when the particulate component $\kappa_p(r)$ is known. The corresponding weighted variable, $\kappa'_W(r)$, is then defined as

$$\kappa'_w(r) = \frac{\kappa_p(r)}{a(r)} + \kappa_m(r) \tag{5.38}$$

This variant of the transformation may be useful in some specific situations, for example, in a combination of aerosol and DIAL measurements, when molecular absorption must be considered.

Both methods to transform the lidar signal described above are based on a modification of the exponential term in the lidar equation. Another method of transforming the signal $P(r)$ into the function $Z(r)$ is based on transformation of the backscatter term of the lidar signal. To transform the original lidar equation to the corresponding function $Z(r)$, an iterative procedure is used (Kovalev, 1993). This variant of transformation is considered in Chapter 7.

Apart from that, the original lidar equation may be transformed into a normalized equation in which the total backscatter coefficient is a new variable. Here the new variable, $y(r)$ in Eq. (5.21), is defined as

$$y(r) = \beta_{\pi,p}(r) + \beta_{\pi,m}(r) \tag{5.39}$$

This type of transformation was made for the lidar signals obtained during extensive tropospheric and stratospheric measurements in the presence of high-altitude clouds (Sassen and Cho, 1992). The transformation allow derivation of the particulate backscatter term rather than the extinction coefficient. Such method made possible the clarification of some atmospheric processes, for example, in periods after excessive volcano eruptions (Hayashida and Sasano, 1993; Kent and Hansen, 1998). The principles underlying such a transformation are discussed in Section 8.1.

5.3. LIDAR EQUATION SOLUTION FOR A SINGLE-COMPONENT HETEROGENEOUS ATMOSPHERE

The assumption of a single-component atmosphere may be used when light scattering created by one atmospheric component significantly dominates over the scattering created by other components. For example, in a heavy fog or a cloudy layer, the light scattering by aerosols is generally much larger than the molecular scattering. Therefore, when processing the lidar data, the molecular scattering can be ignored, so that only the aerosol contribution to scattering is considered. Similarly, the use of an ultraviolet lidar for examining the clear troposphere, especially at high altitudes, may allow consideration of only the molecular contribution. This is especially true when a large molecular absorption is involved in the extinction process.

In this section, a lidar equation solution is considered for a turbid heterogeneous atmosphere that is comprised of aerosol particulates only. For such a single-component atmosphere, one can rewrite Eq. (5.20) in the form

$$P(r) = C_0 T_0^2 \frac{\Pi_p(r)\kappa_p(r)}{r^2} \exp\left[-2\int_{r_0}^{r} \kappa_p(r')dr'\right] \tag{5.40}$$

The equation constant in Eq. (5.40) is comprised of the lidar constant C_0 and the unknown two-way transmittance T_0^2 over the range from $r = 0$ to r_0. Apart from the constants, the equation includes the unknown function $\Pi_p(r)$. To extract $\kappa_p(r)$ from the signal $P(r)$, all of these parameters must be somehow measured or estimated.

Despite the difficulties in determining the equation constants, the main problem is determining the atmospheric backscatter-to-extinction ratio $\Pi_p(r)$, which, in the general case, may be not constant. A variable $\Pi_p(r)$ over the measurement range presents the greatest source of difficulties in inverting elastic lidar measurements. The simplest assumption, which makes it possible to find $\kappa_p(r)$, assumes that the backscatter-to-extinction ratio is range independent, that is,

$$\Pi_p(r) = \Pi_p = const. \tag{5.41}$$

Such an assumption may be considered to be acceptable if its application does not result in an intolerable error for the extracted extinction-coefficient profile.

The validity of the assumption of a constant particulate backscatter-to-extinction ratio depends on the particular atmospheric situation. The backscatter-to-extinction ratio depends on the type, shape, composition, and size distribution of the atmospheric particulates. If these parameters do not significantly change along the examined path, this assumption is reasonable, even if these parameters vary slightly because of small-scale fluctuations.

With Eq. (5.41) and the initial condition of single-component atmosphere $[\kappa_m(r) = 0]$, the transformation function $Y(r)$ in Eq. (5.27) reduces to

$$Y(r) = \frac{C_Y}{\Pi_p} = const. \tag{5.42}$$

As mentioned in Section 5.2, any arbitrary constant value for C_Y may be used. When Π_p is assumed constant, it is convenient to choose the arbitrary constant C_Y to be equal to the backscatter-to-extinction ratio. Note that it is not necessary to know the numerical value of the backscatter-to-extinction ratio to apply the equality $C_Y = \Pi_p$.

In a single-component atmosphere with $\Pi_p = const.$, the extinction coefficient can be found without having to establish the numerical value of the backscatter-to-extinction ratio.

When the transformation function $Y(r) = 1$, no special signal transformation is required. The condition (5.42) allows one to perform the inversion using the range-corrected signal $Z_r(r)$ obtained by multiplying the initial lidar signal $P(r)$ in Eq. (5.40) by the square of range r

$$Z_r(r) = P(r)r^2 = C_r\kappa_p(r)\exp\left[-2\int_{r_0}^{r}\kappa_p(r')dr'\right] \tag{5.43}$$

where

$$C_r = C_0 T_0^2 \Pi_p \tag{5.44}$$

The general solution for the extinction coefficient [Eq. (5.33)] can be reduced and written as (Barrett and Ben-Dov, 1967)

$$\kappa_p(r) = \frac{Z_r(r)}{C_r - 2I_r(r_0, r)} \tag{5.45}$$

where the function $I_r(r_0, r)$ is the range-corrected signal $Z_r(r)$ integrated over the range from r_0 to r

$$I_r(r_0, r) = \int_{r_0}^{r} Z_r(r')\mathrm{d}r' \tag{5.46}$$

At the beginning of the lidar era, the solution given in Eq. (5.45) was developed and analyzed by Barrett and Ben-Dov (1967), Collis (1969), Davis (1969), Zege et al. (1971), and Fernald et al. (1972). During this early period (approximately from 1967 to 1972), this type of straightforward method was commonly considered for lidar signal processing. The approach was based on the idea that the lidar constant might be easily determined through the absolute calibration of the lidar.

However, a number of shortcomings inherent in this method were soon revealed. First, the constant C_r includes not only the lidar instrumental parameter C_0 but also the factors T_0^2 and Π_p. The direct determination of C_r requires knowledge of all of the individual terms. Unlike the constant C_0, the last two terms can be determined during the experiment event only. In clear atmospheres, T_0^2 may be assumed to be unity if the range r_0 is not large. Another option is to estimate in some way the value of the extinction coefficient in an area of the lidar site and then calculate T_0^2 assuming a homogeneous atmosphere in the range from $r = 0$ to r_0 (Ferguson and Stephens, 1983; Marenco et al., 1997). Large uncertainties may arise when relating backscatter and extinction coefficients, that is, when selecting an a priori value of Π_p (Hughes et al., 1985). As will be shown later, the method described above uses an unstable solution, similar to the so-called near-end solution. The poor stability of Eq. (5.45) is due to the subtraction operation in the denominator of the equation. As the range r increases, the denominator decreases. If an error exists in the estimated constant C_r, or if the signal-to-noise ratio significantly worsens, the denominator may become negative, yielding erroneous negative values of the derived extinction coefficient. Also, an absolute calibration must be performed to determine the constant C_0, which in turn, is a product of some instrumental constants, as shown in Section 3.2.1. Attempts to calibrate lidars have revealed that the absolute calibration required a refined technique and was not accomplished simply (Spinhirne et al., 1980). Thus the solution, based on separate determination of the individual instrumentation and atmospheric factors in C_r, is not practical.

5.3.1. Boundary Point Solution

To find the unknown $\kappa_p(r)$ with Eq. (5.45), one must know the constant C_r, that is, the product of $C_0\, T_0^2\, \Pi_p$. Note that it is not necessary to know the individual terms C_0, T_0^2, and Π_p in order to extract the extinction coefficient. It is sufficient to know only the resulting product of these three values. This can be achieved without an absolute calibration. The simplest way to determine the constant C_r is to establish a boundary condition of the equation at some point of the lidar measurement range. This makes it possible to find the constant C_r and then to use it to determine the profile of $\kappa_p(r)$ over the total measurement range. Specifically, the constant can be determined if a point r_b exists within the lidar measurement range at which the extinction coefficient, $\kappa_p(r_b)$ is known, or at least may be accurately estimated or taken a priori. Such methods of solving the lidar equation are known as "boundary point solutions." This solution can be derived in the following way. Solving Eq. (5.43) for the selected boundary point r_b at which the extinction coefficient is known, one can define the constant C_r as

$$C_r = \frac{Z_r(r_b)}{\kappa_p(r_b)\exp\left[-2\int_{r_0}^{r_b}\kappa_p(r')dr'\right]} \tag{5.47}$$

Substituting C_r as defined in Eq. (5.47) into the original lidar equation Eq. (5.43), one can obtain the following equality

$$\frac{Z_r(r_b)}{\kappa_p(r_b)} = \frac{Z_r(r)}{\kappa_p(r)}\exp\left[-2\int_r^{r_b}\kappa_p(r')dr'\right] \tag{5.48}$$

After taking the integral of $Z_r(r)$ in the range from r to r_b, the exponential term in Eq. (5.48) can be derived in the form

$$\exp\left[-2\int_r^{r_b}\kappa_p(r')dr'\right] = 1 - \frac{\kappa_p(r)}{Z_r(r)}\left[2\int_r^{r_b}Z_r(r')dr'\right] \tag{5.49}$$

Substituting the exponent term in Eq. (5.49) into Eq. (5.48), one can obtain the boundary point solution in its conventional form

$$\kappa_p(r) = \frac{Z_r(r)}{\dfrac{Z_r(r_b)}{\kappa_p(r_b)} + 2\int_r^{r_b}Z_r(r')dr'}. \tag{5.50}$$

Thus the boundary point solution makes it possible to avoid a direct calculation of the constant $C_r = C_0 T_0^2 \Pi_p$ in Eq. (5.45) by using some equivalent reference quantity instead of C_r. Such a method is sometimes called the reference calibration. The boundary point may be chosen to be at the near end ($r_b < r$) or the far end ($r_b > r$) of the measurement range [Fig. 5.4, (a) and (b), respectively]. The corresponding solution is defined as the near-end or far-end solution, respectively. Note that when the boundary point r_b is selected at the

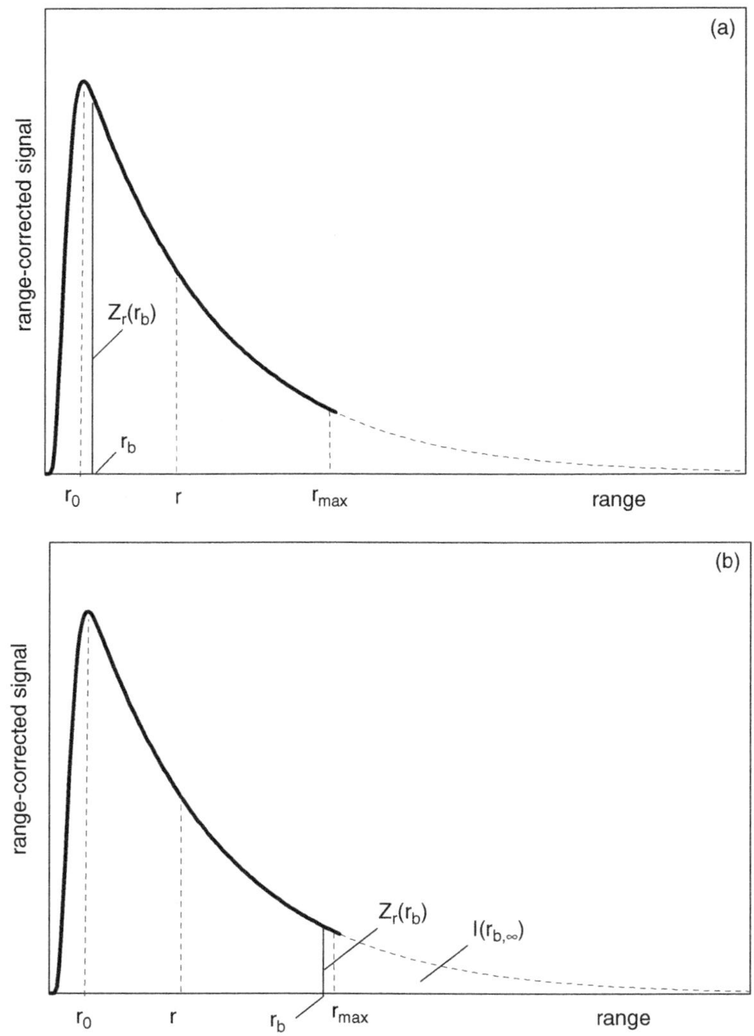

Fig. 5.4. Illustration of the the near end and far-end boundary point solutions. (a) The range r_b, where an assumed (or determined) extinction coefficient $\kappa_p(r_b)$ is defined, is chosen close to the near end of the lidar operating range, r_0. (b) Same as (a) but the point r_b is chosen close to the far end of the lidar operating range, r_{max}.

near end of the measurement range [Fig. 5.4 (a)], the integration limits in Eq. (5.50) are interchanged, so that the summation in the denominator of the equation is replaced by a subtraction

$$\kappa_p(r) = \frac{Z_r(r)}{\dfrac{Z_r(r_b)}{\kappa_p(r_b)} - 2\int_{r_b}^{r} Z_r(r')dr'}. \tag{5.51}$$

When both terms in the denominator become comparable in magnitude, the solution in Eq. (5.51) becomes unstable and can even yield negative values of the measured extinction coefficient (Viezee et al., 1969). The most stable solution for the extinction coefficient is obtained when the boundary point r_b is chosen close to the far end of the lidar measurement range [Fig 5.4 (b)]. Such a solution, given in Eq. (5.50), is widely known as Klett's far-end solution (Klett, 1981).

In comparison, the far-end boundary point solution is much more stable than the near-end solution, at least, in turbid atmospheres. It yields only positive values of the derived extinction coefficient, κ_t, even if the signal-to-noise ratio is poor. However in clear atmospheres, it has no significant advantages as compared to the near-end solution.

The advantage of the far-end boundary point solution in comparison to the near-end solution in turbid atmospheres was first shown by Kaul (1977) and in a later collaborative study by Zuev et al. (1978a). Unfortunately, these studies were not accessible to western readers. In 1981, Klett published his famous study (Klett, 1981), and since then, the far-end solution has been known to western readers as Klett's solution. It would be rightly to refer to this solution as the Kaul–Klett solution, which gives more proper credit.

The far-end solution is always cited as the most practical solution. It is, indeed, a remarkably stable solution in turbid atmospheres (see Section 5.2). Omitting for the moment some specific limitations of this solution, which will be considered later, the basic problem with this solution is the need to establish an accurate value for the local extinction coefficient $\kappa_p(r_b)$ at a distant range of the lidar measurement path, which may be kilometers away from the lidar location. No significant problem in determining $\kappa_p(r_b)$ (except multiple scattering) appears if such a point is selected within a cloud, for which a sensible extinction coefficient can be assumed (Carnuth and Reiter, 1986). Similarly, the problem can be avoided for a remote particulate-free region in which the extinction can be assumed to be purely molecular. For that case, the lidar signal can be processed with an estimate of the molecular extinction as the boundary point (see Section 8.1). However, the most common situation lies between these two extremes, and generally there are no practical methods to establish a boundary value that is accurate enough to obtain acceptable measurement results.

5.3.2. Optical Depth Solution

Another way to solve Eq. (5.43) is to use total path transmittance over the lidar operating range as a boundary value. Similar to the previous case, the optical depth solution is generally applied with the assumption that the backscatter-to-extinction ratio is range independent, that is, $\Pi_p = const.$ over the measurement range. In clear and moderately turbid atmospheres, the total atmospheric transmittance (or the optical depth) may be found from an independent measurement, for example, with a solar radiometer, as proposed by Fernald et al. (1972). In highly turbid, foggy and cloudy atmospheres, the boundary value may be found from the signal $Z_r(r)$ integrated over the maximum operating range (Kovalev, 1973). The optical depth solution has been successfully used both in clear and polluted atmospheres (see e.g., Cook et al., 1972; Uthe and Livingston, 1986; Rybakov et al., 1991; Marenco et al., 1997; Kovalev, 2003).

It is necessary to define the idea of the total path transmittance used as a boundary value. Any lidar system has a particular operating range, where lidar signals may be measured and recorded. We use here the term "operating range" instead of the "measurement range," because with lidar measurements, these two ranges may differ significantly. The measurement range is the range over which the unknown atmospheric quantity can be measured with some acceptable accuracy. However, the lidar operating range generally comprises areas with poor signal-to-noise ratios at the far end of the range, where accurate measurement data cannot be extracted from the signals. However, even these "useless" signals are generally recorded and processed because of at least three reasons. First, neither the operating nor the measurement range can be established before the act of the lidar measurement. Second, the lidar data points over the distant ranges, where the backscatter signal is small and cannot be used for accurate determining extinction profiles due to a poor signal-to-noise ratio, may be used for determining the maximal integral, $I_{r,\max}$ [Eq. (5.53)]. Third, the lidar data points over a distant range, where the signal backscatter component vanishes to zero, are often used to determine the signal background component.

All other conditions being equal, the length of the lidar operating range depends on the atmospheric transparency and the lidar geometry. As shown in Section 5.1, the near end of the lidar measurement range depends on the length of the zone of incomplete overlap. The minimum lidar range $r_{\min}$ is normally taken at or beyond the far end of the incomplete lidar overlap, that is, at $r_{\min} \geq r_0$. The upper lidar measurement limit $r_{\max}$ is restricted because of the reduction of the lidar signal with the range. The magnitude of the useful signal, $P(r)$, decreases with range because of atmospheric extinction and the divergence of the returning scattered light, whereas the background (additive) noise generally has no significant change with the time, it only fluctuates about its mean value. Accordingly, the most significant relative increase of the noise contribution occurs at distant ranges where the backscattered signal vanishes

(Section 3.4). The upper lidar measurement limit r_{max} is commonly taken as the range at which the signal-to-noise ratio reaches a certain threshold value. This maximum range depends both on the extinction coefficient profile along the lidar line of sight and on lidar instrument characteristics, such as the emitted light power and the aperture of receiving optics. Thus the upper limit is variable, whereas the lower range, r_{min}, is a constant value, which depends only on parameters of lidar transmitter and receiver optics.

In the optical depth solution, the two-way transmittance T_{max}^2 over the lidar maximum range from r_0 to r_{max}

$$T_{max}^2 = e^{-2\int_{r_0}^{r_{max}} \kappa_p(r')dr'} \tag{5.52}$$

is used as a solution boundary value. Just as with the boundary point solution, the use of T_{max}^2 as a boundary value makes it possible to avoid direct calculation of the constant C_r. The optical depth solution is derived by estimating T_{max}^2 and calculating the integral of the range-corrected signal $Z_r(r)$ over the maximum range from r_0 to r_{max}. The integral can be found by substituting $r = r_{max}$ in Eq. (5.32)

$$I_{r,max} = \int_{r_0}^{r_{max}} Z_r(r')dr' = \frac{1}{2}C_r\left(1 - T_{max}^2\right) \tag{5.53}$$

The unknown constant in Eq. (5.45) may be found as the function of T_{max}^2 and $I_{r,max}$

$$C_r = \frac{2I_{r,max}}{1 - T_{max}^2} \tag{5.54}$$

By substituting C_r in Eq. (5.54) to Eq. (5.45), one can obtain the optical depth solution for the single-component aerosol atmosphere in the form

$$\kappa_p(r) = \frac{0.5Z_r(r)}{\dfrac{I_{r,max}}{1 - T_{max}^2} - I_r(r_0, r)} \tag{5.55}$$

where the two-way total transmittance T_{max}^2 is the value that must be in some way estimated to determine $\kappa_p(r)$.

For real atmospheric situations, T_{max}^2 is a finite positive value ($0 < T_{max}^2 < 1$), so that the denominator in Eq. (5.55) is also always positive. Therefore, the optical depth solution is quite stable. Like the far-end boundary point solution, it always yields positive values of the derived extinction coefficient.

In studies by Kaul (1977) and Zuev et al. (1978a), a unique relationship was given between the lidar equation constant and the integral of the range-corrected signal measured in a single-component particulate atmosphere. Following these studies, let us consider the integral in Eq. (5.53) with an infinite upper integration limit, that is, when $r_{max} \Rightarrow \infty$. It follows from Eq. (5.53) that the integral with an infinite upper level

$$I(r_0, \infty) = \int_{r_0}^{\infty} Z_r(r')\mathrm{d}r'$$

has a finite value. Indeed, the integral over the range from r_0 to infinity is formally defined as

$$I(r_0, \infty) = \frac{1}{2} C_r \left[(1 - T(r_0, \infty))^2 \right] \tag{5.56}$$

For any real scattering medium with $\kappa_p > 0$, the path transmittance over infinite range, $T(r_0, \infty)$, tends toward zero, thus

$$I(r_0, \infty) = \frac{1}{2} C_r \tag{5.57}$$

There is an interesting application of the theoretical equations above. Note that T_{max}^2 [Eq. (5.52)] differs insignificantly from $T(r_0, \infty)^2$ when the lidar optical depth $\tau(r_0, r_{max})$ is large. For example, if the optical depth $\tau(r_0, r_{max}) = 2$, one can obtain from Eqs. (5.53) and (5.57) that $I(r_0, r_{max}) = 0.98\ I(r_0, \infty)$. Accordingly, the integral $I(r_0, \infty)$ in Eq. (5.57) may be replaced by the integral with a finite upper range r_{max}. Such a replacement will incur only a small error, on the order of 2%. If the lidar constant C_0 is known, that is, is determined by the absolute calibration, and the optical depth of the incomplete overlap zone $(0, r_0)$ is small, so that $T_0^2 \approx 1$, the integral $I(r_0, r_{max})$ may be directly related to the backscatter-to-extinction ratio. Under the above conditions, the backscatter-to-extinction ratio can be found from Eqs. (5.44) and (5.57) as

$$\Pi_p = \frac{2I(r_0, r_{max})}{C_0} \tag{5.58}$$

Eq. (5.58) makes it possible to determine the backscatter-to-extinction ratio with the range-corrected signal after it is integrated over the measurement range with a relevant optical depth. The concept, originally proposed by Kovalev (1973), was later used in studies of high-altitude clouds (Platt, 1979) and artificial smoke clouds (Roy, 1993). The principal shortcoming of this method is the presence of an additional multiple-scattering component when the optical depth is large. To use Eq. (5.58), a multiple scattering must be estimated in some way and removed before Π_p is calculated (Kovalev, 2003a).

It should be noted that, in principle, the optical depth solution can be used with either the total or local path transmittance taken as a boundary value. In other words, the known (or somehow estimated) transmittance of a local zone Δr_b can also be used as a boundary value. If such a zone is at the range from r_b to $[r_b+\Delta r_b]$, the solution in Eq. (5.55) may be transformed into

$$\kappa_t(r) = \frac{Z_r(r)}{\dfrac{2I_r(\Delta r_b)}{1-[T(\Delta r_b)]^2} - 2I_r(r_b, r)} \tag{5.59}$$

It should be pointed out, however, that unlike the basic solution given in Eq. (5.55), the solution in Eq. (5.59) may be not stable for ranges beyond the zone Δr_b.

Some additional comments should be made here concerning the application of range-dependent backscatter-to-extinction ratios in single-component atmospheres. These comments apply to both boundary point and optical depth solutions. With a variable $\Pi_p(r)$, the condition in Eq. (5.42) is invalid. In this case, the profile of $\Pi_p(r)$ along the lidar line of view should be in some way determined, for example, by using data of combined elastic-inelastic lidar measurements. The function $Y(r)$ can be then found as the reciprocal of $\Pi_p(r)$. Note that to determine $Y(r)$, one should know only the relative changes in the backscatter-to-extinction ratio rather than the absolute values. There is a simple explanation of this observation. The relative value of the backscatter-to-extinction ratio can formally be defined as the product $[A_\pi \Pi_p(r)]$, where A_π is an unknown constant. If this function $[A_\pi \Pi_p(r)]$ is known, the transformation function $Y(r)$ can be defined as

$$Y(r) = \frac{1}{[A_\pi \Pi_p(r)]} \tag{5.60}$$

then the lidar solution constant in Eq. (5.44) transforms to

$$C_r = \frac{C_0 T_0^2}{A_\pi} \tag{5.61}$$

Now the backscatter-to-extinction ratio is excluded from C_r, and only constant factors are present in the solution constant, which may be found by either the boundary point or the optical depth solution.

In a single-component atmosphere, the extinction coefficient can be found without having to establish the numerical value of the backscatter-to-extinction ratio. This is true for both Π_p = *const.* and Π_p (r) = *var.* To determine $\kappa_p(r)$, it is only necessary to know the relative change in the backscatter-to-extinction ratio. This is valid for both solutions presented in Sections 5.3.1 and 5.3.2.

To summarize the general points concerning the boundary point and optical depth solutions for a single-component atmosphere:

1. In both solutions, no absolute calibration of the lidar is needed. The constant factor in the equation is determined indirectly, by using a relative rather than absolute calibration.
2. The most stable solution of the lidar equation may be obtained with the far-end boundary point solution or by the optical depth solution with the maximum path transmittance over the lidar range as a boundary value.
3. In both solutions, one can extract the extinction-coefficient profile without the necessity of having to establish a numerical value for the backscatter-to-extinction ratio. The only condition is that this ratio be constant along the measured distance. This condition is practical even if the backscatter-to-extinction ratio varies slightly around a mean value but has no significant monotonic change within the range. Otherwise, at least relative changes in the range-dependent backscatter-to-extinction ratio must be established to obtain accurate measurement results.
4. Both solutions are practical for the extraction of extinction-coefficient profiles in the lower atmosphere, in both horizontal and slope directions. The solutions can be used in various atmospheric conditions: in haze or fog, in moderate snowfall or rain; in clear and cloudy atmospheres, etc. The problem to be solved is the accurate estimate of a boundary parameter, that is, the numerical value of $\kappa_p(r_b)$ or T_{max}^2. Quite often these values are not determined by independent measurements but are assumed a priori.
5. To obtain acceptable inversion data, the boundary conditions should be estimated by analyzing the measurement conditions and the recorded signals rather than taken as a guess. However, it is impossible to give particular recommendations for such estimates for different atmospheric conditions. The only acceptable approach to this problem is to assess the particular atmospheric situation and select the most appropriate algorithm.
6. The boundary point and optical depth solutions are always referenced to two discrete values. In the former, these values are the extinction coefficient $\kappa_p(r_b)$ and the lidar signal $Z_r(r_b)$ [Eqs. (5.50) and (5.51)]. The signal is generally taken at the far end of the measurement range. For the spatially extended measurement range, the signal $Z_r(r_b)$ may be significantly distorted by a poor signal-to-noise ratio and an inaccurate choice for the background offset. Any inaccuracy in the signal $S_r(r_b)$ influences the accuracy of the measurement result in a manner similar to an inaccuracy in the estimated $\kappa_p(r_b)$. The optical depth solution uses

the quantity related to the path-integrated extinction coefficient as a boundary value and the integral of $Z_r(r)$ over an extended range [Eq. (5.55)]. Because of integrat, the latter value is less sensitive to random errors in the lidar signal. Numerous estimates of the measurement errors confirm this point (Zuev et al., 1978; Ignatenko and Kovalev, 1985; Balin et al., 1987; Kunz, 1996).

5.3.3. Solution Based on a Power-Law Relationship Between Backscatter and Extinction

In the late 1950s, Curcio and Knestric (1958) and then Barteneva (1960) investigated the relationship between atmospheric extinction and backscattering and established the famous power-law relationship between the total backscatter and extinction coefficients

$$\beta_\pi = B_1 \kappa_t^{b_1} \tag{5.62}$$

where exponent b_1 and factor B_1 were taken as constants. Although the relationship between β_π and κ_t in Eq. (5.62) is purely empirical and has no theoretical grounds, Fenn (1966) stated that such a dependence was valid to within 20–30% over a broad spectral range of extinction coefficients, between 0.01 and $1\,\text{km}^{-1}$. It was established later that such an approximation may be considered to be valid only for ground-surface measurements and under a restricted set of atmospheric conditions. Fitzgerald (1984) showed that the relationship is dependent on the air mass characteristics and, moreover, is only valid for relative humidities greater than ~80%. Mulders (1984) concluded that the relationship is also sensitive to the chemical composition of the particulates. Thorough investigations have confirmed that the approximation is not universally applicable (see Chapter 7). Nevertheless, in the 1970s and even 1980s, the power-law relationship was considered to be an acceptable approximation for use in lidar equation solutions (Viezee et al., 1969; Fernald et al., 1972; Klett, 1981 and 1985; Uthe and Livingston, 1986; Carnuth and Reiter, 1986, etc.). When using the power-law relationship in lidar measurements, it is assumed that the atmosphere is comprised of a single component and that B_1 and b_1 are constant over the measured range. This dependence makes it possible to derive a simple analytical solution of the lidar equation, similar to that derived in Section 5.3.1. With the relationship in Eq. (5.62), the range-corrected signal [Eq. (5.43)] can be written as

$$Z_r(r) = C_0 T_0^2 B_1 [\kappa_p(r)]^{b_1} \exp\left[-2\int_{r_0}^{r} \kappa_p(r')dr'\right] \tag{5.63}$$

The lidar equation solution can be obtained after transforming Eq. (5.63) into the form

$$[Z_r(r)]^{\frac{1}{b_1}} = [C_0 B_1 T_0^2]^{\frac{1}{b_1}} [\kappa_p(r)] \exp\left[-\frac{2}{b_1}\int_{r_0}^{r} \kappa_p(r')dr'\right] \tag{5.64}$$

With Eq. (5.64), the basic solution in Eq. (5.45) can be rewritten as (Collis, 1969; Viezee et al., 1969)

$$\kappa_p(r) = \frac{[Z_r(r)]^{\frac{1}{b_1}}}{[C_0 B_1 T_0^2]^{\frac{1}{b_1}} - \frac{2}{b_1}\int_{r_0}^{r}[Z_r(x)]^{\frac{1}{b_1}}dx} \tag{5.65}$$

As pointed out by Kohl (1978), the proper choice of the constants b_1 and B_1 is a critical problem when processing lidar returns with Eq. (5.65). Nevertheless, some attempts have been made to use this solution in practical lidar applications. Fergusson and Stephens (1983) proposed an iterative scheme of data processing based on the assumption that the lidar equation is normalized beforehand, specifically, the product $C_0B_1 = 1$. Another simplified version of this method was developed by Mulders (1984). However, Hughes et al. (1985) showed that these methods are extremely sensitive to the selection of both constants relating backscatter and extinction coefficients in Eq. (5.62). Meanwhile, here solutions may be used that do not require an estimate of B_1. In the same way as shown in Section 5.3.1, Eq. (5.65) may be transformed into the boundary point solution. Accordingly, the far-end solution can be written as (Klett, 1981),

$$\kappa_p(r) = \frac{[Z_r(r)]^{\frac{1}{b_1}}}{\frac{[Z_r(r_b)]^{\frac{1}{b_1}}}{\kappa_t(r_b)} + \frac{2}{b_1}\int_{r}^{r_b}[Z_r(r')]^{\frac{1}{b_1}}dr'} \tag{5.66}$$

where r_b is a boundary point within the lidar operating range and $r < r_b$. In the above solution, only the constant b_1 must be known or be selected a priori, whereas the constant B_1 is not required.

Although the solution in Eq. (5.66) has been used widely for both horizontal and slant direction measurements (Lindberg et al., 1984; Uthe and Livingston, 1986; Carnuth and Reiter, 1986; Kovalev et al., 1991; Mitev et al., 1992), the critical problem of the proper choice of the constant b_1 has remained unsolved. For simplicity, most researchers have assumed this constant to be unity, thus reducing Eq. (5.66) to the ordinary boundary point solution [Eq. (5.50)]. Meanwhile, as pointed by Klett as long ago as 1985, the parameter b_1 cannot be considered to be constant in real atmospheres, at least for a wide range of atmospheric turbidity. Numerous experimental and theoretical investigations have confirmed that b_1 may have different numerical values under different measurement conditions, so that the relationship in Eq. (5.62) cannot be considered as practical in lidar applications.

5.4. LIDAR EQUATION SOLUTION FOR A TWO-COMPONENT ATMOSPHERE

In the earth's atmosphere, light extinction is caused by two basic atmospheric components, molecules and particulates. The idea of a two-component atmosphere assumes an atmosphere in which neither the first nor the second component can be ignored when evaluating optical propagation. Such an atmospheric situation is typical, for example, when examining a clear or moderately turbid atmosphere. Here the assumption of a single-component atmosphere as done in Section 5.3 is clearly poor.

The general principles of lidar examination of such atmospheres were based on ideas developed in early searchlight studies of the upper atmosphere (Stevens et al., 1957; Elterman, 1962 and 1963). The principal point of these studies was that for high-altitude measurements the particulates and molecules must be considered as two distinct classes of scatterers, which must be treated separately. Moreover, these early studies proposed the practical idea of using the data from particulate-free areas as reference data when processing the signals at other altitudes. Elterman's method of determining the particulate contribution, based on an iterative procedure, was later modified and used successfully in many lidar studies. The first lidar observations of tropospheric particulates where such an approach was used were reported by Gambling and Bartusek (1972) and Fernald et al. (1972). In the latter study, a general solution for the elastic lidar equation for a two-component atmosphere was given. The authors proposed to use solar radiometer measurements to determine the total transmittance within the lidar operating range. Later, in 1984, Fernald modified the solution. In that study, he proposed a calculation method based on the application of a priori information on the particulate and molecular scattering characteristics at some specific range. Instead of using the data from a standard atmosphere, he proposed to determine the molecular altitude profile from the best available meteorological data. This would allow an improvement in the accuracy in the retrieved particulate extinction obtained after subtracting the molecular contribution. A computational difficulty with Fernald's solution lay in the application of the transcendental equations. To find the unknown quantity, either an iterative procedure or a numerical integration had to be used. Klett (1985) and Browell et al. (1985) proposed an alternative solution for a two-component atmosphere. They developed a boundary point solution based on an analytical formulation. This made it possible to avoid the difficulties associated with the inversion of the transcendental equations in Fernald's (1984) method. Weinman (1988) and Kovalev (1993) developed optical depth solutions for two-component atmospheres, both based on iterative procedures. Later, Kovalev (1995) proposed a simpler version of the optical depth solution based on a transformation of the exponential term, which does not require an iterative procedure. In this chapter, the optical depth solution given is based generally on the latter study.

For a two-component atmosphere composed of particles and molecules, the lidar equation is written in the form [Eq. (5.20)]

$$P(r) = C_0 T_0^2 \frac{\Pi_p(r)\kappa_p(r) + \Pi_m(r)\kappa_m(r)}{r^2} \exp\left\{-2\int_{r_0}^{r} [\kappa_p(r') + \kappa_m(r')]dr'\right\}$$

As explained in Section 5.2, to extract the extinction coefficient, the signal $P(r)$ should first be transformed into the function $Z(r)$, which may be obtained by multiplying the range-corrected signal by the transformation function $Y(r)$. However, for two-component atmospheres, such a transformation may become problematic. To calculate the function $Y(r)$ [Eq. (5.27)], it is necessary to estimate the backscatter-to-extinction ratios $\Pi_p(r)$ and $\Pi_m(r)$ and then calculate the ratio $a(r)$ [Eq. (5.26)]. In the general case, the problem of making such an estimate is related to the need to determine both ratios rather than only the ratio for the particulate contribution, $\Pi_p(r)$. Indeed, the molecular backscatter-to-extinction ratio depends both on scattering and any absorption from molecular compounds that may be present [Eq. (5.18)], that is,

$$\Pi_m(r) = \frac{\beta_{\pi,m}(r)}{\beta_m(r) + \kappa_{A,m}(r)}$$

If the molecular absorption takes place at the wavelength of the lidar, the molecular backscatter-to-extinction ratio cannot be calculated until the profile of the molecular absorption coefficient, $\kappa_{A,m}(r)$, is determined. However in practice, only the scattering term of the molecular extinction is generally available, which can be determined either from a standard atmosphere or from balloon measurements. Therefore, the transformation above is practical only for the wavelengths at which no significant molecular absorption exists. Here $\kappa_m(r) = \beta_m(r)$, and $\Pi_m(r)$ reduces to a range-independent quantity, $\Pi_m(r) = P_{\pi,m} = 3/8\pi$.

Theoretically, the lidar equation transformation for two-component atmospheres can be made when both scattering and absorbing molecular components have nonzero values. However, to accomplish this, the profile of the molecular absorption coefficient should be known. Thus the transformation is practical if no molecular absorption occurs at the wavelength of the measurement.

When no molecular absorption takes place, the transformation function $Y(r)$ in Eq. (5.27) reduces to a form useful for practical applications

$$Y(r) = \frac{C_Y}{\Pi_{\pi,p}(r)} \exp\left\{-2\int_{r_0}^{r} [a(r') - 1]\beta_m(r')dr'\right\} \tag{5.67}$$

where

$$a(r) = \frac{3/8\pi}{\Pi_p(r)}$$

To determine the transformation function $Y(r)$, the numerical value of the backscatter-to-extinction ratio $\Pi_p(r)$ and the molecular scattering coefficient profile $\beta_m(r)$ over the examined path must be known. The simplest assumption is that the particulate backscatter-to-extinction ratio is range independent, that is, $\Pi_p(r) = \Pi_p = const.$; then $a(r) = a = const.$ This chapter assumes a constant particulate backscatter-to-extinction ratio. Data processing with range-dependent $\Pi_p(r)$ is discussed further in Section 7.3.

Unlike the solution for the single-component atmosphere, the solution for the two-component inhomogeneous atmosphere can be only obtained if the numerical value of Π_p is established or taken *a priori*. Moreover, this statement remains true even if the particulate backscatter-to-extinction ratio is a constant, range-independent value.

After the transformation function $Y(r)$ is determined, the corresponding function $Z(r)$ can be found, which has a form similar to that in Eq. (5.28)

$$Z(r) = C[\kappa_p(r) + a\beta_m(r)]\exp\left\{-2\int_{r_0}^{r} [\kappa_p(r') + a\beta_m(r')dr']\right\} \tag{5.68}$$

where C is defined by Eq. (5.29)

$$C = C_Y C_0 T_0^2$$

The new variable for a two-component atmosphere is

$$\kappa_w(r) = \kappa_p(r) + a\beta_m(r) \tag{5.69}$$

where

$$a = \frac{3/8\pi}{\Pi_p} \tag{5.70}$$

The solution for $\kappa_W(r)$ has the same form as that given in Eq. (5.33),

$$\kappa_w(r) = \frac{Z(r)}{C - 2\int_{r_0}^{r} Z(r')dr'}$$

Note that, unlike the constant C_r in the solution for the single-component atmosphere [Eq. (5.44)], here the constant C does not include the backscatter-to-extinction ratio Π_p. In some cases, it is more convenient to have the range-independent term Π_p as a factor of the transformed lidar signal, for example, to have the opportunity to monitor temporal changes in the backscatter-to-extinction ratio. To have the signal intensity be proportional to Π_p, a reduced transformation function $Y_r(r)$ can be used instead of the function $Y(r)$ given in Eq. (5.67). The reduced function is defined as

$$Y_r(r) = \exp\left[-2(a-1)\int_{r_0}^{r} \beta_m(r')dr'\right] \tag{5.67a}$$

With the reduced function, only the exponential term of the original lidar equation is corrected when the transformed function $Z(r) = P(r)r^2Y_r(r)$ is calculated. Accordingly, the constant C is now reduced to C_r as defined in Eq. (5.44), that is, $C_r = C_0T_0^2\Pi_p$. For simplicity, the factor C_Y is taken to be unity.

As with a single-component atmosphere, the most practical algorithms for a two-component atmosphere can be derived by using the boundary point or optical depth solutions. Here the boundary point solution can be used if there is a point r_b within the measurement range where the numerical value of $\kappa_W(r_b)$ is known or can be specified a priori. Because the molecular extinction profile is assumed to be known, this requirement reduces to a sensible selection of the numerical values for the particulate extinction coefficient $\kappa_p(r_b)$ and the backscatter-to-extinction ratio Π_p. The latter value is required to find the ratio a, which must be known to calculate $Y(r)$ with Eq. (5.67) or $Y_r(r)$ with Eq. (5.67a). For uniformity, all of the formulas given below are based on the most general transformation with the function $Y(r)$ defined in Eq. (5.67).

After the boundary point r_b has been selected, the constant C, defined in Eq. (5.35), can be rewritten in the form

$$C = 2\int_{r_0}^{\infty} Z(r')dr' = 2\left[\int_{r_0}^{r} Z(r')dr' + \int_{r}^{r_b} Z(r')dr' + \int_{r_b}^{\infty} Z(r')dr'\right]$$

In the formulas below, the integration limits are written for the far-end solution, when $r < r_b$ (For the near-end solution, the second term in the equation has limits from r_b to r, i.e., it is subtracted rather than added). Substituting the constant C in Eq. (5.33), one obtains the latter in the form

$$\kappa_w(r) = \frac{0.5Z(r)}{I(r_b, \infty) + \int_r^{r_b} Z(r')dr'} \tag{5.71}$$

where $I(r_b, \infty)$ is

$$I(r_b, \infty) = \int_{r_b}^{\infty} Z(r')dr' \tag{5.72}$$

As mentioned in Section 5.2, the integral of $Z(r)$ with an infinite upper limit of integration has a *finite* numerical value when $\kappa_W(r) > 0$. This term may be determined with either the boundary point or the optical depth solution. The first solution may be obtained by substituting $r = r_b$ in Eq. (5.36). The substitution gives the formula

$$\kappa_w(r_b) = \frac{Z(r_b)}{2\int_{r_b}^{\infty} Z(r')dr'} \tag{5.73}$$

With Eqs. (5.72) and (5.73), the integral with the infinite upper limit is then defined as

$$I(r_b, \infty) = \frac{0.5Z(r_b)}{\kappa_w(r_b)} \tag{5.74}$$

After substituting Eq. (5.74) in Eq. (5.71), the far-end boundary point solution for a two-component atmosphere becomes

$$\kappa_w(r) = \frac{Z(r)}{\frac{Z(r_b)}{\kappa_w(r_b)} + 2\int_r^{r_b} Z(r')dr'} \tag{5.75}$$

Eq. (5.75) can be used both for the far- and near-end solutions, depending on the location selected for the boundary point r_b. If $r_b < r$, the near-end solution is obtained; the summation in the denominator is transformed into a subtraction because of the reversal of the integration limits.

After determining the weighted extinction coefficient $\kappa_W(r)$ with Eq. (5.75), the particulate extinction coefficient, $\kappa_p(r)$, can be calculated as the difference between $\kappa_W(r)$ and the product $[a\kappa_W(r)]$ [Eq. (5.34)]. Clearly, to extract the profile of the particulate extinction coefficient, the same values of the molecular profile and the particulate backscatter-to-extinction ratio are used as were used for the calculation of $Y(r)$. Note also that the simplest variant of the boundary depth solution in the two-component atmosphere is achieved when pure molecular scattering takes place at the point r_b. In that case, $\kappa_p(r_b) = 0$, and $\kappa_W(r_b) = a\beta_m(r_b)$, so that the boundary value of the molecular extinction coefficient can be obtained from the available meteorological data or from the appropriate standard atmosphere (see Chapter 8).

Similarly, an optical depth solution may be obtained for the two-component atmosphere, which applies the known (or assumed) atmospheric transmittance over the total range as the boundary value. To derive this solution, Eq. (5.71) is rewritten, selecting the range $r_b = r_0$, that is, moving the point r_b into the near end of the measurement range. For all ranges, $r > r_0$. Eq. (5.71) is now written as

$$\kappa_w(r) = \frac{0.5Z(r)}{I(r_0, \infty) - \int_{r_0}^{r} Z(r')dr'} \tag{5.76}$$

where

$$I(r_0, \infty) = \int_{r_0}^{\infty} Z(r')dr' \tag{5.77}$$

Note that for any $r > r_0$, the inequality $I(r_0, \infty) > I(r_0, r)$ is valid; therefore, the denominator in Eq. (5.76) is always positive. Thus the solution in Eq. (5.76) is stable, as is the boundary point far-end solution. Similar to Eq. (5.57), the integral $I(r_0, \infty)$ is equal to the corresponding equation constant divided by two

$$I(r_0, \infty) = \frac{C}{2} \tag{5.78}$$

For the real signals, the maximum integral can only be calculated within the finite limits of the lidar operating range $[r_0, r_{max}]$, where the function $Z(r)$ is available. This maximum integral over the range $I_{max} = I(r_0, r_{max})$, is related to the integrated value of $\kappa_W(r)$ in a manner similar to that in Eq. (5.32)

$$I_{max} = \int_{r_0}^{r_{max}} Z(r')dr' = \frac{C}{2}\left\{1 - \exp\left[-2\int_{r_0}^{r_{max}} \kappa_w(r')dr'\right]\right\} \tag{5.79}$$

The maximum integral defined here is similar to that for the single-component atmosphere [Eq. (5.53)]. The difference is that here the weighted extinction coefficient $\kappa_W(r)$ rather than the particulate extinction coefficient is the integrand in the exponent of the equation. Denoting the exponent in Eq. (5.79) as

$$V_{max} = V(r_0, r_{max}) = \exp\left[-\int_{r_0}^{r_{max}} \kappa_w(r')dr'\right] \tag{5.80}$$

Eq. (5.79) can be rewritten in a form similar to Eq. (5.53), where the parameter $V_{max} = V(r_0, r_{max})$ is used instead of the path transmittance $T_{max} = T(r_0, r_{max})$. The term V_{max} may be formally considered as the path transmittance over the total measurement range (r_0, r_{max}) for the weighted coefficient $\kappa_W(r)$. In the general form, this parameter is correlated with the actual transmittance of the total range in the following way

$$V_{max} = T_{max} \exp\left[-(a-1)\int_{r_0}^{r_{max}} \kappa_m(r')dr'\right] \tag{5.80a}$$

where T_{max} for the two-component atmosphere is

$$T_{max} = \exp\left\{-\int_{r_0}^{r_{max}} [\kappa_m(r') + \kappa_p(r')]dr'\right\}$$

In terms of the molecular and particulate transmittance, $T_{m,max}$ and $T_{p,max}$, the term V_{max} is correlated with the ratio (a) as

$$V_{max} = T_{p,max}(T_{m,max})^a \tag{5.81}$$

The relationship between the integrals $I(r_0, \infty)$ and I_{max} can be found from Eqs. (5.78) and (5.79) as

$$I(r_0, \infty) = \frac{I_{max}}{1 - V_{max}^2} \tag{5.82}$$

Finally, the most general form of the optical depth solution for a two-component atmosphere can be obtained by substituting Eq. (5.82) into Eq. (5.76). It can be written in the form

$$\kappa_w(r) = \frac{0.5Z(r)}{\dfrac{I_{max}}{1 - V_{max}^2} - \displaystyle\int_{r_0}^{r} Z(r')dr'} \tag{5.83}$$

SUMMARY: In clear atmospheres, for visible or near-visible wavelengths, the particulate and molecular extinction components are, generally, comparable in magnitude. Therefore, for accurate lidar data processing, both components should be considered. To extract the unknown particulate extinction coefficient, the lidar signal is transformed into a function in which the weighted extinction

coefficient, $\kappa_W(r)$ is introduced as a new variable. The general procedure to determine the profile of the particulate extinction coefficient in a two-component atmosphere is as follows: (1) calculation of the profile of function $Y(r)$ with Eq. (5.67); (2) transformation of the recorded lidar signal $P(r)$ into function $Z(r)$; (3) determination of the profile of the weighted extinction coefficient, $\kappa_W(r)$ with either the boundary point or optical depth solution [Eqs. (5.75) and (5.83), respectively]; and (4) determination of the unknown particulate extinction coefficient, $\kappa_p(r)$ [Eq. (5.34)].

Finally, an approximate solution is given that is valid for a two-component homogeneous atmosphere. This solution does not require determination of the transformation function $Y(r)$. The solution may be practical when lidar measurements are made in clear or slightly polluted homogeneous atmospheres, in which all of the involved values, Π_p, Π_m, κ_m, and κ_t can be considered to be range-independent. This solution can be considered as an alternative to the slope method. It may be useful, for example, for routine measurements of horizontal visibility, for pollution monitoring, etc., that is, where a mean value of the atmospheric turbidity should be established. To derive the solution, the lidar signal, $P(r)$, is range-corrected, and the product $P(r) \cdot r^2$ is

$$Z_r(r) = P(r)r^2 = C_0 T_0^2 (\Pi_p \kappa_p + \Pi_m \kappa_m) \exp\left[-2\int_{r_0}^{r} (\kappa_p + \kappa_m) dr'\right] \tag{5.84}$$

After a simple transformation, the equation can be rewritten into the form

$$Z_r(r) = C^* \kappa_t \exp[-2\kappa_t (r - r_0)] \tag{5.85}$$

where

$$C^* = C_0 T_0^2 \Lambda \tag{5.86}$$

and

$$\Lambda = \frac{\Pi_p \kappa_w}{\kappa_t} \tag{5.87}$$

In a horizontally homogeneous atmosphere, where only slight variations of the atmospheric scatterers are assumed, factor Λ, and accordingly, C^* can be assumed to be approximately range independent. Thus the same solutions as in Eqs. (5.75) and (5.83) can be applied for the retrieval of κ_t. No transformation function $Y(r)$ needs to be determined to apply the solution and no individual term in Eq. (5.86) or (5.87) should be known. Therefore, there is no need to evaluate the particulate backscatter-to-extinction ratio Π_p. Practical algorithms based on this transformation, are generally applied to different zones along the same line of sight. Such measurements are considered in Section 12.1.2.

5.5. WHICH SOLUTION IS BEST?

The different solutions considered in this chapter are differently sensitive to different sources of error in the selected constants, to signal random noise and systematic distortions, etc. (see Chapter 6). Therefore, the question posed by the title of this subsection itself is ill-defined. Any certain reply to this simple question may be misleading.

To explain this statement, consider any method analyzed in this chapter; for example, the far-end boundary solution. After publication of the famous study by Klett (1985), in which the author pointed out reliability of the solution, a large number of studies were published concerning the method. It is quite illuminating now to read the early, rapturous remarks followed some years later by far more pessimistic conclusions concerning the same method. Meanwhile, there are no doubts that the method works well, especially, in appropriate atmospheric conditions. The last remark must be stressed: *in appropriate atmospheric conditions*. The question then becomes. What are these appropriate conditions for which this method will work properly? As shown in Chapter 6, generally the method yields good results when the measurement is made in a single-component turbid atmosphere. The method yields only positive values of the extinction coefficient, whereas the alternative near-end boundary method may give nonphysical negative values. Moreover, when the optical depth of the measurement range is restricted by reasonable limits, the former method can yield an extremely accurate result. This can be achieved even with an inaccurately selected far-end boundary value. On the other hand, most of the advantages of the method are lost (1) if the measurement is made in a clear atmosphere, in which the molecular and particulate contributions to scattering are comparable (especially when the extinction coefficient or backscatter-or-extinction ratio changes monotonically over the range); (2) when the optical depth of the atmospheric layer between the lidar and far-end boundary range is too large; (3) when the optical depth of the atmospheric layer between the lidar and far-end boundary range is too small; (4) when the lidar signal over distant ranges is corrupted by systematic distortions.

The acceptable form of the question given in the title of this subsection should be formulated in following way: Which lidar-equation solution is the best for a *particular* type of measurement made in *particular* atmospheric conditions? Obviously, for any individual case, the algorithm must be used that best corresponds with the measurement requirements. To determine this, the goal of the measurement must first be clearly established and the particular atmospheric conditions should be estimated for which the lidar measurement was made. One should thoroughly estimate which algorithm is the best for the *particular* measurement conditions. Before such selection is made, a number

Method	Solution Advantages	Disadvantages	Variables Determined	Variables or Assumption Required	Equation	References
Slope	Simple, no a priori selected quantities are required	Works only in homogeneous atmosphere	Mean κ_t or κ_p over the range	$\kappa_t = const.$ $\beta_\pi = const.$	Eq. (5.11)	Kunz and Leeuw, 1993
Absolute calibration-based solution		Requires sophisticated methodology to calibrate	Range-resolved $\kappa_p(r)$	Π_p and T_0^2 $\Pi_p = const.$	Eq. (5.33), Eq. (5.45)	Hall and Ageno, 1970; Spinhirne et al., 1980
Boundary point far-end solution for single-component atmosphere	Good in turbid atmospheres Π_p need not be selected	Selection of value of $\kappa_p(r_b)$ is a challenge Not accurate enough in clear atmospheres	Range-resolved $\kappa_p(r)$	$\kappa_p(r_b)$ at the far end Π_p = const.	Eq. (5.50)	Klett, 1981; Carnuth and Reiter, 1986
Boundary point near-end solution for single-component atmosphere	Good in clear and moderately turbid atmosphere Π_p need not be selected	Unstable in turbid atmospheres	Range-resolved $\kappa_p(r)$	$\kappa_p(r_b)$ at the near end $\Pi_p = const.$	Eq. (5.51)	Viezee et al., 1969; Ferguson and Stephens, 1983
Boundary point far-end solution for two-component atmosphere	Good with the assumption of a local aerosol-free zone at r_b	$\kappa_p(r_b)$ at the distant range lidar is selected a priori Not practical for moderately turbid atmospheres	Range-resolved $\kappa_p(r)$	$\kappa_p(r_b)$ at the far end and Π_p $\Pi_p = const.$	Eq. (5.75) ($r_b > r$)	Klett, 1981; Fernald, 1984; Browell et al., 1985; Kovalev and Moosmüller, 1994
Boundary point near-end solution for two-component atmosphere	Good in clear atmospheres	Unstable in turbid atmospheres	Range-resolved $\kappa_p(r)$	$\kappa_p(r_b)$ at the near end and Π_p $\Pi_p = const.$	Eq. (5.75) ($r > r_b$)	Fernald, 1984; Kovalev and Moosmüller, 1994
Optical depth solution for single-component atmosphere	Good in turbid atmospheres with $(T_{max})^2 < 0.05$	Solution constant may be estimated from integrated lidar signal	Range-resolved $\kappa_p(r)$	$(T_{max})^2$ $\Pi_p = const.$	Eq. (5.55)	Weinman, 1988; Kovalev, 1993; Kunz, 1996.
Optical depth solution for two-component atmosphere	Good for combined measurements with sun photometer	Not practical without independent estimates of $(T_{max})^2$	Range-resolved $\kappa_p(r)$	$(T_{max})^2$ and Π_p $\Pi_p = const.$	Eq. (5.83)	Fernald et al., 1972; Platt, 1979; Weinman, 1988; Kovalev, 1995.

of questions must be answered. These questions include: (1) Will the measurements be made in a single- or in a two-component atmosphere? (2) Is the atmosphere homogeneous enough to use (or try to use) a solution based on atmospheric homogeneity? (3) Is any independent information available that can help to overcome the lidar equation indeterminacy? (4) What additional information can be obtained from the lidar signals themselves? (5) Is it possible to use reference signals of the same lidar measured, for example, in another azimuthal or zenith direction? (6) What are the most reasonable particular assumptions that can be taken a priori? (7) How sensitive is the assumed lidar equation solution to these assumptions?

There can be no resolution to the question of which lidar solution may be the best until the questions above are answered. The optimum lidar equation solution is that which under other conditions being equal yields the best measurement accuracy of the quantity under investigation. Generally, this is the solution that is least sensitive to the uncertainty of parameters that need to be chosen a priori, such as an assumed backscatter-to-extinction ratio. The table on page 182 summarizes the methods discussed in this chapter. Note that here only the atmospheres are considered where the condition $\Pi_p = \text{const.}$ is valid. Also, a single-component atmosphere is assumed here to be a polluted atmosphere in which particulate scattering dominates, so that the molecular constituent can be ignored. In a two-component atmosphere, the accurate molecular extinction coefficient is assumed to be known as a function of the lidar measurement range.

6

UNCERTAINTY ESTIMATION FOR LIDAR MEASUREMENTS

All experimental data are subject to measurement uncertainty. The uncertainty is the result of two components. The first is due to systematic errors related to the measurement method itself, from the assumptions made in developing an inversion scheme and from uncertainties related to the assumption of required values, such as the backscatter-to-extinction ratio. The second component of the uncertainty is the result of random errors in the measurement. The total uncertainty for lidar measurements depends on many factors, including (1) the measurement accuracy of the signal, (2) the level of the random noise and the relative size of the signal with respect to the noise component (the signal-to-noise ratio), (3) the accuracy of the estimated lidar solution constants, (4) the accuracy of the range-resolved molecular profile used in the inversion procedure in two-component atmospheres, and (5) the relative contribution of the molecular and particulate components to scattering and attenuation. Because the actual lidar signal-to-noise ratio is usually range dependent, the uncertainty of the measurement also depends on the range from the lidar to the scattering volume from which the signal is obtained. The total measurement uncertainty depends on these and others factors in a way that is complicated and unpredictable.

Uncertainty analyses based on standard error propagation principles have been discussed in many lidar studies (see, for example, Russel et al., 1979; Megie and Menzies, 1980; Measures, 1984). However, practical estimates of the

Elastic Lidar: Theory, Practice, and Analysis Methods, by Vladimir A. Kovalev and William E. Eichinger.
ISBN 0-471-20171-5

accuracy of lidar measurements remain quite difficult. What is more, conventional estimates do not necessarily provide a thorough understanding of how different sources of error behave in different atmospheric conditions and, accordingly, how optimal measurement techniques may be developed.

It is well known that to make accurate uncertainty estimates, knowledge of the statistical behavior of the measured variables and their nature is required (see, for example, Taylor, 1982; Bevington and Robinson, 1992). Most practical uncertainty estimate methods are based on simple statistical models, which, unfortunately, are often inappropriate for lidar applications. The conventional theoretical basis for random error estimates puts many restrictions on its practical application. For example, it assumes that (1) the error constituents are small, so that only the first term of a Taylor series expansion is necessary for an acceptable approximation of error propagation; (2) that random errors can be described by some typical (e.g., Gaussian or Poisson) distribution; and (3) that measurement conditions are stationary. This means that the measured quantity does not change its value during the time required to make the measurement. Most practical formulas for making uncertainty estimates are developed with the assumption that the measured or estimated quantities are uncorrelated. Using this assumption avoids problems related to the determination of the covariance terms in the error propagation formulas.

These kinds of conditions are not often realistic for lidar measurements. The quantities used in lidar data processing are often correlated, the level of correlation often changes with range, and no applicable methods exist to determine the actual correlation. Apart from that, the magnitudes of uncertainties are sometimes quite large, preventing the conventional transformation from differentials to the finite differences used in standard error propagation. The measured atmospheric parameters may not be constant during the measurement period because of atmospheric turbulence, particularly during the averaging times used by deep atmospheric sounders. Finally, the total measurement uncertainty includes not only a random (noise) constituent but also a number of systematic errors, which may cause large distortions in the retrieved profiles.

When processing the lidar signal, at least three basic sources of systematic error must be considered. The first is an inaccurate selection of the solution boundary value. The second is an inaccurate selection of the particulate backscatter-to-extinction ratio, and a third may be a signal offset remaining after subtraction of the background component of the lidar signal. These systematic errors may be large, so that standard uncertainty propagation procedures may actually underestimate the actual measurement uncertainty.

Fortunately, apart from the standard error propagation procedure, two alternative ways exist to investigate the effects of systematic errors. The first is a sensitivity study in which expected uncertainties are used in simulated measurements to evaluate the change in the parameter of interest (see, e.g., Russel et al., 1979; Weinman, 1988; Rocadenbosh et al., 1998). The other method may be used when investigating the influence of uncertainty of a particular parameter (especially, one taken a priori). This method is best used, for example, to understand how over- or underestimated backscatter-to-extinction ratios influence the accuracy of the extracted extinction-coefficient profile. To use this method, an analytical dependence is obtained by solving two equations. The first equation is the "true" formula, and the second is that "distorted" by the presence of the error in the parameter of interest. This type of analytical approach is useful when making an uncertainty analysis where large sources of error are involved (Kunz and Leeuw, 1993; Kunz, 1998; Matsumoto and Takeuchi, 1994; Kovalev and Moosmüller, 1994; Kovalev, 1995).

In this chapter, methods of uncertainty analysis are discussed that provide an understanding of the uncertainty associated with the various inversion methods given in Chapter 5. The main purpose of the analysis in this section is to give to the reader a basic understanding of how measurement errors influence the measurement results rather than simply providing formulas for uncertainty estimates. The goal is (1) to explain the behavior of the uncertainty under different measurement conditions; (2) to show the relationship between measurement accuracy and atmospheric turbidity; (3) to explain how the measurement accuracy depends on the particular inversion method used for data processing; and (4) to provide suggestions for what can be done in particular situations to avoid the collection of unreliable lidar data. It is important to understand the physical processes that underlie the formulas as well as which quantities in a formula strongly influence the result and which do not. An extensive list of references on the subject of error propagation is given, and the interested reader is referred to these publications for more detailed studies.

To begin, several terms must be defined. The absolute error of a quantity x is denoted as Δx, that is,

$$\Delta x = \langle x \rangle - x$$

where $\langle x \rangle$ is an estimate or measurement of a true value x (or its best estimate). Accordingly, the relative uncertainty, δx, is

$$\delta x = \frac{\langle x \rangle - x}{x}$$

6.1. UNCERTAINTY FOR THE SLOPE METHOD

As shown in Chapter 5, the mean value of the extinction coefficient over the range Δr may be obtained with the slope method [Eq. (5.11)]

$$\kappa_t(\Delta r) = \frac{-1}{2\Delta r}[\ln Z_r(r + \Delta r) - \ln Z_r(r)]$$

where $Z_r(r)$ and $Z_r(r + \Delta r)$ are the lidar range-corrected signal values measured at ranges r and $(r + \Delta r)$, respectively. Obviously, lidar signals are always corrupted with some error and cannot be measured exactly. When processing the lidar signal, the total measurement uncertainty is the result of both random and systematic errors. The primary sources of random error are electronic noise, originated by the background component, F_{bgr}, and the discrete nature of a digitized signal. Systematic errors may occur for many reasons. They may be caused by incomplete removal of the background light component, F_{bgr}, or by a zero-line shift in the digitizer caused, for example, by low-frequency noise induced in the electrical circuits of the receiver. Thus experimentally determined quantities $Z_r(r)$ and $Z_r(r + \Delta r)$ include uncertainties ΔZ_r and $\Delta Z_{r+\Delta r}$, respectively. Using conventional error analysis techniques, errors may be propagated to find the resulting uncertainty in the measured extinction coefficient $\kappa_t(\Delta r)$. It is important to keep in mind that the uncertainties ΔZ_r and $\Delta Z_{r+\Delta r}$ are highly correlated when the range Δr is small. Therefore, a complete error propagation equation should include covariance terms between these variables (Bevington and Robertson, 1992). For sake of simplicity, we present here a formula for the upper limit of the uncertainty in measured $\kappa_t(\Delta r)$ rather than its standard deviation. Assuming that $\Delta Z_r \ll Z_r(r)$ and $\Delta Z_{r+\Delta r} \ll Z_r(r + \Delta r)$, one can obtain an estimate of the upper limit of the absolute value of uncertainty in $\kappa_t(\Delta r)$ in Eq. (5.11) as

$$\Delta\kappa_t \approx \frac{1}{2\Delta r}\left[\left|\frac{\Delta Z_r}{Z_r(r)}\right| + \left|\frac{\Delta Z_{r+\Delta r}}{Z_r(r+\Delta r)}\right|\right] \tag{6.1}$$

In lidar measurements, it is a conventional practice to use a sum (or an average of the sum) of multiple lidar returns rather than a single laser pulse. This is done to improve the signal-to-noise ratio before data processing is done. If an error component in a single signal is randomly distributed, after signal averaging it is reduced by a factor of $N^{-1/2}$, where N is the number of averaged signals (Bevington and Robertson, 1992). Thus, by increasing the averaged number of pulses, one can reduce the best-fit signal random error, theoretically, to any desired level. However, because of the presence of systematic errors in the measurement, some finite limit to error reduction exists. Below this limit, which is related to the level of the systematic error, no further accuracy improvement can be obtained by an increase in the number of summed pulses, N.

The relationship between the uncertainty ΔZ_r in Eq. (6.1) and the errors in the measured backscattered signals $P(r)$ is

$$\frac{\Delta Z_r(r)}{Z_r(r)} = \frac{\sum_{i=1}^{N} \Delta P_i(r)}{\sum_{i=1}^{N} P_i(r)} \tag{6.2}$$

here $\Delta P(r)$ is the absolute error of the measured lidar signal $P(r)$. Dividing both sides of Eq. (6.1) by $\kappa_t(\Delta r)$ and using Eq. (6.2), the upper limit to the fractional uncertainty of the extinction-coefficient can be written as

$$\delta\kappa_t \approx \frac{1}{2\kappa_t \Delta r}[|\delta P(r)| + |\delta P(r+\Delta r)|] \tag{6.3}$$

where $\delta\kappa_t$ is the fractional uncertainty of the extinction coefficient $\kappa_t(\Delta r)$. For simplicity, the term $\kappa_t(\Delta r)$ is denoted here and below as κ_t. The fractional errors, $\delta P(r)$ and $\delta P(r + \Delta r)$ are

$$\delta P(r) = \frac{\sum_{i=1}^{N} \Delta P_i(r)}{\sum_{i=1}^{N} P_i(r)}$$

and

$$\delta P(r+\Delta r) = \frac{\sum_{i=1}^{N} \Delta P_i(r+\Delta r)}{\sum_{i=1}^{N} P_i(r+\Delta r)}$$

Note that the product $\kappa_t\Delta r$ in the denominator of Eq. (6.3) is the optical depth over the selected measurement range Δr. Thus the fractional uncertainty in the extinction coefficient, $\delta\kappa_t$, is inversely proportional to the optical depth over the measurement range Δr.

An inverse proportion of this nature may result in large uncertainties in the derived extinction coefficient over short ranges in a relatively clear atmosphere (where κ_t is small). This is because the difference between $Z_r(r)$ and $Z_r(r+\Delta r)$ is small. For such a situation, the fractional uncertainty in the derived extinction coefficient $\delta\kappa_t$ may be as much as a hundred times the fractional uncertainty of the measured δP. For example, if Δr = 30 m and visibility is approximately 20 km (this corresponds to $\kappa_t \approx 0.2\,\text{km}^{-1}$ in the visual portion of the spectrum), the optical depth of this range interval is $\tau(\Delta r) = \kappa_t\Delta r = 0.006$. When Δr is small, it may be assumed that $\delta P(r) = \delta P(r + \Delta r)$. It follows from Eq. (6.3) that the fractional error, $\delta\kappa_t$, is related to the original measurement error, $\delta P(r)$, through a magnification factor

$$\delta\kappa_t \approx 167\,\delta P(r)$$

Clearly, the slope method is not appropriate for use with small range intervals Δr.

The uncertainty estimate above is obtained for an ideal case, that is, when no changes take place in the backscatter coefficient β_π. If even slight changes in β_π occur over the range interval Δr, the logarithm of the product $C_0\beta_\pi$ in Eq. (5.7) (Section 5.1) is not constant. Thus an additional error component is present in the retrieved extinction coefficient. The contribution of a change in backscatter coefficient to the uncertainty in the extinction coefficient is

$$\delta\kappa_{t,\beta} = \frac{\ln\beta_\pi(r+\Delta r) - \ln\beta_\pi(r)}{2\kappa_t\Delta r} \tag{6.4}$$

and has the same weighting factor, $(2\kappa_t\Delta r)^{-1}$, as the error in Eq. (6.3).

Thus the use of the slope method for a short spatial range Δr results in large measurement errors. This is why the application of the slope method to small successive range intervals as proposed by Brown (1973) proved to be impractical. However, this method works properly when determining the mean extinction coefficient within an extended range. In other words, to have acceptable measurement accuracy, the length of the lidar signal range interval used in processing should be as long as possible.

It is not possible to specify, in advance, a requirement for the selection of the length of the range increment Δr for slope method measurements. Some recommendations were presented in Chapter 5; however, these cannot be considered universal. It follows from those recommendations that little reliance should be placed on a retrieved extinction coefficient if the slope method measurement interval in a clear atmosphere is less than 2–5 km or if the *a posteriori* estimated optical depth over the selected range is less than about 0.5–1. Note that the values given here are only approximate and can change significantly depending on the specifics of lidar site location.

The uncertainty in the extinction coefficient, as given in Eqs. (6.3) and (6.4), may actually overestimate the uncertainty because the correlation coefficient between the signals $Z_r(r)$ and $Z_r(r + \Delta r)$ is not equal to zero. When an accurate uncertainty estimate is desired, an error covariance component should also be included in the uncertainty estimate. Unfortunately, this is not achievable in practice because of the complexity of determining the covariance component. Ignoring this term is often the only reasonable approximation, especially when the intent is to analyze the general behavior of the error. The basis for such a statement is that the behavior of the error is generally the same, even if the covariance component is ignored. In the slope method, the signals become less correlated as the range Δr becomes large. In that case, ignoring the covariance component can be considered to be a reasonable approximation. With this approximation, a simple formula can be derived for the likely error of the mean extinction-coefficient value measured with the slope method over an extended range from r_1 to r_2

$$\delta\kappa_t = \frac{1}{2\kappa_t(r_2 - r_1)}\sqrt{\delta Z_r(r_1)^2 + \delta Z_r(r_2)^2} = \delta P(r_1)\, F_\tau(r_1, r_2)\left(\frac{r_2}{r_1}\right)^2 \tag{6.5}$$

where

$$F_\tau(r_1, r_2) = \frac{e^{2\tau(r_1, r_2)}}{2\tau(r_1, r_2)} \tag{6.6}$$

The term $\tau(r_1, r_2)$ in Eq. (6.6)

$$\tau(r_1, r_2) = \kappa_t(r_2 - r_1)$$

is the total optical depth of the range interval (r_1, r_2). Unlike measurements made with short range intervals Δr, the assumption of equal relative error δP in signals $P(r_1)$ and $P(r_2)$ may not be valid for extended ranges. This is because the measured signal magnitude changes dramatically when the range interval $(r_2 - r_1)$ is large while the background noise component remains approximately constant. Therefore, in Eq. (6.5), a more practical assumption is used that the absolute error ΔP rather than the relative error is approximately constant within the range interval. In this case, the relative error δP increases with the range, so that the additional factor $[r_2/r_1]^2$ appears in Eq. (6.5).

In contrast to the estimate $\delta\kappa_t$ for a short range interval measurement [Eq. (6.3)], the measurement uncertainty of the extinction coefficient for an extended range (r_1, r_2) depends significantly on the exponential term exp $[2\tau(r_1, r_2)]$, especially in turbid atmospheres. This term becomes a central factor that noticeably increases the measurement uncertainty as the optical depth of the range (r_1, r_2) increases and becomes large. For example, for an optical depth $\tau(r_1, r_2) = 1$, the factor $F_\tau(r_1, r_2)$ in Eq. (6.6) is equal to 3.7; for $\tau(r_1, r_2) = 1.5$, it becomes equal to 6.7, etc. On the other hand, the factor also increases for small values of the optical depth. This occurs because of small values of the denominator in Eq. (6.6). Thus the measurement uncertainty $\delta\kappa_t$ depends on the factor $F_\tau(r_1, r_2)$ that increases for both small and large values of the optical depth (Fig. 6.1). The method is most precise when $\tau(r_1, r_2) \approx 0.3$–$1.0$ A typical dependence of the relative uncertainty in $\kappa_t(r)$ on the measurement range (r_1, r_2), calculated for different values of r_1, is shown in Fig. 6.2. Here the relative signal error at r_1 is taken as $\delta P(r_1) = 0.5\%$ and the extinction coefficient is assumed to be $\kappa_t = 0.3\,\text{km}^{-1}$. It is assumed also that β_π = const. so that no fluctuations in β_π take place.

The dependence of the extinction-coefficient uncertainty on the range interval has a typical U-shaped appearance: The uncertainty increases for both short and long-range intervals (r_1, r_2) and has a minimum uncertainty value within a restricted intermediate area.

The first attempts to apply the slope method in practice were made in the late 1960s, when lidar signals were recorded by photographing the analog trace

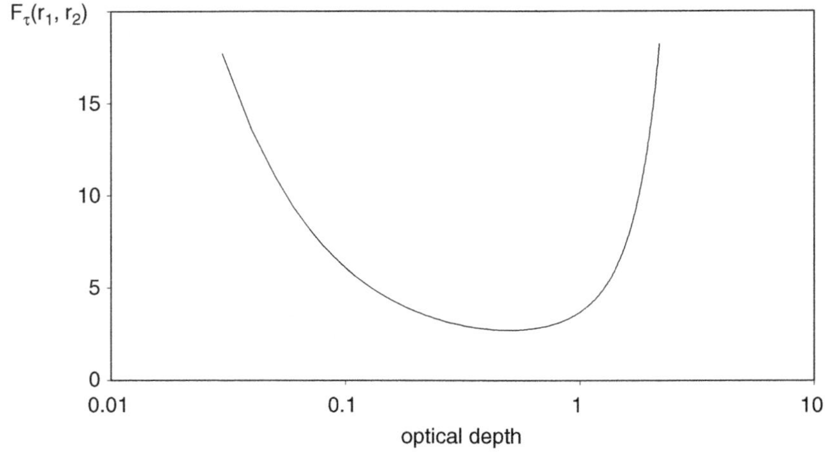

Fig. 6.1. Dependence of the factor $F_\tau(r_1, r_2)$ on the measurement optical depth.

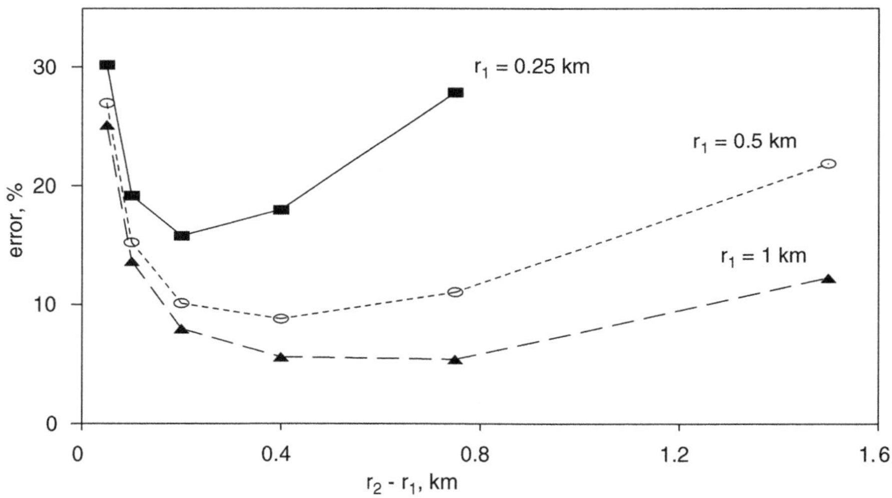

Fig. 6.2. Typical dependence of the relative uncertainty of the extinction coefficient on the measurement range for two-point measurement.

of the signal on an oscilloscope (Viezee et al., 1969). With the advent of the transient signal digitizer and modern computer technology, the conventional application of the slope method has increasingly used least-squares fitting techniques. Generally, the slope method works best when a large number of consecutive, discrete signals (bins) are available (Ignatenko et al., 1988; Kunz and Leeuw, 1993). With the least squares technique, a linear approximation of $\ln Z_r(r)$ inside the range interval can be found and the coefficients κ_t and A

established for a linear fit [Eq. (5.8)]. The appropriate formulas for κ_t and A can be derived by using an estimate of the minimum of the function (Bevington and Robinson, 1992)

$$F = \sum_{j=1}^{M} \frac{[\Phi(r_j) - A + 2\kappa_t r_j]^2}{\sigma_j^2}$$

where M is the total number of data points within the range interval considered, σ_j^2 is a weighting factor related to the dispersion of $\ln Z_r(r_j)$, and $\Phi(r) = \ln Z_r(r)$. The minimum of the function F can be found by minimizing the partial derivatives with respect to the two unknowns, A and κ_t. This yields the following expression for κ_t

$$\kappa_t = \frac{\left(\sum_{j=1}^{M} r_j\right)\left(\sum_{j=1}^{M} \Phi_j\right) - M\left(\sum_{j=1}^{M} r_j \Phi_j\right)}{2\varepsilon} \tag{6.7}$$

where

$$\varepsilon = M\left(\sum_{i=1}^{M} r_i^2\right) - \left(\sum_{i=1}^{M} r_i\right)^2$$

The uncertainty in the measured extinction coefficient (root mean square value, rms) determined with the least-squares method is

$$\Delta\kappa_t = \sqrt{\frac{M\sum_{j=1}^{M}[\Phi(r_j) - A + 2\kappa_t r_j]^2}{4\varepsilon\,(M-1)}} \tag{6.8}$$

The dependence of the relative uncertainty, $\delta\kappa_t = \Delta\kappa_t/\kappa_t$, on the optical depth of the range interval used for determining the linear fit is not obvious from Eqs. (6.7) and (6.8). However, the U-shaped appearance of the relative uncertainty, similar to that in Fig. 6.2, is also found in the least-squares technique. However, the uncertainty in the extinction coefficient with a least-squares technique is considerably less than that of the two-point variant, particularly for long range intervals. It provides a significant improvement in the slope-method measurement accuracy and, in addition, provides criteria by which the degree of atmospheric homogeneity may be estimated. All principal points made concerning the behavior of the measurement uncertainty remain valid for an analysis over any number of range bins. The consideration of the simplest two-bin variant is a simple way to show the general behavior of uncertainty in the slope method.

The dependence of the relative uncertainty of the measured extinction coefficient on the length of the measurement range interval is shown in Fig. 6.3 (Ignatenko et al., 1988). The dependence is determined for different

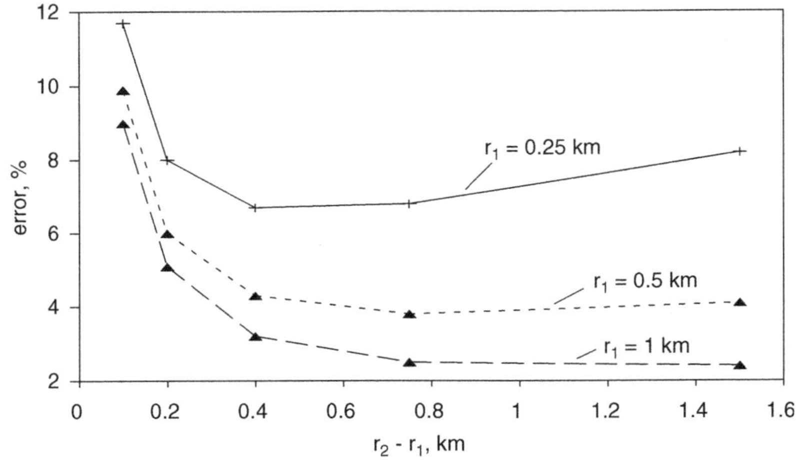

Fig. 6.3. Dependence of the relative uncertainty of the extinction coefficient on the measurement range when derived with the least-squares method for the atmosphere with no atmospheric fluctuations in β_π.

locations of the near-end range, r_1. The total number of equidistant points (discrete signal readings) selected over the range interval (r_1, r_2) is equal to M = 11. To make the variants comparable, the same conditions are used here as in the two-point slope method shown in Fig. 6.2, that is, $\kappa_t = 0.3\,\text{km}^{-1}$ and $\delta P(r_1) = 0.5\%$. The measurement uncertainty for the least-squares method is much less than that for the two-point method. The difference is especially significant for long range intervals. The uncertainty also increases at long range intervals; however, for the lowest two curves, this increase occurs for range intervals ($r_2 - r_1$) longer than the maximum range (1.6 km) presented in Fig. 6.3.

Increasing the number of points used in the least-squares calculations decreases the measurement uncertainty of the derived κ_t. However, the technique significantly reduces the measurement uncertainty compared with the two-point solution only if the quantities used for the regression are normally distributed. Note also that the technique improves the measurement accuracy only if no significant systematic errors occur in the measured set of signals.

In addition to determining κ_t, the least-squares technique makes it possible to estimate the degree of atmospheric homogeneity. As follows from Eq. (6.8), the standard deviation of the linear fit $\Delta\kappa_t$ is proportional to

$$\left[\sum_{j=1}^{M}[Z(r_j) - A + 2\kappa_t r_j]^2\right]^{\frac{1}{2}}$$

and is thus related to the degree of linearity of the function $\Phi = \ln Z_r(r)$. This observation means that the level of $\Delta\kappa_t$ can be considered to be a measure of the degree of atmospheric homogeneity. What is more, the standard deviation may be found for both the total range interval and for separate subintervals within the operating range. In practice, the atmosphere is often considered as homogeneous within an extended interval if the standard deviation $\Delta\kappa_t$ is less than some established, empirical value. If a prominence on the curve exists, such as that for *curve a* in Fig. 5.2, the standard deviation of the linear fit is larger than that for *curve b* in the figure. Heterogeneous areas, such as those shown in *curve a*, should be excluded before the application of the slope method. Similarly, far-end signals with poor signal-to-noise ratios should be excluded.

The standard deviation of $\ln Z_r(r)$ from its linear approximation is often used as an estimate of the degree of atmospheric homogeneity within the selected measurement range. However, this estimate is not enough reliable.

Determining the standard deviation for different subintervals, one can specify a range interval in which the function $\ln Z_r(r)$ may be treated as linear, instead of applying an established criteria to the total range interval. Obviously, such subintervals must be long enough to obtain more or less reliable measurement results. The use of such criteria for atmospheric homogeneity for short-length spatial ranges should be done with great caution.

The practical application of the slope method requires the following: (1) a numerical estimate of the level of the atmospheric homogeneity over the measurement range or extended subintervals, achieved through calculation of the corresponding standard deviation, $\Delta\kappa_t$; (2) exclusion of heterogeneous zones where $\Delta\kappa_t$ is large and the selection of usable range intervals over which the slope method may be applied; and (3) determination of a linear least-squares fit of the logarithm of $Z_r(r)$ over the selected range intervals and the corresponding values of κ_t and $\Delta\kappa_t$. However the calculated absolute uncertainty $\Delta\kappa_t$ (and, accordingly, $\Delta\kappa_t/\kappa_t$) may have nothing common with the actual uncertainty in the retrieved κ_t. This is because the slope-method technique assumes no systematic changes in β_π over the range used for the determination of the extinction coefficient, and this may be not true. Comparisons to other a posteriori estimates of the optical attenuations are strongly recommended, particularly if additional relevant data are available.

The maximum effective range of a lidar is related to the signal-to-noise ratio (Measures, 1984; Kunz and Leeuw, 1993). Accordingly, an acceptable level of noise and the corresponding lidar maximum measurement range should be established. Generally, the random error in the measured lidar signal is taken as the basic error that defines the lidar measurement range. It is common practice to establish the lidar maximum range as the range where the decreasing lidar signal becomes equal to the estimated rms noise level. With this approach, Kunz and de Leeuw (1993) investigated the influence of random noise on the lidar maximum range and the accuracy of backscatter and extinction coefficients inverted with the slope method. The estimates were made by a quantitative analysis of the influence of range-independent white noise; it was implicitly assumed that no systematic offset takes place. The authors assumed also that (1) the shot noise is induced only by background radiation and noise from the electronic circuits and (2) no atmospheric fluctuations in backscatter coefficient occur along the measurement path. The maximum signal-to-noise ratio defined at the point of the complete overlap, r_0, varied in their calculations from 10 to 10^6. The minimum signal-to-noise ratio was kept at a fixed level with an rms value of 1. As was stated, both the extinction coefficient κ_t and the backscatter term β_π can be found from a linear fit [Eq. (5.8)]. However, the errors in obtained κ_t and β_π are different. The authors concluded that the backscatter coefficient in a moderately clear atmosphere ($\kappa_t < 1\,km^{-1}$) can be determined with at least a 10% accuracy. However, this can only be achieved if the signal-to-noise ratio at the starting point is better than ~1000. For turbid atmospheres with $\kappa_t > 1\,km^{-1}$, an accuracy of ~10% in the extinction coefficient can only be achieved if the signal-to-noise ratio is better than ~2000. The authors concluded that this level of signal-to-noise ratio cannot be achieved at least with digitizers that have only 12-bit discrimination, allowing for 4096 different measurement levels. Even a well-adjusted digitizer with no offset, no electronic noise at all, and only with a single-bit digitizing error can record the real (not range corrected) lidar signals over a limited range of values. The basic conclusion of the authors is that, in practice, it is not possible to determine the extinction coefficient with an accuracy better than ~10% in both clear and turbid atmospheres.

Some comments are necessary about these conclusions by Kunz and de Leeuw (1993). First, these conclusions were made for a particular lidar system, with the fixed starting point at $r_0 = 0.05\,km$ and a maximum effective range of 10 km. Different ways exist to reduce problems related to the restricted dynamic range of the digitizer and poor signal-to-noise ratios. It is possible to increase the spatial region of the analog signals recorded by a given digitizer. This can be done by letting the near-end signals saturate the digitizer. The other option is to increase the distance to the complete overlap point, r_0, by reducing the telescope field of view or increasing the offset between the telescope and laser beam and selecting a more realistic starting point at $r_0 = 0.3$–0.5 km instead of 0.05 km. Another option is to use two simultaneously operating digitizers, one for near and the other for far measurement ranges. The signal-to-noise ratio can be improved significantly by increasing the number of averaged shots. Some additional opportunities appear if the measurements are made with photon-counting techniques (albeit with a loss of temporal and spatial resolution). On the other hand, the authors restricted the scope of their study to the analysis of only the influence of the random error. In clear atmospheric conditions, the lidar signal at distant ranges is relatively small, so that even a small zero-line offset remaining after the background component subtraction may produce large systematic errors that can severely reduce the measurement accuracy. Nevertheless, the overall conclusions of the study by Kunz and de Leeuw (1993) remain: (1) Better accuracy is achieved in situations of moderate atmospheric extinction when the lidar operates over its maximum range. (2) The slope method measurement results are less accurate both for large and small extinction coefficients, where the maximum range is limited by the atmospheric transmittance losses or by small backscatter coefficients, respectively.

Some attempts have been made to increase measurement accuracy when the signal-to-noise ratio is low. Instead of the linear approximation of the logarithm of $Z_r(r)$, Rocadenbosh et al. (1998) used direct fitting of the range-corrected signal $Z_r(r)$ to an exponential curve. The authors maintain that this method decreases the influence of large high-frequency noise peaks at the far end of the range-corrected signal, which appear in the conventional slope method. Thus a nonlinear fit may improve the accuracy of the extinction coefficient extracted with the slope method. This observation contradicts the study described above by Kunz and de Leeuw (1993), who concluded that results obtained with an exponential fit are less accurate than those obtained with a linear fit. In their next study (Rocadenbosh et al., 2000), the authors revised their conclusion and agreed that the nonlinear fit has no advantage compared with the conventional slope method, at least when an optimal inversion length is used. In any case, the practical value of a nonlinear fit to the slope method is always questionable, because such conclusions are based on numerical simulations that as a rule ignore all nonrandom sources of error. Unfortunately, it is general practice to assume that the random error component is the dominant source of error, whereas any systematic components and low-frequency offsets can be ignored. Such assumptions may only be relevant when making a general analysis of sources of error to understand which ones are most influential and which ones may be ignored. However, such approximations are inappropriate when comparing, for example, minor differences between linear and nonlinear fit, especially at the far end of the measurement range.

To summarize the uncertainty analysis of the slope method:

1. The slope method is a practical method for measurements of mean extinction coefficients in homogeneous atmospheres. The use of the slope method makes it possible to find the unknown particulate extinction coefficient without the need to estimate the numerical value of the particulate backscatter-to-extinction ratio. This is true for both single- and two-component atmospheres.
2. Under favorable conditions, the application of the least-squares technique to the slope method yields accurate extinction coefficients and provides practical estimates of the degree of atmospheric homogeneity.
3. The dependence of the uncertainty in the extracted extinction coefficient on the optical depth of the measurement range has a U-shaped appearance. The uncertainty increases both for short and long range intervals,

(r_1, r_2), having the smallest values within a restricted intermediate zone.

4. The standard deviation of the linear fit of the logarithm of the range-corrected signal can be used as an estimate of the degree of atmospheric homogeneity. On the other hand, the linearity of the logarithm of $Z_r(r)$ cannot be considered to be absolute evidence of atmospheric homogeneity. This is especially important when short range intervals are analyzed, or when lidar signals are measured in nonhorizontal directions. Note also, that poor optical alignment of the lidar optics may also produce a systematic incline in log of $Z_r(r)$, which may be nicely approximated by a linear fit, giving the researcher a false sense that the system is perfectly aligned.
5. The slope method should not be used for extinction coefficient measurements over range intervals with small optical depths. In this case, the slope of $\ln Z_r(\mathrm{r})$ with respect to the horizontal axis is small, so that the extinction coefficient cannot be accurately estimated. However, such atmospheric conditions are quite favorable for lidar field tests. They allow the application of the slope method to estimate the lidar system performance before routine measurements are made.

6.2. LIDAR MEASUREMENT UNCERTAINTY IN A TWO-COMPONENT ATMOSPHERE

6.2.1. General Formula

The estimates above of the uncertainty associated with the slope method show the strong dependence of the measurement error on the optical depth of the examined range interval. The optical depth of the range interval, or the path transmittance related to it, is the key parameter that influences lidar measurement accuracy. The optical depth generally acts as a factor or exponent in most uncertainty formulations [see, for example, Eqs. (6.3), (6.4), and (6.5)]. In other words, a factor similar to $F_\tau(r_1, r_2)$ [Eq. (6.6)] is introduced and acts as a magnification factor in most uncertainty formulations related to range-resolved extinction, scattering, or absorption coefficient measurements. When the lidar equation transformation is made as described in Section 5.2, this factor is also transformed. It becomes related to the optical depth of the weighted extinction coefficient $\kappa_W(r)$. Similarly, in a differential absorption lidar (DIAL) inversion technique, the measurement accuracy depends on the differential optical depth (Chapter 10). Clearly, to provide an acceptable measurement accuracy, the selection of the optimum optical depths is required.

The determination of the range-resolved profile of an atmospheric parameter is usually much less accurate than the determination of its mean value over an extended interval. There are several specific issues associated with the measurement of the local values of the particulate extinction coefficient $\kappa_p(r)$, that require detailed consideration. As shown in Chapter 5, the extraction of the extinction coefficient $\kappa_p(r)$ from the initial lidar signal, measured in a two-component atmosphere, requires transformation of the signal $P(r)$ into the function $Z(r)$. The general procedure to obtain an unknown $\kappa_p(r)$ may be divided into three steps (Section 5.2). In the first step, the transformation function $Y(r)$ is calculated and the lidar signal $P(r)$ is transformed into the function $Z(r)$ with Eq. (5.28). The transformed equation is solved in the second step, in which the weighted extinction coefficient $\kappa_W(r)$ is determined. In the third step, the inverse transformation is applied to the weighted extinction coefficient to obtain the particulate extinction profile [Eq. (5.34)]. Every step of the transformation can introduce errors. The first step can introduce and transform errors in the signal $P(r)$, inaccurately measured or corrupted by noise and in the function, $Y(r)$, that is used to transform $P(r)$ into the function $Z(r)$. The second step can introduce errors (1) by using incorrect values of $Z(r)$ in Eq. (5.75) to determine the weighted function $\kappa_W(r)$, and (2) in the conversion from the original boundary value of the extinction coefficient $\kappa_p(r_b)$ (or the total transmittance, T_{max}, in the optical depth solution) to the normalized form, $\kappa_W(r_b)$ or V_{max}, respectively. The third step can introduce and transform errors by the incorrect conversion of $\kappa_W(r)$ to the particulate extinction profile $\kappa_p(r)$, which is the parameter of interest.

It was stated above (Section 5.4) that in the two-component atmosphere the lidar equation solution for $\kappa_p(r)$ can be obtained under the following conditions: (1) The molecular extinction coefficient $\kappa_m(r)$ and the particulate backscatter-to-extinction ratio $\Pi_p(r)$ are known or somehow estimated, and (2) no molecular absorption exists, thus $\kappa_m(r) = \beta_m(r)$. The latter condition means that the molecular backscatter-to-extinction ratio $\Pi_m(r)$ is reduced to a constant phase function, $P_{\pi,m} = 3/8\pi$. The above conditions permit the determination of the transformation function $Y(r)$, the transformation of the lidar signal $P(r)$ into the function $Z(r)$ at the first step, and the derivation of the extinction coefficient $\kappa_p(r)$ from $\kappa_W(r)$ at the third step of data processing.

To simplify the following uncertainty analysis, two additional assumptions are made. First, it is assumed that the molecular extinction-coefficient profile $\kappa_m(r)$ is exactly known along the lidar line of sight, that is, the relative uncertainty of the molecular extinction

$$\delta\kappa_m(r) = 0$$

The second condition is that the particulate backscatter-to-extinction ratio Π_p is exactly known and has a constant value over the measurement range,

$$\Pi_p(r) = \Pi_p = const.$$

Such assumptions are necessary to separate the different sources of error and to investigate them separately. The uncertainty caused by an inaccurate selection of the backscatter-to-extinction ratio Π_p is analyzed in Section 7.2. With $\Pi_p = const.$, the weighting function $\kappa_W(r)$ is

$$\kappa_W(r) = \kappa_p(r) + a\kappa_m(r) \tag{6.9}$$

where $\kappa_m(r) = \beta_m(r)$ and

$$a = \left(\frac{3/8\pi}{\Pi_p}\right) = const.$$

As follows from the definition of $Y(r)$ [Eq. (5.27)], the above assumptions yield $\delta Y(r) = 0$, so that no errors are introduced into the transformation function $Y(r)$. Thus, step 1 does not introduce any additional error into the calculated $Z(r)$. Because the transformation from $P(r)$ to $Z(r)$ is multiplicative, $\delta Z(r) = \delta P(r)$. Similarly, no errors are introduced in the transformed boundary values $\kappa_W(r_b)$ or V_{max} when transforming the original boundary values $\kappa_p(r_b)$ or T_{max}, respectively.

In the second step, the general lidar equation solution is used to calculate the function $\kappa_W(r)$. For the uncertainty analysis that follows, the solution given in Eq. (5.71) is used. The solution for $\kappa_W(r)$ is obtained with the use of three different terms: (1) the lidar signal transformed into the function $Z(r)$; (2) the integral of $Z(r)$ calculated in the range from r to r_b, and (3) the lidar solution constant, defined as $I(r_b, \infty)$, which must be estimated in some way, generally by applying boundary conditions. This integral can be considered as the most general form of the lidar solution constant. As shown in Chapter 5, the boundary point and optical depth solutions use, in fact, different ways for determining the integral $I(r_b, \infty)$. For a general uncertainty analysis, it is convenient to use the lidar equation solution of Eq. (5.71) rewritten for $r > r_b$, i.e.,

$$\kappa_W(r) = \frac{0.5Z(r)}{I(r_b, \infty) - I(r_b, r)} \tag{6.10}$$

where

$$I(r_b, r) = \int_{r_b}^{r} Z(r')\, dr' \tag{6.11}$$

Obviously, the terms $Z(r)$, $I(r_b, \infty)$, and $I(r_b, r)$ in Eq. (6.10) are always determined with some degree of uncertainty, $\delta Z(r)$, $\delta I(r_b, \infty)$, and $\delta I(r_b, r)$, respectively, that influence accuracy of the unknown $\kappa_W(r)$. The uncertainty of the lidar solution is generally not symmetric with respect to large positive and negative errors of the parameters involved. The uncertainty may depend significantly on whether the estimated boundary value, $I(r_b, \infty)$, used for the solution is over- or underestimated. For example, if $I(r_b, \infty)$ in Eq. (6.10) is underestimated, the solution may yield not physical negative values of $\kappa_W(r)$, whereas an overestimated $I(r_b, \infty)$ will yield only positive values. To have a comprehensive understanding of the error behavior, the signs of the error components cannot be ignored, as is done with conventional uncertainty analysis. With this observation, the uncertainty of the weighted extinction coefficient $\kappa_W(r)$ can be derived as a function of the three errors components above as (Kovalev and Moosmüller, 1994)

$$\delta\kappa_W(r) = \frac{\delta Z(r)\, V^2(r_b, r) - \delta I(r_b, \infty) + \delta I(r_b, r)[1 - V^2(r_b, r)]}{V^2(r_b, r) + \delta I(r_b, \infty) - \delta I(r_b, r)[1 - V^2(r_b, r)]} \tag{6.12}$$

The function $V^2(r_b, r)$ in Eq. (6.12) is the two-way atmospheric transmittance of the range interval (r_b, r) calculated with the weighted extinction coefficient $\kappa_W(r)$

$$V^2(r_b, r) = \exp[-2\tau_W(r_b, r)] = \exp\left\{-2\int_{r_b}^{r} [\kappa_p(r') + a\kappa_m(r')]dr'\right\} \tag{6.13}$$

where the function $\tau_W(r_b, r)$ is the optical depth of the weighted extinction coefficient $\kappa_W(r)$ over the range interval from r_b to r

$$\tau_W(r_b, r) = \int_{r_b}^{r} \kappa_W(r')\, dr' \tag{6.14}$$

In the next sections of the chapter, the uncertainty analysis is given restricted to boundary point solutions. The uncertainties inherent to the optical depth solution are analyzed in Sections 12.1 and 12.2.

6.2.2. Boundary Point Solution: Influence of Uncertainty and Location of the Specified Boundary Value on the Uncertainty $\delta\kappa_W(r)$

To determine the influence of the uncertainty and location of the boundary value on the solution accuracy, only terms related to the boundary values in Eq. (6.12) will be considered. In other words, all other contributions to the uncertainty in Eq. (6.12) are assumed to be negligibly small and can be ignored. If $\delta Z(r) = 0$, and $\delta I(r_b, r) = 0$, the only uncertainty introduced in step *2* of the inversion stems from the uncertainty of the boundary value estimate, so that Eq. (6.12) is reduced to

$$\delta\kappa_W(r) = \frac{-\delta I(r_b, \infty)}{V^2(r_b, r) + \delta I(r_b, \infty)} \tag{6.15}$$

In the boundary point solution, the integral $I(r_b, \infty)$ is found by using either an assumed or in some way determined value of the particulate extinction coefficient at the boundary point, $\kappa_p(r_b)$. With this value, the corresponding value of $\kappa_W(r_b)$, is calculated with Eq. (6.9). After that, the integral $I(r_b, \infty)$ is determined with Eq. (5.74)

$$I(r_b, \infty) = \frac{0.5Z(r_b)}{\kappa_W(r_b)}$$

and together with Eq. (6.10) yields the solution in Eq. (5.75).

An incorrectly determined value of the weighted extinction coefficient $\kappa_W(r_b)$ introduces an uncertainty in the estimate of the integral $I(r_b, \infty)$. The relative error $\delta\kappa_W(r_b)$ may be quite large, especially when the value of $\kappa_p(r_b)$ is taken a priori. Assuming for simplicity that $\Delta I(r_b, \infty)$ is the absolute uncertainty of the integral $I(r_b, \infty)$ due to uncertainty $\Delta\kappa_W(r_b)$, and that the uncertainty in $Z(r_b)$ is small and can be ignored, one can write the above equation as

$$I(r_b, \infty) + \Delta I(r_b, \infty) = \frac{0.5S(r_b)}{\kappa_W(r_b) + \Delta\kappa_W(r_b)} \tag{6.16}$$

Solving Eqs. (5.74) and (6.16), an expression for the relative uncertainty $\delta I(r_b, \infty)$ is obtained:

$$\delta I(r_b, \infty) = \frac{-\delta\kappa_W(r_b)}{1 + \delta\kappa_W(r_b)} \tag{6.17}$$

where $\delta I(r_b, \infty) = \Delta I(r_b, \infty)/I(r_b, \infty)$ and $\delta\kappa_W(r_b) = \Delta\kappa_W(r_b)/\kappa_W(r_b)$. It should be noted that the uncertainties $\delta I(r_b, \infty)$ and $\delta\kappa_W(r_b)$ have opposite signs. This means that an overestimated $\kappa_W(r_b)$ yields an underestimated integral $I(r_b, \infty)$ in Eq. (6.10), and vice versa. Note that when $\delta\kappa_W(r_b) \ll 1$, Eq. (6.17) reduces to $|\delta I(r_b, \infty)| = |\delta\kappa_W(r_b)|$, which may also be obtained with a conventional uncertainty propagation. After substitution of Eq. (6.17) into Eq. (6.15), the latter is reduced to (Kovalev and Moosmüller, 1994)

$$\delta\kappa_W(r) = \left[V^2(r_b, r) + \frac{V^2(r_b, r)}{\delta\kappa_W(r_b)} - 1\right]^{-1} \tag{6.18}$$

Thus the uncertainty in $\kappa_W(r)$ is related to the uncertainty of $\kappa_W(r_b)$ and the two-way path transmission, $V(r_b, r)^2$. The latter is related to the optical depth $\tau_W(r_b, r)$ of the variable $\kappa_W(r)$ in the range interval from r_b to r [Eq. (6.13)]. In Fig. 6.4, the uncertainty $\delta\kappa_W(r)$ is shown as a function of the optical depth $\tau_W(r_b, r)$ for different uncertainties in the assumed boundary value $\kappa_W(r_b)$. At the location of the boundary point itself, for $r = r_b$, the relative uncertainty in $\kappa_W(r)$ is equal to the uncertainty in the specified boundary value, $\delta\kappa_W(r_b)$. The boundary points $\delta\kappa_W(r_b)$ are shown as black squares. Moving away from these points, the uncertainty changes monotonically as a function of the variable $\tau_W(r_b, r)$. It can be seen that the optical depth rather than the geometric length of the range (r_b, r) influences the uncertainty in the measurement. For the near-end

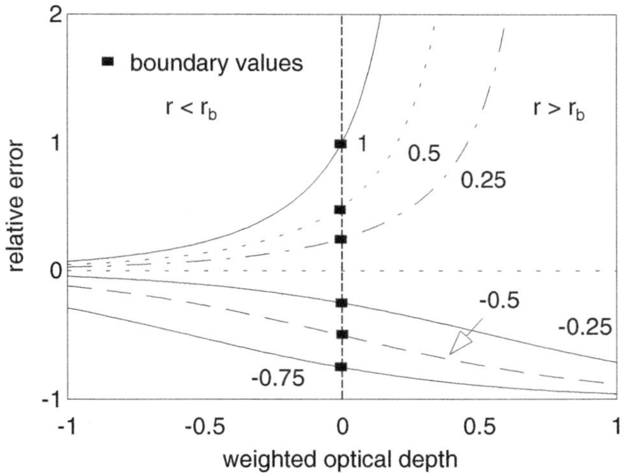

Fig. 6.4. The uncertainty $\delta\kappa_W(r)$ as a function of the optical depth $\tau_W(r_b, r)$ for different uncertainties in the boundary value $\delta\kappa_W(r_b)$. The numbers are the specified values of $\delta\kappa_W(r_b)$ (Kovalev and Moosmüller, 1992).

solution ($r > r_b$), the absolute value of the relative uncertainty increases with the increase of the optical depth, $\tau_W(r_b, r)$, as shown on to the right side of Fig. 6.4, where values of $\tau_W(r_b, r)$ are shown as positive. When the boundary point is selected at the far end, the operating measurement range extends to the left side of Fig. 6.4, where values of $\tau_W(r_b, r)$ are shown as negative. Note that the uncertainties in this case are always less than the uncertainty in the assumed boundary value $\kappa_W(r_b)$. The most accurate result is achieved close to and at the near end of the measurement range (Kaul, 1977; Zuev et al., 1978a; Klett, 1981).

The uncertainty in $\delta\kappa_W(r)$ decreases monotonically as a function of $\tau_W(r_b, r)$ in the direction toward the lidar system, that is, to the left border of Fig. 6.4, whereas it increases in the opposite direction. Thus improved measurement accuracy is attained when the location of the boundary point is selected to be as far as possible from the lidar site, as shown in Fig. 5.4 (b). Generally, it is selected as close to the far end of the lidar operating range as possible while maintaining an acceptable signal-to-noise ratio.

The statement above applies when the particulate backscatter-to-extinction ratio, Π_p, has a constant value and is accurately estimated. As shown in Section 7.2, the far-end solution may yield an inaccurate measurement result if the assumed backscatter-to-extinction ratio is taken incorrectly, especially if the extinction coefficient has monotonic changes with range. Note also that in turbid atmospheres where a single-component particulate atmosphere assumption is valid, the optical depth $\tau_W(r_b, r)$ reduces to the optical depth of the particulate atmosphere, $\tau_p(r_b, r)$. Here the uncertainty $\delta\kappa_W(r)$ is strongly related to the total particulate depth (Balin et al., 1987, Jinhuan, 1988).

As can be seen in Fig. 6.4, the behavior of the uncertainty $\delta\kappa_W(r)$ depends significantly on the accuracy of the assumed boundary value, that is, on the value and the sign of the error in $\kappa_W(r_b)$. For the far-end solution, a positive error in $\delta\kappa_W(r_b)$, that is, overestimated $\kappa_W(r_b)$, is preferable because it provides a smaller measurement error. The larger the optical depth τ_W between r and r_b, the more accurate the measurement result that is obtained. On the other hand, when the boundary point r_b is selected at the near end of the measurement range ($r > r_b$), an underestimated $\kappa_W(r_b)$ is preferable. Here overestimated $\kappa_W(r_b)$ yields a measurement error that increases monotonically toward a pole at

$$\tau_{W,\text{pole}}(r_b, r) = -0.5 \ln\left[\frac{\delta\kappa_W(r_b)}{1 + \delta\kappa_W(r_b)}\right] \tag{6.19}$$

where the value of $\kappa_W(r) \Rightarrow \infty$ toward the pole. This occurs when the denominator in Eq. (6.10) becomes equal to zero because of an incorrectly established $I(r_b, \infty)$.

The behavior of the uncertainty of the measured extinction coefficient $\delta\kappa_W(r)$ in Fig. 6.4 clearly shows that the near-end solution is generally inaccurate, because the measurement uncertainty may increase significantly at long distances from the lidar when the boundary condition $\kappa_W(r_b)$ is inaccurate.

For negative values of $\delta\kappa_W(r_b)$, that is, for an underestimation of the boundary value $\kappa_W(r_b)$, the uncertainty $\delta\kappa_W(r)$ is also negative. In this case, the increase in the uncertainty in the near-end solution is not so rapid as for an overestimated $\kappa_W(r_b)$ (Fig. 6.4). Therefore, for the near-end solution, an underestimate of the boundary value is preferable to an overestimate of $\kappa_W(r_b)$. Note also that in clear atmospheres, where the optical depth over the lidar operating range is small, the near-end solution becomes more stable. In this case, the location of the boundary point is less important than the uncertainty in the specified boundary value (Bissonnette, 1986). This observation is most often the case for lidar systems operating in clear atmospheres in the visible or infrared, where the optical depth of the measured range is small. Examples of the $\kappa_p(r)$ profiles calculated for a clear atmosphere are shown in Fig. 6.5. The profiles are calculated for a homogeneous atmosphere with $\kappa_p = 0.05\,\text{km}^{-1}$, $\kappa_m = 0.0116\,\text{km}^{-1}$, and $\Pi_p = 0.05\,\text{sr}^{-1}$. The boundary values of $\kappa_p(r_b)$ are specified at three different locations: at the near end ($r_b = 1\,\text{km}$), at the far end ($r_b = 4\,\text{km}$), and at an intermediate point ($r_b = 2.5\,\text{km}$) in the measurement range for both positive [$\delta\kappa_p(r_b) = 0.5$] and negative [$\delta\kappa_p(r_b) = -0.5$] relative uncertainty. The uncertainties $\delta I(r_b, r)$ and $\delta P(r_b, r)$ are ignored. It can be seen that the influence of the boundary-point location is relatively small. The slope of the uncertainty with range, shown in Fig. 6.5, will increase if a lidar with a shorter wavelength is used. This is because, for shorter wavelengths, larger molecular scattering increases the optical depth τ_W over the same range intervals. In the

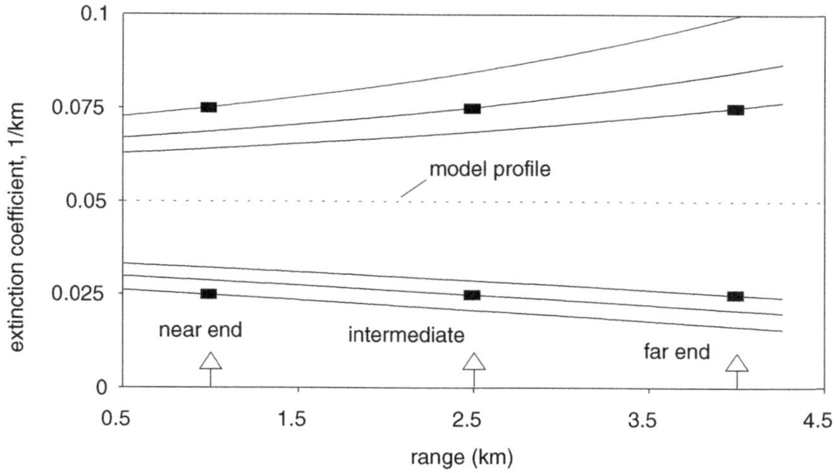

Fig. 6.5. Example of the particulate extinction profiles derived with different boundary point locations in a clear atmosphere. The model profile of the homogeneous atmosphere is used with $\kappa_p = 0.05\,km^{-1}$. Boundary values, shown as black squares, are specified at the near end ($r_b = 1\,km$), at the far end ($r_b = 4\,km$), and at an intermediate point ($r_b = 2.5\,km$) of the measurement range with both positive [$\delta\kappa_p(r_b) = 0.5$] and negative [$\delta\kappa_p(r_b) = -0.5$] relative uncertainties (Kovalev and Moosmüller, 1992).

ultraviolet region, even a clear unpolluted atmosphere can result in an increased optical depth $\tau_W(r_b, r)$ because of the λ^{-4} increase in the molecular extinction.

The application of the near-end solution [Eq. (5.75), $r > r_b$] requires attention to even small errors that may generally be ignored in the far-end solution. One can easily demonstrate the sensitivity of the near-end solution to even minor processing errors. For example, noticeable errors in the extracted extinction coefficient may even be caused by errors introduced by numerical integration. Such errors occur when a small number of discrete points (range bins) are available, especially in areas of thin layering where the backscatter coefficient changes rapidly. Similar errors in the retrieved profile may also occur in clear atmospheres if a significant change in the extinction coefficient occurs near the selected boundary point, r_b. In the simulated data in Fig. 6.6 (a–d), a conventional trapezoidal method is used to numerically integrate a signal recorded with a range resolution of 30 m. The atmospheric situation can be interpreted as a thin turbid layer moving along the lidar measurement range. It is assumed that no other sources of error exist, that is, the backscatter-to-extinction ratio is constant and precisely known and the correct boundary values $\kappa_W(r_b)$ are used. The latter values are shown in Fig. 6.6 as black rectangles. The discrepancies between the model and inverted profiles, shown in the figure as dotted and solid lines, respectively, are due solely to errors from the numerical integration method used. Although these integration errors are normally dwarfed by signal and transformation errors, their influence

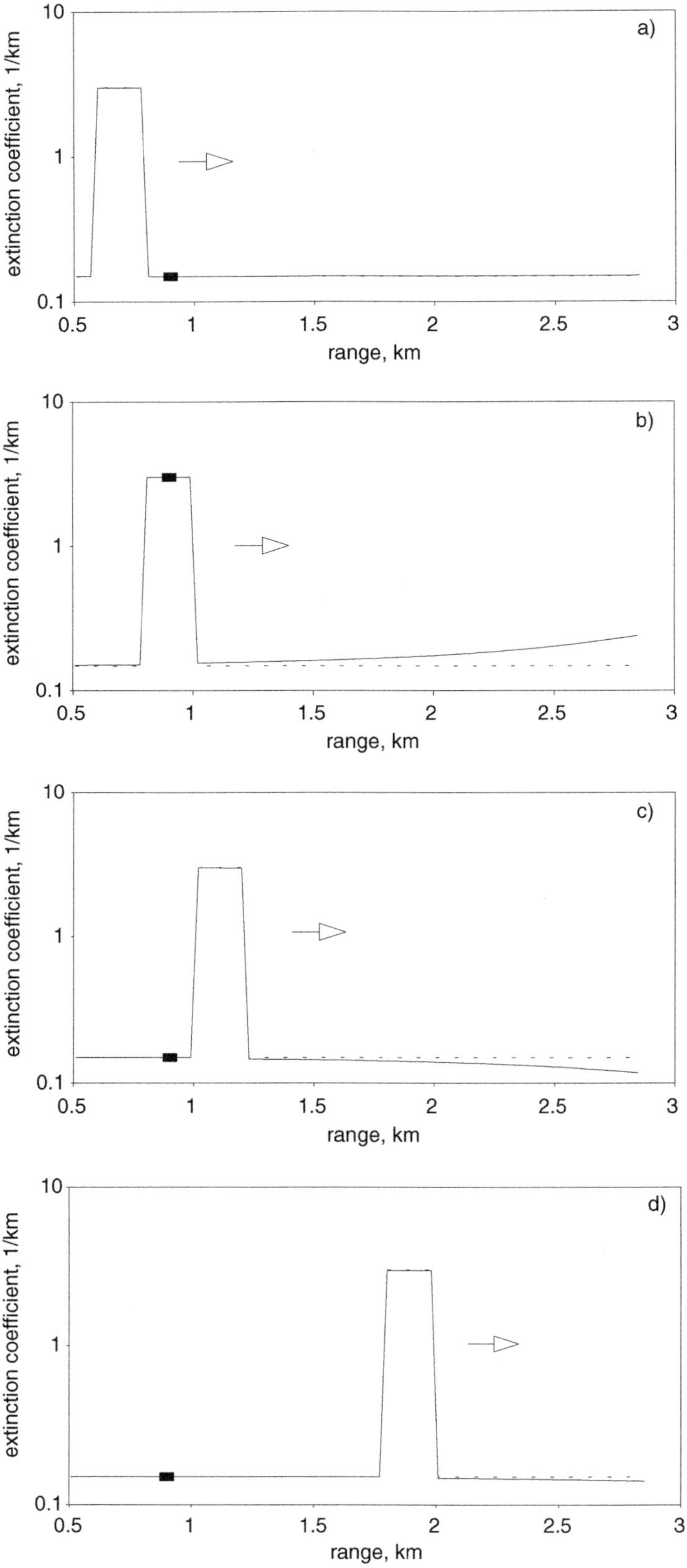

a)
b)
c)
d)
extinction coefficient, 1/km
range, km
10
1
0.1
0.5
1
1.5
2
2.5
3

demonstrates the sensitivity of the near-end solution in heterogeneous atmospheres to minor distortions of the parameters involved. To improve the stability of the near-end solution, a combination of the near-end and optical depth solutions can be used, as shown in Section 8.1.4.

6.2.3. Boundary-Point Solution: Influence of the Particulate Backscatter-to-Extinction Ratio and the Ratio Between $\kappa_p(r)$ and $\kappa_m(r)$ on Measurement Accuracy

After solving Eq. (5.75), the weighted extinction coefficient $\kappa_W(r)$ is determined. The coefficient $\kappa_W(r)$ is only an intermediate function, from which the quantity of interest, namely, the particulate extinction coefficient profile, is then obtained. The particulate extinction coefficient is found from Eq. (6.9) as

$$\kappa_p(r) = \kappa_W(r) - a\kappa_m(r)$$

Considering the relationship between $\kappa_p(r)$ and $\kappa_W(r)$, the relative uncertainties in these values can be written as

$$\delta\kappa_p(r) = \left[1 + a\frac{\kappa_m(r)}{\kappa_p(r)}\right]\delta\kappa_W(r) \tag{6.20}$$

Eq. (6.20) is obtained by conventional error propagation (Bevington and Robinson, 1992). This equation is derived assuming that only the error in $\kappa_W(r)$ contributes to the uncertainty in retrieved $\kappa_p(r)$. Using the relationship between the extinction and backscatter coefficients given in Section 5.2 [Eqs. (5.17) and (5.18)], Eq. (6.20) can also be rewritten as

$$\delta\kappa_p(r) = \left[1 + \frac{\beta_{\pi,m}(r)}{\beta_{\pi,p}(r)}\right]\delta\kappa_W(r) \tag{6.21}$$

where $\beta_{\pi,m}(r)$ and $\beta_{\pi,p}(r)$ are the molecular and particulate backscatter coefficients, respectively. Thus the uncertainties $\delta\kappa_p(r)$ and $\delta\kappa_W(r)$ are a function of the ratio of the molecular and particulate backscatter coefficients. However,

Fig. 6.6. (a)–(d) Inversion example of an extinction coefficient profile where a relatively thin turbid layer is moving through the lidar measurement range. The location of the boundary point ($r_b = 0.9$ km) is the same for (a)–(d). Correct boundary values are used for calculations, and only the error in the numerical integration influences measurement accuracy. The particulate backscatter-to-extinction ratio and the molecular extinction coefficient are $\Pi_p = 0.015\,\text{sr}^{-1}$ and $\kappa_m = 0.067\,\text{km}^{-1}$, respectively (Kovalev and Moosmüller, 1992).

in performing an uncertainty analysis, it is useful to separate the contribution to the uncertainty caused by different proportions between the particulate and molecular extinction constituents and the contribution due to an uncertainty in the backscatter-to-extinction ratio. In most cases, Eq. (6.20) is preferable when making error analysis.

The molecular extinction-coefficient profile and the particulate backscatter-to-extinction ratio are assumed to be precisely known, so that the uncertainty in $\kappa_p(r)$ is the result of inaccuracies in the function $Z(r)$ and the assumed boundary value used in processing. However, the uncertainty $\delta\kappa_p(r)$ is highly dependent on the proportion between the atmospheric particulate and molecular scattering components and the parameter a. Defining the ratio of the particulate and molecular extinction coefficients as

$$R(r) = \frac{\kappa_p(r)}{\kappa_m(r)} \tag{6.22}$$

one can rewrite the uncertainty in the derived particulate extinction-coefficient profile in Eq. (6.20) as

$$\delta\kappa_p(r) = \left[1 + \frac{a}{R(r)}\right]\delta\kappa_W(r) \tag{6.23}$$

The proportion between the atmospheric particulate and molecular extinction coefficients significantly influences the accuracy of the derived profile of the particulate extinction coefficient. This is true even if the molecular extinction coefficient and particulate backscatter-to-extinction ratio used in the solution are precisely established.

In clear atmospheres, particulate extinction may be only a few percent of the molecular extinction. In this case, the problem is to accurately separate the particulate and molecular components. This problem is inherent in high-altitude measurements at visible and infrared wavelengths, where the scattering from particulates can be less than 1% of the total scattering. Substituting Eq. (6.18) into Eq. (6.23) transforms the latter into

$$\delta\kappa_p(r) = \frac{1 + \dfrac{a}{R(r)}}{\left[V^2(r_b, r) - 1 + \dfrac{V^2(r_b, r)}{\delta\kappa_W(r_b)}\right]} \tag{6.24}$$

With Eq. (6.24), the influence of the uncertainty in the boundary value, $\delta\kappa_W(r_b)$, on the accuracy of the derived particulate extinction-coefficient profile $\kappa_p(r)$ can be determined. Note that the selected boundary value of the particulate extinction coefficient, $\kappa_p(r_b)$, is transformed to the boundary value of the weighted extinction coefficient, $\kappa_W(r_b)$, and only then used in Eq. (5.75). Because the relationship between $\kappa_W(r_b)$ and $\kappa_p(r_b)$ is

$$\kappa_W(r_b) = \kappa_p(r_b) + a\kappa_m(r_b)$$

the uncertainty in the calculated value of $\kappa_W(r_b)$ in Eq. (6.24) differs from the uncertainty in the selected value of $\kappa_p(r_b)$ that was estimated or taken a priori. The relationship between these values obeys Eq. (6.23); thus

$$\delta\kappa_W(r_b) = \frac{\delta\kappa_p(r_b)}{1 + \dfrac{a}{R(r_b)}} \tag{6.25}$$

where $\delta\kappa_p(r_b)$ is the relative uncertainty in the specified boundary value $\kappa_p(r_b)$. After substituting Eq. (6.25) into Eq. (6.24), the uncertainty in the calculated extinction-coefficient profile $\kappa_p(r)$ can be determined as

$$\delta\kappa_p(r) = \frac{1 + \dfrac{a}{R(r)}}{\left[V^2(r_b, r) - 1 + \dfrac{V^2(r_b, r)}{\delta\kappa_p(r_b)}\left(1 + \dfrac{a}{R(r_b)}\right)\right]} \tag{6.26}$$

The relative uncertainty of the measured profile of $\kappa_p(r)$ depends not only on the uncertainty in the selected value of $\kappa_p(r_b)$ but also on the ratio of a to $R(r_b)$. Note that the function $V^2(r_b, r)$, defined in Eq. (6.13), may also be presented as a function of the ratio $a/R(r)$

$$V^2(r_b, r) = \exp\left[-2\int_{r_b}^{r} \kappa_p(r')\left(1 + \frac{a}{R(r')}\right)dr'\right] \tag{6.27}$$

When the molecular contribution to extinction at the reference point becomes small compared with the particulate contribution, it can be ignored, and the ratio $a/R(r_b)$ tends toward zero. For such an atmosphere, the term $[1 + a/R(r_b)] \approx 1$. Then the uncertainty of the boundary value no longer depends on the value of a, so that $\kappa_W(r_b) \approx \kappa_p(r_b)$.

Some additional comments here may be helpful to provide a more comprehensive understanding of the relationships between the uncertainties. The transformation of the original lidar signal into the function $Z(r)$ changes the original proportions between the particulate and molecular contributions in the new variable, $\kappa_W(r)$. These new proportions are also maintained in the corresponding dependent values, such as the optical depth and path transmission, which now become the functions defined as $\tau_W(r_b, r)$ and $V(r_b, r)$, respectively. The transformed optical depth, $\tau_W(r_b, r)$, can be expressed as a total of the particulate and weighted molecular optical depths, $\tau_p(r_b, r)$ and $\tau_m(r_b, r)$, as

$$\tau_W(r_b, r) = \tau_p(r_b, r) + a\tau_m(r_b, r) \tag{6.28}$$

Similarly to Eq. (5.81), the function $V(r_b, r)$ in Eq. (6.26) may be defined with the molecular and particulate transmission over the range (r_b, r) and the ratio a as

$$V(r_b, r) = T_p(r_b, r)[T_m(r_b, r)]^a \tag{6.29}$$

Thus the molecular contribution to the new quantities is weighted by a factor of a, that is, by the ratio of $3/8\pi$ to Π_p [Eq. (5.70)]. Generally, the molecular phase function is twice (or even more) as much as the particulate backscatter-to-extinction ratio, Π_p. Therefore, a is usually larger than 1. This feature increases the weight of the molecular component compared with the particulate component when determining the new variable $\kappa_W(r)$ and the related terms $\tau_W(r_b, r)$ and $V(r_b, r)$. This may result in two opposing effects in clear atmospheres where $R(r)$ is small. First, as follows from Eq. (6.25), a decrease in the uncertainty in the boundary value $\kappa_W(r_b)$ occurs relative to that in the assumed value of $\kappa_p(r_b)$. Second, an increase of the uncertainty in the measured particulate component occurs when extracting a profile from an inaccurately obtained $\kappa_W(r)$ with Eq. (6.23). Generally, these effects compensate each other, at least to some extent.

In Fig. 6.7, the relative error in the retrieved extinction coefficient $\kappa_p(r)$ is shown as a function of the total (particulate and molecular) optical depth, $\tau(r_b, r) = \tau_p(r_b, r) + \tau_m(r_b, r)$. Here the positive values of $\tau(r_b, r)$ correspond to the near-end solution, and the negative values correspond to the far-end solution [i.e., $-\tau(r_b, r) = \tau(r, r_b)$]. The relative uncertainties in the specified boundary values of $\kappa_p(r_b)$ are $\delta\kappa_p(r_b) = -0.5$ and $\delta\kappa_p(r_b) = 0.5$; the boundary values are shown as black rectangles. The uncertainty relationships are shown for different ratios a/R, and the bold lines show the case of a single-component particulate atmosphere ($a/R = 0$). In all cases, the uncertainty in the measured extinction coefficient increases when the near-end solution is applied. For the far-end solution, the relative uncertainty of the derived particulate extinction coefficient is smaller when the ratio a/R and, accordingly, the molecular extinction coefficient, become larger. Thus, when the far-end solution is used for a moderately turbid atmosphere, better measurement accuracy might be achieved when the measurement is made in the visible portion of the spectrum rather than in the infrared. One should keep in mind, however, that this might be only true if the molecular extinction coefficient profile and the ratio a are precisely known. The uncertainty in these values and especially in the measured signals will implement additional errors in $\kappa_p(r)$, which can be large when the lidar operates in visual or ultraviolet spectra.

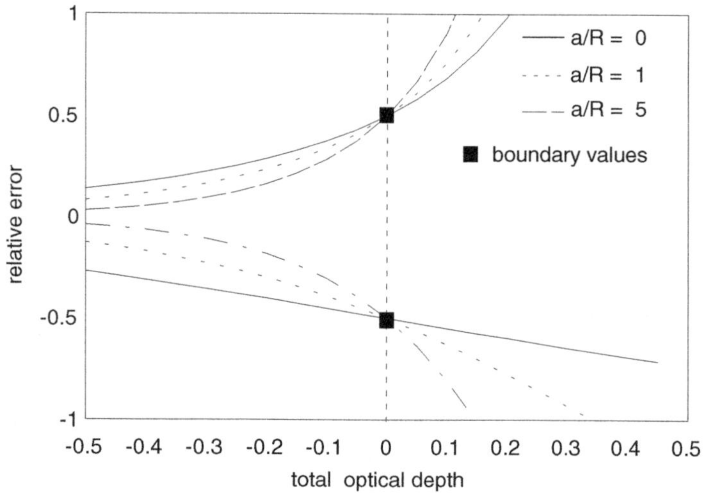

Fig. 6.7. Relative uncertainty in the derived $\kappa_p(r)$ profile as a function of the total optical depth for different ratios of a/R and both positive $[\delta\kappa_p(r_b) = 0.5]$ and negative $[\delta\kappa_p(r_b) = -0.5]$ errors in the specified boundary value $\kappa_p(r_b)$ (adapted from Kovalev and Moosmüller, 1992).

For better understanding of the above relationships, one can differentiate between the influence of the values of R and a. The influence of these parameters are shown in Figs. 6.8 and 6.9, respectively. As above, here the boundary values $\kappa_p(r_b)$ are shown as black rectangles. Figure 6.8 shows that the same uncertainty in the assumed $\kappa_p(r_b)$ may result in different errors in the retrieved extinction coefficient if different proportions occur between the particulate and molecular components. For the far-end solution, the measurement errors are less when the ratio of the particular-to-molecular extinction coefficient R is small, and vice versa. The explanation of this effect is similar to that given above. When R is small, smaller uncertainties result in the weighted extinction coefficient $\kappa_W(r_b)$ [Eq. (6.25)]. Obviously, the least amount of measurement error can be expected when the pure molecular scattering takes place at the boundary point r_b. This specific condition is widely used in lidar examination of clear and moderately turbid atmospheres (see Chapter 8). In Fig. 6.9, the uncertainty relationships are shown for different particulate backscatter-to-extinction ratios and, accordingly, for different a. Here the ratio R is taken as constant and equal to 1, that is, the particulate and molecular extinction coefficients are assumed to be equal. The figure shows the same tendency in the behavior of the uncertainty as that in Fig. 6.8, for both the near- and far-end solutions. For the latter solution, larger particulate backscatter-to-extinction ratios result in an increase in the measurement uncertainty.

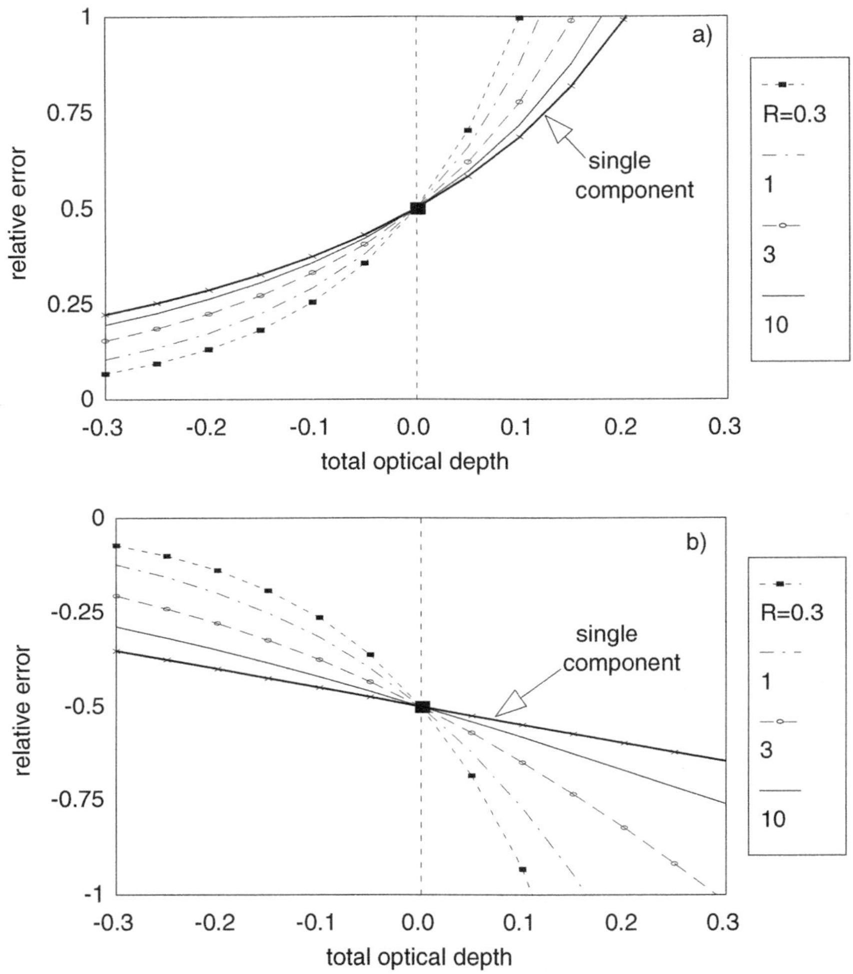

Fig. 6.8. Relative uncertainty in the derived $\kappa_p(r)$ profile as a function of the total optical depth calculated for (a) the positive $[\delta\kappa_p(r_b) = 0.5]$ and (b) negative $[\delta\kappa_p(r_b) = -0.5]$ errors in the specified boundary value $\kappa_p(r_b)$. The bold curves show the limiting case of a single-component particulate atmosphere (adapted from Kovalev and Moosmüller, 1992).

In the two-component atmospheres, the gain in the accuracy in the far-end boundary solution is related to the optical depth $\tau_W(r, r_b)$ of the weighted extinction coefficient $\kappa_W(r)$ rather than the total optical depth $\tau(r, r_b) = \tau_p(r, r_b) + \tau_m(r, r_b)$.

It is generally accepted that the far-end solution works best when the optical depth $\tau_W(r, r_b)$ is large. However, this statement should be taken only as a general conclusion. The assumptions made in this section regarding accurate

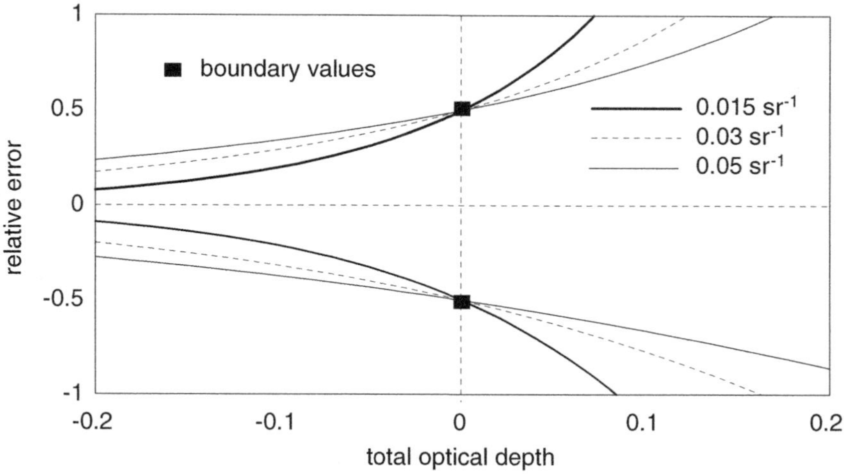

Fig. 6.9. Relative uncertainty in the derived $\kappa_p(r)$ profile as a function of the total optical depth for different particulate backscatter-to-extinction ratios and the positive $[\delta\kappa_p(r_b) = 0.5]$ and negative $[\delta\kappa_p(r_b) = -0.5]$ errors of the specified boundary value (adapted from Kovalev and Moosmüller, 1992).

knowledge of the particulate backscatter-to-extinction ratio and molecular extinction-coefficient profile are quite restrictive. Meanwhile, to estimate the total measurement uncertainty, all of the error sources must be taken into consideration, including even the uncertainty in the calculated $Z(r_b)$ at the far end of the range, where the signal-to-noise ratio may be poor. Atmospheric heterogeneity may also be a factor that exacerbates the problem. For a heterogeneous atmosphere, where local layering (plumes, cloud) exists, the most stable far-end solution can yield incorrect, even negative, particulate extinction coefficients. This can occur, for example, if a turbid layer (a cloud) is found at the far end of the measured range and the specified boundary value is underestimated. An example of such an optical situation is shown in Fig. 6.10. Here the boundary value at the far end of the measured range, $r_b = 3.5$ km, is specified as $\kappa_p(r_b) = 0.15\,\text{km}^{-1}$, whereas the actual value is $\kappa_p(r_b) = 0.3\,\text{km}^{-1}$. An incorrect estimate of the boundary value results in negative particulate extinction coefficients near the turbid area. As shown in Section 7.2, similar incorrect results for the far-end solution can also be obtained when lidar measurements are made in a clear atmosphere in which the vertical extinction coefficient profile has a monotonic change.

It is generally assumed that the influence of uncertainties in the integral $I(r_b, r)$ in Eq. (6.12) can be neglected because they are much smaller than those of the boundary value, that is, $\delta I(r_b, r) \ll \delta I(r_b, \infty)$. However, it can be shown that even a small error in $I(r_b, r)$ can at times result in an appreciable difference between the actual and derived extinction-coefficient profiles. The uncertainty in $\delta I(r_b, r)$ may be the result of (1) an uncertainty, $\delta P(r)$, in the measured

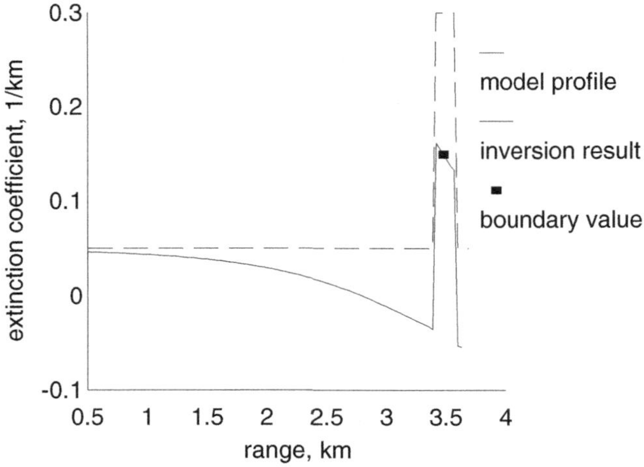

Fig. 6.10. Example of an inversion where the far-end solution yields negative values for the particulate extinction coefficient. The boundary value is specified as $\kappa_p(r_b) = 0.15\,\text{km}^{-1}$, whereas the actual value is $\kappa_p(r_b) = 0.3\,\text{km}^{-1}$. The inversion result is obtained with $\Pi_p = 0.015\,\text{sr}^{-1}$ (adapted from Kovalev and Moosmüller, 1992).

lidar signal, (2) an incorrectly estimated background offset, (3) an uncertainty in the function $Y(r)$, and (4) an error in the numerical integration, as shown in Fig. 6.6. The error $\delta I(r_b, r)$ is equivalent to a change in the specified boundary value as a function of the range. Indeed, if the function $Z(r)$ contains an offset $\Delta Z(r)$, then the integral in the range from r_b to r can be written as the sum of two terms

$$\langle I(r_b, r)\rangle = \int_{r_b}^{r} Z(r')\,dr' + \int_{r_b}^{r} \Delta Z(r')\,dr' \tag{6.30}$$

where $\Delta Z(r)$ can be either positive or negative. This term can be considered as an additional constituent of the integral $I(r_b, \infty)$ in Eq. (6.10). After substitution of Eq. (6.30) into Eq. (6.10), the general solution for $\kappa_W(r)$ can be written as

$$\kappa_W(r) = \frac{0.5\,[Z(r) + \Delta Z(r)]}{\left[I(r_b, \infty) - \int_{r_b}^{r} \Delta Z(r')\,dr'\right] - \int_{r_b}^{r} Z(r')\,dr'} \tag{6.31}$$

The integral $\int \Delta Z(r')dr'$ in the square brackets can be treated as a range-dependent error in the boundary value $I(r_b, \infty)$. Note that the offset $\Delta Z(r)$, being accumulated in any local range from r_b to r_j, worsens the measurement accuracy for all points beyond this range. Examples of the influence of the uncertainty $\delta I(r_b, r)$ on the measurement accuracy for the near- and far-end solutions are shown in Fig. 6.11 (a) and (b), respectively. The model particulate extinction profiles are shown as *curves 1*, whereas the inversion results are shown as *curves 2*. Here the shift ΔZ is assumed to exist only within the range of the turbid region. Such a shift can be introduced, for example, by uncompensated multiple scattering within the cloud or can be due to a difference between the actual backscatter-to-extinction ratio within the cloud and that used for inversion. The distortion of the extracted profile is similar to that caused by an incorrect estimate of the boundary value. The discrepancies between the actual and retrieved $\kappa_p(r)$ profiles are generally larger for relatively small values of the particulate backscatter-to-extinction ratios ($\Pi_p = 0.01$–$0.02\,\mathrm{sr}^{-1}$) and for increased values of *a/R*.

6.3. BACKGROUND CONSTITUENT IN THE ORIGINAL LIDAR SIGNAL AND LIDAR SIGNAL AVERAGING

When recorded during the day, lidar signals may contain a large offset because of background solar radiation. The recorded signal is the sum of two terms

$$P_\Sigma(r) = P(r) + P_{bgr} \tag{6.32}$$

where $P(r)$ is the true backscatter signal and P_{bgr} is the signal offset (Fig. 4.12). Generally, two major contributions to the offset may exist. The first is the residual skylight that passes a narrow optical bandpass filter, and the second is an electrical offset generated in the receiver electronics. The former component is mostly dominant. After substituting $P(r)$ [Eq. (5.2)] into Eq. (6.32), the recorded signal can be rewritten as

$$P_\Sigma(r) = P(r)\left\{1 + P_{bgr}\left[\frac{r^2 e^{2\tau(0,r)}}{C_0\beta_\pi}\right]\right\} \tag{6.33}$$

where $\tau(0, r)$ is the optical depth of the range from $r = 0$ to r. It can be seen that the weight of the offset term, P_{bgr}, in the recorded signal, $P_\Sigma(r)$, rapidly increases with an increase in the range r and the optical depth $\tau(0, r)$. To obtain accurate measurement data, the value of the background component must be precisely estimated and subtracted from the recorded signal before data processing is done. It is common practice to estimate the signal offset by recording the background level at the photoreceiver either before the light pulse is emitted or at long times after its emission. For the latter method, the time used to determine the background level must be long enough to ensure that the backscattered signal has completely decayed away. In Fig. 4.12, this time corresponds to a range of more than 2.5–3 km. In this range, $P(r)$ is indistinguishable from zero, so that the remaining signal magnitude $P_\Sigma(r)$ can be

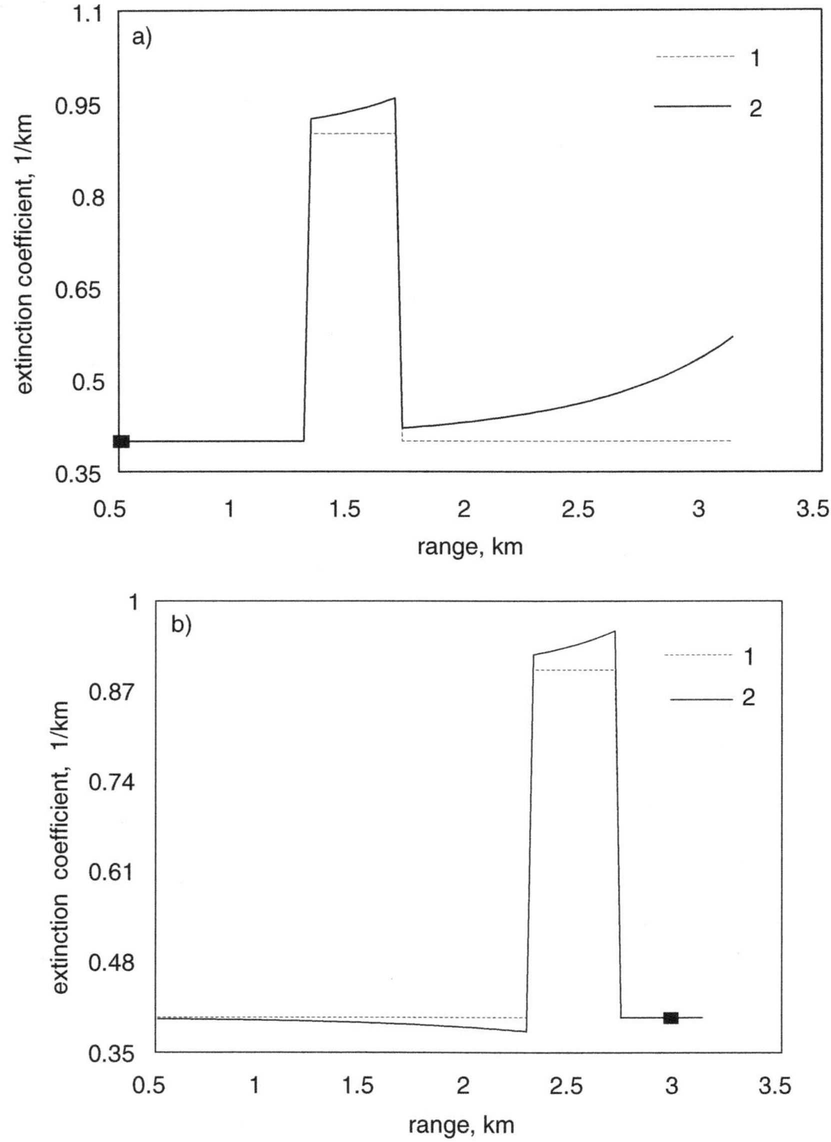

Fig. 6.11. (a) Example of a near-end solution where the measurement error is due only to $\delta I(r_b, r) \neq 0$ in the turbid area between 1.3 and 1.7 km. The signal shift in this region is $\Delta P = 0.02\ P(r)$, and the particulate backscatter-to-extinction ratio is $\Pi_p = 0.03\ \text{sr}^{-1}$. (b) Example of the far-end solution where the measurement error is due only to $\delta I(r_b, r) \neq 0$ in the turbid area. The signal shift in this region is $\Delta P = 0.05\ P(r)$, and the particulate backscatter-to-extinction ratio is $\Pi_p = 0.03\ \text{sr}^{-1}$ (Kovalev and Moosmüller, 1992).

assumed to represent only the background component. Note that such a method assumes that the value of the background P_{bgr} remains constant during the recording time.

The accurate estimate of the background constituent is extremely difficult for two basic reasons (Milton and Woods, 1987). The first arises when the background constituent is relatively large, when $P_{bgr} \gg P(r)$. In Fig. 4.12, this takes place at the ranges from 1 to 2 km, where the accuracy of the measured signal $P(r)$ becomes poor. The signal $P(r)$ is found here as a small difference of two large quantities, $P_\Sigma(r)$ and P_{bgr}. A subtraction inaccuracy results in a shift, ΔP, which may remain in the signal $P(r)$ after subtracting the background constituent P_{bgr}. The failure to subtract all of the background signal may significantly increase the calculated value of the signal $P(r)$ and, accordingly, artificially increase the estimated signal-to noise ratio. Generally, this results in a systematic shift in the retrieved extinction coefficient that is especially noticeable at the far end of the measurement range. The second problem is that both P_{bgr} and $P(r)$ are subject to statistical fluctuations caused by noise. If at long distances from the lidar, the subtracted background constituent becomes greater than $P_\Sigma(r)$, then the estimated backscatter signal $P(r)$ may have nonphysical negative values. All the above observations result in certain restrictions on the lidar measurement range and measurement accuracy. The accuracy at distant ranges cannot be significantly improved by the increase of the number of shots that are averaged. This is because variations in the remaining shot-to-shot shift mostly have both random and systematic components.

The signal offsets remaining after the background subtraction are generally small and are mostly ignored in measurement uncertainty estimates. Meanwhile, lidar signals measured in clear atmospheres can only be inverted accurately if the systematic signal distortions are excluded or compensated. To give to the reader some feelings how such an apparently insignificant offset can distort profiles of the derived extinction coefficient, we present in Figs. 6.12 and 6.13 simulated inversion results obtained for a clear homogeneous atmosphere with the particulate extinction coefficient $\kappa_p = 0.01\,\text{km}^{-1}$. Here it is assumed that the lidar operates at 532 nm, the extinction coefficient profile is retrieved over the range from $r_{min} = 500$ m to $r_{max} = 5000$ m, the maximal signal at the range 500 m is approximately 4000 bins, and the actual background offset is 200 bins. The inversions of the simulated signal are made with both the near-end and the far-end solution, i.e., by using the forward and backward inversion algorithms. In these simulations it is assumed that no signal noise exists and the boundary values for the solutions are precisely known, so that the retrieved extinction-coefficient profile distortion occurs only due to a small offset 2 bins remaining after background subtraction. As compared with the maximum value of the lidar signal (~4000 bins), the offset, 2 bins, seems to be insignificant (~0.05%). However, in clear atmospheres even such a small shift can yield large measurement errors. In Fig. 6.12 the inversion results are shown when the offset is equal −2 bins, i.e., the signal used for the inversion is less that the actual one. The dependencies for the offset equal to +2 bins are shown in Fig. 6.13. One can see that in such clear atmospheres, the measurement error becomes significant, for both the far and near end solutions. However, in the near zone (500 m–3000 m), the near-end solution provides a more accurate inversion result than that by the far-end solution. Particularly the near-end

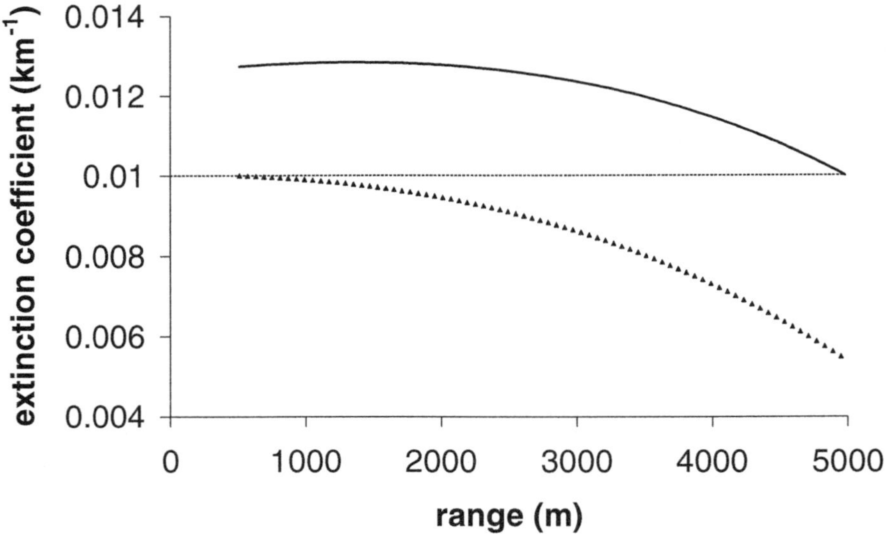

Fig. 6.12. Simulated inversion results obtained for a clear homogeneous atmosphere with the particulate extinction coefficient, $\kappa_p = 0.01\ \text{km}^{-1}$ (dotted line). The inversion results, obtained with the far and near-end solutions are shown as a bold curve and that with black triangles, respectively. The zero-line offset is −2 bins.

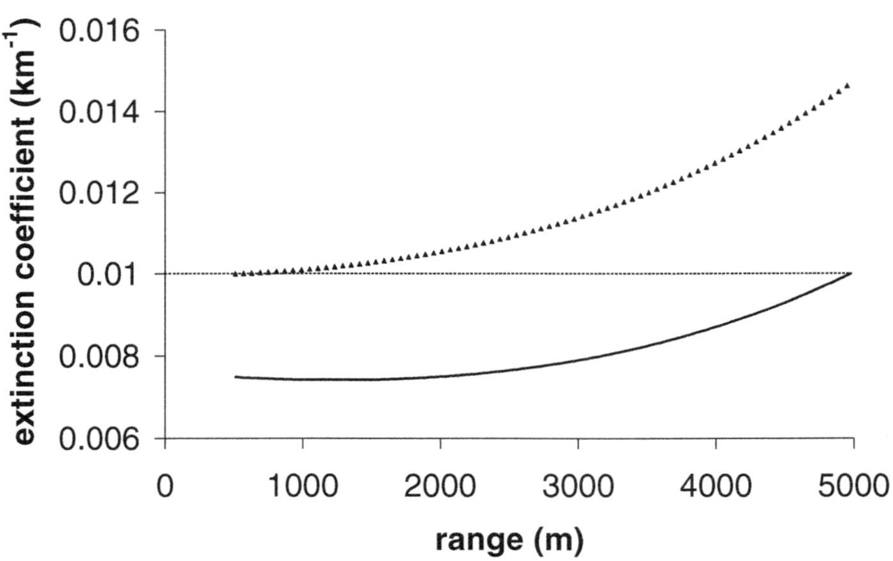

Fig. 6.13. Same as in Fig. 6.12, except that the zero-line offset is +2 bins.

solution results in systematic shifts in the derived κ_p of less than 14%, whereas the far-end solution yields profiles where systematic shifts over this zone range from 21 to 28%. Note also that in the near-end solution, the zones of minimum systematic and minimum random errors coincide, so that for real signals with a zero-line offset, this solution may often be preferable as compared to the stable far-end solution.

Thus, a zero-line offset remaining after the subtraction of an inaccurately determined value of the signal background component may cause significant distortions in the derived extinction-coefficient profiles. A similar effect can be caused by a far-end incomplete overlap due to poor adjustment of the lidar-system optics. These systematic distortions of lidar signals can dramatically increase errors in the measured extinction coefficient profile, especially when measured in clear atmospheres. In such atmospheres the near end solution may often be more accurate than the far-end solution, at least, over the ranges adjacent to the near incomplete-overlap zone, where the relative weight of the lidar-signal systematic offset is small and does not distort significantly the inversion result. On the other hand, the far-end solution can yield strongly shifted extinction coefficient profiles. This is due to the fact that the boundary value is estimated at distant ranges where the relative weight of even a small systematic offset is large.

The accuracy of extinction coefficient measurements may be significantly influenced by minor instrument defects that often seem negligible.

The return from a single laser pulse is usually too weak to be accurately processed. Any atmospheric parameter calculated from a single shot is noisy. Theoretically, the greatest sensitivity is achieved when the lidar minimum detectable energy is limited only by the quantum fluctuations of the signal itself (the signal shot noise limit) (Measures, 1983). However, lidar operations are often influenced by strong daylight background illumination. This is because most lidars operate at wavelengths within the spectral range of the solar spectrum. The background may be so great that it may even saturate the detector. Usually, the researcher is faced with an intermediate situation and is forced to take this problem as inevitable.

To make an accurate quantitative measurement, any remote-sensing technique must distinguish between signal variations due to changes in the parameter of interest and changes due to signal noise. Temporal averaging may be a simple and effective way to improve the signal-to-noise ratio. It follows from the general uncertainty theory that the measurement uncertainty of the averaged quantity is proportional to $N^{-1/2}$ when N independent measurements are made (Bevington and Robinson, 1992). However, this is only true when the errors are independent and randomly distributed. If this condition is met for the lidar signals, the measurement error may be reduced significantly by increasing the number of averaged shots and processing the mean rather than a single signal. The first lidar measurements revealed, however, that strong departures from $N^{-1/2}$ may be observed for lidar returns from turbid atmospheres. Experimental studies have shown that in the lower troposphere, departures from $N^{-1/2}$ are actually quite common. The studies included measurements of lidar signals from topographic and diffusely reflecting targets (Killinger and Menyuk, 1981; Menyuk and Killinger, 1983; Menyuk et al., 1985) and the signal backscattered from the atmosphere (Durieux and Fiorani, 1998). The authors explained this effect by the temporal correlation of the successive lidar signals. According to the general theory, the result of smoothing is worse than $N^{-1/2}$ when a positive correlation exists between the data points. On the other hand, for a negative correlation between points, the effect of smoothing will be better than $N^{-1/2}$. The common point among the authors cited above is that the temporal autocorrelation is a direct consequence of the fact that the atmospheric transmission varies during the time it takes to make the measurement. As shown by Elbaum and Diament (1976), for a photon-counting system, the standard deviation of p backscattered photons detected during the response time of the detector is

$$\Delta_s p = \left[\left(\frac{\eta_e \lambda}{h_p c}\Delta_s W\right)^2 + p + p_{dgr} + p_{dc}\right]^{1/2} \tag{6.34}$$

where η_e is the quantum efficiency of the detector, λ is the wavelength, c is the velocity of light, and h_p is Plank's constant. The term $\Delta_s W$ defines the standard deviation of the backscatter energy that reaches the detector during the response time. The value of $\Delta_s W$ includes fluctuations caused by atmospheric turbulence. The values of p, p_{bgr}, and p_{dc} are the numbers of photons detected during the response time and originate from the backscattered signal, the sky background, and the dark current photons, respectively. It is assumed that these contributions to the noise may be regarded as random, independent, and distributed according to Poisson statistics.

Departures from $N^{-1/2}$, observed in the lower troposphere, may severely limit the amount of improvement achievable through signal averaging. On the other hand, Grant et al. (1988) have shown experimentally that backscattered returns can be averaged with an $N^{-1/2}$ reduction in the standard deviation for N in the range, at least, of several hundred to a thousand. According to this study, deviations from $N^{-1/2}$ behavior are due to the influence of the background noise constituent, changes in the atmospheric differential backscatter, and/or the absorption of the lidar signals. A similar conclusion about the absence of significant temporal correlation in experimental lidar data was made in a study by Milton and Woods (1987). The validity of the $N^{-1/2}$ law, at least when processing the lidar data with acceptable signal-to-noise ratios, seemed to be confirmed. However later, new investigations were made that again confronted the validity of the $N^{-1/2}$ law. At the Swiss Federal Institute of Technology, Durieux and Fiorani (1998) carried out the measurement of the signal noise with a shot-per-shot lidar. The authors revealed significant discrepancies between the experimental results and the estimates based on a simple $N^{-1/2}$ dependence. The ratio of the standard deviation $\Delta_s N$ to $N^{-1/2}$ was much higher than unity, the value expected according to the $N^{-1/2}$ law. The authors concluded that atmospheric turbulence was responsible for the fluctuations observed, so that the optimal averaging level depends significantly on the particular atmospheric conditions. Such controversial results require additional studies. It appears that both positions have good grounds. The proposal made by Durieux and Fiorani (1998) that the noise behavior should be estimated with atmospheric turbulence taken into account seems reasonable. Unfortunately, the question arises as to how corrections to the $N^{-1/2}$ law can be made in a practical sense to determine the actual limits for optimal averaging. Because the application of shot averaging remains the most practical option to increase the signal-to-noise ratio, the amount of averaging should be limited to shorter periods, especially if the particulate loading is changing rapidly in the area of interest (Grant et al., 1988). With measurements made in the lower troposphere, one must be cautious when estimating the uncertainty of lidar measurements with long-period averages.

It is necessary to distinguish between the operating range and the measurement range of the lidar. Generally, the lidar maximum operating range is defined as the range where the decreasing lidar signal $P(r)$ becomes equal to the standard deviation of noise constituent. For practical convenience, systematic offset is generally ignored, so that the maximum operating range is related only to the signal-to-noise ratio. With real lidar measurements, the actual measurement range may be significantly less than the lidar operating range. This is because the general definition of measurement range is related to the measurement accuracy of the retrieved quantity of interest rather than the accuracy of the lidar signal. In particular, the measurement range is an area over which a quantity of interest is measured with some acceptable accuracy. Meanwhile, as shown above, the accuracy of the measured lidar signal worsens with increase in the range. Accordingly, the accuracy of any atmospheric parameter obtained by lidar signal inversion (such as the extinction or the absorption coefficient) will also become worse as the range increases. Thus, at distant ranges, the measurement uncertainty of the retrieved quantity may be unacceptable. In lidar measurements, it is quite common that the range over which the atmospheric parameter of interest can be measured is significantly less than the lidar operating range, where the signal-to-noise ratio exceeds unity.

Finally, the uncertainty in the molecular scattering profile should be mentioned. In two-component atmospheres, knowledge of the real profile of the atmospheric molecular density is required to differentiate between the particulate and molecular contributions. The molecular density can be retrieved either from balloon measurements or from models of the local atmosphere. In both cases, the measurement uncertainty in aerosol loading will be influenced by accuracy of the molecular profile used in lidar data processing. This uncertainty may significantly distort the retrieved particulate extinction coefficient profile, especially in an atmosphere in which the particulate contribution is relatively small, so that the ratio a/R is large. The uncertainty in the molecular extinction coefficient at the boundary point may significantly worsen the accuracy of the boundary value $\kappa_W(r_b)$ in the boundary point solution. The requirements for the accuracy of the molecular density profiles are surprisingly exacting. According to a study by Kent and Hansen (1998), when the molecular density at the assumed aerosol-free altitude is known to an accuracy of 1–2%, a potential 20–40% error in the particulate extinction-coefficient profile can be expected. When the molecular density is obtained from the average of several density profiles, the standard deviation of the density profile must be considered as an additional component of the uncertainty in the derived particulate extinction coefficient profile (Del Guasta, 1998).

7

BACKSCATTER-TO-EXTINCTION RATIO

7.1. EXPLORATION OF THE BACKSCATTER-TO-EXTINCTION RATIOS: BRIEF REVIEW

The problem of selecting an appropriate backscatter-to-extinction ratio for lidar data processing in different atmospheres has been widely discussed in the scientific literature. In this section we present a brief overview of investigations in this area, concerning only the characteristics for spherical particles. The relationship between backscatter and extinction for nonspherical particulates, such as ice particles or mixed-phase clouds, is beyond the scope of this consideration. The reader is directed to more specialized studies, such as Van de Hulst (1957) or Bohren and Huffman (1983), where these questions are addressed in detail.

As shown in previous chapters, an analytical solution of the elastic lidar equation requires knowledge of the backscatter-to-extinction ratios along the line of sight examined by the lidar. Meanwhile, the particulate backscatter-to-extinction ratio depends on many factors, such as the laser wavelength, the aerosol particle chemical composition, particulate size distribution, and the atmospheric index of refraction (see Chapter 2). Because of the large variability of actual aerosols or particulates in the atmosphere, it is generally difficult to establish credible backscatter-to-extinction ratios for use in specific measurement conditions.

Elastic Lidar: Theory, Practice, and Analysis Methods, by Vladimir A. Kovalev and William E. Eichinger.
ISBN 0-471-20171-5

The selection of a relevant value of the backscatter-to-extinction ratio for a particular atmospheric situation is a painful problem for practical elastic lidar measurements. The real atmosphere is always filled with polydisperse scatterers of different sizes, origins, and compositions, so that the particulate backscatter-to-extinction ratio varies, at least slightly, along any examined path. Scatterers of different size have differently shaped phase functions (see Chapter 2). The scattering of the ensemble of particulates is the sum of the scattering due to all of the scatterers in the examined volume. Therefore the total amount of atmospheric backscattering and, accordingly, the backscatter-to-extinction ratios represent integrated parameters that vary considerably less than those of the individual particles found in the examined volume. This is why the particulate backscatter-to-extinction ratios measured in the atmosphere mostly vary over a factor of only 10 to 20, whereas the measured total scattering or backscattering coefficients may vary by factors of $\sim 10^4$ to 10^6, and even more.

To achieve the most accurate inversion of the measured lidar signal, the range variations of the backscatter-to-extinction ratio along the examined atmospheric path should be considered. As discussed in Chapter 11, the most practical way to obtain such information is a combination of elastic and inelastic lidar measurements along the same line of sight. The combination of the elastic and Raman techniques may noticeably improve the measured data quality (Ansmann et al., 1992; Reichardt et al., 1996; Donovan and Carswell, 1997). However, there are many difficulties in the practical application of such combined techniques. When such a combination is not available, the most common way of the lidar signal inversion is to select a priori some constant value for the backscatter-to-extinction ratio. Such a selection may be based on information about the ratios for the aerosols found in the literature for similar optical situations.

Numerous experimental investigations have shown that large variations in the backscatter-to-extinction ratio occur in both time and space. For mixed-layer aerosols, this value may vary, approximately, from $0.01\,sr^{-1}$ to $0.11\,sr^{-1}$ and may even be as large as $0.2\,sr^{-1}$ (Reagan et al., 1988; Sasano and Browell, 1989). On the other hand, backscatter-to-extinction ratios may often be considered to be constant in unmixed atmospheres, for example, in some clear atmospheres or in water clouds. It has been established, for example, that the ratio is nearly the same in water clouds, at least for wavelengths up to $1\,\mu m$. This follows from both experimental and theoretical studies (Sassen and Lou, 1979; Pinnick et al., 1983; Dubinsky et al., 1985; Del Guasta et al., 1993). Theoretical studies have also revealed that the backscatter-to-extinction ratio may remain almost constant in cloud layers even when the particle density and size distribution are varied (Carrier et al., 1996; Derr, 1980).

It has been found in most studies, for example, by Pinnick et al. (1980), Dubinsky et al. (1985), and Parameswaran et al. (1991), that values for backscatter-to-extinction ratio less than $0.05\,sr^{-1}$ are the most common in the atmosphere. Such values correspond to scattering from particles whose size is larger than or close to the wavelength of the scattered light, a condition also common with stratospheric aerosols. Reagan et al. (1988) investigated the backscatter-to-extinction ratio by slant-path lidar observations at a wavelength of 694 nm. These observations yielded values of the ratio from 0.01 to 0.2 sr^{-1}, with the majority of the data in the range from approximately 0.02 to 0.1 sr^{-1}. In fact, this range of values could be obtained from any of the commonly assumed size distributions and refractive indices. The authors pointed out that large values of the backscatter-to-extinction ratio (0.05–0.1 sr^{-1}) corresponded to scattering from particles with large real refractive indices and with imaginary indices close to zero. The corresponding size distributions contained significant coarse-mode concentrations. For particles with small real indices and larger imaginary components, the backscatter-to-extinction ratios had lower values (~0.02 sr^{-1} and less).

It is, unfortunately, not possible to establish a general dependence of the backscatter-to-extinction ratio with particular aerosol types in a way that could be practical in real atmospheres. Numerous studies, both theoretical and experimental, show that the backscatter-to-extinction ratio is related to many parameters. In 1967, Carrier et al. made theoretical computations of backscatter-to-extinction ratios for the wavelengths 488 and 1060 nm, varying the density and size distribution of the particles. The backscatter-to-extinction ratios obtained ranged between 0.0625 and 0.045 sr^{-1}, respectively. In the theoretical computations of Derr (1980), the backscatter-to-extinction ratio was determined for a set of different water clouds types for two wavelengths, 275 and 1060 nm. The mean ratios were 0.061 and 0.056 sr^{-1}, respectively, with a variance of 15%. In the experimental studies of Sassen and Liou (1979) and Pinnick et al. (1983), the relationship between extinction and backscattering was investigated at 632 nm. In the former study the established values of the backscatter-to-extinction ratios were 0.033–0.05 sr^{-1}, and in the latter the mean value was 0.0565 sr^{-1}. In a study by Dubinsky et al. (1985), a linear relationship was established between the cloud extinction coefficient and the backscatter coefficient at a wavelength of 514 nm. However, the backscatter-to-extinction ratio for different clouds varied from 0.02 to 0.05 sr^{-1}, depending on the droplet size distribution. Spinhirne et al. (1980) made lidar measurements at a wavelength of 694.3 nm within the lower mixed layer of the atmosphere and found that the backscatter-to-extinction ratio varied generally in a range near 0.05 sr^{-1}. However, the standard deviation was large (0.021 sr^{-1}). In the aerosol corrections to the DIAL measurements made at 286 and 300 nm, Browell et al. (1985) used different values of the backscatter-to-extinction ratio for urban, rural, and maritime aerosols. These values were 0.01 sr^{-1} for urban aerosols, 0.028 sr^{-1} for rural continental aerosols, and 0.05 sr^{-1} for maritime aerosols.

Relative humidity plays an important role in particulate properties and thus in the backscatter-to-extinction ratio. In response to changes in relative humidity, particulates absorb or release water. During this process, their physical and chemical properties change, including their size and index of refraction. In turn, these changes can significantly influence the optical parameters of the particulates, such as scattering, backscattering and absorption. The chemical composition of the particulates, especially close to urban areas, may vary significantly in space and time. Although the aerosol chemical composition varies in a wide range, inorganic salts and acidic forms of sulfate may compose a substantial fraction of the aerosol mass. Because these species are water soluble, they are commonly found in atmospheric aerosols. On the other hand, hydrophilic organic carbon compounds should also be considered to be a significant component of atmospheric aerosols. For example, investigations made at some tens of sites throughout the United States revealed that organic carbon compounds may contribute up to 60% of the fine aerosol mass (Sisler, 1996). Atmospheric aerosols can be composed of different mixtures of organic and inorganic compounds, and therefore the particulate scattering characteristics may be quite different. This is the major factor that explains why experimental studies often reveal such different values of the backscatter-to-extinction ratio under similar atmospheric conditions.

Takamura and Sasano (1987) examined wavelength and relative humidity dependence on the backscatter-to-extinction ratio at four wavelengths with the Mie scattering theory. Their analysis showed that for the shortest wavelength, 355 nm, the ratios increase with relative humidity within the range ~0.01–0.02 sr^{-1}, whereas the ratios show a weak dependence on humidity for wavelengths between 532 and 1064 nm. In this wavelength range, the backscatter-to-extinction ratio ranged from ~0.01 to 0.025 sr^{-1}. The difference in the backscatter-to-extinction ratios between the wavelengths is reduced under high humidity. In a study by Leeuw et al. (1986), the variations of the backscatter-to-extinction ratio with relative humidity were analyzed with lidar experimental data and Mie calculations. The database contained nearly 500 validated lidar measurements over a near-horizontal path made at the wavelengths 694 and 1064 nm over a 2-year period. In these studies, no distinct statistical relationship was observed between the backscatter-to-extinction ratio and humidity. The experimental plots presented by the authors showed an extremely large range of the ratio variations, which varied, approximately, more than one order of magnitude. Anderson et al. (2000) obtained similar large variations using a 180° backscatter nephelometer. However in the study by Chazette (2003) the dependence of the backscatter-to-extinction ratio on humidity does not have such large variations; it decreases slightly within the range, from 0.02 sr^{-1} to approximately 0.12–0.15 sr^{-1} when the relative humidity increase from 55 to 95%.

In the experimental study by Day et al. (2000), scattering from the same particulate types was investigated under different relative humidities. The measurements were made with an integrating nephelometer at a wavelength of 530 nm. The range of the relative humidity was changed from 5% to 95% when sampled aerosol passed an array of drying tubes that allowed control of sample relative humidity and temperature. The ratio of the scattering coefficients of "wet" particulates at relative humidities from 20% to 95% to the scattering coefficients for the "dry" aerosol was calculated. The latter was defined as an aerosol with a relative humidity less than 15%. The authors established that the scattering ratio smoothly and continuously increased as the "wet" sampling air humidity increased and vice versa. Results of the study did not reveal any discontinuities in the ratio, so the authors concluded that the particulates were never completely dried, even when humidity decreased below 10%.

Extensive in situ ground surface measurements and a detailed data analysis were made by Anderson et al. (2000). In this study, the experimental investigations were made with an integrating nephelometer at 450 and 550 nm and a backscattering nephelometer at 532 nm, described in the study by Doherty et al. (1999). Nearly continuous measurements were made in 1999 over 4 weeks in central Illinois. In addition, the data were analyzed obtained with the same instrumentation at a coastal station in 1998. Some relationships were found between the backscatter-to-extinction ratio and humidity; however, this explained only a small portion of the variations of the ratio. The authors concluded that most of the variations were associated with changes between two dominant air mass types, which were defined as rapid transfer from the northwest and regional stagnation. For the former, the backscatter-to-extinction ratios were mostly higher than ~0.02 sr^{-1}, whereas for the latter, the values were generally smaller. Averages for these situations were 0.025 and 0.0156 sr^{-1}, respectively. The authors also presented a plot of the extinction-to backscatter ratio versus extinction coefficient. In fact, no correlation was found between these values for clear atmospheres. The backscatter-to-extinction ratios varied chaotically over the range from ~0.01 to 0.1 sr^{-1}. The authors did not comment such large scattering in clear atmospheres. It is not clear whether these variations are real or due to instrumental noise, which may significantly worsen the signal-to-noise ratio, especially when measuring weak scattering and backscattering in clear atmospheres. The data presented show also that high-pollution events have, generally, a much narrower range of variations in the ratio compared with clear atmospheres. Moreover, the range of the variations in polluted atmospheres proved to be the same for both the coastal station and central Illinois. The authors concluded that the extinction levels may provide approximate predictions of the expected backscatter-to-extinction ratios, but only within a pollution source region rather than outside it, so that no general relationship between extinction and backscattering can be expected.

Evans (1988) made measurements of the aerosol size distribution simultaneously with an experimental determination of the backscatter-to-extinction ratio at visible wavelengths and at 694 nm. He established that the backscatter-to-extinction ratio varied from 0.02 to 0.08 sr^{-1}, but 67% of these values fell in the narrow range from 0.05 to 0.06 sr^{-1}. Ansmann et al. (1992a) measured the backscatter-to-extinction ratio for the lower troposphere over northern Germany using a Raman lidar at 308 nm. The average value of the backscatter-to-extinction ratio in cloudless atmosphere at the altitude range 1.3–3 km was 0.03 sr^{-1}. In a study by Del Guasta et al. (1993), the statistics are given for 1 year of ground-based lidar measurements. The measurements of tropospheric clouds were made in the coastal Antarctic at a wavelength of 532 nm. The data on the extinction, optical depth, and backscatter-to-extinction ratio of the clouds revealed an extremely wide data dispersion, which might reflect changes in the macrophysical and optical parameters of the clouds. In a study by Takamura et al. (1994), tropospheric aerosols were simultaneously observed with a multiangle lidar and a sun photometer. The comparison between the optical depth obtained from the lidar and sun photometer data made it possible to estimate a mean columnar value of backscatter-to-extinction ratios. These values were in a range from 0.014 to 0.05 sr^{-1}. Daily means of the backscatter-to-extinction ratios for the measurements carried out over the Aegean Sea in June 1996 were close to 0.051 sr^{-1} (Marenco et al., 1997). Aerosol backscatter-to-extinction profiles at 351 nm over a lower troposphere, at altitudes up to 4.5 km, were measured in the study by Ferrare et al. (1998). The values varied in a wide range between 0.012 and 0.05 sr^{-1}. Doherty et al. (1999) made measurements of atmospheric backscattering of continental and marine aerosol and determined the backscatter-to-extinction ratio at wavelength 532 nm. For these measurements, a backscatter nephelometer was used in which the light was measured scattered over the angular range from 176° to 178°. This study confirmed that the coarse-mode marine air has much higher values for the backscatter-to-extinction ratio than fine-mode-dominated continental air, what is consistent with Mie theory. For marine aerosols, the mean backscatter-to-extinction ratio was established to be 0.047 sr^{-1}, whereas for continental air it was, approximately, in the range from 0.015 to 0.017 sr^{-1}. For the former, the backscatter-to-extinction ratio remained relatively constant. The variability of the ratio was less than 20%, which the authors explained by instrumental noise rather than by actual variation of the backscatter-to-extinction ratios.

Table 7.1 presents a summary of backscatter-to-extinction ratios for different atmospheric and measurement conditions based on both theoretical and experimental studies. A brief review of studies of the backscatter-to-extinction ratios for tropospheric aerosols is presented also in the study by Anderson et al. (1999).

Even this short review shows that the principal question concerning the determination or estimation of the backscatter-to-extinction-ratio to be used in the lidar data inversion is unsolved. The most common approach used to invert elastic lidar signals is based on the use of a constant, range-independent backscatter-to-extinction ratio. This assumption is often made because it is the simplest way to invert the lidar equation and because there is little basis on which to predict how the ratio might vary along a given line of sight. The use of a constant backscatter-to-extinction ratio significantly simplifies the computations, especially if the measurement is made in a single-component atmosphere. As shown in Chapter 5, it is not necessary to establish a numerical value for the backscatter-to-extinction ratio for measurements in a single-

TABLE 7.1. Backscatter-to-Extinction Ratios in the Real Atmospheres

Aerosol Type	Value, sr^{-1}	Wavelength, nm	Source
Arizona ABL	0.051	694	Spinherne et al., 1980
Water droplet clouds	0.02–0.05	514	Dubinsky et al., 1985
Maritime (Mie calculations)	0.015	355	Takamura and Sasano, 1987
	0.017	532	
	0.019	694	
	0.024	1064	
Continental	0.028	300	Sasano and Browell, 1989
Maritime	0.052–0.02	300	
Saharan dust	0.017–0.02	300	
	0.017–0.066	600	
Rain forest	0.029	300	
	0.017–0.023	600	
Lower troposphere	0.05–0.06	Visible, 694	Evans, 1988
Arizona, ABL	0.022–0.1	694	Reagan et al., 1988
Lower troposphere	0.015–0.03	532	Takamura and Sasano, 1990
Lower troposphere	0.03	308	Ansmann et al. (1992a)
Tsukuba (Japan) Troposphere	0.014–0.05	532	Takamura et al., 1994
Maritime	0.04–0.05	355	Marenco et al., 1997
SW ABL	0.024	490	Rosen et al., 1997
Lower troposphere	0.013–0.033	351	Ferrare et al., 1998
Maritime	0.02–0.04	1064	Ackerman, 1998
Desert	0.021–0.024	355	
Desert	0.04–0.059	532–1064	
Marine	0.047	532	Doherty et al., 1999
Continental	0.015–0.017		

component atmosphere. Such a situation is often met, for example, in turbid atmospheres where particulates dominate the scattering process and molecular scattering can be ignored. In this case, the determination of the extinction coefficient requires only a knowledge of the relative behavior of the backscatter-to-extinction ratio along the examined path rather than its numerical value. In relatively clean and moderately turbid atmospheres, which are considered to be two-component atmospheres, the inversion procedure requires knowledge of the numerical value of the backscatter-to-extinction ratio.

Unlike a single-component atmosphere, the extraction of the particulate extinction coefficient in a two-component atmosphere cannot be made without selection of a particular numerical value for the particulate backscatter-to-extinction ratio.

7.2. INFLUENCE OF UNCERTAINTY IN THE BACKSCATTER-TO-EXTINCTION RATIO ON THE INVERSION RESULT

In Chapter 6, the amount of distortion in the derived extinction coefficient profile that occurs because of an incorrect selection of the boundary value for the lidar equation was analyzed. The analysis was made with an assumption that the particulate backscatter-to-extinction ratio is known accurately. However, the backscatter-to-extinction ratio is usually known either poorly or not at all. Its value is generally chosen a priori; therefore, it may significantly differ from the actual value. As a result, an additional error may occur in the extracted extinction coefficient profile. The uncertainty due to an inaccurate selection of the backscatter-to-extinction ratio depends on how the boundary conditions are determinated. The question of interest is whether the accuracy of the retrieved extinction coefficient may be improved by using some optimal lidar solution, particularly if independent measurement data are available. The problem is quite real for slant-angle measurements, especially when these are made in directions close to vertical (Ferrare et al., 1998). In this case, the selection of an appropriate backscatter-to-extinction ratio is difficult because of atmospheric vertical heterogeneity. On the other hand, vertical and near-vertical lines of sight are most advantageous when high-altitude atmospheric aerosols and gases are to be remotely investigated.

In this section, estimates of uncertainty are presented for the two basic methods of extinction coefficient retrieval, the boundary point and optical depth solutions. Unfortunately, such estimates are quite difficult, because none of the simple models is universally true. The error in the selected backscatter-to-extinction ratio, Π_p, may include a large systematic component of unknown sign. The difference between the actual Π_p and that taken a priori to invert measured lidar signals may be as large as 100% and even more. Meanwhile, as mentioned in Section 6.1, the conventional theoretical basis for the error estimate assumes that the error constituents are small, so that only the first term of a Taylor series expansion is necessary for error propagation. When the errors may be large, this approach is not applicable. An extremely large systematic uncertainty may be implemented in the assumed Π_p, forcing the use of a more sophisticated method of error analysis in this section.

As shown in Chapter 5, to obtain a lidar equation solution for a two-component atmosphere, the measured signal and its integrated profile must be transformed with an auxiliary function $Y(r)$ [Eq. (5.67)]. It was shown in Chapter 6 that three steps in the calculation of the extinction coefficient profile must be made and that different errors are introduced at the different steps. These three-step transformations impede the analysis of the uncertainty due to an incorrect selection of the particulate backscatter-to-extinction ratio. The general method used here is as follows. If the assumed aerosol backscatter-to-extinction ratio $[\Pi_p(r)]_{as}$ is inaccurate, then an incorrect ratio

$$a_{as}(r) = \frac{3/8\pi}{[\Pi_p(r)]_{as}} \tag{7.1}$$

is used for the calculation of the auxiliary function $Y(r)$ in Eq. (5.67). This distorted function is determined as

$$\langle Y(r)\rangle = C' a_{as}(r) \exp\left\{-2\int_{r_0}^{r} [a_{as}(r') - 1]\, \kappa_m(r')\, dr'\right\} \tag{7.2}$$

If no molecular absorption occurs, $\kappa_m(r) = \beta_m(r)$ and $C' = C_Y\, 8\pi/3$. The incorrect function $\langle Y(r)\rangle$ is then used for transformation of the original lidar signal into the function $Z(r)$ with Eq. (5.28). With the incorrect transformation function, a distorted function $\langle Z(r)\rangle$ is obtained with the formula

$$\langle Z(r)\rangle = P(r)\langle Y(r)\rangle r^2 \tag{7.3}$$

When the inversion procedure is applied to this distorted function $\langle Z(r)\rangle$, a distorted value of the weighted extinction coefficient $\kappa_W(r)$ is obtained. Using single algebraic transformation, one can present Eq. (7.3) as

$$\langle Z(r)\rangle = C'' D(r)[\kappa_W(r)]_{est} \exp\left\{-2\int_{r_0}^{r} [\kappa_W(r')]_{est}\, dr'\right\} \tag{7.4}$$

Here C'' is an arbitrary constant and $[\kappa_W(r)]_{est}$ is the weighted extinction coefficient estimated with the assumed ratio $a_{as}(r)$. With Eq. (5.30), the extinction coefficient can be presented in the form

$$[\kappa_W(r)]_{est} = \kappa_m(r)[a_{as}(r) + R(r)] \tag{7.5}$$

where $R(r)$ is the ratio of the particulate-to-molecular extinction coefficient. The function $D(r)$ in Eq. (7.4) may be considered as a range-dependent distortion factor defined as

$$D(r) = \frac{1 + \dfrac{R(r)}{a(r)}}{1 + \dfrac{R(r)}{a(r)} \dfrac{[\Pi_p(r)]_{as}}{\Pi_p(r)}} \tag{7.6}$$

If a point r_b exists in which the particular and molecular extinction coefficients are known, the boundary point solution can be used to find the weighted extinction coefficient. However, if an incorrect selection of the particulate backscatter-to-extinction ratio is made, an error is also introduced into this boundary value, even if both molecular and particulate extinction coefficients at r_b are known precisely. This is because of the use of the incorrect ratio $a_{as}(r_b)$ instead of a correct $a(r_b)$. The estimated boundary value of the weighted extinction coefficient can be written as

$$[\kappa_W(r_b)]_{est} = \kappa_m(r_b)[a_{as}(r_b) + R(r_b)] \tag{7.7}$$

When the distorted function $\langle Z(r)\rangle$ and the inaccurate boundary value $[\kappa_W(r_b)]_{est}$ are substituted into the lidar equation solution [Eq. (5.75)], the distorted profile $\langle\kappa_W(r)\rangle$ is obtained. With Eqs. (5.75) and (7.4), the ratio of the function extracted from $\langle Z(r)\rangle$ to $[\kappa_W(r)]_{est}$ defined in Eq. (7.5) can be written in the form

$$\frac{\langle\kappa_W(r)\rangle}{[\kappa_W(r)]_{est}} = \frac{D(r)V_c^2(r_b, r)}{D(r_b) - 2\int_{r_b}^{r} D(r')[\kappa_W(r')]_{est} V_c^2(r_b, r')\,dr'} \tag{7.8}$$

where function $V_c^2(r_b, r)$ defines the two-way transmittance for $[\kappa_W(r)]_{est}$,

$$V_c^2(r_b, r) = \exp\left[-2\int_{r_b}^{r} [\kappa_W(r')]_{est}\,dr'\right] \tag{7.9}$$

The relative uncertainty of the retrieved particulate extinction coefficient can be determined via the ratio in Eq. (7.8) as

$$\delta\kappa_p(r) = \left(1 + \frac{a_{as}(r)}{R(r)}\right)\left[\frac{\langle\kappa_W(r)\rangle}{[\kappa_W(r)]_{est}} - 1\right] \tag{7.10}$$

As follows from Eq. (7.8), the ratio of $\langle\kappa_W(r)\rangle$ to $[\kappa_W(r)]_{est}$ is equal to unity if the distortion factor $D(r) = D = const.$ in the range from r_b to r. Under this condition, the uncertainty in the calculated particulate extinction coefficient is equal to zero. In other words, the retrieved extinction coefficient does not depend on the assumed backscatter-to-extinction ratio if the two ratios, $[\Pi_p(r)_{as}]/\Pi_p(r)$ and $R(r)/a(r)$ in Eq. (7.6), are range independent. Unfortunately, in the lower troposphere, large changes in the aerosol extinction coefficient generally occur (McCartney, 1977; Zuev and Krekov, 1986; Sasano, 1996, Ferrare et al., 1998), so the actual factor, $D(r)$, is not constant. Therefore, the measurement uncertainty caused by an incorrectly chosen $\Pi_p(r)$ may increase from point r_b, where the boundary condition is specified, in both directions. This, in turn, means that even the far-end solution may yield large errors in the particulate extinction coefficient.

With similar transformations with Eqs. (5.83) and (7.4), the optical depth solution can be obtained in the form

$$\frac{\langle \kappa_W(r) \rangle}{[\kappa_W(r)]_{est}} = \frac{D(r)V_c^2(r_0, r)}{\frac{2}{1 - V_c^2(r_0, r_{max})} \int_{r_0}^{r_{max}} D(r')[\kappa_W(r')]_{est} V_c^2(r_0, r')\,dr' - 2\int_{r_0}^{r} D(r')[\kappa_W(r')]_{est} V_c^2(r_0, r')\,dr'} \tag{7.11}$$

where the values $V_c^2(r_0, r)$ and $V_c^2(r_0, r_{max})$ are determined similarly to those in Eq. (5.80) but with integration ranges from r_0 to r and from r_0 to r_{max}, respectively. In the optical depth solution, the retrieved extinction coefficient also does not depend on assumed $[\Pi_p(r)]_{as}$ if the ratio of the assumed to the actual backscatter-to-extinction ratios and the ratio $R(r)/a(r)$ are constant over the measurement range. The conclusion is only true if an accurate boundary value $T^2(r_0, r_{max})$ is used.

The accuracy of a lidar signal inversion depends on whether $[\Pi_p(r)]_{as}$ is over- or underestimated. This can easily be shown by relating the uncertainties in $\Pi_p(r)$ and $a(r)$. Defining the assumed value of $a(r)$ as $a_{as}(r) = a(r) + \Delta a(r)$, where $\Delta a(r)$ is the absolute error in $a(r)$, the relative uncertainty of $a(r)$ can be determined as

$$\frac{\Delta a(r)}{a(r)} = \frac{-\Delta\Pi_p(r)}{\Pi_p(r) + \Delta\Pi_p(r)} \tag{7.12}$$

where $\Delta\Pi_p(r)$ is the absolute uncertainty of the assumed particulate backscatter-to-extinction ratio. As follows from Eq. (7.12), the uncertainty in the assumed ratio $a_{as}(r)$, which influences measurement accuracy [Eq. (7.10)], is not symmetric with respect to a positive or negative error in the backscatter-to-extinction ratio. Therefore, for both lidar equation solutions, different uncertainties occur in the measured extinction coefficient for an underestimated and an overestimated particulate backscatter-to-extinction ratio.

In a two-component atmosphere, the accuracy in the derived particulate extinction coefficient is generally worse when smaller (underestimated) values of the specified backscatter-to-extinction ratio are used.

For a single-component particulate atmosphere, in which ratio $R(r)/a(r) >> 1$, Eq. (7.6) reduces to

$$D(r) = \frac{\Pi_p(r)}{[\Pi_p(r)]_{as}}$$

In such an atmosphere, the uncertainty in the retrieved extinction coefficient does not depend on the profile of the particulate extinction coefficient when the ratio of the actual $\Pi_p(r)$ to the assumed $[\Pi_p(r)]_{as}$ is constant and, accordingly, $D(r) = D = const.$ In other words, in a single-component particulate atmosphere, knowledge of the relative change in the backscatter-to-extinction ratio rather than its absolute value is preferable to obtain an accurate inversion result (Kovalev et al., 1991). This observation confirms the advantage of the use of variable backscatter-to-extinction ratios for single-component atmospheres, at least in some specific situations. The sensitivity of lidar inversion algorithms to the accuracy of the assumed backscatter-to-extinction ratio has been analyzed in many studies (see Kovalev and Ignatenko, 1980; Sasano and Nakane, 1984; Klett, 1985; Sasano et al., 1985; Hudhes et al., 1985; Kovalev, 1995 among others.) It has been shown that the far-end solution generally reduces the influence of an inaccurately selected backscatter-to-extinction ratio (Sasano et al., 1985). However, this remains true only when there is no significant gradient in the particulate extinction coefficient along the lidar line of sight (Hudhes et al., 1985), especially when a two-component atmosphere is examined (Ansmann et al., 1992; Kovalev, 1995). Although the far-end solution usually yields a more accurate measurement result, this may be not true for clear areas containing large gradients in $\kappa_p(r)$. Here the derived extinction coefficient may not converge to the true value at the near end if an incorrect aerosol backscatter-to-extinction ratio is assumed. It may even result in unrealistic negative values for the particulate extinction coefficient close to lidar location. Note that this is true even for atmospheres where Π_p = const.

To illustrate this observation, in Figs. 7.1 and 7.2, two sets of retrieved extinction-coefficient profiles are shown, in which incorrect values of the backscatter-to-extinction ratio were used for the inversion. The initial model profiles of the particulate extinction coefficients used for the simulations are shown in both figures as *curve 1*. These profiles incorporate a mildly turbid layer at ranges from 1.3 to 1.7 km from the lidar. The synthetic lidar signals corresponding to these profiles were calculated with an "actual" backscatter-to-extinction ratio and then inverted with an "incorrect" (assumed) $[\Pi_p(r)]_{as}$. For simplicity, the "actual" backscatter-to-extinction ratio is taken to be range independent, having the same value of $\Pi_p = 0.03\,sr^{-1}$ for both turbid and clear areas. The molecular extinction coefficient is also constant over the range ($\kappa_m = 0.067\,km^{-1}$). It is also assumed that no other errors exist and that the correct boundary value of $\kappa_p(r_b)$ is known at the far end, $r_b = 2.5$ km. *Curves 2–5* in both figures are extracted from the synthetic signals by means of the far-end solution with incorrect backscatter-to-extinction ratios. It can be seen that the retrieved extinction coefficient does not depend on assumed backscatter-to-extinction ratios only for a restricted homogeneous area near the far end, where the boundary value is specified. For this area (1.7–2.5 km), the measurement error is equal to zero, although the assumed Π_p are specified incorrectly.

The explanation of such error behavior was given in Section 6.4. In a homogeneous turbid layer, all derived extinction coefficient profiles tend to converge to the true value when the range decreases, as is typical for the far-end

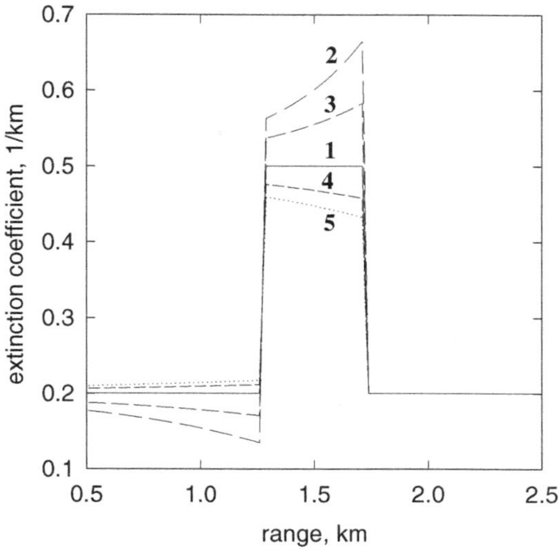

Fig. 7.1. Dependence of the retrieved $\kappa_p(r)$ profiles on assumed aerosol backscatter-to-extinction ratios. The model $\kappa_p(r)$ profile is shown as *curve 1*. *Curves 2–5* show the $\kappa_p(r)$ profiles retrieved with $\Pi_p = 0.015\,\text{sr}^{-1}$, $\Pi_p = 0.02\,\text{sr}^{-1}$, $\Pi_p = 0.04\,\text{sr}^{-1}$, and $\Pi_p = 0.05\,\text{sr}^{-1}$, respectively, whereas the model backscatter-to-extinction ratio is $\Pi_p = 0.03\,\text{sr}^{-1}$. The correct boundary value of $\kappa_p(r_b)$ is specified at $r_b = 2\,\text{km}$ (Kovalev, 1995).

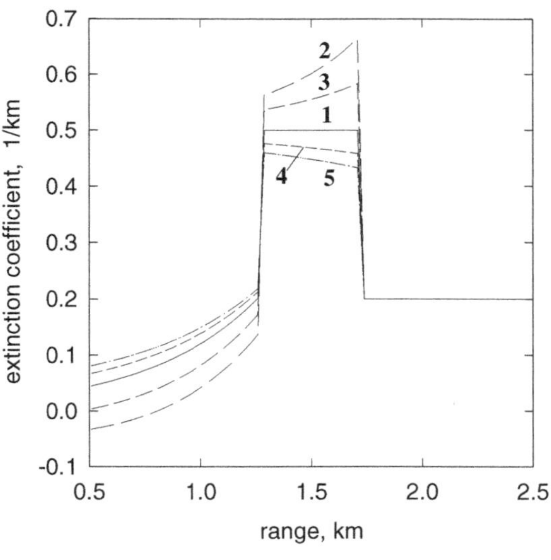

Fig. 7.2. Conditions are the same as in Fig. 7.1 except that the model $\kappa_p(r)$ profile changes monotonically at the near end, within the range from 0.5 to 1.3 km (Kovalev, 1995).

solution. The behavior of the retrieved extinction coefficient at the near end of the measurement range (0.5–1.3 km) is different for both figures. In Fig. 7.1, the particulate extinction coefficient has a tendency to converge into the true value over the homogeneous area, just as in the turbid area. This is not true for the retrieved extinction coefficient profiles shown in Fig. 7.2. The reason is that here the initial synthetic profile (*curve 1*) has a monotonic change in the extinction coefficient $\kappa_p(r)$ at the near end. This monotonic change results in a corresponding change of the ratio $R(r)/a(r)$ and, accordingly, in the factor $D(r)$ in Eq. (7.6). Despite the same retrieval conditions as in Fig. 7.1, the extracted extinction coefficients do not converge to the true value at the near end.

In two-component atmospheres, atmospheric heterogeneity is the dominant factor when estimating the measurement uncertainty caused by errors in the assumed backscatter-to-extinction ratio. A monotonic change in $\kappa_p(r)$ may result in large measurement errors even if the far-end solution is used with the correct boundary value.

Typical distortions of the derived $\kappa_p(h)$ altitude profiles, caused by incorrectly selected particulate backscatter-to-extinction ratios $[\Pi_p]_{as}$ are shown in the study by Kovalev (1995). The distortions are found for an atmosphere where $\kappa_p(h)$ changes monotonically with altitude (Fig. 7.3). The particulate extinction coefficient profile $\kappa_p(h)$ is taken from the study by Zuev and Krekov (1986, p. 145–157). This type of profile for a wavelength of 350 nm is typical for very clear atmospheres in which ground-level visibility is high, not less than

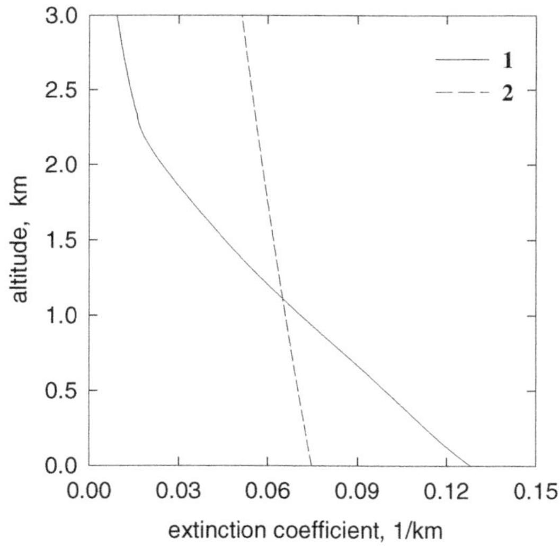

Fig. 7.3. $\kappa_p(h)$ and $\kappa_m(h)$ altitude profiles (*curves 1* and *2*, respectively) used for the numerical experiments shown in Figs. 7.4–7.7 below (Kovalev, 1995).

30–40 km. The numerical experiment is done both for a ground-based vertically staring lidar and for an airborne down-looking lidar with a minimum range for complete lidar overlap, $r_0 = 0.3$ km. In the simulations, it is assumed for simplicity that the backscatter-to-extinction ratio $\Pi_p = 0.03\,sr^{-1}$ is constant at all altitudes. The results of the inversions made for the ground-based and airborne lidars are shown in Figs. 7.4 and 7.5, respectively. All curves in the figures are extracted with the far-end solution in which the precise boundary values were used. The distortion in the retrieved $\kappa_p(h)$ profiles is due only to incorrectly assumed backscatter-to-extinction ratios Π_p (the subscript "*as*" here and below is omitted for brevity). In both figures, *curve 1* is the model $\kappa_p(h)$ profile given in Fig. 7.3. The retrieved $\kappa_p(h)$ profiles (*curves 2–5*) are calculated with constant values of Π_p, which differ from the initial value, $0.03\,sr^{-1}$. The curves show the profiles retrieved with $\Pi_p = 0.01\,sr^{-1}$, $\Pi_p = 0.02\,sr^{-1}$, $\Pi_p = 0.04\,sr^{-1}$, and $\Pi_p = 0.05\,sr^{-1}$, respectively. It can be seen that an incorrect value in the assumed Π_p can even result in an unrealistic negative extinction coefficient profile (*curve 5* in Fig. 7.5). The occurrence of such unrealistic results may allow restriction of the range of likely backscatter-to-extinction ratios and thus may put additional limitations on possible solutions to the lidar equation.

The atmospheric profile obtained under the same retriving conditions as that in Figs. 7.4 and 7.5, but inverted with an optical depth solution, are given in Figs. 7.6 and 7.7. Here, the precise value of the two-way total transmittance,

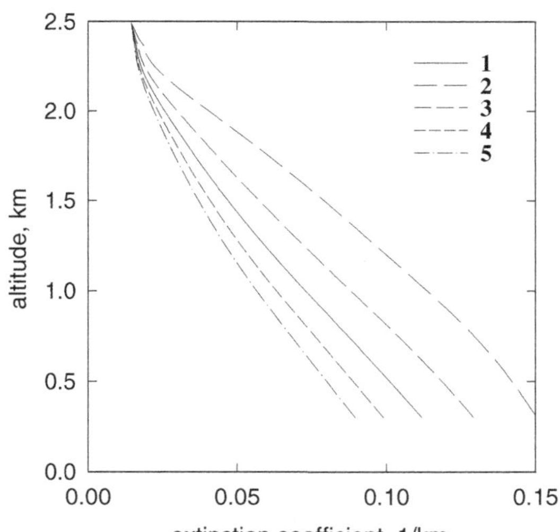

Fig. 7.4. $\kappa_p(h)$ profiles retrieved with incorrect Π_p values. The model $\kappa_p(h)$ and $\kappa_m(h)$ altitude profiles are shown in Fig. 7.3. The numerical experiment is made for a ground-based up-looking lidar, and the correct boundary value of $\kappa_p(h_b)$ is specified at the altitude of 2.5 km (Kovalev, 1995).

$[T(r_0, r_{max})]^2$ is taken as the boundary value. Just as before, the error in the solution stems only from the error in the incorrectly assumed backscatter-to-extinction ratio. Unlike the boundary point solution, in this case, a limited region exists within the operating range in which the retrieved extinction coef-

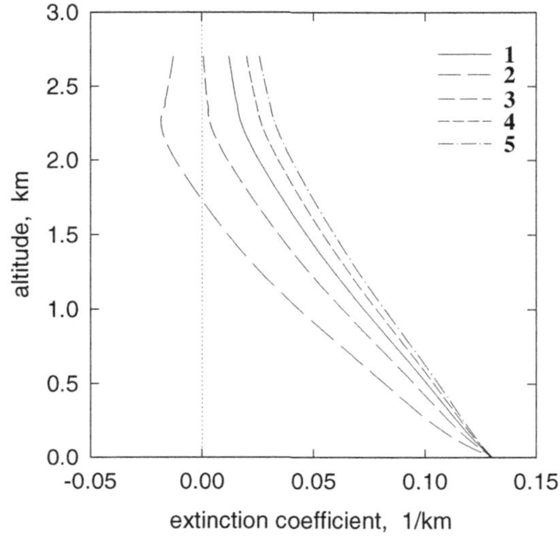

Fig. 7.5. Conditions are the same as in Fig. 7.4, but with the numerical experiment made for an airborne down-looking lidar. The plane altitude is 3 km, and the correct boundary value of $\kappa_p(h_b)$ is specified near the ground surface (Kovalev, 1995).

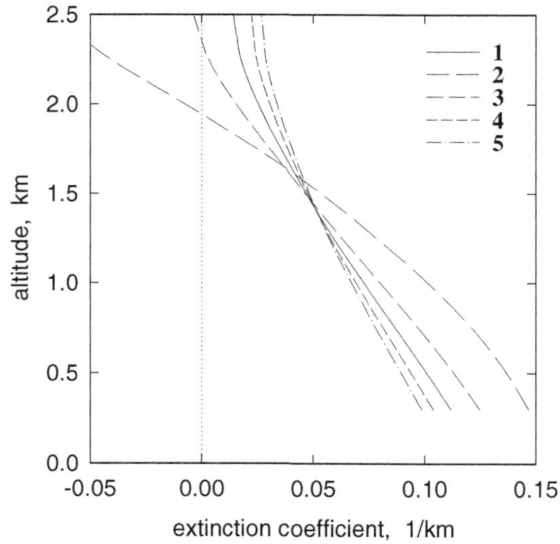

Fig. 7.6. $\kappa_p(h)$ profiles retrieved with the optical depth solution. The model $\kappa_p(h)$ profile is shown as *curve 1*, and retrieving conditions are the same as in Fig. 7.4 (Kovalev, 1995).

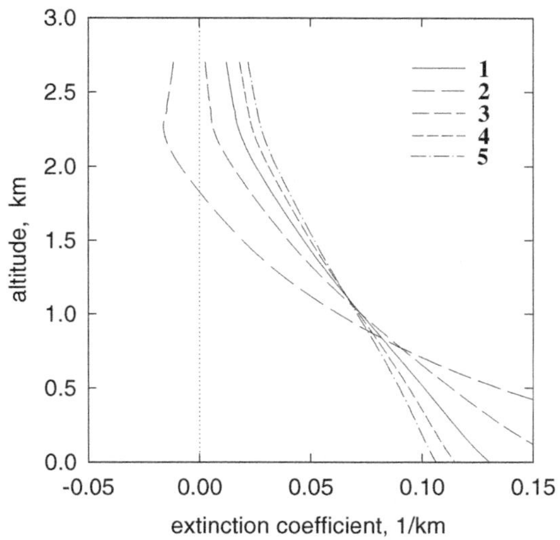

Fig. 7.7. $\kappa_p(h)$ profiles retrieved with the optical depth solution. The model $\kappa_p(h)$ profile is shown as *curve 1*, and retrieving conditions are the same as in Fig. 7.5 (Kovalev, 1995).

ficients are close to the actual value of $\kappa_p(h)$ regardless of the assumed value for Π_p. The extinction coefficient values obtained in such regions can be considered to be the most reliable data and used as reference values for an additional correction to the retrieved profile. However, this effect is generally inherent only in monotonically changing extinction coefficient profiles, such as those shown in Fig. 7.3. Furthermore, to achieve this result, an accurate value of the total atmospheric transmittance $[T(r_0, r_{max})]^2$ over the range from r_0 to r_{max} must be initially determined. This can be accomplished, for example, through the use of an independent measurement of total transmittance through the atmosphere made with a sun photometer (see Section 8.1.3). Note also that the worst profiles in all figures (Figs. 7.4–7.7) are obtained with $\Pi_p = 0.01\,sr^{-1}$, that is, when the backscatter-to-extinction ratio is the most severely underestimated with respect to the real values, $0.03\,sr^{-1}$.

To summarize the results of the measurement uncertainty caused by an incorrectly determined backscatter-to-extinction ratio in atmospheres with a large monotonic change in the extinction coefficient, the distortion of the derived profile $\kappa_p(h)$ depends both on the accuracy of the assumed Π_p and on the method by which the signal inversion is made. For the boundary point solution, the uncertainty in the derived $\kappa_p(h)$ profile may increase in both directions from the point at which the boundary condition is specified. When optical depth solution is used with precise value $[T(r_0, r_{max})]^2$, a restricted zone exists within the range $r_0 - r_{max}$ where measurement uncertainty is minimal. In both cases, the uncertainties are generally larger when the backscatter-to-extinction ratios are underestimated.

7.3. PROBLEM OF A RANGE-DEPENDENT BACKSCATTER-TO-EXTINCTION RATIO

In an atmosphere filled with aerosols, the lidar equation always contains two unknown quantities related to particulate loading, the backscattering term, $\beta_{\pi,p}(r)$, and the extinction term, $\kappa_p(r)$. Both quantities may vary in an extremely wide range, a million times and even more, whereas the ratio of the two values, $\Pi_p(r)$, changes over a much smaller range, typically from 0.01 to 0.05 sr^{-1}. When attempting to invert the lidar signal, it is logical to apply an analytical relationship between the values $\beta_\pi(r)$ and $\kappa_p(r)$. This makes it possible to replace the backscattering term $\beta_{\pi,p}(r)$ by the more slowly varying function, $\Pi_p(r)$. Obviously, for such a replacement, some relationship between the extinction and backscatter coefficients must be chosen for any particular measurement. The conventional approximation for the backscatter-to-extinction ratio assumes a linear dependence between the backscatter and total scattering (or total extinction). Such an approximation does not stem directly from Mie theory, at least for polydisperse aerosols. Nevertheless, this assumption may be practical in many optical situations (Derr, 1980; Pinnick et al., 1983; Dubinsky et al., 1985). On the other hand, this approximation is often not adequate to describe actual atmospheric conditions. This is especially true in atmospheres in which the particulate size distribution and, accordingly, the particulate extinction coefficient vary significantly along the lidar measurement range. Clearly, the application of a variable backscatter-to-extinction ratio in an inhomogeneous, especially, multilayer atmosphere is preferable to using an inflexible constant value that is chosen a priori.

As shown in Chapter 5, the lidar equation solution for a single-component atmosphere requires knowledge of the relative change of the backscatter-to-extinction ratio $\Pi_p(r)$ along the lidar line of sight. Here the relative change in $\Pi_p(r)$ rather than its numerical value is a major factor that determines the measurement accuracy. Ignoring such changes may result in large measurement errors. The largest distortions in the retrieved extinction coefficient profiles occur either in layered atmospheres or in the atmosphere where a systematic change of $\Pi_p(r)$ with range takes place. The latter may occur, for example, when a ground-based lidar measurements are made in slope directions in low-cloudy atmospheres. In the region below the cloud, backscattering results from moderately turbid or even clear air. In the region of the cloudy layer, the backscattering is originated by large cloud aerosols. The use of a range-invariant backscatter-to-extinction ratio for the signal inversion creates systematic shifts in the derived profiles, which are related to the elevation angle of the lidar line of sight when low stratus are investigated (Kovalev et al., 1991). The only way to avoid such distortions is the use of a nonlinear dependence between extinction and backscattering.

There are two ways to implement a range-dependent backscatter-to-extinction ratio in the lidar data processing technique. The first method makes use of additional instrumentation to determine this function directly along the lidar line of sight. The second method is to establish and apply approximate analytical relationships between the extinction and backscattering coefficients. Such an established dependence could be substituted into the lidar equation, thus removing the unknown backscattering term, that is, transforming this equation into a function of the extinction coefficient only. Unfortunately, both methods have significant drawbacks.

The first method may be achieved by a combination of elastic and inelastic lidar measurements. Fairly recent developments in inelastic remote-sensing techniques make it possible to estimate backscatter-to-extinction ratios and improve the accuracy of elastic lidar measurements. The idea of such a combination, which has become quite popular, proved to be fruitful (Ansmann et al., 1992 and 1992a; Donovan and Carswell, 1997; Ferrare et al., 1998; Müller et al., 1998 and 2001). A combined elastic-Raman lidar system can provide the information on both the backscattering and extinction coefficients along the searched path (see Chapter 11). The basic problem with this method is the large difference between the Raman and elastic scattering cross sections and, accordingly, the large difference in the intensity of the measured signals. Raman signals are about three orders of magnitude weaker than the signals due to elastic scattering. This may result in quite different measurement ranges or averaging times for the elastic and inelastic signals. To equalize the measurement capabilities for elastic and Raman returns, recording the Raman signals is generally made using the photon-counting mode, and the time of photon counting is selected much larger than the averaging time required for elastic signals; for distant ranges the time may be of 10–15 minutes and more (Section 11.1). Such averaging is mostly applied in stratospheric measurements. For low-tropospheric measurements, the combined processing the data of elastic and Raman lidars may be an issue, because generally these measurements cannot cover the same range interval (r_0, r_{max}), especially, in nonstationary atmospheres and daytime conditions. Although a lot of lidars for combined elastic-inelastic measurements are built, the problem of their accurate data inversion still remains.

Such difficulties do not occur if an analytical dependence between backscattering and extinction is somehow established. The analytical dependence may be practical for many specific tasks or particular situations. As shown further in Section 7.3.2, such an approach may be practical for slope measurements of extinction profiles in cloudy atmospheres or when correcting the backscatter-to-extinction ratio in thin layering, where multiple scattering cannot be ignored. As follows from the analysis in Section 7.1, the most obvious problems for the use of a analytical dependence between the backscatter and the extinction coefficient are as follows. First, the backscatter-to-extinction ratio is different for different types of aerosol, size distributions, refraction indices, etc. Second, it depends on atmospheric conditions, such as humidity, temperature, etc. Third, for the same atmospheric conditions and types of aerosols, the ratio is different for different wavelengths. Thus any general dependence, such as the power-law relationship, has, in fact, no physical basis. It is impossible to define the relationship between backscattering and extinction without some initial knowledge of the aerosol origins, their type, etc. This follows from numerous studies, such as those by Fymat and Mease (1978), Pinnick et al. (1983), Evans (1985), Leeuw et al. (1986), Takamura and Sasano (1987), Sasano and Browell (1989), Parameswaran et al. (1991), Anderson et al. (2000), and others.

An alternative way is a combination of two above methods. To our best knowledge, such a combination, i.e., the use of an analytical dependence between backscattering and extinction when processing data of a combined elastic-Raman lidar, has never been considered. At a glance, there is no reason to apply such an analytical dependence for the backscatter-to-extinction ratio, $\Pi_p(r)$, because the Raman-lidar system can determine both backscattering and total extinction coefficients. One can agree that there is no need for such a dependence when advanced multiwavelength elastic-Raman systems are used which operates simultaneously on 3–5 or more wavelengths (Ansmann, 1991, 1992, and 1992a; Ferrare et al., 1998 and 1998a; Müller et al., 1998, 2000, 2001, and 2001a). Such systems allow applying most sophisticated data-processing methods and algorithms and make it possible to extract vast information on particulate properties in the upper troposphere and stratosphere, including the particulate albedo, refraction indices, particulate size distribution, etc. (Zuev and Naats, 1983; Donovan an Carswell, 1997; Müller et al., 1999 and 1999a; Ligon et al., 2000; Veselovskii et al., 2002). However such advanced technologies are not applicable for simplest elastic-Raman lidars, for example for a lidar that uses one elastic and one Raman channel. In fact, there is no alternative processing method that could be actually practical for such simple systems. The application of the best-fit analytical dependence between backscattering and extinction, found with the same system during a preliminary calibration procedure, that would preceded the atmospheric measurement, might be helpful for such systems.

Thus the latter method requires an initial calibration procedure made before the measurements of atmospheric extinction, during which a preliminary set of the inelastic and elastic lidar measurement data is first obtained. These data are used to determine the particular relationship between the backscattering and extinction for the searched atmosphere. An analytical fit for this relationship is found and then used to invert the elastic lidar signals from areas both within and beyond the overlap of Raman and elastic lidar measurement ranges.

It should be noted that for elastic signal inversion with variable backscatter-to-extinction ratios, the use of an analytical fit of the obtained relationship is preferable to the use of a numerical look-up table relating extinction and backscattering. The reason for this observation is that the inversion algorithms often use iterative procedures, in which the actual value of the extinction coefficient is only obtained after some number of iterations. The values of the extinction coefficient obtained during the first cycles of iteration can significantly differ from the final values, and, moreover, these intermediate values can be outside the actual range of values. Clearly, the elastic-Raman measurements may not provide backscatter-to-extinction ratios for all of the possible intermediate values for the extinction coefficient that could appear during iteration. The iteration may not converge if all intermediate values for the backscatter-to-extinction ratios are not available. The use of an expanded analytical dependence allows avoid this. What is more, it will allow to obtain accurate inversion results for the full measurement range of the elastically scattered signal, including distant ranges, where the Raman signal is too week to be accurately measured.

The above data processing procedure for the elastic-Raman lidar system can be shortly described as follows. Before atmospheric measurements, an initial calibration procedure is made, in which the elastic and Raman lidar data are processed and the backscatter and extinction profiles are determined in the range where both elastic and inelastic signals have acceptable signal-to-noise ratios. With a subset of the measurements, a numerical relationship between the backscatter-to-extinction ratio and extinction coefficient is established (or renewed). An analytical fit is then found for this relationship. The fit can be based on some generalized dependence, so that only the fitting constants of this dependence are varied when a new adjustment to the dependence shape is made. This analytical dependence is then used in all elastic lidar measurements until the next calibration is made.

7.3.1. Application of the Power-Law Relationship Between Backscattering and Total Scattering in Real Atmospheres: Overview

The simplest variant, which assumes a range-independent backscatter-to-extinction ratio, may yield large errors in lidar signal inversion when the lidar measurement range comprises regions including both clear areas and turbid layers (Sasano et al., 1985; Kovalev et al., 1991). As mentioned in Section 5.3.3, some attempts have been made to establish a practical nonlinear relationship between backscatter and extinction. Nonlinear correlations were first developed by atmospheric researchers in experimental studies in the 1960s and 1970s. In 1958, Curcio and Knestric established that, in their experimental data, the linear relationship took place between the logarithms of κ_t and β_π rather than between the values of backscatter and total scattering. The dependence can be written in the form

$$\log \beta_\pi = a_1 + b_1 \log \kappa_t \tag{7.13}$$

where a_1 and b_1 are constants. In the lidar equation, this approximation was generally applied as the power-law relationship between the backscatter and extinction coefficients, with a fixed exponent and constant of proportionality,

$$\beta_\pi = B_1 \kappa_t^{b_1} \tag{7.14}$$

so that $a_1 = \log B_1$. As shown in Section 5.3.3, for single-component turbid atmospheres, only the exponent b_1 must be known to solve the lidar equation and determine κ_t. In studies made during 1960–1980, the relationship in Eq. (7.13) was investigated mostly in the visible range of the spectrum. The studies were made in a wide range of atmospheric turbidity, and both B_1 and b_1 were assumed to be constant. In the moderately turbid atmospheres under investigation, the small amount of molecular scattering does not significantly influence the constants B_1 and b_1. Therefore, in the early studies, the molecular term was just ignored when determining the linear fit of $\log \beta_\pi$ versus $\log \kappa_t$ in Eq. (7.13). In the above pioneering study of Curcio and Knestric (1958), the constant b_1 in Eq. (7.13) was found to be 0.66. In later experimental studies by Barteneva (1960), Gavrilov (1966), Barteneva et al. (1967), Stepanenko (1973), and Gorchakov and Isakov (1976), the linear correlation between the logarithms of the backscatter and total scattering coefficients was also confirmed with a b_1 value close to 0.7. According to the analysis made by Tonna (1991), a power-law relationship can be used, at least in the wavelength range from 250 to 500 nm. On the other hand, studies have been published in which the dependence between logarithms of the backscatter and total scattering was found to be nonlinear (Foitzik and Zschaeck, 1953; Golberg, 1968 and 1971; Lyscev, 1978). According to the latter, the relationship between $\log \beta_\pi$ and $\log \kappa_t$ could be considered to be linear only within a restricted range of atmospheric turbidity. The numerical value of constant b_1 in these studies was related to the turbidity range, and under bad visibility conditions b_1 was generally larger than that (0.66–0.7) established in the earlier studies. Both experimental and theoretical published data for the relationship between backscatter and total scattering coefficients were analyzed by Kovalev et al. (1987). In this study, the values of constant b_1 were compiled from the studies made during 1953–1978, information on which was available to the authors. The result of this compilation is given in Table 7.2.

The relationships between backscatter and extinction compiled in the study by Kovalev et al. (1987) are shown in Fig. 7.8. *Curves 1–6* show the relationships between β_π and κ_t obtained from the different studies. The bold vertical lines are taken from the study by Hinkley (1976). These lines show the likely range of the backscatter coefficient values for discrete ranges of the extinction coefficient at 550 nm. A specific feature of the curves shown in Fig. 7.8 is the noticeable increase in the slope when κ_t becomes more than 1 km^{-1}. This effect is clearly seen when the average of the curves is considered (Fig. 7.9). As follows from the figure, the average relationship can be approximated by two different straight lines. For relatively clear atmospheres, with extinction coefficients up to 1 km^{-1}, the constant b_1 is, approximately 0.7, whereas for more turbid atmospheres with κ_t greater than 1 km^{-1}, constant b_1 becomes equal to 1.3. Note that the latter value is close to that determined for stratus in a study by Klett (1985), where b_1 was established to be 1.34. The values of 0.7 and 1.3 must be considered to be average estimates for small and large κ_t. As follows from Table 7.2, for specific optical situations and restricted ranges,

TABLE 7.2. Constant b_1 in the Linear Relationship Between the Logarithms of the Backscatter and Extinction Coefficients Determined Close to the Ground Surface

	Wavelength, nm	κ_t, km^{-1}	b_1
Curcio and Knestric (1958)	350–680	0.06–40	0.66
Barteneva (1960)	White light	0.02–0.4	0.7
Barteneva et al. (1967)	"	0.02–15	0.66*
Stepanenko (1973)	"	0.2–6	0.66
Gorchakov and Isakov (1976)	550	0.02–1	0.69
Golberg (1968)	White light	0.4–20	1.2*
Golberg (1971)	"	0.2–0.4	0.5
		0.56–7.8	1.0
		>7.8	1.2
Lyscev (1978)	920	0.7–7	1.5–2.5
Foitzik and Zschaeck (1953)	White light	0.8–4	1.2*
		0.08–0.5	0.12*
Toropova et al. (1974)	630		1.02
Panchenko et al. (1978)	546	0.05–0.5	0.71
Pavlova (1977)	630	>20	1.4

* Based on analysis of the experimental date published in the cited study.

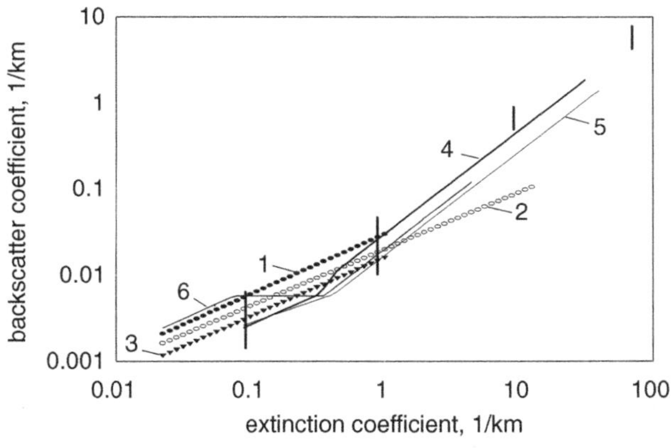

Fig. 7.8. Typical relationships between the backscatter and extinction coefficients at the wavelength 550 nm and for achromatic light. The curves are derived from published theoretical and experimental data, obtained near the ground surface. *Curves 1* and *2* are based on the studies by Barteneva (1960) and Barteneva et al. (1967); *curve 3* on the study by Gorchakov and Isakov (1976); *curves 4 and 5* on the study by Golberg (1968 and 1971); and *curve 6* on the study by Foitzik and Zschaeck (1953). The bold vertical segments show the backscatter coefficient range for the discrete ranges of κ_t as estimated in the study by Hinkley (1976) (Adapted from Kovalev et al., 1987).

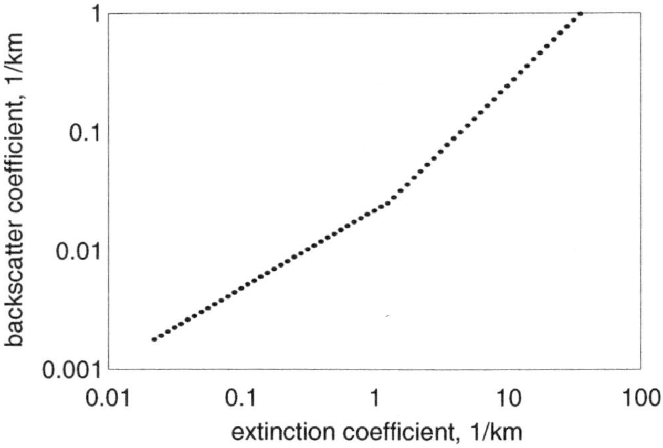

Fig. 7.9. Mean dependence between the backscatter and extinction coefficients as estimated from data in Fig. 7.8 (Adapted from Kovalev et al., 1987).

the value of constant b_1 may vary, at least in the range from 0.5 to approximately 2–2.5. These large uncertainties in the constant b_1 are the reason why most investigators, accepting in principle the power-law relationship, generally applied $b_1 = 1$ when analyzing results of lidar measurements (see Viezee et al., 1969; Lindberg et al., 1984; Carnuth and Reiter, 1986, etc.).

Klett (1985) was the first to recognize that the most realistic approach was to consider the relationship between the total scattering and backscattering in a more complicated form than that given in Eq. (7.14). Direct Mie scattering theory calculations yielded a similar conclusion (Takamura and Sasano, 1987; Parameswaran et al., 1991). In a study by Parameswaran et al. (1991), the relationship between particulate backscattering and the extinction coefficient at a ruby laser wavelength of 694.3 nm was examined with Mie theory. The validity of the power-law dependence in Eq. (7.14) was examined for particulates with different size distributions and indices of refraction. The authors concluded that in the general case, the constants in the power-law dependence are correlated with the total-to-molecular backscatter coefficient ratio, so that the use of a power-law solution with fixed constants is not physical. A similar conclusion also follows from Fig. 7.8, which shows that the backscatter coefficients increase abruptly when the total scattering coefficient increases and becomes more than $1\,\text{km}^{-1}$. Thus the dependence between the logarithms of the backscatter and total extinction coefficients cannot be treated as linear over an extended range of extinction coefficients, from clear air to heavy haze. The numerical value of $b_1 \approx 0.7$ proposed in the early studies by Curcio and Knestric (1958) and Barteneva (1960) may only be typical at the ground level in moderately turbid atmospheres. However, this value is not appropriate for clouds and fogs, where larger values of b_1 seem to be more realistic. Note that in dense layering, an additional signal component may occur because of multiple scattering. It stands to reason that for large κ_t, some relationship may exist between the increase of the constant b_1 and the increase in signal due to multiple scattering. However, to our knowledge, this relationship has never been properly investigated. The lidar community remains skeptical to the application of analytical dependencies between backscatter-to-extinction ratio and extinction coefficient in practical measurements. Large data-pont scattering in the dependencies between these values experimentally established from lidar data (see, for example, the studies by Leeuw et al., 1986; Del Guasta et al., 1993; Anderson et al., 2000) can only discourage researchers, because under such conditions no analytical dependence seems to be sensible. However, the question always emerges what is real accuracy of all such measurements; It is difficult to believe that the revealed data-point scattering is only due to actual fluctuations in Π_p and neither systematic nor random measurement errors influence the measurement results. Meanwhile the estimated standard deviations in experimentally derived Π_p, when these are determined (see for example, Ferrare et al. 1998; Voss et al., 2001), show that accuracy of such estimates may be rather poor. Anyway, as will be shown in the next section, in many real atmospheric situations the use of the approximation of a constant backscatter-to-extinction ratio is not the best inversion variant.

7.3.2. Application of a Range-Dependent Backscatter-to-Extinction Ratio in Two-Layer Atmospheres

The analysis by Kovalev et al. (1991) showed that significant discrepancies in the retrieved extinction coefficient profiles may occur when multiangle lidar data, measured in a two-layer cloudy atmosphere, are processed with a range-invariant backscatter-to-extinction ratio. The use of a constant ratio may result in systematic shifts in the extinction coefficient profiles at the far end of the measured range. This systematic shift is also related to the elevation angle of the lidar. This is because the changes in the elevation angle change the relative lengths of two adjacent areas with different backscattering. An analysis confirmed that the shifts disappeared when different constants b_1 were used for the cloudy layer and the layer below it. Particularly, the use of $b_1 = 1.3$–1.4 for extracting optical characteristics from the cloudy area and $b_1 = 0.7$ for extracting the extinction coefficient below the cloud completely eliminated the above shifts. Thus, for situations when the lidar operating range (r_0, r_{max}) is comprised of two stratified zones with significantly different backscattering, the first step in the data processing is to establish the ranges for these zones, (r_0, r_b), and (r_b, r_{max}), respectively. In the nearest zone from r_0 to r_b, the lidar beam propagates through a relatively clear atmosphere, whereas in the remote area from r_b to r_{max}, it propagates through a more turbid, cloudy layer. Values of b_1 used for these areas are further denoted as b_n for the nearest relatively clear area, and as b_c for the cloudy area. The point r_b is taken as the boundary point, and the value of the extinction coefficient in this point is estimated with the signals obtained from the cloudy area (r_b, r_{max}). With the power-law relationship [Eq. (7.14)], the solution in Eq. (5.66) may be rewritten as

$$\kappa_p(r_b) = \frac{b_c[S_r(r)]^{\frac{1}{b_c}}}{2\int_{r_b}^{\infty}[S_r(r')]^{\frac{1}{b_c}}dr'} \tag{7.15}$$

The integral with the infinite upper limit in the denominator of Eq. (7.15) can be estimated with the integrated lidar signal over cloudy area, from r_b to r_{max}

$$\int_{r_b}^{\infty}[S_r(r')]^{\frac{1}{b_c}}dr' = \eta(1+\varepsilon)\int_{r_b}^{r_{max}}[S_r(r')]^{\frac{1}{b_c}}dr' \tag{7.16}$$

where η is a multiple scattering factor (see Section 3.2.2), and the correction factor ε can be estimated with the ratio $S_r(r_{max})/S_r(r_b)$ (see Section 12.2). As $\varepsilon > 0$, and $\eta < 1$, the product $\eta(1 + \varepsilon)$ can be assumed to be unity if no additional information is available. With this approximation, one can obtain the value of $\kappa_p(r_b)$ with Eq. (7.15) in which the upper (infinite) integration limit is replaced by r_{max}. The profile of the extinction coefficient over the near range from r_0 to r_b can then be found with the value $\kappa_p(r_b)$ and the appropriate constant b_n

$$\kappa_p(r) = \frac{\left[\frac{S_r(r)}{S_r(r_b)}\right]^{\frac{1}{b_n}}}{\frac{1}{\kappa_p(r_b)} + \frac{2}{b_n}\int_{r}^{r_b}\left[\frac{S_r(r')}{S_r(r_b)}\right]^{\frac{1}{b_n}}dr'} \tag{7.17}$$

Eq. (7.17) is the stable far-end boundary solution for a single-component atmosphere; therefore, in moderately turbid atmospheres, a possible uncertainty in the boundary value, $\kappa_p(r_b)$, does not result in large errors in the profile $\kappa_p(r)$ over the range (r_0, r_b). The determination of the extinction coefficient profile in the cloudy layer, from r_b to r_{max}, is more problematic. In principle, the profile of the extinction coefficient in this range can be found by using the same value of $\kappa_p(r_b)$, but this time the near-end solution must be used. However, the near-end solution is here quite inaccurate, because of uncertainties in both ε and η. The signals measured in the cloud area may only be relevant to estimate the total optical depth over the range (r_b, r_{max}). Whereas such a method is not enough accurate for determining range-resolved extinction coefficient profiles, its application is sensible for determining the total transmission and optical depths of aerosol layers of the atmosphere (see Section 12.2).

There is a more straightforward solution for the lidar signal inversion in atmospheres that comprise two or more layers with well-defined boundaries between the layers. Such situations, for example, may be found when making plume dispersion experiments (Eberhard et al., 1987), investigating aerosols from biomass fires (Kovalev et al., 2002) or screening military smokes (Roy et al., 1993), or when examining the plumes from launch vehicles powered by rocket motors (Gelbwachs, 1996). In such situations, the lidar measurement range includes at least two adjacent zones with significantly different optical properties. Generally within a near not polluted zone, over some range up to $r < r_b$, the backscatter signals are associated with background aerosol scattering. Smoke plumes are dispersed at distant ranges, $r > r_b$, generally at distances of 1 km or more from the lidar.

The lidar signal inversion may be based on a simple approximation, which assumes that the particulate backscatter-to-extinction ratios over the near (background aerosol) and distant (smoky) zones, $\Pi_{p,cl}$ and $\Pi_{p,sm}$, respectively, are constant over each zone but not equal, that is, $\Pi_{p,cl} \neq \Pi_{p,sm}$. To obtain the solution for a two-layered atmosphere where the particulate backscatter-to-extinction ratios are significantly different over the adjacent zones, one should first determine the ranges of these zones, $[r_0, r_b]$ and $[r_b, r_{max}]$, respectively. The zones where significantly different backscatter-to-extinction ratios occur can be established from a preliminary examination of the lidar signal intensity. The above inversion principle may be applied for three and more zones, but here, for simplicity, it is assumed that the backscattered signal vanishes in the second zone, at some range r_{max}. The procedure to transform the lidar signal is the same as that described in Section 5.2, namely, the signal transformation is done by means of multiplying the range-corrected lidar signal by a transformation function $Y(r)$. To determine $Y(r)$, one needs to know the molecular extinction coefficient profile $\kappa_m(r)$ and the backscatter-to-extinction ratios along the lidar searching path (Section 5.2). For the first zone, $r_0 < r < r_b$, the transformation function $Y_{cl}(r)$ is defined with the backscatter-to-extinction ratio $\Pi_{p,cl}$

$$Y_{cl}(r) = (\Pi_{p,cl})^{-1} \exp\left[-2\int_{r_0}^{r} (a_{cl} - 1)\,\kappa_m(x)\,dx\right] \tag{7.18}$$

where $a_{cl} = 3/[8\pi\,\Pi_{p,cl}]$ and $\kappa_m(r)$ is the molecular extinction coefficient profile, which is assumed to be known. It is assumed also that no molecular absorption takes place, so that $\kappa_m(r) = \beta_m(r)$.

For the second zone, $r_b < r < r_{max}$, the transformation function $Y_{sm}(r)$ is

$$Y_{sm}(r) = (\Pi_{p,sm})^{-1} \exp\left[-2\int_{r_0}^{r_b} (a_{cl} - 1)\,\kappa_m(x)\,dx\right] \exp\left[-2\int_{r_b}^{r} (a_{sm} - 1)\,\kappa_m(x)\,dx\right] \tag{7.19}$$

where $a_{sm} = 3/[8\pi\,\Pi_{p,sm}]$. The function $Z(r) = P(r)\,Y(r)\,r^2$ over the range from r_0 to r_b is defined as

$$Z(r) = C_0 T_0^2 [\kappa_p(r) + a_{cl}\kappa_m(r)] \exp\left\{-2\int_{r_0}^{r} [\kappa_p(x) + a_{cl}\kappa_m(x)]\,dx\right\}$$
$$= C_0 T_0^2 \kappa_W(r)[T_p(r_0, r)]^2 [T_m(r_0, r)]^{2a_{cl}} \tag{7.20}$$

The terms $T_p(r_0, r)$ and $T_m(r_0, r)$ are the total path transmittance over the range from r_0 to r for the particular and molecular constituents, respectively. Over the smoky area, that is, over the range from r_b to r_{max}, the function $Z(r)$ is found as

$$Z(r) = C_0 T_0^2 [\kappa_p(r) + a_{sm}\kappa_m(r)] \exp\left\{-2\int_{r_0}^{r_b} [\kappa_p(x) + a_{cl}\kappa_m(x)]\,dx\right\}$$
$$\exp\left\{-2\int_{r_b}^{r} [\kappa_p(x) + a_{sm}\kappa_m(x)]\,dx\right\} \tag{7.21}$$

The product of the exponent terms in Eq. (7.21) can be defined through the two-way path transmittance $[V(r_0, r)]^2$ for the particulate and molecular constituents as

$$[V(r_0, r)]^2 = [T_p(r_0, r)]^2 [T_m(r_0, r_b)]^{2a_{cl}} [T_m(r_b, r)]^{2a_{sm}} \tag{7.22}$$

where the first term in the right side of Eq. (7.22) is the total path transmittance over the range from to r_0 to r for the particular constituent, and two others are related to the molecular transmittance over the ranges (r_0, r_b) and (r_b, r), respectively.

7.3.3. Lidar Signal Inversion with an Iterative Procedure

The application of different constants b_1 or different fixed backscatter-to-extinction ratios $\Pi_{p,i}$ for different zones with the method discussed in the previous section may be helpful for a two-layer atmosphere that has a well-defined boundary between a smoke plume or a cloud (subcloud) and moderately turbid air below it. However, it is difficult to do this when the layer boundaries are not clearly defined, so that the extinction coefficient changes monotonically over some extended range between the cloud and the clear air below it. In this case, the alternative approach can be used based on the application of some analytical dependence between the extinction and backscatter coefficients.

There are two ways to apply this approach to practical lidar measurements. The first approximation may be done similarly to that discussed in the previous section, when aerosols with significantly different backscattering intensity (for example, smokes and clear-air background particulates) are found at extended areas within the lidar measurement range. To avoid the need to establish geometric boundaries for these areas by analyzing the signal profiles, as discussed in the previous section, one can establish some threshold level of the backscatter or the extinction coefficient to separate the smokes from the clear air. During the iteration procedure, the lidar signal inversion is made with two different backscatter-to-extinction ratios, $\Pi_{p,sm}$ and $\Pi_{p,cl}$, selected (in the worst case, a priori) for the smoky and clear areas. The second way, described below in this section is to transform some experimental dependence of β_π on the extinction coefficient, for example, such as shown in Figs. 7.8 and 7.9, or that derived from simultaneous elastic and inelastic measurements, into an analytical dependence of $\Pi_p(r)$ on $\kappa_p(r)$. Such an analytical dependence would make it possible to apply a range-dependent backscatter-to-extinction ratio directly for the lidar signal inversion. This could be done without a preliminary examination of the elastic signal profile and determination of the boundaries between aerosols of different nature.

As was stated, the inversion procedure may be applied to the combined elastic-inelastic lidar measurements even if a concrete dependence between the extinction and backscattering is only established over some restricted range. To apply this dependence for the elastic lidar measurements, the experimental dependence of $\Pi_p(r)$ on $\kappa_p(r)$ must be fit to an analytical formula and then applied to the signal-processing algorithm. To see how this can be done, consider the application of the dependence shown in Fig. 7.9 for such a procedure. The analytical dependence of the curve shown in the figure was obtained in the study by Kovalev (1993). In fact, this dependence is a sophisticated form of Eq. (7.13). However, the exponent term b_1 is treated here as a function of the particulate extinction coefficient rather than a constant. Accordingly, Eq. (7.13) is rewritten as

$$\log \beta_{\pi,p} = a_2 + b(\kappa_p) \log \kappa_p \tag{7.23}$$

or in the exponential form

$$\beta_{\pi,p} = C_2 \kappa_p^{b(\kappa_p)} \tag{7.24}$$

where $a_2 = \log C_2$, and the exponent $b(\kappa_p)$ is considered to be a function of the particulate extinction coefficient. It follows from Eq. (7.24) that

$$\Pi_p = C_2 \kappa_p^{b(\kappa_p)-1} \tag{7.25}$$

In the study by Kovalev (1993), $b(\kappa_p)$ is defined by the formula

$$b(\kappa_p) = b_0 + C_3 \kappa_p^b \tag{7.26}$$

where b, b_0, and C_3 are constants. The best analytical fit for the mean dependence shown in Fig. 7.9 was obtained with $C_2 = 0.021$, $b_0 = -0.3$, and $b = 0.5$. The initial data, used to calculate the analytical dependence, were established within a restricted range of turbidities, in which the extinction coefficient ranged approximately from 0.02 to 30 km^{-1} (Fig. 7.9).

Note that by changing the value C_3 the behavior of the function Π_p for large extinction coefficients can be adjusted. Particularly by increasing the value of C_3, a significant increase in Π_p can be obtained. Thus the selection of a relevant value of C_3 can to some degree compensate for the contribution of multiple scattering and, accordingly, improve inversion accuracy. This kind of method, which can be considered to be an alternative to the approach by Platt (1973) and Sassen et al. (1989) (Chapter 8), is based on a simple approximation of the lidar equation. Considering the total backscattering at the range r to be the sum of the single-scattering components $\beta_{\pi,p}(r)$ and the multiple-scattering components $\beta_{ms}(r)$ the range-corrected signal for the particulate single-component atmosphere can be rewritten as (Bissonnette and Roy, 2000)

$$Z_r(r) = C_0 T_0^2 [\beta_{\pi,p}(r) + \beta_{ms}(r)] \exp\left[-2\int_{r_0}^{r} \kappa_p(r')dr'\right] \tag{7.27}$$

Eq. (7.27) is easily transformed to

$$Z_r(r) = C_0 T_0^2 \Pi_{p,eff}(r) \kappa_p(r) \exp\left[-2\int_{r_0}^{r} \kappa_p(r')dr'\right] \tag{7.28}$$

where

$$\Pi_{p,eff}(r) = \Pi_p(r)\left[1 + \frac{\beta_{ms}(r)}{\beta_{\pi,p}(r)}\right] \tag{7.29}$$

Note that in areas where multiple scattering does not occur, namely, $\beta_{ms}(r) = 0$, $\Pi_{p,eff}(r) = \Pi_p(r)$, and Eq. (7.28) automatically reduces to the conventional single-component lidar equation.

This approach, proposed in the study by Bissonnette and Roy (2000), was used for the inversion of lidar signals containing a multiple scattering component by Kovalev (2003a). For the transformation of the lidar signal, a special transformation function $Y_\delta(r)$ was used, which included the multiple-to-single scattering ratio, $\delta(\tau)$, defined as a function of the optical depth. For the two-component atmosphere, the transformation function is defined as

$$Y_\delta(r) = \frac{1}{\Pi_p(r)[1+\delta(\tau)]} \exp\left\{-2\int_{r_1}^{r}\left[\frac{3/(8\pi)}{\Pi_p(r)}\frac{1}{[1+\delta(\tau)]} - 1\right]\beta_m(r')dr'\right\}$$

where r_1 is the measurement near-end range, and $\beta_m(r)$ is the molecular scattering coefficient. After multiplying the range-corrected signal by this transformation function, $Y_\delta(r)$, the original lidar is transformed into the same form as that in Eq. (5.21). The new variable of the solution is

$$\kappa_\delta(r) = \kappa_p(r) + \frac{3\beta_m(r)}{8\pi\Pi_p(r)[1+\delta(\tau)]}$$

The inversion of the lidar signal with a variable backscatter-to-extinction ratio differs from that described in Section 5.2. Signal normalization, described in Section 5.2, transforms the shape of the range-corrected lidar signal into the function $Z(r)$ by correcting the exponential term in the original lidar equation. Despite some differences in the computational techniques, this or a similar transformation has been used in many studies, for example, by Klett (1985), Browell et al. (1985), Kaestner (1986), Weinman (1988), etc. However, when using a variable backscatter-to-extinction ratio that is a function of the extinction coefficient, another variant of lidar signal transformation should preferably be used. Here the backscatter term of the lidar equation is transformed rather than the exponential portion of the equation. In this variant, either a constant or a variable particulate backscatter-to-extinction ratio, $\Pi_p(r)$, can be used to invert the signal. Moreover, the ratio can either be determined as a function of the particulate extinction coefficient profile or be taken as a function of the distance from the lidar.

To better understand this variant, we present the basic elements of the iteration procedure. Similar to the signal transformation described in Section 5.2, the iteration procedure makes it possible to transform the original lidar signal into the same form as that in Eq. (5.21)

$$Z(x) = Cy(x)\exp\left[-2\int y(x)\,dx\right]$$

However, now the conversion is made without transforming the exponential term of the original lidar equation. The iteration procedure transforms the backscattering term $\beta_\pi(r)$ of the original lidar signal in Eq. (5.2) rather than the extinction coefficient $\kappa_t(r)$ in the exponential term. This is the basic difference between the transformations. The total backscatter coefficient, $\beta_\pi(r) = \beta_{\pi,p}(r) + \beta_{\pi,m}(r)$, in the lidar equation may be considered as the weighted sum of the particulate and molecular extinction coefficients, that is,

$$\beta_\pi(r) = \Pi_p(r)\,\kappa_p(r) + \frac{3}{8\Pi}\kappa_m(r) \qquad (7.30)$$

In Eq. (7.30) the particulate backscatter-to-extinction ratio $\Pi_p(r)$ may be considered as a weighted function of particulate component $\kappa_p(r)$, whereas the molecular phase function $3/(8\pi)$ is the weight of the molecular component $\kappa_m(r)$. The purpose of the given below iteration procedure is to equalize the weights of the particulate and molecular components. After completion of the iteration procedure, the original lidar signal is transformed into a function in which such an equivalence is made, so that its structure is similar to that in the above function $Z(x)$. In other words, in the function $Z^{(n)}(r)$ obtained after the final, nth, iteration, the weights of the molecular and particulate extinction constituents in Eq. (7.30) are equalized. This allows us to define a new variable $y(r)$ as the total extinction coefficient

$$y(r) = \kappa_m(r) + \kappa_p(r) \tag{7.31}$$

Several issues are associated with this type of transformation. Unlike the solution in Section 5.2, here the iteration also changes the transformation term $Y(r)$ at each iteration cycle. To distinguish the transformation term $Y(r)$ in Eq. (5.27) from that in the formulas below, the latter is denoted as $Y^{(i)}(r)$, where the superscript (i) defines the iterative cycle at which this value was determined. Accordingly, the normalized signal, defined as the product of the range corrected signal $Z_r(r)$ and the transformation function $Y^{(i)}(r)$ is denoted here as $Z^{(i)}(r)$, so that $Z^{(i)}(r) = Z_r(r)Y^{(i)}(r)$. In the solution below, either the boundary point or the optical depth solution can be used. The only difference is that in the boundary point solution, the function $Z^{(i)}(r_b)$ changes at each cycle of iteration. In the optical depth solution, which is described here, the value of the maximal integral [Eq. (5.53)] is recalculated at each cycle of iteration. The sequence of the iteration calculations is as follows (Kovalev, 1993):

(1) In the first cycle of the iteration, the initial transformation function $Y^{(1)}(r)$ is taken to be $Y^{(1)}(r) = 1$. The normalized signal $Z^{(1)}(r)$ is now equal to the range-corrected signal, $Z^{(1)}(r) = Z_r(r) = P(r)r^2$. To start the iteration, the initial particulate backscatter-to-extinction ratio $\Pi_p{}^{(1)}(r)$ is assumed to be equal to the molecular backscatter-to-extinction ratio, so that the ratio $a^{(1)} = 1$. With these conditions, the initial extinction-coefficient profile $\kappa_p^{(1)}(r)$ determined with the solution in Eq. (5.83) is reduced to

$$\kappa_p^{(1)}(r) = \frac{0.5Z^{(1)}(r)}{\dfrac{I_{max}^{(1)}}{1 - T_{max}^2} - I^{(1)}(r_0, r)} - \kappa_m(r) \tag{7.32}$$

where $I_{max}^{(1)}$ is the integral of $Z^{(1)}(r)$ over the range from r_0 to r_{max} and $\kappa_m(r)$ is the molecular extinction coefficient, which is assumed be known. T_{max}^2 is the assumed total transmittance over the lidar measurement range, that is, the boundary value. Note that the value of T_{max}^2 remains the same for all iterations.

(2) The next step depends on whether a constant or a variable backscatter-to-extinction ratio is used for the solution. Let us assume that the particulate backscatter-to-extinction ratio is related to the extinction coefficient over the measurement range by Eq. (7.25). With the profile $\kappa_p^{(1)}(r)$ obtained in Eq. (7.32), the profile of the backscatter-to-extinction ratio for the next iteration is found as

$$\Pi_p^{(2)}(r) = C_2[\kappa_p^{(1)}(r)]^{b\kappa_p^{(1)}(r)-1} \tag{7.33}$$

and the corresponding ratio $a^{(2)}(r)$ is

$$a^{(2)}(r) = \frac{3/8\pi}{\Pi_p^{(2)}(r)} \tag{7.34}$$

If a constant backscatter-to-extinction ratio is assumed to be valid, the calculation in Eq. (7.33) is omitted. The initially assumed constant Π_p and the corresponding constant ratio a are then used in all further iterations.

(3) Using the profiles $\kappa_p^{(1)}(r)$ and $a^{(2)}(r)$, the corresponding correction function $Y^{(2)}(r)$ is determined by means of the formula

$$Y^{(2)}(r) = \frac{\kappa_m(r) + \kappa_p^{(1)}(r)}{\kappa_m(r) + a^{(2)}(r)\,\kappa_p^{(1)}(r)} \tag{7.35}$$

(4) The new transformation function $Z^{(2)}(r)$ is then calculated as

$$Z^{(2)}(r) = Z_r(r)\,Y^{(2)}(r) \tag{7.36}$$

Note that the same initial range-corrected signal $Z_r(r)$ used in Eq. (7.36) is then applied in all next iterations, whereas the values $Y^{(i)}(r)$, $\kappa_p^{(i)}(r)$, and $a^{(i)}(r)$ are recalculated (updated) for each iteration.

(5) The next step of the iteration is to determine a new extinction coefficient profile, $\kappa_p^{(2)}(r)$. To accomplish this, the function $Z^{(2)}(r)$ and two integrals of this function, $I_{max}^{(2)}$ and $I^{(2)}(r_0, r)$, are used. The integrals are calculated over the ranges (r_0, r_{max}) and (r_0, r), respectively. The extinction coefficient $\kappa_p^{(2)}(r)$ is found with a formula similar to that in *step 1*

$$\kappa_p^{(2)}(r) = \frac{0.5Z^{(2)}(r)}{\dfrac{I_{max}^{(2)}}{1-T_{max}^2} - I^{(2)}(r_0, r)} - \kappa_m(r) \tag{7.37}$$

Steps 2–5 are then repeated until the iteration procedure converges to a stable shape of the updated extinction-coefficient profile $\kappa_p^{(i)}(r)$.

It is useful to repeat here that to apply this kind of retrieval method with variable backscatter-to-extinction ratios, the dependence between Π_p and κ_p for an extended extinction coefficient range should be established. In other words, at least an approximate dependence should be known beyond the actual range of the measured extinction coefficient. It is very likely that at some step of the iteration, an intermediate value of the retrieved extinction coefficient $\kappa_p^{(i)}(r)$ may be far beyond the range of the actual values. To ensure the convergence of an automated analysis program, it is necessary to have corresponding values of $\Pi_p^{(i)}(r)$ even for outlying values of the extinction coefficient.

To summarize, in order to effectively invert elastic lidar signals, some particular relationship between extinction and backscattering must be used. However, the use of a constant backscatter-to-extinction ratio in strongly heterogeneous atmospheres is a major issue that precludes obtaining accurate values for the extinction coefficient from an elastic lidar measurements. In mixed atmospheres, the application of a range-dependent backscatter-to-extinction ratio is far preferable to the use of a constant value. A combination of Raman or high-spectral-resolution lidar measurements with elastic lidar measurements is the first step toward the practical use of range-dependent ratios in elastic lidar measurements.

8

LIDAR EXAMINATION OF CLEAR AND MODERATELY TURBID ATMOSPHERES

8.1. ONE-DIRECTIONAL LIDAR MEASUREMENTS: METHODS AND PROBLEMS

In this section, one-directional measurement methods are analyzed. These methods assume that the lidar data set to be processed is obtained with a fixed spatial orientation of the lidar line of sight during the measurements. The data could be obtained, for example, by an airborne lidar, in which a laser beam is constantly directed to either the nadir or the zenith during the measurement. The data could also be from a ground-based lidar system, operating with fixed azimuth and elevation angles.

The data processing methods considered here are generally used to determine particulate extinction coefficient profiles in clear and moderately turbid atmospheres. In addition to the common problems of determining the lidar solution boundary value and selecting a reasonable backscatter-to-extinction ratio, in clear atmospheres further difficulties occur when separating the molecular and particulate scattering components. For this type of situation, the particulate extinction may be only a few percent of the weighted sum, κ_W, so that differentiating between the particulate and molecular contributions is a difficult task. Moreover, it requires an accurate evaluation of the particulate backscatter-to-extinction ratio.

Nevertheless, establishing the boundary value for the solution is the first problem that must be solved while processing the data. With lidar measure-

Elastic Lidar: Theory, Practice, and Analysis Methods, by Vladimir A. Kovalev and William E. Eichinger.
ISBN 0-471-20171-5

ments made along one direction in clear and moderately turbid atmospheres, the determination of the unknown particulate loading may be achieved by using the boundary point or optical depth solutions of the lidar equation. The details of the methods as applied to clear atmospheres are examined further below.

8.1.1. Application of a Particulate-Free Zone Approach

In 1972, Fernald et al. developed practical algorithms for lidar signal processing in a two-component atmosphere. The key point of this study is that to invert lidar data, the scattering characteristics of the aerosols and molecules should be determined separately. A similar approach was used earlier by Elterman (1966) in his atmospheric searchlight studies and later in a lidar study by Gambling and Bartusek (1972). However, the study by Fernald et al. (1972) was the first in which it was clearly stated that in two-component atmospheres the extinction coefficient profile may be obtained without an absolute calibration of the lidar. To determine the lidar solution constant, the authors proposed to use the known vertical molecular backscattering profile. In this work, the idea of the optical depth solution was formulated. However, the initial version of the lidar equation solution, proposed by the authors, was based on an iterative solution of a transcendental equation. Later, Fernald (1984) summarized a general approach for the analysis of measurements in clear and moderately turbid atmospheres, an approach that is still used in most lidar measurements. This approach is based on the following principal elements: (i) the molecular scattering profile is determined from available meteorological data or is approximated from an appropriate standard atmosphere, and (ii) a priori information is used to specify the boundary value of the particulate extinction coefficient at a specific range within the measured region. These principles have been widely used in lidar measurements in clear atmospheres. The main problem that limits the application of this method in clear and moderately turbid atmospheres is related to the uncertainty of the particulate backscatter-to-extinction ratio. In such atmospheres, the accuracy of the retrieved particulate extinction coefficient is extremely dependent on the accuracy of the backscatter-to-extinction ratio used for inversion.

The most straightforward approach to lidar data processing can be used when the lidar is operating in a permanently staring mode. Such a mode assumes that the lidar data are collected over some extended time without any realignment or adjustment to the lidar system. When a long series of these measurements are made, data obtained during different weather conditions can be compared and the best data can be used to correct the rest. Such an approach may be especially effective when relevant data from independent atmospheric measurements are available for the analysis. If such data are not available, the lidar signals measured during the cleanest days may be used as reference data. This approach was used, for example, by Hoff et al. in 1996 during an aerosol and optical experiment in Ontario, Canada. A monostatic lidar at 1.064 μm operated in a permanent upward staring mode over a long period. This allowed a check of the lidar calibration with lidar data obtained during the cleanest days. At a selected altitude range, the profile measured on clear days was assumed to be the result of purely molecular scattering. The data obtained during other days were processed by referencing the signal to the pure Rayleigh scattering. A typical calibration procedure was used in which the ratio of the lidar signal obtained in the presence of particulate loading to that obtained on the clear days was calculated. Clearly, it is difficult to estimate the accuracy of the retrieved data based on such an assumption unless relevant atmospheric information is available. Nevertheless, this type of straightforward approach is quite useful when investigating the characteristics and dynamics of atmospheric processes in time.

The assumption of the existence of an aerosol-free region within the lidar operating range is often used in analyzing tropospheric and stratospheric measurements. The lidar returns from such an area may be considered as a reference signal to determine the solution constant. This, in turn, makes it possible to determine the particulate extinction coefficient profile in all other areas, that is, in regions of nonzero particulate loading. Historically, the method that applies lidar signals from aerosol-free areas was proposed by Davis (1969) for the investigation of cirrus clouds. Later it was widely used for studies of any weakly scattering atmospheric layers, especially layering that is invisible to the unaided eye. This was a time when the scientific community was focused on possible climatic effects associated with thin aerosol layers, especially cirrus clouds. The problem initiated a large number of lidar programs. Extended observations of cirrus clouds were made with a set of instruments including different lidar systems (Platt, 1973 and 1979; Hall et al., 1988; Sassen et al., 1989; Grund and Eloranta, 1990; Sassen and Cho, 1992, Ansmann et al., 1992, etc.) In these and other studies, different versions of the algorithms were developed. However, in the main, they used lidar reference signals obtained from areas assumed to be aerosol free as references.

Before data processing formulas are presented, several remarks should be made concerning multiple-scattering effects in measurements of optically thin clouds. Multiply scattered light from cloud particulates is a source of the most significant difficulties in lidar signal inversion. There currently are no reliable and accurate methods to estimate the effects of multiple scattering or to adjust the signal to remove these effects. Researchers in practical situations tend to avoid using awkward and complicated theoretical formulas to calculate and compensate for multiple-scattering components in backscattered light. Instead, it is more common to make a simple correction to the transmission term of the lidar equation. The basis for this is as follows. When the lidar signal is contaminated by multiple scattering, the use of the conventional lidar equation [Eq. (5.14)] to determine the cloud extinction will distort the retrieved extinction coefficient profile within the cloud. This distortion is caused by strong forward scattering of the light from large-size cloud particles. The most common approach to compensate this effect is to apply an additional constant factor in the transmission term of the lidar equation (Platt, 1979).

One can consider the reduced optical depth obtained with the conventional single-scattering lidar equation as effective optical depth, $\tau_{p,eff}(r)$. To restore the actual optical depth within the cloud, which is larger than $\tau_{p,eff}(r)$, an artificial factor $\eta(r)$ is introduced, which is assumed to be less than 1. The actual optical depth $\tau_p(r)$ is related to $\tau_{p,eff}(r)$ by the simple formula (Section 3.2.2),

$$\tau_{p,eff}(r) = \eta(r)\tau_p(r) \tag{8.1}$$

With the multiple-scattering factor η, the original lidar equation [Eq. (5.14)] for a vertically staring lidar can be rewritten in the form

$$P(h)h^2 = C_0 T_0^2 [\beta_{\pi,p}(h) + \beta_{\pi,m}(h)] \exp\left\{-2\int_{h_0}^{h} [\eta(h')\kappa_p(h') + \kappa_m(h')]\, dh'\right\} \tag{8.2}$$

where h is the altitude above the ground surface. In the exponential term of the equation, an effective extinction coefficient is used, defined as $[\eta(h)\ \kappa_p(h) + \kappa_m(h)]$, rather than the simple sum of the particulate and molecular components, $[\kappa_p(h) + \kappa_m(h)]$. In other words, when combining the particulate and molecular extinction coefficients in the cloud, the former component must weighted by the factor $\eta(h)$. As follows from multiple-scattering theory, this factor is a function not only of the cloud microphysics but also of the lidar geometry, especially the field of view of the photoreceiver. It depends as well on the distance from the lidar to the scattering volume, the optical depth of the layer between it and the lidar, and the geometry of the cloud. However, there are no simple analytical formulas to calculate $\eta(h)$. Therefore, a variable factor $\eta(h)$ is not practical, so that the simplified condition that $\eta(h) = \eta =$ *const.* is most commonly used.

Consider a lidar equation solution based on the assumption of pure molecular scattering in some area within the measurement range used by Sassen et al. (1989) and Sassen and Cho (1992). Measurements were made with a ground-based, vertically staring lidar. The molecular profile was calculated from air density profiles obtained from local sounding data. The optical characteristics of the cirrus cloud aerosols were assumed to be invariant with height, so that the backscatter-to-extinction ratio in the cloud could also be assumed to be constant. The lidar signal was normalized to the signal at a reference point chosen to correspond with a local minimum in the lidar signal. To avoid issues related to poor signal-to-noise ratios, the aerosol-free area was chosen to be below rather than above the cirrus cloud base. If, at some altitude h_b located just below the cloud base, pure molecular scattering exists, that is, the particulate constituent $\kappa_p(h_b) = 0$, the ratio of the range-corrected signal from the cloud area, at the altitude $h > h_b$ and the reference altitude, h_b, can be written as

$$Z_r^*(h) = \frac{P(h)h^2}{P(h_b)h_b^2} = \left[\frac{\beta_{\pi,p}(h)+\beta_{\pi,m}(h)}{\beta_{\pi,m}(h_b)}\right]\exp\left\{-2\int_{h_b}^{h}[\eta\kappa_p(h')+\kappa_m(h')]\,dh'\right\} \quad (8.3)$$

where the factor η is assumed to be constant. In the study by Sassen et al. (1992), the factor η was taken as $\eta = 0.75$. Note that the use of the assumption of the pure molecular atmosphere at h_b removes the lidar equation constants C_0 and T_0^2 from the equation, that is, it eliminates the need to determine these constants.

As discussed in Section 5.2, the lidar signal must be transformed before an inversion can be made. The procedure must transform the lidar signal into a function that has a structure similar to that defined in Eq. (5.21). In this case, the authors transformed the function $Z_r^*(h)$ in Eq. (8.3) in the form

$$Z^*(x) = y(x)\exp\left[-2C\int y(x)dx\right] \quad (8.4)$$

thus the difference is that now the constant C is in the exponent.

A feature of the particular solution obtained by this method is that the aerosol backscatter coefficient $\beta_{\pi,p}$, rather than the extinction coefficient κ_p, is directly derived from the measured lidar return. Accordingly, the independent solution variable is

$$y(x) = \beta_\pi(x) = \beta_{\pi,p}(x) + \beta_{\pi,m}(x) \quad (8.5)$$

To transform Eq. (8.3) into the form in Eq. (8.4), a transformation function $Y^*(h)$ must be found that allows to one to obtain the product of the functions $Z^*(h)$ and $Y^*(h)$ in the form

$$Z^*(h) = Z_r^*(h)Y^*(h) = [\beta_{\pi,p}(h)+\beta_{\pi,m}(h)]\exp\left\{-2C\int_{h_b}^{h}[\beta_{\pi,p}(h')+\beta_{\pi,m}(h')]\,dh'\right\}$$

(8.6)

The transformation function $Y^*(h)$ can be found from Eqs. (8.3) and (8.6) as

$$Y^*(h) = \frac{Z^*(h)}{Z_r^*(h)}$$

$$= \beta_{\pi,m}(h_b)\exp\left\{-2\int_{h_b}^{h}[C\beta_{\pi,p}(h')+C\beta_{\pi,m}(h')-\eta\kappa_p(h')-\kappa_m(h')]\,dh'\right\} \quad (8.7)$$

Using the relationship between extinction and backscattering [Eqs. (5.17) and (5.18)], Eq. (8.7) can be reduced to

$$Y^*(h) = \beta_{\pi,m}(h_b)\exp\left(-2\left[C - \frac{\eta}{\Pi_p}\right]\int_{h_b}^{h} \beta_{\pi,p}(h')dh'\right)$$
$$\exp\left(-2\left[C - \frac{8\pi}{3}\right]\int_{h_b}^{h} \beta_{\pi,m}(h')dh'\right) \tag{8.8}$$

and by setting

$$C = \frac{\eta}{\Pi_p}$$

the transformation function is obtained as

$$Y^*(h) = \beta_{\pi,m}(h_b)\exp\left[-2\left(\frac{\eta}{\Pi_p} - \frac{8\pi}{3}\right)\int_{h_b}^{h} \beta_{\pi,m}(h')\,dh'\right] \tag{8.9}$$

To calculate the transformation function, it is necessary to establish or assume the molecular scattering profile with altitude, the backscatter-to-extinction ratio of the cloud aerosols, and the multiple scattering factor η. Note that the two latter quantities are assumed to be constant within the cloud. The solution for $y(x)$ is the sum of the particulate and molecular backscattering coefficients [Eq. (8.5)] and can be written in the form (Sassen and Cho, 1992)

$$\beta_{\pi,p}(h) + \beta_{\pi,m}(h) = \frac{Z^*(h)}{1 - \dfrac{2\eta}{\Pi_p}\displaystyle\int_{h_b}^{h} Z^*(h')\,dh'} \tag{8.10}$$

The formula above is notable for the presence of the ratio η/Π_p in the integral term of the denominator. Note that for a single-scattering atmosphere, where $\eta = 1$, the ratio reduces to the reciprocal of Π_p. The selection of the multiple-scattering factor $\eta < 1$ is, in fact, equivalent to the use of a corrected value of the backscatter-to-extinction ratio. This characteristic makes it possible to apply a slightly modified form of the conventional lidar equation in areas where multiple scattering cannot be ignored.

Thus, according to the cited studies, to find the vertical profile of the aerosol backscattering coefficients in high-altitude cirrus clouds, it is necessary to perform the following operations and procedures:

(1) Determine the vertical molecular scattering profile, ideally from an air density profile obtained from local sounding data;

(2) Determine a point below the cloud base at which a local minimum in the measured lidar signal occurs, and then calculate the normalized function $Z_r^*(h)$ with Eq. (8.3);

(3) Select a reasonable particulate backscatter-to-extinction ratio Π_p and a multiple-scattering factor η for use in the cloud, and calculate the transformation function $Y^*(h)$ with Eq. (8.9) and $Z^*(h) = Z_z^*(h) \cdot Y^*(h)$;

(4) Determine the profile of the total backscattering coefficient with Eq. (8.10);

(5) Determine the profile of the particulate backscattering coefficient by subtracting the molecular contribution.

Using this method, Sassen and Cho (1992) normalized their lidar signals, averaged vertically and temporally, to the signal at a point just below the cloud base. In addition to the normalization, an iterative procedure was used to adjust the derived profile. In their iteration procedure, different ratios of $2\eta/\Pi_p$ were used to find the best agreement between particulate and molecular backscattering above the cirrus cloud.

The approach described above is quite typical for measurements in clear atmospheres (see Platt, 1979; Browell et al., 1985; Sasano and Nakano, 1987; Hall et al., 1988; Chaikovsky and Shcherbakov (1989); Sassen et al., 1989 and 1992, etc.) The differences between the methods stem, generally, from the details of the methods used to normalize the lidar equation when different locations for the assumed particulate-free area are specified. For example, Hall et al. (1988) selected a reference point above the cirrus cloud. However, the method was not applicable after the 1991 eruption of Mt. Pinatubo in the Philippines. After the eruption, a long-lived particulate layer appeared that overlaid the high tropical cirrus clouds.

When estimating the accuracy of such measurements, the principal question becomes the measurement error that may occur because of ignorance of the amount of aerosol loading in the areas assumed to have purely molecular scattering. As demonstrated by Del Guasta (1998), an inaccurate assumption of a completely aerosol-free area may result an erroneous measurement result. In general, the presence of aerosol loading cannot be ignored even in regions where the lidar signal is a minimum. Such situations when no aerosol-free areas exist within the lidar measurement range were considered in studies by Kovalev (1993), Young (1995), Kovalev et al. (1996), and Del Guasta (1998). To reduce the amount of error due to incorrectly selected particulate loading at the reference point, two boundary values may be used. One boundary value is selected above the cloud layer and the other below it, so that two separated reference areas are used. This approach is analyzed further in Section 8.2.2.

At times, the lidar signal at distant ranges may be excessively noisy, so that selecting a point where the calibration is to be made becomes extremely difficult. Clearly, fitting the signal over some extended area is preferable to normalization at a point. Such a method was used, for example, in DIAL measurements made by Browell et al. (1985). Here the lidar signal was calibrated with a molecular backscatter profile determined within an extended area below the aerosol layer.

A comprehensive analysis of different methods that may be used to estimate the true minimum from a signal profile corrupted by noise is given by Russell et al. (1979). The authors pointed out that no rigorous solution for this problem is known. In a noisy profile, an estimate of the true minimum made by choosing the smallest signals may provide unsatisfactory results. This is because these signals may be corrupted by distortions that reduce the size of the signal. Choosing the minimum of a lidar signal as the best estimate of the true minimum of the atmospheric loading may introduce a significant underestimate of the aerosol loading. Such methods are especially unsatisfactory if large signal variations occur in the area of interest. Generally, the best methods are based on a normal distribution approximation for the lidar signal in the region of interest. The simplest version assumes that each deviation, Δx_i, in the profile of interest is assumed to obey a normal distribution with a mean deviation of zero. In other words, the estimate of the minimum, x_{min}, for the profile of interest may be made with a best estimate $\hat{x}$ and its standard deviation $\Delta_s x$. For example, to determine x_{min}, small groups of adjacent lidar data points are averaged together. Because the errors within the groups are likely differ in sign, their averages tend to zero. Such smoothing may significantly improve the signal-to-noise ratio in the area of interest. This, in turn, reduces the possibility that the minimum value will be corrupted by a large negative value. With a running mean, a coarse-resolution profile is then obtained and the minimum of this profile is taken as the best estimate of x_{min}. An obvious shortcoming of such a simple method is that errors over a limited averaging distance may be correlated, so that the error in the coarse profile does not approach zero. In another method, analyzed by Russell et al. (1979), the best estimate of x_{min} is taken to be the weighted mean of data points in a limited set of data. The best estimate is found as

$$x_{min} = \frac{\sum x_i w_i}{w_i} \tag{8.11}$$

where each point is weighted by the inverse standard deviation, that is,

$$w_i \equiv [\Delta_s x]^{-2} \tag{8.12}$$

The authors in the above-cited study proposed another best-estimate method. In this method, the estimate of the profile minimum is taken as a weighted mean of the data points, where the weight of each point $\hat{x}_i$ is the conditional probability $[P(\hat{x}_i - \hat{x}_m \mid x_i \leq x_m]$. The latter term is the probability of obtaining the difference $\hat{x}_i - \hat{x}_m$ under the condition that the true value x_i is less than or equal to the true value x_m. Thus the best estimate of $x_{\min}$ is found with the same formula as in Eq. (8.11), but where

$$w_i \equiv P(\hat{x}_i - \hat{x}_m \mid x_i \leq x_m) \tag{8.13}$$

Unfortunately, as stated in the study by Russell et al. (1979), none of the methods has been rigorously tested to determine the best. Thus the selection of an optimum method to determine the best fit of $x_{\min}$ for a noisy profile remains empirical, or based on numerical simulations.

It should be noted that significant errors in the retrieved particulate profile may also arise from errors in the vertical molecular extinction profiles used for the signal inversion (Donovan and Carswell, 1997). These errors may arise from uncertainties in the density profile used to determine the molecular backscatter or extinction coefficients. This is especially critical if a large error in the density profile occurs in the region that is used to normalize the lidar signal. The influence of density profile errors may be greatly reduced when simultaneous Raman lidar data are available. The Raman signal from atmospheric nitrogen can be used as a proxy for density.

It should be noted that the assumption of an aerosol- or particulate-free area can easily be applied to the formulas for a two-component atmosphere given in Chapter 5. For such an aerosol-free area in a range interval from $r_1 > r_0$ to r, Eq. (5.20) is reduced to

$$P(r) = C_0 T_0^2 T(r_0, r_1)^2 \frac{\Pi_{\mathrm{m}}(r)\kappa_{\mathrm{m}}(r)}{r^2} \exp\left[-2\int_{r_1}^{r} \kappa_{\mathrm{m}}(r')\, dr'\right] \tag{8.14}$$

where $T(r_0, r_1)^2$ is the total two-way transmittance over the range interval (r_0, r_1). For an atmosphere with purely molecular scattering, $\kappa_{\mathrm{m}}(r) = \beta_{\mathrm{m}}(r)$ and $\Pi_m(r) = 3/8\pi = const.$ Accordingly, after multiplying Eq. (8.14) by r^2 and with $Y(r)$ defined in Eq. (5.67), the function $Z(r)$ may be obtained as

$$Z(r) = C_0 C_Y T_0^2 T(r_0, r_1)^2 \left(\frac{3/8\pi}{\Pi_{\mathrm{p}}}\right) \beta_{\mathrm{m}}(r) \exp\left[-2\left(\frac{3/8\pi}{\Pi_{\mathrm{p}}}\right)\int_{r_1}^{r} \beta_{\mathrm{m}}(r')\, dr'\right] \tag{8.15}$$

Eq. (8.15) has the same structure as Eq. (5.68). The only difference is that the function $\kappa_{\mathrm{W}}(r)$ in the aerosol free area is reduced to

$$\kappa_{\mathrm{W}}(r) = \frac{3/8\pi}{\Pi_{\mathrm{p}}} \beta_{\mathrm{m}}(r) \tag{8.16}$$

Note that the constant Π_{p} in the above formulas no longer has a physical meaning. It is now only a mathematical factor selected to enable the calculation of the transformation function $Y(r)$. It does not matter what numerical value is used for Π_p in the areas where $\kappa_p(r) = 0$. The only requirement is that the same positive value must be used both for the transformation function $Y(r)$ and for determining $\kappa_W(r)$ in Eq. (8.16).

8.1.2. Iterative Method to Determine the Location of Clear Zones

In moderately clear atmospheres, an area with minimal aerosol loading within the lidar operating range may be established by an iterative procedure (Kovalev, 1993). As in the methods considered above, a vertical molecular extinction profile must be known to extract the profile of the unknown particulate component. The initial assumption is that, within the lidar operating range, a restricted area exists where the relative particulate loading is least. After this area is determined, the ratio of the particulate to molecular extinction coefficients [Eq. (6.22)]

$$R(r) = \frac{\kappa_p(r)}{\kappa_m(r)}$$

is chosen and used for this area as a boundary value. Thus the determination of the boundary condition is reduced to the choice of a reasonable value for the ratio $R(r)$ in the clearest part of the lidar operating range. For a particulate-free area, the ratio $R(r) = 0$. The more general approach assumes that no aerosol-free area exists within the lidar operating range, so that at any point, $R(r) > 0$. In this case, some area exists where the ratio $R(r)$ is least. Note that here the idea of a relative rather than absolute particulate loading is used, that is, the clearest area is one in which the ratio $R(r)$ is a minimum. An important feature in this approach is the use of an iterative procedure that makes it possible to examine the signal profile and find a least aerosol-loaded area. In this range interval, the boundary value of $R(r)$ is then specified. However, the minimum value of $R(r)$, which is taken as the boundary value of the lidar solution, must generally be established or taken a priori. This method may be most useful with measurements made by a ground-based lidar in a cloudless atmosphere, when the least polluted air is mostly at the far end of the lidar operating range. Here, the stable far-end boundary solution is applied. Note also that the iterative method makes it possible to use either a constant or a range-dependent backscatter-to-extinction ratio.

Consider the method for determining the location of the area with the least aerosol loading. The iteration procedure used here is similar to that described in Section 7.3.3. However, in this case, the total extinction coefficient is rewritten as

$$\kappa_t(r) = \kappa_m(r)[1 + R(r)] \tag{8.17}$$

With Eq. (8.17), the basic solution used for the iteration [Eq. (7.32)] can be rewritten in the form

$$\kappa_m(r)[1+R^{(i)}(r)]=\frac{0.5Z^{(i)}(r)}{\dfrac{I^{(i)}_{max}}{1-T^2_{max}}-I^{(i)}(r_0,r)} \tag{8.18}$$

From Eq. (8.18), the two-way transmittance T^2_{max} can formally be written as

$$T^2_{max}=1-\frac{2I^{(i)}_{max}}{\dfrac{Z^{(i)}(r)}{\kappa_m(r)[1+R^{(i)}(r)]}+2I^{(i)}(r_0,r)} \tag{8.19}$$

which is valid for any range r within the range $r_0 \le r \le r_{max}$. In the measurement range, the ratio $R(r)$ may vary within some interval between minimum and maximum values. Because the quantity T^2_{max} is always a positive value, this also limits the possible values of $R(r)$ in Eq. (8.19). Accordingly,

$$R^{(i)}(r)<\frac{Z^{(i)}(r)}{2\kappa_m(r)[I^{(i)}_{max}-I^{(i)}(r_0,r)]}-1 \tag{8.20}$$

For any given molecular profile, Eq. (8.20) establishes the largest values that the ratio $R(r)$ may assume for any range r, that is, it also puts some restrictions on the lidar equation solution from above. In other words, when $\kappa_p(r) = 0$, the value of the ratio $R(r)$ may only range from 0 to the value defined in Eq. (8.20).

To obtain the profile $R(r)$, it is necessary to establish the location of the distant area with the least particulate loading. An iteration procedure may be used to determine this location. The most stable results are generally obtained for situations in which the particulate loading decreases toward the far end of the measurement range. To determine the least polluted area, that is, the area where $R(r)$ is minimum, an auxiliary function must be initially determined over the range from r_0 to r_{max}. The function γ is found with a formula similar to Eq. (8.19). The only difference is that here the minimum ratio, $R_{min,b}$ is used instead of a variable $R(r)$, that is

$$Y^{(i)}(r,R_{min,b})=1-\frac{2I^{(i)}_{max}}{\dfrac{Z^{(i)}(r)}{\kappa_m(r)(1-R_{min,b})}+2I^{(i)}(r_0,r)} \tag{8.21}$$

A practical procedure for lidar signal inversion includes at least two series of iterations. First, a value for the minimum of the ratio $R(r_b) = R_{min,b}$ is specified in the clearest area of the examined range, at r_b, to initiate the iteration process. The best initial assumption is that $R_{min,b} = 0$. This initial assumption assumes the existence of some zone (or even a point) within the lidar operating range where only molecular scattering takes place. With this assumption, the iteration is triggered as described in Section 7.3.3. Note that the initial iteration with $R_{\min,b} = 0$ must be made even if $R_{\min,b}$ is obviously not equal to 0. The reason is that an iteration with $R_{\min,b} = 0$ produces an initial profile with the minimum possible positive values of the particulate extinction coefficient profile.

Thus, for the first iteration series, the profile of $\gamma(r, R_{\min,b})$ is calculated with $R_{\min,b} = 0$. After that, the minimum value of the function $\gamma_{\min}(r, R_{\min,b} = 0)$ is determined within the range $(r_0, r_{\max})$. Then the iteration cycle is executed in the same way as shown in Section 7.3.3. With the calculated value of $\gamma_{\min}(r, R_{\min,b} = 0)$ used instead of $T^2_{\max}$, the extinction coefficient $\kappa_p^{(1)}(r)$ is found as

$$\kappa_p^{(1)}(r) = \frac{Z_r(r)}{\dfrac{2I_{r,\max}}{1-\gamma_{\min}(r, R_{\min,b} = 0)} - 2I_r(r_0, r)} - \kappa_m(r) \tag{8.22}$$

Just as with Eq. (7.32) in Section 7.3.3, $Z_r(r)$ is the range-corrected signal $Z_r(r) = P(r)r^2$ and $I_{r,\max}$ is the integral of $Z_r(r)$ over the range from r_0 to $r_{\max}$. After determining $\kappa_p^{(1)}(r)$, the correction function, $Y^{(2)}(r)$ is obtained with Eq. (7.35). If a range-dependent backscatter-to-extinction ratio $\Pi_p(r)$ is used, the latter must be established before the iteration and the corresponding ratio $a^{(2)}(r)$ must be calculated. After the correction function $Y^{(2)}(r)$ is obtained, the normalized profile $Z^{(2)}(r)$ is found with Eq. (7.36). With the values of $Z^{(2)}(r)$, the iteration procedure is repeated, and the following values are calculated in succession: the new profile $\gamma^{(2)}(r, R_{\min,b} = 0)$ and its minimum value; the corrected extinction coefficient profile $\kappa_p^{(2)}(r)$; the profile $Y^{(3)}(r)$ and a new normalized profile $Z^{(3)}(r)$. Note that all profiles $Z^{(2)}(r), Z^{(3)}(r) \ldots Z^{(n)}(r)$ are found by using the same original range-corrected signal $Z_r(r)$, whereas the other functions are new with each iteration. The first series of iterations is repeated until subsequent profiles of $\kappa_p^{(i)}(r)$ and $Z^{(i)}(r)$ converge. Typically from 5 to 10 iterations are needed. This completes the first series of iterations. The inversion results thus obtained apply to the condition $R_{\min,b} = 0$, that is, for the initial assumption of an aerosol-free area in the least polluted area.

In those situations in which the assumption of nonzero aerosol loading in the clearest area is believed to be more realistic, so that actual $R_{\min,b} > 0$, a second series of iterations is made. The particulate extinction coefficient at the boundary point r_b is related to the selected $R_{\min,b}$ as

$$\kappa_{p,\min}(r_b) = \kappa_m(r_b) R_{\min,b} \tag{8.23}$$

Note that this new value of $R_{\min,b}$ must be consistent with the condition given in Eq. (8.20). Otherwise, the iteration will not converge, and an unrealistic negative or infinite value of the extinction coefficient may be obtained. The chosen value of $R_{\min,b}$ must always be consistent with the condition

$$0 \le R_{\min,b} \le (R_{\min,b})_{\text{upper}}$$

that is, it is restricted both from below and from above. Here the quantity $(R_{min,b})_{upper}$ is obtained with Eq. (8.20). The upper restriction is because the transmittance T^2_{max} of the lidar operating range is also restricted ($0 < T^2_{max} < 1$). If this value can be somehow estimated, for example, by sun photometer measurements of the total atmospheric transmission, T_{total}, then $(R_{min,b})_{upper}$ can be found as the minimum value of the profile

$$[R(r)]_{upper} \le \frac{\dfrac{0.5Z(r)}{\kappa_m(r)}}{\dfrac{I_{max}}{1-T^2_{total}} - I(r_0, r)} - 1 \tag{8.24}$$

The range from $R_{min,b} = 0$ to the maximum value, $(R_{min,b})_{upper}$, defines a range over which a realistic set of lidar equation solutions with not negative $\kappa_p(r)$ may be obtained. The simplest version with $R_{min,b} = 0$, yields a robust estimate of the extinction profile in clear atmospheres, where a local region involving only molecular scattering may be reliably assumed.

8.1.3. Two-Boundary-Point and Optical Depth Solutions

As shown in the previous sections, the main problem of elastic lidar measurements along a fixed line of sight is the uncertainty in the measurement accuracy of the retrieved extinction coefficient. The key problem is that to invert the lidar return, some reference signal must be specified, such as that obtained from an aerosol-free area. The question will always remain of whether purely molecular scattering actually exists in the range where the range-corrected lidar signal is a minimum. If this assumption is wrong, it may yield large measurement errors. This problem is especially important in measurements where the area with the scattering minimum is located at the near end of the lidar operating range. Such a situation, for example, may take place in a clear atmosphere if the measurement is made by a nadir-directed airborne or satelite lidar. Here, the least polluted atmosphere is, generally, close to the lidar carrier. Accordingly, an aerosol-free area approach leads to the use of the near-end solution, which may be unstable in many situations (Chapter 5). Moreover, the presence of particulate loading in the area assumed to be particulate free, or any other irregularity in the assumed boundary conditions, may yield large systematic distortions in the derived extinction coefficient profile. With the near-end solution, these distortions may be especially large at the distant end of the measurement range. In Fig. 8.1 (a), inversion results from an actual lidar signal are shown. The data, which are typical for cloudless conditions, were obtained by a nadir-looking airborne lidar at a wavelength of 360 nm. With the method discussed in the previous subsection, the area between the altitudes ~1.9–2 km was established as the region in which the ratio $R(r)$ is a minimum. The aircraft altitude was 2.5 km, so that this area

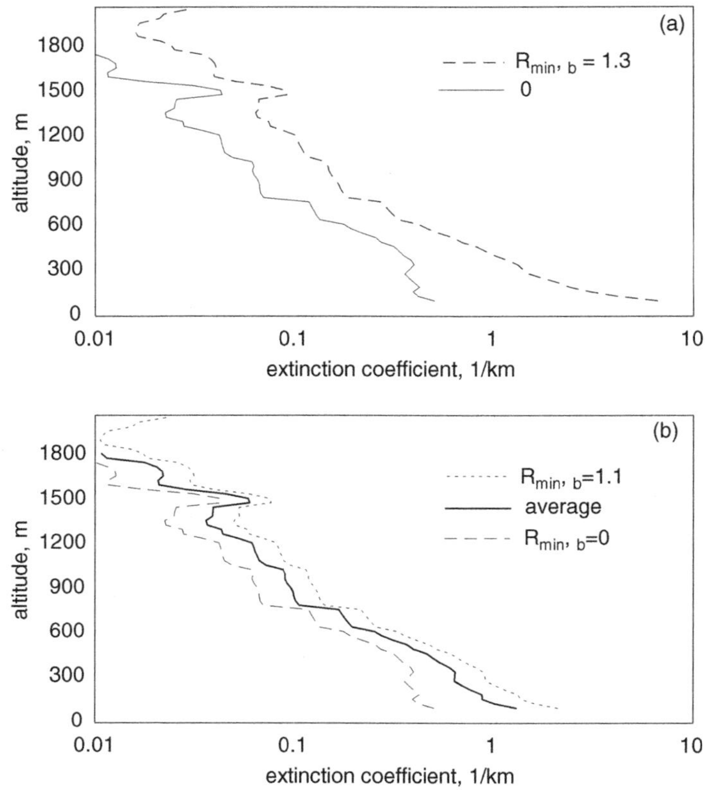

Fig. 8.1. (a) An example of the inversion of experimental data obtained with a nadir-looking airborne lidar. The curves are the particulate extinction coefficient profiles derived with extreme values of $R_{min,b}$. (b) Particulate extinction coefficient profiles obtained with the data in (a) but within a restricted range of $R_{min,b}$ from 0 to 1.1.

was located approximately 600 m below the aircraft. Thus the near-end solution with the boundary range $r_b \approx 0.6$ km was used for the signal inversion, and the anticipated increase in the particulate extinction coefficient was obtained for the lower heights, when approaching the ground surface. For the solution, the inversion procedure with different $R_{min,b}$ was used, which provided different profiles; the ratios $R_{min,b}$, that yielded sensible (positive) extinction coefficients over the whole measurement range ranged from 0 to 1.3. As expected, the retrieved extinction coefficient at the distant end of the measured range was extremely dependent on the specified boundary value, $R_{min,b}$. This becomes especially noticeable when $R_{min,b}$ is larger than 1. In such situations, the application of some restrictions for the far-end range may be helpful to narrow the possible range of the lidar equation solutions. When no independent atmospheric data are available, the application of reasonable criteria and knowledge of typical behaviors for extinction coefficient profiles in the lower troposphere can noticeably improve the quality of the retrieved data. In particular, some realistic minimum and maximum values for the extinction coefficients near the ground surface, related to the ground visibility conditions can be used as restricting criteria. These values will determine the range of possible lidar equation solutions, restricting them from below and from above. An obvious criterion that restricts the set of possible lidar equation solutions from below is that $\kappa_p(r) \geq 0$ for all points within the lidar measurement range. To constrain values from above, a restriction on the maximum value of the extinction coefficient profile is established with some reasonable maximum value of $\kappa_p(r)$ within the measurement range. Generally the maximum value may be assumed at the most distant range, that is, close to the ground surface. In the case shown in Fig. 8.1 (a), the measurements were made in clear atmospheric conditions, the lower value of visibility at the ground surface was estimated as 10–20 km. Even if the lower limit is chosen to be 10 times smaller (i.e., ~2 km), it results in a maximum boundary value of $R_{min,b} \approx 1.1$. The particulate extinction coefficient profiles, restricted by boundary values $R_{min,b} \approx 0$ and $R_{min,b} \approx 1.1$, are shown in Fig. 8.1 (b) as dashed and dotted lines, respectively. The bold curve shows the average profile.

Unfortunately, it is impossible to give a unique rule for the selection of a boundary value when using a small portion of one-directional measurement data and having no other independent data. In any case, some a posteriori analysis may be quite helpful, which includes an examination of the inversion results and checks to ensure that the data obtained are consistent with the particular optical situation. An analysis can also be made to establish whether the calculated extinction coefficient profile is reasonable at specific locations. The examination would involve determining the location of the least aerosol-polluted atmospheric areas and whether the initially specified boundary value is reasonable for these altitudes. Note also that even a moderate increase in $R_{min,b}$ in the near-end solution may cause a large increase in the extinction coefficient at the distant end of the range. Accordingly, a reasonable extinction coefficient gradient at the far end of the measurement range may be used as another restricting parameter. Reducing the indeterminacy of the lidar solution requires the rejection of uninformed guesses when estimating the boundary value. Such guesses must be replaced by a comprehensive estimate of the possible range of these values, by logical treatment of the lidar signal and an a posteriori analysis.

The advantage of the optical depth solution is that in this solution a range-integrated value is used as the reference parameter. Here, the total transmittance (or optical depth) of the atmospheric layer examined by lidar is chosen as the boundary value instead of a local extinction coefficient at a specified point or a zone. The optical depth solution uniquely restricts the solution set simultaneously from below and from above. This is because here the integrated extinction over the measurement range is fixed by the selected boundary value used for the inversion. If the total optical depth is accurately defined, the errors in the other parameters, including errors in the assumed backscatter-to-extinction ratio, are generally less influential than in the boundary point solution. This is why the optical depth solution often is used to determine profiles of the extinction coefficient in thin atmospheric layering. The boundary value, that is, the total optical depth of the layer, may be determined from the lidar signals measured above and below the layering boundaries. This technique is discussed further in Section 8.2.2.

The optical depth solution may be most useful in the following situations. First, it may be used when the atmospheric transmission can be obtained with an independent measurement. For extended tropospheric or stratospheric measurements made with ground-based lidars, a sun photometer (solar radiometer) may be used as an independent measurement of total atmospheric turbidity. In a clear, cloudless atmosphere, this instrument often allows an accurate estimation of the boundary value of the atmospheric transmittance (Fernald et al., 1972). The combination of lidar and solar measurements in clear atmospheres has been used in one-directional and multiangle measurements by Spinhirne et al. (1980), Takamura et al. (1994), and Marenco et al. (1997). Second, the optical depth solution can be used in situations in which targets, such as cloud layers or beam stops, are available in the lidar path. Such an approach was used in studies by Cook et al. (1972), Uthe and Livingston (1986), and Weinman (1988). In these studies, lidar system performance was tested by using synthetic targets of known reflectance. Finally, an optical depth solution is possible when the measurements are made in turbid atmospheres. When the optical depth of the total operating range of the lidar is 1.5 or more, the lidar signal, integrated over the total operative range, can be used as the solution boundary value (Kovalev, 1973 and 1973a; Roy et al., 1993).

There are advantages and disadvantages to the optical depth solution with a boundary value obtained with an independent photometric technique. The obvious restriction of this method is that it requires a clear line of sight to the sun as the light source. In addition, the method requires the solution of several issues. First, the maximum effective range of the lidar is always restricted by an acceptable signal-to-noise ratio, whereas the sun photometer measures the total atmospheric transmittance (or the total-column optical depth) over the entire depth of the atmosphere. Therefore, an optical depth derived from a sun photometer measurement is the sum of contributions from both the troposphere and the stratosphere. However, nearly all of the aerosol loading is concentrated in the troposphere, and only small fraction is spread over the stratosphere (volcanic events being a notable exception). Thus sun photometer data may be helpful to evaluate the boundary values for ground-based tropospheric lidars. However, after volcanic eruptions, the stratospheric particulate content may be significant, so that the optical depth of the stratospheric particulates may be noticeably increased (Hayashida and Sasano, 1993). Before the eruption of Mt. Pinatubo, the Philippines, measurements with a lidar and the sun photometer made by Takamura et al. (1994) showed almost the same optical depth. After the eruption, the optical depth obtained with the sun photometer systematically showed larger values than those obtained with the lidar. Under such circumstances, the application of sun photometer data for the determination of lidar boundary values becomes impractical, at least in clear atmospheric conditions. Because of the lack of mixing between the troposphere and stratosphere, an increase in the amount of stratospheric particulates may last for years. Another problem with the application of the optical depth solution deals with estimating the extinction coefficient in the lowest layer of the atmosphere. Ground-based lidars for upper tropospheric or stratospheric measurements have total measurement ranges of tens of kilometers. Such a lidar, generally pointed in the vertical direction, usually has a large zone of incomplete overlap between the laser beam and the field of view of the receiving telescope. In this area, the length of which is from several hundred meters to kilometers, no accurate lidar data are available. Thus a vertically staring lidar cannot provide a measurement data for the lowest, most polluted portion of the surface layer. This causes a disparity between the lidar and sun photometer measurements, which significantly complicates the use of the sun photometer data when processing lidar data. In some specific situations, for example, in a hilly region, a sun photometer measurement can be made at the elevation of the lidar overlap. However, this is not generally practical. Thus, in the general case, corrections to sun photometer data are necessary to remove the portion of the optical depth from a zone near the surface and from above the lidar measurement range. Such a correction is not a trivial task. Practically, it requires an estimate of the atmospheric turbidity at ground level (Marenco et al., 1997). For this, additional instrumentation (for example, a nephelometer) may be used to obtain reference data at the ground surface (see Section 8.1.4).

It should be noted that no additional information used for lidar signal processing can completely eliminate uncertainty associated with lidar data interpretation. In fact, lidar data inversion always requires the use of some set of assumptions, even when data from independent atmospheric measurements are available. To illustrate this statement, take for example the comprehensive experimental study by Platt (1979). In this study, the visible and infrared properties of high ice clouds were determined with a ground-based lidar and an infrared radiometer. The data from the radiometer were applied to evaluate the optical depth of the clouds and thus to accurately determine the boundary conditions for the lidar equation solution. To invert the lidar data, a set of additional assumptions had to be used. The basic assumptions used for that inversion included: (1) the backscatter-to-extinction ratio is constant within the cloud; (2) the ratio of the extinction coefficient in the visible to the infrared absorption coefficient is constant; (3) multiple scattering can accurately be determined and compensated when making the signal inversion; and (4) the ice crystals in the cloud are isotropic scatterers in the backscatter direction. Note that the latter is equivalent to the assumption that the backscatter-to-extinction ratio is independent of crystal shape. Clearly, all of these assumptions may only be approximately true. Therefore, each of them is a source of additional uncertainty in the measurement results. What is the worse, the measurement uncertainty of the retrieved data cannot be reliably evaluated.

The problems that arise in any practical lidar measurement are related to the number and type of assumptions (often made implicitly) used to invert the lidar signal. Many straightforward attempts have failed to achieve a unique lidar equation solution that would miraculously improve the quality of inverted lidar data. Even the most convoluted solutions [such as Klett's (1985) far-end solution] have not resulted in a noticeable improvement of practical lidar measurements. It appears that the only way to obtain a real improvement in inverted elastic lidar measurements is to revise in some way the general approach, that is, to apply new principles to the approach by which lidar data are processed. In particular, the combination of different lidar techniques (elastic, Raman, and high-resolution lidars) has produced quite promising results. The most significant problems related to such a combination are discussed briefly below.

A common feature of conventional single-directional lidar inversion methods is the "lack of memory." Even when processing a set of consecutive returns, each measured signal is considered to be independent and in no way related to the others. Every inversion is made independently, and the lidar equation constant is determined individually for each inverted profile. Meanwhile, it is reasonable to assume that, in the same set of consecutive measurements, the solution constants are at least highly correlated, if not the same value. The same observation is valid for the scattering parameters of the atmosphere, at least in adjacent areas. However, neither the statistics of the signals nor the uncertainties in the boundary values are taken into account in commonly used computational techniques. To overcome this limitation of lidar inversion methods, Kalman filtering may be helpful. The application of this technique was analyzed in studies by Warren (1987), Rue and Hardesty (1989), Brown and Hwang (1992), Grewal and Andrews, (1993), and Rocadenbosch et al. (1999). In this technique, the information obtained from previous inversions is taken into account when inverting the current signals. Having new incoming signals, the Kalman filter updates itself by estimating the inconsistencies between the parameters taken a priori and those obtained during current inversions. At every step of the process, a new, improved a posteriori estimate is made. The key point of any such technique is that to perform the computations, some set of criteria must be used, for example, a statistical minimum-variance criterion (Rocadenbosch et al., 1999). In other words, to use a Kalman filter for lidar data inversion, an a priori assumption on the signal noise characteristics is necessary in addition to the general assumptions such as the behavior of the backscatter-to-extinction ratio. If these characteristics are accurately established, even atmospheric nonstationarity effects can be overcome. On the other hand, if reliable a priori knowledge is not available, the advantage of Kalman filtering is lost. In that case, its estimates have no particular advantages compared with the conventional estimators. This latter drawback is the main reason why, until now, these methods are rarely used in practical measurements.

Simple conventional estimators, such as the standard deviation, have also been used to interrelate consecutively obtained returns when processing. As shown in Chapter 7, the unknown spatial variation of the backscatter-to-extinction ratio of the particulate scatterers is a dominant factor that causes ambiguity in the lidar equation solution. This is why the reliability of lidar measurement data is often open to question. In highly heterogeneous atmospheres, an accurate elastic lidar inversion may be made only when the spatial behavior of the ratio along the lidar line of sight is adequately estimated. If no information on the backscatter-to-extinction ratio is available, the commonly used approximation is a range-independent ratio. However, as shown in Chapter 7, this assumption is often too restrictive, so that it is generally true in horizontal-direction measurements, and then only in a highly averaged sense. The backscatter-to-extinction ratio may be assumed invariant over uniform and flat ground surfaces when no local sources of particulate heterogeneity exist such as, for example, a dusty road. The spatial behavior of the backscatter-to-extinction ratio in sloped or vertical directions is essentially unknown, and the assumption of an altitude-independent ratio may yield inaccurate measurement results. Therefore, an inelastic lidar technique, such as the use of Raman scattering or high-spectral-resolution lidars, may be helpful to estimate the spatial behavior of the backscatter-to-extinction ratio. The combination of the elastic and inelastic scattering measurements appears promising (Ansmann et al., 1992a; Reichard et al., 1992; Donovan and Carswell, 1997). It should be stressed, however, that the inaccuracies of inelastic measurements must be considered when estimating the merits of such a combination. Inaccurate measurement results obtained with inelastic lidar techniques may significantly reduce the gain of this instrument combination. Currently all of the inelastic methods are short ranged or require the use of photon counting, which requires long averaging times. Large measurement uncertainties may occur because of a nonstationary atmosphere and the nonlinear nature of averaging (Ansmann et al., 1992) or because of the influence of multiple scattering (Wandinger, 1998). In regions of local aerosol heterogeneity, the errors in inelastic lidar measurements are generally increased. Therefore, the areas of aerosol heterogeneity must be established when data processing is performed.

8.1.4. Combination of the Boundary Point and Optical Depth Solutions

As shown in the previous section, in situ measurements of atmospheric optical properties, made independently during lidar examination of the atmosphere, may be helpful for lidar signal inversion. Such measurements allow one to avoid, or at least to minimize, the need for a priori assumptions when lidar data are processed. This, in turn, may significantly improve the reliability and accuracy of the retrieved data. Nephelometer, sun photometer, and radiometer are the instruments most commonly used simultaneously with lidar (Platt, 1979; Hoff et al., 1996; Marenco et al., 1997; Takamura et al., 1994; Sasano, 1996; Brock et al., 1990; Ferrare et al., 1998; Flamant et al., 2000; Voss et al., 2001). However, the practical application of such additional information meets some difficulties. To date, no generally accepted lidar data processing technique is available that applies the data obtained independently with such instruments. This is primarily because of the quite different measurement volumes of lidars, nephelometers, and sun photometers or because of poor correlation between lidar backscatter returns and the scattered radiation intensity measured by radiometer.

The problems related with the application of independent data obtained with a sun photometer for lidar signal inversion procedure were discussed in previous section. Inversion of lidar data with the use of nephelometer data also makes it possible to avoid a purely a priori selection of the solution boundary value. Moreover, unlike a sun photometer or radiometer, the use of a nephelometer adds fewer complications, and therefore this instrument often yields more relevant and useful reference data for lidar inversion. However, the practical application of the nephelometer data is an issue. The near-end boundary solution is most relevant to the measurement scheme used when the nephelometer is located close to the lidar measurement site. However, this solution is known to be unstable. In addition, the application of the near-end solution is also exacerbated by the presence of an extended "dead zone" near the lidar caused by incomplete overlap.

Despite these difficulties, the nephelometer is the instrument most widely used with lidar, particularly during long-term lidar studies to investigate aerosol regimes in different regions. For example, such observations were made during the Aerosols 99 cruise, which crossed the Atlantic Ocean from the U.S. to South Africa (Voss et al., 2001). Here extensive comparisons were made between integrating nephelometer readings and data of a vertically oriented micropulse lidar system. Brock et al. (1999) investigated Arctic haze with airborne lidar measurements of aerosol backscattering along with nephelometer measurements of the total scattering. Extensive airborne lidar measurements were made over the Atlantic Ocean during a European pollution outbreak during ACE-2 (Flamant et al., 2000). Here the aerosol spatial distribution and its optical properties were analyzed with data of an airborne lidar, an on-board nephelometer, and a sun photometer.

In the studies by Kovalev et al. (2002), an inversion algorithm was presented for combined measurements with lidar and nephelometer in clear and moderately turbid atmospheres. The inversion algorithm is based on the use of near-end reference data obtained with a nephelometer. The combination of the near-end boundary point and optical depth solutions seems to be practical for measurements in clear atmospheres. Such a combination allows one to obtain a stable solution without the use of the assumption of an aerosol-free area within the lidar measurement range. For data retrieval, the conventional optical depth solution algorithm [Eq. (5.83)] is used, which in the most general form can be written as

$$\kappa_p(r) = \frac{Z(r)}{\dfrac{2I_{max}}{1 - V_{max}^2} - 2\int\limits_{r_0}^{r} Z(r')\,dr'} - a(r)\kappa_m(r) \tag{8.25}$$

To determine $\kappa_p(r)$, it is necessary to know the molecular extinction coefficient profile $\kappa_m(r)$ and the backscatter-to-extinction ratio along the lidar examination path to calculate the ratio of the molecular to particulate backscatter-to-extinction ratio, $a(r)$. Note that, depending on the atmospheric conditions, the particulate backscatter-to-extinction ratio may be either range independent or range dependent, for example, stepped over the measurement range. The key point of the use of the solution is that here $(V_{max})^2$ is estimated from nephelometer rather than sun photometer data. This is achieved by a procedure that matches the extinction coefficient retrieved from the lidar data in the near zone to the extinction coefficient obtained from nephelometer measurements. Because of the lidar incomplete overlap zone, the value of the extinction coefficient $\kappa_p(r)$ cannot be retrieved with Eq. (8.25) at the point $r = 0$, where the nephelometer is most easily located. Therefore, a more sophisticated procedure is proposed to combine the lidar and nephelometer measurements. This is based on the assumption that the extinction coefficient over the lidar near-field zone changes monotonically or remains constant. Accordingly, the boundary condition is reduced to the assumption that a linear or a nonlinear fit to the extinction coefficient profile, found for a near-field range interval from r_0 to $r_0 + \Delta r$ (i.e., over a range interval just beyond the incomplete overlap area) can be extrapolated to the lidar zone of incomplete overlap $(0, r_0)$. In the simplest case of a linear change in $\kappa_p(r)$, the extinction coefficient at the lidar location, $\kappa_p(r = 0)$, can be found from the linear fit for $\kappa_p(r)$ over the zone Δr just beyond the incomplete overlap zone

$$\kappa_p(r) = [\kappa_p(r = 0)] + br \tag{8.26}$$

where b depends on the slope of the extinction coefficient profile over the zone Δr. Obviously, b can be positive or negative, and its value becomes zero for a range-independent $\kappa_p(r)$. If the retrieved extinction coefficient profile shows a significant nonlinear change over this range Δr, a nonlinear fit may be used. The simplest variant is the application of an exponential approximation for the extinction coefficient over the range of interest. In this case, the dependence in Eq. (8.26) may be transformed into the form

$$\ln \kappa_p(r) = \ln[\kappa_p(r = 0)] + b_1 r \tag{8.27}$$

The best initial value of $V^2_{max,init}$ that allows starting the procedure of equalizing the nephelometer and lidar data is obtained by matching the reference data obtained by the nephelometer to a nearest available bin of the lidar signal. In particular, the value of $V^2_{max,init}$ may be found from Eq. (8.25) by taking $r = r_0$ to obtain

$$V^2_{max,init} = 1 - \frac{2\kappa_W(r_0)}{Z(r_0)} \int_{r_0}^{r_{max}} Z(x)\,dx$$

where $\kappa_W(r_0)$ is the total of the nephelometer reference value, $\kappa_p(r_0)$, and the product $a\kappa_m(r_0)$. The latter term can be ignored when measuring in the infrared, where the inequality $\kappa_p(r_0) \gg a\kappa_m(r_0)$ is generally true, at least on and near the ground. Note that a negative value of $V^2_{max,init}$ obtained with this formula means that an unrealistic value of $\kappa_W(r_0)$ or Π_p was used for the inversion. The presence of a large multiple-scattering component in the signal, especially at the far end of the measurement range, may also yield a negative value of $V^2_{max,init}$ (Kovalev, 2003a).

Unlike the conventional near-end solution, which may yield erroneous negative or even infinite values for the extinction coefficient, the combination of near-end and optical depth solutions yields most realistic inversion data. The method "refuses" to work if the boundary conditions or assumed backscatter-to-extinction ratios are unrealistic, that is, these do not match to the measured lidar signal. One can easily understand this by comparing the solution in Eq. (8.25) with the conventional near-end solution. As follows from Eqs. (5.75) and (5.34), the latter can be written as

$$\kappa_p(r) = \frac{Z(r)}{\dfrac{Z(r_b)}{\kappa_W(r_b)} - 2\displaystyle\int_{r_b}^{r} Z(x)\,dx} - a\kappa_m(r) \tag{8.28}$$

where r_b is a near-end range for which the reference value of the extinction coefficient, $\kappa_p(r_b)$ must be known to be transformed to the boundary value $\kappa_W(r_b)$. Thus the only (and fundamental) difference between Eqs. (8.25) and (8.28) is that the first terms in the denominator of the right-hand side differ. In Eq. (8.28) the two terms in the denominator are nearly independent, at least when r is large compared to r_b, whereas the two integrals in the denominator of Eq. (8.25) are highly correlated. Moreover, the level of the correlation between the integrals in Eq. (8.25) increases with the increase of range r toward r_b. As follows from general error analysis theory, the covariance becomes large in such situations, and it will significantly influence the measurement accuracy. Unlike the solution in Eq. (8.28), an overestimation of the boundary value in Eq. (8.25) cannot result in a dramatic increase of the measurement error with divergence of $\kappa_p(r) \to \infty$ toward a pole (see Section 6.2.2). Simply speaking, with Eq. (8.28), one can obtain infinite and negative $\kappa_p(r)$

[for example, if $\kappa_W(r_b)$ is underestimated], whereas the difference between the integrals in the denominator of Eq. (8.25) is always positive. If the atmospheric optical properties have not been assumed with sufficient accuracy, for example, if the backscatter-to-extinction ratio is badly underestimated, matching the extinction coefficients retrieved from lidar data with Eq. (8.25) and nephelometer becomes impossible due to the constraint of $0 < V_{max}^2 < 1$. In this case, the extinction coefficient at $r = 0$, obtained from the linear fit over the regression range Δr, is always less than the reference extinction coefficient obtained with the nephelometer.

Another advantage of the method deals with the relationship between nephelometer data from a location near the lidar with lidar data from beyond the lidar incomplete overlap area. For example, in the study by Voss et al. (2001) aerosol was probed with a nephelometer at 19-m altitude, and these extinction measurements were related to the extinction coefficient retrieved from lidar signal inversion at the lowest altitude level (75 m). The authors found that in some cases, the lidar data underestimate the extinction coefficient in the lowest layer. The likely reason was assumed to be a bias due to the difference in sampling heights between the nephelometer and that of the lowest lidar bin available for processing. The solution described here decreases or even eliminates such bias.

Finally, an additional advantage of the method arises when measuring strong backscatter signals from distant layers, for example, from cirrus clouds. The most common lidar signal inversion approach for such cases is based on the use of reference data points measured in an assumed aerosol-free area beyond and close to the layer boundaries (for example, Hall et al., 1988; Sassen and Cho, 1992; Young, 1995). Generally, the signals at the far-end area of the measurement range, above the layer, have a poor signal-to-noise ratio. Therefore, the aerosol-free area is mostly assumed below the layer and is often a dubious assumption. In the method considered in this section, neither the assumption of an aerosol-free area nor a reference point outside the layer is required for the inversion. Moreover, the inversion of signals from distant aerosol formations with strong backscattering is achievable even when the lidar returns outside the boundaries of the formation under investigation are indiscernible from noise (Kovalev, 2003). Such a real case is given in Fig. 8.2 (a–c), where an experimental signal measured in a very clear atmosphere and its inversion results are shown. The signal [Fig. 8.2 (a)] comprises three different constituents: (i) the backscattered signal from the clear atmosphere near the lidar, which extends approximately up to 1200 m, (ii) the pure background component of the signal (~170 bins), and (iii) a distant smoke plume over the range from approximately 4100 to 4500 m. Note that the backscatter signal beyond (outside) this layer is not discernible from high-frequency fluctuations of the background compoent [Fig. 8.2 (b)]. In this case, no reliable data points can be found outside, close to the layer that could be used as references. However, the extinction coefficient profile of the layer may be retrieved by using the reference data from the nephelometer located at the

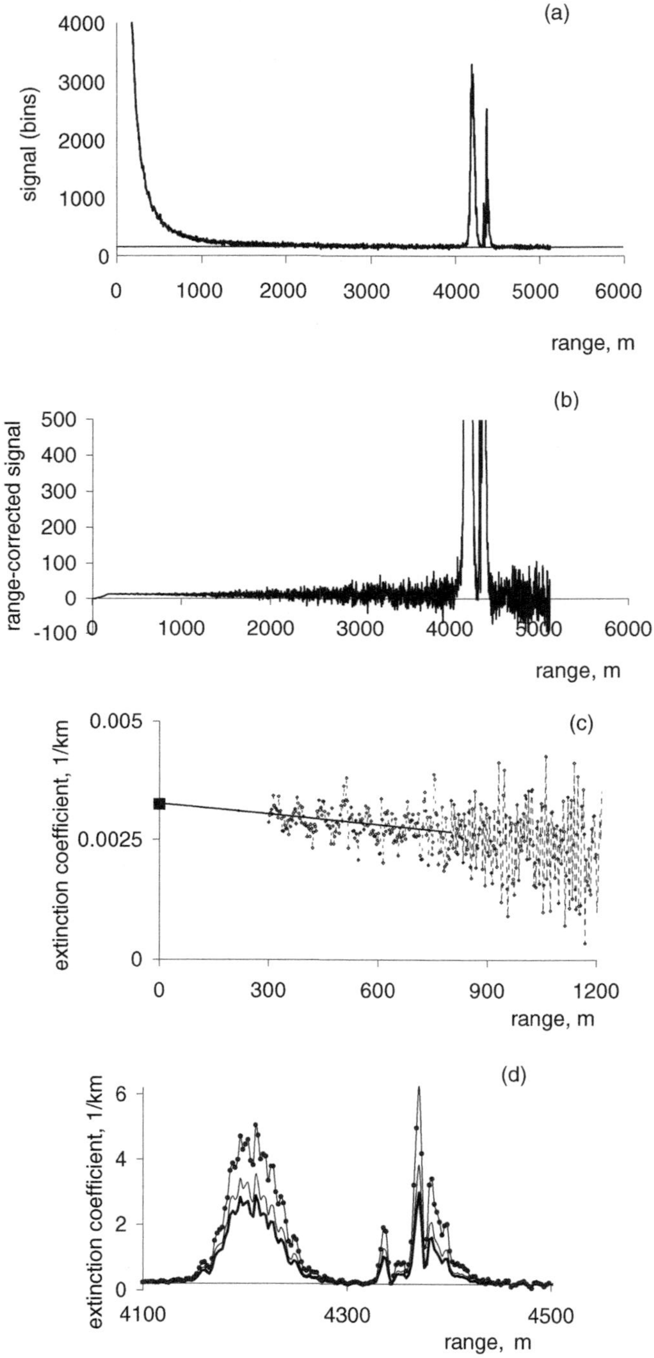
(a)
signal (bins)
4000
3000
2000
1000
0
0
1000
2000
3000
4000
5000
6000
range, m
(b)
range-corrected signal
500
400
300
200
100
0
-100
0
1000
2000
3000
4000
5000
6000
range, m
(c)
extinction coefficient, 1/km
0.005
0.0025
0
0
300
600
900
1200
range, m
(d)
extinction coefficient, 1/km
6
4
2
0
4100
4300
4500
range, m

lidar measurement site. In the above case, the nephelometer reading measured at 530 nm is 0.013 km^{-1}, and the corresponding matching value for the lidar wavelength 1064 nm is estimated to be 0.0033 km^{-1}. In Fig. 8.2 (c), this reference value is shown as a black rectangular mark. The extinction coefficient over the near area (300–1200 m) is here shown as a dashed curve, and the linear fit, found with Eq. (8.26) over the range 300–800 m, is shown as a solid line. The extinction coefficient profile derived from the signal is shown in Fig. 8.2 (d). The backscatter-to-extinction ratios for the clear and smoky areas are selected a priori. For the clear air, $\Pi_{p,cl} = 0.05\,sr^{-1}$. To show the influence of the selected backscatter-to-extinction ratio in the smoky areas, the extinction coefficients are calculated with $\Pi_{p,sm} = 0.05\,sr^{-1}$ (bold curve), $\Pi_{p,sm} = 0.04\,sr^{-1}$ (solid curve), and $\Pi_{p,sm} = 0.03\,sr^{-1}$ (solid curve with black circles).

Thus, when an appropriate algorithm is used, the near-end solution of the lidar equation may provide a stable inversion equivalent to the far-end Klett solution (Klett, 1981). The use of this stable near-end boundary solution allows one to take advantage of the optical depth algorithm, in which the boundary value is estimated by using independent data from a nephelometer at the lidar measurement site. For the inversion, a simple procedure is used that matches the extinction coefficient retrieved from the lidar data over the near-end range with the extinction coefficient obtained from the nephelometer readings. To avoid a bias due to the difference between the nephelometer sampling location and nearest available bins of the lidar returns, a regression procedure is applied to estimate the extinction coefficient behavior in a lidar near area. The signal inversion is based on the assumption that the particulate extinction coefficient in a restricted area close to the lidar is either range independent or changes monotonically with the same slope over that near area. Accordingly, the estimated behavior of the extinction coefficient profile retrieved from a set of the nearest bins of the lidar signal (within the zone of complete overlap) may be extrapolated over the zone of the incomplete lidar overlap.

The solution presented here has significant advantages in comparison to the conventional near-end boundary solution. First, it is stable, equivalent to the conventional optical depth solution. It simply refuses to work if the involved data are not compatible. Second, the inversion of signals from distant aerosol formations with strong backscattering is achievable even when an extended zone exists between the distant formation and the lidar near range in which the lidar returns are indiscernible from noise. The solution can be used for

Fig. 8.2. Inversion of the signal from a distant smoke plume. (a) The lidar signal (bold curve) that comprises near-end backscatter return from the clear air and that from the distant smoke. The solid line shows the background offset. (b) The same signal as in (a) but after subtraction of the background offset and the range correction. To show the weak near-end signal, the scale is enlarged, so that the distant smoke plume signal is out of scale. (c) The extinction coefficient in the nearest zone and its linear fit. (d) Smoke extinction coefficient profiles calculated with different backscatter-to-extinction ratios, 0.05 sr^{-1} (bold curve), 0.04 sr^{-1} (solid curve), and 0.03 sr^{-1} (solid curve with black circles).

two-layered atmospheres with significantly different backscatter-to-extinction ratios. Unlike conventional solutions, the solution given here does not require the determination of backscattered signals beyond the aerosol layer, as with the assumption of an aerosol-free atmosphere. Finally, the method considered here may decrease or even eliminate the bias of the retrieved profile due to the difference in sampling height between the nephelometer and the near bins of the lidar.

8.2. INVERSION TECHNIQUES FOR A "SPOTTED" ATMOSPHERE

If the use of lidars has accomplished anything, it has established that, in general, the atmosphere is neither homogeneous nor stationary. This observation makes accurate lidar data inversion quite difficult. The application of conventional assumptions of range-invariant backscatter-to-extinction ratios is often inappropriate and clearly wrong when heterogeneous layering occurs. Second, in turbid heterogeneous areas, multiple scattering may sometimes be considerable. The effects of multiple scattering must be corrected during or before data processing to obtain acceptable measurement results. Third, because of nonstationary spatial variations of the atmospheric scatterers, lidar signal averaging may not provide the correct mean values. Signal averaging is only useful in conditions when the temporal change in the scattering intensity at any averaged point is small and is approximately normally distributed. Because the particulate density influences two terms in the lidar equation, simple summing of lidar signals does not necessarily result in a correctly averaged condition. The presence of quite different aerosol loading is real and can clearly be seen when plotting multidimensional lidar scans like those shown in Chapter 2.

Lidar one-directional measurements generally comprise a set of signals measured during some time period. However, even then lidar signal inversion is often accomplished without interrelating the data inside the collection set. Data processing methodologies based on the straightforward use of the independent inversions for individual short-time signal averages have obvious deficiencies. Such methods are based on the dubious assumption that a reasonable boundary value may be established independently for any and every individual signal profile. Meanwhile, when applying this approach, the only way to establish such a solution boundary value is by using either an a priori assumption or information somehow extracted from the profile of the examined signal. It is worth keeping in mind that when the measurements are made during some extended time and the measurement conditions significantly vary, the best lidar data may be found and used as reference data in an a posteriori analysis.

A two-dimensional image of the set of lidar shot profiles contains much more information than a one-directional lidar signal or a pair of signals in the two-angle method. Obviously, with multiangle measurements, independent processing of the data in each line of sight is not productive. The inversion solutions made in adjacent angular directions independently may be inconsistent if the boundary conditions are not accurately estimated. In other words, the data of the adjacent lines of sight are related to each other, and the atmosphere can often be considered to be locally homogeneous.

The multiangle or two-angle methods, which are considered in the next section, allow estimation of the boundary conditions using overall information from different lines of sight. To achieve an improved lidar signal inversion result, a set of lidar shots, rather than the signals from each separate line of sight should be processed. However, before inversion of these signals, analyzed in Chapter 9, those angles or segments must be identified and excluded where the assumptions of horizontal homogeneity and constant backscatter-to-extinction ratio are obviously wrong. Such areas can be identified by examining two-dimensional images of the range-corrected lidar signals.

8.2.1. General Principles of Localization of Atmospheric "Spots"

The inversion formulas given in Chapter 5 are based on rigid assumptions that often are not true for local areas that are nonstationary. When local nonstationary heterogeneities are found within the volume examined by the lidar, it is reasonable to exclude such areas before using conventional inversion formulas. Moreover, it can be stated with certainty that an improvement in the accuracy of the measurements requires that the lidar data processing procedure include the separation of the signal data points from local aerosol layers and plumes from the signals from the background aerosols and molecules. This can be done by using the information contained in the lidar signal profiles themselves. Lidars can easily detect the boundaries between different atmospheric layers, and one can easily visualize the location and boundaries of heterogeneous areas. Two-dimensional images of the lidar backscatter signals are especially useful for this purpose. Different methodologies to process such data have been proposed (Platt, 1979; Sassen et al., 1989 and 1992; Kovalev and McElroy, 1994; Piironen and Eloranta, 1995; Young, 1995; Kovalev et al., 1996a). The general purpose of these methods is to separate the regions with large levels of backscattering variance or gradient.

Historically, the basic principles of localizing the areas of nonstationary particulate concentrations were developed in studies of atmospheric boundary layer dynamics and its evolution with visualizations of lidar data. Because the boundary layer has an elevated particulate concentration relative to that in the free atmosphere above, the dynamics of this layer are easily observed with lidar remote sensing. The convective boundary layer is generally marked by sharp temporal and spatial changes of the particulate concentration at the layer boundaries (Chapter 1). These spatial fluctuations and temporal evolution can be easily monitored with a lidar. For this, different data processing algorithms have been developed that make it possible to discriminate the atmospheric layering from clear air (Melfi et al., 1985; Hooper and Eloranta,

1986; Piironen and Eloranta, 1995; Menut et al., 1999). The discrimination methods are based on large spatial or time variations of the lidar signal intensity from the layering relative to that in clear areas. Generally, two methods are applied to localize the layer. In the first method, the shape of the lidar signal is analyzed and the spikes in the signal intensity are considered to be aerosol plumes. This method can be applied both to single and averaged lidar signals. The second method deals with the variance in the lidar signal intensity.

The first method has been used in lidar studies of atmospheric boundary layer dynamics and height evolution for almost 20 years. In the early studies, the presence and location of heterogeneous layers were determined with simple empirical criteria. For example, Melfi et al. (1985) determined the height of the atmospheric boundary layer as a point where the backscatter intensity exceeds that of the free atmosphere, at least by 25% or more. Later, such areas of the boundary layer were localized through the determination of the derivative of the lidar signal profiles with respect to altitude. This makes it possible to detect the gradient change at the transition zone from clear air to the layer. Using this approach, Pal et al. (1992) developed an automated method for the determination of the cloud base height and vertical extent by analyzing the behavior of the lidar signal derivative. Similarly, Del Guasta et al. (1993) determined the cloud base, top, and peak heights by using the derivative of the raw signal with respect to the altitude. Flamant et al. (1997) determined the height of the boundary layer by analyzing the change of the first derivative of the range-corrected signal and its standard deviation with height. The height of the boundary layer was defined as the distance at which the standard deviation reaches an established threshold value. This value was empirically established to be to three times the standard deviation in the free atmosphere. A similar approach was used by Spinhirne et al. (1997) to exclude the signals measured from the clouds in multiangle lidar measurements. The authors identified cloud presence by means of a threshold analysis of the lidar signals and their derivatives. One should note that because of the large degree of variability of real atmospheric situations, the shape of the signals may be significantly different. This makes it quite difficult to establish simple criteria for discriminating clouds with an automated method. The practice revealed that any such automatic method will sometimes fail, so that the data must always be checked by a human operator. A somewhat different approach was used in a study of urban boundary layer height dynamics over the Paris area made by Menut et al. (1999). Here the filtered second-order derivative of the averaged and range-corrected lidar signal with respect to the altitude was analyzed. The authors processed a large set of lidar data and made the conclusion that the minimum of the second derivative provides a better measure of the height of the boundary layer than the first-order derivative.

Another method that allows localization of the boundary layer is described in studies of Hooper and Eloranta (1986) and Piironen and Eloranta (1995).

The authors developed automatic methods to obtain convective boundary layer depths, cloud-base height, and associated characteristics. The method was based on the evaluation of the signal variance at each altitude. The lowest altitude with a local maximum in the variance profile was taken to be the mean height of the convective boundary layer. To avoid spurious maxima of the variance caused by signal noise or atypical signal shapes, the authors checked the behavior of the points on both sides of the maximum and also those next adjacent. Thus, to find the unknown altitude, the behavior of the variance at five consecutive altitudes was specified.

In the above studies of boundary layer dynamics, localizing the heterogeneous areas rather than lidar signal inversion was a primary purpose of the investigation. The extraction of quantitative scattering characteristics from the lidar signals in these areas is fraught with difficulty. Because of extremely large fluctuations in the backscattered signal in time and space, caused by the movement of the plumes, averaging procedures may be not practical; no normal distribution can be expected in the measured signals. However, as follows from the above-cited studies, specific criteria can be used to separate the "spotted" and clear areas, for example, by calculating a running average and the standard deviation of the signal in a two-dimensional image. It allows one to discriminate and exclude locally heterogeneous areas before determining the extinction coefficients in the background areas. For these background areas, conventional methods can be used that are based, for example, on the assumptions of an invariant backscatter-to-extinction ratio or horizontal homogeneity. The exclusion of the heterogeneous particulate "spots" before performing the inversion may significantly reduce the errors of the inverted data. Note also that, for convenience of data processing, the heterogeneous areas may be considered as independent aerosol formations that are superimposed over a background level of scattering.

On the basis of theoretical and experimental studies by Platt (1979), Sassen et al. (1989 and 1992), Piironen and Eloranta (1995), Young (1995), Kovalev et al. (1996a), and Spinhirne et al. (1997), a practical methodology for lidar data processing in "spotted" atmospheres may be suggested:

(1) Before the unknown atmospheric characteristic is extracted from a prerecorded set of lidar returns, a corresponding two-dimensional image of the lidar signal is analyzed to separate the clear or stationary zones, in which no significant plumes or aerosol layering exists, from zones of large aerosol heterogeneity.

(2) The particulate component in the stationary or background areas is found. For these areas, in which no significant particulate heterogeneity has been established, the conventional assumptions concerning the behavior of the atmospheric characteristics may be used for the signal inversion. In other words, the absence of significant heterogeneity in these zones makes it possible to apply conventional inversion algo-

rithms (Chapter 5) or to use the algorithms for two-angle or multiple angle measurements as discussed in Chapter 9.

(3) The particulate extinction coefficients found in the background areas are then used as reference values to determine the scattering characteristics of the heterogeneous layers and "spots." In other words, the extinction coefficients calculated for the stationary particulate loading are used as the boundary values for the signal inversion in the nonstationary areas. In the latter areas, the influence of multiple scattering must often be taken into consideration.

(4) The data obtained for the heterogeneous areas are superimposed on the two-dimensional image of the atmospheric background component. If no inversion method proves to be reliable to determine the extinction parameters in the nonstationary areas, these areas can be considered as blank spaces.

With these methods, a significant improvement in the accuracy of the lidar data and its reliability can be expected. This is particularly useful in studies of boundary layer dynamics, in environmental and toxicology studies, in monitoring and mapping the sources of pollution, the transport and dilution of contaminants, etc. Note that the similar approach may be used for different lidar measurement technologies, including DIAL measurements of trace gases in the atmosphere (such as ozone), for example, when examining the real accuracy of the retrieved concentration profiles.

8.2.2. Lidar-Inversion Techniques for Monitoring and Mapping Particulate Plumes and Thin Clouds

In this section, lidar inversion techniques are described for determining the extinction coefficient profiles that have spatially restricted areas of particulate heterogeneity, such as plumes, smokes, or cloudy layers. The techniques may also be applied to measurements of aerosol layering higher up in the troposphere, such as contrails or cirrus clouds.

As stated above, areas with stable atmospheric conditions and areas with nonstationary aerosol content should be analyzed separately, with different processing methodologies. For nonstationary areas, for example, when measuring the optical characteristics of optically thin cloud or dust plumes, significant problems arise when inverting the lidar data. Generally, the information that can be extracted from lidar signals from such heterogeneous areas is quite limited and not accurate. The lidar signals obtained from these areas must be processed with caution, because even the effectiveness of signal averaging in these regions becomes problematic. It is also difficult to select a reasonable value for the solution boundary value within the nonstationary area. Therefore, the boundary values for the inversion of signals in such areas are generally determined outside these areas, in the adjacent stationary (preferably in

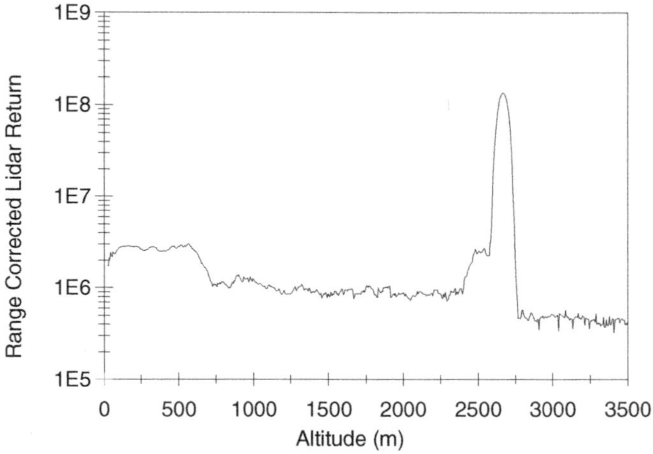

Fig. 8.3. An example of the lidar return from a cloud in which the signal below the cloud is noticeably larger than that above the cloud. The difference may be used to determine the optical depth of the cloud. The wavelength is 532 nm. Note the sharp drop in signal magnitude at 600 m, the top of the boundary layer.

aerosol free) area. This principle was used in the lidar methods beginning with the early study by Cook et al. (1972). Here, the transmittance of a smoke plume was obtained by comparing the clear air lidar return at the near side of the plume with that at the far side (Fig. 8.3). However, the difference may only be used to determine the optical depth of the cloud if the backscattering outside the cloud boundaries are the same values. More accurate results will be obtained when the air around the heterogeneous aerosol or particulate areas contains no particulates, so that it may be assumed that only purely molecular scattering takes place in the nearby region (see Browell et al., 1985; Sassen et al., 1989, etc.)

Before inversion methods for inhomogeneous thin layers are considered, the concept of an optically thin layer used below should be established. As defined by Young (1995), an optically thin cloud or any other local layer refers to an area that can be penetrated by the lidar light pulse. This means that measurable signals are present from the atmosphere on both near and far sides of the cloud and that each signal has an acceptable signal-to-noise ratio. This definition assumes a small optical depth rather than a small geometric thickness in the distant layer.

A theoretically elegant solution for determining particulate extinction coefficient for a thin aerosol layer located within an extended area of the aerosol-free atmosphere was proposed by Young (1995). Following this study, consider an ideal situation, when outside the boundaries of the thin aerosol layer, h_1 and h_2 (Fig. 8.4), only molecular scattering exists, or at least the aerosol scattering is small enough to be ignored. In this case, the clear regions below and

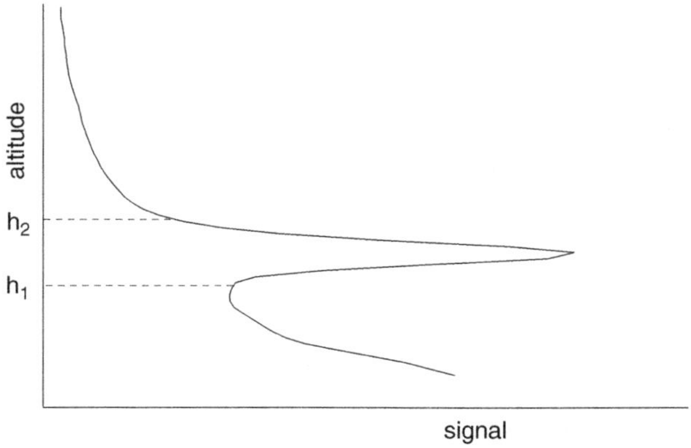

Fig. 8.4. The backscatter signal measured from a ground-based and vertically directed lidar in an atmosphere with an optically thin aerosol layer.

above the cloud can be used as the areas of the reference molecular profile. For a ground-based, vertically staring lidar, the lidar signal measured at height h above the cloud, for the altitude $h > h_2$, can be written as

$$P(h) = C_0 \frac{\beta_{\pi,m}(h)}{h^2} T_m^2(0, h) T_{cl,eff}^2(h_1, h_2) + \Delta P_0 \tag{8.29}$$

where $T_m(0, h)$ is the molecular transmittance of the layer $(0, h)$ and $T_{cl,eff}(h_1, h_2)$ is the vertical transmittance of the cloud. Because the signal may be distorted by multiple scattering, this quantity should be considered to be "effective" path transmittance. Note also that a signal offset, ΔP_0, is included in the equation.

To perform the signal inversion, a synthetic lidar signal profile for molecular scattering is first calculated as a function of altitude. Such a calculation can be based, for example, on data from a molecular density profile obtained either from local radiosonde ascents or by using mean profiles. The synthetic lidar signal profile for the molecular component may be written as

$$P_m(h) = \frac{\beta_{\pi,m}(h)}{h^2} T_m^2(0, h) \tag{8.30}$$

where the lidar signal has been normalized so that the lidar constant is unity. If only molecular scattering exists for heights above the cloud ($h > h_2$), the lidar signal can be written as

$$P(h > h_2) = [C_0 T_{cl,eff}^2(h_1, h_2) P_m(h) + \Delta P_0] \tag{8.31}$$

Eq. (8.31) can be treated as a linear equation in which $P_m(h)$ is an independent variable. With a conventional linear regression of the measured signal $P(h > h_2)$ against $P_m(h)$, both unknown constants, the product $C_0[T_{cl,eff}(h_1, h_2)]^2$ and the offset ΔP_0 can be found. On the other hand, for the heights below the cloud, that is, for $h < h_1$, another linear equation can be obtained

$$P(h < h_1) = C_0 P_m(h) + \Delta P_0 \tag{8.32}$$

Here the regression of the measured signal $P(h < h_2)$ against $P_m(h)$ determines both the unknown offset ΔP_0 and the constant C_0. With these constants, the total cloud transmittance $T_{cl,eff}(h_1, h_2)$ can be determined. With a constant multiple-scattering factor η in the cloud transmission term, as proposed by Platt (1979), this term now becomes

$$T_{cl,eff}(h_1, h_2) = \exp\left[-\eta \int_{h_1}^{h_2} \kappa_p(h')\, dh'\right] \tag{8.33}$$

Formally, once the boundary conditions are established, the particulate extinction coefficient $\kappa_p(h)$ within the thin cloud can be found. However, the result may be not reliable because of the unknown behavior of term η, which may change rather than remaining constant as the light pulse penetrates the cloud. The multiple scattering factor is the main source of the uncertainty for $\kappa_p(h)$ because it can vary with the cloud microphysics, the lidar geometry, the distance from the lidar, etc. A number of other assumptions used in this method may also be a source of errors in the retrieved profile of $\kappa_p(h)$. Thus only the transmission term, $T_{cl,eff}(h_1, h_2)$, and the total optical depth of the layer can more or less accurately be obtained if the molecular extinction coefficient and, accordingly, $P_m(h)$, are accurately estimated. This is because the use of two-boundary algorithms significantly constrains the lidar equation solution (Kovalev and Moosmüller, 1994; Young, 1995; Del Guasta, 1998).

The method proposed by Young (1995) is extended to optical situations when purely molecular scattering can be assumed either below or above the cloud layer, but not both. In such a situation, an additional backscattering profile must be measured from cloud-free sky to obtain a reference signal. The measurement schematic is shown in Fig. 8.5. The lidar at the point *L* measures the signals in two directions, *I* and *II*. When measured in direction *I*, the signal contains backscattering from a local aerosol layer, *P*, under investigation. The second measurement is made with the same (preferably) elevation angle, but in a slightly shifted azimuthal direction *II*. The signal is obtained from a cloud-free sky, and it may be used as the source for the background (reference) signal. The reference profile is found by averaging many cloud-free signals in direction *II*. Then the particular lidar signal, measured in direction *I*, is fitted to the reference signal in the corresponding region. In the simplest case of an overlying aerosol loading, purely molecular scattering is assumed below the

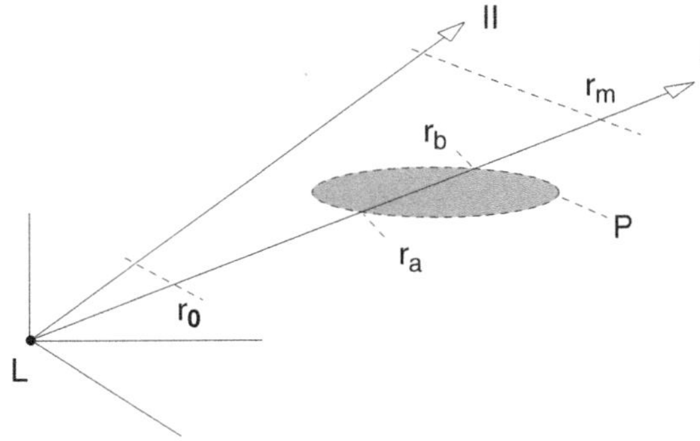

Fig. 8.5. Schematic of the lidar measurement in a "spotted" atmosphere.

aerosol layer *P*. The averaged signal profile in direction *II* is fitted and rescaled to the molecular profile in the lower area. With the assumption of the aerosol-free zone below the layer *P*, the solution constant and the extinction coefficient profiles for direction *II* can be determined and then used to calculate a reference signal as

$$W(r) = r^{-2}[\beta_{\pi,m}(r) + \beta_{\pi,p}(r)]T_m^2(0, r)T_p^2(0, r) \tag{8.34}$$

whereas the signal measured in direction *I*, at $r \geq r_b$ (Fig. 8.5), is

$$P(r \geq r_b) = C_0 r^{-2}[\beta_{\pi,m}(r) + \beta_{\pi,p}(r)]T_m^2(0, r)T_p^2(0, r)T_{cl,eff}^2(r_a, r_b) + \Delta P_0 \tag{8.35}$$

where the subscripts "*cl*" and "*p*" denote the terms related to the particulate extinction in the cloud *P* and outside it, respectively. Note that the ranges r_a and r_b are selected so as to be close but beyond the layer *P*. As follows from Eqs. (8.34) and (8.35), the signal $P(r)$ below the layer *P* then may be written as

$$P(r \leq r_a) = C_0 W(r) + \Delta P_0 \tag{8.36}$$

On the other hand, above the cloud, the signal $P(r)$ is

$$P(r \geq r_b) = C_0 T_{cl,eff}^2(r_a, r_b)W(r) + \Delta P_0 \tag{8.37}$$

With a linear fit for the dependence of $P(r)$ on $W(r)$ in Eq. (8.36), the constant C_0 and the offset ΔP_0 can be determined. After that, the effective two-way transmittance $[T_{cl,eff}(r_a, r_b)]^2$ can be found from Eq. (8.37). Just as with the previous method, an accurate determination of the extinction coefficient profile within the cloud from the term $T_{cl,eff}(r_a, r_b)$ can be made only when the contribution of multiple scattering to the signal is negligible.

For the case of an underlying aerosol or particulate layer, a solution can be found with the assumption of purely molecular scattering above the layer P. Even from purely theoretical considerations, this solution looks less practical. This is because additional assumptions and, accordingly, additional uncertainties are involved in the inversion. The thorough analysis made in the study by Del Guasta (1996) confirmed the principal advantages of the application of the two-boundary algorithms. It should be kept in mind, however, that the signals at the far end of the measured range, at $r \geq r_b$, generally have a poor signal-to-noise ratio, so that the application of such algorithms is practical only for relative thin aerosol layering.

The atmospheric "spots" and plumes often have an anthropogenic origin. Anthropogenic emissions, such as urban chimney plumes, smog spots near the highways, or stratospheric particles injected during a spacecraft launch, can be considered to be an independent particulate formation that is superimposed on the background aerosols. Similarly, some natural aerosol formations such as dusty clouds can be treated in the same way. The principle of superimposition assumes that the presence of the local spot or plume does not influence the optical characteristics of the background aerosols. Obviously, this approximation may be not valid when some physical processes take place, for example, when particles absorb moisture because of high humidity at a particular height (this typically occurs at the top of the boundary layer). Nevertheless, the assumption of independent aerosol formations, superimposed on background aerosol levels, may be fruitful for lidar data inversion. A variant of the two-boundary solution for determining the transmittance of such spots and plumes was proposed by Kovalev et al. (1996a). Here the local plume or spot under consideration was considered as a formation of particulates that is superimposed on background aerosols and molecules. Just as with the study by Young (1995), the approach assumes that a reference signal is available from an adjacent spot-free region. A set of plume-free profiles is averaged, and this average profile is used as a reference. Unlike Young's (1995) method, in the method by Kovalev et al. (1996a), the atmosphere beyond the plume is not considered to be free of aerosol loading, either above or below the plume. Second, data processing is based on an analysis of the ratio of the signals measured along directions I and II (Fig. 8.5), rather than on the regression technique.

With the multiple scattering factor η defined in Eq. (8.1), the lidar signal measured along direction I at $r_a < r < r_b$ can be written as

$$P^{(I)}(r) = C_0 T_0^2 r^{-2} [\beta_{\pi,p}^{(I)}(r) + \beta_{\pi,m}(r) + \beta_{\pi,pl}(r)] \exp\left\{-2\int_{r_0}^{r} [\kappa_p^{(I)}(r') + \kappa_m(r')]\, dr'\right\} \exp\left\{-2\int_{r_b}^{r} \eta(r')\, \kappa_{pl}(r')\, dr'\right\} \quad (8.38)$$

where $\beta_{\pi,pl}(r)$ and $\kappa_{pl}(r)$ are the volume backscatter and extinction coefficients of the plume P and the superscript (I) denotes the signal, the extinction, and the backscatter coefficients measured in direction *I*. The lidar reference signal measured along direction *II* is

$$P^{(II)}(r) = C_0 T_0^2 r^{-2} [\beta_{\pi,p}^{(II)}(r) + \beta_{\pi,m}(r)] \exp\left\{-2\int_{r_0}^{r} [\kappa_p^{(II)}(r') + \kappa_m(r')]\, dr'\right\} \quad (8.39)$$

where the superscript (*II*) denotes the extinction and backscatter coefficients measured in direction *II*. It is assumed here that any temporal instability in the emitted laser energy while measuring the signals $P^{(I)}(r)$ and $P^{(II)}(r)$ is compensated, so that C_0 does not vary during the measurement. Denoting the differences between the background backscatter and extinction coefficients in directions *I* and *II* as

$$\Delta\beta_{\pi,p}(r) = \beta_{\pi,p}^{(I)}(r) - \beta_{\pi,p}^{(II)}(r) \quad (8.40)$$

and

$$\Delta\kappa_{\pi,p}(r) = \kappa_{\pi,p}^{(I)}(r) - \kappa_{\pi,p}^{(II)}(r) \quad (8.41)$$

the ratio of the signals is written in the form

$$U(r) = \frac{P^{(I)}(r)}{P^{(II)}(r)} = \left[1 + \frac{\beta_{\pi,pl}(r) + \Delta\beta_{\pi,pl}(r)}{\beta_{\pi,p}^{(II)}(r) + \beta_{\pi,m}(r)}\right] \exp\left\{-2\int_{r_a}^{r_b} [\eta(r')\kappa_{pl}(r') + \Delta\kappa_p(r')]\, dr'\right\} \quad (8.42)$$

As the ranges r_a and r_b are selected so as to be beyond the boundaries of the plume (Fig. 8.4), $\beta_{\pi,pl}(r)$ at these points is zero, and the logarithm of the ratio of $U(r_b)$ to $U(r_a)$ is

$$\ln\frac{U(r_b)}{U(r_a)} = \Delta B(r_a, r_b) - 2\int_{r_a}^{r_b} [\eta(r')\kappa_{pl}(r') + \Delta\kappa_p(r')]\, dr' \quad (8.43)$$

where

$$\Delta B(r_a, r_b) = \ln\left[1 + \frac{\Delta\beta_{\pi,p}(r_b)}{\beta_{\pi,p}^{(II)}(r_b) + \beta_{\pi,m}(r_b)}\right] - \ln\left[1 + \frac{\Delta\beta_{\pi,p}(r_a)}{\beta_{\pi,p}^{(II)}(r_a) + \beta_{\pi,m}(r_a)}\right] \quad (8.44)$$

The terms $\Delta\beta_{\pi,p}(r_a)$ and $\Delta\beta_{\pi,p}(r_b)$ are the differences between the backscatter coefficients in the clear regions in directions *I* and *II*. If the differences are small enough, the term $\Delta B(r_a, r_b)$ may be ignored. Then the integrand in Eq. (8.43), which is related to the total optical depth of the plume, can be obtained as

$$\int_{r_a}^{r_b} [\eta(r')\kappa_{pl}(r') + \Delta\kappa_p(r')]\, dr' = -0.5 \ln \frac{U(r_b)}{U(r_a)} \tag{8.45}$$

The integral in the left side of Eq. (8.45) can be considered to be an estimate of the optical depth of the plume. It can be used as a boundary value to determine the extinction coefficient $\kappa_{pl}(r)$ within the area P. An iterative method to obtain the profile of $\kappa_{pl}(r)$ is given in study by Kovalev et al. (1996a). To determine the extinction coefficient of the plume, the backscatter-to-extinction ratio and the extinction coefficient of the background profile $\kappa_p^{(II)}(r)$ must be known, at least approximately. The analysis made by the authors of the study revealed that the solution, being constrained from above and from below by Eq. (8.45), is rather insensitive to the accuracy of both the background extinction coefficient and the backscatter-to-extinction ratio. When multiple scattering can be ignored, that is, $\eta(r) = 1$, the method yields an acceptable measurement result even if the a priori information used for data processing, is somewhat uncertain. Moreover, the method makes it possible to estimate a posteriori the reliability of the retrieved extinction coefficient profile. However, the uncertainty in the solution due to the likely presence of multiple scattering can significantly worsen the inversion results, especially the derived profile of $\kappa_{pl}(r)$.

A similar two-boundary solution for remote sensing of ozone density was proposed by Gelbwachs (1996). The ozone concentration had to be measured within the exhaust plumes of Titan IV launch vehicles. The application of the conventional DIAL methods was made particularly challenging by the injection of a large quantity (50–80 tons) of aluminum oxide particles into the stratosphere during the launch. The method proposed by the author was based on the comparison of DIAL on- and off-line signals before passage of the launch vehicle and after it, in the presence of the plume segments. As was done with the methods discussed above, Gelbwachs (1996) also assumed that plume was limited to a well-defined area, so that backscattering in the upper stratosphere, beyond the plume, might be used as a reference value.

9

MULTIANGLE METHODS FOR EXTINCTION COEFFICIENT DETERMINATION

9.1. ANGLE-DEPENDENT LIDAR EQUATION AND ITS BASIC SOLUTION

Under appropriate circumstances, the difficulties in the selection of a boundary value in slant direction measurements can be overcome with multiple-angle measurement approaches. In general case of multiangle measurements, the lidar scans the atmosphere in many angular directions at a constant azimuth, starting from a direction close to horizontal, producing a two-dimensional image known as a range-height indicator (RHI) scan. The original concepts behind multiangle measurements were developed by Sanford (1967, 1967a), Hamilton (1969), and Kano (1969), and later modified and applied in atmospheric investigations by Spinhirne et al. (1980), Rothermel and Jones (1985), Sasano and Nakane (1987), Takamura et al. (1994), Sasano (1996), and Sicard et al. (2002). The general principles of data processing in this approach are based on the assumption of a horizontally uniform atmosphere with constant scattering characteristics at each altitude. The type of horizontal layering implied by this requirement occurs during stable atmospheric conditions, generally at night. Figure 9.1 is an example of such a nocturnal, stable atmosphere at high altitudes. Note that near the surface, the atmosphere is turbulent and heterogeneous.

Under the condition of a horizontally uniform atmosphere, the optical depth of the atmosphere can be found directly from lidar multiangle mea-

Elastic Lidar: Theory, Practice, and Analysis Methods, by Vladimir A. Kovalev and William E. Eichinger.
ISBN 0-471-20171-5

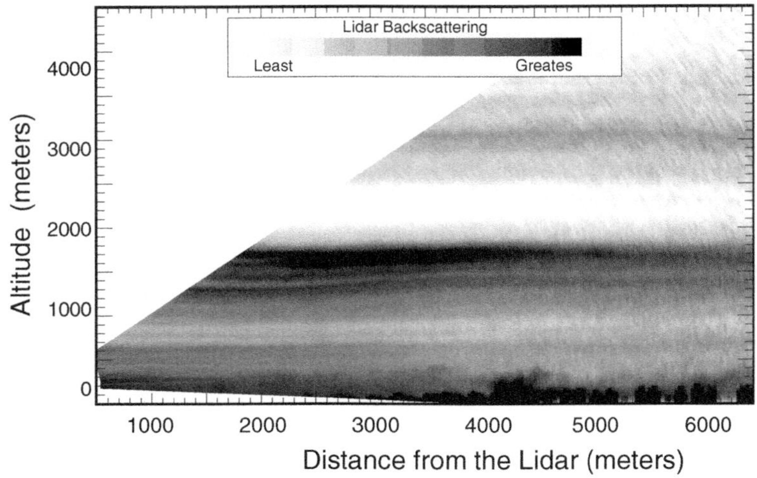

Fig. 9.1. An example of a stably stratified boundary layer over Barcelona, Spain made at 1:30 AM. A stable boundary layer will exhibit the type of horizontal homogeneity required for multiangle analysis methods.

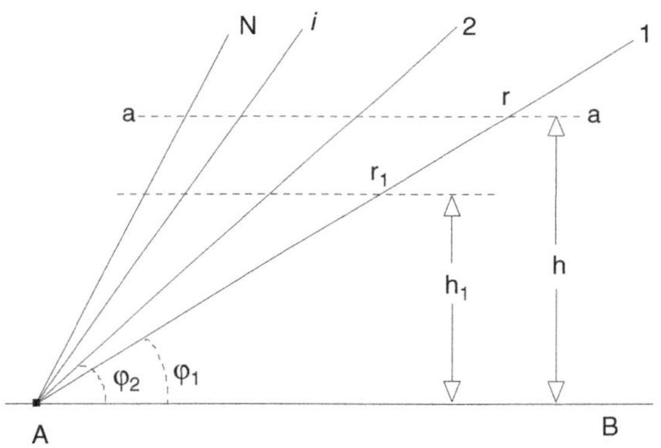

Fig. 9.2. Schematic of lidar multiangle measurements.

surements (Sanford, 1967 and 1967a; Hamilton, 1969; Kano, 1969). The data processing technique, where the atmosphere is considered to be horizontally layered like a puff pastry pie with very thin horizontal slices, is based on two principal conditions. First it is assumed that within the operating area of the lidar, the backscatter coefficient in any thin slice is constant and does not change during the time in which the lidar scans the atmosphere over the selected range of elevation angles. In other words, when the lidar scans along N different slant paths with elevation angles $\phi_1, \phi_2, \ldots \phi_N$ (Fig. 9.2), the backscatter coefficient at the each altitude h remains invariant

$$\beta_\pi(h, \phi_1) = \beta_\pi(h, \phi_2) = \ldots = \beta_\pi(h, \phi_N) = const. \tag{9.1}$$

In the simplest version considered in this section, this horizontal homogeneity is assumed be true within the entire altitude range from the ground surface to the specified maximum altitude h_{max}. If this condition is valid, the optical depth of the layer from the ground level to any fixed height h along different slant paths is inversely proportional to the sine of the elevation angle. For the elevation angles $\phi_1, \phi_2, \ldots \phi_N$, this condition may be written in the form

$$\tau(h, \phi_1)\sin\phi_1 = \tau(h, \phi_2)\sin\phi_2 = \ldots = \tau(h, \phi_N)\sin\phi_N = const. \tag{9.2}$$

where $\tau(h, \phi_i)$ is the optical depth of the atmospheric layer from the ground ($h = 0$) to the height h, measured in the slope direction with the elevation angle ϕ_i

$$\tau(h, \phi_i) = \int_0^r \kappa_t(r')dr' = \frac{1}{\sin\phi_i}\int_0^h \kappa_t(h')dh' \tag{9.3}$$

here $h = r/\sin\phi_i$. It follows from Eq. (9.2) that the optical depth in the vertical direction of the atmospheric layer $(0, h)$ can be calculated from the lidar measurement made in any slope direction and vice versa. Equation (9.3) can be rewritten as

$$\tau(h, \phi_i) = \overline{\kappa_t}(h, \phi_i)\frac{h}{\sin\phi_i} \tag{9.4}$$

where $\overline{\kappa_t}(h, \phi)$ is the mean value of the total (molecular and particulate) extinction coefficient of the layer $(0, h)$. Unlike the optical depth $\tau(h, \phi_i)$, the value $\overline{\kappa_t}(h, \phi)$ measured along any slant path of the sliced atmosphere is an invariant value for any fixed h. By substituting Eq. (9.4) in Eq. (9.2), one obtains

$$\overline{\kappa_t}(h, \phi_1) = \overline{\kappa_t}(h, \phi_2) = \ldots = \overline{\kappa_t}(h, \phi_N) = \overline{\kappa_t}(h) = const. \tag{9.5}$$

Thus, in a horizontally homogeneous atmosphere, the mean extinction coefficient of the fixed layer $(0, h)$ does not change when it is measured at different angles $\phi_1, \phi_2, \ldots \phi_N$. This feature can be used to extract atmospheric parameters from lidar measurement data. To derive a vertical transmission profile or any related parameters, such as the mean extinction coefficient, measurements are made at two or more elevation angles. Actually, the necessary information can be obtained from a two-angle measurement, that is, by making measurements only along two slant paths. Several variants of the two-angle method are considered below in Sections 9.3 and 9.4. In this section, the simplest theoretical variant is examined. This theoretical consideration clearly shows the extreme sensitivity of two-angle and multiangle methods to measurement errors, especially when the angular separation of the lidar lines of sight is small. Consider a lidar pointed alternately along two optical paths with the elevation angles, ϕ and $\phi + \Delta\phi$. To extract information on the examined atmosphere, the lidar returns must be compared at the same height. This is why in two-angle and multiangle measurements, the height h rather than the lidar range r is generally used as the independent variable. Replacing the range r by the corresponding ratios [$h/\sin\phi$] and [$h/\sin(\phi + \Delta\phi)$] in Eq. (3.11), two independent equations can be written in which the lidar signal is presented as a function of the height. For the elevation angles ϕ and $\phi + \Delta\phi$, the following equations are obtained:

$$P(h,\phi) = C_0\beta_\pi(h)\frac{\sin^2\phi}{h^2}\exp\left[\frac{-2h}{\sin\phi}\overline{\kappa_t}(h)\right] \tag{9.6}$$

and

$$P(h,\phi+\Delta\phi) = C_0\beta_\pi(h)\frac{\sin^2(\phi+\Delta\phi)}{h^2}\exp\left[\frac{-2h}{\sin(\phi+\Delta\phi)}\overline{\kappa_t}(h)\right] \tag{9.7}$$

Note that in Eqs. (9.6) and (9.7) the same constant C_0 is used for the different lines of sight, along the slant paths, ϕ and $\phi + \Delta\phi$. This can only be done if the lidar signals are normalized, that is all fluctuations in the intensity of the emitted laser energy are compensated. Such a signal normalization and extended temporal averaging is required for all types of the multiangle measurements which are based on the assumptions of atmospheric horizontal homogeneity.

Combining Eqs. (9.6) and (9.7), the solution for the mean value of the extinction coefficient, $\overline{\kappa_t}(h)$, can be obtained as

$$\overline{\kappa_t}(h) = \frac{1}{2h}\left(\frac{1}{\sin\phi} - \frac{1}{\sin(\phi+\Delta\phi)}\right)^{-1}\ln\left[\frac{P(h,\phi+\Delta\phi)\sin^2\phi}{P(h,\phi)\sin^2(\phi+\Delta\phi)}\right] \tag{9.8}$$

Using conventional methods to propagate the uncertainties in the measured signals $P(h, \phi)$ and $P(h, \phi + \Delta\phi)$ to the uncertainly in the dependent variable (Bevington and Robinson, 1992), and ignoring for simplicity the covariance term, the following formula can be derived for the relative uncertainty in the extinction coefficient, $\overline{\kappa_t}(h)$, derived with Eq. (9.8)

$$\delta\overline{\kappa_t}(h) = \frac{1}{2\tau(0,h)}\left(\frac{1}{\sin\phi} - \frac{1}{\sin(\phi+\Delta\phi)}\right)^{-1}\sqrt{[\delta P(h,\phi)]^2 + [\delta P(h,\phi+\Delta\phi)]^2} \tag{9.9}$$

where $\delta P(h, \phi)$ and $\delta P(h, \phi + \Delta\phi)$ are the relative uncertainties in the measured signal at height h at the elevation angles ϕ and $\phi + \Delta\phi$, respectively; $\tau(0, h)$ is the vertical optical depth of the layer $(0, h)$, defined as

$$\tau(0,h) = \overline{\kappa_t}(h)h \tag{9.10}$$

Note that when the angular separation $\Delta\phi$ tends to zero, the factor in brackets in Eq. (9.9) also tends to zero; accordingly, the uncertainty $\delta\overline{\kappa_t}(h)$ tends to infinity. This means that the two-angle method is extremely sensitive to the measurement errors $\delta P(h, \phi)$ and $\delta P(h, \phi + \Delta\phi)$ when the angular separation $\Delta\phi$ is small. It means that errors originating from signal noise, zero-line offset, receiver nonlinearity, and inaccurate optical adjustment of the system influence the measurement accuracy with an extremely large magnification factor.

A similar formula can be written for the uncertainty caused by the violation of the condition in Eq. (9.1), that is, by a difference in the backscattering coefficients $\beta_\pi(h, \phi_i)$ at altitude h. For the lidar signals, measured along angles ϕ and $\phi + \Delta\phi$, this error is

$$\delta\overline{\overline{\kappa_t}}(h, \Delta\phi) = \frac{\delta\beta_\pi^*(h)}{2\tau(0, h)}\left(\frac{1}{\sin\phi} - \frac{1}{\sin(\phi+\Delta\phi)}\right)^{-1} \tag{9.11}$$

where

$$\delta\beta_\pi^*(h) = \ln\left[\frac{\beta_\pi(h, \phi+\Delta\phi)}{\beta_\pi(h, \phi)}\right]$$

As follows from Eqs. (9.9) and (9.11), the two-angle measurement uncertainties are proportional to the error magnification factor

$$\psi = \left(\frac{1}{\sin\phi} - \frac{1}{\sin(\phi+\Delta\phi)}\right)^{-1}$$

which depends on the angular separation $\Delta\phi$ between the selected slope directions. The dependence of ψ on $\Delta\phi$ is given in Fig. 9.3. It can be seen that the magnification factor tends to infinity when the angle separation between the examined directions tends to zero. Thus the magnification factor ψ and the uncertainty in the derived extinction coefficient [Eq. (9.9)] dramatically increase if $\Delta\phi$ is chosen too small. Note also that the uncertainty increases more rapidly when ϕ is large (Fig. 9.3). To reduce the factor ψ, the angular separation $\Delta\phi$ must be increased. However, an increase in $\Delta\phi$ increases the distance between the measured scattering volumes at height h. This may invalidate or weaken the horizontal homogeneity assumption, $\beta_\pi(h, \phi) = \beta_\pi(h, \phi + \Delta\phi)$, and significantly increase the uncertainty of $\delta\beta_\pi^*(h)$ [Eq. (9.11)]. It stands to reason that the differences in $\beta_\pi(h)$ are smaller when the angular separation is small.

In order the differences in $\beta_\pi(h)$ at the height of interest h be small, the distance along the horizontal line *aa* (Fig. 9.2) connecting the examined *directions 1* and *2* must be as small as possible. On the other hand, to obtain small values for the magnification factor ψ, the angular separation $\Delta\phi$ should be large. Thus the

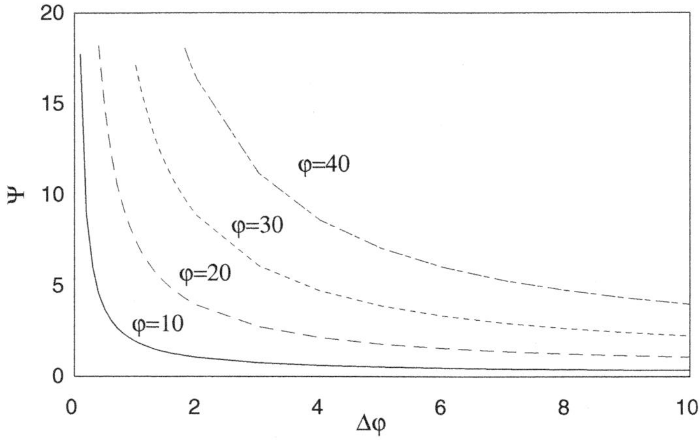

Fig. 9.3. Dependence of the factor Ψ on the separation angle, $\Delta\phi$ between the slope directions.

requirements for the selection of an optimal angular separation in two-angle and multiangle measurements are contradictory.

Thus the measurement uncertainly increases both for small and large increments $\Delta\phi$. Accordingly, the dependence of the measurement uncertainty on the angular separation has the same U shape as that of the slope method, where the error increases when choosing a too-small or too-large range resolution Δr (Section 5.1). This means that with multiangle measurements, the uncertainty has an acceptable value only for some restricted range of angular separations $\Delta\phi$.

The total measurement uncertainty, defined as the sum of the uncertainty components given by Eqs. (9.9) and (9.11), can also be written in the form

$$\delta\overline{\kappa_{t,\Sigma}}(h, \Delta\phi) = \frac{0.5}{\tau(h, \phi) - \tau(h, \phi + \Delta\phi)} \sqrt{[\delta P(h, \phi)]^2 + [\delta P(h, \phi + \Delta\phi)]^2 + [\delta\beta_\pi^*(h)]^2} \tag{9.12}$$

where $\tau(h, \phi)$ and $\tau(h, \phi + \Delta\phi)$ are the optical depths of the layer (0, h) measured along the slope angles ϕ and $\phi + \Delta\phi$, respectively. The measurement uncertainty is large when the difference in these optical depths is small. This is why in clear atmospheres, this approach requires the use of larger angular separations. In such atmospheres, the optical depths $\tau(h, \phi)$ and $\tau(h, \phi + \Delta\phi)$ are small, leading to a small difference between these in the denominator of Eq. (9.12). This may result in an extremely large measurement uncertainty.

To illustrate this, consider two lidar signals measured at 1064 nm in a clear atmosphere over the slant paths, 70° and 90°. Let $\kappa_t = 0.1\,\text{km}^{-1}$, which is a typical value at 1064 nm near the ground in a clear atmosphere. For the atmospheric layer that extends from the ground level to the height, let say, h = 500 m, the corresponding optical depth will be 0.05 for the vertical direction, and 0.0532 for the slope direction of 70°. Accordingly, $0.5[\tau(h, 70°) - \tau(h, 90°)]^{-1} \approx 156$. If the total uncertainty of three terms $\delta P(h, 70°)$, $\delta P(h, 90°)$, and $\delta\beta^*_\pi(h)$ in Eq. (9.12) is 10%, the measurement uncertainty in the derived extinction coefficient will exceed a thousand percent. The use of the multiangle rather than the two-angle data set can significantly reduce the random uncertainty but does not influence the systematic error.

When the measurement data are collected along several lines of sight, the measurement uncertainty that originates from random errors may be reduced. The large number of slant directions used in multiangle measurements provides an opportunity to incorporate a least-squares method. This variant of the multiangle method was initially published by Hamilton (1969). The basic idea of this version is quite similar to the slope method discussed in Chapter 5. The difference is that with multiangle measurements, the independent variable is related to the set of elevation angles at which the measurements were made. If the condition given in Eq. (9.2) is true, the lidar equation for any fixed height h can be written as a function of the sine of angle ϕ

$$P(h,\phi) = C_0\beta_\pi(h)\frac{\sin^2\phi}{h^2}\exp\left[\frac{-2h}{\sin\phi}\overline{\kappa_t}(h)\right] \tag{9.13}$$

here $\overline{\kappa_t}(h)$ is the mean extinction coefficient of the layer (0, h). After taking the logarithm of the range-corrected signal, $Z_r(r, \phi) = P(r, \phi)r^2$, Eq. (9.13) can be rewritten in the form

$$\ln Z_r(h,\phi) = \ln[C_0\beta_\pi(h)] - 2\overline{\kappa_t}(h)\left(\frac{h}{\sin\phi}\right) \tag{9.14}$$

Defining the independent variable as $x = h/\sin\phi$ and the dependent variable as $y = \ln[Z_r(h, \phi)]$, one obtains the linear equation $y = B - 2Ax$. The straight line intersection with the vertical axis is $B = \ln[C_0\beta_\pi(h)]$, and the slope of the fitted line is $A = \overline{\kappa_t}(h)$. By using the set of range-corrected lidar signals $Z_r(r, \phi_1)$, $Z_r(r, \phi_2)$, . . . $Z_r(r, \phi_N)$ at the same height h, the constants A and B can be found through linear regression.

With Hamilton's (1969) method in two-component atmospheres, it is not necessary to know the numerical value of the backscatter-to-extinction ratio to extract the particulate extinction coefficient constituent. Moreover, the backscatter coefficient $\beta_\pi(h')$ can itself be evaluated from the constant B of the linear fit if the calibration constant C_0 is in some way estimated.

Thus the mean value of the extinction coefficient for an extended atmospheric layer can be determined as the slope of the log-transformed, range-corrected lidar signal but, unlike the ordinary slope method, taken here as a function of (h/sin ϕ). Because the mean extinction coefficient (or the optical depth) can be found for all altitudes within the lidar operating range, the local extinction coefficient can then be obtained (at least theoretically) by determining the increments in the optical depth for consecutive layers. However, this possibility is not often realized in practice because the errors in the derived local extinction coefficients are generally too large.

The principal question for the application of a multiangle approach is the question whether the assumption of horizontal homogeneity is appropriate for the examined atmosphere. All of the early lidars and many still today operate only during the hours of darkness, when this atmospheric condition can occur. However, this condition may be not valid during daylight hours (see the discussion in Chapter 1). Thus the method described in this section is not useful for studies of unstable boundary layer. Even when the atmosphere is highly stable, the layers near the surface may still not be horizontally homogeneous. Examination of Fig. 9.1 reveals such an area near the surface.

Analyzing the results of airborne lidar measurements made as part of the Global Backscatter Experiment, Spinhirne et al. (1997) concluded that "horizontal and vertical inhomogeneity is the rule rather than the exception." This is especially true in and above the boundary layer and in areas of cloud formations, where dynamic processes of cloud formation and dissipation change the structure of the ambient atmosphere. To obtain accurate measurement results, a preliminary examination of the available data always must be made. This examination must be considered to be the rule. As a first step, cloud detection and filtering procedures must be constructed so as to exclude heterogeneous layering. Second, restricted spatial regions should be identified where the assumption of atmospheric homogeneity may be considered to be valid. The different multiangle measurement variants have different sensitivity to the violation of the horizontal homogeneity assumption, so that the errors caused by the atmospheric heterogeneity depend on details of the method used. On the other hand, one should have a clear understanding of how accurately examined atmospheric parameters will be estimated if initial assumptions are violated. For example, the assumption that the optical depth of the layer of interest is uniquely related to the sine of the elevation angle may not be good enough to determine the fine atmospheric structure in a clear atmosphere but may be acceptable for determining the total transmittance or visibility in a lower layer of a turbid atmosphere, that is, in situations where the transmission term of the lidar equation dominates the lidar return (see Sections 12.1 and 12.2).

This section has discussed the simplest variant of multiangle analysis, one that was initially proposed for the analysis of elastic lidar measurements. In practice, this variant revealed many limitations. First, the basic requirement for horizontal homogeneity [Eq. (9.1)] in thin spatially extended horizontal layers may often be inappropriate for real atmospheres. To complicate the situation, local heterogeneity at any height h_{in}, will also influence the measurement accuracy for all higher altitudes, that is, for all $h > h_{in}$ (Fig. 9.4). Second, to have acceptable accuracy, a large number of data points should be

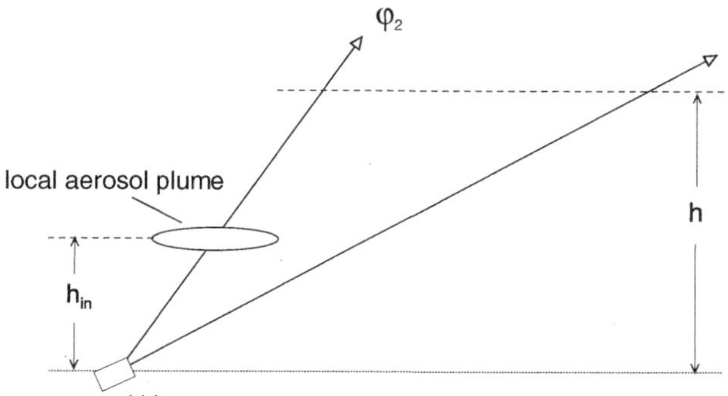

Fig. 9.4. Local inhomogeneity that distorts the retrieved profiles for all altitudes $h > h_{in}$.

used to determine $\overline{\kappa_t}(h)$ with the least-squares method. This means that a large number of sloped paths (ϕ_1, $\phi_2 \ldots \phi_N$) should be used where the signals $P(h, \phi_1)$, $P(h, \phi_2) \ldots P(h, \phi_N)$ should be determined for the same height h, so that the distances from the lidar to height h increase proportionally to $1/\sin\phi$. Obviously, the signal-to-noise ratios of the lidar signal worsen when the selected elevation angles become small. This significantly restricts the lines of sight that can be used to determine the slope with Eq. (9.14).

The restrictions in the application of the horizontal homogeneity assumption in the multiangle method are quite similar to those for the slope method discussed in Section 5.1. To avoid processing lidar data from areas inconsistent with the restrictions of the multiangle method, the computer program must first determine the spatial location of the heterogeneous areas or "spots" and select only relevant data for inversion. It should be mentioned that the use of the method, especially in a clear atmosphere, requires a properly tested and adjusted instrument. In other words, to avoid disenchantment with multiangle measurements, all of the systematic distortions that may occur in the lidar signal, caused by optical misalignment, receiver nonlinearity, or zero-line offsets, should be preliminarily investigated and either eliminated or compensated. Our practice has revealed that even a slight monotonic change in the overlap function with the range, when not taken into consideration, can destructively influence the measurement result when doing multiangle data inversion. Finally, an additional deficiency of the multiangle method should be mentioned. It lies in the assumption of a "frozen" atmosphere during the entire period of the multiangle measurement. Generally, local heterogeneities are evolving in time and moving in space; thus even an increase or change in the wind speed devalues the data obtained. All these shortcomings restrict the use of this analysis method.

Practical investigations of the multiangle approach have shown that the most significant errors occur because of horizontal heterogeneity in the backscatter coefficients, systematic distortions, and signal noise associated with measured lidar signal power. As follows from the study of Spinhirne et al.

(1980), the standard deviation of the horizontal variations in the backscatter cross section within the mixing layer typically ranges from 0.05 to 0.15. Large errors in the values of the mean extinction coefficient obtained by this method complicate the subsequent extraction of extinction coefficients by height differentiation. However, despite the obvious shortcomings of this version of multiangle measurement analysis, it may be applied in practice (Rothermel and Jones, 1985; Sicard et al., 2002).

9.2. SOLUTION FOR THE LAYER-INTEGRATED FORM OF THE ANGLE-DEPENDENT LIDAR EQUATION

The requirements for horizontal homogeneity given in Eqs. (9.1) and (9.2) are quite restrictive. Spinhirne et al. (1980) developed a variant that does not require homogeneity within the thin horizontal layers. The method is based on the use of the slant-angle lidar equation integrated over some extended atmospheric layer between heights h_1 and h (Fig. 9.2). The authors considered vertically extended rather than thin atmospheric layers, for which two basic assumptions are made. Similar to the method described in Section 9.1, it was assumed that the vertical optical depth of any such a layer $\Delta h = h - h_1$ (Fig. 9.2) can be determined as the product of the slant optical depth by the sine of the elevation angle

$$\tau(\Delta h, \phi_1)\sin\phi_1 = \tau(\Delta h, \phi_2)\sin\phi_2 = \ldots = \tau(\Delta h, \phi = 90°) \tag{9.15}$$

As shown in the previous section, this assumption is equivalent to the assumption that the mean extinction coefficient of the layer Δh does not depend on the elevation angle [Eq. (9.5)]. Second, Spinhirne et al. (1980) assumed that the particulate backscatter-to-extinction ratio is constant throughout the extended atmospheric layer under consideration. Thus, within the layer Δh, the backscatter-to-extinction ratio is an altitude-independent value

$$\Pi_p(\Delta h, \phi) = const. \tag{9.16}$$

This condition must be valid for any slope direction (i), that is, for all elevation angles $\phi_1, \phi_2 \ldots \phi_N$ used in the measurement. Note that this assumption significantly differs from the assumption of atmospheric horizontal homogeneity in Eq. (9.1). The latter assumes horizontal homogeneity in thin horizontal layers, whereas the assumption in Eq. (9.16) is considered as applicable for an extended layer Δh. When applying the method, some averaging of the backscatter coefficients takes place over a sufficiently thick layer. This results in some smoothing of the local heterogeneities.

The theoretical foundation of the method is as follows. As follows from Eq. (5.31), with the scale constant $C_Y = 1$, the function $Z(r)$ can be written in the form

$$Z(r) = C_0[\kappa_W(r)]\exp\left\{-2\int_0^r [\kappa_W(r')]dr'\right\} \tag{9.17}$$

where $\kappa_W(r)$ is the weighted extinction coefficient, defined as [Eq. (5.30)]

$$\kappa_W(r) = \kappa_p(r) + a\kappa_m(r)$$

The lower limit of integration for $Z(r)$ in Eq. (9.17) is taken as zero; accordingly, the term T_0^2 here is excluded. Note also that according to Eq. (9.16), the ratio $a(\Delta h, \phi) = a = const.$ The additional condition is that no molecular absorption occurs, so that $\kappa_m = \beta_m$.

To obtain the solution for the angle-dependent lidar equation, the relationship between the integrals of $Z(r)$ and $\kappa_W(r)$ should be first established. As shown in Chapter 5, the integration of $Z(r)$ may be made by implementing a new variable $y(r) = \int \kappa_W(r')dr'$. Then $dy = \kappa_W(r')dr'$, so that the integration of $Z(r)$ from a fixed range $r_1 > 0$ to r gives the formula

$$\int_{r_1}^{r} Z(r')dr' = \frac{C_0}{2}\exp\left[-2\int_0^{r_1} \kappa_W(r')dr'\right] - \frac{C_0}{2}\exp\left[-2\int_0^{r} \kappa_W(r')dr'\right] \tag{9.18}$$

Using the function $V(0, r)$, defined through the integral of $\kappa_W(r)$, similar to that in Eq. (5.80)

$$V(0,r) = \exp\left[-\int_0^r \kappa_W(r')dr'\right] \tag{9.19}$$

Eq. (9.18) is rewritten as

$$\int_{r_1}^{r} Z(r')dr' = \frac{C_0}{2}\left[V(0,r_1)^2 - V(0,r)^2\right] \tag{9.20}$$

With the relationship $r = h/\sin\phi$, Eq. (9.20) can be rewritten as

$$[V(0,h)]^{\gamma} = [V(0,h_1)]^{\gamma} - \frac{\gamma}{C_0}\int_{h_1}^{h} Z(h')dh' \tag{9.21}$$

where

$$\gamma = \frac{2}{\sin\phi}$$

The function $V(0, h)$ can be defined in terms of the particulate and molecular transmissions, $T_p(0, h)$ and $T_m(0, h)$, in a manner similar to that in Eq. (5.81)

$$V(0,h) = T_p(0,h)[T_m(0,h)]^{a} \tag{9.22}$$

to transform Eq. (9.21) into the form [Spinhirne et al. (1980)]

$$[T_p(0,h)]^{\gamma}[T_m(0,h)]^{a\gamma} = [T_p(0,h_1)]^{\gamma}[T_m(0,h_1)]^{a\gamma} - \frac{\gamma}{C_0}\int_{h_1}^{h} Z(h')dh' \quad (9.23)$$

where $Z(h)$ can be found as

$$Z(h) = \frac{P(h)h^2}{\Pi_p \sin^2\phi}\exp\left[-\gamma\int_{h_1}^{h}\kappa_m(h')[a-1]dh'\right] \quad (9.24)$$

The molecular terms in Eqs. (9.23) and (9.24) may be obtained from the atmospheric pressure and temperature profiles. Thus four unknown quantities must be determined, namely, the constant C_0, the assumed constant Π_p (and accordingly, the exponent a), and the particulate transmission terms $T_p(0, h)$ and $T_p(0, h_1)$. In the study, the constant C_0 was determined by the preliminary calibration of the lidar with a flat target of known reflectance. The transmission in the bottom layer, $T_p(0, h_1)$, which is unity at the surface, is obtainable by consecutive derivation of the transmission in the lower layers. In clear atmospheres, $T_p(0, h_1)$ may be assumed unity even for an extended range of the heights h_1. Two other unknowns in Eq. (9.23), $T_p(0, h)$ and Π_p, can be found by using data obtained from measurements at different angles. With Eq. (9.23), a nonlinear system of equations with two unknowns is obtained. An iterative technique can be used to find the optimum solution for the system of equations. Note that the transmission terms $T_p(0, h)$ and $T_p(0,h_1)$ are generally only intermediate values, from which the particulate extinction coefficient must then be extracted. By taking the logarithm of these functions, the corresponding optical depths $\tau_p(0, h)$ and $\tau_p(0, h_1)$ are determined. The total extinction coefficient can then be calculated as the change in the optical depth for small height increments Δh_i.

Thus just as with the method by Hamilton (1969), the method by Spinhirne et al. (1980) directly yields only the transmission term of the lidar equation [Eq. (9.23)], whereas the extinction coefficient profile is, generally, the main subject of interest. In both methods, the extinction coefficient may be calculated as the change in the optical depth for small height increments. Unfortunately, the determination of the extinction coefficient from changes in the optical depth is a procedure that is fraught with large measurement uncertainty. The second problem, inherent to most methods of multiangle measurements, is related to the determination of the atmospheric parameters close to ground surface, particularly the term $T_p(0, h_1)$. To provide this information, additional measurements can be made at low elevation angles, beginning from directions close to horizontal. Such an approach, for example, was used in the study of tropospheric profiles by Sasano (1996). When the least elevation angle available for examination significantly differs from zero, information near the ground is not obtainable because of incomplete overlap in the lidar near-field area. In this case, the transmission in the lower layers can be estimated from independent measurements or taken a priori. Note also that lidar measurements close to the horizon, which might help solve the problem, may be impossible because of eye safety requirements or the presence of buildings, trees, or other obstacles in the vicinity of the measurement site. This often makes multiangle solutions inapplicable for atmospheric layers close to the ground surface. In practice, acceptable multiangle data are generally available only for some restricted altitude range from h_{min} to h_{max}. The minimum height is $h_{min} = r_0 \sin\phi_{min}$, where r_0 is the minimum range of complete overlap and ϕ_{min} is the least elevation angle that can be used for atmospheric examination at the lidar measurement site. The maximum height is restricted by the acceptable signal-to-noise ratio of the measured lidar signals. In the above study by Spinhirne et al. (1980), this issue significantly impeded the application of the method above the atmospheric boundary layer. Obviously, for the same height, the signal-to-noise ratio is poorer when the signal is measured at a smaller elevation angle. Therefore, high altitudes in the troposphere can usually be reached only in near-vertical directions. In general, the maximum range of the multiangle technique ultimately depends on the lidar dynamic range, the accuracy of the subtraction of the background component, the signal-to-noise ratio, the existence of signal systematic distortions, and the linearity of the receiver system.

It should also be kept in mind that the accuracy of the solution for the angle-dependent equation significantly depends on the validity of the assumption that the optical depth of the atmospheric layer of interest is uniquely related to the elevation angle. If a local inhomogeneity with an optical depth $\Delta\tau_{inh}$ appears at some low height h_{in} (Fig. 9.4), the assumption is violated for all heights above it. This is because for the slope path ϕ_2, the value $\Delta\tau_{inh}$ will now be added to the optical depths at all higher levels. The second assumption used by Spinhirne et al. (1980) is the assumption of a constant backscatter-to-extinction ratio. It allows one to apply a constant value of the ratio a in Eqs. (9.23) and (9.24). Note that the general solution of the angle-dependent lidar equation is valid for both the constant and the range-dependent backscatter-to-extinction ratios. Thus the second assumption might be avoided if the behavior of the altitude-dependent backscatter-to-extinction ratio might in some way be estimated. However, to apply the latter variant in practice, a mean profile of the particulate backscatter-to-extinction ratio $\Pi_p(h)$ over the examined layer (h_1, h) must be known.

There are other problems and drawbacks of the solution for the angle-dependent lidar equation to consider. Among these, the requirement of an absolute calibration is an issue because it significantly impedes the practical application of this approach. The calibration of a lidar is a delicate operation that requires solving a number of attendant problems.

It is worthwhile to outline the basic conclusions made by Spinhirne et al. (1980) about multiangle lidar measurements. According to the study, this methodology is applicable when applied within the lower mixed layer of the atmosphere. However, to obtain acceptable accuracy in the measurement results, the total aerosol optical depth of the examined layer should not be less than approximately 0.04. The reason is that the measurement error is large when the difference in the optical depths measured at adjacent elevation angles is small (Section 9.1). The limitations of the lidar system used in the investigation did not permit the direct application of the multiangle analysis in the upper troposphere. There, the particulate scattering was small in comparison to that within the boundary layer. At times, it was only a few percent of the molecular scattering. Therefore, even small errors in the assumed value of the particulate backscatter-to-extinction ratio would result large errors when differentiating the molecular and particulate contributions.

The assumption that the optical depth of the atmospheric layer of interest is uniquely related to the cosine of the zenith angle was used in the study by Gutkowicz-Krusin (1993). Here, a multiangle method was analyzed in which a realistic presumption is included concerning the presence of local aerosol heterogeneities. The homogeneous areas are found through the examination of the behavior of the derivative of $d[\ln Z_r(h, \theta)]/d\theta$ with a formula similar to Eq. (9.14). A function dependent on the zenith angle is introduced to establish the locations of the homogeneous areas. In general, this approach is similar to the slope method and, unfortunately, has similar uncertainties. Although the function introduced by the author remains constant in homogeneous areas, the inverse assertion may be not true. In other words, the invariability of the function is not sufficient evidence of atmospheric homogeneity at a fixed altitude.

As noted in a study by Takamura et al. (1994), the multiangle approach has great advantages in comparison to single-angle measurements, but only if particular assumptions about atmospheric spatial and temporal characteristics are valid. One key assumption made implicitly is atmospheric stationarity. To obtain accurate measurement results, the atmosphere must be temporarily stationary, so that the large-scale heterogeneities should not significantly change location during the scanning period and their boundaries could be accurately determined. On the other hand, it is well known that turbid atmospheres can often be treated as statistically homogeneous if a sufficiently large set of lidar signals is being averaged; thus the signal average can be treated as a single signal measured in a homogeneous medium. Presumably, the longer periods used to accumulate the measured data allow smoothing to reduce noise and small-scale aerosol fluctuations. For example, in the study by Spinhirne et al. (1980), the measurement period was approximately 10 min; in the study by Sicard et al. (2002), the data were acquired during 5-minute periods at each line of sight. Obviously, a method, which requires only a pair of slope directions, might be most practical when the data are assumed to be averaged. This method would simplify many problems that arise when the measurements are made along many slant paths. The first advantage of such a method would be a significantly smaller volume of data to be processed. The second advantage is that the measurement time for two slant paths is proportionally less than that for a multiangle measurement, so that the requirement of the atmospheric stationarity can be more easily satisfied.

9.3. SOLUTION FOR THE TWO-ANGLE LAYER-INTEGRATED FORM OF THE LIDAR EQUATION

The version of the two-angle method presented in this section was proposed by Kovalev and Ignatenko (1985) for slant visibility measurements in turbid atmospheres. A schematic of the method is shown in Fig. 9.5. The lidar at point A measures the backscattered signal in two slope directions, at the elevation angles ϕ_1 and ϕ_2, where $\phi_1 < \phi_2$. The lidar altitude range is restricted to the height range from h_1 to h_2. The minimum measurement height h_1 is restricted by the length of the incomplete-overlap zone, r_0, of the lidar and the elevation angle ϕ_2

$$h_1 \geqslant r_0 \sin \phi_2$$

and the maximum height h_2 is determined by the lidar maximum range r_2 and the elevation angle ϕ_1

$$h_2 = r_2 \sin \phi_1$$

The two assumptions used to solve the lidar equation are basically similar to the assumptions used by Spinhirne et al. (1980). The first is that the optical depth of any layer measured in the slope direction is unequivocally related to the elevation angle and to the vertical optical depth of the examined layer.

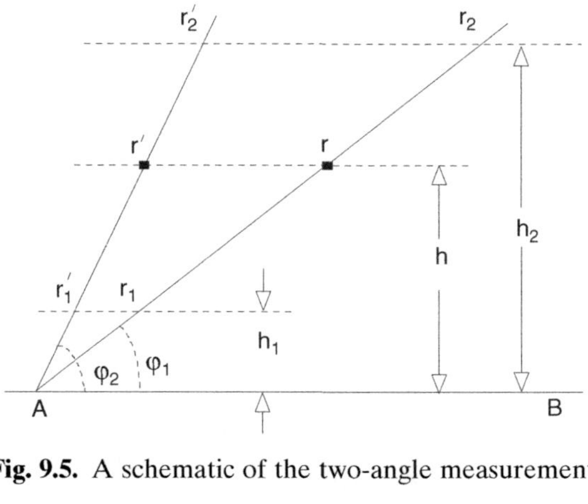

Fig. 9.5. A schematic of the two-angle measurement.

For the two-angle method, this gives the following formulas for the optical depths in adjacent layers (h_1, h) and (h, h_2) (Fig. 9.5)

$$\tau_{\phi,1}(r_1, r)\sin\phi_1 = \tau_{\phi,2}(r_1', r')\sin\phi_2 \tag{9.25}$$

and

$$\tau_{\phi,1}(r, r_2)\sin\phi_1 = \tau_{\phi,2}(r', r_2')\sin\phi_2 \tag{9.26}$$

Here $\tau_{\phi,1}$ and $\tau_{\phi,2}$ are the optical depths of the layers (h_1, h) and (h, h_2) measured in the corresponding slope direction. The second assumption is that the particulate backscatter-to-extinction ratio is constant over both atmospheric layers (h_1, h) and (h, h_2) in any slope direction. Note that, as with the approach by Spinhirne et al. (1980), a two-angle solution can be derived for both constant and range-dependent backscatter-to-extinction ratios. The latter can be accomplished when elastic and inelastic lidar measurements are made simultaneously. Otherwise, the assumption of a constant backscatter-to-extinction ratio is the only option.

Just as with the previous variants, the lidar signal must be range corrected and transformed into the function $Z_\phi(r)$ by multiplying it by the correction function $Y(r)$. This operation transforms the original lidar signal into a function of the variable $\kappa_W(r)$. The function can be written in the form

$$Z_\phi(r) = C_\phi \kappa_W(r) V_\phi(r_1, r)^2 \tag{9.27}$$

where C_ϕ is the solution constant and the term $V_\phi(r_1, r)$ is related to the particulate and molecular path transmittance along the slope through the layer (h_1, h), similar to Eq. (9.22)

$$V_\phi(r_1, r) = T_{p,\phi}(r_1, r)[T_{m,\phi}(r_1, r)]^a \tag{9.28}$$

Note that the term $V_\phi(r_1, r)$ in Eq. (9.28) is written for a constant backscatter-to-extinction ratio and, accordingly, with the constant ratio, a. Clearly, these relationships are similar for both slope directions ϕ_1 and ϕ_2.

Simple mathematical transformations show that the ratio of the functions $Z_\phi(r)$ integrated over the altitude range (h_1, h) and (h_1, h_2) is related to the path transmittance of these layers. As follows from Eq. (9.20), these ratios, defined for slope directions ϕ_1 and ϕ_2 as $J_{\phi,1}$ and $J_{\phi,2}$, can be written as

$$J_{\phi,1}(h) = \frac{\int_r^{r_2} Z_{\phi,1}(x)dx}{\int_{r_1}^{r_2} Z_{\phi,1}(x)dx} = \frac{V(r_1, r)^2 - V(r_1, r_2)^2}{1 - V(r_1, r_2)^2} \tag{9.29}$$

and

$$J_{\phi,2}(h) = \frac{\int_{r'_1}^{r'} Z_{\phi,2}(x)\mathrm{d}x}{\int_{r'_1}^{r'_2} Z_{\phi,2}(x)\mathrm{d}x} = \frac{V(r'_1, r')^2 - V(r'_1, r'_2)^2}{1 - V(r'_1, r'_2)^2} \tag{9.30}$$

where the lidar range r_j, and the corresponding height h_j, are related through the sine of the elevation angle ϕ_i. Denoting for brevity $V_1 = V(r_1, r)$ and $V_2 = V(r_1, r_2)$ and using the condition in Eqs. (9.25) and (9.26), one can rewrite Eqs. (9.29) and (9.30) as

$$J_{\phi,1}(h) = \frac{V_1^2 - V_2^2}{1 - V_2^2} \tag{9.31}$$

and

$$J_{\phi,2}(h) = \frac{V_1^{\frac{2}{m}} - V_2^{\frac{2}{m}}}{1 - V_2^{\frac{2}{m}}} \tag{9.32}$$

where

$$m = \frac{\sin\phi_2}{\sin\phi_1} \tag{9.33}$$

Thus, for any height h, the system of two equations [Eqs. (9.31) and (9.32)] is written with two unknown parameters V_1 and V_2. After solving these equations, the transmittance and the mean extinction coefficients for the corresponding layers (h_1, h) and (h_1, h_2) are found. To determine the particulate path transmittance or the particulate extinction coefficients in these layers, it is necessary to know the molecular extinction profile. As with the other multiangle methods, the molecular extinction coefficient profile may be calculated with vertical profiles of the atmospheric pressure and temperature obtained from balloons or a standard atmosphere.

The simplest solution for Eqs. (9.31) and (9.32) can be obtained if the ratio m is selected to be $m = 2$. Then Eq. (9.32) is reduced to

$$J_{\phi,2}(h) = \frac{V_1 - V_2}{1 - V_2} \tag{9.34}$$

and the following formula can be derived from Eqs. (9.31) and (9.34):

$$\frac{J_{\phi,1}(h)}{J_{\phi,2}(h)} = \frac{V_1 + V_2}{1 + V_2} \tag{9.35}$$

Solving Eqs. (9.34) and (9.35), one can obtain the relationship

$$\frac{J_{\phi,1}(h)}{J_{\phi,2}(h)} = 1 - \left(\frac{1 - V_2}{1 + V_2}\right)[1 - J_{\phi,1}(h)] \tag{9.36}$$

which can be treated as a linear equation

$$y(h) = 1 - cx(h) \tag{9.37}$$

with the dependent variable

$$y(h) = \frac{J_{\phi,1}(h)}{J_{\phi,2}(h)} \tag{9.38}$$

and the independent variable

$$x(h) = 1 - J_{\phi,1}(h) \tag{9.39}$$

The equation constant can be presented as the function of V_2

$$c = \frac{1 - V_2}{1 + V_2} \tag{9.40}$$

Thus a linear relationship exists between the functions $y(h)$ and $x(h)$, in which the slope of the straight line is uniquely related to the unknown function V_2 (Fig. 9.6). This function, in turn, is related to the total transmittance of the layer (h_1, h_2) at the angle ϕ_1, that is,

$$V_2 = V_{\phi 1}(r_1, r_2) = T_{\mathrm{p},\phi 1}(r_1, r_2)[T_{\mathrm{m},\phi 1}(r_1, r_2)]^a$$

Selecting different heights h within the measurement range (h_1, h_2), one can determine a set of the related pairs $y(h)$ and $x(h)$ with Eqs. (9.38) and (9.39) and then apply a least-squares method to find the constant c in Eq. (9.37). After the constant is determined, the particulate path transmittance can be determined by separating the molecular component $T_{\mathrm{m},\phi 1}(r_1, r_2)$. In turbid atmospheres, this procedure can be omitted, and the approximate equality $V_2 \approx T_{\mathrm{p},\phi 1}(r_1, r_2)$ can be used.

The methods based on the assumption of atmospheric horizontal homogeneity require that at least two signals be processed simultaneously to obtain the data of interest [Eq. (9.8)]. These signals must always be chosen at the same height and, accordingly, at different ranges. Therefore, any disturbance in the assumed measurement conditions will result in different, asymmetric

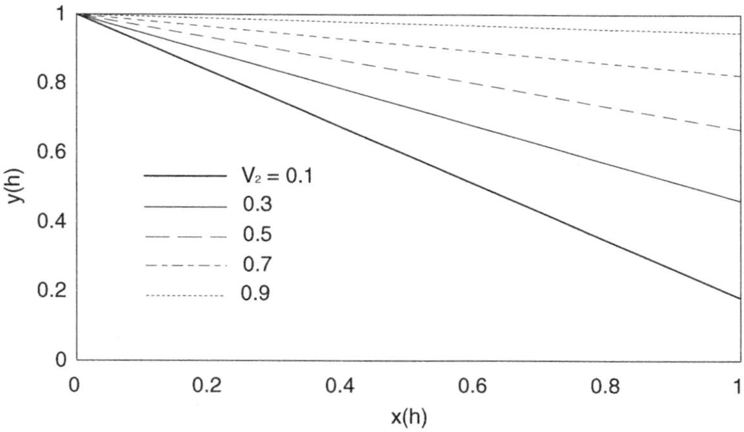

Fig. 9.6. Relationship between functions $y(h)$ and $x(h)$ for different V_2.

signal distortions when performing the signal inversion. In other words, the inversion result depends on which one of two signals is distorted. This is especially inherent in the solutions for the layer-integrated form of the lidar equation, that is, where the assumption given in Eq. (9.15) is applied. If a local heterogeneity with a vertical optical depth $\Delta\tau$ intersects the line of sight along the direction ϕ_2, as shown in Fig. 9.4, the condition in Eq. (9.15) [the same as in Eqs. (9.25) and (9.26)] is no longer true for any height $h > h_{in}$. The actual dependence between the optical depth $\tau(h)$ in the areas not spoiled by the local heterogeneity and the value, $\langle\tau(h)\rangle$ retrieved with the layer-integrated form of the lidar equation is (Pahlow, 2002)

$$\frac{\langle\tau(h)\rangle}{\tau(h)} = \frac{\dfrac{1}{\sin\phi_1} - \dfrac{1}{\sin\phi_2}[1 + \Delta\tau(h)/\tau(h)]}{\dfrac{1}{\sin\phi_1} - \dfrac{1}{\sin\phi_2}} \tag{9.41}$$

Thus the retrieved value of the optical depth $\langle\tau(h)\rangle$ depends on the ratio of the term $[1 + \Delta\tau(h)/\tau(h)]$ to $\sin\phi_2$. If the same heterogeneous formation intersects direction ϕ_1, the measured optical depth will depend on $\sin\phi_1$. One should also point out that, in real inhomogeneous atmospheres, these distortions accumulate with increasing height h. In the next two sections, methods that use an angle-independent lidar equation are considered.

9.4. TWO-ANGLE SOLUTION FOR THE ANGLE-INDEPENDENT LIDAR EQUATION

As shown in Section 9.1, the direct multiangle measurement of the extinction coefficient in a clear atmosphere is an extremely difficult task. This is not only because of the atmospheric inhomogeneity, but also due to extremely harsh requirements to the lidar measurement accuracy, that is, to the accuracy of determining the light backscatter intensity versus time. In some cases, the multiangle approach may be more efficient for lidar relative calibrations, that is, for determining the lidar-equation constant, rather than for direct calculations of extinction profiles. Such a constant determined for the whole two-dimensional lidar scan can be then used for the determination of the extinction-coefficient profiles along individual lines of sight without using now the restrictive atmospheric homogeneity assumptions. Two-angle methods might be most effective for such a variant.

In this section, a two-angle method is presented that applies an angle-independent lidar equation. The method is based on the study by Ignatenko (1991). It can be used either in an independent mode or in multiangle measurements to determine the solution constants. In the latter case, two-angle subsets are selected in some "background" or "reference" aerosol area (see Section 8.2). The method can also be used for long-term unattended lidar operation in a permanent upward-looking, two-angle mode. An advantage of the method is that it may include a posteriori estimates of the validity of the signal inversion result and allow corrections in the initial profiles with these estimates under favorable conditions.

The basic concepts behind the method follow. As with the previous method, the lidar signals $P_1(r)$ and $P_2(r)$ are measured at two relevant angles to the horizon, ϕ_1 and ϕ_2. Before the signal inversion is made, the signals are transformed into the functions $Z_1(r)$ and $Z_2(r)$. This operation is made in the same way as described in Section 9.3. To transform the signals, they are range corrected and multiplied by the correction functions $Y_1(r)$ and $Y_2(r)$. For the same altitude h and two slope paths ϕ_1 and ϕ_2, the transformed functions are

$$Z_1(h) = P_1(h)Y_1(h)\left(\frac{h}{\sin\phi_1}\right)^2 \tag{9.42}$$

and

$$Z_2(h) = P_2(h)Y_2(h)\left(\frac{h}{\sin\phi_2}\right)^2 \tag{9.43}$$

To find the transformation functions $Y_1(r)$ and $Y_2(r)$, the vertical molecular extinction coefficient profile $\kappa_m(h)$ and the particulate backscatter-to-extinction ratio $\Pi_p(h)$ should be known. As above, the latter quantity is assumed range independent, that is, $\Pi_p(\phi) = \Pi_p = const.$, so that $a = const.$ Using the general lidar equation solution for the variable $\kappa_W(h)$ [Eq. (5.33)], one can write the solutions for directions ϕ_1 and ϕ_2 as

$$\kappa_{W,1}(h) = \frac{Z_1(h)}{C_1 - 2I_1(h_1, h)} \tag{9.44}$$

and

$$\kappa_{W,2}(h) = \frac{Z_2(h)}{C_2 - 2I_2(h_1, h)} \tag{9.45}$$

where C_1 and C_2 are lidar equation constants. The integrals $I_1(h_1, h)$ and $I_2(h_1, h)$ are determined as

$$I_1(h_1, h) = \int_{h_1}^{h} Z_1(h')dh' \tag{9.46}$$

and

$$I_2(h_1, h) = \int_{h_1}^{h} Z_2(h')dh' \tag{9.47}$$

where the height h_1 is a fixed height in the lidar operating range, above which the atmospheric layer of interest is located (Fig. 9.5). Equations (9.44) and (9.45) were obtained with the assumption that the particulate backscatter-to-extinction ratio and, accordingly, $a(h)$ are constants over the altitude range from h_1 to h. Note that here, as in Section 9.3, the height h_1 is chosen as the lower limit of integration in the integrals $I_1(h_1, h)$ and $I_2(h_1, h)$ and when determining $Y(r)$. The constants C_1 and C_2 may differ from each other. As shown in Section 4.2, the lidar equation constant is the product of several factors. Because, for simplicity, C_Y is taken to be unity, the constants C_1 and C_2 are the products of two factors [Eq. (5.29)]. These are the constant C_0, and the two-way transmittance T_1^2 over the altitude range $(0, h_1)$, that is, $C = C_0 T_1^2$. The latter term, T_1^2, depends on the elevation angle and may be different for each of the slant paths ϕ_1 and ϕ_2. Accordingly, the constants C_1 and C_2 may also differ from each other. In clear atmospheres, the difference may be not significant if the energy emitted by the lidar is sufficiently stable and h_1 is not too high. Note that the term T_1^2 is the function of the extinction coefficient $\kappa_t(h)$ rather than of $\kappa_W(h)$. This is because the lower integration limit was set as h_1 when determining the transformation function $Y(r)$. If the limit is kept as 0, the term T_1^2 must be replaced by V_1^2 defined similar to Eq. (9.19) over the altitude range $(0, h_1)$.

To find the functions $\kappa_W(h)$ over the range from h_1 to h, the solution constants C_1 and C_2 are first established. The basic assumption that is used to solve the system of Eqs. (9.44) and (9.45) is related to atmospheric horizontal homogeneity. The assumption is that the weighted extinction coefficient κ_W is invariant in horizontal directions, that is, it does not depend on the selected angle of the lidar line of sight. This condition, which is similar to that given in Eq. (9.1), is written in the form

$$\kappa_{W,1}(h) = \kappa_{W,2}(h) = \kappa_W(h) \tag{9.48}$$

In clear atmospheres, where both constituents of $\kappa_W(h)$, namely, the terms $\kappa_p(h)$ and $a\kappa_m(h)$ [Eq. (5.30)], are of comparable of value, the assumption made in Eq. (9.48) may be less restrictive because of the larger weight of the molecular component. As shown in Chapter 7, the range of typical particulate backscatter-to-extinction ratios is ~0.02–0.05 sr^{-1}. The molecular phase function is a constant value of $3/8\pi$. Thus the typical range of the function a varies, approximately, from 2.4 to 6. This means that the contribution of the molecular component in $\kappa_W(h)$ is generally larger than that in the total extinction component, $\kappa_t(h) = \kappa_p(h) + \kappa_m(h)$. This is a favorable factor for the assumption of horizontal homogeneity in clear atmospheres, especially in the *UV* range. Molecular extinction coefficients are related to the temperature (density) and are generally horizontally homogeneous. The difference in the weight function of the molecular and particulate components reduces to some extent the influence of horizontal heterogeneity in the aerosol concentration or composition.

Three unknowns remain in the system of equations above, namely, C_1, C_2, and $\kappa_W(h)$. The system can be solved by excluding $\kappa_W(h)$, so that the least-squares method can then be applied to determine C_1 and C_2. With the assumption in Eq. (9.48), the following formula can be obtained from Eqs. (9.44) and (9.45)

$$\left[\frac{Z_1(h)}{Z_2(h)}\right]\left[\frac{C_2 - 2I_2(h_1, h)}{C_1 - 2I_1(h_1, h)}\right] = 1 \tag{9.49}$$

This can then be transformed into the form

$$2I_1(h_1, h) - 2I_2(h_1, h)\frac{Z_1(h)}{Z_2(h)} = C_1 - C_2\frac{Z_1(h)}{Z_2(h)} \tag{9.50}$$

Eq. (9.50) can be considered as a linear equation (Ignatenko, 1991)

$$y(h) = C_1 - C_2 z(h) \tag{9.51}$$

where the independent variable is

$$y(h) = 2I_1(h_1, h) - 2I_2(h_1, h)\frac{Z_1(h)}{Z_2(h)} \tag{9.52}$$

and the dependent variable is

$$z(h) = \frac{Z_1(h)}{Z_2(h)} \tag{9.53}$$

Equation (9.50) is a linear equation in which the dependent and independent variables, defined as $y(h)$ and $z(h)$, are known functions of altitude whereas the constant terms are unknown lidar solution constants, C_1 and C_2. The variables $y(h)$ and $z(h)$ can be found with Eqs. (9.52) and (9.53) for any altitude h using only the functions $Z(h)$ and these integrals. Applying the least-squares fit for the left-side term in Eq. (9.50), the constants of the regression line, C_1 and C_2, can be found that correspond to the slant paths to ϕ_1 and ϕ_2, respectively. After determining C_1 and C_2, two corresponding profiles of $\kappa_W(h)$ can be determined with Eqs. (9.44) and (9.45), and then the particulate extinction coefficient profiles $\kappa_p(h)$ may be found. This is done by subtracting the weighted molecular contribution, $a\kappa_m(h)$, from the calculated $\kappa_w(h)$ [Eq. (5.30)].

With this method, two assumptions are used to determine the constants C_1 and C_2. The first assumption is atmospheric horizontal homogeneity, that is, the assumption of an invariant backscattering and, accordingly, constant $\kappa_W(h)$ at each altitude [Eq. (9.48)]. The other assumption is a constant backscatter-to-extinction ratio $\Pi_p(\Delta h, \phi)$ within the layer of interest along any slant path ϕ [Eq. (9.16)]. Despite the seeming similarity of this two-angle solution to that given in previous sections, these solutions are significantly different. The differences between the methods are subtle, so that some explanation is in order. The first major difference in this two-angle method is that the assumption in Eq. (9.15) is not used here. No relationship is assumed between the optical depth of the atmospheric layer of interest and the slope of the lidar line of sight. Thus the basic assumption of the conventional multiangle variants (Hamilton, 1969; Spinhirne et al., 1980; Sicard et al., 2002), given in Eq. (9.15), is not required for the inversion. Therefore, for any height h, the validity of the basic equation of the two-angle method [Eq. (9.49)] depends on the atmospheric parameters at this altitude only. The heterogeneities at the heights below h do not violate Eq. (9.49). This is a considerable advantage of the two-angle method, which makes it possible to obtain an acceptable solution even when local heterogeneity occurs below the altitude range of the aerosol layer of interest.

The second difference between the methods is that the most restrictive condition in Eq. (9.48) applied in the method is not directly used to determine the profiles of the extinction coefficient but only for determining the solution constants.

Unlike the methods considered in the previous sections, in the two-angle method, the condition of horizontal homogeneity is applied only when determining the solution constants C_1 and C_2. This condition is not used for calculations of the particular profiles $\kappa_{W,1}(h)$ and $\kappa_{W,2}(h)$.

The extinction coefficient profiles are determined for each slope direction separately only after the constants C_1 and C_2, are established. The constants C_1 and C_2 may be found with a restricted altitude range of the horizontal homogeneity $[h_1, h_2]$ and within some restricted angular sector $[\phi_{min}, \phi_{max}]$. However, the extinction coefficient profiles $\kappa_{W,1}(h)$ and $\kappa_{W,2}(h)$ can then be calculated far beyond the area where these constants were determined. Clearly, a violation of the requirement for horizontal homogeneity will result in significantly different errors when determining the solution constants and when determining the extinction coefficient profiles.

Originally, the method by Ignatenko (1991) was used in relatively polluted, one-component atmospheres. Tests of the method made in clear atmospheres revealed some characteristics of the method (Pahlow, 2002). First, the lidar equation transmission term in clear atmospheres generally remains very close to unity over the entire range of interest. Accordingly, the ratio of the signals, that is, the variable $y(h)$, varies only slightly close unity. In this case, it is more practical to swap the variables $y(h)$ and $z(h)$ and use for the regression Eq. (9.51) transformed into the form

$$z(h) = \frac{C_1}{C_2} - \frac{1}{C_2} y(h) \tag{9.54}$$

To estimate a real value and the prospects for the method, more realistic situations should be analyzed, particularly, the atmospheric heterogeineity and likely signal distortions should be considered. First of all, real lidar signals are always corrupted by noise, so that one can obtain only approximate extinction coefficient profiles. In other words, using real signals in Eqs. (9.44) and (9.45), one will derive from the functions $Z_1(h)$ and $Z_2(h)$ the corrupted profiles $\kappa_w(h)[1 + \delta\kappa_1(h)]$ and $\kappa_W(h)[1 + \delta\kappa_2(h)]$, where the terms $\delta\kappa_1(h)$ and $\delta\kappa_2(h)$ are the relative errors in the retrieved extinction coefficient caused by signal noise in $Z_1(h)$ and $Z_2(h)$, respectively. This distortion of the retrieved profiles will occur even when the basic condition, $\kappa_{w,1}(h) = \kappa_{w,2}(h) = \kappa_w(h)$, is valid. Second, the assumption of atmospheric horizontal homogeneity is also only an approximation of reality. For real atmospheres, the extinction coefficient along a horizontal layer at a fixed height h can be considered, at best, to be a value that fluctuates close to some mean value, so that the ratio of $\kappa_{w,1}(h)$ to $\kappa_{w,2}(h)$ cannot be omitted, at least until some averaging is performed. Accordingly, Eq. (9.49) should be rewritten in the more general form

$$\left[\frac{S_1(h)}{S_2(h)}\right]\left[\frac{C_2 - 2I_2(h_1, h)}{C_1 - 2I_1(h_1, h)}\right] = \frac{\kappa_{W,1}(h)}{\kappa_{W,2}(h)} \tag{9.55}$$

Equation (9.50) should now be rewritten as

$$2I_1(h_1, h) - 2I_2(h_1, h)z(h)\frac{\kappa_{W,2}(h)}{\kappa_{W,1}(h)} = C_1 - C_2 z(h)\frac{\kappa_{W,2}(h)}{\kappa_{W,1}(h)} \tag{9.56}$$

As explained above, the variations in the ratio of $\kappa_{w,1}(h)$ to $\kappa_{w,2}(h)$ originated from horizontal atmospheric heterogeneity are enhanced by signal noise. After some simple transformations, the following equation may be obtained from Eq. (9.56):

$$z(h) = \frac{1}{1 - \psi(h)[V_2(h_1, h)]^2}\left[\frac{C_1}{C_2} - \frac{1}{C_2} y(h)\right] \tag{9.57}$$

where

$$\psi(h) = 1 - \frac{\kappa_{W,2}(h)}{\kappa_{W,1}(h)} \tag{9.58}$$

and

$$[V_2(h_1, h_2)]^2 = \exp\left[-2 \int_{\frac{h_1}{\sin\phi_2}}^{\frac{h}{\sin\phi_2}} \kappa_{W,2}(x)dx\right] \tag{9.59}$$

One can see that in turbid atmospheres, where the term $[V_2(h_1, h)]^2$ is much less than 1, fluctuations in $\kappa_w(h)$ are significantly damped, and if the approximation is valid that

$$\psi(h)[V_2(h_1, h_2)]^2 << 1$$

then Eq. (9.57) reduces to

$$z(h) \approx \left[\frac{C_1}{C_2} - \frac{1}{C_2} y(h)\right] \tag{9.60}$$

In this case, small-scale fluctuations in $\kappa_w(h)$ do not destroy the linear dependence from which the constants C_1 and C_2 are found. However, in clear atmospheres, the term $[V_2(h_1, h)]^2$ may be close to unity, so that

$$\psi(h)[V_2(h_1, h_2)]^2 \approx \psi(h)$$

In this case, Eq. (9.57) transforms to

$$z(h) \approx \frac{\kappa_{W,1}(h)}{\kappa_{W,2}(h)}\left[\frac{C_1}{C_2} - \frac{1}{C_2} y(h)\right] \tag{9.61}$$

and the fluctuations in $\kappa_w(h)$ become influential and may significantly change the slope of the linear fit for $z(h)$ in Eq. (9.54). To compound the problem, the solution in Eq. (9.61) is asymmetric. If the equality $\kappa_{W,1}(h) \approx \kappa_{W,2}(h)$ is significantly violated, the parameter $z(h)$ will depend on which one of the $\kappa_{w,j}(h)$ is larger. For example, if the equality is violated because of the presence of a local particulate layer in the direction ϕ_1, so that $\kappa_{w,1}(h) = 2\kappa_{w,2}(h)$, the first ratio in Eq. (9.61) becomes 2. However, if the same layer crosses the direction ϕ_2, then $\kappa_{w,2}(h) = 2\kappa_{w,1}(h)$, and the first term becomes 0.5, so that the mean value is 1.25 rather than 1. This shift can significantly distort the inversion result when the set of ratios $Z_1(h)/Z_2(h)$ are averaged. This drawback can be avoided if a logarithmic variant of the two-angle method is used, that is, if Eq. (9.55) is transformed to the logarithmic form, so that

$$\ln\left[\frac{Z_1(h)}{Z_2(h)}\right] = \ln\left[\frac{\kappa_{W,1}(h)}{\kappa_{W,2}(h)}\right] + \ln\left[\frac{C_1 - 2I_1(h_1, h)}{C_2 - 2I_2(h_1, h)}\right] \tag{9.62}$$

and the logarithm of the ratio $Z_1(h)/Z_2(h)$ is then used as the regression variable (Kovalev et al., 2002). In this case the first term on the right-hand side becomes symmetric about zero, and no systematic shift occurs as the result of the local heterogeneities when determining an average of the logarithm ratio in the left side of Eq. (9.62).

Thus, with the present method, the lidar equation constant is found with a regression procedure using lidar data from two-angle measurements. This approach significantly simplifies the measurement of atmospheric parameters, making it possible to use a permanent two-angle mode for routine atmospheric monitoring. The two-angle method can also be used in combination with a multiangle technique. In particular, having a set of multiangle measurement data, one can select from these the slant paths that may provide the highest quality data, that is, those that are not contaminated by heterogeneous areas. These data can be used to determine boundary conditions for "background" regions in the examined two-dimensional image (see Section 8.2). If necessary, the latter procedure can be repeated by using a different set of the signal pairs. This makes it possible to estimate the actual level of measurement uncertainty. With this variant, one can obtain an accurate average value for the solution constant for the whole two-dimensional image. Small angular separations in each pair reduce the influence of horizontal heterogeneity, whereas averaging of a large number of variables may reduce the influence of random noise. However, any systematic distortions of the measured signals caused, for example, by poor optical adjustment, may result in a systematic change in the overlap function and even make a solution impossible.

In Table 9.1, the characteristics of the different methods, considered in Sections 9.1–9.4, are compared.

9.5. HIGH-ALTITUDE TROPOSPHERIC MEASUREMENTS WITH LIDAR

Despite many difficulties in practical application, multiangle measurements have been used in many scientific investigations, particularly when the optical characteristics over the depth of the troposphere satisfy the required conditions. In the method presented in this section, the boundary conditions are inferred from an assumption of the existence of aerosol-free zones at high altitudes. For lidar measurements, the idea was proposed by Fernald (1972) and used in many studies (Platt, 1973 and 1979; Fernald, 1984; Sasano and Nakane, 1987; Sassen et al., 1989; Sassen and Cho, 1992).

As with the two-angle method in Section 9.4, the use of the assumption of an aerosol-free zone makes it possible to invert lidar data without using the

TABLE 9.1. Comparison of the Lidar Signal Inversion Methods of Multiangle Measurement Based on the Assumption of a Horizontally Structured Atmosphere

	Classic Approach	Integrated Form Solution	Two-Angle Method (TAM)	Two-Angle Logarithmic Method (TALM)
	Kano (1968); Hamilton (1969); Sicard et al. (2002)	Spinhirne et al. (1980); Kovalev and Ignatenko (1985; Kovalev et al. (1991)	Ignatenko (1991)	Kovalev et al. (2002)
Invariant backscattering along horizontal layers	yes	no	yes	yes
Unique relationship between the searching slope and the optical depth of the searched layer	yes	yes	no	no
Invariable backscatter-to extinction ratio in slope directions	no	yes	yes	yes
Local aerosol layers worsen the measurement accuracy at all altitudes above these layers	yes	yes	no	no
Asymmetric lidar equation solution	yes	yes	yes	no
Poor lidar optics adjustment or systematic signal distortions in receiver's channel do not allow performance of the signal inversion	yes	yes	yes	yes
Time or spatial averaging of signal ratio (or the log of the signal ratio) allow improvement of measurement accuracy	used only in the study by Sicard et al. (2002)	yes	in moderately turbid atmospheres	in clear atmospheres*
The method is practical for the atmospheric long-term monitoring in a permanent two-angle mode.	no	no	in moderately turbid atmospheres	in clear atmospheres*

*The conclusion follows from theoretical analyses.

assumption of a unique relationship between the elevation angle and the optical depth of the atmospheric layer of interest.

For tropospheric studies, this approach was applied by Takamura et al. (1994) and Sasano (1996). The initial methodology was proposed in the study by Sasano and Nakane (1987). A variant of the multiangle measurement technique was presented in which the measurement scheme was used with constant distances for the maximum lidar measurement range for all elevation angles. This scheme is quite practical, especially in clear-sky atmospheres. The basic assumption that enables processing of the data from the multiangle measurements is the existence of an aerosol-free zone at some altitude within the measurement range of the lidar. This assumption is most likely to occur at high altitudes, so that the initial signal used in processing is the one measured closest to the vertical direction. With this assumption, the extinction coefficient profile is found for the lidar maximum elevation angle, ϕ_{max}. The profile is found over an altitude range from $h_{0,1} = r_0/\sin\phi_{max}$, defined by the lidar incomplete-overlap range r_0, to the maximum height, $h_{max,1} = r_{max,1}/\sin\phi_{max}$ (Fig. 9.7). The lidar elevation angle is then decreased, so that the new operating range is within a smaller altitude range, from $h_{0,2}$ to $h_{max,2}$, where $h_{0,2} < h_{0,1}$ and $h_{max,2} < h_{max,1}$. This measurement range covers a part of the altitude range below $h_{0,1}$, which was within the lidar blind zone when making the previous measurement. From the second line of sight, the boundary conditions are determined from the extinction coefficient profile obtained with the previous line of sight. After that, the lidar elevation angle is again decreased, so that now the lidar operating range is within the altitude range from $h_{0,3} < h_{0,2}$ to $h_{max,3} < h_{max,2}$, and so on. The other requirement in the study by Sasano and Nakane

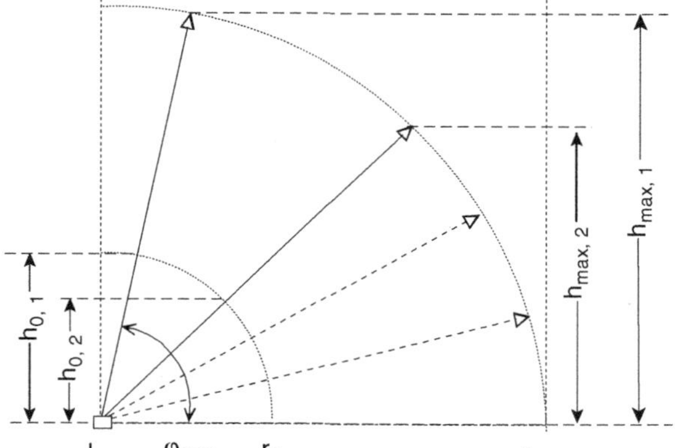

Fig. 9.7. Schematic of a multiangle measurement with the assumption of an aerosol-free area at high altitudes. The lidar is located at point *L*.

(1987) is the application of an iterative method. Initially, the selection of the boundary value for the far-end solution must be made for every line of sight. After that, a mean vertical profile may be obtained and refined boundary conditions are assigned for the next iteration. The procedures are repeated until some criterion is satisfied for convergence. These principles were implemented in tropospheric studies made with a scanning lidar over Tsukuba, Japan, for 3 years from 1990 to 1993. The purpose of the study was to analyze the variations and trends in aerosol optical thickness from the winter of 1990–1991 to the spring of 1992 and to investigate the loading of aerosols from Mt. Pinatubo's eruption in June 1991. These lidar measurements covered the altitude range from the ground level up to the altitude 12 km. Tropospheric aerosol characteristics were investigated with a complex instrumental setup, which, in addition to the multiangle lidar, included a sun photometer and an optical particle counter. Analysis of the measurement data was made by Takamura et al. (1994) and Sasano (1996). Because the principles used in data processing in these studies are slightly different, they are considered separately.

In the study by Takamura et al. (1994), the measurement scheme above was used where the vertical distribution of particulates from the highest altitude down to the lidar level was retrieved. This study used the following assumptions: (1) The backscatter-to-extinction ratio of the particulates are assumed to be the same in both the horizontal and vertical directions, that is, $\Pi_p(\phi) =$ *const.* [Eq. (9.16)]. (2) At each altitude, the particulate concentration and, accordingly, the extinction coefficient is assumed to fluctuate about a constant value in horizontal direction. (3) A particulate-free zone is assumed to exist within the lidar measurement range. This means that within some altitude range (h_b, h_c), generally found near the lidar maximum altitude, the condition is valid

$$\kappa_p(h_b \le h \le h_c) = 0 \tag{9.63}$$

With the last assumption, which is critical to the method, the boundary conditions can be easily inferred in the manner that is discussed in Chapter 8. To find the location of the assumed particulate-free zone an iterative process was used, based on the so-called matching method (Russell et al., 1979). The lidar data were analyzed with different backscatter-to-extinction ratios Π_p, which were allowed to vary from approximately 0.01 to 0.1 sr^{-1}. The particulate optical depth was determined independently by the lidar and from direct solar radiation measurements with a sun photometer. A comparison of these optical depths makes it possible to estimate a mean value of Π_p. According to estimates made by the authors of the study, the values Π_p generally ranged from 0.015 to 0.05 sr^{-1}. Obviously, the accuracy of these estimates depends on the validity of the initial assumption that $\Pi_p(\phi) = const.$ The other assumption that influences the accuracy of the obtained Π_p is the assumption in Eq. (9.63) that the contribution of the particulate loading near the maximum lidar measurement altitude (12 km) is negligible and can be ignored. The data analysis revealed that before Mt. Pinatubo's eruption, the measurements of the optical depth from the lidar and the sun photometer showed almost the same value. However, after the eruption, the optical depths obtained with the sun photometer were larger than those from the lidar. This is because the assumption of a particulate-free atmosphere might be not accurate enough to properly process the data obtained after the eruption. Therefore, the matching method might underestimate the particulate loading after the eruption.

Basically the same methodology was later applied by Sasano (1996) to obtain seasonal profiles of the particulate extinction coefficient. For this, the same observations made at Tsukuba were used, obtained from 1990 to 1993. However, the author of the latter work did not use sun photometer data to estimate the value of the backscatter-to-extinction ratio. He stated that this technique requires an extremely accurate determination of the particulate optical depth from sun photometer data obtained during the lidar measurements. For clear atmospheres, the accuracy of the optical depth obtained from sun photometer data is poor. Therefore, in the study by Sasano (1996), a constant value for the backscatter-to-extinction ratio, $\Pi_p = 0.2\,\text{sr}^{-1}$, was chosen a priori. The iterative procedure used to determine the particulate extinction coefficient was as follows. First, the lidar measurement range r_{min}–r_{max} was established. The minimum range, $r_{min} = 5\,\text{km}$, was selected to avoid current saturation in the photomultiplier of the lidar receiver. The maximum range, $r_{max} = 12\,\text{km}$, was selected to yield an acceptable signal-to-noise ratio. These ranges were the same for all of the lines of sight at different angles, from ϕ_{min} to ϕ_{max}. At all elevation angles, the maximum distances r_b were established close to r_{max} ($r_b \le r_{max}$), where the boundary values were iterated. For the first iteration cycle, the boundary values $\kappa_p(r_b)$ were chosen to be zero for all of the lines of sight from ϕ_{min} to ϕ_{max}. Thus some of the particulate-free zones were assumed to be in directions close to horizontal. The corresponding extinction coefficient profiles $\kappa_p(r)$ were calculated for each slope direction. For this, Fernald's (1984) solution was used with signal integration from the farthest point back toward the lidar, which works similar to the conventional far-end solution. Then a two-dimensional image y versus x was built. On this image, a grid with a spatial resolution Δx and Δy was applied. The mean value of the particulate extinction coefficient was determined for every subgrid cell. All extinction coefficients located within a cell were averaged to yield a single value for each cell. After that, a mean vertical profile was calculated by horizontally averaging the two-dimensional gridded data. Now these averaged extinction coefficients could be used to find new boundary values for each altitude level. The process was repeated until the difference between the latest and previous averaged extinction coefficients $\kappa_p(h)$ became less than some established criterion.

Potentially, this iteration method is a powerful tool when processing a large set of experimental data in which the quantities are in some way related. However, two difficulties must be overcome. First, the iteration may or may not converge with the particular data set of interest. Second, the quality of the iteration result is strongly dependent on both the atmospheric conditions and the accuracy of the initial data used to start the iteration (Russel et al., 1979; Ferguson and Stephens, 1983; Sasano and Nakane 1987; Rocadenbosh et al., 1998). There is no reason to believe that when using inappropriate initial assumptions [for example, an assumption of purely molecular scattering at altitudes where actual $\kappa_p(h) \neq 0$] the set of lidar equation solutions will converge to the true values. The selection of the relevant boundary value in clear atmospheres is always problematic, the same as the selection of the particulate backscatter-to-extinction ratio. When using such a priori values, it is not possible to make grounded estimates of the actual uncertainty in the retrieved extinction coefficient profile unless relevant independent data are available.

9.6. WHICH METHOD IS THE BEST?

The question of which method is best should be formulated as the question of what particular assumptions yield the most reliable results and least measurement errors when used for multiangle measurements. The obvious reply is that the best set of assumptions is that which most accurately describes the particular atmospheric conditions. This statement requires some additional comments. As follows from this chapter, there are two alternative methods of signal inversion for multiangle measurements: (a) application of the assumption of a horizontally layered atmosphere, and (b) the use of the *a priori* assumption of an aerosol-free area within the lidar measurement range (Fig. 9.8). Note that on occasion, for example, in the studies by Takamura et al. (1994) and Sasano (1996), both assumptions, that is, the assumptions of horizontal homogeneity and an aerosol-free area, are used. Nevertheless, the difference between the alternative methods lies in the assumption hierarchy, namely, which one is required for the inversion and which one is supplementary. The solution stability, retrieved data accuracy, and reliability significantly depend on which one of the two assumptions is fundamental when performing the inversion. The characteristics of the options are briefly compared in Table 9.2.

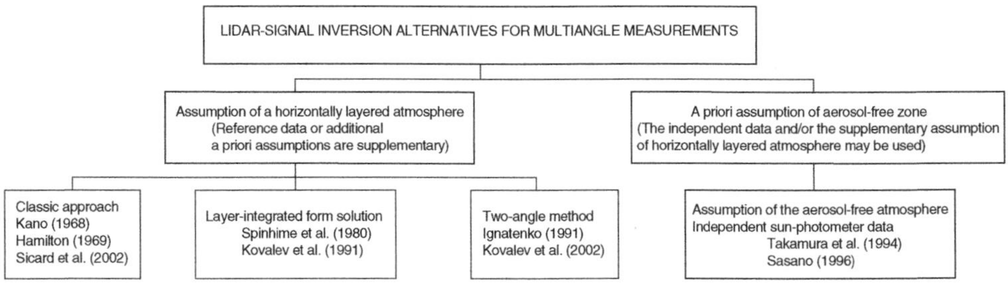

Fig. 9.8. Lidar signal inversion alternatives for multiangle measurements.

TABLE 9.2. Comparison of Alternative Methods to Invert Lidar Signals in Multiangle Measurements

The Assumption of a Horizontally Structured Atmosphere as a Basis for Inversion	An a priori Assumption of an Aerosol-Free Zone or Independent Reference Data as a Basis for Inversion
Preferable for measurements in the lower troposphere using simple lidar systems with measurement range 3–5 km.	Preferable for tropospheric and stratospheric investigations with a lidar measurement range ~10 km and more.
No reference data are required for the inversion. The zones of a horizontally structured atmosphere can be established from two-dimensional images of the range-corrected signals. The validity of the inversion results can be checked a posteriori by an analysis of the retrieved profiles.	Requires either an a priori assumption of an aerosol-free zone or independent reference data. The accuracy and validity of the measurement results generally cannot be checked a posteriori without having additional independent information.
Allows both day- and nighttime measurements. High-altitude clouds do not influence the measurements in the lower troposphere.	Requires clear-sky conditions to obtain measurable signals from high altitudes. Application of sun photometer reference data is restricted by daytime measurements.
Requires application of thoroughly adjusted and properly tested lidar system. All systematic shifts in the lidar signal must be eliminated or compensated before the inversion can be performed.	Poor lidar optics adjustment and/or systematic distortions of the measured lidar signal do not prevent obtaining seeming reasonable (plausible) inversion results with the unknown actual uncertainty.
A poor signal-to-noise ratio in the backscatter signals, especially in clear atmospheres, does not allow the signal inversion.	A poor signal-to-noise ratio in the backscatter signals results in noisy profiles of the measured quantity.
If the examined atmosphere is not horizontally structured, the measurement uncertainty can be reduced by time or spatial averaging.	The use of inaccurate reference data results in hidden and unknown systematic shifts in the retrieved profiles. Data averaging does not reduce measurement uncertainty.
Can be used for atmospheric long-term monitoring in a permanent two-angle mode.	Cannot be used for long-term measurements in a permanent two-angle mode.

The method based on the assumption of an aerosol-free zone is easier to work with. Neither systematic signal distortion due to optical alignment, zero-line offset, or receiver nonlinearity nor poor signal-to-noise ratio can prevent inversion results that are plausible and difficult if not impossible to verify. In fact, the accuracy and reliability of such data are difficult to establish even with independent sun photometer data. Under the assumption of an aerosol-free atmosphere, the auxiliary assumption of a horizontally stratified atmosphere becomes less restrictive even in heterogeneous atmospheres. The first method that does not assume an aerosol-free zone is more difficult to implement in practice. In this case, both systematic distortions and signal noise can make the lidar data impossible to invert, just as will atmospheric heterogeneity. This is especially true for measurements in clear atmospheres, where the requirements for system linearity, precise optical adjustment, and noise level become restrictive. Using this method, one can obtain either no inversion results or good results that are easily checked by a posteriori analysis.

Multiangle measurement techniques based on the assumption of a horizontally structured atmosphere require several assumptions concerning the nature of the lidar returns from the same height but measured in different slope directions. The number of likely assumptions is restricted because the lidar equation includes only two unknown parameters, both related to the degree of atmospheric turbidity. These parameters are the backscatter coefficient, $\beta_\pi(h, \phi)$, and the transmission term, $\exp[-2\tau(\Delta h, \phi)]$, where $\tau(\Delta h, \phi)$ is the optical depth of the atmospheric layer from the ground surface (or from some fixed height) to the height of interest, h, measured at the elevation angle ϕ. Because $\beta_\pi(h, \phi)$ is related to the extinction coefficient through the backscatter-to-extinction ratio Π_p, the set of useful assumptions is limited to those that relate the backscatter coefficient $\beta_\pi(h, \phi)$, optical depth $\tau(\Delta h, \phi)$, the backscatter-to-extinction ratio $\Pi_p(h)$ at the fixed altitude h, or the ratios $\Pi_p(\phi)$ along the slant paths ϕ.

The basic advantages and drawbacks of multiangle measurement methods based on the assumption of a horizontally structured atmosphere are summarized in Table 9.3. As follows from the table, the first, most basic method has the most restrictive assumptions. It is assumed that for any fixed altitude h that the backscatter coefficient $\beta_\pi(h, \phi) = const.$ and the optical depth $\tau(\Delta h, \phi)$ is uniquely related to the sine of the elevation angle. Here the ground surface is taken as the lower boundary of the layer Δh when determining $\tau(\Delta h, \phi)$. The method is sensitive to atmospheric heterogeneities both at the altitude of interest and below it. The assumption of horizontal homogeneity in thin horizontal layers may be not true, particularly for unstable atmospheric conditions found during daylight hours. Apart from that, aerosol heterogeneities at low altitudes influence the measurement accuracy for higher altitudes. The method of Spinhirne et al. (1980) uses the assumption of a constant backscatter-to-extinction ratio $\Pi_p(\phi)$ along any slant path ϕ within the layer of interest, Δh. The other assumption is the same unique relationship between the optical depth of an extended atmospheric layer $\tau(\Delta h, \phi)$ and the angular

TABLE 9.3. Advantages and Drawbacks of the Methods Used with Multiangle Measurements that Use an Assumption of a Horizontally Structured Atmosphere

Method	Assumptions Used	Advantages	Drawbacks	Reference
Basic	Eqs. (9.1) and (9.2)	No estimates of Π_p are needed to determine the atmospheric transmission	Large measurement errors, especially in clear atmospheres.	Sanford (1967); Hamilton (1969); Kano (1969); Sicard et al. (2002)
Layer-integrated form solution	Eqs. (9.15) and (9.16)	Works well in moderately turbid atmospheres to determine the atmospheric transmission	An estimate of the Π_p value is required Large measurement errors in clear atmospheres.	Spinhirne et al. (1980); Kovalev, et al. (1991)
Two-angle variant of the layer-integrated form solution	Eqs. (9.15) and (9.16)	Good in moderately turbid atmospheres to determine the atmospheric transmission	An estimate of Π_p is required. Large measurement errors in clear atmospheres.	Kovalev and Ignatenko (1985)
Two-angle method of Ignatenko	Eqs. (9.16) and (9.48)	Most practical for determining the constants in the lidar equation in moderately turbid, atmospheres.	An estimate of Π_p is required when using in two-component atmospheres. Large measurement errors in clear atmospheres.	Ignatenko (1991)

direction of the lidar line of sight as that used in the previous variant. Accordingly, the method is sensitive to horizontal atmospheric heterogeneities in the layer Δh, especially in clear atmospheres, where the differential optical depth of the layer is small. The method is most practical when the transmission term of the lidar equation is found in turbid or cloudy atmospheres, for example, when determining the slant visibility (Kovalev et al., 1991). However, it is difficult to obtain acceptable measurement accuracy when the local extinction coefficients are obtained through the increment change in the optical depth derived from the above transmission term. The methods of Ignatenko (1991) and Pahlow (2002) also use the assumption of a constant backscatter-to-extinction ratio $\Pi_p(\phi)$ within the layer of interest along the slant path ϕ. The other assumption concerns the horizontal homogeneity of the extinction coefficient, or in a more general form, the homogeneity of the weighted extinction coefficient, $\kappa_W(h)$ [Eq. (9.48)]. No relationship is assumed between the optical depth of the atmospheric layer and the direction of the lidar line of sight. Therefore, for any height h, the basic two-angle equation [Eq. (9.49)] depends only on atmospheric parameters at this altitude and does not depend on particulate heterogeneity at lower altitudes. This is a basic property of the two-angle method that makes it possible to obtain acceptable solution constants even when local heterogeneities occur along the examined direction. However, because of the asymmetry of the basic solution, the method becomes unstable in clear atmospheres [Eq. (9.61)]. A variant of the two-angle method has been proposed in which the asymmetry is eliminated (Kovalev et al., 2002).

It should be emphasized that methods based on an assumption of a horizontally structured atmosphere can only be applied to signals from a thoroughly adjusted and properly tested lidar system. Any systematic shift in the lidar signal must be eliminated or compensated before an inversion can be performed. Even then, every real lidar has a lower limit of the atmospheric attenuation where it can still be used, that is, where its instrumental characteristics still provide the required measurement accuracy of the atmospheric parameter under investigation. The use of a lidar that does not meet the measurement accuracy requirements may only bring disenchanting results. The multiangle approach, which is extremely sensitive to the lidar system distortions, may be more valuable for lidar-system tests and relative calibrations than for direct calculations of vertical extinction profiles. It looks like a combination of the multiangle approach for determining the lidar-equation constant for a whole two-dimensional scan with the next determination of the extinction-coefficient profiles under individual lines of sight might be the most efficient method for processing the two-dimensional (RHI) lidar scans.

10

DIFFERENTIAL ABSORPTION LIDAR TECHNIQUE (DIAL)

The ability of differential absorption lidar (DIAL) measurements to determine and map the concentrations of selected molecular species in ambient air is one of the most powerful and useful. With the DIAL technique, one can investigate the most important man-made pollutants in both the free atmosphere and in polluted areas, such as cities or near industrial plants. The differential absorption technique can be extremely sensitive and is able to detect gas concentrations as low as a few hundred parts per billion (ppb). This makes it possible to measure trace pollutants in the ambient atmosphere and monitor stack emissions in the parts per million range. Range-resolved DIAL systems are sensitive enough to measure the ambient air concentrations and distribution of most of the important polluting gases, including SO_2, NO_2, NO, and ozone. This technique makes it possible to obtain vertical profiles of the atmospheric gas concentrations from ground, airborne, or space platforms. A set of DIAL systems for these measurements has been built in different countries, and the systems are now widely used for routine monitoring throughout the world (Ancellet et al., 1989; Stefanutti et al., 1992; Kempfer et al., 1994; Sunesson et al., 1994; Reichardt et al., 1996; Fiorani et al., 1998; Carnuth et al., 2002).

Elastic Lidar: Theory, Practice, and Analysis Methods, by Vladimir A. Kovalev and William E. Eichinger.
ISBN 0-471-20171-5

10.1. DIAL PROCESSING TECHNIQUE: FUNDAMENTALS

10.1.1. General Theory

The DIAL technique uses the idea of differential-absorption measurement. Two light pulses of different wavelengths are launched along the same path into the atmosphere, and two corresponding backscattered profiles are simultaneously measured. The DIAL wavelengths are selected so that the light at the one wavelength, λ_{on}, is strongly absorbed by the absorbing species under investigation, whereas the light at the second wavelength, λ_{off}, is absorbed not at all or at least much less. In such absorbing media, the total (molecular and particulate) extinction coefficients at λ_{on} and λ_{off} are the sums of the scattering and absorbing constituents

$$\kappa_{t,on}(r) = \beta_{on}(r) + \sigma_{on}\, n(r) \tag{10.1}$$

and

$$\kappa_{t,off}(r) = \beta_{off}(r) + \sigma_{off}\, n(r) \tag{10.2}$$

Here $\beta_{on}(r)$ and $\beta_{off}(r)$ are the total scattering coefficients, and σ_{on} and σ_{off} the absorption cross sections of the species under investigation; n is the number of the absorbing molecules in a unit volume at range r. Here and before, the subscripts "on" and "off" denote parameters at the wavelengths λ_{on} and λ_{off}, respectively, and the subscripts (A) in the absorption terms are omitted for brevity. Note that often several, rather than one, gaseous absorbing compounds may influence light propagation in the same portion of the spectrum. The principal requirement for the selection of the wavelength pair λ_{on} and λ_{off} for DIAL measurements is that at these wavelengths the absorption of the species under consideration is significantly greater than any other, so that the absorption by other species is negligible and can be ignored. The corresponding lidar equation [Eq. (5.2)] for wavelength λ includes the absorption term and may then be rewritten as

$$P_\lambda(r) = C_\lambda \frac{\beta_{\pi,\lambda}(r)}{r^2} \exp\left[-2\int_{r_1}^{r} [\beta_\lambda(r') + \sigma_\lambda n(r')]\, dr'\right] \tag{10.3}$$

where

$$C_\lambda = C_0[T_\lambda(0, r_1)]^2 \tag{10.4}$$

The range r_1 denotes the starting point of the examined path along which the unknown gas concentration is measured. It is assumed that $r_1 \geq r_0$, where r_0 is the incomplete overlap zone. The term $[T_\lambda(0, r_1)]^2$ is the two-way atmospheric transmission over the range from the lidar to r_1

$$[T_\lambda(0,r_1)]^2 = \exp\left\{-2\int_0^{r_1}[\beta_\lambda(r') + \sigma_\lambda n(r')]\,dr'\right\}$$

With DIAL measurements, the unknown gas concentration profile $n(r)$ is derived from the ratio of the signals $P_{on}(r)$ and $P_{off}(r)$ measured at the selected wavelengths within and outside the region of strong absorption of the gas, λ_{on} and λ_{off}

$$\frac{P_{on}(r)}{P_{off}(r)} = \left[\frac{C_{on,1}}{C_{off,1}}\right]\frac{\beta_{\pi,on}(r)}{\beta_{\pi,off}(r)}\exp\left\{-2\int_{r_1}^{r}[(\sigma_{on} - \sigma_{off})\,n(r') + \beta_{on}(r') - \beta_{off}(r')]\,dr'\right\} \tag{10.5}$$

where $C_{on,1}$ and $C_{off,1}$ are the lidar equation constants defined with Eq. (10.4) and $\beta_{\pi,on}(r)$ and $\beta_{\pi,off}(r)$ are the total (i.e., molecular and particulate) backscatter coefficients.

The selection of a relevant wavelength pair $\lambda_{on} - \lambda_{off}$ is an important aspect of DIAL measurements. On the one hand, the DIAL system wavelengths are selected so that the difference in the absorption cross sections of the species under investigation, σ_{on} and σ_{off}, is large. In this case, the term $(\sigma_{on} - \sigma_{off})n(r)$ in the exponent of Eq. (10.5) is also large, and $n(r)$ could be accurately extracted. On the other hand, the difference in the scattering coefficients $\beta_{on}(r)$ and $\beta_{off}(r)$ must be small, so that the exponential term in Eq. (10.5) is primarily related to differential absorption and not differential scattering. The absorbing gas concentration $n(r)$ can be determined from Eq. (10.5) as

$$n(r) = \frac{-1}{2\Delta\sigma}\left\{\frac{d}{dr}\left[\ln\frac{P_{on}(r)}{P_{off}(r)}\right]\right\} + \frac{1}{2\Delta\sigma}\left\{\frac{d}{dr}\left[\ln\frac{\beta_{\pi,on}(r)}{\beta_{\pi,off}(r)}\right]\right\} - \frac{1}{\Delta\sigma}[\beta_{on}(r) - \beta_{off}(r)] \tag{10.6}$$

where $\Delta\sigma = \sigma_{on} - \sigma_{off}$ is the differential absorption cross section of the measured gas. As follows from Eq. (10.6), three terms must be known to obtain the concentration $n(r)$: the derivative of the logarithm of the ratio $P_{on}(r)/P_{off}(r)$; the so-called backscatter correction term, which is related to the derivative of the logarithm of $\beta_{\pi,on}(r)/\beta_{\pi,off}(r)$; and the extinction correction term, which is a function of the differential scattering $\beta_{on}(r) - \beta_{off}(r)$. Accordingly, Eq. (10.6) can be rewritten as the sum of three terms

$$n(r) = n'(r) + \Delta n_\beta(r) + \Delta n_e(r) \tag{10.7}$$

The first term, $n'(r)$, is the basic term. Being determined directly from the ratio of the signals,

$$n'(r) = \frac{-1}{2\Delta\sigma}\frac{d}{dr}\left[\ln\frac{P_{on}(r)}{P_{off}(r)}\right] \tag{10.8}$$

it is quite handy for operational data analysis to provide an initial estimate of the gas concentration distribution in space and time. In this kind of estimate, the terms $\Delta n_\beta(r)$ and $\Delta n_e(r)$ are considered to be unknown systematic errors in the measured concentration $n(r)$.

The basic term $n'(r)$ can provide operational estimates of the quality of the obtained data. The term is directly influenced by the signal-to-noise ratio of the off and on signals. The other terms, Δn_β and Δn_e, may be considered as correction terms, which must in some way be determined to determine a more accurate chemical species concentration, $n(r)$, when making the final analysis. Unfortunately, both parameters depend on atmospheric conditions and often cannot be accurately established, especially in the lower troposphere.

Note that integration of the transformed Eq. (10.8) results in the formula

$$\ln \frac{P_{\mathrm{off}}(r)}{P_{\mathrm{on}}(r)} = 2\Delta\sigma \int n'(r) \mathrm{d}r + const. \tag{10.9}$$

Thus the logarithm of the signal ratio is proportional to the two-way differential absorption optical depth, that is, to the gas concentration column content for the constituent $n'(r)$. This parameter is often used in the analysis of the accuracy of DIAL measurements.

Basic Solution. The initial estimate of the absorbing gas concentration profile $n'(r)$ is a key procedure in DIAL signal inversion. If this term cannot be accurately obtained from the measured signals, for example, because of a poor signal-to-noise ratio, the remaining corrections are useless. This feature requires that special attention be paid to practical methods to determine the term $n'(r)$.

To obtain $n'(r)$ with Eq. (10.8), a numerical differentiation of the logarithm of $[P_{\mathrm{on}}(r)/P_{\mathrm{off}}(r)]$ must be performed. The numerical differentiation of experimental data is always a challenge (Wylie and Barret, 1982; Zuev et al., 1983; Godin et al., 1999; Beyerle and McDermid, 1999). Generally, the range-resolved gas concentration profile is derived by calculating the logarithmic differences in the $P_{\mathrm{on}}(r)$-to-$P_{\mathrm{off}}(r)$ ratio for range increments Δr that are large with respect to the lidar range resolution. In the simplest theoretical variant, the logarithmic differences can be defined with four discrete lidar data points. These signals are measured at both wavelengths, at two ranges, r and $(r + \Delta r)$. The mean value $n'(r, r + \Delta r)$, defined for brevity as $n'(r)$, can be derived from Eq. (10.8) as

$$n'(r) = \frac{1}{2\Delta\sigma\Delta r}\left[\ln \frac{P_{\mathrm{on}}(r)}{P_{\mathrm{on}}(r+\Delta r)} - \ln \frac{P_{\mathrm{off}}(r)}{P_{\mathrm{off}}(r+\Delta r)}\right] \tag{10.10}$$

By calculating the logarithmic differences of the $P_{\mathrm{on}}(r)$-to-$P_{\mathrm{off}}(r)$ ratio for successive range elements Δr, one can calculate the absorbing gas concentration profile $n'(r)$ over the total measured range. In fact, this numerical differentiation determines an average gas number density in a finite range interval Δr, rather than a finely resolved profile of $n'(r)$. Accordingly, with real DIAL measurements, the average value of the gas concentration for an extended range interval Δr (generally, tens or even hundreds of meters) can be obtained.

Note that Eq. (10.10) has a structure similar to that used to determine the extinction coefficient with the slope method [Eq. (5.11)]. Therefore, as with the slope method, the error in the measured gas concentration is sensitive to the length of the range element Δr. Assuming for simplicity that the error in the off signal can be ignored in comparison with that in the on signal, one can obtain a simple formula for the relative error of the chemical species concentration. The error, $\delta n' = \Delta n'/n$ can be derived from Eq. (10.10) by conventional error propagation. The formula is

$$\delta n' = \frac{1}{2\Delta\tau_{A,\mathrm{dif}}}\sqrt{\left[\frac{\Delta P_{\mathrm{on}}(r)}{P_{\mathrm{on}}(r)}\right]^2 + \left[\frac{\Delta P_{\mathrm{on}}(r+\Delta r)}{P_{\mathrm{on}}(r+\Delta r)}\right]^2 \pm [\mathrm{COV}(P_r, P_{r+\Delta r})]^2} \qquad (10.11)$$

where $\Delta\tau_{A,\mathrm{dif}}$ is the differential optical depth, that is, the difference between the optical depths $\tau_{A,\mathrm{on}}$ and $\tau_{A,\mathrm{off}}$ over the range Δr. The value can be written as

$$\Delta\tau_{A,\mathrm{dif}} = \tau_{A,\mathrm{on}} - \tau_{A,\mathrm{off}} = n(r)\Delta\sigma\,\Delta r \qquad (10.12)$$

When Δr is small, the quantities $P_{\mathrm{on}}(r)$ and $P_{\mathrm{on}}(r + \Delta r)$ may be highly correlated; therefore, the covariance term of the signals, $\mathrm{COV}(P_r, P_{r+\Delta r})$ is included in Eq. (10.11).

As follows from Eq. (10.11), the relative error in the measured gas concentration is inversely proportional to the difference in the optical depths at on and off wavelengths, $\tau_{A,\mathrm{on}}(\Delta r)$ and $\tau_{A,\mathrm{off}}(\Delta r)$. Following the terminology of Measures (1984), we denote this difference $\Delta\tau_{A,\mathrm{dif}}$ as the local differential absorption optical depth. As the length Δr tends to zero, the optical depth also tends to zero, and according to Eq. (10.11), the relative error tends to infinity.

As with the slope method used to determine the extinction coefficient (Section 5.1), the range element Δr in DIAL measurements must be long enough to provide acceptable accuracy in the retrieved chemical species concentration. Thus the local differential absorption optical depth is the most important factor that influences accuracy of the measured data.

As with the slope method (Section 5.1), a least-squares technique is commonly used in DIAL measurements rather than a two-point variant. However, a consideration of the simplest two-bin variant is the simplest way to show the dependence of the measured error on the differential optical depth. The use of a least-squares technique reduces the uncertainty but does not change the general dependence of the measurement uncertainty on $\Delta\tau_{A,\mathrm{dif}}$.

During the last decades, comprehensive experiments with differential absorption lidar systems have been conducted, with measurements in both the troposphere and the stratosphere. The development of lasers that can emit light pulses at several wavelengths simultaneously has significantly improved the potential capabilities of DIAL systems. Measurements of atmospheric ozone concentrations has proven to be the most widespread application of the DIAL technique. Ground-based and airborne differential absorption lidars, particularly in the ultraviolet portion of the spectrum, have been used successfully in many regional studies of ozone in the lower troposphere. At this time, ozone concentration measurements are perhaps the most advanced as compared with other DIAL applications. Therefore, all further analysis in this chapter is developed for this application. The importance of ozone measurements is related to its importance in the chemistry of atmospheric pollution in the boundary layer and in the troposphere as an attenuator of harmful ultraviolet light from the sun. The problem specifically concerns ozone distributions in the lowest layers of the troposphere, which are inhabited by humans. Important issues with respect to the ozone problem include regional transport, air quality control, and predicting unexpectedly high ground-level concentrations.

Backscatter and Extinction Correction Term Estimates. The initial version of the DIAL method (Hinkley, 1976) assumed that the differences in aerosol scattering and backscattering between the two DIAL wavelengths can be neglected. Accordingly, the approximation given in Eq. (10.10) was considered to be the basic DIAL solution. Later, it was established that this approximation can cause large errors in the measured gas concentration. Therefore, a more advanced variant was developed that recommended applying corrections for systematic errors that originate from aerosol and molecular differential scattering. An examination of these errors can be found in many studies (e.g., Schotland, 1974; Megie and Menzies, 1980; Pelon and Megie, 1982; Menyuk and Killinger, 1983; Megie et al., 1985; Browell et al., 1985; Papayannis et al., 1990; Kovalev and McElroy, 1996).

The correction terms $\Delta n_\beta(r)$ and $\Delta n_e(r)$ include contributions to the bias from both molecules and particulates. For the molecular differential correction, an appropriate standard molecular profile can be used, so that no additional measurements are required. For the particulate differential correction, some estimates of the aerosol profile are required. To provide a basis for an estimate of the backscatter correction term, information is required on the variations in the particulate backscatter intensity along the searched direction. If the absorption coefficient at the off-line is negligible, the off signal can be used as a reference signal to determine the aerosol corrections. Otherwise, an independent lidar measurement must be made at a reference wavelength λ_{ref} taken somewhere outside the absorption region. The drawback of this technique is that data at λ_{ref} are helpful to determine the terms $\Delta n_e(r)$ and $\Delta n_\beta(r)$ only if some specific conditions are true. The conditions are that unique relationships exist between $\beta_{p,ref}(r)$, $\beta_{p,on}(r)$, and $\beta_{p,off}(r)$ and also between $\kappa_{p,ref}(r)$, $\kappa_{p,off}(r)$, and $\kappa_{p,on}(r)$, and that these relationships are known. In practice, a priori assumptions about the wavelength dependence between the scattering characteristics are usually chosen. Generally, it is assumed that the particulate extinction and backscattering coefficients vary inversely with the wavelength over a wavelength range that includes the wavelengths λ_{on} and λ_{off} (and λ_{ref} if it differs from λ_{off}).

The particulate extinction correction may be evaluated with a power law dependence for the particulate component (see Chapter 2). It is commonly assumed that the aerosol optical attenuation (scattering) has a power law dependence with a constant Angstrom coefficient u as the exponent

$$\beta_p = \frac{const.}{\lambda^u}$$

Generally, the wavelength difference $\Delta\lambda$ between λ_{on} and λ_{off} is small. Therefore, the approximate relationship between $\kappa_{p,off}(r)$ and $\lambda_{p,off}(r)$ may be written as

$$\beta_{p,on}(r) \approx \beta_{p,off}(r)\left[1 + u\frac{\Delta\lambda}{\lambda_{off}}\right]$$

For molecular scattering, the exponent $u = 4$, thus

$$\beta_{m,on}(r) \approx \beta_{m,off}(r)\left[1 + 4\frac{\Delta\lambda}{\lambda_{off}}\right]$$

The formula to determine the total particulate and molecular extinction correction term in Eq. (10.6) can then be written as (Megie and Menzies, 1980, Browell et al., 1985)

$$\Delta n_e(r) = \frac{-1}{\Delta\sigma}[\beta_{on}(r) - \beta_{off}(r)] \approx -B_\lambda[u\beta_{p,off}(r) + 4\beta_{m,off}(r)] \qquad (10.13)$$

where the spectrum factor B_λ is

$$B_\lambda = \frac{1}{\lambda_{off}\left[\dfrac{\Delta\sigma}{\Delta\lambda}\right]} \qquad (10.14)$$

Note that the error $\Delta n_e(r)$ is directly proportional to the factor B_λ (Kovalev and Bristow, 1996; Simeonov et. al., 2002). This factor, in turn, is inversely proportional to the ratio $\Delta\sigma/\Delta\lambda$, which determines the sensitivity of the differential method in the particular spectrum range. Obviously, the ratio $\Delta\sigma/\Delta\lambda$ is small if the difference between σ_{on} and σ_{off} is small. In this case, the spectrum factor B_λ and, accordingly, the systematic error $\Delta n_e(r)$, defined by Eq. (10.13), may be very large.

The ozone concentration in the lower troposphere varies, generally, from 30–50 to 100–150 ppb (1 ppb is equal to 1 part ozone in 10^9 parts air by number density). To give an idea of the magnitude of these correction values in typical atmospheric conditions, we present the estimates made for the NASA airborne DIAL system during the EPA 1980 PEPE/NEROS Field Experiment (Browell et al., 1985). For the DIAL system, which operated at $\lambda_{on} = 286$ nm and $\lambda_{off} = 298.3$ nm, the aerosol extinction correction varied from 2 to 10 ppb with an average of approximately 5 ppb. To evaluate these corrections, the particulate extinction coefficient profile at the off (or reference) wavelength should be first determined, and this is the basic difficulty. Because of the uncertainty in both the particulate extinction coefficient and the Angstrom coefficient u, DIAL measurements are often corrected only for the molecular component of $\Delta n_e(r)$. The molecular extinction coefficients are known from the molecular scattering (Rayleigh) theory. The component is independent of altitude. In the above experiment, the value of the molecular correction was 6.7 ppb.

The estimation of the backscatter correction term $\Delta n_\beta(r)$ is the most difficult problem. This value depends on the gradient of the particulate extinction coefficient profile and is found by taking the derivative of the backscatter ratio [Eq. (10.6)]. The backscatter correction for small range differences Δr can be calculated with logarithmic differences of the $\beta_{on}(r)$ and $\beta_{off}(r)$, similar to that in Eq. (10.10)

$$\Delta n_\beta(r) = \frac{-1}{2\Delta\sigma\,\Delta r}\left[\ln\frac{\beta_{\pi,on}(r)}{\beta_{\pi,on}(r+\Delta r)} - \ln\frac{\beta_{\pi,off}(r)}{\beta_{\pi,off}(r+\Delta r)}\right] \tag{10.15}$$

The backscatter relative error can be defined from Eq. (10.15) by the ratio of $\Delta n_\beta(r)$ to the gas concentration, that is, $\delta n_\beta(r) = \Delta n_\beta(r)/n(r)$. After dividing both sides of Eq. (10.15) by $n(r)$, the factor ahead of the square brackets becomes equal to $[2n(r)\Delta\sigma\Delta r]^{-1} = [2\Delta\tau_{A,dif}]^{-1}$. Thus, similar to the error $\delta n'$ in Eq. (10.11), the error $\delta n_\beta(r)$ is proportional to the reciprocal of the local differential absorption optical depth, $\Delta\tau_{A,dif}$. When a small Δr is used, the error may become large, especially in areas with sharp spatial changes in the particulate backscattering, for example, in clouds, and the values $\Delta n_\beta(r)$ may also be large in these areas, up to tens of ppb. The systematic error caused by aerosol differential backscattering is a key problem with DIAL measurements in heterogeneous atmospheres. In areas where no significant heterogeneity exists, the ozone profile correction can be made with an approximate method developed by Browell et al. (1985). The method is based on the introduction of a power law relationship for backscattering in the operating wavelength range. If the particulate backscattering coefficients vary inversely with wavelength to the power $\xi = const.$, the aerosol-to-molecular backscatter ratio defined as

$$Q(r,\lambda) = \frac{\beta_{\pi,p}(r,\lambda)}{\beta_{\pi,m}(r,\lambda)} = \frac{\Pi_{\pi,p}(r,\lambda)\,\kappa_p(r,\lambda)}{\frac{3}{8\pi}\beta_m(r,\lambda)} \tag{10.16}$$

can be rewritten as the wavelength-dependent function ($\lambda < \lambda_{ref}$), that is

$$Q(r,\lambda) = Q(r,\lambda_{ref})\left(\frac{\lambda}{\lambda_{ref}}\right)^{4-\xi} \tag{10.17}$$

If the wavelength difference is small, the following approximate relationship applies

$$\left(\frac{\lambda}{\lambda_{ref}}\right)^{4-\xi} \approx 1-(4-\xi)\frac{\Delta\lambda}{\lambda_{ref}} \tag{10.18}$$

so that the backscatter correction term can be found as (Browell et al., 1985)

$$\Delta n_\beta(r) = \frac{(4-\xi)B_\lambda}{2\Delta r}\left[\frac{Q_{off}(r)}{1+Q_{off}(r)} - \frac{Q_{off}(r+\Delta r)}{1+Q_{off}(r+\Delta r)}\right] \tag{10.19}$$

here the term λ in $Q_{off}(r)$ is omitted for brevity. The use of a power law approximation makes it possible to find the backscatter correction term by calculating $Q_{off}(r)$ and selecting some value of ξ. Note that formally no determination of the derivatives of $\beta_p(r)$ is made in Eq. (10.19). However, the quantity Q_{off} is found at both ends of the range Δr. Thus the operation in Eq. (10.19) is equivalent to a conventional numerical differentiation. Note also that the backscatter correction term is directly proportional to the spectrum factor, B_λ, the same as for the extinction correction term $\Delta n_e(r)$ in Eq. (10.13).

In the studies by Schotland (1974), Menyuk and Killinger (1983), and Browell et al. (1985), the following correction procedure was assumed for the DIAL measurements:

(1) During DIAL measurements, a lidar signal is recorded at a wavelength at which no absorption takes place. If the off wavelength of the DIAL meets this requirement, the off wavelength signal may also be used as the reference signal. Otherwise, the extinction coefficient measurement must be made at some additional reference wavelength, λ_{ref}.

(2) The profile of the particulate extinction coefficient is found at the reference wavelength, and the corresponding profile of the aerosol-to-molecular backscatter ratio $Q(r, \lambda_{ref})$ is calculated. Here the common problem is the indeterminacy of the elastic lidar equation. When no independent inelastic measurements are available, an a priori backscatter-to-extinction ratio must be chosen. Also, a solution boundary value should be established.

(3) The differential extinction correction for the ozone profile is found with Eq. (10.13) and an Angstrom coefficient u that is assumed to be true over the operational wavelength range.
(4) The differential backscatter correction is made with a similar approach. For the backscatter-spectral dependence, the same power law dependence is assumed to be valid, but with another constant exponent, ξ. Both power law exponents, u and ξ, are usually chosen a priori.

Let us make a short summary. The determination of gas concentration profiles with DIAL must include the following operations (Browell, 1985):

(1) Measurement of the elastic lidar signals at the on and off wavelengths. An additional lidar signal measurement may also be made at a reference wavelength, λ_{ref}, that allows determination of the backscattering and extinction corrections.
(2) Calculation of the first raw estimate of the absorbing gas concentration profile $n'(r)$ with Eq. (10.10). This makes it possible to estimate the data quality and the achieved measurement range.
(3) Calculation of the particulate extinction coefficient profile at the reference wavelength and determination of the backscatter and extinction corrections for the ozone concentration.
(4) Calculation of the final absorbing gas concentration profile by using the backscatter and extinction corrections. Note that the backscatter and extinction corrections can be made either after taking the derivative of the signal ratio logarithm or before this operation. One can avoid additional numerical differentiation when determining the backscatter correction term, making the corrections before the ozone concentration is extracted (Kovalev and McElroy, 1994; Kovalev et al., 1996).

10.1.2. Uncertainty of the Backscatter Corrections in Atmospheres with Large Gradients of Aerosol Backscattering

An accurate backscattering correction can only be obtained for atmospheres in which the backscatter coefficients are constant or vary slightly with range. The validity of the dependence in Eq. (10.17) is questionable when the measurements are made in a heterogeneous atmosphere, in which large changes in the aerosol concentration and/or particle size distribution occur. It is even possible to worsen rather than improve the accuracy of the derived concentration when using Eq. (10.17) for such conditions. The sensitivity of the ozone concentration backscatter correction, $\Delta n_\beta(r)$, to the constant ξ in heterogeneous atmospheres was shown by Kovalev and McElroy (1994). The authors calculated the backscatter corrections for $\lambda_{on} = 276.9\,\text{nm}$ and $\lambda_{off} = 312.9\,\text{nm}$ with an extinction coefficient profile measured at a reference wavelength

$\lambda_{ref} = 359.6$ nm. Some results from this study are given below. The particulate extinction coefficient profiles used for the numerical experiment are shown in Fig. 10.1. *Curve 1* represents an artificial extinction coefficient profile, the shape of which is typical for altitude profiles obtained by an airborne down-looking DIAL system. *Curve 2* is the same profile, but two artificial turbid atmospheric layers have been added. The backscatter correction term has been estimated for an ideal atmosphere, where the power law relation [Eq. (10.17)] holds with the same constant value of ξ at all wavelengths λ_{on}, λ_{off} and λ_{ref}. It was also assumed that no measurement errors exist in the measured extinction coefficient $\kappa_p(r, \lambda_{ref})$ and, accordingly, in the corresponding profile of $Q(r, \lambda_{ref})$. The ozone corrections $\Delta n_\beta(r)$ that correspond to the monotonic profile (*curve 1* in Fig. 10.1) are presented in Fig. 10.2 (a). It can be seen that the backscatter corrections here are small. This is because the spatial changes in the initial particulate extinction coefficient are small. The correction values become much larger in heterogeneous regions with strong aerosol gradients [Fig. 10.2 (b)]. In this case there is a significant difference in the calculated correction term $\Delta n_\beta(r)$ when using a different constant ξ. In these locations, an a priori selection of the constant is fraught with the possibility of large errors. A decrease in the uncertainty of $\Delta n_\beta(r)$ in areas of strong aerosol layering can only be achieved by worsening the measurement range resolution. This results in some smoothing and reduces spikes in the backscatter corrections. However, when this is done, the distortion range expands into adjacent areas, outside the actual layering. In Fig. 10.3, the absolute errors in the function $\Delta n_\beta(r)$ that are due to the difference between the assumed and actual values of ξ are shown for range resolutions of 120 m (*curves 1* and *2*) and

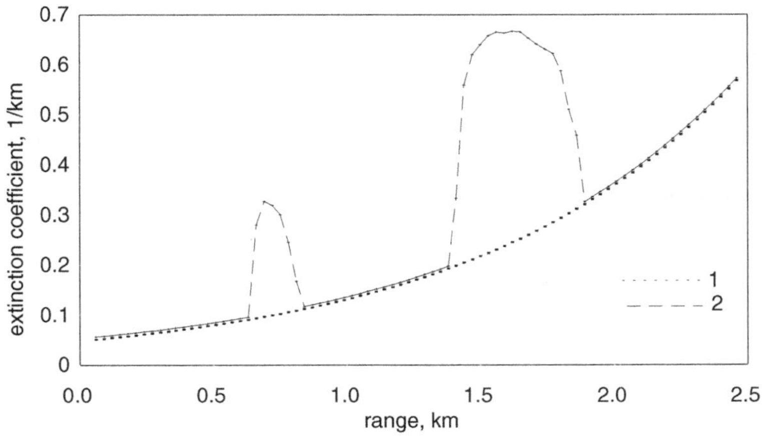

Fig. 10.1. Model particulate extinction coefficient profiles at λ_{ref} used for the calculation of the backscatter corrections in Figs. 10.2 (a) and (b). *Curve 1* is an artificial vertical profile of κ_p for a cloudless atmosphere. *Curve 2* is the same profile but where additional turbid atmospheric layers exist (Kovalev and McElroy, 1994).

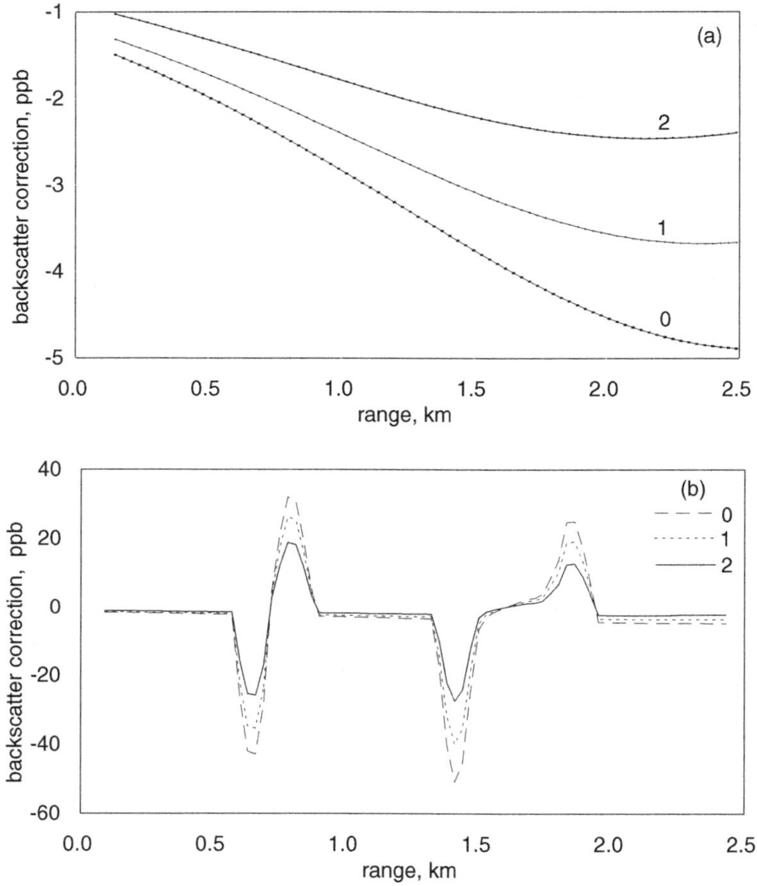

Fig. 10.2. (a) Backscatter correction functions $\Delta n_\beta(r)$ calculated for the smooth extinction coefficient profile shown as *curve 1* in Fig. 10.1. The values of ξ used for the calculation are shown as the numbers of the corresponding curves, and $\Pi_p = 0.03\,\text{sr}^{-1}$. The functions are obtained with the conventional regression procedure using a five-point running mean, with a cell size of 120 m. (b) Same as in (a) but for the extinction coefficient profile shown as *curve 2* in Fig. 10.1 (Kovalev and McElroy, 1994).

300 m (*curves 3* and *4*). The concentration calculation is made with $\xi = 1$, whereas the actual values of ξ are assumed to be 0 and 2. Note the difference in the sign of the correction term $\Delta n_\beta(r)$ for $\xi = 0$ and $\xi = 2$.

To clarify the origin of the large uncertainty obtained in heterogeneous layers, note that the dependence in Eq. (10.17) is extremely restrictive for real atmospheres in which heterogeneous areas exist. In fact, with ξ = const., Eq. (10.17) implies that

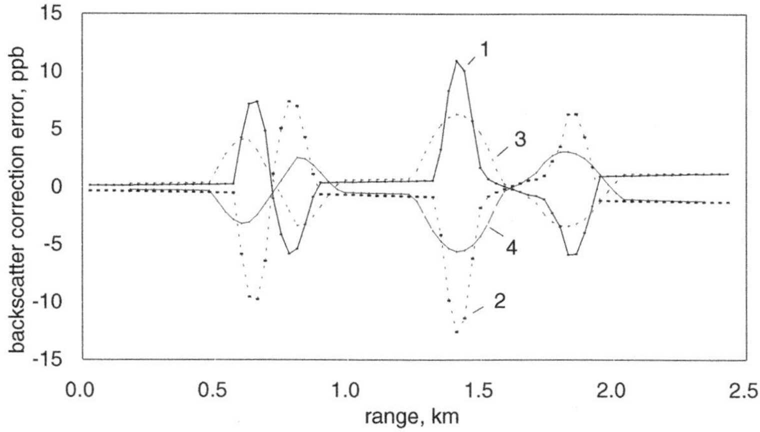

Fig. 10.3. Absolute errors in backscatter correction function $\Delta n_\beta(r)$ caused by inaccurate specification of ξ for the range cell length of 120 m (*curves 1* and *2*) and 300 m (*curves 3* and *4*). The errors are calculated with the extinction coefficient profile shown as *curve 2* in Fig. 10.1. The constant ξ is chosen as unity, whereas its actual value is 0 (*curves 1* and *3*) and 2 (*curves 2* and *4*) (Kovalev and McElroy, 1994).

$$\frac{Q(r,\lambda)}{Q(r,\lambda_{ref})} = const.$$

and, accordingly, the ratio of the backscatter particulate coefficients at the two wavelengths (λ_{on} and λ_{off}) is also a constant, range-independent value,

$$\frac{\Pi_p(r,\lambda_{on})\,\kappa_p(r,\lambda_{on})}{\Pi_p(r,\lambda_{off})\,\kappa_p(r,\lambda_{off})} = \frac{\beta_p(r,\lambda_{on})}{\beta_p(r,\lambda_{off})} = const. \tag{10.20}$$

The relationship in Eq. (10.20) assumes that, within the measurement range, the ratio remains invariant both in the clear atmosphere and within the areosol/cloud layers. In other words, the particulate backscattering ratio at λ_{on} and λ_{off} does not change regardless of any changes in the atmospheric aerosol characteristics along the lidar measurement range, such as the concentration or size distribution. This is an unrealistic assumption. A more realistic presumption would be that the ratio $Q_{on}(r)/Q_{off}(r)$ varies, at least slightly, over the measurement range. Accordingly, instead of the rigid condition expressed by Eq. (10.17), a more flexible relationship between $Q_{on}(r)$ and $Q_{off}(r)$ is

$$\frac{Q_{on}(r)}{Q_{off}(r)} = [1 + \delta Q^*(r)]\left[\frac{\lambda_{on}}{\lambda_{off}}\right]^{4-\xi} \tag{10.21}$$

where a variable term $[1 + \delta Q^*(r)]$ is included, which allows the possibility of local spatial variations in the value of the ratio $Q_{on}(r)/Q_{off}(r)$. Formally, these variations in the ratio can be considered to be variations in ξ. With this observation, Eq. (10.21) can be rewritten in the form

$$Q_{on}(r) = Q_{off}(r)\left[\frac{\lambda_{on}}{\lambda_{off}}\right]^{4-[\xi+\Delta\xi(r)]} \tag{10.22}$$

where the value $\Delta\xi(r)$ is uniquely related to the term $\delta Q^*(r)$. The relationship between $\delta Q^*(r)$ and $\Delta\xi(r)$ can be found from Eqs. (10.21) and (10.22) as

$$\Delta\xi(r) = \frac{\ln[1+\delta Q^*(r)]}{\ln\left[\frac{\lambda_{off}}{\lambda_{on}}\right]} \tag{10.23}$$

It follows from Eq. (10.23) that any slight fluctuation in $\delta Q^*(r)$ is equivalent to a significant change in ξ. For example, for a KrF DIAL system with $\lambda_{on} = 268.4$ nm and $\lambda_{off} = 291.8$ nm, a fluctuation $\delta Q^*(r) = 0.05$ is equivalent to $\Delta\xi(r) = 0.58$; for $\delta Q^*(r) = 0.1$, the value of $\Delta\xi(r) = 1.14$, etc. Therefore, the change in the correction term $\Delta n_\beta(r)$ can be large even if $Q_{off}(r)/Q_{on}(r)$ varies only slightly. To illustrate this, in Fig. 10.4, the backscatter correction functions $\Delta n_\beta(r)$ are shown calculated for the same model extinction coefficient profile shown as *curve 2* in Fig. 10.1, but for different $\delta Q(r)$. One can see large dif-

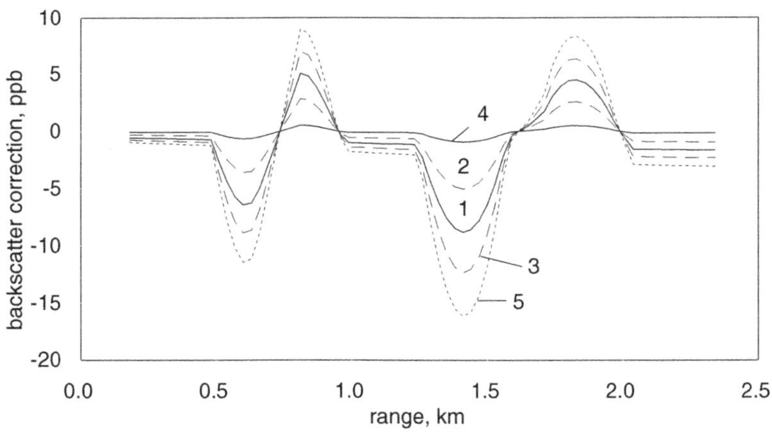

Fig. 10.4. Backscatter correction function $\Delta n_\beta(r)$ for the wavelength pair λ_{on} = 268.4 nm and $\lambda_{off} = 291.8$ nm calculated with the extinction coefficient profile shown as *curve 2* in Fig. 10.1. The reference wavelength is $\lambda_{ref} = 359.6$ nm. *Curve 1* shows $\Delta n_\beta(r)$ calculated with the assumption that the aerosol-to-backscatter ratio is constant over the measurement range and $\xi = 1$. *Curves 2–5* are the same as *curve 1* but with $\delta Q(r, \lambda_{off}) = 0.1$, $\delta Q(r, \lambda_{on}) = 0.1$, $\delta Q(r, \lambda_{off}) = 0.2$, and $\delta Q(r, \lambda_{on}) = 0.2$, respectively. The range cell size is 300 m (Kovalev and McElroy, 1994).

ferences in the calculated backscatter correction term in the heterogeneous regions.

Thus the uncertainty in an a prior selected ξ is the first source of significant errors in $Q_{off}(r)$ and, accordingly, in the backscatter correction term [Eq. (10.19)]. An error in the retrieved extinction coefficient profile is another source of uncertainty in $Q_{off}(r)$. First, this type of error occurs from inaccuracies in the lidar solution constant and in the assumed aerosol backscatter-to-extinction ratio, Π_p (Chapter 7). Second, lidar signal averaging can introduce a significant error due to temporal variations in the aerosol optical properties in areas of local heterogeneity (Hooper and Eloranta, 1986; Menut et al., 1999). This is a significant source of errors in the DIAL measurements. To have acceptable signal statistics, the minimum averaging time for the ozone DIAL measurements in the troposphere must be 5–10 s or even more (Browell et al., 1985; Papayannis et al., 1990; Uthe et al., 1992; Moosmüller et al., 1993). The same average time should be used to determine the reference extinction coefficient profile for aerosol corrections. If such measurements are made from a moving airborne platform, an error may originate from spatial variations at the top or bottom boundaries of turbid layers, or from local, spatially restricted aerosol tatters. Such errors are critical for data that are averaged and processed with numerical differentiation and can severely limit the amount of improvement achievable with lidar signal averaging. To illustrate this, several experimental ozone concentration profiles calculated for the lower troposphere are shown in Fig. 10.5. The profiles were measured by an airborne down-looking multiwavelength UV-DIAL system developed by the EPA (Moosmüller et al., 1991). The plane altitude is approximately 2500 m. In the figure, consecutive ozone concentration profiles are shown, obtained in the presence of local thin-cloudy, heterogeneous layering at altitudes of 1000–1300 m. Each curve is obtained with a 20 single-shot (1 s) average, and six con-

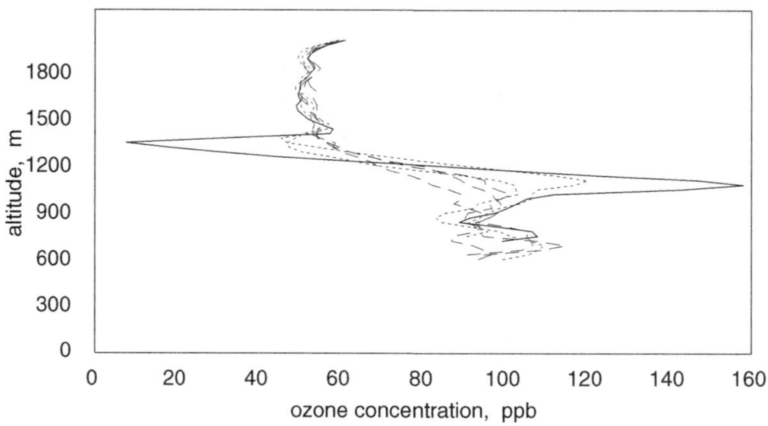

Fig. 10.5. Experimental ozone concentration profiles $n'(h)$ obtained with the down-looking airborne DIAL in the lower troposphere.

secutive profiles are shown, obtained with a running mean of an 11-point linear regression. There is an enormous amount of scattering and false fluctuations in the measured ozone concentration profiles that appear in areas of thin heterogeneous layering.

An improvement of DIAL measurement accuracy with estimation of the aerosol backscatter correction is only achievable in areas with no strong heterogeneous layering. Backscatter corrections are generally not practical in heterogeneous zones. The basic difficulty in determining the backscatter correction term in such areas is the extremely high sensitivity of the backscatter correction term to the gradient of the ratio $[\beta_{p,on}(r)/\beta_{p,off}(r)]$.

10.1.3. Dependence of the DIAL Equation Correction Terms on the Spectral Range Interval Between the On and Off Wavelengths

DIAL systems, including those for ozone concentration measurements, can operate in the region of the spectrum from the ultraviolet (UV) to the infrared (IR). However, with IR measurements, the amount of molecular differential scattering is small because of the λ^{-4} dependence. Therefore, a relatively high concentration is necessary to obtain an acceptable differential optical depth, $\Delta\tau_{A,dif}$, which is the difference between the optical depths $\tau_{A,on}$ and $\tau_{A,off}$ over the range Δr [Eq. (10.12)]. Small values of $\Delta\tau_{A,dif}$ result in a large measurement uncertainty [Eq. (10.11). Thus the UV region is preferred because of the large absorption cross sections, and the spectral range between 260 and 290 nm is most practical for ozone measurements.

As shown in the study of Megie and Menzies (1980), the difference in the nature of the UV and IR transitions involved in the absorption process results in different optimization of the lidar parameters. In the UV portion of the spectrum, absorption is independent of altitude. This effect significantly simplifies UV differential measurements and is due to the characteristics of the UV absorption lines. Unlike the IR portion of the spectrum, where the absorption transitions involve the vibrational rotation levels of a single electronic state, in UV the absorption lines correspond to vibrational/rotational transitions between different electronic states (Bohren and Huffman, 1983). The latter lines overlap at all altitude levels, making the absorption spectra independent of the background pressure. Although there is a temperature dependence on the absorption intensity in the UV, it is insignificant and can be ignored (Megie et al., 1985). With IR spectra, the absorption line widths are altitude dependent. Therefore, the ratio of the local optical depth, as defined by the measurement range resolution, to the integrated optical depth may depend on the altitude. This significantly complicates processing DIAL data in the IR. On the other hand, a global ozone distribution cannot be examined with any single wavelength pair of laser lines. For global ozone measurements with DIAL, the UV and IR spectral measurements can be considered to be complementary rather than competitive. This is because the maximum altitude achievable in the UV portion of the spectrum is restricted to the altitudes of the ozone concentration maximum (about 23 km). This barrier exists in both directions, from space and ground-based systems. To measure such high concentrations, IR DIAL measurements may be preferable.

In Fig. 10.6, the absorption spectrum of ozone in the UV region is shown for normal atmospheric pressure and T = 298 K. The graph is built with the data of Molina and Molina (1986). Using this spectra, one can select optimum on-off wavelength pairs for ozone concentration measurement in both the troposphere and the stratosphere. Considering aerosol heterogeneous loading to be the most serious problem for DIAL measurements, we concentrate our analysis on UV DIAL measurements in the lower troposphere, where the loading is maximum. The choice of optimal wavelengths allows one to obtain the highest accuracy achievable for particular conditions of interest. According to the estimates of Papayannis et al. (1990), the wavelength separation between the off- and on-line should be not less than 5–15 nm. This allows the establishment of an optimal balance between the errors that are caused by spatial variations of particulate scattering and by statistical error. The optimization of DIAL system requires some numerical computation and a joint analysis of both the experimental parameters of the DIAL system and parameters related to ozone absorption cross sections and typical particulate loading. Obviously, the altitude where the ozone concentration measurement has to be made is the most important characteristic. In the study by Papayannis et al. (1990), two different altitude regions in the troposphere are considered for which the DIAL systems must have significantly different characteristics. These regions are the planetary boundary layer (0–2 km) and the free troposphere (2–15 km). In the planetary boundary layer, the error $\Delta n_\beta(r)$ is a predominant factor. This is the result of the large particulate loading and good signal-to-noise ratios in this altitude range for ground-based DIAL.

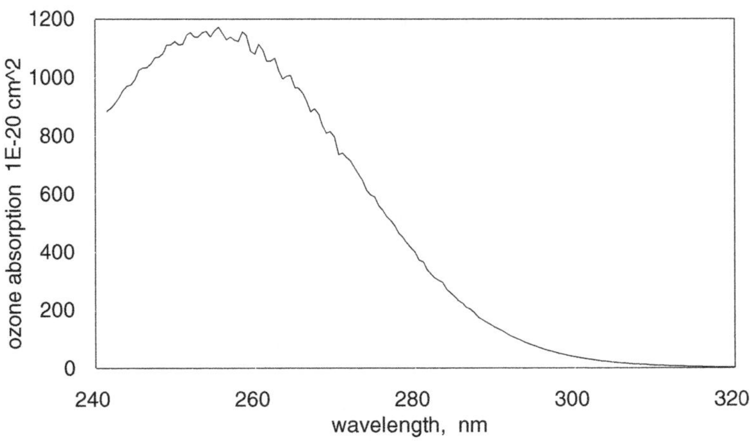

Fig. 10.6. Ozone absorption spectra in the ultraviolet spectrum at 298 K (Molina and Molina, 1986).

The maximum values of the error $\Delta n_\beta(r)$ are generally observed at the area near the top of the planetary boundary layer, where large gradients exist in the vertical profile of particulate backscattering. In addition to aerosol loading, high concentrations of some pollutants such as SO_2 and NO_2 may also become a source of systematic error in this layer. Because of the high UV scattering coefficients, the total extinction is large in the first few kilometers but then the signal-to-noise ratio rapidly worsens with altitude. Therefore, in the free troposphere, the signals are significantly reduced. Here, the statistical error becomes a key factor, whereas the systematic error becomes less important because of low aerosol loading and pollution content. Accordingly, for measurements in the upper troposphere, the wavelength λ_{on} must be shifted to the upper level of the UV region, closer to 280–290 nm. At these wavelengths, the absorption cross section is smaller and absorption optical depth is not too large. This makes it possible to obtain an acceptable signal-to-noise ratio at the on wavelength when examining altitudes up to 10–15 km.

Difficulties with the particulate backscatter corrections to DIAL measurements lead to attempts to reduce the problem by reducing the spectral interval, $\Delta\lambda$, between the on and off wavelengths. As follows from Eq. (10.22), the ratio $Q_{on}(r)/Q_{off}(r)$ is rather insensitive to variations of $\Delta\xi(r)$ when λ_{on} is close enough to λ_{off}. In some studies, it was even assumed that the correction terms can be ignored if the wavelength separation between λ_{on} and λ_{off} is small. On this basis, DIAL systems were built in which the wavelength separation was only a few nanometers (Proffitt and Langford, 1997). Unfortunately, the reduction of the spectral interval $\Delta\lambda$ can improve the measurement accuracy only within quite modest limits. The decrease of the wavelength separation between λ_{on} and λ_{off} below some optimum may worsen rather than improve the measurement accuracy. This statement stems from an analysis of Eqs. (10.13) and (10.19). The extinction and backscatter errors are proportional to the spectrum factor B_λ, which, in turn, is proportional to the reciprocal of the ratio of the ozone differential absorption cross section, $\Delta\sigma$, to the on and off wavelength separation [Eq. (10.14)]. If the wavelength separation $\Delta\lambda$ tends to zero, $\Delta\sigma$ also tends to zero and the absolute value of the factor B_λ becomes proportional to the reciprocal of the derivative $(d\sigma/d\lambda)$. Therefore, when the wavelength separation decreases, the errors Δn_β and Δn_e can increase, remain constant, or fall, depending on the slope and values of the absorption coefficient in the particular range $\Delta\lambda$. Usually (but not always), the error is reduced when λ_{off} is shifted toward a fixed λ_{on}. However, the factor B_λ rather than the wavelength separation $\Delta\lambda$ is a key factor that determines the ozone measurement accuracy. For example, consider the tropospheric DIAL system developed in the Netherlands (Sunesson et al., 1994) that measures the backscattered signals at the wavelengths 266, 289, and 299 nm. The factor B_λ for the different pairs of this DIAL is significantly different. For the 289/299 nm pair, B_λ is a maximum ($B_\lambda = 2.96 \times 10^{16}$ mol·cm^{-2}). Note that the pair with the worst B_λ has the smallest wavelength separation (10 nm). For the other pair, 266/289 nm, in which the wavelength separation is much larger (23 nm),

$B_\lambda = 1.01 \times 10^{16}$ mol·cm^{-2}, so that the backscatter correction is the least. It is not surprising then that the analysis of DIAL experimental data made by Sunesson et al. (1994) revealed that the 266/289-nm pair is much more suitable for measurements in the mixed layer than the 289/299-nm pair. In fact, most tropospheric UV DIAL systems with $\lambda_{on} < 300$ nm have wavelength separations of 10–30 nm (see, for example, Browell et al. (1985); McDermid et al., 1990; Moosmüller et al., 1991; Zhao et al., 1993; Ancellet and Ravetta (1998); Carnuth et al., 2002). This wavelength separation can be considered to be close to optimum for UV ozone measurements.

To clarify this point, one can consider the behavior of B_λ within the absorption spectra given in Fig. 10.6. Using this spectrum, one can compare different on-off wavelength pairs used for ozone concentration retrieval, and determine the pair in which the factor B_λ is the least. For such a wavelength pair, particulate corrections are also the least. The behavior of the factor B_λ for different separation $\Delta\lambda$, calculated with the UV absorption spectra in Fig. 10.6, is shown in Fig. 10.7. As in Eqs. (10.13) and (10.19), all of the atmospheric parameters are determined for the off wavelengths; the factor B_λ in Fig. 10.7 is also given as the function of this wavelength. One can see that a decrease in the wavelength separation generally increases B_λ and, accordingly, the corresponding extinction and backscatter uncertainty. Clearly, other parameters must also be considered when making a comprehensive analysis of a optimal DIAL system, especially when evaluating it for use in the lower stratosphere. In some situations, the systematic error can be independent of $\Delta\lambda$ or can even increase with $\Delta\lambda$ (Pelon and Megie, 1982). However, to make a preliminary estimate of the efficiency of a DIAL system, the dependencies like that shown in Fig. 10.7 can be quite helpful.

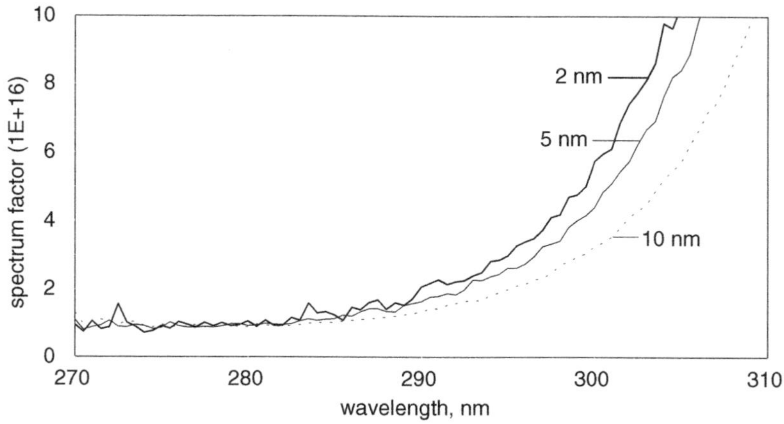

Fig. 10.7. Factor B_λ as a function of the wavelength λ_{off} for different wavelength separation $\Delta\lambda$, calculated with the ozone absorption spectrum given in Fig. 10.6.

The selection of optimum values of λ_{on} and the separation $\Delta\lambda$ between the on and off-line wavelengths must be based on the particular requirements of the DIAL system. The optimum values depend primarily on the desired operating altitude, on the range of expected ozone concentrations, and, correspondingly, on the signal attenuation. For example, the strong attenuation of the UV DIAL signal intensity in the upper troposphere is the most important factor for ground measurements of stratospheric ozone; however, this factor is not important for spacecraft stratospheric measurements.

The contribution of $\Delta n_\beta(r)$ cannot be considered to be a purely systematic error in atmospheres where heterogeneous layering exists. In such atmospheres, $\Delta n_\beta(r)$ may add both systematic and random contributions in addition to the conventional signal noise contribution. In Fig. 10.8, the experimental altitude ozone concentration profiles, $n'(h)$, are shown for different pairs of λ_{off} and λ_{off}, measured simultaneously by a UV DIAL system (Moosmüller et al., 1991 and 1993). The wavelength pairs used for the profile calculation are 276.9/312.9 nm (bold curve), 276.9/291.6 nm (solid curve), and 291.6/319.4 nm (dotted curve). All of the range-resolved profiles are obtained with 11-point linear regression. Thus any point in the profile is, in fact, a running average with an altitude resolution $\Delta h = 300$ m. No backscatter and extinction corrections are made; therefore, all profiles include some systematic shifts. The values of the spectrum factor are $B_\lambda = 2.3E + 16\,mol\cdot cm^{-2}$ for the 276.9/312.9-nm pair; $B_\lambda = 1.3E + 16\,mol\cdot cm^{-2}$ for the 276.9/291.6-nm pair, and $B_\lambda = 7.5E + 16\,mol\cdot cm^{-2}$ for the 291.6/319.4-nm pair. One can note increased ozone fluctuations obtained for the pair 291.6/319.4 nm, for which the factor B_λ is the largest.

DIAL measurement uncertainty has been thoroughly investigated in many studies pioneered by the study by Schotland (1974). The conventional formula

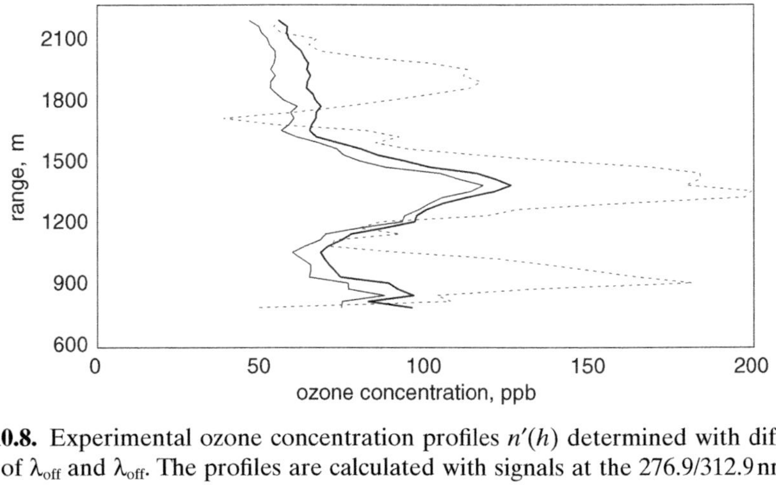

Fig. 10.8. Experimental ozone concentration profiles $n'(h)$ determined with different pairs of λ_{off} and λ_{off}. The profiles are calculated with signals at the 276.9/312.9 nm pair (bold curve), at the 276.9/291.6 nm pair (solid curve), and at the 291.6/319.4 nm pair (dotted curve).

for the estimate of the statistical error in the retrieved ozone concentration, used in most practical estimates, is based on standard error propagation for uncorrelated terms (Megie and Menzies, 1980; Pelon and Megie, 1982; Megie et al., 1985; Papayannis et al., 1990; Godin et al., 1999)

$$\delta n_s = \frac{1}{2\Delta\tau_{A,\mathrm{dif}} N^{1/2}} \sqrt{\sum_{i,k} [\mathrm{SNR}_{i,k}]^{-2}} \tag{10.24}$$

where $\mathrm{SNR}_{i,k}$ is the signal-to-noise ratio for the measurement at λ_i and the range r_k; N is the number of averaged shots. Note that the random measurement error is also proportional to the reciprocal of the two-way differential absorption optical depth $\Delta\tau_{A,\mathrm{dif}}$ defined in Eq. (10.12). Obviously, the measurement error may be large if either $\Delta\sigma$ or Δr in Eq. (10.12) is small. Thus, small values of $\Delta\sigma$ in DIAL measurements result in both large systematic errors, Δn_β and Δn_e, and a large random error δn_s. It should be noted that these values are related. Consider the random errors caused by signal noise in high-altitude measurements made by a ground-based lidar. As stated above, at these altitudes the signal is significantly smaller, so that the statistical error becomes dominant. Such DIAL measurements made in the upper troposphere and stratosphere generally use photon-counting techniques, in which only zero and positive integral values are obtained (see Chapter 4). Because of this characteristic, the estimate of the statistical error is made on the assumption that the uncertainty in the number of photons counted by the photodetector current is governed by Poisson statistics. With this assumption, the signal-to-noise ratio is

$$\mathrm{SNR}_\lambda(r) = \frac{p_\lambda(r)}{\sqrt{p_\lambda(r) + p_{\mathrm{bgr}} + p_{\mathrm{dc}}}}$$

where $p_\lambda(r)$ is the lidar signal (the number of photons) at the wavelength λ and p_{bgr} is the background contribution. The term p_{dc} denotes the photodetector dark current fluctuation within the time gate interval $2\Delta r/c$, where c is the velocity of light. It is assumed that no changes occur in the backscattering intensity that reaches the detector during the measurement interval. However, it should be mentioned that Poisson statistics, as with any theoretical model, should be used cautiously when applied to a particular practical measurement. Donovan et al. (1993) found that Poisson statistics overestimated the error compared with that obtained from a more thorough analysis, in which the parameters involved (the sampling time, widths of the pulse, and the count rate) are taken into consideration. The statistics may also be invalidated by photomultiplier saturation, afterpulsing, the presence of systematic errors due to temporal variability of atmospheric backscattering, and so on. The presence of a discriminator that attempts to separate noise and signal pulses in the photocounting receiver system may be an additional reason for the statistical distortion (Baray et al., 1999). Nevertheless, Poisson statistic is considered to be the most practical approximation when estimating the uncertainty of photon-counting systems.

Some comments must be added concerning DIAL signal averaging. As follows from Eq. (10.24), the random error is proportional to $N^{-1/2}$. Therefore, the uncertainty may be significantly reduced by increasing the number of averaged shots. However, this is only true under the condition that atmospheric turbulence can be ignored. This condition may often be inappropriate when the measurements are made in the lower troposphere (Elbaum and Diament, 1976; Killinger and Menyuk, 1981; Menyuk et al., 1982; Durieux and Fiorani, 1998). Apart from that, some random errors do not follow a Poisson distribution. These are the errors caused by baseline uncertainty caused, for example, by signal induced noise. Such errors cannot be reduced by increasing the number of averaged shots. On the other hand, these errors remain proportional to the reciprocal of the optical depth $\Delta\tau_{A,dif}$, and they also increase when the wavelength separation between the on and off wavelengths is small.

10.2. DIAL PROCESSING TECHNIQUE: PROBLEMS

In Section 10.1, the theoretical basis for DIAL measurements was presented. In this section, some problems of practical DIAL measurements are discussed that may significantly influence DIAL measurement accuracy. The equations for the error estimate for a column content from the ozone concentration profile are considered, as are questions related to numerical differentiation and smoothing techniques. Also discussed is the amount of reduction in the random noise that is achieved by temporal and spatial averaging.

10.2.1. Uncertainty of the DIAL Solution for Column Content of the Ozone Concentration

There are several practical variants of the technique for ozone concentration determination. The concentration may be obtained by the following ways: (a) with the difference of the derivatives calculated separately for each DIAL signal (Pelon and Megie, 1982; McDermid et al., 1990); (b) by numerical differentiation of the logarithm of the on-to-off signal ratio, making the aerosol and backscatter corrections after determining the raw concentration profile (Measures, 1984; Browell et al., 1985; Steinbrecht et al., 1989: Uchino and Tabata, 1991; Carnuth et al., 2002); and (c) as in (b) but making the corrections before determining the derivative of the logarithmic ratio (Kovalev and McElroy, 1994). Considering these variants from the point of pure mathematics, all of the variants are equivalent. In practice, the validity of the particular assumptions used in these approaches is always questionable. This is why no variant is generally accepted. In this section, version (c) is analyzed because of its relative simplicity when making backscatter corrections. This variant does not require the calculation of the second derivative in Eq. (10.6), that is, the derivative of the logarithm of the ratio $\beta_{\pi,on}(r)/\beta_{\pi,off}(r)$.

For the uncertainty analysis, it is often convenient to consider primarily the uncertainty in a column content of the ozone concentration, which is found directly from the logarithm of the off-to-on signal ratio. The initial DIAL equation [Eq. (10.5)] can be rewritten as

$$\ln\frac{P_{off}(r)}{P_{on}(r)} = \ln\frac{C_{off}}{C_{on}} + \ln\frac{\beta_{\pi,off}(r)}{\beta_{\pi,on}(r)} + 2\int_{r_0}^{r}\kappa_{dif}(r')\,dr' \tag{10.25}$$

where the constants C_{off} and C_{on} are determined at the starting point r_0 rather than r_1 as in Eq. (10.5), $\kappa_{dif}(r)$ is the total differential extinction coefficient, and its integral is directly related to the columnar ozone content over the range from r_0 to r. This term may be considered to be the sum of two components. The first component originates from the differential absorption of ozone (or another gas), the concentration of which is the subject of interest. The second term in $\kappa_{dif}(r)$ comprises the remaining differential extinction that is not due to the presence of ozone. In terms of the differential optical depths of the column (r_0, r), this can be written as

$$\int_{r_0}^{r}\kappa_{dif}(r)\,dr = \tau_{A,dif}(r_0, r) + \tau_{e,dif}(r_0, r) \tag{10.26}$$

where $\tau_{A,dif}(r_0, r)$ is the differential absorption optical depth of ozone, that is,

$$\tau_{A,dif}(r_0, r) = \int_{r_0}^{r}\kappa_{A,dif}(r')\,dr' = \Delta\sigma\int_{r_0}^{r}n(r')\,dr' \tag{10.27}$$

and $\tau_{e,dif}(r_0, r)$ is the remaining differential extinction optical depth, which takes into account the effect of particulate and molecular scattering and, if any, differential absorption by constituents other than ozone. The contribution of these constituents acts as systematic uncertainty and must be removed before an accurate ozone concentration can be extracted from the integral in Eq. (10.26). If the differential absorption by other gases at λ_{off} and λ_{on} is negligible and can be ignored, the term $\tau_{e,dif}(r_0, r)$ is due only to particulate and molecular differential scattering and can be written as

$$\tau_{e,dif}(r_0, r) = \int_{r_0}^{r}\Delta\beta_{dif}(r')\,dr' \tag{10.28}$$

where $\Delta\beta_{dif}(r) = \beta_{on}(r) - \beta_{off}(r)$.

As shown in Chapter 2, the total flux on the lidar photoreceiver is the sum of several components. Usually, the lidar signal is comprised of three constituents, a constituent originating from single scattering of the laser light, a constituent from multiple scattering, and a constituent originating from the solar background. In clear atmospheres, analyzed here, the multiple-scattering contribution may be ignored. Then the backscattered signal on the photoreceiver, $P_\Sigma(r)$, is the total of the singly scattered constituent, $P(r)$, and an additive background constituent, P_{bgr}. In the measured on and off signals, the background constituent should be estimated and subtracted before further processing, as

$$P(r) = P_\Sigma(r) - P_{bgr} \tag{10.29}$$

Using the definitions in Eqs. (10.27) and (10.28), one can determine the column differential optical depth of the ozone from Eq. (10.25) as

$$\tau_{A,dif}(r_0, r) = 0.5\ln\left[\frac{P_{\Sigma,off}(r) - P_{bgr,off}}{P_{\Sigma,on}(r) - P_{bgr,on}}\right] - 0.5\ln\left[\frac{\beta_{\pi,off}(r)}{\beta_{\pi,on}(r)}\right] - 0.5\ln\left[\frac{C_{off}}{C_{on}}\right] - \tau_{e,dif}(r_0, r) \tag{10.30}$$

The column optical depth $\tau_{A,dif}(r_0, r)$ may be considered to be the most convenient value in the DIAL error analysis because there is no contribution to the uncertainty from signal differentiation. Accordingly, the uncertainty of $\tau_{A,dif}(r_0, r)$ can be accurately estimated with conventional error propagation techniques, using the estimated uncertainties in the values of the equation rather than their gradients.

Eq. (10.30) can be rewritten in the form

$$\tau_{A,dif}(r_0, r) = R_{dif}(r) - B_\pi^*(r) - \tau_{e,dif}(r_0, r) - C^* \tag{10.31}$$

where

$$R_{dif}(r) = 0.5\ln\left[\frac{P_{\Sigma,off}(r) - P_{bgr,off}}{P_{\Sigma,on}(r) - P_{bgr,on}}\right] \tag{10.32}$$

$$B_\pi^*(r) = 0.5\ln\left[\frac{\beta_{\pi,off}(r)}{\beta_{\pi,on}(r)}\right] \tag{10.33}$$

and

$$C^* = 0.5\ln\left(\frac{C_{off}}{C_{on}}\right) \tag{10.34}$$

To estimate the achievable accuracy in the extracted column ozone concentration, the uncertainty of each of the terms in Eq. (10.31) must be determined. The first term of the error components, $R_{dif}(r)$, is caused by uncertainty in the measured signals, $P_{\Sigma,off}(r)$ and $P_{\Sigma,on}(r)$. These signals may be corrupted by both random noise and a systematic offset of unknown sign. The random noise can be caused by speckle effects, shot noise, etc. The systematic error may have a different origin. It may be introduced, for example, by signal averaging or by the so-called signal-induced noise, which causes a significant curvature in the background level (McDermid et al., 1990; Sunesson et al., 1994). An incorrect estimate of P_{bgr} is the other reason for a systematic error when separating two noise-corrupted constituents, $P_{\Sigma}(r)$ and P_{bgr}. The uncertainty in $R_{dif}(r)$ can be found by conventional procedures for error propagation for uncorrelated quantities. The absolute uncertainty caused by the errors in the measured signals is

$$\Delta R_{dif}(r) = 0.5\sqrt{\begin{aligned}&\left[\frac{\Delta P_{\Sigma,off}(r)}{P_{\Sigma,off}(r) - P_{bgr,off}}\right]^2 + \left[\frac{\Delta P_{bgr,off}}{P_{\Sigma,off}(r) - P_{bgr,off}}\right]^2\\ &+\left[\frac{\Delta P_{\Sigma,on}(r)}{P_{\Sigma,on}(r) - P_{bgr,on}}\right]^2 + \left[\frac{\Delta P_{bgr,on}}{P_{\Sigma,on}(r) - P_{bgr,on}}\right]^2\end{aligned}} \tag{10.35}$$

where $\Delta P_{\Sigma,off}(r)$, $\Delta P_{bgr,off}(r)$, $\Delta P_{\Sigma,on}(r)$, and $\Delta P_{bgr,on}(r)$ are the absolute errors of $P_{\Sigma,off}(r)$, $P_{bgr,off}(r)$, $P_{\Sigma,on}(r)$, and $P_{bgr,on}(r)$, respectively. Note that in practice, the on and off signals are always averaged before the inversion. This means that two averaged values are always used in the first term of Eq. (10.32) rather than two single shots. Accordingly, the error contributions in Eq. (10.35) must be estimated for the averaged values, the same way as described in Section 6.1.

The backscatter and extinction corrections can be calculated in a way similar to that in Section 10.1.1. This means that the same assumption of a power law relationship between the backscatter coefficients at on and off wavelengths is used to determine the ratio $\beta_{\pi,off}(r)/\beta_{\pi,on}(r)$. Note that, unlike the backscatter corrections in the previous sections, the ratio above is used to correct the column ozone concentration over the range (r_0, r) rather than the range-resolved concentration. If no strong heterogeneous layers exist along the examined path, the spectral dependencies of the particulate backscattering can be taken as range independent. Under such an assumption, the profile of the backscatter ratio can be found with Eq. (10.16) as

$$\frac{\beta_{\pi,off}(r)}{\beta_{\pi,on}(r)} = \frac{\beta_{m,off}(r)}{\beta_{m,off}(r)}\left[\frac{1+Q_{off}(r)}{1+Q_{on}(r)}\right] \tag{10.36}$$

Because the ratio of the molecular scattering at λ_{on} and λ_{off} is a known range-independent value, one can find the absolute error of $\Delta B^*_{\pi}(r)$ as a function of relative errors in $Q_{off}(r)$ and $Q_{on}(r)$

$$\Delta B^*_{\pi}(r) = 0.5\sqrt{\left[\frac{Q_{on}(r)}{1+Q_{on}(r)}\right]^2 \delta Q^2_{on} + \left[\frac{Q_{off}(r)}{1+Q_{off}(r)}\right]^2 \delta Q^2_{off}} \tag{10.37}$$

If the reference signal is measured at the wavelength λ_{ref}, which differs from λ_{off} ($\lambda_{ref} > \lambda_{off}$), the ratio in Eq. (10.36) is transformed into the formula

$$\frac{\beta_{\pi,off}(r)}{\beta_{\pi,on}(r)} = \frac{\beta_{m,off}(r)}{\beta_{m,on}(r)} \left[\frac{1 + Q_{ref}(r)\left(\dfrac{\lambda_{off}}{\lambda_{ref}}\right)^{4-\xi}}{1 + Q_{ref}(r)\left(\dfrac{\lambda_{on}}{\lambda_{ref}}\right)^{4-\xi}} \right] \tag{10.38}$$

To find the backscattering ratio in Eqs. (10.36) or (10.38), $Q(r)$ at λ_{off} or λ_{ref} and the constant ξ must be known. Accordingly, the error $\Delta B^*_\pi(r)$ in Eq. (10.31) depends on the uncertainly in the calculated profile $Q(r)$ and on the accuracy of ξ, generally chosen a priori.

Finally, the last range-dependent term in Eq. (10.31), that is, the optical depth of the differential extinction coefficient, must be found. This term can be calculated by multiplying both sides of Eq. (10.13) by $\Delta\sigma$ and integrating the result over the range from r_0 to r

$$\tau_{e,dif}(r_0, r) = \frac{\Delta\lambda}{\lambda} \int_{r_0}^{r} [u\beta_{p,off}(r') + 4\beta_{m,off}(r')]\, dr' \tag{10.39}$$

The uncertainty in the term $\tau_{e,dif}(r_0, r)$ can originate from errors in the calculated particulate and molecular scattering coefficients and from an inaccurately selected Angstrom coefficient u. The absolute uncertainty $\Delta\tau_{e,dif}(r_0, r)$ is generally smaller than that for the backscattering correction, at least in heterogeneous atmospheres. Nevertheless, it should be considered, especially when the DIAL wavelength separation $\Delta\lambda$ is large.

It is not necessary to know the constant term C^* in Eq. (10.31) when determining the range-resolved ozone concentration profile. This is because the derivative of $\tau_{A,dif}(r_0, r)$ does not depend on the constant term. However, if necessary, the constant term can easily be excluded from the equation by putting $r = r_0$. At this point $\tau_{A,dif} = \tau_{e,dif} = 0$, and Eq. (10.31) is reduced to

$$R_{dif}(r_0) - B^*_\pi(r_0) - C^* = 0$$

from which the constant C^* can easily be found. The uncertainty in C^* can be considered as a constant offset in the function $\tau_{A,dif}(r_0, r)$, which can be omitted from consideration. After determining the terms in Eq. (10.31) and making the backscatter and extinction corrections, the total uncertainty remaining in the calculated optical depth $\tau_{A,dif}(r_0, r)$ is

$$\Delta\tau_{A,dif}(r_0, r) = \sqrt{[\Delta R_{dif}(r)]^2 + [\Delta\tau_{e,dif}(r_0, r)]^2 + [\Delta B^*_\pi(r)]^2} \tag{10.40}$$

The uncertainty in the differential optical depth, $\Delta\tau_{A,dif}(r_0, r)$, can be estimated through the uncertainties in the terms $R_{dif}(r)$, $\tau_{e,dif}(r_0, r)$, and $B^*_\pi(r)$.

An additional source of measurement uncertainty exists, which should also be mentioned. In the estimates above, it was assumed that only ozone differential absorption occurs at the on-off wavelengths selected for ozone measurements. However, in the UV, additional absorption may occur, mainly due to O_2, SO_2, and NO_2 absorption. In some measurements, especially in urban or industrial areas, interference from these compounds must also be taken into account when estimating the total extinction correction term, $\tau_{e,dif}(r_0, r)$. The influence of absorbing species, other than ozone, in UV spectra is analyzed in many studies, for example, by Bass et al., 1976; Brassington, 1981; Trakhovsky et al., 1989; Papayannis et al., 1990; Sunnesson et al., 1994.

10.2.2. Transition from Integrated to Range-Resolved Ozone Concentration: Problems of Numerical Differentiation and Data Smoothing

The range-resolved ozone concentration profile is found from the derivative of $\tau_{A,dif}(r_0, r)$. If the backscatter and extinction corrections are made before the differentiation, the final value of the concentration $n(r)$ is directly obtained

$$n(r) = \frac{1}{\Delta\sigma}\frac{d}{dr}[\tau_{A,dif}(r_0, r)] \tag{10.41}$$

The range-resolved ozone concentration is related to the local gradient of $\tau_{A,dif}(r_0, r)$. Accordingly, the measurement error originates from an inaccurate determination of the slope rather than by an inaccuracy in the calculated value of $\tau_{A,dif}(r_0, r)$ itself. Meanwhile, conventional error propagation techniques are based on estimates of the uncertainty in the numerical values of the quantities involved. No practical relationships exist between the uncertainties in the function value and in its slope. This is the issue that prevents a reliable estimation of the actual DIAL measurement uncertainty. There is no practical way to obtain accurate estimates of the local slope variations of the function $\tau_{A,dif}(r_0, r)$ even if the uncertainty boundaries of its value is known.

Differentiation is similar to applying a high-pass filter to the signal (Zuev et al., 1983; Beyerle and McDermid, 1999). It amplifies noise, increases the distortions, and thus significantly exacerbates the problem of accurate measurements of range-resolved ozone concentration. Note that this issue is the principal difficulty for both DIAL and Raman measurements. To compensate for the increase in the high-frequency components a low-pass smoothing filter is generally used. The question always arises of how much detail in the ozone concentration profile can be extracted from a particular noisy signal and, accordingly, what type of filtering will result in the minimum uncertainty. After differentiation, the question remains whether the small-scale changes in the ozone concentration profile are real or are the result of noise or aerosol loading.

The derivative in Eq. (10.41) is generally approximated by a differential quotient in a given range interval. In practical measurements, a least-squares technique is generally applied for numerical differentiation. Accordingly, some number of consecutive data points are used to determine an averaged slope of the function of interest over the resolved range. The first-level filtering is chosen by selecting the length of the resolution range $\Delta r = r_{(j+n)} - r_j$ for numerical differentiation, that is, by selecting the number of the discrete ranges r_j, $r_{(j+1)} \ldots r_{(j+n)}$ at which data points of $\tau_{A,dif}(r_0, r)$ are taken to calculate the local ozone concentration. After that, some algorithm for numerical differentiation is applied to the discrete data points of $\tau_{A,dif}(r_0, r)$. Sometimes such an algorithm is applied individually to the off and on signals rather than to $\tau_{A,dif}(r_0, r)$ (Pelon and Megie, 1982, Godin et al., 1999)

The use of different range resolution lengths Δr is equivalent to application of different low-pass filters for high-frequency noise components. The length Δr determines the frequency cut-off parameters of the low-pass filter. To explain the influence of different range resolutions on the noise suppression, the basic principles of digital filtering are outlined here. Digital filtering theory is applied to consecutive numerical quantities with a temporal resolution, Δt, that is related to the sampling frequency of the receiver. With DIAL measurements, this time interval determines the spatial resolution of the digital recording system, that is, $\Delta r_d = (c\Delta t)/2$. The highest spatial frequency that can be extracted from the recorded data, the Nyquist frequency, f_N, is equal to $f_N = 1/(2\Delta t)$ (Hamming, 1989). Clearly, the highest frequencies in the signal may only be obtained if no filtering is applied to the recorded signal. This occurs if the range resolution Δr, used for numerical differentiation, is chosen be equal to the sampling resolution, Δr_d. Commonly, the length of the range Δr is larger than Δr_d, that is, $\Delta r = m\Delta r_d$, where m may be equal to 2, 4, etc. (see Chapter 4). The increase of Δr is equivalent to the use of a narrower low-pass filter, which reduces the high-frequency components in the signal spectrum that are considered to be noise. Conventional digital filters are defined by the linear formula (Rabiner and Gold, 1975; Hamming, 1989; Godin et al., 1999)

$$Y_k = \sum_{i=-M}^{M} c_i X_{i-k}$$

where Y_k is the output signal of the filter, X_{i-k} is the input signal, and c_i are the weighting coefficients of the filter. The number of coefficients M is the filter order that determines the so-called cutoff frequency, that is, the highest spatial frequency component that will pass through the filter. It is the filter order, uniquely related to the range resolution Δr that establishes how much of the detail in the measured profile may be extracted after application of the filter. If the range resolution selected is too long, useful details in the retrieved profile, which could have been determined, are lost. On the other hand, if the order of the filter selected is too small, that is, Δr is too short, high-frequency noise contributions will be considered to be details in the profile of interest.

It should be stressed that no amount of filtering can, by itself, separate the noise and the signal. The basic question that has to be answered is whether a detail in the retrieved profile is real or whether it is just noise. There is no digital filtering theory that provides a certain answer to this question. The question must be answered by the researcher on the basis of a thorough analysis.

To illustrate the importance of the selection of the length of the range resolution interval, in Fig. 10.9, two vertical ozone concentration profiles are shown extracted from the same DIAL signals. The signals are measured at 276.9 nm (the on-line) and at 312.9 nm (the off-line), and the ozone concentration profiles are obtained with conventional regression procedures. The profiles are extracted with a running least-squares linear fit with an altitude resolution of 120 and 300 m. In Fig. 10.9, these profiles are shown by the dotted and solid lines, respectively. No aerosol corrections are made, so the profiles are initial raw estimates, $n'(r)$. Significant differences in the range-resolved ozone concentration are caused only by the difference in the applied range resolution. Increasing the range resolution significantly smooths the retrieved profile.

When making measurements of the ozone concentration in the upper troposphere and stratosphere with a ground-based, upward-staring lidar system, the signal-to-noise ratio rapidly worsens with altitude. Therefore, in high-altitude areas, the signals may be significantly distorted by noise. To compensate for this effect and to equalize the data quality over the whole altitude range, one can increase the resolution range Δr when calculating ozone concentrations at distant ranges. This type of filtering is a general practice with high-altitude measurements (see, for example, Megie and Menzies, 1980; Measures, 1984; Godin et al., 1999; Beyerle and McDermid, 1999; Carnuth et al., 2002).

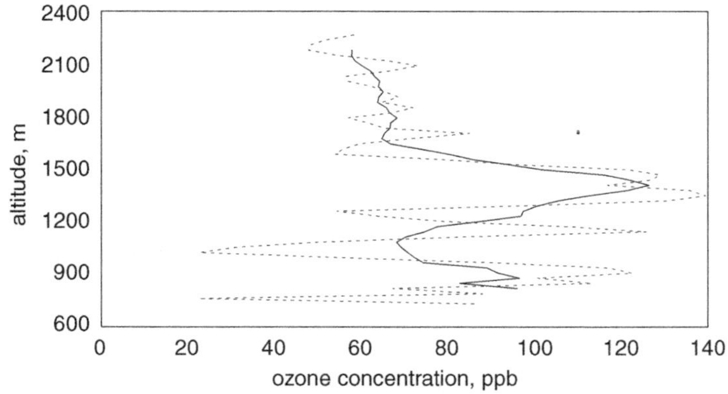

Fig. 10.9. Experimental ozone concentration profiles $n'(h)$ obtained with the numerical regression procedure. The dotted and solid curves show the ozone concentration profiles derived from the same on- and off-signal pair the 5 and 11-point linear regression (120- and 300-m resolution range), respectively.

Thus there are several conflicting requirements when selecting optimal filtering of DIAL data. The most relevant way to detect an actual ozone perturbation from a noise fluctuation may be based on some knowledge of the spatial ozone field parameters. In other words, to use the proper filtering to extract the ozone concentration, it is necessary to estimate the scale of the actual spatial heterogeneity in the concentration. Such estimates are quite difficult. In practice, the main purpose of filtering is to compensate for a decreasing signal-to-noise ratio at distant ranges. The most common method is to use a digital filter in which the range resolution Δr, that is, the number of data points used to determine the linear (or nonlinear) fit, increases with range (Godin et al., 1999). Accordingly, the greatest amount of filtering is done on the most distant ranges, where the signal-to-noise ratio is poorest. Unfortunately, this straightforward approach does not take into consideration a possible increase in the systematic error at distant ranges. It must always be kept in mind that no amount of filtering can compensate for systematic errors at the far end of the profile. Therefore, the real improvement in accuracy, achieved by filtering at distant ranges, is actually quite moderate.

No commonly accepted methods exist to estimate the adequacy of a given filter. The standard deviation in the measured concentration profiles as a function of the range is, in fact, the only criterion. The most difficult question remains whether the details of the spatial structure of the extracted ozone concentration profile are an accurate representation of the real ozone profile or are due to noise and unknown systematic distortions.

On the other hand, the selection of the length of the range resolution and the algorithm (linear or nonlinear fit) is equivalent to the selection of some model of the assumed ozone concentration behavior within this range resolution. The model is uniquely related both to the selected range and to the algorithm used for numerical differentiation. The last statement requires some additional explanation. When different range resolutions $[r_j, r_{(j+n)}]$ are used for the same data, different concentration profiles are retrieved. This occurs not only because of the different level of noise smoothing, but also because of discrepancies in the computational models used. The effect of the use of different lengths for the range resolution $[r_j, r_{(j+n)}]$ for numerical differentiation is shown in Fig. 10.10. Here *curve 1* is an artificial ozone concentration profile used for the simulation. In the range from 1500 to 1800 m, an increased ozone concentration, 70 ppb, takes place, whereas beyond this region the ozone concentration is only 30 ppb. The boundaries of this change are sharp and clearly defined. For the profile, the corresponding column-integrated ozone concentration was calculated, and after that the integrated profile was inverted with a conventional numerical differentiation. In this procedure, the moving means were calculated by a linear fit with the range resolution 120 and 300 m. The inverted ozone concentration profiles are shown as *curves 2* and *3*, respectively. No noise or measurement error is assumed when calculating the on and off signals for the above profiles. The distortions in *curves 2* and *3* are generated

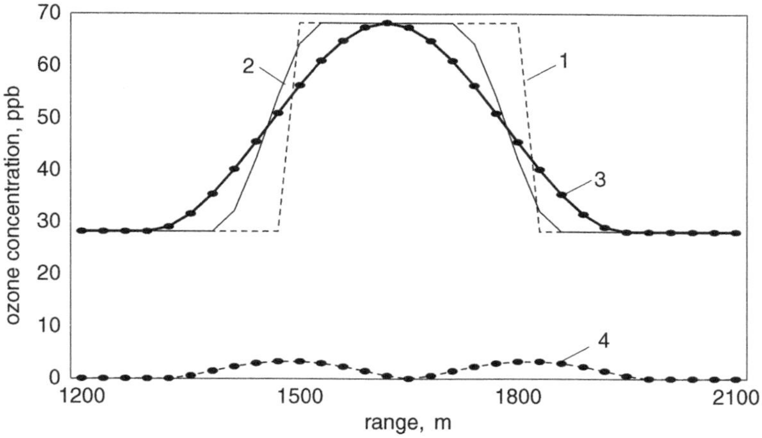

Fig. 10.10. Synthetic ozone concentration profile used for the inversion (*curve 1*) and the inverted profiles obtained with the numerical regression. The ozone concentration profiles determined with 5- and 11-point linear fit are shown as *curves 2* and *3*, respectively. *Curve 4* shows the standard deviation for *curve 3*.

only by the error in the differentiation model. The difference between the original and restored profiles in Fig. 10.10 is caused by the inconsistency between the inversion model and the actual profile in areas with sharp changes in the ozone concentration. The inversion model assumes ozone homogeneity in each local zone within the resolved range. Such an assumption is not valid at the boundaries of a layer with an increased ozone concentration (~1500 and 1800 m). This is why large systematic discrepancies between model and retrieved profiles occur in these areas. With the selected range resolution, distorted profiles are obtained in which the high-frequency components are lost. Note that the distortion of the inverted profiles is followed by an increase in the standard deviation of the linear fit (*curve 4*) in the corresponding zones. Note also that, beyond the areas of the systematic distortions, the standard deviation is zero.

The amount of smoothing in the retrieved ozone concentration profile is established by the length of the range resolution Δr and by the order of the polynomial fit used for numerical differentiation. As follows from Taylor's theorem (Wylie and Barret, 1982), the term $\tau_{A,dif}(r_0, r)$ in Eq. (10.31) for the range from r to $r + \Delta r$ can be written as the series representation

$$\tau_{A,dif}(r+\Delta r) = \tau_{A,dif}(r) + \frac{d}{dr}[\tau_{A,dif}(r)]\,\Delta r + \sum\left\{\frac{d^{(i)}}{dr^{(i)}}\left[\tau_{A,dif}(r)\frac{\Delta r^i}{i!}\right]\right\} \quad (10.42)$$

where

$$\sum\left[\frac{d^{(i)}}{dr^{(i)}}[\tau_{A,dif}(r)]\frac{\Delta r^i}{i!}\right] = \frac{d^{(2)}}{dr^{(2)}}[\tau_{A,dif}(r)]\frac{\Delta r^2}{2!} + \cdots + \frac{d^{(n)}}{dr^{(n)}}[\tau_{A,dif}(r)]\frac{\Delta r^n}{n!} + R_{n+1}$$

is the sum of the higher-order terms in the Taylor series. Denoting this sum for brevity as Σ, one can write the precise formula for the first-order derivative in the form

$$\frac{d}{dr}[\tau_{A,dif}(r)] = \frac{\tau_{A,dif}(r+\Delta r) - \tau_{A,dif}(r) - \Sigma}{\Delta r} \tag{10.43}$$

When calculating numerical derivatives from experimental data, we omit the higher-order terms in the Taylor series, retaining only the first-order derivative term. After that, the numerical derivative is found from Eq. (10.43) reduced to

$$\frac{d}{dr}[\tau_{A,dif}(r)]_{num} \approx \frac{\tau_{A,dif}(r+\Delta r) - \tau_{A,dif}(r)}{\Delta r} \tag{10.44}$$

which generally is accurate enough only for small Δr. The distortions of the inverted functions, shown in Fig. 10.10, are just due to the omission of the higher-order terms in the Taylor series. The relationship between the numerical and actual derivatives is

$$\frac{d}{dr}[\tau_{A,dif}(r)] = \frac{d}{dr}[\tau_{A,dif}(r)]_{num} - \frac{\Sigma}{\Delta r} \tag{10.45}$$

To summarize, each of the variants used for numerical differentiation is based on some particular approximation for the parameters involved. The simplest and the most common way to compute the numerical derivative is to use only the first term in the Taylor series. This is equivalent to the assumption that the ozone concentration within the local range resolution Δr is (or can be treated as) constant. If this is true, the logarithm of the on and off signal ratio is linear over the range Δr and a straight line is an adequate fit for $\tau_{A,dif}(r_0, r)$ over this range.

The evaluation of a numerical derivative through a least-squares fit to a straight line means that the extracted quantity is assumed approximately constant over the selected range interval.

Thus the retrieved concentrations may be systematically distorted because of the difference between an approximate numerical differentiation and a strictly analytical differentiation. Note that the numerical differentiation of DIAL measurements can be made with either a linear or a nonlinear polynomial fit. The linear approximation is the most simple and straightforward. There are two basic reasons for the common use of this approximation. First, the linear fit has the simplest mathematical formulation. Second, no evidence exists that any nonlinear fit yields more accurate results when processing noisy signals. Nevertheless, higher-order polynomial fitting is sometimes used, for example, by Pelon and Megie (1982), Kempfer et al. (1994), and Fujimoto et al. (1994). It should also be mentioned that a polynomial fit of a specific order can be applied on either a local or a global scale and used for an analytical approximation of different functions. For example, in the study by Pelon and Megie (1982), the ozone concentration was found from the difference of the range derivatives for the on and off DIAL signals rather than from $\tau_{A,dif}(r_0, r)$. The derivative at each of the range intervals was determined by fitting the range-corrected signal to a second-order polynomial. This was made by using a nonlinear least-squares method. A similar approach was used by McDermid et al. (1990).

Each particular algorithm to determine the numerical derivative has its own smoothing characteristics. The algorithm yields the most accurate result when the assumed statistical model is relevant to the data points involved. This means that the numerical differentiation is always based on some implicitly assumed behavior of the measured gas concentration over the local range of interest.

In the research by Godin et al. (1999), the results of different fitting methods are compared using synthetic lidar signals. The objective of this study was to test different numerical differentiating techniques used in DIAL measurements. A synthetic lidar signal set was computed at different wavelengths with three models of assumed ozone altitude profiles. These synthetic profiles were smooth but contained small scale perturbations that were put in regions both low and high in the atmosphere. The perturbations were included to test the vertical resolution of 10 algorithms used by various lidar groups to invert DIAL data. Most teams used similar techniques, based on the fit of the logarithm of the signal ratio, $R_{dif}(r)$, to a first- or second-order polynomial. Particularly, in four algorithms, the logarithm of the on and off signal ratio was fitted to a straight line and the ozone concentration was derived from a linear fit. In three other algorithms, the ozone concentration was derived from the difference in the derivative of a second-order polynomial, fitted to the logarithm of each lidar signal. Only two algorithms used a higher-order polynomial to fit the logarithm of the signal ratio. Thus the data processing technique used by most researchers was mostly based on a simple linear or parabolic fit. However, even this unique test did not reveal what type algorithm can be considered to be optimal. The comparison revealed that the simplest technique most often provided the best inversion results. The results of the test showed that all of the methods, including ones that applied a high-order polynomial fit, revealed a large bias in the inverted profile at high altitudes. In fact, no technique showed acceptable results over the whole altitude range from 10 to 50 km. The simulations revealed the obvious fact that the DIAL technique could potentially detect the perturbations only by using reduced range resolution. The unsolved problem remains of how to discriminate real changes in concentration from fluctuations due to noise. Obviously, profile perturbations can be reliably detected only in areas where the signal-to-noise ratio is high. However, even in these areas, the discrepancies obtained by the methods proved to be on the order of several hundred percent. Moreover, higher vertical resolution did not always correspond to the best response to an ozone perturbation. It was also established that the use of a high-order polynomial fit can result in large additional perturbations. Finally, the results showed some inconsistencies in the definition of the range resolution.

It can be expected that, in a real atmosphere, such a comparison would reveal considerably more bias and overshoots. In the study by Godin et al. (1999), attention was concentrated on the comparison of the filters used to differentiate. In fact, for these simulations, quite favorable measurement conditions were assumed. First, it was assumed that the return signals were obtained from a single-component atmosphere, free of aerosols. Only Gaussian random noise was added to the signals, and no systematic errors were involved. Nevertheless, even for such favorable conditions, the principal result of the study is that no unique algorithm exists that could be recommended as most acceptable. As stressed in the study by Beyerle and McDermid (1999), different calculational models and empirical definitions yield different results even when these are based on clear geophysical interpretations. Unfortunately, the particular measurement conditions may often be far from the conditions presumed by the interpretations.

Two additional error sources with DIAL measurement must also be mentioned. First, the least-squares technique assumes that the data used for the regression are normally distributed. However, as pointed out by Whiteman (1999), the quantities that are usually used in the regression procedure are, in fact, not normally distributed. As with the extinction coefficient calculation for the slope method, some particular error distribution is assumed when the DIAL data are processed. The other error that may be involved in DIAL (and Raman) measurements is caused by averaging of the lidar returns. This procedure requires that the spatial distribution of the scatterers remain constant along the examined path, that is, the atmosphere must be "frozen" while recording the signals that are then averaged. Keeping all these likely errors in mind, one can conclude that it is no evidence that a nonlinear fit can actually produce a significant improvement in the quality of the retrieved data. Moreover, such a nonlinear fit can generate false variations in the retrieved profile, and there is no basis on which to determine whether these variations are false or real.

Let us summarize the issues associated with the determination of the range-resolved ozone concentration. First, the use of any particular (linear or nonlinear) fitting for numerical differentiation is accompanied by tacit presumptions on the behavior of the quantity of interest over the resolved range. This, in turn, may create a significant error through the use of an inappropriate model for differentiation. To improve DIAL measurement accuracy, data filtering must be based not only on estimates of the signal-to-noise ratio, but also on estimates (or at least on reasonable assumptions) of the spatial scales of the heterogeneity in the quantity of interest. Unfortunately, this characteristic is often omitted from consideration, and an invariable distribution model for the quantity of interest is generally assumed to be valid for any location within the differentiation range. A nonlinear fit may decrease (at least in principle) these systematic distortions. However, the amount of gain that may be obtained is quite questionable.

Second, the likely systematic distortions in the signals, particularly when measured at long distances, should be taken into consideration. Meanwhile, when making error estimates, the most common tacit assumption is that no systematic error occurs in the $\tau_{A,dif}(r_0, r)$ profile except that related to $\Delta n_\beta(r)$ and $\Delta n_e(r)$. This assumption becomes questionable in distant areas where the remaining background offset becomes comparable to the backscatter signal. Here the weight of instrumental systematic errors, not related to $\Delta n_\beta(r)$ and $\Delta n_e(r)$, may become significant, and the amount of gain obtained by increasing the differentiation range resolution is questionable. What is more, the random-error accuracy improvement, achieved by increasing the number of laser pulses, N, for signal averaging, may also differ significantly from the $N^{-\frac{1}{2}}$ law.

Finally, the temporal and spatial variability of aerosol scattering in the lower troposphere, which exacerbates all of the above problems, should be stressed. The ratio of the on and off signals at the edges of a heterogeneous layer frequently exhibits large local fluctuations caused by spatial and temporal variations in aerosol layers, by the variability of the backscatter ratio, etc. (Kovalev and McElroy, 1994). When conventional numerical differentiation is used to retrieve an ozone profile, local fluctuations in the calculated $\tau_{A,dif}(r_0, r)$, such as bulges and concavities, result in erroneous fluctuations in the retrieved ozone concentration. Negative concentration values in the retrieved ozone profile can even be obtained in such areas. One can see this effect in Fig. 10.5, where typical experimental data were shown. The spikes in the DIAL signals at λ_{on} and λ_{off}, obtained from local aerosol layers at altitudes of ~1000–1300 m are caused by the variations in the layer edge altitudes and corresponding changes in the local backscatter-to-extinction ratios during the signal averaging. This creates a local concavity in the function $\tau_{A,dif}(r_0, r)$, which, in turn, results in large variations in the retrieved ozone concentration. This effect is described in studies by Kovalev and McElroy (1994) and Godin et al. (1999).

10.3. OTHER TECHNIQUES FOR DIAL DATA PROCESSING

10.3.1. DIAL Nonlinear Approximation Technique for Determining Ozone Concentration Profiles

As stated in the previous section, it is reasonable to analyze the achievable discrimination in the retrieved ozone concentration through the analysis of the uncertainty in the optical depth $\tau_{A,dif}(r_0, r)$ rather than in $n(r)$. In this case, the application of an overall analytical approximation for $\tau_{A,dif}(r_0, r)$ rather than that for local zones might be beneficial. Such an analytical approximation makes it possible to use analytical differentiation over the entire measurement range. This in turn reduces the filtering selection problem to the selection of an adequate polynomial order (or the number of terms in a Fourier series) used to approximate the function $\tau_{A,dif}(r_0, r)$. However, at least three problems must be overcome to make such an approximation practical. First, the analytical fit must accurately approximate the slope rather than the magnitude of the function $\tau_{A,dif}(r_0, r)$ over the entire distance from r_0 to r_{max}, and this is an issue. To obtain an accurate range-resolved concentration profile from such an approximation, it must be accurate at all local ranges. This means that the requirements for the accuracy of the analytical approximation are quite severe. The second problem is that all of the corrupted areas of $\tau_{A,dif}(r_0, r)$ over the range from r_0 to r_{max} should be excluded before an overall approximation is made. The third problem is related to the selection of a functional form for the approximate function $\tau_{A,dif}(r_0, r)$. The problem is that this function has a monotonic change over the range (r_0, r_{max}). To provide an accurate overall approximation for such a monotonic curve, complicated analytical function must be used, even after strong concavities and bulges in $\tau_{A,dif}(r_0, r)$ are removed. High-order polynomial fits are not reasonable when the approximated function is corrupted with high-frequency noise. On the other hand, a low-order polynomial fit can yield an inaccurate approximation in regions with large ozone concentration gradients, which result an inaccurate values of the range-resolved concentration in these areas. The problem can be overcome if different levels of approximation are used consecutively.

A variant of the two-step analytical approximation was developed in studies by Kovalev and McElroy (1994) and Kovalev et al. (1996). The approximation procedure is as follows. First, the function $\tau_{A,dif}(r_0, r)$ is found as described in Section 10.2.1. Then the differential path transmission is calculated as

$$T_{A,dif}(r_0, r) = e^{-\tau_{A,dif}(r_0, r)} = \exp\left[-\int_{r_0}^{r} \kappa_{A,dif}(r')\, dr'\right] \tag{10.46}$$

where $\kappa_{A,dif}(z)$ is the differential absorption coefficient. It is uniquely related with the ozone concentration $n(r)$ as $\kappa_{A,dif}(r) = n(r)\Delta\sigma$. For simplicity, it is assumed that the aerosol and molecular corrections have been made and the constant C^* in Eq. (10.31) has been determined so that the differential path transmission $T_{A,dif}(r_0, r)$ is normalized to unity at the starting point, $r = r_0$. For rough estimates, a function similar to that in Eq. (10.46) can be determined directly from the on-off signal ratio.

Then an intermediate function, $\kappa_{est}(r)$ is introduced that is related to the differential transmission $T_{A,dif}(r_0, r)$ as (Kovalev et al., 1996)

$$T_{A,dif}(r_0, r) = B[\kappa_{est}(r)]^{p} \exp\left[-\int_{r_0}^{r} \kappa_{est}(r')\, dr'\right] \tag{10.47}$$

where B_i and p are constants. Note that an infinite number of corresponding functions $\kappa_{est}(r)$ can be obtained for the same $T_{A,dif}(r_0, r)$ by selecting different constants B and p in Eq. (10.47). For further processing, any such function can be used. Because $T_{A,dif}(r_0, r) = 1$ at $r = r_0$, one can find B from Eq. (10.47) as

$$B = [\kappa_{est}(r_0)]^{1/p}$$

The selection of a particular value of B is equivalent to the selection of a particular $[\kappa_{est}(r_0)]^{1/p}$ at the starting point, r_0. Equation (10.47) can then be rewritten as

$$T_{A,dif}(r_0, r) = \left[\frac{\kappa_{est}(r)}{\kappa_{est}(r_0)}\right]^p \exp\left[-\int_{r_0}^{r} \kappa_{est}(r')\,dr'\right] \tag{10.48}$$

To determine the relationship that relates the function $\kappa_{est}(r)$ to the differential transmission term $T_{A,dif}(r_0, r)$, Eq. (10.48) is rewritten as

$$[T_{A,dif}(r_0, r)]^{1/p} = \frac{\kappa_{est}(r)}{\kappa_{est}(r_0)} \exp\left[-\frac{1}{p}\int_{r_0}^{r} \kappa_{est}(r')\,dr'\right] \tag{10.49}$$

The relationship between $\kappa_{est}(r)$ and $T_{A,dif}(r_0, r)$ may be derived by integration of the terms on both sides of Eq. (10.49). By introducing a new variable, $\zeta = \int \kappa_{est}(r')dr'$ (so that $d\zeta = \kappa_{est}(r)dr$), the relationship between the integrals may be obtained in the form

$$\int_{r_0}^{r} [T_{A,dif}(r_0, r')]^{1/p}\,dr' = \frac{p}{\kappa_{est}(r_0)}\left\{1 - \exp\left[-\frac{1}{p}\int_{r_0}^{r} \kappa_{est}(r')\,dr'\right]\right\} \tag{10.50}$$

The function $\kappa_{est}(r)$ may be found from Eqs. (10.48) and (10.50) as

$$\kappa_{est}(r) = \frac{[T_{A,dif}(r_0, r)]^{1/p}}{\dfrac{1}{\kappa_{est}(r_0)} - \dfrac{1}{p}\displaystyle\int_{r_0}^{r} [T_{A,dif}(r_0, r')]^{1/p}\,dr'} \tag{10.51}$$

On the other hand, taking the logarithm of Eq. (10.48) and rearranging the terms, one can obtain

$$\int_{r_0}^{r} \kappa_{est}(r')dr' - \int_{r_0}^{r} \kappa_{A,dif}(r')dr' = p \ln \frac{\kappa_{est}(r)}{\kappa_{est}(r_0)} \tag{10.52}$$

The solution for $\kappa_{A,dif}(r)$ can be obtained by differentiating Eq. (10.52). Taking the derivative, a simple formula that relates $\kappa_{A,dif}(r)$ to $\kappa_{est}(r)$ can be found:

$$\kappa_{A,dif}(r) = \kappa_{est}(r) - p\frac{d}{dr}\ln \kappa_{est}(r) \tag{10.53}$$

As follows from Eq. (10.53), the introduction of the function $\kappa_{est}(r)$ makes it possible to represent the unknown function $\kappa_{A,dif}(r)$ as the algebraic sum of two components that can be determined separately.

Before the approximation technique is presented, consider how the constants p and $\kappa_{est}(r_0)$ in Eq. (10.51) influence the behavior of the introduced function $\kappa_{est}(r)$. As pointed out above, different shapes for the function $\kappa_{est}(r)$ are obtained when different constants p and $\kappa_{est}(r_0)$ are used in Eq. (10.51). If the constant p is chosen small enough, the second term in the right side of Eq. (10.53) becomes much less than the first term, $\kappa_{est}(r)$. This is true, at least in areas with moderate gradients in the logarithm of $\kappa_{est}(r)$. For such areas, where

$$p\frac{d}{dr}[\ln \kappa_{est}(r)] << \kappa_{est}(r) \tag{10.54}$$

Eq. (10.53) reduces to the simple equality

$$\kappa_{A,dif}(r) \approx \kappa_{est}(r)$$

In other words, by using a small value of p in Eq. (10.51), a function $\kappa_{est}(r)$ can be found that is sufficiently close to the value of interest $\kappa_{A,dif}(r)$ over the entire measurement range under consideration. This is true at all locations where Eq. (10.54) is valid. In these areas, the function $\kappa_{est}(r)$ can be considered to be an estimate of $\kappa_{A,dif}(r)$, that is, as the derivative of $\tau_{A,dif}(r_0, r)$ but obtained analytically. In other words, with small values of p, the solution in Eq. (10.51) can be treated as an algorithm for analytical differentiation. To show how Eq. (10.51) works when p is small, an example of analytical differentiation is shown in Fig. 10.11. Here the simulated rectangular profile of $\kappa_{A,dif}(r)$, used for the inversion, is shown as the solid curve. With this function, the corresponding differential path transmission, $T_{A,dif}(r_0, r)$, was calculated with Eq. (10.46). Then the profile of $\kappa_{est}(r)$ was calculated with Eq. (10.51), using a small constant p ($p = 0.04$). The function is shown as the bold curve. It can be seen that Eq. (10.51) restores the initial function $\kappa_{A,dif}(r)$ fairly accurately when the constant p is small enough.

Unfortunately, the above simple differentiation method, based on the use of extremely small constants p in Eq. (10.51), is not practical when real noisy experimental data are processed. This is because the noise level in the calculated $\kappa_{est}(r)$ depends significantly on the value of the selected value of p. When data corrupted with noise are inverted with Eq. (10.51) using a small value of p, a significant noise contribution appears in the calculated function $\kappa_{est}(r)$. This effect is similar to using small increments Δr for numerical differentiation. Therefore, to apply Eq. (10.51) for analytical differentiation of real

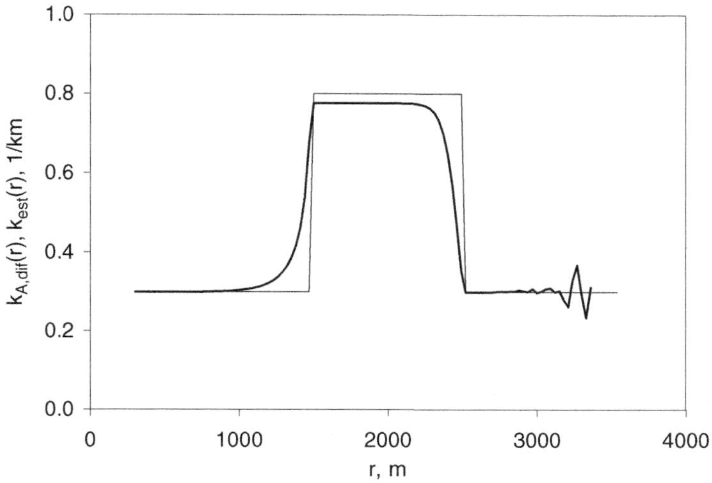

Fig. 10.11. Simulated function $\kappa_{A,dif}(r)$ (solid line) and function $\kappa_{est}(r)$ obtained with Eq. (10.51) (bold line).

signals, some optimal range for p must be established that provides an acceptable noise level in the function $\kappa_{est}(r)$ and, accordingly, in the restored concentration profile. In Fig. 10.12 (a)–(c), some results of simulations are shown that illustrate the influence of the selected value of p on the noise level in the retrieved function $\kappa_{est}(r)$. In all of the panels, the same original synthetic profile is used as that in Fig. 10.11. This model profile is shown in Fig. 10.12 (a)–(c) by solid lines. As before, the corresponding function $T_{A,dif}(r_0, r)$ was calculated by the integration of the model profile, but then it was artificially distorted by quasi-random noise. This noise-contaminated profile was then used to calculate $\kappa_{est}(r)$ with Eq. (10.51) and different values of p. The profiles of $\kappa_{est}(r)$ shown in Fig. 10.12 (a–c) are derived with $p = 0.04$, $p = 0.08$, and $p = 0.25$, respectively. It can be seen that very small values of p result in an increased noise level, similar to a short range resolution Δr in a numerical differentiation. The use of larger values of p produces a smaller level of the high-frequency noise variations but simultaneously increases the low-frequency distortions in the derived $\kappa_{est}(r)$. This effect is similar to the use of a large resolution range Δr in conventional numerical differentiation.

A practical method of the above analytical approximation should take into consideration both components of the right side of Eq. (10.53). Note that here only the second term, which contains a derivative of the logarithm of $\kappa_{est}(r)$, requires the use of differentiation. To avoid numerical differentiation of this term, an analytical fit for the function $\kappa_{A,dif}(r)$ (or its logarithm) is first found. Note that an analytical fit for $\kappa_{est}(r)$ can be made much more accurately than that for the initial function, $\tau_{A,dif}(r_0, r)$ or $T_{A,dif}(r_0, r)$. This is because of the differences in the shape of these functions. By selecting an optimal constant

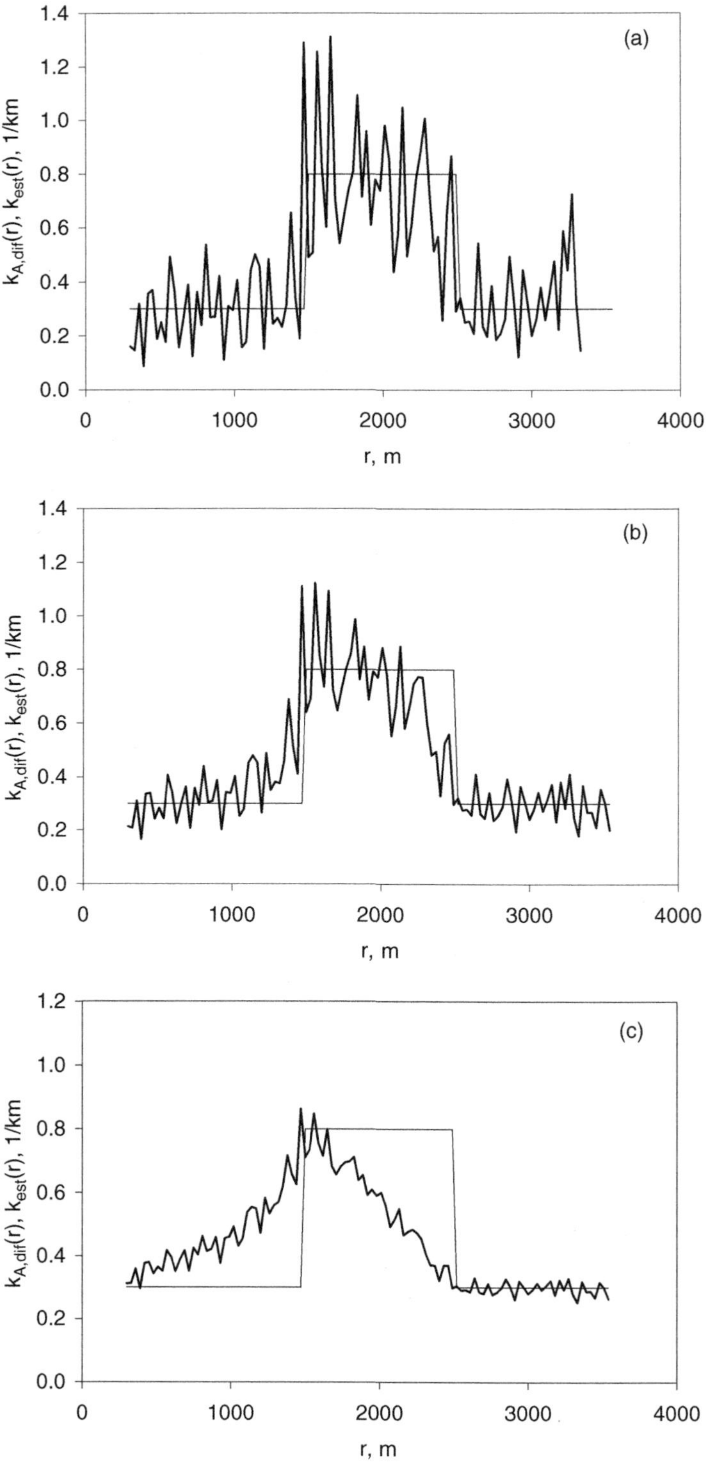
(a)
(b)
(c)
$k_{A,dif}(r)$, $k_{est}(r)$, 1/km
r, m
0
1000
2000
3000
4000
0.0
0.2
0.4
0.6
0.8
1.0
1.2
1.4

$\kappa_{est}(r_0)$ in Eq. (10.51), a function $\kappa_{est}(r)$ may be obtained that has a minimal slope within the total measurement range and, accordingly, a small change over the total distance of interest. Unlike the initial function, $\tau_{A,dif}(r_0, r)$, the function $\kappa_{est}(r)$ may be accurately approximated by a low-order polynomial fit if the constants p and $\kappa_{est}(r_0)$ in Eq. (10.51) were properly selected.

The introduction of the function $\kappa_{est}(r)$ allows one to split the unknown range-dependent function, $\kappa_{A,dif}(r)$, which is directly related to the measured concentration, into two parts, only one of which requires differentiation. Moreover, properly selected constants p and $\kappa_{est}(r_0)$ in Eq. (10.51) allow one to obtain an accurate analytical fit for the logarithm of $\kappa_{est}(r)$ in Eq. (10.53) with a minimal-order polynomial. This is a key point for this method for analytical differentiation.

After selecting an optimal p and $\kappa_{est}(r_0)$ and obtaining the corresponding function $\kappa_{est}(r)$, the following operations are made: (1) The determination of a low-order polynomial fit for $\kappa_{est}(r)$ or for its logarithm; (2) the separation of the calculated polynomial constituent and the remaining high-frequency fluctuations in $\kappa_{est}(r)$; (3) the determination of a low-order trigonometric fit for the remaining function obtained in item (2); (4) the analytical differentiation of the polynomial and the trigonometric functions; and (5) determination of the ozone concentration profile as the sum of the corresponding constituents.

As a first step in the procedure, a low-order polynomial fit is found for $\kappa_{est}(r)$ or its logarithm. Note that the approximation is made for the entire operating range (r_0, r_{max}). The dependence between the function $\kappa_{est}(r)$ and its low-order polynomial fit, $\kappa_{appr}(r)$, can be written as

$$\kappa_{est}(r) = \kappa_{appr}(r)[1 + \delta\kappa_{est}(r)] \tag{10.55}$$

where the term in the brackets is a factor that contains the remaining medium- and high-frequency constituents in $\kappa_{est}(r)$. With the latter formula, Eq. (10.53) can be rewritten as

$$\kappa_{A,dif}(r) = \kappa_{est}(r) - p\frac{d}{dr}\ln[\kappa_{appr}(r)] - \frac{d}{dr}\Delta\tau_1(r) \tag{10.56}$$

where

$$\Delta\tau_1(r) = p\ln[1 + \delta\kappa_{est}(r)] \tag{10.57}$$

Fig. 10.12. (a). The model profile (solid line) and that obtained with Eq. (10.51) (bold line) for the noise-corrupted data. Here $p = 0.04$; (b) Same as in (a) but with $p = 0.08$; (c) Same as in (a) but with $p = 0.25$.

The term $\Delta\tau_1(r)$ in Eq. (10.56) can be considered to be a remaining uncertainty term, when the first solution for $\kappa_{A,dif}(r)$ is found as the algebraic sum of the two other terms, that is

$$\kappa^{(1)}_{A,dif}(r) = \kappa_{est}(r) - p\frac{d}{dr}\ln[\kappa_{appr}(r)] \tag{10.58}$$

Eq. (10.58) can be considered as a first solution for $\kappa_{A,dif}(r)$. Note that $\kappa_{appr}(r)$ is the analytical function, thus the solution for $\kappa^{(1)}_{A,dif}(r)$ may be obtained analytically, without using numerical differentiation.

The function $\kappa^{(1)}_{A,dif}(r)$ is a first-approximation profile for the unknown $\kappa_{A,dif}(r)$, in which high-frequency components are not included. Specifically, $\kappa^{(1)}_{A,dif}(r)$ is a low-order polynomial fit of $\kappa_{A,dif}(r)$, in which some part of the high-frequency components may contribute to $\kappa_{est}(r)$. The amount of contribution from the high-frequency components depends on the selected constant p in Eq. (10.51) and is larger for smaller p [Fig. 10.12 (a–c)].

The next step is to extract the high-frequency concentration components, if any, from the third term in the right side of Eq. (10.56). To find the areas where such an operation is required, it is necessary to analyze the term $\Delta\tau_1(r)$ and establish whether this term contains ozone concentration components or whether it contains only noise. To determine this, it is necessary first to identify and separate regions where the difference between $\Delta\tau_1(r)$ and the estimated uncertainty $\Delta\tau_{A,dif}(r_0, r)$ may be caused by aerosol interference (such as a local turbid aerosol layer). As shown in previous sections, no accurate ozone concentration can be extracted from these regions. To avoid confusion, such regions must be identified and excluded before the additional ozone concentration is extracted from $\Delta\tau_1(r)$. The methods by which these regions may be identified may be different. It can be done, for example, by determining the amount of increased variance in the lidar signal, as in study by Hooper and Eloranta (1986) and Piironen and Eloranta (1995). In the study by Kovalev and McElroy (1994), local regions were found where the deviations in $\Delta\tau_1(r)$ were much greater than its standard deviation over the total operating range. This simple criterion enables demarcation of the regions in which the $\Delta\tau_1(r)$ profile is unusable for ozone concentration retrieval. The values of $\Delta\tau_1(r)$ in such regions were considered as outliers and excluded before the second-step approximation was made. Similarly, areas were excluded with large fluctuations in $\Delta\tau_1(r)$, caused by a poor signal-to-noise ratio at distant ranges. These procedures avoid short-range distortions in the retrieved ozone concentration when performing the analytical approximation of $\Delta\tau_1(r)$.

After the determination and exclusion of the outliers, one must decide whether the remaining term $\Delta\tau_1(r)$ still contains changes that can be attributed to ozone absorption rather than to signal noise. In these regions, an additional component to the ozone concentration may be extracted. For this, an analytical approximation for $\Delta\tau_1(r)$ can be determined in the same way as was done for $\kappa_{est}(r)$. However, this time, trigonometric (Fourier) series are more appropriate to determine the best fit for $\Delta\tau_1(r)$. The number of the terms in the Fourier series used for the approximation can be based, in principle, on the assumed scale of the ozone concentration heterogeneity. In fact, selecting the number of the terms in the Fourier series for the approximation is equivalent to selecting the low-pass filtering parameters discussed in Section 10.2.2, that is, this operation establishes the level of filtering for the constituent $\Delta\tau_1(r)$. Denoting the trigonometric fit for $\Delta\tau_1(r)$ as $W_{appr}(r)$, this term can be represented as a sum of two components:

$$\Delta\tau_1(r) = W_{appr}(r) + \Delta W(r) \tag{10.59}$$

where $\Delta W(r)$ is the difference between the actual $\Delta\tau_1(r)$ and its trigonometric fit, $W_{appr}(r)$. After $W_{appr}(r)$ is determined, the second solution for the unknown $\kappa_{A,dif}(r)$ is found as the algebraic sum of $\kappa^{(1)}_{A,dif}(r)$ in Eq. (10.58) and the analytical derivative of the trigonometric fit $W_{appr}(r)$. The final profile, $\kappa^{(2)}_{A,dif}(r)$, and the residual range-dependent noise components, $\Delta\tau_2(r)$, are now defined as

$$\kappa^{(2)}_{A,dif}(r) = \kappa^{(1)}_{A,dif}(r) - \frac{d}{dr} W_{appr}(r) \tag{10.60}$$

and

$$\Delta\tau_2(r) = \Delta\tau_1(r) - W_{appr}(r) \tag{10.61}$$

respectively. Similar to the first approximation procedure, the resultant function $\Delta\tau_2(r)$ may be compared with the uncertainty $\Delta\tau_{A,dif}(r_0, r)$. In addition, an analysis of the derivative $d/dr[\Delta\tau_2(r)]$ may be recommended, which may reveal the presence of a large-scale systematic error.

Note that with this method, no filtering is done by changing the range resolution at the far end of the operating range. As was previously mentioned, the gain due to such filtering may be quite dubious because it ignores systematic errors in the initial data caused by remaining offsets, signal-induced noise, inaccurate background subtraction, and so on. With the method considered here, removing the outliers excludes poor data points, including those at distant ranges with a poor signal-to-noise ratio. Finally, one can recommend that the retrieved concentration profile be checked close to the near end of the measurement range, where the effect of different overlap zones for the on and off channels may produce additional systematic distortions.

When the final function $\kappa^{(2)}_{A,dif}(r)$ is found, the ozone concentration profile, $n(r)$, is determined by substituting it into Eq. (10.41)

$$n(r) = \frac{\kappa^{(2)}_{A,dif}(r)}{\Delta\sigma} \tag{10.62}$$

The remaining noise component, $\Delta n(r)$, which was excluded from the calculated $n(r)$ profile, can be checked by examining the term $\Delta\tau_2(r)$ in Eq. (10.61). This is especially recommended when two-dimensional ozone concentration images are analyzed simultaneously with corresponding two-dimensional images of the noise component extracted from $\Delta\tau_2(r)$. The noise component can be obtained by conventional numerical differentiation of $\Delta\tau_2(r)$, that is

$$\Delta n(r) \approx \frac{1}{\Delta\sigma}\frac{\mathrm{d}}{\mathrm{d}r}[\Delta\tau_2(r)] \tag{10.63}$$

In Fig. 10.13 (a)–(d), typical functions $\tau_{A,\mathrm{dif}}(r_0, r)$, $\kappa_{\mathrm{est}}(r)$, $\kappa_{\mathrm{appr}}(r)$, etc. are shown, extracted from experimental DIAL data. The functions are obtained from

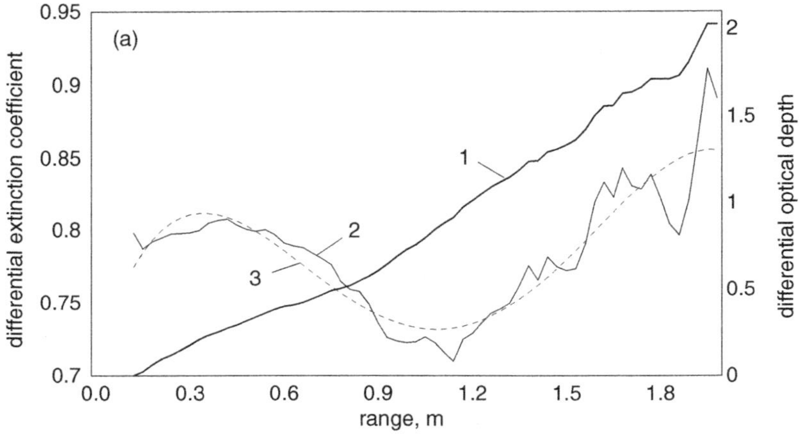

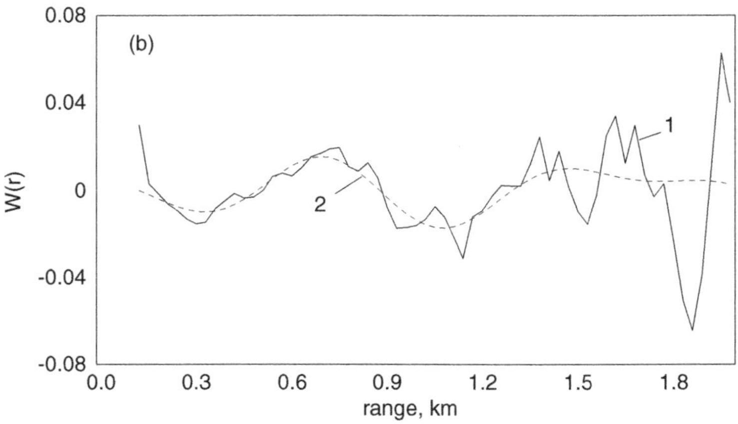

Fig. 10.13. (a) Typical functions $\tau_{A,\mathrm{dif}}(r_0, r)$, $\kappa_{\mathrm{est}}(r)$, and $\kappa_{\mathrm{appr}}(r)$, shown as *curves 1, 2,* and *3*, respectively, obtained from experimental data during the first approximation procedure. (b) Corresponding function $\Delta\tau_1(r)$ (*curve 1*) and its trigonometric fit, $W_{\mathrm{appr}}(r)$ (*curve 2*).

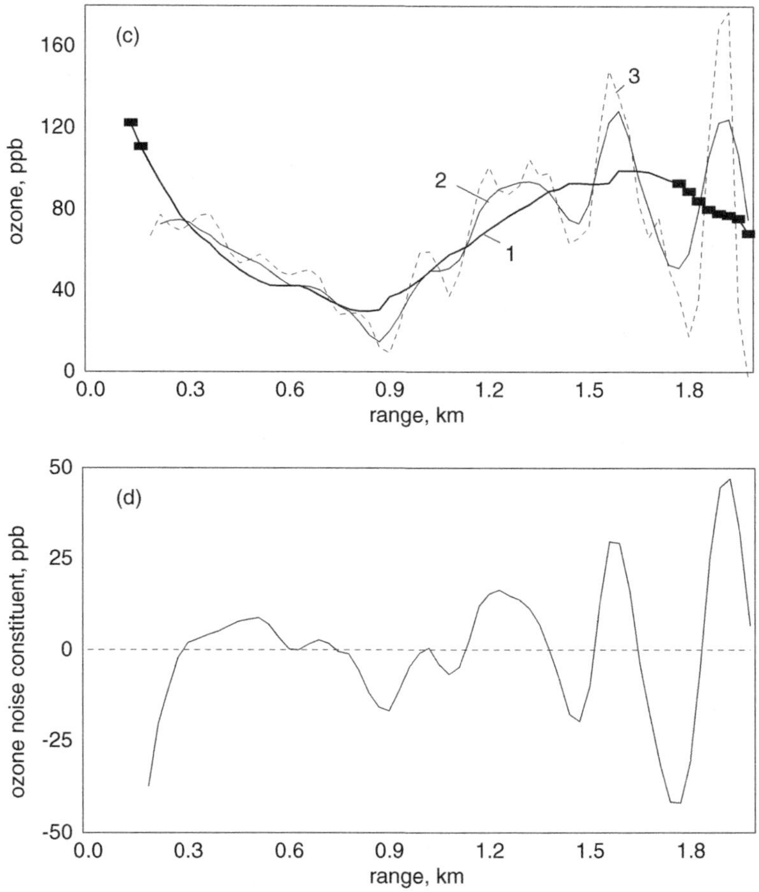

Fig. 10.13. (c) Ozone concentration profile $n(r)$ obtained with the nonlinear approximation method (*curve 1*). *Curve 2* is the same profile obtained without excluding the estimated noise constituent. The ozone concentration profile obtained with the conventional numerical differentiation is shown as *curve 3*. (d) Noise constituent $\Delta n(r)$ corresponding to the ozone concentration profile shown in (c) as *curve 1*.

signals measured in the lower troposphere with a down-looking airborne UV-DIAL system (Kovalev et al., 1996). In all of the panels, the range r is the nadir-viewed distance from the lidar. For the calculations, a 1-s set of DIAL signals is used, involving the average of 20 individual lidar returns, measured simultaneously at the off and on wavelengths, 312.9 nm and 276.9 nm, respectively. The aerosol extinction and backscatter corrections were made with an extinction coefficient profile measured at the reference wavelength, 359.6 nm. In Fig. 10.13 (a), the functions $\tau_{A,dif}(r_0, r)$, $\kappa_{est}(r)$, and $\kappa_{appr}(r)$ are shown as *curves 1, 2*, and *3*, respectively. The corresponding function $\Delta\tau_1(r)$ and its trigonometric fit, $W_{appr}(r)$ are shown in Fig. 10.13 (b) as *curves 1* and *2*, respectively. In Fig. 10.13 (c) the extracted ozone concentration profiles are shown. The profile $n(r)$

obtained with the above analytical approximation is shown as *curve 1*. The black squares show the $n(r)$ values considered as dubious according to the criteria used. *Curve 2* in Fig. 10.13 (c) shows the ozone profile that would be obtained from the same function $\tau_{A,dif}(r_0, r)$ given in (a) without excluding the noise component from $\Delta\tau_2(r)$. For comparison, the ozone concentration profile is also shown that was obtained with conventional numerical differentiation of $\tau_{A,dif}(r_0, r)$ (*curve 3*). As expected, the high-frequency fluctuations in *curves 2* and *3* are similar. The differences in these oscillations are caused by different smoothing when calculating the profiles. The remaining noise component $\Delta n(r)$, extracted from $\Delta\tau_2(r)$ [Eq. (10.63)] is shown in Fig. 10.13 (d); here the moving average is calculated with a range resolution $\Delta r = 120\,\text{m}$.

The nonlinear approximation technique can yield some improvement in DIAL measurement data in comparison to the results obtained with conventional techniques based on numerical differentiation. The application of the approximation technique to the experimental data shows that the erroneous fluctuations induced in the retrieved profile by noise and aerosol inhomogeneity are significantly reduced. Moreover, these can be separated and presented in a two-dimensional image for an a posteriori analysis. There are also other ways to avoid direct calculations of small logarithmic increments in DIAL measurements, for example, as discussed in studies by Zuev et al. (1983) and Stelmaszczyk et al. (2000). In the study by Kovalev (2002), a variant of analytical differentiation is proposed based on minimizing the second term in Eq. (10.53). This can be done by selecting different constants $\kappa_{est}(r_0)$ in Eq. (10.52) for every region of interest.

Summary. The basic procedures of the method are as follows. The on-off signal ratio is transformed into an intermediate analytical function. The ozone concentration retrieval is then made by processing this transformed function rather than the initial on-off signal ratio. Consecutive fits are applied to the transformed function, and a separation of low- and high-frequency components is made. This enables the consecutive retrieval of ozone concentration over areas with small and large gradients. The derivative of the transformed function is found by analytical differentiation. The ozone concentration is obtained by subsequent separation of the low- and high-frequency components. For this, some criteria are established to discriminate the signal from the noise. This is achieved by the selection of the number of terms in a Fourier series. The nonlinear approximation technique makes it possible to use simple criteria to determine the length of the actual measurement range and to determine the details of the extracted profile. The method of nonlinear approximation is general and can be applied to both DIAL and Raman data processing, which have the same numerical differentiation issues.

10.3.2. Compensational Three-Wavelength DIAL Technique

As discussed in Section 10.2.2, the aerosol corrections to DIAL measurements are based on some assumptions and theoretical approximations that may often not be valid, especially in the lower troposphere in regions with strong aerosol layering. Generally, these corrections are accurate if no significant changes in the aerosol concentration and particle size distribution occur. However, this can only be achieved if the assumptions taken a priori, such as in Eqs. (10.17) and (10.20), are true. Otherwise, the corrections can worsen rather than improve the accuracy of the derived ozone concentration profile.

Alternative methods for DIAL measurements that may reduce the influence of aerosol differential scattering were analyzed in studies by Wang et al. (1994), Kovalev and Bristow (1996), and Wang et al. (1997). The principal advantages of these compensational techniques are that no corrections for aerosol differential extinction and backscatter effects are needed for a good first approximation. This avoids having to obtain a particulate extinction coefficient profile at a reference wavelength and having to invoke assumptions regarding the spectral dependence of the aerosol scattering coefficients. Another potential advantage of the compensational technique is the reduction of errors caused by absorption from chemical species other than ozone. This may be achieved by a sensible selection of the operating wavelengths.

With conventional DIAL techniques, the ozone concentration is calculated from a pair of signals measured at the on and off wavelengths. When two different pairs of the signals are available, the ozone concentration can be obtained either by each pair processed separately or by using a four-wavelength differential method. The four-wavelength method was widely used at the world network monitoring stations for spectrophotometric measurements of the total ozone in the atmosphere. The advantage of the four-wavelength differential method as compared to the conventional two-wavelength method is the reduced influence of aerosol differential scattering on the measurement accuracy. For lidar measurements, this method was first proposed by Wang et al. (1994). In the method, two pair wavelengths, $\lambda_{on,1} - \lambda_{off,1}$ and $\lambda_{on,2} - \lambda_{off,2}$, are used. To reduce the influence of aerosol scattering, these two spectral bands must overlap. In the UV spectra with $\lambda > 260$ nm, the ozone absorption cross section reduces with the increase of the wavelength (Fig. 10.6), so that the wavelength sequence in the two-pair method must be $\lambda_{on,1} < \lambda_{on,2} < \lambda_{off,1} < \lambda_{off,2}$. In the reduced three-wavelength technique, two medium wavelengths are selected to be equal, that is, $\lambda_{on,2} = \lambda_{off,1}$ (Kovalev and Bristow, 1996; Wang et al., 1997). Accordingly, the ozone concentration is determined from the signals measured concurrently at wavelengths $\lambda_{on,1}$, $\lambda_{on,2} = \lambda_{off,1}$, and $\lambda_{off,2}$, which correspond to a high, medium, and low absorption of ozone, respectively. Accordingly, the DIAL solution is transformed, so that the differential optical depth is determined for three rather than two wavelengths. This reduces the aerosol differential scattering without having to introduce the corrections $\Delta n_\beta(r)$ and $\Delta n_e(r)$ and use of a priori assumptions. Unlike the variants of the three-wavelength techniques given by Sasano (1985) and Jinhuan (1994), no a priori information regarding the aerosol characteristics is involved in data processing with the compensational technique given below.

The ozone concentration is determined by using DIAL signals $P(r, \lambda_i)$ measured at three wavelengths denoted further as λ_1, λ_2, and λ_3, where $\lambda_1 < \lambda_2 <$

λ_3. The measurement is assumed to be made in the ultraviolet with $\lambda >$ 260 nm; thus the absorption is a maximum at λ_1 and least at λ_3. The basic function used for ozone concentration retrieval is related to the three signals $P(r, \lambda_i)$ as

$$H(r) = \frac{[P(r, \lambda_2)]^2}{P(r, \lambda_1) P(r, \lambda_3)} \tag{10.64}$$

The logarithm of $H(r)$ is related to the three-wavelength differential optical depth, from which the integrated ozone concentration is determined. Unlike the differential optical depth for a two-wavelength DIAL measurement (Section 10.1), this term is determined from Eqs. (10.3) and (10.64) as

$$\ln H(r) = const_3 + \ln \frac{[\beta_\pi(r, \lambda_2)]^2}{\beta_\pi(r, \lambda_1)\, \beta_\pi(r, \lambda_3)} + 2\int_{r_1}^{r} [\Delta\sigma_{(3)} n(r') + \Delta\kappa_{A(3)}(r') + \Delta\beta_{(3)}(r')]\, dr' \tag{10.65}$$

where $\beta_\pi(r, \lambda_i)$ is the backscatter coefficient at wavelength λ_i, and $n(r)$ is the unknown ozone concentration at range r. The three-wavelength differential absorption cross section for ozone, $\Delta\sigma_{(3)}$ is

$$\Delta\sigma_{(3)} = \sigma(\lambda_1) + \sigma(\lambda_3) - 2\sigma(\lambda_2) \tag{10.66}$$

where $\sigma(\lambda_i)$ is the ozone absorption cross section at the wavelength λ_i. The three-wavelength differential absorption coefficient $\Delta\kappa_{A(3)}(r)$ for other (interfering) absorbing species (e.g., SO_2) and the three-wavelength total differential scattering coefficient $\Delta\beta_{(3)}(r)$ are defined similar to $\Delta\sigma_{(3)}$:

$$\Delta\kappa_{A(3)}(r) = \kappa_A(r, \lambda_1) + \kappa_A(r, \lambda_3) - 2\kappa_A(r, \lambda_2) \tag{10.67}$$

and

$$\Delta\beta_{(3)}(r) = \beta(r, \lambda_1) + \beta(r, \lambda_3) - 2\beta(r, \lambda_2) \tag{10.68}$$

where $\beta(r, \lambda_i)$ is the total (particulate and molecular) scattering coefficient at λ_i

$$\beta(r, \lambda_i) = \beta_p(r, \lambda_i) + \beta_m(r, \lambda_i)$$

The differential scattering coefficient $\Delta\beta_{(3)}(r)$ can also be rewritten as the sum of the particulate and molecular scattering constituents

$$\Delta\beta_{(3)}(r) = \Delta\beta_{p(3)}(r) + \Delta\beta_{m(3)}(r) \tag{10.69}$$

The column optical depth of the ozone can be obtained from Eq. (10.65) as

$$\tau_{a,dif(3)}(r_1, r) = 0.5\left[\ln H(r) - \ln\frac{[\beta_\pi(r,\lambda_2)]^2}{\beta_\pi(r,\lambda_1)\beta_\pi(r,\lambda_3)} - const_3\right] - \int_{r_1}^{r}[\Delta\kappa_{A(3)}(r') + \Delta\beta_{(3)}(r')]\,dr' \tag{10.70}$$

The molecular scattering constituent $\Delta\beta_{m(3)}(r)$ can be calculated and excluded from Eq. (10.69). After that, $\Delta\beta_{(3)}(r) = \Delta\beta_{p(3)}(r)$, so that it is necessary to consider only the particulate component in the term $\Delta\beta_{(3)}(r)$. The optimal selection of the wavelengths makes it possible to reduce the systematic errors caused by both the particulate scattering and the interfering absorbing species. This can be achieved by proper selection of the wavelengths λ_1, λ_2, and λ_3. The optimal selection of the wavelengths is reached when $\Delta\kappa_{A(3)}(r)$ and $\Delta\beta_{p(3)}(r)$ are close to zero, whereas $\Delta\sigma_{(3)}$ remains large enough to provide acceptable measurement sensitivity for ozone.

The optimization variant, when the wavelength intervals $\Delta\lambda_{1,2}$ and $\Delta\lambda_{2,3}$ are significantly different is analyzed in the study by Wang et al. (1997). Here the ratio of $\Delta\lambda_{1,2}$ to $\Delta\lambda_{2,3}$ may be calculated and used as an additional constant factor in the data processing algorithms.

As shown in the previous sections, the systematic error caused by the differential aerosol backscatter term is often dominant in tropospheric measurements. Moreover, this error is most difficult to correct in a conventional DIAL measurement, where spatial variability in the backscattering causes a large error in the derived ozone concentration. This is why the analysis below concentrates on the uncertainties caused by backscatter gradients. Specifically, the two- and three-wavelength techniques are compared. Using a transformation similar to that in Section 10.1, one can rewrite the logarithmic term in the right side of Eq. (10.65) as

$$\ln\frac{[\beta_\pi(r,\lambda_2)]^2}{\beta_\pi(r,\lambda_1)\beta_\pi(r,\lambda_3)} = const_3 + \ln\frac{[1+Q(r,\lambda_2)]^2}{[1+Q(r,\lambda_1)][1+Q(r,\lambda_3)]} \tag{10.71}$$

where $Q(r, \lambda_i)$ is the aerosol backscatter ratio at wavelength λ_i defined in Eq. (10.16). The formula for the conventional two-wavelength DIAL, which operates at the wavelengths λ_1 and λ_2, can be derived from Eq. (10.36) as

$$\ln\frac{\beta_\pi(r,\lambda_2)}{\beta_\pi(r,\lambda_1)} = const_2 + \ln\frac{1+Q(r,\lambda_2)}{1+Q(r,\lambda_1)} \tag{10.72}$$

The comparison of the incremental changes for the logarithmic terms in Eqs. (10.71) and (10.72) is a good opportunity to show the behavior of the systematic error Δn_β caused by particulate backscattering in these methods. To calculate the incremental changes, a spectral dependence of the aerosol backscatter coefficient over the spectral range $\Delta\lambda = \lambda_3 - \lambda_1$ must be taken. It is sensible to use for the analysis the same assumptions on the scattering spectral dependencies as in the previous sections. The assumptions are that the particulate backscatter coefficient $\beta_{\pi,p}(r)$ for wavelengths λ_i and λ_j vary inversely with the wavelength to the power of $\xi_{i,j}$ (Section 10.1). In the real atmospheres, the exponent $\xi_{i,j}$ may be range dependent, that is, $\xi_{i,j} = \xi_{i,j}(r)$. Accordingly, the ratio of $\beta_\pi(r, \lambda_i)$ to $\beta_\pi(r, \lambda_j)$ can be range dependent, that is

$$\frac{\beta_{\pi,p}(r,\lambda_i)}{\beta_{\pi,p}(r,\lambda_j)} = \left(\frac{\lambda_i}{\lambda_j}\right)^{-\xi_{i,j}(r)} \tag{10.73}$$

On the other hand, the spectral dependence of the aerosol backscatter coefficient can be different for different spectral intervals, so that the terms $\xi_{1,2}$ and $\xi_{2,3}$ in adjacent intervals $(\lambda_1 - \lambda_2)$ and $(\lambda_2 - \lambda_3)$ may also be different. Therefore, this dependence may be more accurately approximated by different exponents. Taking into consideration that the molecular volume backscattering coefficients for λ_i and λ_j vary inversely with the wavelength to the fourth power, and assuming that the relative separation between the adjacent wavelengths λ_i and λ_j is small,

$$\delta\lambda_{i,j} = \frac{|\lambda_j - \lambda_i|}{\lambda_j} << 1$$

one can write the ratio of $Q(r, \lambda_i)$ to $Q(r, \lambda_j)$ in a form similar to that in Eqs. (10.17) and (10.18):

$$\frac{Q(r,\lambda_i)}{Q(r,\lambda_j)} = \left(\frac{\lambda_i}{\lambda_j}\right)^{4-\xi_{i,j}(r)} \approx 1 - g_{i,j}(r) \tag{10.74}$$

where

$$g_{i,j}(r) = [4 - \xi_{i,j}(r)]\,\delta\lambda_{i,j} \tag{10.75}$$

The systematic error Δn_β, caused by particulate differential backscattering is different for the two- and three-wavelength techniques. These errors can be found from the corresponding values of the logarithmic differences, as in Eq. (10.15). As shown in Section 10.2, such logarithmic differences can be treated either as correction terms (when these can be in some way estimated) or as systematic errors (when such an estimate is impossible). In many situations, the compensational technique might be more robust than the conventional one, at least under conditions where $\xi_{1,2}$ and $\xi_{2,3}$ do not differ dramatically. In the analysis below, we consider situations in which the correction terms cannot be determined accurately. Accordingly, these must be considered as sources of systematic error $\Delta\tau$ in the calculated differential optical depth $\tau_{A,dif(3)}(r_0, r)$. These errors for the two- and three-wavelength technique, $\Delta\tau_{(2)}$ and $\Delta\tau_{(3)}$, may be obtained by substituting Eqs. (10.74) and (10.75) into Eq. (10.72) for the conventional technique and into Eq. (10.71) for the compensational technique, respectively. By calculating the increments of the logarithmic terms in Eqs. (10.71) and (10.72), one can obtain the following formulas for $\Delta\tau_{(2)}$ and $\Delta\tau_{(3)}$:

$$\Delta\tau_{(2)}(r, r+\Delta r) = -0.5\ln\frac{1-\dfrac{Q(r+\Delta r,\lambda_2)\,g_{1,2}(r+\Delta r)}{1+Q(r+\Delta r,\lambda_2)}}{1-\dfrac{Q(r,\lambda_2)\,g_{1,2}(r)}{1+Q(r,\lambda_2)}} \tag{10.76}$$

and

$$\Delta\tau_{(3)}(r, r+\Delta r) = \Delta\tau_{(2)}(r, r+\Delta r) - 0.5\ln\frac{1+\dfrac{Q(r+\Delta r,\lambda_2)}{1+Q(r+\Delta r,\lambda_2)}\left[\dfrac{g_{2,3}(r+\Delta r)}{1-g_{2,3}(r+\Delta r)}\right]}{1+\dfrac{Q(r,\lambda_2)}{1+Q(r,\lambda_2)}\left[\dfrac{g_{2,3}(r)}{1-g_{2,3}(r)}\right]} \tag{10.77}$$

where Δr is a selected resolution range. The corresponding systematic errors of the ozone concentration for the conventional and compensational technique, $\Delta n_{\beta(2)}$ and $\Delta n_{\beta(3)}$, can be found as

$$\Delta n_{\beta,(2)}(r, r+\Delta r) = \frac{\Delta\tau_{(2)}(r, r+\Delta r)}{\Delta\sigma_{(2)}\Delta r} \tag{10.78}$$

and

$$\Delta n_{\beta,(3)}(r, r+\Delta r) = \frac{\Delta\tau_{(3)}(r, r+\Delta r)}{\Delta\sigma_{(3)}\Delta r} \tag{10.79}$$

here $\Delta\sigma_{(3)}$ is given by Eq. (10.66), and

$$\Delta\sigma_{(2)} = \sigma(\lambda_1) - \sigma(\lambda_2) \tag{10.80}$$

The numerical experiments made by Kovalev and Bristow (1996) showed that in many cases, the three-wavelength technique can significantly improve differential absorption measurement accuracy. Some results of these calculations are given in Figs. 10.14–10.17. All calculations are made for a multiwavelength UV-DIAL system described in the study by Moosmüller et al. (1991). The wavelengths used for the analysis of the three-wavelengths technique are $\lambda_1 = 276.9$ nm, $\lambda_2 = 291.6$ nm, and $\lambda_3 = 312.9$ nm. When calculating the

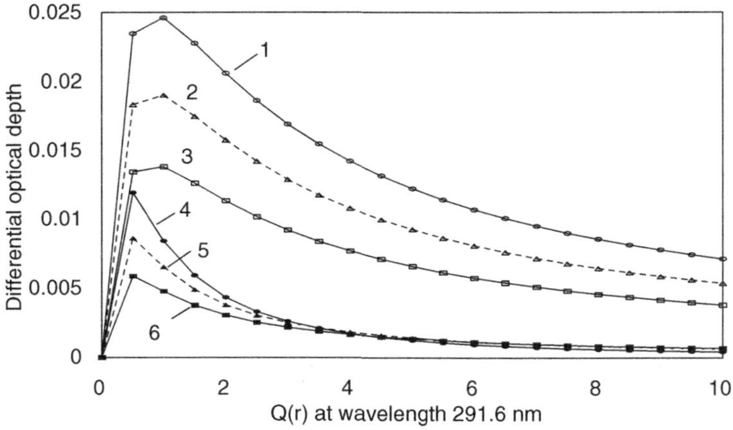

Fig. 10.14. $\Delta\tau_{(2)}$ as a function of the aerosol backscattering ratio $Q(r, \lambda_2)$ determined for the wavelength pair $\lambda_1 = 276.9$ nm and $\lambda_2 = 291.6$ nm (curves with the open data points, i.e., circles, triangles, etc.) and $\Delta\tau_{(3)}$ calculated for the wavelengths $\lambda_1 = 276.9$ nm, $\lambda_2 = 291.6$ nm, and $\lambda_3 = 312.9$ nm (curves with the solid data points). Exponent values were $\xi = -1$ for *curves 1* and *4*, $\xi = 0$ for *curves 2* and *5*, and $\xi = 1$ for *curves 3* and *6* (Kovalev and Bristow, 1996).

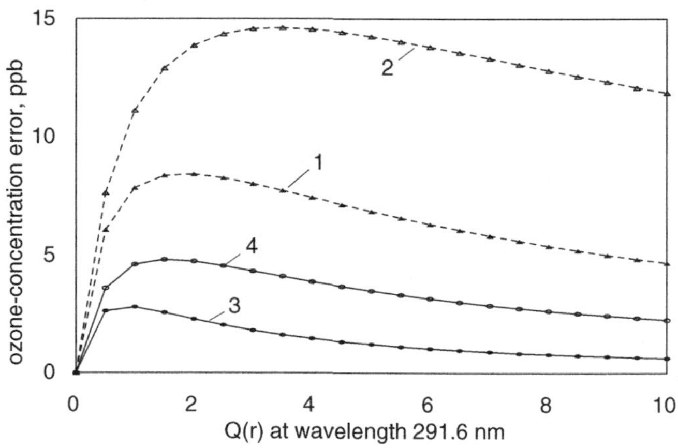

Fig. 10.15. The systematic error in the ozone concentration, $\Delta n_\beta(r, r + \Delta r)$, caused by aerosol differential backscattering versus $Q(r, \lambda_2)$. The error is calculated for the conventional two-wavelength technique (curves 2 and 4) and for the three-wavelength compensational technique (curves 1 and 3). The extinction coefficient ratio at ranges $(r + \Delta r)$ and r is set to 0.1 (*curves 1* and *2*) and 0.5 (*curves 3* and *4*) (Kovalev and Bristow, 1996).

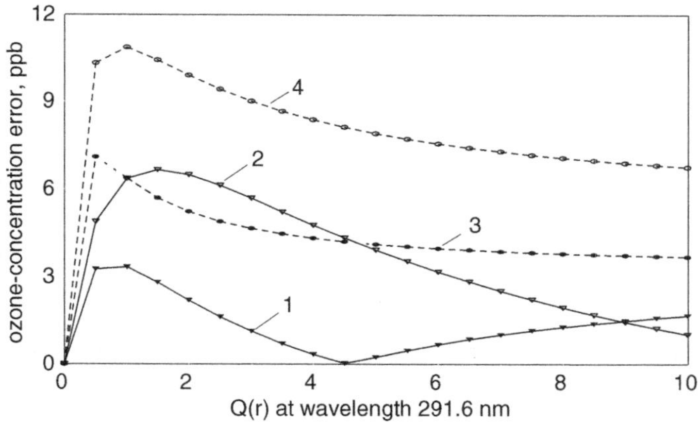

Fig. 10.16. Same as plot shown in Fig. 10.15, except that the exponent ξ now varies as $\xi_{i,j}(r + \Delta r)/\xi_{i,j}(r) = 0.5$. Also, the extinction coefficient ratio at ranges $(r + \Delta r)$ and r is now set equal to 0.3 for *curves 1* and *2* and to 3 for *curves 3* and *4* (Kovalev and Bristow, 1996).

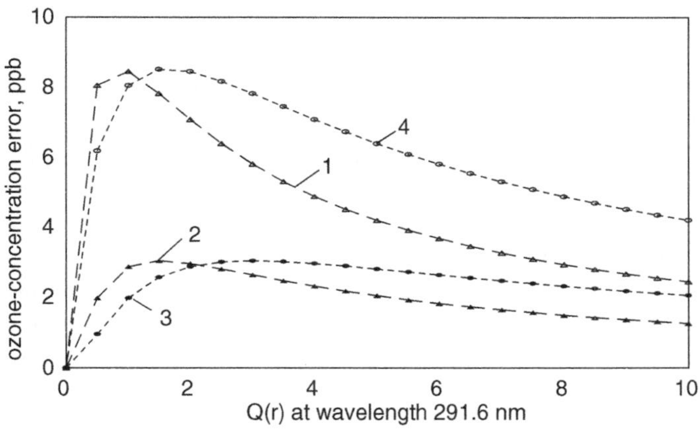

Fig. 10.17. Systematic error $\Delta n_\beta(r, r + \Delta r)$ for the two-wavelength technique (*curves 1* and *4*) and for the three-wavelength compensational technique (*curves 2* and *3*) where $\xi_{1,2} = -1$ and $\xi_{2,3} = 1$. The extinction coefficient ratio is set to 2 for *curves 1* and *2* and to 0.5 for *curves 3* and *4* (Kovalev and Bristow, 1996).

errors for the conventional DIAL, the signals are assumed to be measured at wavelengths 276.9 and 291.6 nm. In Fig. 10.14, the simplest case is shown, when the dependence given by Eq. (10.74) holds between wavelengths λ_1, λ_2, and λ_3 with $\xi_{i,j}(r) = \xi = const.$ Here the $\Delta\tau$ values are calculated for the exponents $\xi = -1$ (*curves 1* and *4*), $\xi = 0$ (*curves 2* and *5*), and $\xi = 1$ (*curves 3* and *6*). The particulate extinction coefficient is set to increase with the range, where the particulate scattering coefficient ratio, $\beta_p(r + \Delta r, \lambda_2)/\beta_p(r, \lambda_2)$ is set equal to 2.

Comparing $\Delta\tau_{(2)}$ and $\Delta\tau_{(3)}$, one can see the significant advantage of the compensational technique. The values of $\Delta\tau_{(3)}$ are much less than those for the conventional technique. The errors in the ozone concentration caused by not correcting for particulate differential backscattering are shown in Figs. 10.15–10.17. In Fig. 10.15, *curves 1* and *3* show the errors for the compensational technique and *curves 2* and *4* show the corresponding errors for the conventional technique. Here the aerosol extinction coefficient is set to decrease with the range. The exponent ξ is set to 1 and is assumed to be constant over the total wavelength spectrum from 276.9 to 312.9 nm; the spatial range interval is $\Delta r = 300$ m.

As stated above, the assumption that the exponent $\xi_{i,j}(r)$ is constant over the measurement range may often be invalid. In real atmospheres, significant variations in $\xi_{i,j}(r)$ are quite likely, especially, if large intervals between wavelengths are used. The analysis shows that, generally, the compensational technique provides more accurate results even if the spectral dependencies for aerosol backscattering are range dependent, that is, if $\xi_{i,j}(r)$ is not constant within the range interval Δr. In Fig. 10.16, systematic errors in the ozone concentration are shown caused by particulate differential backscattering when $\xi_{i,j}(r)$ is variable. One can see that the compensational technique (*curves 1* and *3*) yields reduced systematic errors as compared with the conventional method (*curves 2* and *4*).

In practice, it is very likely that the exponent $\xi_i(r)$ may have different values for adjacent spectral intervals $(\lambda_1 - \lambda_2)$ and $(\lambda_2 - \lambda_3)$. An example of systematic errors in the ozone concentration obtained for an atmosphere where $\xi_{1,2} \neq \xi_{2,3}$ is shown in Fig. 10.17. The ratio $\beta_p(r + \Delta r, \lambda_2)/\beta_p(r, \lambda_2)$ is set to 2 for *curves 1* and *2* and 0.5 for *curves 3* and *4*. Note that, as above, the three-wavelength method significantly reduces the systematic error caused by particulate loading. However, the estimates by Kovalev and Bristow (1995) revealed that the compensational method is more sensitive to signal noise than the conventional two-wavelength method. This is primarily because the additional signal, corrupted by noise, is involved in calculations [Eq. (10.64)]. The second reason is a relative decrease of the differential absorption cross section [Eq. (10.66)] as compared with the conventional DIAL technique [Eq. (10.80)]. Therefore, to obtain more accurate measurements, the ozone fluctuations caused by random noise must be thoroughly suppressed. The experimental tests also showed that the compensational method is much more effective if it is used in combination with an approximation technique, such as that given in Section 10.3.1.

A similar compensational approach was used in the study of Wang et al. (1997). In this study, the analysis of the three-wavelength technique was made for stratospheric ozone measurements. On the base of a theoretical analysis and experimental data, the authors concluded that the three-wavelength technique provides much more accurate concentration profiles than the conventional method even after making backscatter corrections. It was pointed out that the method greatly reduces the effect of volcanic aerosols on the accuracy of the ozone concentration measurements. According to the authors' estimates, the statistical error of the three-wavelength DIAL proved to be slightly larger than that for the conventional DIAL. As mentioned above, this increase occurs because the three-wavelength method incorporates an additional error from the third signal. Wang et al. (1997) estimated that this increase in error is only 2% at a height of 30 km. The systematic error caused by aerosol backscattering is estimated by the authors to be reduced as much as 10 times, compared with conventional DIAL. The authors concluded that the method is almost insensitive both to spatial inhomogeneity of aerosol loading and to the wavelength dependence of the aerosol backscatter and its spatial change. However, an unbiased consideration of the experimental data presented in this study shows that the accuracy estimates in the study may be too optimistic. One should be cautious when estimating small measurement errors, especially in new methods. It is necessary to consider thoroughly the validity of all of the assumptions that were used (such as absence of systematic errors or zero-line offsets) and make a thorough analysis of the experimental data, comparing results obtained by the new and old methods.

To summarize, the compensational technique can be very helpful in many situations, including these, where more than one absorbing species absorbs the light in the same spectral range. The most significant merit of the compensational DIAL technique is that the aerosol corrections may be omitted. Thus no a priori information is needed regarding the spectral dependencies of the aerosol scattering properties along the searched paths. However, the gain in accuracy obtained with the compensational technique depends on particular atmospheric conditions. Extremely large gradients in aerosol backscattering, or large changes in aerosol spectral dependencies, can significantly reduce the benefits of the compensational technique. The technique is more sensitive to the signal noise and distortions compared with the conventional two-wavelength technique. The differential absorption coefficient for the three-wavelength method is, generally, less than that for the conventional DIAL, that is, $\Delta\sigma_{(3)} < \Delta\sigma_{(2)}$. Therefore, for the same range resolution, the local differential optical depth is always less when the three-wavelength method is used instead of the two-wavelength technique. This, in turn, increases the error constituents, which are proportional to the reciprocal of $\Delta\sigma$.

11

HARDWARE SOLUTIONS TO THE INVERSION PROBLEM

As discussed previously, the power received by a monostatic lidar with time is proportional to the backscatter coefficient in a sampled volume at some distance from the lidar and the two-way attenuation along the path from the lidar to the sampled volume and back. For single-wavelength lidars, only one piece of new data (the lidar return) is available at each range element whereas two new unknowns present themselves (the backscatter and attenuation coefficients). Without other information, the problem is fundamentally insoluble. To solve the problem, many different hardware solutions have been examined by a wide variety of researchers. The principal goal of these studies was to develop a lidar system that provides enough information to uniquely and unambiguously determine the unknowns in each range element. The solution of this problem is far from complete; however, serious accomplishments in this direction have been made.

In Sections 11.1 and 11.2, two techniques are presented. Both of the techniques use the scattering from molecules (as opposed to particulate scattering) as the basis of a method that determines the particulate attenuation at each range element. Because the molecular density profile of the atmosphere can be determined with pressure and temperature data from balloons or climatological information, the profile of the backscatter coefficients for molecular scattering can be determined. Under certain conditions, this allows an unambiguous determination of the particulate attenuation coefficients. Unfor-

Elastic Lidar: Theory, Practice, and Analysis Methods, by Vladimir A. Kovalev and William E. Eichinger.
ISBN 0-471-20171-5

tunately, as will be shown, these systems do have some limitations. Multiple-wavelength lidars are discussed in Section 11.3.

11.1. USE OF N_2 RAMAN SCATTERING FOR EXTINCTION MEASUREMENT

11.1.1. Method

Raman lidars use a technique originally pioneered by Fiocco and Smullins (1963), who first detected scattering layers in the upper atmosphere at Raman-shifted wavelengths. For these observations, a ruby lidar system at 694.3 nm was used with two telescopes directed to the zenith. The backscattered photons were detected by a photomultiplier, displayed on an oscilloscope, and photographed. The signal was quite weak, so the individual photons were counted in 10-km (66 μs) intervals. Nevertheless, altitudes were examined up to 180 km. The return signal not only showed Rayleigh scattering up to 50–60 km, but also revealed echoes from altitude ranges 60–90 and 110–140 km; the latter were assumed to originate in dust layers. Leonard (1967) first reported the observation of the Raman scattering from atmospheric nitrogen. Then in 1968 Cooney presented results of Raman scattering from atmospheric nitrogen with a ruby laser and a year later showed that this could be used to determine atmospheric attenuation coefficients uniquely (Cooney et al., 1969). Melfi et al. (1969) and Cooney (1970) later reported the observation of not only nitrogen and oxygen but also Raman scattering by atmospheric water vapor. Most if not all Raman lidars have been built to measure water vapor concentrations as well as particulate properties. Additionally, some are used to measure temperature, a capability discussed in Chapter 12.

Similar to an elastic lidar, a Raman lidar operates by emitting a pulsed laser beam, usually in the ultraviolet or near ultraviolet, into the atmosphere. Atmospheric gases, such as nitrogen, oxygen, and water vapor interact with this light via the Raman scattering process, causing light of longer wavelengths to be scattered. Thus, in addition to the elastically backscattered light, molecules in the atmosphere also scatter a wavelength-shifted component (Chapter 2). The amount of the wavelength shift is unique to each molecule. For example, the Raman-shifted nitrogen returns are shifted by 2331 wavenumbers from the laser line. Therefore, atmospheric gaseous species can be distinguished by this technique. Figure 11.1 is a plot of the spectrum of light returning from an emitted 248-nm pulse, showing the peaks from the various atmospheric constituents. Note that elastic scattering at the emitted wavelength, 248 nm, is off of the scale in this figure. Because of the small value for the cross sections for Raman scattering and, accordingly, the small backscatter coefficients, the number of photons returning to the lidar is small. Because the probability of Raman scattering is proportional to $1/\lambda^4$, that is, the same as for molecular elastic scattering (see Chapter 2), the use of short wavelengths

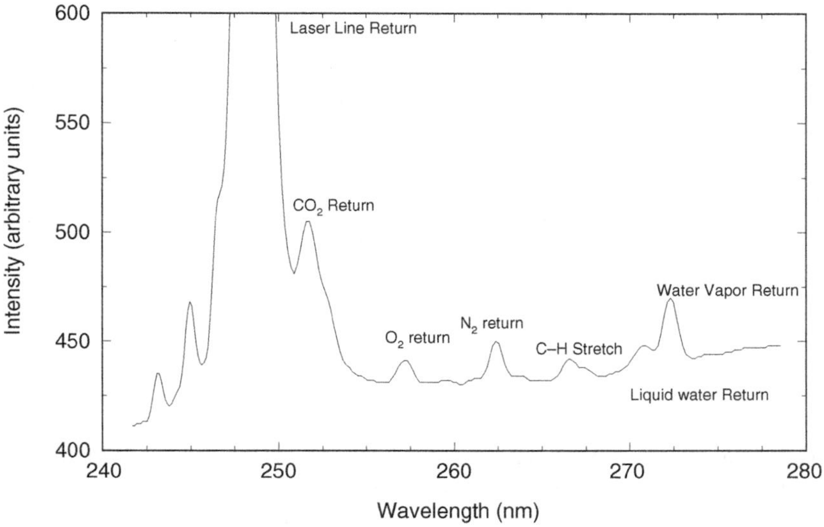

Fig. 11.1. A plot of the spectrum of light returning from a 248-nm Raman lidar showing the peaks from the various atmospheric constituents.

increases the magnitude of the signal. Thus more modern Raman lidars are found at ultraviolet wavelengths, particularly at 248 nm (KrF excimer), 266 nm (quadrupled Nd:YAG), 308 nm (XeCl excimer), 351 nm (XeF excimer), and 355 nm (tripled Nd:YAG). Figures 11.2 and 11.3 are diagrams showing the layout of the Los Alamos Raman lidar and the optics used to separate the wavelengths of light behind the telescope. This lidar and the separation optics are typical of those used in Raman lidars. In this lidar, the laser is mounted below the telescope. A series of mirrors and lenses is used to expand the beam to make it eye safe and collinear with the telescope. A 45° angled mirror is used to change the optical direction to vertical, allowing the system to make vertical soundings. With the scanning mirror mounted, the system can perform three-dimensional scanning near the surface. At the back of the telescope, a series of dichroic beam splitters are used to separate the elastically scattered light from the light at the two Raman-shifted wavelengths from nitrogen and water vapor. Narrow band interference filters block unwanted wavelengths in each channel. Occasionally, a cuvette of an organic liquid is used to provide an additional level of rejection of the elastic scattered light. For rejection of ultraviolet light at 248 nm, ethyl formate or butyl acetate may be used.

Because of the small cross section for Raman scattering, the number of photons returning to the lidar is small, so that photon counting is required to achieve meaningful signals at long ranges. The discrimination of these photons from background light is another issue that must be addressed. To work during

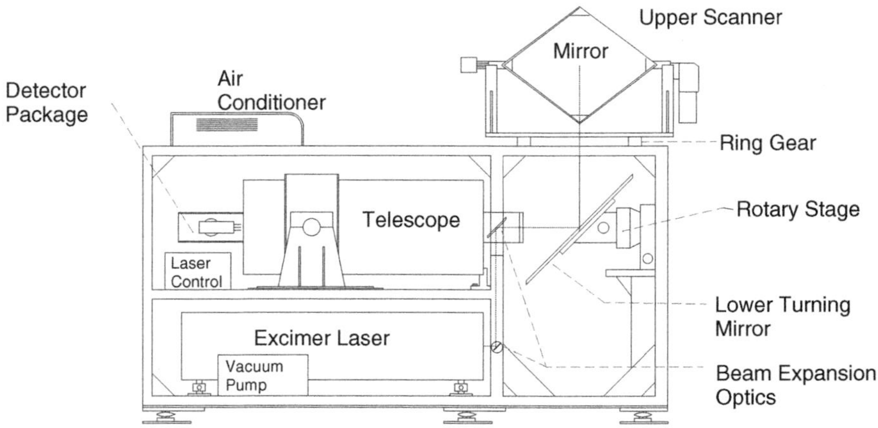

Fig. 11.2. Diagram showing the layout of the Los Alamos Raman lidar. With the exception of the scanning mirror, the arrangement is typical for Raman lidars.

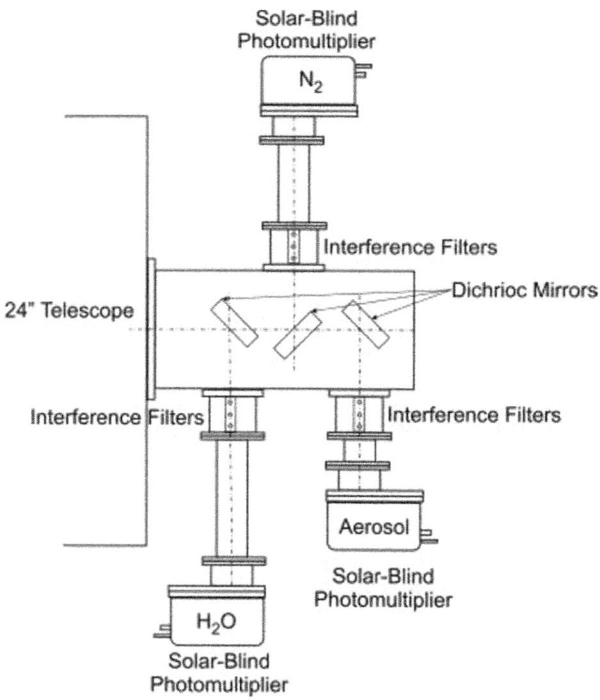

Fig. 11.3. Diagram showing the layout of the beam splitters, filters, and lenses that separate the light at the back of the telescope in the Los Alamos Raman lidar. Three wavelengths of light are separated, an elastic scattered wavelength (generally 248 nm), a nitrogen Raman-scattered wavelength (generally 263 nm), and a water vapor Raman-scattered wavelength (generally 273 nm).

the day, many systems operate in the region of the spectrum below about 300 nm, where ozone and oxygen strongly absorb sunlight, and are thus "blind" to solar photons. Daytime solar-blind operation for Raman lidars was developed by Renault et al. (1980), Cooney et al. (1985), and Renault and Capitini (1988). Solar-blind operation requires the use of a laser near 250–260 nm so that the Raman-shifted lines will be below 300 nm. The use of a laser with a wavelength longer than 266 nm will have contamination from sunlight at the Raman-shifted wavelengths. Laser wavelengths shorter than 248 nm will be so strongly absorbed at both the emission and Raman-shifted wavelengths that the maximum range of the system will be severely restricted.

Because wavelengths longer than 300 nm are not strongly absorbed by atmospheric ozone and molecular scattering is reduced at longer wavelengths, much greater range is possible with the use of longer wavelengths. However, because of the small Raman cross section, discrimination of Raman-scattered photons from sunlight is an issue requiring special measures. If the system is expected to operate during the day, the use of an extremely narrow field of view in the receiving telescope is required. Several of these systems have been built, with mixed results (Cooney, 1983; Annsman et al., 1992; Goldsmith et al., 1998). Because of the limited amount of returning light, Raman lidar systems tend to use large, powerful lasers and large telescopes. Therefore, they are unusually large. Figure 11.4 is a photograph of the Los Alamos Scanning Raman lidar mounted on its trailer.

At least in part because of the limitations of photon counting, most Raman lidars operate in a vertically staring mode. The NASA Goddard Space Flight Center (GSFC) Raman lidar (Ferrare et al., 1998), which can scan in a verti-

Fig. 11.4. Photograph of the Los Alamos scanning Raman Lidar. The size of this system is smaller than the typical Raman lidar. Many are mounted in semitrailers or large shipping containers.

cal plane, and the Los Alamos Scanning Raman Lidar (Eichinger et al., 1999), which can scan in three dimensions, are currently the two exceptions. The GSFC Raman lidar operates primarily in a staring mode along different azimuthal angles. Operating at various angles to the ground enables the system to achieve higher spatial resolution at lower altitudes.

Because Raman lidars are most often used as vertical sounders, in all of the equations below, the height above ground, h, is used as the lidar equation independent variable. The backscattered signals from the elastic, nitrogen, and water vapor channels are given by the following equations:

$$P_{\mathrm{elastic}}(h) = \frac{C_1 E[\beta_{\pi,\mathrm{p}}(h,\lambda) + \beta_{\pi,\mathrm{m}}(h,\lambda)]\exp\left[-2\int_0^h (\kappa_{\mathrm{t}}(h',\lambda)dh')\right]}{h^2} \tag{11.1}$$

$$P_{\mathrm{N2}}(h) = \frac{C_{\mathrm{N2}} E n_{\mathrm{N2}}(h)\sigma_{\mathrm{N2}} \exp\left[-\int_0^h [\kappa_{\mathrm{t}}(h',\lambda) + \kappa_{\mathrm{t}}(h',\lambda_{\mathrm{N2,R}})dh']\right]}{h^2} \tag{11.2}$$

and

$$P_{\mathrm{H2O}}(h) = \frac{C_2 E n_{\mathrm{H2O}}(h)\sigma_{\mathrm{H2O}} \exp\left[-\int_0^h [\kappa_{\mathrm{t}}(h',\lambda) + \kappa_{\mathrm{t}}(h',\lambda_{\mathrm{H2O,R}})]dh'\right]}{h^2} \tag{11.3}$$

where λ, $\lambda_{\mathrm{N2,R}}$, and $\lambda_{\mathrm{H2O,R}}$ are the laser and Raman N_2 and H_2O scattered wavelengths, respectively. Note that the Raman wavelength in the above equations has different values, $\lambda_{\mathrm{N2,R}}$ for nitrogen and $\lambda_{\mathrm{H2O,R}}$ for water vapor. Functions $P_{\mathrm{elastic}}(h)$, $P_{\mathrm{N2}}(h)$, and $P_{\mathrm{H2O}}(h)$ are the received signals in the elastic, nitrogen, and water vapor channels, E is the laser energy per pulse; $\beta_{\pi,\mathrm{m}}$ and $\beta_{\pi,\mathrm{p}}$ are the molecular and particulate scattering coefficients at 180° at the wavelength λ, emitted by the laser; σ_{N2} and σ_{H2O} are the Raman backscatter cross sections for the laser wavelength; $n_{\mathrm{N2}}(h)$ and $n_{\mathrm{H2O}}(h)$ are the number density of nitrogen and water molecules at height h; $\kappa_{\mathrm{t}}(h, \lambda)$, $\kappa_{\mathrm{t}}(h, \lambda_{\mathrm{N2,R}})$, and $\kappa_{\mathrm{t}}(h, \lambda_{\mathrm{H2O,R}})$ are the total attenuation coefficients at the laser wavelength λ and at the Raman-shifted wavelengths of nitrogen and water vapor molecules; and C_1, C_{N2} and C_{H2O} are the system coefficients, which take into account the effective area of the telescope, the transmission efficiency of the optical train, and the detector quantum efficiency at the elastic and Raman-shifted wavelengths.

The principal advantage of the Raman lidar technique lies in having an additional signal from atmospheric gases (specifically nitrogen or oxygen) in addition to the conventional elastic signal. The backscatter coefficient from a particular molecule is proportional to the gas density with altitude. The nitrogen and oxygen density is known or can be calculated from temperature and pressure measurements, which in turn can be obtained from meteorological balloons or climatological data. The extinction coefficients at the emitted and Raman-scattered wavelengths in the exponent term of Eqs. (11.1), (11.2), and

(11.3), $\kappa_t(h, \lambda)$, $\kappa_t(h, \lambda_{N2,R})$, and $\kappa_t(h, \lambda_{H2O,R})$, are nearly the same if the Raman shift is not too large.

Although the discussion here primarily concerns the use of scattering from atmospheric nitrogen to determine attenuation coefficients, it is possible to use scattering from oxygen as well. Oxygen is well mixed and has constant concentration throughout the troposphere. The frequency shift from oxygen is two-thirds that of nitrogen, so the effects of differential attenuation are less important than for nitrogen. The GSFC Raman lidar has the ability to monitor the Raman-shifted signals from oxygen or nitrogen (Ferrare et al., 1998). Because the density of oxygen in the atmosphere is about one-fourth that of nitrogen and the cross section for oxygen is only 30% larger, the signal from oxygen is significantly smaller than the nitrogen signal. Because the signal quality and maximum range for a Raman lidar are limited by the low intensity of scattered light, using the oxygen signal limits the system capability further.

Similar to elastic lidar measurements, the signals from laser pulses are summed to increase the statistical significance of the measurements and to improve the signal-to-noise ratio. Because the signals in the Raman channel are extremely weak, photon counting is nearly always used and may require long summing times (commonly 5–10 min) to accumulate returns from high altitudes. One consequence of photon counting is the requirement of correcting the count rate in the near field for photons that are missed during the finite time (dead time) required to count each individual photon. When recording the first photon, the scalar is effectively "dead" or incapable of recording the second photon (Chapter 4). Near the Raman lidar (from 500 m to 1 km), where high counting rates occur, the corrections can be quite large. At long distances from the Raman lidar, this correction is negligible. There are well-established techniques for dead time correction developed by the nuclear instrumentation community. A summary of dead time correction techniques can be found in Knoll (1979), and detailed discussions of dead time and the necessary corrections can be found in Funck (1986) and Donovan et al. (1993).

The Raman technique makes it possible to make quantitative measurements of the spatial distribution of atmospheric molecular gases. The mixing ratio of any gas is the mass of gas divided by the mass of the dry air in a given volume. A combination of Eqs. (11.2) and (11.3) allows determination of the water vapor mixing ratio as a function of distance from the lidar. The value can be obtained from the ratio of the signal magnitude in the water vapor channel, $P_{H2O}(h)$, to the magnitude of the signal in the nitrogen channel, $P_{N2}(h)$, with the formula (Melfi, 1972)

$$q_w(h) = \frac{P_{H2O}(h)}{P_{N2}(h)} \times \left[\frac{C_{N2}\sigma_{N2}n_{N2}}{C_{H2O}\sigma_{H2O}n_{H2O}fr_{N2}} \right] \times \exp\left\{ \int_0^h [\kappa_t(h', \lambda_{N2,R}) - \kappa_t(h', \lambda_{H2O,R})] dh' \right\} \tag{11.4}$$

where fr_{N2} is the fractional N_2 content of the atmosphere (0.78084). Thus the water vapor mixing ratio at any altitude is given by the ratio of the magnitude of the signal in the water vapor channel to the magnitude of the signal in the nitrogen channel, a multiplicative constant (the part in square brackets), and an exponential correction due to difference in extinction between the nitrogen-shifted and water vapor-shifted wavelengths. The multiplicative constant can be determined by comparison of the lidar signal with radiosondes or by aiming the lidar horizontally and comparing to calibrated water vapor point sensors at various distances from the lidar. The technique can be applied to any molecular constituent.

In the study by Melfi (1972), comparison and calibration were accomplished by a weighted least-squares fit of the lidar mixing ratio to that from a balloon measurement. In early Raman studies, the exponential term was often ignored. To reduce the uncertainty due to the differential attenuation term, one can correct the data for molecular scattering. For this, a calculation has to be made of the molecular scattering transmission ratio, using either a standard-atmosphere model or radiosonde data for temperature and pressure. The effect of different particulate optical depth at two Raman-shifted wavelengths can be reduced by the method proposed later in studies by Ansmann et al. (1990, 1992), which is considered below.

The Raman signal in Eq. (11.2) can be inverted to obtain the total particulate extinction coefficients at the emitted and corresponding Raman wavelengths

$$\kappa_p(h,\lambda)+\kappa_p(h,\lambda_{N2,R})=\frac{d}{dh}\left\{\ln\left[\frac{n_{N2}(h)}{h^2 P_{N2}(h)}\right]\right\}-\kappa_m(h,\lambda)-\kappa_m(h,\lambda_{N2,R}) \tag{11.5}$$

Here $\kappa_m(h, \lambda)$ and $\kappa_m(h, \lambda_{N2,R})$, are the molecular extinction coefficients due to absorption and Rayleigh scattering at the laser wavelength and at the N_2 Raman scattered wavelength, and $\kappa_p(h, \lambda)$ and $\kappa_p(h, \lambda_{N2,R})$ are the same for the particulate coefficients. The inversion can be made only because the fractional amount of nitrogen at all points in the atmosphere is constant. Assuming an analytical dependence between $\kappa_p(h, \lambda)$ and $\kappa_p(h, \lambda_{N2,R})$ in the same manner as in the DIAL analysis technique, one can uniquely extract the particulate extinction coefficient (Ansmann et al., 1990 and 1992a; Ferrare et al., 1992)

$$\kappa_p(h,\lambda)=\frac{\dfrac{d}{dh}\left\{\ln\left[\dfrac{n_{N2}(h)}{h^2 P_{N2}(h)}\right]\right\}-\kappa_m(h,\lambda)-\kappa_m(h,\lambda_{N2,R})}{1+\left(\dfrac{\lambda}{\lambda_{N2,R}}\right)^{u}} \tag{11.6}$$

The molecular scattering coefficients are well known and can be found from Rayleigh scattering theory. For Raman lidars operating in the ultraviolet portion of the spectrum, the attenuation coefficients, $\kappa_m(h, \lambda)$ and $\kappa_m(h, \lambda_{N2,R})$, must include molecular absorption from ozone and possibly oxygen, depend-

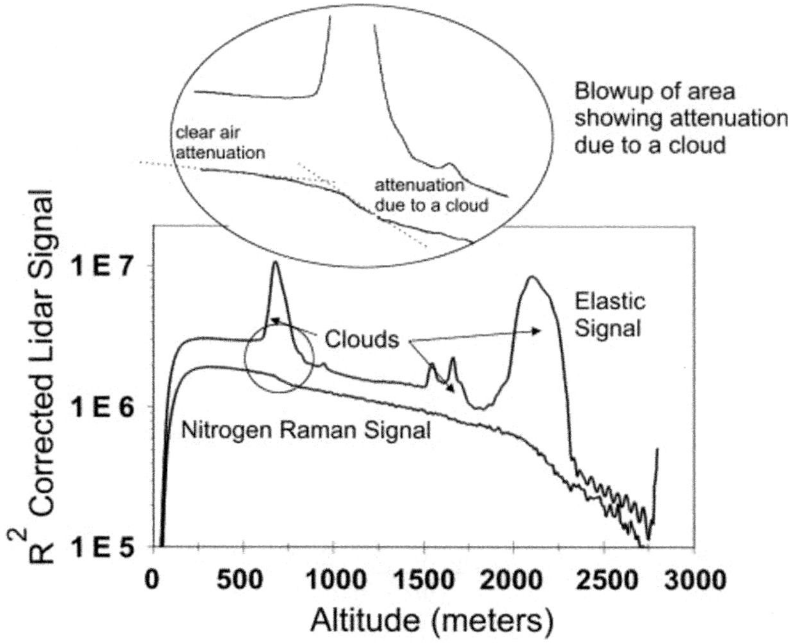

Fig. 11.5. An example of a profile showing the particulate and nitrogen signals in the presence of clouds. Although the N_2 signal clearly shows attenuation in the clouds, it also shows the limitation of the method. There is a small noise component that makes any use of a derivative method difficult.

ing on the wavelength. The ozone concentration profile must be measured or assumed from a standard atmosphere to apply Eq. (11.6) to determine the molecular extinction coefficients in the ultraviolet region. Figure 11.5 shows an example of a profile of the particulate and nitrogen signals in the presence of clouds.

The λ/λ_{N2R} term in the denominator of Eq. (11.6) corrects for the small difference in particulate attenuation between the laser and Raman-scattered wavelengths. This form of the equation assumes an exponential relationship for attenuation coefficients as a function of the wavelength (see Chapter 10). Over small wavelength differences, the relationship is commonly assumed to be linear, resulting in the constant u long being taken as unity. Atmospheric scattering theory establishes the range of the constant from 0 to 4 (Van de Hulst, 1957; McCartney, 1977). For particulates and water droplets with sizes comparable to the laser wavelength, larger values are valid, that is, for large droplets or ice particles, $u = 0$ is considered most appropriate. In horizontal directions under visibility from 3 to 120 km and for the total atmospheric column it is close to 1.3 (Curcio and Durbin, 1959; Angstrom, 1951; Müller et al. 2003). Measurements at the Department of Energy Southern Great Plains site have shown that u can vary from 0 to 2 (von der Gathen, 1995; Ferrare et al., 1998). Even negative values for u have been found (Valero and Pilewski, 1992). It should be stressed that an accurate estimate of the constant u from lidar experimental data requires extremely accurate lidar data. Moreover, an accurate estimate of u can only be made in a stationary scattering field. Inhomogeneous particulate or cloud layers within the lidar measurement range may significantly distort the data so as to obtain an erroneous value of u (Kovalev and McElroy, 1994). As was shown with the particulate corrections for DIAL measurements discussed in Chapter 10, the application of a wavelength dependence correction is questionable within any areas of particulate heterogeneity.

It is useful to demonstrate how cautious one must be when estimating values of u with experimental data. Assuming that the aerosol attenuation has a power law dependence with a constant Angstrom coefficient u as the exponent (Chapter 2), and the particulate extinction coefficients at two wavelengths, $\kappa_p(\lambda)$ and $\kappa_p(\lambda_1)$ are somehow determined, one can formally write the solution for u as,

$$u = \frac{\ln \kappa(\lambda_1) - \ln \kappa(\lambda)}{\ln\lambda - \ln \lambda_1} \tag{11.7}$$

The absolute uncertainty in the calculated parameter u can be easily derived through conventional uncertainty analysis methods. The estimate for the uncertainty is

$$\Delta u = \frac{1}{\ln\lambda - \ln \lambda_1}\sqrt{[\delta\kappa(\lambda)]^2 + [\delta\kappa(\lambda_1)]^2} \tag{11.8}$$

where $\delta\kappa_p(\lambda,)$ and $\delta\kappa_p(\lambda_1)$ are the relative uncertainties in the particulate extinction coefficients at the two wavelengths. Because the difference between λ, and λ_1 is small, the denominator of the term is small, resulting in a large value of the uncertainty in the calculated u. For example, if $\lambda = 248$ nm and $\lambda_{N2,R} = 262$ nm, then

$$\Delta u \approx 18\sqrt{[\delta\kappa(\lambda)]^2 + [\delta\kappa(\lambda_1)]^2} \tag{11.9}$$

Thus, if the extinction coefficients are determined with the accuracy of 5%, the absolute uncertainty $\Delta u = 1.26$, that is, the relative uncertainty is 126% for $u = 1$ and 64% for $u = 2$. This simple numerical example nicely shows that even accurate algorithms do not guarantee the practical usefulness of the corresponding measurement method unless it is backed up with a comprehensive uncertainty analysis.

The contribution to the uncertainty in the extinction coefficients determined with Eq. (11.6) can be especially significant for measurements in the near ultraviolet, where the wavelength difference between the laser wavelength and the Raman-scattered wavelength may be large. For a somewhat extreme example, for a XeF lidar at 351 nm, the shift is 32 nm and an uncertainty in u of ±1 results in a systematic uncertainty in the derived extinction coefficients on the order of 5%.

When the Raman technique is used, the particulate backscatter coefficient can be determined by reference to the signal at some height. The conventional assumption is made of the existence of a particulate-free area somewhere within the lidar measurement range. This is similar to the practice for elastic lidars that was considered in Section 8.1. The reference height, h_{ref}, is chosen to be a clear-air condition in which the particulate density is insignificant. Then the particulate scattering is negligible compared with molecular scattering, so that $\beta_{\pi,m} \gg \beta_{\pi,p}$. Combining Eqs. (11.1) and (11.2), and eliminating $\beta_{\pi,p}(h_{ref})$ as insignificant, one can obtain the following formula:

$$\begin{aligned}\beta_{\pi,p}(h,\lambda) = -\beta_{\pi,m}(h,\lambda) + \\ \beta_{\pi,m}(h_{ref},\lambda)\left[\frac{P_{N2}(h_{ref})P_{elastic}(h)n_{N2}(h)}{P_{elastic}(h_{ref})P_{N2}(h)n_{N2}(h_{ref})}\right] \\ \exp\left[-\int_{h}^{h_{ref}}[\kappa_t(h',\lambda)-\kappa_t(h',\lambda_{N2,R})]\,dh'\right]\end{aligned} \tag{11.10}$$

To solve Eq. (11.10) with experimental Raman lidar data, one should know or estimate the vertical profiles of the following parameters: (1) the air density; (2) the molecular scattering (backscattering) and absorption at wavelength λ; (3) the molecular scattering and absorption at wavelength λ_R; (4) the particulate scattering and absorption at λ; and (5) the constant term u that corrects for the difference in the particulate extinction at the Raman wavelength, thus making it possible to determine the term $\kappa_p(h, \lambda_{N2,R})$. The molecular backscattering term and air density can be estimated if the temperature and pressure are available at each altitude or can be estimated from a standard atmosphere. The basic difficulty is in obtaining accurate enough extinction coefficient profiles $\kappa_p(h, \lambda)$. This can be achieved by numerical differentiation using Eq. (11.6). The technique is often plagued by large errors, especially in areas of heterogeneous aerosol loading. Another difficulty is the uncertainty associated with choosing a particular altitude as the reference height. If all the above problems are successfully resolved, the profile of the particulate backscatter-to-extinction ratio can then be determined.

11.1.2. Limitations of the Method

Although the Raman method is significantly simpler in terms of the hardware and data processing than that for high-spectral-resolution lidars, discussed in Section 11.2, there are several limitations, all of which are a result of the small Raman scattering cross sections. Raman scattering cross sections are on the order of 10^3 smaller than for molecular scattering. This leads to the use of large-diameter receiving telescopes and, accordingly, to large-size lidar systems. All of the systems in use today are semitrailer sized. The large lasers and chillers used demand a great deal of power. Their size and power requirements limit their use in many situations where information on particulate properties might be of value, for example, in pollution control.

Also because of the small cross sections, Raman lidars are forced to resort to photon counting to achieve long ranges. The requirement for photon counting limits the use of these systems during hours of daylight. The use of laser and Raman-shifted wavelengths below 300 nm allows solar-blind operation, but the strong attenuation found in the ultraviolet severely limits the maximum range, requiring long averaging times to sound the entire troposphere. Daylight operation is essential for a method to be successful for use as a long-term operational method. This is because particulate and pollutant emissions from the surface are at a maximum during daylight hours and long-range transport both vertically and horizontally is maximized. In addition, because much of the development of Raman lidars is driven by the need to determine cloud properties and their effects on climate forcing, measurement of cloud properties during the day, and when other instruments can support, is required.

In the visible and near-visible portion of the spectrum, there are essentially two methods that can be used for optical filtration. The Raman spectrum has three major parts (Measures, 1984); a central line called the *Q* branch that contains the bulk of the signal, and two wings, the *O* and *S* branches. The shape of the *O* and *S* branches is sensitive to temperature, whereas the *Q* branch is insensitive to temperature. Different methods of filtration with respect to these wings can used. The first method of filtration uses a narrow band laser and an extremely narrow filter (0.3 nm) that isolates the *Q* branch (see, for example, Whiteman et al., 1992). The second method uses a filter that passes the entire rotational spectrum, so that temperature effects in the shape of the *O* and *S* branches are minimized (Whiteman et al., 1993). The spectral widths of the Raman *O* and *S* branches are relatively broad so that wide filters (on the order of 3–5 nm) must be used. Filters that do not either exclude the *O* and *S* branches or include all of the *O* and *S* branches will be temperature dependent. For daylight operation, the use of the narrow filter technique limits the amount of sunlight contamination of the signal. However, because this method requires the use of a spectrally narrow laser line, this limits the use of excimer lasers because of their wide lasing linewidths or multiple lasing lines.

Even with the use of a narrow filter, the field of view of the telescope must be narrowed to minimize the solar background. The intensity of the background solar radiation is proportional to the square of the telescope divergence angle. Reducing the telescope field of view, particularly in photon-counting systems, makes aligning the laser and telescope difficult. It also reduces, but does not eliminate, solar photons. Subtraction of these background photons becomes difficult as the number of photons becomes small because of statistical or counting errors. Narrow field of view systems have been built, but this technique is difficult to use in practice, even for elastic systems that have a factor of at least 1000 times more photons available to use than the Raman method.

11.1.3. Uncertainty

The uncertainty in the extinction coefficient obtained with the Raman technique may be very large under unfavorable conditions. To our knowledge, there has never been a rigorous presentation of the uncertainty associated with the Raman measurements based on a comprehensive theoretical analysis similar to that made by Russel et al. (1979) for elastic lidar measurements. Some exceptions can be mentioned, for example, the studies by Ansmann et al. (1992) and Whiteman (1999). In the analysis by Ansmann et al. (1992), three sources of uncertainties were presented that determine the uncertainty of the particulate properties calculated with the Raman technique. These are: a statistical uncertainty caused by photon or signal noise, a systematic uncertainty from errors in the input parameters, and uncertainty associated with procedures such as signal averaging. Statistical uncertainty associated with the use of a finite number of photon counts is estimated with Poisson statistics in which the standard error in the estimate is the square root of the number of photon counts. In the study by Whiteman (1999), an analysis of uncertainty specific to DIAL and Raman measurements was made with statistical analysis techniques. One should stress that, similar to the DIAL measurement technique, in Raman data processing large errors may occur when the derivative of the logarithm of the ratio of two quantities is calculated with Eq. (11.6). As shown in Chapter 10, no generally accepted method exists for numerical differentiation of lidar data. The evaluation of the derivative of the experimental data corrupted with random noise and unknown systematic distortions may produce a significant measurement uncertainty. In other words, the quantities regressed are often not normally distributed and no rigorous or accepted methods exist to evaluate the actual measurement uncertainty.

Sources of systematic uncertainty are primarily associated with uncertainties in the estimates of supporting parameters such as the temperature, pressure, and ozone density at a given altitude and the value of the wavelength parameter u [Eq. (11.6)]. Of these, the most significant is the uncertainty associated with the temperature gradient. Uncertainty in the temperature gradient affects the derivative of the molecular number density term in Eq. (11.6), $d/dh[\ln n_{N2}(h)]$. In the absence of strong temperature gradients, the amount of uncertainty due to pressure and temperature uncertainties is small. Ansmann et al. (1992) estimated the uncertainty for a combined error of 10 K and 1 kPa. The uncertainty in the extinction coefficient proved to be on the order of 5%. However, the uncertainty in the extinction coefficients can approach 50% in regions where a sharp temperature gradient occurs, such as those usually associated with inversion layers. The magnitude of the uncertainty decreases as the smoothing window used to calculate the attenuation coefficient increases. If the estimated ozone concentration is in error by a factor of 2, the resulting uncertainty in the attenuation coefficient is about 7% at 308 nm, being larger at shorter wavelengths and smaller at longer. Uncertainties in the wavelength parameter *u* are normally not large for short wavelength systems where the laser and Raman-scattered wavelengths are nearly the same. At 308 nm, a 100% uncertainty in *u* may result in an uncertainty in the extinction coefficient of less than 4%. But for systems at 350 nm and longer, it may become significant. For normal operations, a total systematic uncertainty is estimated to be on the order of 5–10%.

The uncertainties in measurements from standard meteorological instruments should not be underestimated or minimized. The uncertainty of measurements of water vapor from balloons has long been a source of concern for water vapor Raman measurements because the lidars commonly calibrate to them. Capacitance hygrometers are now known to give erroneous results when the relative humidity is less than 20% (see, for example, Ferrare et al., 1995). Errors in standard radiosondes have been studied and relatively well quantified by the meteorological community (Luers, 1990; Wade, 1994; Connell and Miller, 1995). The uncertainty from radiosonde measurements propagates into the lidar measurements through the calculation of molecular density and scattering.

Signal averaging is necessary in photon-counting lidars to obtain sufficient statistical significance of the signal at some desired altitude. Significant errors can be introduced if the optical properties of the cloud change during the measurement period, which may be long. In the study by Ansmann et al. (1992), the measurement time was 12 and 26 min for the profiles shown as examples. The correct extinction coefficient in a given range element is obtained only if the optical properties are constant inside the range element over the measurement period and the changes in optical depth between the lidar and the range element are small. The magnitude of the fractional uncertainty increases with increasing optical depth. For cirrus clouds, uncertainties on the order of 10% in the lower portion of the cloud and 30% in the upper portion of the cloud should be expected. Uncertainties of this type can be reduced by recording the data with high temporal resolution and then separating the data into intervals with similar extinction conditions, evaluating the extinction coefficients for each of these intervals, and then averaging the result. Theoretically, averaging the logarithms of the derivatives of the signals should lead to the correct average extinction coefficient. However, because the individual signals are noisy and derivatives of noisy signals tend to emphasize the noise, new types of uncertainty are introduced and little improvement is obtained by this method (Theopold and Bosenberg, 1988).

The methods described above assume single scattering of the returning photons. However, in perhaps the most useful situations, the examination of dense particulate concentrations and clouds, the effects of multiple scattering must be considered. Similar to elastic scattering situations, multiply scattered photons reentering the telescope field of view artificially increase the magnitude of the signal received by the telescope. The degree of influence of multiple scattering is related to the size of the volume being examined, distance to the scattering volume, and optical density and particle size in the volume. Although the divergence of the lasers and fields of view of the receiver optics are generally narrowed to reduce the examined volume, little can be done about the distance to the scatterer and optical depth of the volume. Wandinger (1998) studied the effects of multiple scattering in high-spectral-resolution and Raman lidars and concluded that the effects of multiple scattering can implement large measurement errors. Although this observation is true for both the elastic and Raman-shifted signals, it is more significant for the latter. Obviously, multiple-scattering effects are most significant in the presence of heterogeneous particulate layers, such as cirrus clouds. The largest errors are found in the extinction coefficients at the base of clouds and are as large as 50%.

Although the uncertainty of the extinction coefficients determined with the Raman method may be significant in some situations, the Raman method has been shown to be superior to conventional elastic inversion methods like the Klett method (Ansmann et al., 1992; Mitev et al., 1992; Ansmann et al., 1991). However, one must be cautious with such general conclusions because the elastic and Raman methods have different situations in which they may be favorably applied. What is more, different methodologies can be used to process elastic lidar data, each of which may yield different accuracies for the measured extinction coefficients. Obviously, the results of such comparisons strongly depend on the measurement objectives, particular atmospheric conditions, the method of elastic data processing, and the investigator's skill.

11.1.4. Alternate Methods

The requirement to link the extinction coefficients between two different wavelengths is considered to be a weak point in the analysis of particulate properties with the Raman technique. The problem is that, unlike the elastic signal that contains two one-way transmission terms at the laser wavelength, the Raman signal has two different one-way transmission terms, one at the laser wavelength, and one at the Raman-shifted wavelength. The value of the exponent u in Eq. (11.6) is generally unknown and has to be selected a priori.

Perhaps the simplest and most easily achievable (at least, theoretically) method to overcome this obstacle was presented by Cooney (1986, 1987), who suggested that systems be built that can detect Raman scattering from atmospheric oxygen simultaneously with that from nitrogen. Two equations can be written using the two Raman-scattered signals, based on Eq. (11.5):

$$\begin{aligned}
\kappa_p(h,\lambda)+\kappa_p(h,\lambda_{N2,R}) &= \frac{d}{dh}\left\{\ln\left(\frac{n_{N2}(h)}{h^2 P_{N2}(h)}\right)\right\} - \kappa_m(h,\lambda) - \kappa_m(h,\lambda_{N2,R}) \\
\kappa_p(h,\lambda)+\kappa_p(h,\lambda_{O2,R}) &= \frac{d}{dh}\left\{\ln\left(\frac{n_{O2}(h)}{h^2 P_{O2}(h)}\right)\right\} - \kappa_m(h,\lambda) - \kappa_m(h,\lambda_{O2,R})
\end{aligned} \tag{11.11}$$

Now assuming that u is constant in the wavelength range that includes all three wavelengths, laser wavelength λ, and Raman shifted, λ_{N2R} and λ_{O2R}, the relationship can be written

$$\kappa_p(\lambda_j)\lambda_j^u = const. \tag{11.12}$$

which is valid for the above range. Now there are three wavelengths, all related by the relations

$$\kappa_p(h,\lambda)\lambda^u = \kappa_p(h,\lambda_{N2,R})\lambda_{N2,R}^u = \kappa_p(h,\lambda_{O2,R})\lambda_{O2,R}^u \tag{11.13}$$

which provides two more equations, so that the four unknowns, $\kappa_p(z, \lambda_l)$, $\kappa_p(z, \lambda_{N2,R})$, $\kappa_p(z, \lambda_{O2,R})$, and u have unique solutions. The reliability of this solution depends on the validity of the assumption given in Eq. (11.12), which requires that $u = const.$ over the entire distance being examined and over the entire range of values that the extinction coefficient may assume. Unfortunately, the value of the coefficient u may not be constant throughout the particular measurement conditions, and there is currently no method to check the validity of Eq. (11.13) without additional measurements. The second limitation of this method is the degree to which the molecular extinction, especially the molecular absorption coefficients, is known. The presence of ozone in the near-ultraviolet region is the biggest contributor to this uncertainty.

Adding a capability for simultaneous detection of oxygen adds some degree of complexity to the system but in comparison to any of the other methods is simple and cost effective. On the other hand, photon count rates from oxygen are even lower than for nitrogen and require longer averaging periods. Lower count rates and longer averaging times have a number of uncertainty sources associated with them. Any additional information retrieved from an additional instrumental measurement is always followed by an additional uncertainty contribution. In the above case, the additional signal measured at the oxygen-shifted wavelength λ_{O2R} has some nonzero noise component, which influences the total measurement accuracy.

The signal-to-noise ratio, related to the magnitude of the oxygen signal, is range dependent. Accordingly, at some range from the lidar, the benefit from this additional information is overwhelmed by the uncertainty contribution. The long-term question is whether the expected theoretical improvement of the measurement accuracy exceeds the accuracy worsening due to presence of the additional uncertainty source. At least, an estimate of the range over which the actual improvement is achieved is required.

A number of methods have been proposed for the unambiguous determination of the extinction coefficient by combining simultaneous Raman and elastic component measurements. The most typical approach was first outlined by Cooney (1987) as a means of determining the ozone concentration. Then

Mitchenkov and Solodukhin (1990) proposed to build a simple Raman system that emits one additional laser beam at the nitrogen Raman scattering wavelength. In addition to the Raman-shifted signal from the primary laser, an elastic lidar signal can also be obtained that has the same transmission term at the Raman-scattered wavelength. Such a system provides simultaneous measurement of the Raman- and two elastically scattered returns. The idea was developed later by Moosmüller and Wilkerson (1997), who proposed diverting the laser beam every other pulse through a nitrogen cell to generate a beam at the Raman-scattered wavelength. For ground-based, vertically pointing lidars, the equations for the elastically scattered signals at laser wavelength λ and Raman-shifted wavelength, denoted here as λ_R, are

$$P_{\text{elastic}}(h) = \frac{C_1 E[\beta_{\pi,p}(h,\lambda) + \beta_{\pi,m}(h,\lambda)]\exp\left[-2\int_0^h \kappa_t(h',\lambda)\,dh'\right]}{h^2} \tag{11.14}$$

and

$$P_{\text{N2elastic}}(h) = \frac{C_3 E[\beta_{\pi,p}(h,\lambda_R) + \beta_{\pi,m}(h,\lambda_R)]\exp\left[-2\int_0^h \kappa_t(h',\lambda_R)\,dh'\right]}{h^2} \tag{11.15}$$

The Raman-shifted signal at λ_R is

$$P_{\text{N2R}}(h) = \frac{C_2 E n_{\text{N2}}(h)\,\sigma_{\text{N2}} \exp\left\{-\int_0^h [\kappa_t(h',\lambda) + \kappa_t(h',\lambda_R)]\,dh'\right\}}{h^2} \tag{11.16}$$

Multiplying the two elastic signals and dividing by the square of the nitrogen Raman-scattered signal removes all of the transmission terms, leaving

$$\frac{P_{\text{elastic}}(h)P_{\text{N2elastic}}(h)}{[P_{\text{N2,R}}(h)]^2} = \frac{C_1[\beta_{\pi,p}(h,\lambda) + \beta_{\pi,m}(h,\lambda)]\,C_3[\beta_{\pi,p}(h,\lambda_R) + \beta_{\pi,m}(h,\lambda_R)]}{[C_2\beta_{\text{N2,R}}(h)]^2} \tag{11.17}$$

where $\beta_{\text{N2,R}}(h) = n_{\text{N2}}(h)\sigma_{\text{N2}}$. Equation (11.17) can be rearranged to obtain

$$\begin{aligned}&[\beta_{\pi,p}(h,\lambda) + \beta_{\pi,m}(h,\lambda)][\beta_{\pi,p}(h,\lambda_R) + \beta_{\pi,m}(h,\lambda_R)] \\ &\quad = [\beta_{\text{N2,R}}(h)]^2\left[\frac{C_2^2}{C_1 C_3}\right]\left[\frac{P_{\text{elastic}}(h)P_{\text{N2elastic}}(h)}{[P_{\text{N2,R}}(h)]^2}\right]\end{aligned} \tag{11.18}$$

The square root of the left side of Eq. (11.18) can be viewed as the geometric mean of the backscatter coefficients at the laser and the nitrogen Raman wavelengths. Although this parameter is not a normal characteristic used to categorize particulates or clouds, Moosmüller and Wilkerson (1997) show that profiles of this parameter can be determined with good accuracy. Using a power law wavelength dependence for backscattering, as proposed by Browell et al. (1985) (see Chapter 10), one can transform Eq. (11.18) into a form that makes it possible to determine the backscatter coefficients at both wavelengths. The advantage of this method is that all of the values on the right side are known or can be determined. The product of the three lidar coefficients, C_1, C_2, and C_3, form a constant coefficient that needs to be determined only once and can be found by using the assumption of the existence of a region of the atmosphere free of particulates. The problem remains of the validity of the assumed altitude-independent power law wavelength dependence for backscattering.

To avoid the necessity of assuming that $u = const.$, some curious (even exotic) methods have been proposed. Generally, these methods have a simple theoretical basis. However, their practical value is often questionable because of the complexity of the required hardware and because of a lack of appropriate estimates of the amount of measurement accuracy gained by their use. The principle of a unambiguous determination of the extinction coefficient outlined by Cooney (1987) was developed by Van der Gathen (1995), who suggested a lidar system that emits three collimated beams at different wavelengths. The method is based on the fact that the Raman shift from oxygen is almost exactly two-thirds of the spectral shift from nitrogen Raman scattering. Accordingly, three laser wavelengths are used such that the second wavelength produces a nitrogen signal at the same wavelength as the oxygen signal from the first. Similarly, the third laser line is chosen so that the Raman oxygen signal is at the same wavelength as the first laser. This creates a chain of six wavelengths separated by $\Delta v = 777.5\,\text{cm}^{-1}$, of which three wavelengths have two signals, elastic and Raman shifted (Fig. 11.6). The vertical profile of the total extinction coefficient at the wavelength λ_v, obtained from the elastic-channel signal, can be written as the function of the altitude h in the form

$$\kappa_t(h, \lambda_v) = \kappa_p(h, \lambda_v) + \kappa_m(h, \lambda_v) = \frac{1}{2\beta_\pi(h, v)} \frac{d\beta_\pi(h, v)}{dh} - \frac{1}{2} \frac{d\Phi(h, v)}{dh} \tag{11.19}$$

where $\Phi(h) = \ln[P(h)h^2]$. The extinction coefficient defined with Eq. (11.19) is obtained at three wavelengths. There are also six extinction coefficients at four Raman-shifted wavelengths that can be written as

$$\kappa_t(h, \lambda_{v-3}) + \kappa_t(h, \lambda_v) = \frac{1}{n(h)} \frac{dn(h)}{dh} - \frac{d\Phi(h, v-3)}{dh} \tag{11.20}$$

Subtraction of these equations eliminates the terms having to do with changes in molecular density, so that the differences are related only to changes in the

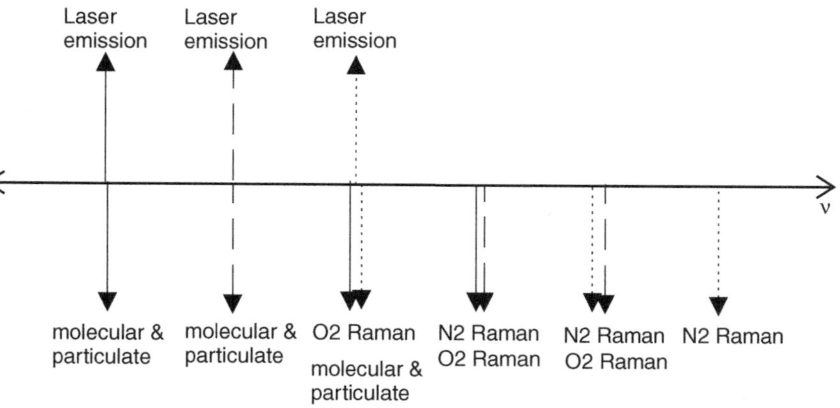

Fig. 11.6. The elastic and Raman-shifted returns from three laser emissions spaced 777.5 cm^{-1} apart creates a chain of returns spaced at the same interval. Because of the overlap of the elastic and Raman returns, the extinction coefficients at the wavelengths can be determined uniquely (Gathen, 1995).

particulate densities. This set of equations can be solved with matrix methods to give the extinction coefficients at the four wavelengths. One of these wavelengths is also one of the elastic wavelengths, so that the backscatter coefficients can be determined.

It should be noted that the practical application of this method is difficult. Laser wavelength shifting in the ultraviolet portion of the spectrum is difficult and inefficient. However, despite all the problems, the conclusion can be made that the combination of elastic and inelastic Raman measurements may provide a notable improvement in the accuracy of the measured atmospheric parameters (Donovan and Carswell, 1997). Fundamental limitations apply that limit the accuracy and usefulness of the technique. Perhaps the most serious is the requirement for long averaging times and the natural variability of the atmosphere, which challenge the homogeneity assumptions inherent in the technique.

11.1.5. Determination of Water Content in Clouds

A significant contribution was made by Whiteman and Melfi (1999) with the addition of a capability to determine the liquid water content, the mean droplet radius, and the number density of cloud water droplets. This ability stems from the fact that Raman scattering from a collection of water droplets is proportional to the total amount of water present. The Raman spectrum from liquid water is shifted in the range of 2800–3800 cm^{-1} from the exciting wavelength. This overlaps the region in which water vapor is detected (a shift of 3420–4140 cm^{-1}) with Raman lidar. Thus it is not surprising that excess

Raman scattering was noticed in clouds in the form of lidar returns that indicated water vapor concentrations in excess of saturation (Melfi et al., 1997).

The determination of the liquid water concentration is difficult because of the overlap with the water vapor Raman shift and because of the temperature dependence of the liquid water Raman cross section. Whiteman and Melfi (1999) determined the liquid water content as that amount of water vapor in excess of the saturation amount. Whiteman et al. (1999) identified an isobestic point in the Raman liquid water spectrum. This is a wavelength at which the amplitude of the Raman cross section is constant with temperature; a measurement made at that wavelength will not be temperature dependent. This wavelength is located at a shift of 3425 cm^{-1} from the laser line. A narrow filter of about 100 cm^{-1} full-width half-maximum will isolate this portion of the spectrum with a negligible contribution from water vapor. The particulate backscatter ratio defined as

$$R_\beta(h) = \frac{\beta_{\pi,p}(h) + \beta_{\pi,m}(h)}{\beta_{\pi,m}(h)}$$

can be determined from the elastic and inelastic lidar signals as

$$R_\beta(h) = C_4\left[\frac{P_{\text{elastic}}(h)}{P_{\text{N2,R}}(h)}\right]\exp\left\{2\int_0^h [\kappa_t(h',\lambda) - \kappa_t(h',\lambda_{\text{N2,R}})]dh'\right\} \quad (11.21)$$

where C_4 is a constant that can be determined in a region of the atmosphere free of particulates. Accordingly, the particulate backscatter coefficient in a cloud may be found with $R_\beta(h)$ as

$$\beta_{\pi,p}(h) = \beta_{\pi,m}(h)[R_\beta(h) - 1] \quad (11.22)$$

The droplet size distribution in a cloud can be assumed to be described by a gamma distribution. Specifically, a Khrgian–Mazin distribution (Khrgian, 1963; Pruppacher and Klett, 1997) has been shown to be a good representation for real clouds. It can be written as

$$n(a) = \frac{27}{2}\frac{N}{\bar{a}^3}a^2\exp\left(-3\frac{a}{\bar{a}}\right) \quad (11.23)$$

where N is the total number of droplets per cubic centimeter, and $\bar{a}$ is the average droplet radius. Combining the droplet distribution with the volume of each droplet and water density, the cloud liquid water content can be written as

$$w_L(g/m^3) = 10^6\left(\frac{4\pi}{3}\right)\rho_w\int_0^\infty a^3 n(a)da \quad (11.24)$$

where w_L is the liquid water content of the cloud, ρ_w is the density of water, and $n(a)da$ is the number of droplets per cubic centimeter with radii ranging between a and $a + da$.

Performing the integration and solving for the total number of droplets, N, one obtains

$$N = \frac{27}{80\pi} 10^{-6} \frac{w_L}{\rho_w \bar{a}^3} \tag{11.25}$$

Once the lidar has determined the value of w_L, then the product $(N\bar{a})$ is known. This provides one constraint on the problem. The other constraint is provided by the backscatter intensity from the cloud droplets. The backscatter coefficient for the cloud droplets can be found from Mie scattering theory as

$$\beta_{\pi,p} = \int_0^\infty n(a)\sigma_{\pi,p}(a)da \tag{11.26}$$

where $\sigma_{\pi,p}(a)$ is the cross section for particulate backscattering. Using the expression for the cloud backscatter and the Khrgian–Mazin distribution for the cloud droplets, one can obtain an expression for the backscatter coefficient as a function of the liquid water content and average droplet size

$$\beta_{\pi,p} = \beta_{\pi,m}(R_\beta - 1) = \frac{729}{160\pi} \frac{10^{-6}}{\bar{a}^3} \frac{w_L}{\rho_w} \int_0^\infty a^2 \exp\left(-3\frac{a}{\bar{a}}\right) \sigma_{\pi,p}(a)da \tag{11.27}$$

This equation must be solved at each range increment inside the cloud. This requires an iterative method to determine the value of $\bar{a}$. Once $\bar{a}$ is determined, then the cloud number density can be found with Eq. (11.25). Whiteman and Melfi calculated this integral over the range of 0.06–100 μm with a step size of 0.001 μm. In the Mie calculations, the index of refraction for pure water was used for the droplets.

11.2. RESOLUTION OF PARTICULATE AND MOLECULAR SCATTERING BY FILTRATION

11.2.1. Background

In Chapter 2, elastic scattering was defined as a process in which scattering from molecular and particulate scatterers occurs at the same wavelength of the incident light, that is, at the emitted laser wavelength. However, the actual backscatter from molecules and particulates in the air is always slightly shifted in wavelength from the wavelength of the emitted light. This is because of Doppler broadening of the reflected light caused by the motion of the molecules and particulates. Let us assume that a source of electromagnetic radiation is composed of a single frequency, ν, and the scattering particle is in motion with respect to the source. An observer located with the source will detect the elastic scattered light not at ν, but at a shifted frequency, ν'. For remote-sensing situations, in which the source of light and receiver are collocated and for which the velocity of the scatterer is much less than the speed of light, c, the Doppler shift for a single scatterer becomes

$$\frac{\nu'}{\nu} = 1 \pm 2\frac{V}{c} \tag{11.28}$$

where V is the component of the velocity along the line between the lidar and the scatterer. The plus sign is used when the scatterer is moving toward the lidar and the minus sign when it is receding.

Molecules and particulates in the atmosphere may be assumed to have Maxwellian velocity distribution. It can be shown that this produces a continuous intensity profile as a function of frequency. The scattered light returning to the lidar will have a continuous Gaussian-shaped profile. The width of this profile corresponds to a characteristic frequency shift of $\Delta\nu = \nu - \nu'$, that is proportional to the quantity

$$\Delta\nu \propto \frac{\nu}{c}\left(\frac{2kT}{m}\right)^{1/2} \tag{11.29}$$

where k is the Boltzmann constant, T is the absolute temperature of the scatterers, and m is the mass of the scatterers (the molecules or particulates). In practice, the laser line has a finite width so that the actual intensity distribution is a convolution of the laser intensity profile and a Gaussian profile.

It may be assumed that the molecules and particulates are in thermal equilibrium and have the same temperature T. The scattered light from molecules will be distributed over a spectral width on the order of 2 pm as shown in Fig. 11.7. However, the mass of the particles is sufficiently larger than that of the molecules so that their thermal velocity is small, and thus the scattered light spectrum from particulates is essentially unbroadened. More precisely, the width of the Doppler broadening due to motion of the particulate scatterers is generally smaller than the line width of the laser and is therefore insignificant. The total "elastic" signal is actually the sum of the two components (Fig. 11.7). A more complete discussion of the spectra of scattered light in the atmosphere can be found in the study by Fiocco and DeWolf (1968).

A high-spectral-resolution lidar (HSRL) separates the two components, resolving the contribution from particulates from the contribution from molecules. Therefore, the particulate attenuation coefficient can be determined without the backscatter-to-extinction ratio taken a priori. The technique was first suggested in a paper by Schwiesow and Lading (1981) and first demonstrated by Shipley et al. (1983) and Sroga et al. (1983).

11.2.2. Method

For a two-component atmosphere, the elastic lidar equation can be written as [Eq. (3.11), Chapter 3]

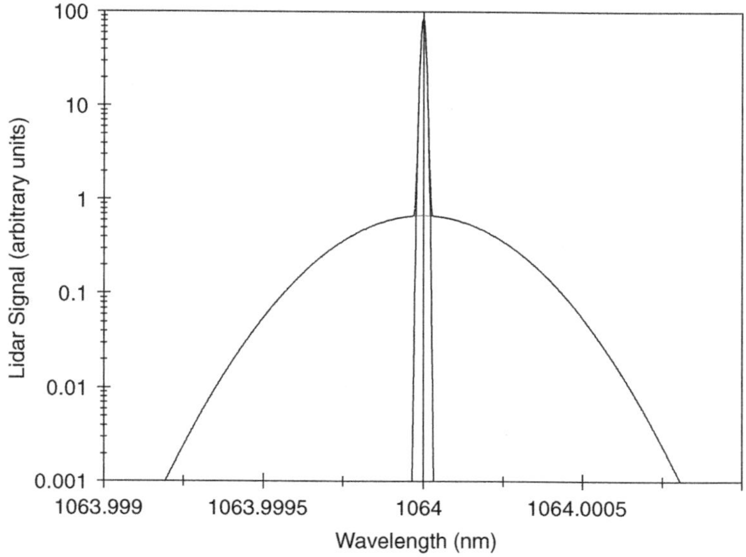

Fig. 11.7. Plot showing the spectral distribution of elastically scattered light from particles (the narrow distribution) and from molecules (the wide distribution).

$$P(r) = \frac{C_0[\beta_{\pi,p}(r) + \beta_{\pi,m}(r)]\exp\left[-2\int_0^r \kappa_t(r')dr'\right]}{r^2} \tag{11.30}$$

where C_0 is a system constant. An HSRL includes hardware elements to enable it to measure molecular and particulate backscatter separately. Thus, instead of Eq. (11.30), two equations for the quantity of light obtained from molecular and from particulate backscattering can be written. These two equations are

$$P_{\text{molecular}}(r) = \frac{C_{0,1}\beta_{\pi,m}(r)\exp\left[-2\int_0^r \kappa_t(r')dr'\right]}{r^2} \tag{11.31}$$

and

$$P_{\text{particulate}}(r) = \frac{C_{0,2}\beta_{\pi,p}(r)\exp\left[-2\int_0^r \kappa_t(r')dr'\right]}{r^2} \tag{11.32}$$

Note that these equations are coupled by the same attenuation term; however, the equation constants are different. This is because of different hardware elements used to discriminate between the molecular and particulate signals. The molecular backscattering coefficient, $\beta_{\pi,m}(r)$, is a function of the air density, which can be calculated with a measured atmospheric temperature profile. Accordingly, Eq. (11.31) can be inverted to obtain a unique value for the total extinction coefficient, $\kappa_t(r)$ at every range r

$$\kappa_t(r) = -\frac{1}{2}\left\{\frac{d}{dr}\ln[r^2 P_{molecular}(r)] - \frac{d}{dr}\ln[\beta_{\pi,m}(r)]\right\} \tag{11.33}$$

At high altitudes, the difference between the signals $P_{molecular}(r)$ at sequential altitudes is small and the uncertainty in $\kappa_t(r)$ may be relatively large. The error may be reduced by taking this derivative over longer distances that may increase with altitude, in a manner similar to that used in the DIAL measurements (Chapter 10). This is also done to increase the magnitude of the differences in the logarithm of the product $[r^2\, P_{molecular}(r)]$. In other words, the changes in the product $[r^2\, P_{molecular}(r)]$ between the two ranges used to estimate the derivative must be measurable and statistically significant. This is necessary to yield a meaningful value of attenuation, especially in the presence of noise in the system. The atmospheric molecular backscatter coefficient can be determined as

$$\beta_{\pi,m}(r)(\text{cm steradian})^{-1} = 3742.8\left[\frac{p(r)(\text{kPa})}{T(r)(\text{K})}\right]\left[\frac{1}{\lambda^4(\text{nm})}\right] \tag{11.34}$$

where $p(r)$ is the atmospheric pressure and $T(r)$ is the atmospheric temperature at a distance r from the lidar. The lidar backscatter ratio, defined to be the ratio between the particulate lidar return and the molecular return, can be found from the ratio of Eqs. (11.32) to (11.31)

$$R^*_\beta(r) = \frac{\beta_{\pi,p}(r)}{\beta_{\pi,m}(r)} = \left[\frac{C_{0,1}}{C_{0,2}}\frac{P_{particulate}(r)}{P_{molecular}(r)}\right] \tag{11.35}$$

From the backscatter ratio, the particulate backscatter coefficient can be found as

$$\beta_{\pi,p}(r) = R^*_\beta(r)\beta_{\pi,m}(r) = \left[\frac{C_{0,1}}{C_{0,2}}\frac{P_{particulate}(r)}{P_{molecular}(r)}\right]\beta_{\pi,m}(r) \tag{11.36}$$

The particulate backscatter-to-extinction ratio can be calculated by substituting Eq. (11.36) in Eq. (5.17)

$$\Pi_p(r) = \frac{\beta_{\pi,p}(r)}{\kappa_p(r)} = \frac{\beta_{\pi,m}(r)}{\kappa_p(r)}\left[\frac{C_{0,1}}{C_{0,2}}\frac{P_{particulate}(r)}{P_{molecular}(r)}\right] \tag{11.37}$$

The analysis above assumes that the molecular and particulate signals have been completely separated. However, what is actually measured by an HSRL is a linear superposition of each of the two signals in two channels. There is always some component of each signal measured in the other channel. Therefore, the measured signal intensities, $MS_{\text{molecular}}(r)$ and $MS_{\text{particulate}}(r)$, in each channel are a linear combination, so that

$$\begin{aligned} MS_{\text{molecular}}(r) &= \gamma[C_{\text{pm}} P_{\text{particulate}}(r) + C_{\text{mm}} P_{\text{molecular}}(r)] + P_{\text{bgr,molecular}} \\ MS_{\text{particulate}}(r) &= \gamma[C_{\text{pp}} P_{\text{particulate}}(r) + C_{\text{mp}} P_{\text{molecular}}(r)] + P_{\text{bgr,particulate}} \end{aligned} \quad (11.38)$$

where C_{mm} is the fraction of the molecular scattering that is detected in the molecular channel, C_{pm} is the fraction of the particulate scattering that penetrated into the molecular channel, C_{pp} is the fraction of the particulate scattering that is detected in the particulate channel, and C_{mp} is the fraction of the molecular scattering that penetrated into the particulate channel. The factor γ is the lidar system photon efficiency, and $P_{\text{bgr,molecular}}$ and $P_{\text{bgr,particulate}}$ are the background signal counts in each channel. The background counts can be determined from the average number of counts found in channels far beyond those in which scattered photons from the laser are expected. Factors C_{pm}, C_{mm}, C_{pp}, and C_{mp} are calibration coefficients that must be measured to determine $P_{\text{molecular}}(r)$ and $P_{\text{particulate}}(r)$. Note that $C_{0,1}$ and $C_{0,2}$ need not be known separately, only the ratio of $C_{0,1}$ to $C_{0,2}$ must be determined to use the data in the analysis method above [Eqs (11.35)–(11.37)]. These coefficients must be determined to accuracies on the order of 0.01% because the magnitude of the particulate-scattering signal detected in the molecular channel can be as much as a thousand times larger than the magnitude of the molecular signal in the molecular channel when examining clouds (Piironen and Eloranta, 1993).

11.2.3. Hardware

An HSRL with a wavelength of 532 nm was developed at the University of Wisconsin (UW) (Grund and Eloranta, 1991). For this wavelength, the single-scattering albedo is close to unity for water and ice clouds and some particulates. The early versions of the HSRL, discussed in this section, used a high-resolution étalon to separate the particulate and molecular backscatter signals. This étalon has a 0.5-pm bandpass. The transmission through the étalon varies as the angle of incidence is changed. Light scattered from different ranges is focused by the receiving telescope at different points and thus enters the étalon at slightly different angles. To reduce this effect, the backscattered light collected by the receiver telescope is sent through a fiber-optic scrambler. This scrambler reduces the range dependence of the étalon transmission due to the angular sensitivity of the étalon (Grund and Eloranta, 1991). Without the fiber-optic scrambler, the calibration coefficients in Eq. (11.38) are range dependent. Figure 11.8 shows the layout of the UW HSRL system. To reduce the background solar radiance in daytime measurements, the incoming light was prefiltered with an interference filter and a pair of low-

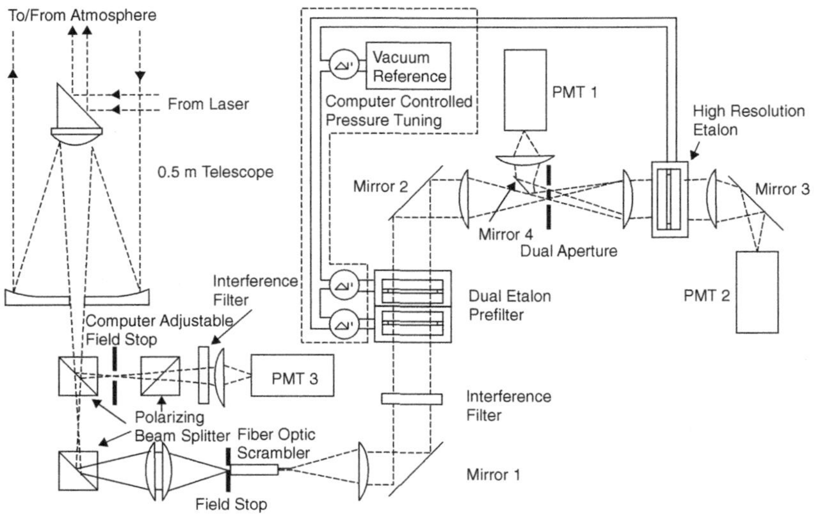

Fig. 11.8. The layout of an étalon-based high-spectral-resolution lidar. The backscatter signal is collected with a telescope. The separation between particulate and molecular backscatter signals is done with the high-resolution étalon. The light passing through the system is detected with PMT1 and PMT2 (Grund and Eloranta, 1991).

resolution étalons. After passing a dual aperture, the light was directed to a pressure-tuned, high-resolution étalon. The high-resolution étalon is tilted with respect to the optical axis so that light that does not pass through the étalon is reflected back through the dual aperture and then to the molecular channel photodetector (PMT1). The light that passes through the high-resolution étalon is directed to the particulate channel photodetector (PMT2). The étalons are not perfect filters and are not able to completely separate the two signals. Thus the signal in the particulate channel contains a contribution from the center of the molecular backscatter spectrum. Likewise, the signal in the molecular channel contains a contribution of light from the part of the particulate backscatter spectrum that did not pass through the high-resolution étalon. Because of the low power output of the laser and relatively low receiver transmission, photon counting is required. However, by using photon counting, signals can be obtained with over four decades of dynamic range. The averaging time required to profile a cloud with an optical depth of 1 at a distance of 8 km is approximately 1 min (Grund and Eloranta, 1991). Basic parameters of the system are presented in Tables 11.1.

The laser used by an HSRL must be line narrowed, which for a Nd:YAG laser generally requires injection seeding. The laser used in the UW HSRL system is tunable over a 124-GHz range with less than 100 MHz/h frequency drift. The laser generates 1 μJ per pulse at a rate of 4 kHz. It should also be noted that because of the temperature and pressure sensitivity of the étalons,

TABLE 11.1. University of Wisconsin High Spectral Resolution Lidar (HSRL)

Transmitter		Receiver	
Wavelength	532 nm	Type	Dall–Kirkham
Pulse Length	~130 ns	Diameter	0.5 m
Pulse Repetition Rate	4 kHz	Focal Length	5.08 m
Frequency Stability	0.09 pm/hr w/o I_2 locking	Filter Bandwidth	0.3 nm (night)
	0.052 pm with I_2 locking		8 pm (daylight)
		Field of View	0.16 to 4.0 mrad adj. (night) 0.16 to 0.5 mrad adj. (daylight)
		Polarization Rejection	$\sim 1 \times 10^{-3}$
		Data Collection Method	Photon Counting 100 ns bin minimum

a considerable degree of effort must be invested in maintaining the stability of these elements (Grund et al., 1988).

The same as with conventional elastic scattering measurements, multiple scattering along with the simultaneous single scattering is measured in the HSRL. The multiply scattered return is a function of the telescope field of view, the particle size, the range from the lidar, and the optical depth of the cloud. Accordingly, cloud particle sizes can be estimated by measuring signal variations as a function of the telescope field of view. A detailed description of the multiple-scattering approximations used for the HSRL measurements is presented in the study by Eloranta and Shipley (1982).

11.2.4. Atomic Absorption Filters

The use of a Fabry–Perot étalon in the HSRL as a filter has limitations. The performance of the étalon is governed by its finesse and the angular distribution of the incoming light. This requires a high degree of control over the state of the étalon. The UW HSRL was required to control the pressure to better than 0.1 mbar and the temperature to better than 0.1°C (Piironen and Eloranta, 1994). Even at this level of control, when the signal from particulates is much larger than the signal from molecules (for example, when examining clouds), there may be insufficient rejection of the particulate scattered light in the molecular channel. Bleed-through of the particulate signal may render the system ineffective. The need for an additional level of filtration has led to the use of atomic absorption filters.

Shimizu et al. (1983) first proposed the use of a narrow-band atomic absorption filter in a high-spectral-resolution lidar. This paper is an excellent summary of the considerations required to use an atomic filter in this way. The concept is to match the wavelength of a strong absorption line of some atom with the laser wavelength and thus the particulate scattered wavelength. Atomic absorption lines are ideal because of their inherently narrow line width. The line width of the filter can be broadened by heating the filter to achieve the desired absorption width. The use of the atomic filter gives an additional level of filtration to remove the strong particulate scattering signal. An atomic filter has the added advantages that the absorption lines are stable and have no angular dependence, so that alignment of the filter in the optical train is not an issue. Also, the amount of absorption can be easily controlled by either varying the concentration of the absorber or reducing the length of the cell. The following elements have been suggested as likely candidates for use as lidar filters: barium (at 553.701 nm), rubidium (at 780.023 nm), cesium (at 388.865 nm), lead (at 283.306 nm) (Shimizu et al., 1983), potassium (at 532 nm) (Yang et al., 1997), and thalium (276.787 nm) (Luckow et al., 1994).

A barium atomic absorption filter in such a lidar was demonstrated by She et al. (1992). The use of barium at a wavelength of 553 nm required the use of a highly tuned dye laser. An improvement was the use of an iodine filter by the UW HSRL (Piironen and Eloranta, 1994) at wavelengths near 532.2 nm. The use of iodine as a narrow-band optical filter was first suggested by Liao and Gupta (1978). The use of an iodine filter allows the use of a frequency-doubled Nd:YAG laser. Injection seeding is required to narrow the line width of the laser, but this also allows tuning the laser over a limited range of wavelengths. Several absorption lines of iodine are accessible within the lasing range of a frequency-doubled Nd:YAG laser (Fig. 11.9). The 1109 line of iodine was chosen because of its strength and isolation. A feedback system with a second iodine cell, through which a small fraction of the emitted laser light is directed, is used to dynamically tune the laser wavelength during measurements to maintain the laser at the center of the iodine absorption line. A second set of optical fibers transmits part of the outgoing light to the receiver system as part of this feedback system. Figure 11.11 is a diagram of the system used in the UW HSRL to stabilize the laser. This system has achieved a rejection ratio of 1:5000 of the scattered light from particulates in the molecular channel. The added rejection offered by atomic filtering is shown graphically in Fig. 11.10. The étalon system is capable of about a 1:2 rejection of the light scattered by particulates, whereas a rejection of about 1:1000 is shown for the atomic filter.

The layout of an HSRL using the molecular filtering technique is shown in Fig. 11.12 (Piironen and Eloranta, 1994). The backscattered light is collected with a telescope and passed through a polarizing beam splitter. The signal is filtered to reduce background light with an interference filter and a pair of low-resolution étalons. A fiber-optic scrambler precedes these filters to reduce the range dependence of the étalons due to the angular sensitivity of the étalon

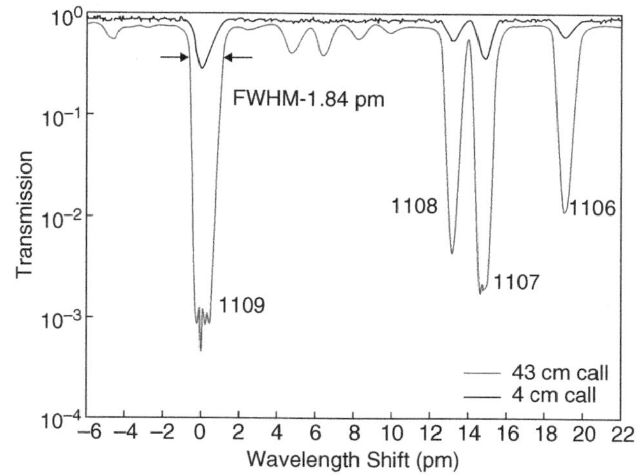

Fig. 11.9. Iodine absorption lines that may be used with a frequency-doubled and seeded Nd:YAG laser.

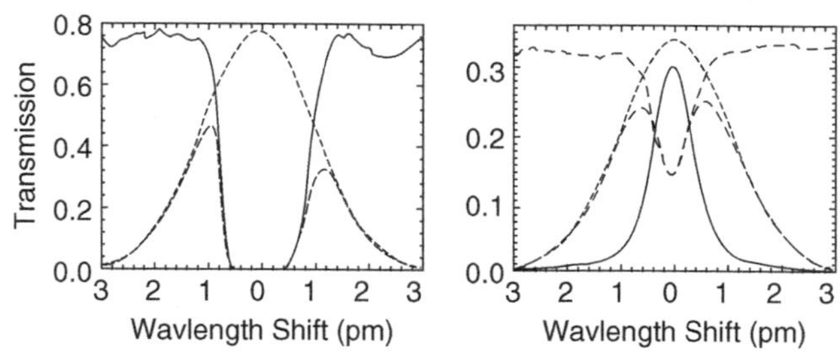

Fig. 11.10. The difference in the blocking afforded by the use of a molecular filter is shown. The transmission of the absorption cell is shown on the left as a solid line. The dashed-dot curve is the molecular spectrum for air at –65°C, and the dashed line is the effective transmission of the molecular spectrum. On the right, the transmission of the high-resolution étalon is shown as a solid line and the transmission as a dashed line. The dashed-dotted curve shows the effective transmission of the molecular spectrum (Piironen and Eloranta, 1994).

transmission. The separation of the particulate from the molecular backscatter signals is accomplished with the filter cell with this signal detected by PMT2. A portion of the total signal is directed to PMT1. This light is a combination of the total particulate and molecular backscatter spectra.

Because the bandwidth of the scattered light from molecules is a function of the temperature of the air, the amount of this signal passing through the filter is also a function of the air temperature. The width of the absorption line becomes important when the line is relatively wide. The amount of light

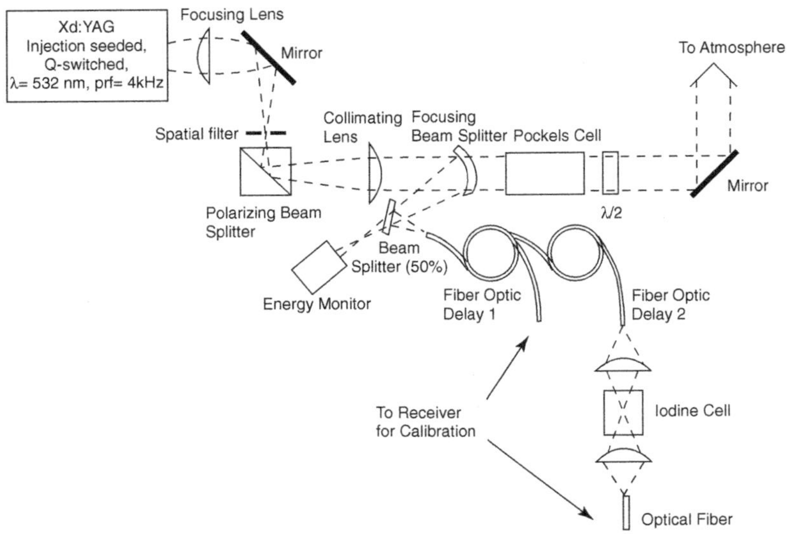

Fig. 11.11. The laser wavelength stabilization system used with the UW HSRL (Piironen and Eloranta, 1994).

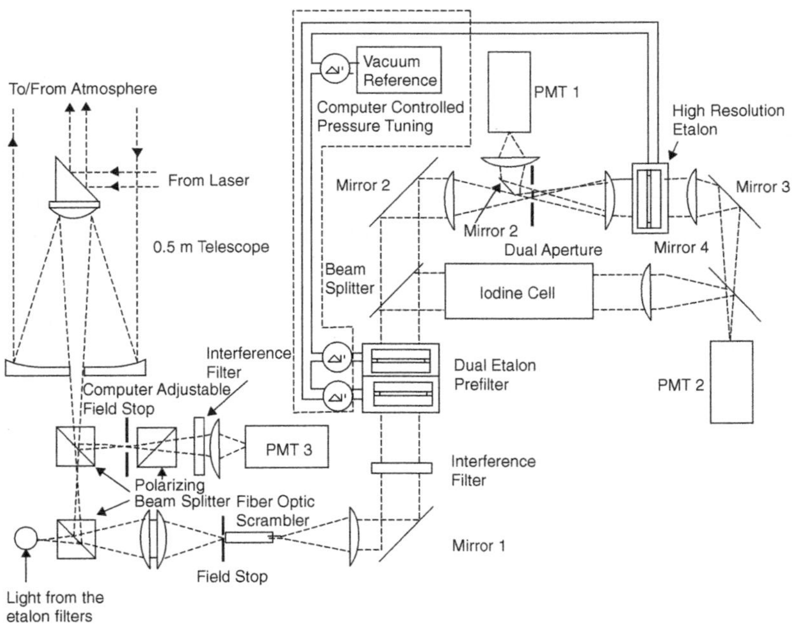

Fig. 11.12. The layout of an molecular filter-based high-spectral-resolution lidar. This layout is used in the UW HSRL (Piironen and Eloranta, 1994).

returning with wavelengths near the center of the distribution does not change a great deal with temperature, but the amount near the edges of the distribution is strongly affected. For a filter that is wide with respect to the width of the particulate line, the signal comes primarily from the edges and is thus strongly affected by the air temperature. Correcting for this requires information on the temperature profile of the atmosphere (obtainable from radiosonde measurements) and detailed information on the characteristics of the system.

The HSRL uses the iodine absorption at a wavelength 532.26 nm that is well isolated from the neighboring lines. The full-width half-maximum width of the line is 1.8 pm. Because of the width of the absorption line, the transmission of molecular scattered light through the iodine filter is more dependent on the air temperature than is an étalon. Although the iodine cell can be used at room temperature, the operating temperature of the cell must be controlled, because the vapor pressure of iodine is temperature sensitive. In the HSRL, the cell temperature is maintained with ±0.1°C accuracy by operating the cell in a temperature-controlled environment. Over a cell temperature range of 27°C to 0°C, the online transmission can be changed from 0.08% to 60%. In a short-term operation, the stability of the absorption characteristics has proven to be so good that system calibration scans from different days can be used for the calculations of the system calibration coefficients.

11.2.5. Sources of Uncertainty

The primary uncertainty sources for this type of lidar result from photon-counting statistics, background subtraction uncertainty, photomultiplier afterpulsing uncertainty, uncertainty in the determination of the calibration coefficients (including misalignment uncertainty), the effects of multiple scattering, molecular density estimation uncertainty (uncertainty in the temperature profile), and wavelength tuning uncertainty. The accuracy of optical depth measurements is limited primarily by photon-counting statistics. Photon-counting statistics are the primary limitation on the accuracy of the background correction as well. HSRL measurements are strongly dependent on the accuracy of the system calibration coefficients, particularly in the case of clouds. The calibration coefficients can be determined to an accuracy of 2–5%.

In performing a detailed analysis of the uncertainty in the UW HSRL, Piironen (1993) estimated that a 3-min averaging time is sufficient for 10% measurement accuracy for backscatter cross section of dense particulates and thin cirrus clouds. Longer averaging times are required to obtain the same accuracy for measurements in clear air. The cloud phase function can be determined to an accuracy of 10–20% when 6-min averaging times are used. Through the use of longer averaging times, more accurate measurements of the phase function can be made, assuming that multiple scattering may be ignored. The determination of the extinction cross section is also dependent on the accuracy of the molecular density profile. Because of the dynamics of clouds, averaging may lead to nonlinear errors as the clouds move and evolve. This effect is not normally accounted for in uncertainty analysis.

Consideration must also be given to the component of the wind velocity in the direction along the lidar field of view. For a zenith- or near-zenith-pointing lidar, the Doppler shift in the particulate spectrum is insignificant. However, this shift may be significant if large zenith angles are used. This consideration would seem to limit this type of lidar to sounding measurements.

11.3. MULTIPLE-WAVELENGTH LIDARS

It usually does not take long time for those attempting to produce quantitative information from data to conclude that a single-wavelength lidar system generally provides insufficient information to reliably invert the data and determine particulate concentrations or properties. Not only is the lidar equation [Eq. (3.12)] indeterminate, having two unknowns at every range location, but for a given extinction coefficient, there are an infinite number of particulate size distributions and indices of refraction that could have produced that value of the extinction. More information is clearly required. Because the scattering properties of particulates are a function of the ratio of the particle diameter and the wavelength of light, the use of multiple wavelengths in the lidar system is a potential way out of the problem. Thus the use of multiple laser wavelengths is a common lidar variation, and many have been built, even in the early years of lidar research. Each wavelength potentially provides an additional piece of information that could be used to determine some desired parameter. For example, the index of refraction having been assumed or measured, a three-color lidar could, in principle, be used to determine the size distribution and concentration for an exponentially distributed collection of particulates. More sophisticated models of the particulates would require more wavelengths. A number of methods have been developed for the inversion of multiple-wavelength systems (Potter, 1987; Girolamo, 1995; Post, 1996; Yoshiyama et al., 1996; Ackermann, 1997; Bockmann et al., 1998; Gobbi, 1998; Rajeev and Parameswaran, 1998; Ackermann, 1999; Kunz, 1999; Müller et al., 2000, 2002, and 2001a). These methods differ radically from one another in the assumptions that are made and the mathematical methods for the inversion. In general, the methods are complex and require human intervention or interpretation to work. In addition, there is still discussion in the literature on issues of completeness and uniqueness (see, for example, Kunz, 1997; Ackermann, 1999; Gimmestad, 2001).

There are two major reasons for using multiple-wavelength lidars. The first is, as stated above, that there exist, at least theoretically, unique solutions for inversions for a given set of assumptions and a set of measurements at a sufficient number of wavelengths. The use of multiple wavelengths reduces the ill-conditioned nature of the lidar solution problem. The problem is highly nonlinear, involving a complex convolution of particulate size distribution, index of refraction, and particulate size-wavelength interactions. In this kind of problem, small errors in the measured quantities may result in large errors in the reconstructed size distribution. The use of measurements at a sufficient number of wavelengths reduces the possible occurrence of false solutions. In this case, the result of the inversion solves the problem in an optimal way, minimizing the effects of measurement errors. Finally, when a larger number of wavelengths is used, fewer assumptions are, at least in principle, required to invert the data.

In its simplest form, the basic laser wavelength used is frequency doubled, tripled, or quadrupled with all of the wavelengths simultaneously transmitted along the same path. This has most commonly been done with Nd:YAG (1.064-μm fundamental) and ruby (0.694-μm fundamental). Other combinations have involved multiple lasers, Raman-shifted wavelengths, or the use of dye lasers. The technique requires that the additional wavelengths be emitted collinearly from the lidar so that all of the wavelengths examine the same volume of space. For systems that are intended to examine high-altitude clouds, the upper troposphere or stratosphere, this does not pose a particular problem. By the time the beams reach high altitudes, the laser beams are sufficiently large that offsets at the surface are small in comparison. But for devices intended to profile particulates in the boundary layer, this requires that the beams be collinear to within centimeters. This also means that all of the detectors have the same field of view. These requirements are difficult to achieve when using more than one laser, but not impossible. Many lasers are made so that the doubled or tripled frequencies are emitted through the same aperture. Separation of the light at the back of the telescope is relatively simple, using dichroic mirrors that reflect a narrow wavelength band and pass all others. Interference filters can be used to reject any light from other wavelengths that may enter the detector area. The alignment of multiple laser wavelengths through the same aperture is a particularly difficult task for which there is no simple or straightforward solution.

As shown in Chapter 2, in particulate scattering theory, two dimensionless parameters are defined. Q_{sc}, the scattering efficiency, is defined as the ratio of particulate scattering cross section σ_p to the geometric cross-sectional area of the scattering particle [Eq. (2.30)],

$$Q_{sc} = \frac{\sigma_p}{\pi\rho^2}$$

where ρ is the particle radius. The second dimensionless parameter is the size parameter, ϕ, defined as [Eq. (2.31)]

$$\phi = \frac{2\pi\rho}{\lambda}$$

where λ is the wavelength of the incident light. The scattering coefficient for a single particle of radius ρ can be written as [Eq. (2.32)]

$$\beta_p = \pi\rho^2 Q_{sc}(\rho)$$

The total scattering or attenuation coefficient in a polydisperse atmosphere with a distribution of particles of various radii is [Eq. (2.36)]

$$\beta_p = \int_{\rho 1}^{\rho 2} \pi\rho^2 Q_{sc}(\rho) n(\rho) d\rho \tag{2.36}$$

There are two fundamentally different ways to pose the multiple-wavelength signal inversion problem. The first is to attempt a solution of Eq. (2.36), a Riemann integral equation. Given the mathematical complexity of $Q_{sc}(\rho)$ and $n(\rho)$, this type of equation is extremely difficult to solve. The advantage of attempting to solve this equation is that it will provide the aerosol number density and size distribution. The downside of solving this equation is that some information must be known concerning the index of refraction (see Chapter 2), a larger number of wavelengths is required, and any solution will be complex. Given that the solution of Eq. (2.36) is difficult, an alternative way to take advantage of data at multiple wavelengths is to write the lidar equation for each wavelength and assume some relationship between the backscatter and/or attenuation coefficients at the various wavelengths. In this way, the ill-posed nature of the lidar equation can be circumvented. The cost for this advantage is a restriction on the information that can be retrieved from this method of multiple-wavelength signal inversion; here only the particulate extinction or backscatter coefficients are obtained. All of the methods discussed in Section 11.3.1 are variations of the latter solution method, although some may take advantage of supplemental information, such as measured particle size distributions. Methods to derive particulate microphysical characteristics from the signals of a multiple-wavelength lidar, such as the particulate concentration or the particulate size distribution, are beyond the scope of this book. A brief outline of such methods is given in Section 11.3.2.

11.3.1. Application of Multiple-Wavelength Lidars for the Extraction of Particulate Optical Parameters

Different algorithms have been proposed to extract particulate optical parameters from multiple-wavelength lidar data. The simplest approach is based on the use of fixed relationships between the same scattering parameters at different wavelengths. This variant for multiple-wavelength data analysis requires that the backscattered signals are simultaneously measured at least at two wavelengths. Some common elements used in processing data from a two-wavelength lidar system are discussed in studies by Krekov and Rakhimov (1986), Potter (1987), and Askermann (1997). Unfortunately, such studies are based primarily on theoretical considerations and are not supported with experimental results. At best, these ideas have been tested by simulated data. Generally, when using a two-wavelength approach, some fixed analytical relationship between the extinction and backscatter coefficients at different wavelengths is assumed. In the variant proposed by Krekov and Rakhimov (1986), a two-wavelength method was proposed for stratospheric measurements. The method was based on the assumption that the backscatter-to-extinction ratio is the same at both wavelengths. In a version proposed by Potter (1987), the assumption is made that the ratio of the extinction coefficients, measured at two wavelengths λ_1 and λ_2, is a constant value independent of range, that is, $\kappa_p(r, \lambda_1)/\kappa_p(r, \lambda_2) = b = const.$ As follows from scattering theory, such a simple assumption is formally true only for a monodisperse aerosol, that is, for particulates with the same composition and size. In some situations, this approximation may be acceptable for nonuniform particulates, at least in relatively homogeneous atmospheres. The applicability of this approximation for inhomogeneous atmospheres is severely restricted. The assumption of a range-independent value of b also assumes that integrated optical characteristics of the different particulates are invariant or vary insignificantly over the lidar measurement range. Such an assumption for inhomogeneous atmospheres is generally impractical (Kunz, 1999). As follows from the Mie theory, the assumption $b = const.$ may be true if the two wavelengths λ_1 and λ_2 are very close to each other. However, the signals from these wavelengths will be nearly identical and the accuracy of the retrieved extinction coefficient will be poor.

To retrieve the optical parameters of particulates with a two-wavelength approach, the assumption that $b = const.$ is insufficient. A related requirement is that the ratio b must be significantly different from unity. This condition is required to obtain acceptable measurement accuracy with the two-wavelength method. Consequently, it is necessary to increase the separation of the wavelengths λ_1 and λ_2 as much as possible. However, this requirement and the assumption $b = const.$ are contradictory for any real atmosphere.

To illustrate the basic features and the problems associated with practical multiple-wavelength measurements and inversions related to the extraction of the particulate optical parameters, we outline here a more sophisticated inversion methodology used in a typical experimental study by Spinhirne et al. (1997) to extract atmospheric backscatter cross section profiles. A major goal of the experiment was to investigate the variability of atmospheric backscatter cross sections across the Pacific region during the Global Backscatter Experiment (1989–1990). Simultaneous lidar measurements at three wavelengths were made in the visible and near infrared, at wavelengths of 0.532, 1.064, and 1.54 μm. For the measurements, an airborne lidar was used that could be pointed in the nadir or zenith directions. The data processing method developed by the authors was based on a combination of a preliminary hard target calibration and a normalization of the lidar signal. To normalize the signals, lidar signals were used obtained from areas assumed to be aerosol free.

A short explanation is necessary to clarify the concept of calibrating a lidar. The lidar constant C_0 in Eq. (3.11) includes two different factors that must be distinguished when an absolute calibration of the lidar is made. The first factor, C_1, depends on the characteristics of the transmitter and receiver optics, the diameter of the receiver telescope, the transmission of the optical system, and so on (Section 3.2.1). The second factor, E, is a product of the energy in each laser pulse and the conversion factor of the input radiant flux into the output lidar signal power. Thus constant C_0 may be defined as the product of two terms

$$C_0 = C_1 F_0 \frac{c\eta_0}{2} \gamma_{an} = C_1 E \tag{11.39}$$

The separation of the terms C_1 and E is required because of likely changes in factor E during the measurement event. This may occur because of a temporal instability in the pulse-to-pulse laser energy F_0 or degradation of the transformation factor γ_{an}. When a calibration constant is used for lidar data processing, the changes in E must be recorded during the measurements to be able to correct the retrieved data for these changes.

For simplicity, the equations below are considered for a ground-based lidar. Consider a lidar system operating at two wavelengths, λ_1 and λ_2, where $\lambda_1 > \lambda_2$. For a vertically staring lidar, the altitude-corrected lidar signal at the wavelength λ_1 measured at the attitude h can be written as

$$P(h, \lambda_1)h^2 = C_1^{(1)} E(\lambda_1)\beta_{\pi,m}(h, \lambda_1)[1 + \delta(h, \lambda_1)](T_{0,1})^2 \tag{11.40}$$

where $C_1^{(1)}$ is the lidar constant at the wavelength λ_1 and $(T_{0,1})^2$ is two-way vertical transmittance of the atmospheric layer from $h = 0$ to h at the wavelength λ_1. The function $\delta(h, \lambda_1)$ is

$$\delta(h, \lambda_1) = \frac{\beta_{\pi,p}(h, \lambda_1)}{\beta_{\pi,m}(h, \lambda_1)} = \frac{R(h, \lambda_1)}{a(h, \lambda_1)} \tag{11.41}$$

If no molecular absorption occurs at λ_1, the molecular backscattering profile $\beta_{\pi,m}(h, \lambda_1)$ can be calculated with a vertical temperature sounding. Defining the range-corrected signal, normalized by the product of $[E(\lambda_1)\beta_{\pi,m}(h, \lambda_1)]$ as (Spinhirne et al., 1997)

$$Z(h, \lambda_1) = \frac{P(h, \lambda_1)h^2}{E(\lambda_1)\beta_{\pi,m}(h, \lambda_1)} \tag{11.42}$$

the normalized signal $Z(h, \lambda_1)$ can be rewritten with Eqs. (11.40) and (11.42) in the form

$$Z(h, \lambda_1) = C_1^{(1)}[1 + \delta(h, \lambda_1)](T_{0,1})^2 \tag{11.43}$$

The lidar system calibration $C_1^{(1)}$ can be obtained by a hard-target measurement procedure. However, as noted by Spinhirne et al. (1997), the relative constant between wavelengths can be determined much more accurately. Accordingly, this type of calibration is preferable when multiple-wavelength measurements are made. If the calibration ratio at two wavelengths λ_1 and λ_2

$$Q_{2,1} = \frac{C_1^{(2)}}{C_1^{(1)}} \tag{11.44}$$

is known, then the lidar equation for the second wavelength λ_2 can be written as

$$Z(h, \lambda_2) = Q_{2,1} C_1^{(1)}[1 + R_{2,1}(h)\,\delta(h, \lambda_1)]\,(T_{0,2})^2 \tag{11.45}$$

where $R_{2,1}(h)$ is the ratio between backscattering terms at λ_1 and λ_2, defined as

$$R_{2,1}(h) = \frac{\delta(h, \lambda_2)}{\delta(h, \lambda_1)} = \frac{\beta_{\pi,m}(h, \lambda_1)\,\beta_{\pi,p}(h, \lambda_2)}{\beta_{\pi,m}(h, \lambda_2)\,\beta_{\pi,p}(h, \lambda_1)} \tag{11.46}$$

An appropriate selection of the wavelengths λ_1 and λ_2 makes it possible to ignore the term in rectangular brackets of Eq. (11.45). The backscattering coefficients for particulate and molecular constituents vary inversely with the wavelength to the power of ~1 and 4, respectively. Therefore, for the wavelengths $\lambda_1 = 1.064\,\mu m$, and $\lambda_2 = 0.532\,\mu m$ used in the experiment, the parameter $R_{2,1}(h)$ has a value of about one-eighth. It can be assumed that for clear-air conditions

$$1 + R_{2,1}(h)\,\delta(h, \lambda_1) \approx 1$$

so that

$$Z(h, \lambda_2) = Q_{2,1} C_1^{(1)}(T_{0,2})^2 \tag{11.47}$$

Thus, with the known calibration factor $Q_{2,1}$, there is a system of two equations, Eqs. (11.43) and (11.47), with four unknowns, $C_1^{(1)}$, $\delta(h, \lambda_1)$, $(T_{0,1})^2$, and $(T_{0,2})^2$. There are different ways to determine the unknowns, depending on the particular optical situation. In clear atmospheres, the particulate component of the total transmission term over the path (h_0, h) is negligible, at least at the longer wavelength, $\lambda_1 = 1.064\,\mu m$. In this case, the term $(T_{0,1})^2$ may be either ignored or reduced to the transmission for molecular scattering

$$(T_{0,1})^2 = \exp\left[-2\int_0^h \beta_{m,1}(h')dh'\right] \tag{11.48}$$

where $\beta_{m,1}(h)$ is the molecular extinction coefficient (scattering) at λ_1. Now only three unknowns remain, $C_1^{(1)}$, $\delta(h, \lambda_1)$, and $(T_{0,2})^2$, so that the solution can be found with an iterative procedure. Initially the transmission at $\lambda_2 = 532\,\mu m$ is taken to be only due to molecular extinction, so that a first estimate of $(T_{0,2})^2$ can be found via the molecular component. Under the initial condition that both transmission terms within the altitude range (h_0, h) are known, the remaining terms can be determined. The value of $C_1^{(1)}$ can be found from Eq. (11.47) and the values of $\delta(h, \lambda_1)$ from Eq. (11.43). After that, improved transmission terms can be found with an iterative procedure, where a simple equation is used,

$$T^2 = T_m^2 T_p^2 = T_m^2 \exp\left[-2\int_0^h \kappa_p(x)dx\right] \tag{11.49}$$

Here all the indexes and variables in brackets are omitted. The particulate extinction term can be found with the use of a backscatter-to-extinction ratio estimated initially.

To improve multiple-wavelength solution accuracy, sensible assumptions and independently measured particulate parameters may be used. In the study by Spinhirne et al. (1997), the solution of Eqs. (11.43) and (11.47) was found with the additional assumption of the aerosol-free upper troposphere. In this region, only the transmission term was updated in the iteration procedure. To calibrate and process the data, the signals at $0.532\,\mu m$ were first normalized to a molecular profile in the region that showed the least backscatter during the flight. The term $R_{j,i}(h)$ and the particulate backscatter-to-extinction ratios were calculated with the Mie theory using particle measurements made by on-board particle samplers. The relative target calibration values, which were corrected for any flight-to-flight variations, were applied to obtain the backscatter profiles at 1.064 and $1.54\,\mu m$. As follows from the authors' estimates, the combination of the relative and absolute calibration made it possible to reduce the backscatter measurement uncertainty to the order of 10^{-9} $(m\,sr)^{-1}$ at wavelengths 1.06 and $1.54\,\mu m$ and to the order of 10^{-8} $(m\,sr)^{-1}$ for the measurement at $0.532\,\mu m$.

Thus, following the study by Spinhirne et al. (1997), the following procedure can be specified for a practical multiple-wavelength methodology: (1) determination of the system calibration ratio between the wavelengths with hard-target measurements and its regular correction; (2) calculation of the vertical molecular profiles with the best available temperature profiles; (3) examination of the lidar signal to determine the clearest areas where particulate loading is least; (4) identification of the presence of clouds by means of a threshold analysis of the signals and their derivative; (5) exclusion of the signals from within the clouds; (6) retrieval of the backscatter profiles with an iterative procedure; and (7) spatial and temporal smoothing of the data. In addition to this, particulate measurements with on-board particle samplers were made, and a calculation of the scattering terms was performed with Mie theory.

To summarize this section, data processing methodologies for the above multiple-wavelength techniques are based on differences between the scattering parameters at different wavelengths. This approach makes it possible to ignore some parameters at marginal wavelengths. This, in turn, decreases the number of the unknown quantities in the equation set. The multiple-wavelength approach may be especially effective when it is combined with methods to establish supporting information (for example, the use of aerosol-free areas, or Mie calculations based on in situ data).

When a multiple-wavelength lidar system is used, the signals measured at the different wavelengths can be used in a different way to obtain optimal lidar equation solutions. The lidar calibration parameters may be determined from aerosol-free areas with data at the shortest operating wavelength where the weight of the particulate constituent in the total signal is least. On the other hand, the unknown particulate extinction coefficient may be determined at the longest operating wavelength of the lidar, where the ratio of the particulate-to-molecular scattering is the largest in value.

The key problem in multiple-wavelength lidar measurements of particulate optical parameters is the unknown relationship between the particulate scattering at different wavelengths. To extract the information contained in the data of a multiple-wavelength lidar, these corresponding relationships must be somehow established or assumed.

It is necessary to point out that multiple-wavelength lidar measurements are uniquely complicated and require a quite delicate computational approach. To complicate matters, a huge volume of raw data is involved in the data processing. The most important point to be made with such measurements is that data collection must be accomplished with extremely high accuracies. This requirement arises because of the fact that all of the data used in the analysis are interrelated. Therefore, even a small inaccuracy in an intermediate result, obtained at one wavelength, will worsen the results extracted from the signal at the other wavelengths. An inaccurate calibration of the lidar system is also inadmissable, because it will cause a systematic error in the retrieved data, generally much larger than for a one-wavelength measurement. A common effect is that the measurement error increases when an increased number of error sources are involved in the data retrieval.

11.3.2. Investigation of Particulate Microphysical Parameters with Multiple-Wavelength Lidars

The main purpose of multiple-wavelength measurements is to investigate the basic characteristics of atmospheric particulates, their microphysical parameters, such as the number and volume concentration, the particulate size distribution, and the index of refraction. Unfortunately, the inversion of multiple-wavelength lidar signals is a complicated task. No simple analytical solution is available to reconstruct the particle parameters from measured data. Existing solutions are generally ill-posed, so that they require complex computational methods, for example, those developed by Tikhonov and Arsenin (1977). Obtaining the characteristics of the atmospheric particulates is more difficult than determination of their extinction or backscatter coefficient. When extracting the extinction coefficient, it is necessary only to determine a solution boundary value and the backscatter-to-extinction ratio. The determination of the particulate characteristics requires knowledge of some other characteristics, such as the refractive index and the particulate size distribution, or knowledge of relationships between particulate characteristics.

A detailed discussion of the techniques of multiple-wavelength inversion is a highly technical topic, closely related to Mie scattering theory and worthy of a book in itself. This question is presented in many theoretical studies (e.g., in studies by Twomey, 1977; Zuev and Naats, 1983; Müller et al., 1999; Liu et al., 1999). In this section, only a brief review of the problem is given without considering details. The purpose is to give the reader a general understanding of the principal concepts and difficulties related to this problem.

As early as 1989, Sasano and Browell practically demonstrated the potential of multiple-wavelength measurements to discriminate between different aerosol types. Using experimental data, they showed that with a multiple-wavelength technique, it is possible to discriminate between maritime, continental, stratospheric, and desert aerosols. This study used an assumption of similarity in the derived profiles of the backscatter coefficients at three wavelengths (300, 600, and 1064 nm). A conventional power law dependence of the particulate backscatter coefficient on the wavelength was assumed to be

$$\beta_{\pi,p}(\lambda_2) = \beta_{\pi,p}(\lambda_1)\left[\frac{\lambda_1}{\lambda_2}\right]^{\xi}$$

In their analysis, aerosol size distribution data were obtained simultaneously with the lidar measurements. With these in situ data, Mie calculations were made, and the results of the calculations were compared with the lidar data. The backscatter coefficients were assumed to be related only to the total number density of the particulates. This means that the size distribution and the refractive index for the aerosol were assumed to be invariant along the lidar line of sight. As often happens in experimental studies, quantitative disagreements were found between the theoretical and empirical results, that is, between the Mie calculations and the lidar data. The authors assumed that the disagreement might be partly due to uncertainties in the lidar data analysis and partly caused by uncertainties in the particulate size distributions and refractive indices. The nonsphericity of the particulates was assumed be an additional reason for the disparity. The authors stated that the parameter ξ in the power law dependence may change depending on the assumed refractive index.

Obviously, a limited number of wavelengths can provide only limited information about scattering properties of particulates. In a numerical study, Müller and Quenzel (1985) investigated the feasibility of determining the particulate size distribution from particulate extinction and backscatter coefficients determined with lidar at four wavelengths, 347, 530, 694, and 1064 nm. It was found that the accuracy of conventional lidar measurements is insufficient to fulfil all of the requirements necessary to obtain accurate inversion results. The authors concluded that a real improvement can only be achieved if the particulate refractive index is determined independently, for example, from particulate sampling. The authors' conclusion was that a lidar alone can only provide qualitative information rather than quantitative determination of the aerosol parameters.

Potentially, the increase in the number of wavelengths used to simultaneously search the atmosphere increases the amount of available information with fewer assumptions. The combination of elastic and Raman measurements in multiple-wavelength measurements can further improve the quality of the extracted information (Müller et al., 2000; Müller et al., 2001 and 2001a).

A large number of theoretical studies on the topic of multiwavelength inversion have been published during the last decade. A comprehensive theoretical analysis and the principles of retrieval of aerosol properties from multiple-wavelength lidars can be found, for example, in studies by Müller et al. (1998, 1999, 1999a, 2000, and 2001). Ligon et al. (2000) proposed an inversion technique based on a Monte Carlo method. The latter can be considered to be an alternative to the traditional regularization technique (Müller et al., 1999). According to the authors, the Monte Carlo method is extremely accurate when estimating the aerosol size distribution. The assumption made here is that the aerosols under investigation are spherical dielectrics, for which the refractive index is known. Rajeev and Parameswaran (1998) proposed a method to invert multiple-wavelength lidar signals without assuming any analytical form for the particulate size distribution. The method requires a lidar system with eight operating wavelengths, a constant, range-independent backscatter-to-extinction ratio, and a priori knowledge of the refractive index at all of the wavelengths. It can be seen even from this brief outline of recent studies that an uncertainty in the aerosol refractive index can significantly reduce the value of any inversion method. This is a general conclusion of most studies, and none of the currently available techniques entirely overcomes this problem. When the refractive index is assumed to be known, the inversion results are (at least, theoretically) stable and accurate even when the data have significant noise (Ligon et al., 2000).

In an experimental study by Müller et al. (1998), two multiple-wavelength lidar systems, a transportable and a stationary Raman lidar, were used to investigate profiles of tropospheric aerosols. With these systems, the particulate backscatter profiles at five wavelengths and extinction profiles at two wavelengths were simultaneously measured. To derive the particulate microphysical parameters, such as the number and volume concentration and the complex refractive index, the regularization described by Tikhonov and Arsenin (1977) was used. The authors of this study pointed out the requirement to obtain accurate values for the particulate backscatter coefficients. To achieve reliable inversion results, the backscatter coefficients must be known with an error of less than 20%. On the basis of their theoretical studies, the authors stated that at least two extinction coefficients and six backscatter coefficients are necessary to obtain accurate information on particulate properties, such as number or volume concentrations, the effective radius, and complex refractive index. The experimental results obtained by researchers from the Institute for Tropospheric Research were published by Müller et al. (2000, 2001 and 2001a).

A method of retrieving atmospheric particulate properties that uses a linear combination of the measured aerosol backscatter at different wavelengths was discussed recently by Donovan and Carswell (1997) and Yue (2000). In the latter study, the author concluded that the size distribution can be reasonably retrieved from backscattering even using only two or three wavelengths. To achieve this, it is sufficient reduce the possible range that some parameters of the particulate size distribution may have. To reduce the range of these parameters, an in situ measurement must be collected close to the lidar measurements.

In a study by Donovan and Carswell (1997), the authors show how a principal component analysis, based on Mie theory, may be used to determine the parameters of stratospheric sulfate aerosols. Unlike the rather pessimistic conclusion made by Müller and Quenzel (1989), the key point of these authors is that many atmospheric particulate parameters can be determined with the information that is available only from multiple-wavelength lidar measurements. According to Donovan and Carswell (1997), principal component analysis allows estimation of the parameters of the integrated particulate size distribution with a linear combination of the measured aerosol backscatter and extinction coefficients. Such an analysis allows an assessment of how much information can be obtained with a given kernel set and, moreover, how sensitive the extracted parameters are to measurement errors. The authors considered situations in which particulate and molecular backscattering is available at different combinations of five wavelengths. Their research states that, for sulfate aerosols, multiple-wavelength lidar data may be inverted without any a priori assumption concerning the aerosol size distribution. This can be achieved, however, only if the assumption of spherical aerosols is valid and the refractive index is known.

In the recent studies of Donovan et al. (2001 and 2001a), a method is presented for inverting simultaneously measured lidar and radar signals that makes it possible to retrieve cloud particle radii and water content profiles. The authors proposed an algorithm that treats the lidar extinction, derived cloud particle effective size, and cloud multiple-scattering effects together in a consistent fashion. According to the authors of this study, the use of the radar and lidar signals together allows one to overcome the lidar problem of the extracting accurate values of atmospheric extinction. The inversion algorithms were experimentally tested and compared with ground-based passive remote-sensing observations and with in situ airborne particle probes. The comparisons showed a good agreement between the lidar/radar results, the in situ measurements, and an independent IR radiometer. The basic problem of such combined measurements lies in the different atmospheric albedo for lidar and radar wavelengths. In optically thick clouds, reliable information can only be obtained in a restricted altitude range, up to heights at which the lidar signal-to-noise ratio is acceptable for the inversion. On the other hand, the method is not applicable when the cloud particles are so small that they are not detected by radar.

To briefly summarize the discussion of the analysis of multiple-wavelength measurements, there are numerous studies devoted to the problem of extracting data from the multiple-wavelength measurements that basically differ only by the particular set of assumptions used for the inversion. This means that the value of the particular theoretical approach often depends on the applicability of the particular assumptions. At best, this means that the particular solution is mainly relevant for some particular set of atmospheric conditions. As pointed out by Donovan and Carswell (1997), many of the methods discussed in the literature contain various unrealistic assumptions. The simplest are that the aerosol properties do not vary with height, that the refractive index along the searching path can be found, that the particulate size distribution has some fixed shape, which is exactly known, for example, a single log-normal mode, etc. Obviously, for a particular optical situation these assumptions may or may not be appropriate.

11.3.3. Limitations of the Method

Not all of the data collected from different wavelengths are effectively independent measurements. And, in a practical sense, only a limited number of different wavelengths are reasonable. Although lasers beyond 1 μm exist, molecular scattering is almost nonexistent in this region of the spectrum, making the lidar signal small. When coupled with the decreased detector response beyond 1 μm, lidars using wavelengths longer than 1 μm are inherently shorter-range instruments than those using wavelengths shorter than 1 μm. Wavelengths shorter than about 0.27–0.3 μm are strongly attenuated by atmospheric ozone and consist primarily of molecular scattering. In short, there is a limited range of wavelengths from which the operating set can be chosen. Wavelengths as long as 10 μm have been used for multiple-wavelength measurements on clouds (Post et al., 1996, 1997), but inversion for these systems is a serious issue. The limitation on the usable wavelength range implies that particulate sizes from roughly 0.1 to 2.5 μm can be effectively measured. Fortunately, this is an important range for pollution measurements, corresponding to the particulate matter, PM2.5 and standards of the U.S. Environmental Protection Agency (EPA). However, because cloud droplets are so much larger than the wavelengths suggested here, it seems unlikely that this range of wavelengths will be effective in measuring cloud drop size distributions. The difference in returns between size parameters ($2\pi r/\lambda$) of 20 and 25 is too subtle and beyond the precision with which lidar measurements can be made.

When measuring in clear atmospheres, the most significant problem is to accurately separate the particulate-scattering component from the molecular-scattering component. In the lidar signal measured at visible spectra, the particulate component of the scattering can be hundreds time of less than the molecular component. In situations when molecular scattering dominates, the aerosol constituent is obtained as a negligible difference between two large numbers, that is, between the total and molecular scattering terms. The only useful factor in this situation is that the signal from the molecular scattering can be used as a lidar calibration source. As shown in the previous sections, this can be achieved if areas may be identified in which the particulate scattering does not take place. The other useful factor in the multiple-wavelength method is a significant difference in the particulate and molecular scattering for different wavelengths. When the frequency λ of the emitted laser pulse is doubled, the molecular scattering, which is proportional to λ^{-4}, changes by a factor of 16, whereas the aerosol scattering generally changes by a factor of 2–4 with the wavelength (see Chapter 2).

12

ATMOSPHERIC PARAMETERS FROM ELASTIC LIDAR DATA

12.1. VISUAL RANGE IN HORIZONTAL DIRECTIONS

There are many reasons to measure atmospheric transmission and visibility. The first stems from the widespread use of ground, water, and air transportation. Poor visibility in areas near airports is a key factor limiting aircraft safety during take off and landing. Poor visibility conditions restrict traffic on highways and contribute to shipping accidents, especially in constricted areas near the shore or along rivers. The second reason deals with the need to monitor the sources and dynamics of atmospheric pollution. This includes monitoring the emissions from burning forests or oil wells, studying the uptake and transport of dust and particulates. The methods developed to measure atmospheric transmission may also be helpful to determine reference (boundary) values for two-dimensional images obtained in "spotted" atmospheres.

12.1.1. Definition of Terms

The most general formulation defines visibility as the ability to discern distant objects by the unaided human eye. Some portion of the atmosphere always lies between the observer and the distant objects. In bad weather conditions, such as haze or fog, the large aerosol contents in the atmosphere may significantly decrease visual perception of distant objects. Generally, atmospheric visibility is limited because of the effects of light scattering and absorption by water droplets,

Elastic Lidar: Theory, Practice, and Analysis Methods, by Vladimir A. Kovalev and William E. Eichinger.
ISBN 0-471-20171-5

dust, microscopic salt crystals, and soot particles that are suspended in the atmosphere near the earth's surface. Mists and fogs are caused by the condensation of water onto microscopic particles (nuclei). In practice, the term "fog" is usually applied if visibility falls below 1000 meters. Limited visibility due to dust or other dry microscopic particles in the atmosphere is called haze. Haze, mist, and fog are the primary causes for severely decreased atmospheric visibility.

The visibility of a distant object depends on the characteristics of the object such as its size, geometric form, and color. It also depends on the background against which the object is observed, the contrast between the object and the background, and the level of illumination. The object is scarcely seen or may even be invisible if any of the following conditions take place: (1) The angular size of the distant object is less than the angular discrimination of the human eye. (2) The difference in color and brightness between the object and the background against which the object is seen is small. In other words, the object becomes invisible if the contrast between the object and the background is so small that it cannot be discriminated by the human eye. (3) The object, which does not shine and is not illuminated, is observed in the dark. An excellent discussion of the practical issues associated with visibility is given by Bohren (1987).

In meteorological practice, the following terminology for atmospheric visibility is generally used:

(1) *Visual range* is the maximum range, usually in a horizontal direction, at which a given light source or object becomes barely visible under a given atmospheric transmittance and background luminance.
(2) *Meteorological visibility range* is a formal characteristic of daytime visibility, defined as the greatest distance at which a black object of a relevant size can be seen when observed against a background of fog or sky.

In a homogeneous atmosphere, the relationship between the meteorological visibility range, L_M, and the extinction coefficient, κ_t, is determined as (Koschmider, 1925; Horwath, 1981),

$$L_M = \frac{-\ln \varepsilon}{\kappa_t} \tag{12.1}$$

The relationship in Eq. (12.1) is known as Koschmider's law. Here, ε is the visual threshold of the luminance contrast. The visual threshold is the least luminance contrast between the object and its background that makes it possible to visually distinguish and identify the object. The object becomes invisible if the luminance contrast of the object against the background is less than the visual threshold of luminance of the human eye. Numerical investigations established that the value of ε mostly ranges between 0.02 and 0.05 to allow the object be distinguished, and it increases at least, up to 0.05–0.08 to allow the object be identified. In the most visibility measurements (except that made in civil airports), the value $\varepsilon = 0.02$ is commonly used. As follows from Eq. (12.1), the optical depth of an atmospheric layer with a visual range L_M is a constant value

$$\tau(L_M) = \kappa_t L_M = -\ln \varepsilon \tag{12.2}$$

With the equations above, the mean value of the extinction coefficient κ_t close to the ground surface can easily be obtained if the horizontal visibility is known. The relationship between κ_t and visibility was used at meteorological network stations to estimate the atmospheric extinction without the use of optical instruments. This type of approximate estimates can also be obtained for light of different wavelengths (Kruse et al., 1963).

In the practice of meteorological support of civil aviation, two basic visibility measures are used: the *meteorological optical range* and the *runway visual range*. The definition of the meteorological optical range is related to light transmittance that, in turn, defines what part of the original luminous flux remains in a light beam after traversing an optical path of a given length (Section 2.1). The meteorological optical range is the length of a path in the atmosphere over which the total transmittance is 0.05. As follows from this definition, the relationship between the meteorological optical range L, transmittance $T(L)$, and the extinction coefficient κ_t can be written as

$$T(L) = \exp\left[-\int_0^L \kappa_t(x)\,dx\right] = 0.05 \tag{12.3}$$

Thus the optical depth of an atmospheric layer of length L will have the value

$$\tau(L) = \int_0^L \kappa_t(x)\,dx = 3 \tag{12.4}$$

It follows from the formulas above that the optical depth of an atmospheric column with length L is a constant value. The same applies to L_M. In a homogeneous atmosphere, the relationship between the extinction coefficient and the meteorological optical range is

$$L = \frac{3}{\kappa_t} \tag{12.5}$$

As follows from Eqs. (12.1) and (12.5), the values of L and L_M are equal if the visual threshold of the luminance contrast in Eq. (12.1) is selected to be $\varepsilon = 0.05$. If the threshold contrast ε is chosen to be different from 0.05, the meteorological visibility range differs from the meteorological optical range. For example, in meteorological practice not related to aviation, a threshold of $\varepsilon = 0.02$ was generally used (Koschmider, 1924; Kruse et al., 1963; Barteneva et al., 1967; Measures, 1984). In this case, the meteorological visibility range must be considered to be the length of an atmospheric column in which optical depth is equal to the logarithm of 0.02, accordingly,

$$L_M = \frac{3.91}{\kappa_t}$$

The values of L_M obtained with different ε differ from each other and from L by a constant factor, so that their ratio does not depend on the extinction coefficient. If the uncertainty in the selected value of ε is ignored, the relative uncertainty of the meteorological optical range L and the meteorological visibility range L_M are equal. Therefore, we will not discriminate between the meteorological optical range and meteorological visibility range in the discussion that follows.

Another atmospheric visibility measure used in meteorological practice in support of civil aviation is the runway visual range. This value is the most important visibility measure used to estimate runway visibility. The main purpose for its use was to provide pilots and air traffic services with specific information on runway visibility conditions during periods of low visibility caused by fog, rain, snow, sandstorms, etc. Knowledge of the runway visual range makes it possible to decide whether the weather conditions are acceptable for plane landing or take off. Formally, information is needed to determine whether the visibility is above or below some specified operating minimum for a particular airport. Based on this (and some additional) information, a decision authorizing plane landings or take offs can be made. The formal definition of the term follows. The runway visual range, L_R, is the distance over which the pilot of an aircraft can see the runway surface markings or the runway lights when moving along the runway. This value depends on whether nonilluminated landing marks or runway lights are used to orient the pilot. In the first case, the runway visual range is estimated through the meteorological optical range L. In the second case, the runway visual range is determined as the visibility range of the runway lights. During hours of darkness, the lights that delineate the runway or identify its center line are always switched on during take off and landing. Note that in bad visibility conditions, that is, in heavy fogs, rains, and snowfalls, the lights are seen better than the daytime markings; therefore, under poor visibility conditions the runway lights are switched on, even in the daytime.

The range L_R, defined as the maximum range at which the runway lights can be seen, can be determined from Allard's law [Eq. (2.11)]. This is a transcendental equation for the unknown L_R

$$E_T = \frac{I_R}{L_R^2} e^{-\kappa_t L_R} \tag{12.6}$$

where I_R is the intensity of the runway edge or runway center-line lights and E_T is the visual threshold of illumination. The visual threshold is the least level of the illumination required to make visible a distant point source (or a small size) light to the naked eye. Note that the visual threshold E_T is related to the background luminance against which the light is observed. Depending on the type of illumination, E_T varies from approximately 10^{-6} lx (for nighttime conditions) to 10^{-3} lx (for daytime conditions).

The visibility range of runway lights changes during the transition period from day to nighttime conditions (and vice versa) even if the atmospheric turbidity does not change.

As follows from its definition, the runway visual range cannot be measured directly on the runway but must be calculated. For this, all of the other terms in Eq. (12.6) must be known. This requires knowledge of several quite disparate pieces of information. These include physical and biological factors such as the visual threshold of illumination, operational factors such as the runway light intensity, and atmospheric factors such as the background illumination E_T and the extinction coefficient of the atmosphere κ_t. At airports, the atmospheric extinction coefficient is determined by a special instrument, a transmissometer.

12.1.2. Standard Instrumentation and Measurement Uncertainties

A transmissometer is considered to be the most accurate instrument for atmospheric transparency measurements. It directly measures atmospheric transmittance over some fixed distance with two spatially separated instrument units. In a conventional double-ended transmissometer, a light projector directs a narrow beam of light to a remote photodetector in a receiver unit. The equation to determine the extinction coefficient may be obtained from Beer's law for a homogeneous atmosphere [Eq. (2.10)]. Denoting the distance between the projector and the receiver units (the transmissometer baseline) as Δr, one can determine the extinction coefficient κ_t as

$$\kappa_t = \frac{-\ln T(\Delta r)}{\Delta r} \tag{12.7}$$

where $T(\Delta r)$ is the atmospheric transmittance over the baseline distance. Transmissometer output data can easily be transformed into values of meteorological visibility range and meteorological optical range and used to calculate the visibility range of a distant light. As follows from Eqs. (12.5) and (12.7), the meteorological optical range L can be calculated as

$$L = \frac{-3\Delta r}{\ln T(\Delta r)} \tag{12.8}$$

Real light beams are always divergent rather than parallel, so that Beer's law written for a parallel light beam [Eq. (2.3)] cannot be used directly in practical calculations. For a real transmissometer with a baseline Δr, Beer's law can be applied in the form

$$F_\lambda = F_{up,\lambda} e^{-\int_0^{\Delta r} \kappa_{t,\lambda}(r)dr} \tag{12.9}$$

where $F_{up,\lambda}$ is the flux measured by the photodetector at the upper scale limit of the transmissometer range. In other words, $F_{up,\lambda}$ is the maximum value of the flux on the photodetector, measured in a very clear atmosphere, when the optical depth of the range Δr is very small, that is

$$\int_0^{\Delta r} \kappa_{t,\lambda}(r)dr \approx 0$$

In this case, light extinction over Δr can be ignored and F_λ becomes equal to $F_{up,\lambda}$. For a homogeneous atmosphere, Eq. (12.9) reduces to

$$F = F_{up} e^{-\kappa_t \Delta r} \tag{12.10}$$

where the subscript λ is omitted for simplicity. The transformation from κ_t to visibility range is multiplicative, so that the fractional uncertainty in the measured extinction coefficient is equal to the fractional uncertainty in the meteorological optical range and in visibility. The uncertainty is defined by Eq. (12.11), obtained by uncertainty propagation applied to Eq. (12.10)

$$\delta\kappa_t = \delta L = \frac{1}{\kappa_t \Delta r}\sqrt{\delta F_{up}^2 + \delta F^2} \tag{12.11}$$

where $\delta\kappa_t$ and δL are the fractional uncertainties of κ_t and the meteorological optical range, respectively. The term δF is the fractional uncertainty of the luminous flux F measured after light beam propagation through the turbid layer Δr. The component δF_{up} is the fractional uncertainty in established F_{up} at the upper scale limit. This parameter is, in fact, the calibration uncertainty. The calibration is generally made in the clearest atmospheric conditions available, when light losses along the transmissometer baseline can be ignored. Assuming for simplicity that the absolute uncertainties ΔF_{up} and ΔF are equal, one can rewrite Eq. (12.11) in the form

$$\delta\kappa_t = \delta L = \frac{1}{\kappa_t \Delta r}\frac{\Delta F}{F_{up}}\sqrt{1 + e^{2\kappa_t \Delta r}} \tag{12.12}$$

As with a lidar, transmissometer measurement accuracy is inversely proportional to the optical depth of the measurement range, that is, to the optical depth of the instrument baseline, $\tau = \kappa_t \Delta r$.

The accuracy of the visibility range, as measured by a transmissometer, is related to the length of the transmissometer baseline Δr. To obtain a general uncertainty relationship for a transmissometer measurement, a nondimensional parameter, z_{tr}, is introduced. This parameter is equal to the ratio of the optical depth of the measured meteorological optical range L to the optical depth over the instrument baseline during the measurement

$$z_{tr} = \frac{\tau(L)}{\tau(\Delta r)} \tag{12.13}$$

where the subscript (tr) denotes "transmissometer." For a homogeneous atmosphere, the parameter z_{tr} reduces to the ratio of the meteorological optical range to the baseline length of the transmissometer

$$z_{tr} = \frac{L}{\Delta r} \tag{12.14}$$

The general dependence of the uncertainty in the meteorological optical range on z_{tr} can be derived from above Eq. (12.12) in the form

$$\delta L = 0.33 z_{tr} \frac{\Delta F}{F_{up}} \sqrt{1 + \exp(6/z_{tr})} \tag{12.15}$$

The main parameters that determine the accuracy of transmissometer measurements are (1) the instrument uncertainty of the transmissometer and (2) the parameter z_{tr}, which is the ratio of the optical depth over the range L to that of the baseline.

In Table 12.1, the dependence of $\delta L\%$ on z_{tr} is given. Here the fractional uncertainty of the instrument is taken to be $\Delta F/F_{up}$ = 1%. Note that the transmissometer measurement uncertainty is a minimum when the transmissometer baseline length and the measured meteorological optical range are nearly the same, or at least, when $L = (1 - 10)\Delta r$. The uncertainty significantly increases if L becomes much larger than Δr ($L > 10\Delta r$).

In Fig. 12.1, the dependence of the relative uncertainty δL in percentage is shown as a function of the measured meteorological optical range. This dependence is calculated for transmissometers with different baseline lengths, $\Delta r = 0.2$ km and $\Delta r = 1$ km (*curve 1* and *curve 2*, respectively). Here the instru-

TABLE 12.1. Dependence of $\delta L\%$ and $\tau(\Delta r)$ on z_{tr}

z_{tr}	1	2	4	6	10	15	20	30
$\tau(\Delta r)$	3	1.5	0.75	0.5	0.3	0.2	0.15	0.1
δL, %	6.6	3.0	3.1	3.8	5.5	7.8	10.1	14.8

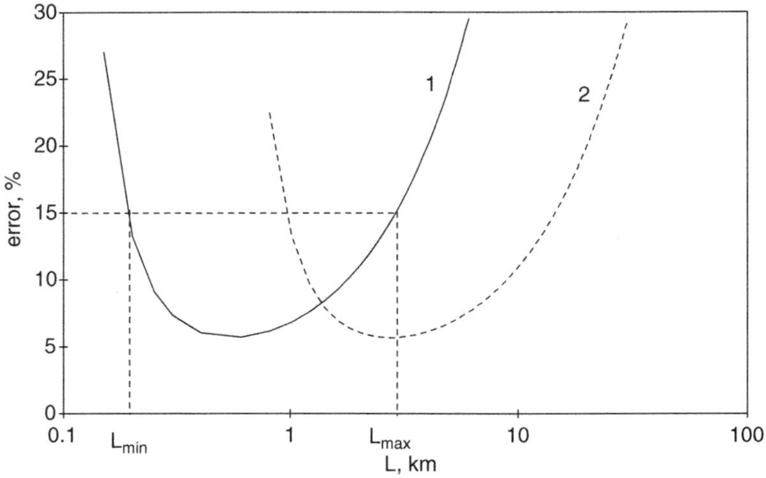

Fig. 12.1. Dependence of uncertainty δl, % on the meteorological optical range for the different baseline length. *Curves 1* and *2* show the uncertainty δL for the baseline length $\Delta r = 0.2$ km and $\Delta r = 1$ km, respectively. The instrumental uncertainty for the both cases is $\Delta F/F_{up} = 2\%$.

mental uncertainty for both instruments is the same, $\Delta F/F_{up} = 2\%$. The dependence of the uncertainty δL on the measured meteorological optical range has the same U-shaped appearance as that for the lidar (Chapter 6). Note that the curves in Fig. 12.1, obtained for different baseline lengths, are shifted relative to each other. Because the acceptable level of measurement uncertainty is always restricted, the range of L that can be measured with a transmissometer with a fixed baseline length is also limited. For example, if the acceptable measurement uncertainty level $\delta L = 15\%$, the optical ranges L that may be measured with a transmissometer with $\Delta r = 0.2$ km extends from $L_{min} = 0.2$ km to only $L_{max} = 3$ km (Fig. 12.1). This is why transmissometers with a baseline length of 0.2 km cannot be used for accurate measurements in clear atmospheres. Similarly, a transmissometer with a baseline length of 1 km cannot be used for measurements in turbid and foggy atmospheres, when the visibility is less than 1 km. In other words, the baseline length Δr of the instrument must be chosen to suit the particular application. It is not possible to measure the meteorological visibility or optical range in high-visibility conditions by using a transmissometer with a short baseline length, and vice versa. Generally, the value of the instrument baseline length should be equal to or a little less than the minimum value of the meteorological visibility (or optical) range that must be measured. Otherwise, the measurement uncertainty at the minimal measurement range may be unacceptable. On the other hand, to measure the meteorological visibility or meteorological optical range in clear atmospheres, a transmissometer with a large baseline length should be used.

A transmissometer may also be used to determine the meteorological visibility range L_M for a specified visual threshold for the luminance contrast ε. Using a simple mathematical transformation, one can obtain the dependence of the measurement uncertainty δL_M on z_{tr} similar to that in Eq. (12.15)

$$\delta L_m = \frac{z_{tr}}{\ln 1/\varepsilon} \frac{\Delta F}{F_{up}} \sqrt{1 + \exp\left(\frac{-2\ln\varepsilon}{z_{tr}}\right)} \tag{12.16}$$

An additional source of uncertainty exists in visibility measurements with transmissometers. In practice, this uncertainty cannot even be accurately estimated and therefore is usually ignored. The source of this uncertainty lies in the difference between the baseline length and the visibility range. The baseline length of the transmissometer is usually much less than the measured meteorological optical (or visual) range. Therefore, the visual range is measured over a restricted transmissometer baseline and extrapolated outside the baseline length. Such an extrapolation assumes that the optical characteristics are identical within and outside the baseline, that it assumes atmospheric optical homogeneity. Atmospheric heterogeneity can significantly increase the actual measurement uncertainty as compared to that determined with Eqs. (12.12), (12.15), or (12.16).

Note that visibility measurements with lidar can significantly reduce the uncertainty caused by atmospheric heterogeneity. This is because the lidar operating range is variable and, under condition of acceptable lidar signal-to-noise ratios, it can increase with an increase of atmospheric visibility.

In practical application, transmissometers suffer from a number of limitations:

(1) Transmissometer consists of two spatially separated pieces and therefore can only determine the visual range close to the ground surface. It is generally capable of measurements only in a fixed horizontal direction. However, many practical applications require measurements in slope or vertical directions.

(2) The instrument baseline of the transmissometer is fixed. It cannot be adjusted or changed during the measurement or analysis process to improve the measurement accuracy when visibility changes.

(3) The transmissometer baseline is, generally, much less than the measured visual range. It means that the transmissometer data are often extrapolated beyond the baseline distance. Therefore, in a heterogeneous atmosphere, for example, during a snowstorm or a dissipating fog, the uncertainty of the calculated meteorological optical range may be enormously large.

(4) Even in homogeneous atmospheres, transmissometer provides an acceptable measurement uncertainty for a relatively restricted range of extinction coefficients. The uncertainty will increase enormously for

measurements outside this range. Thus an instrument with a fixed baseline length provides only a limited spread of measureable visibility ranges.

Until recently, the transmissometer was the only optical instrument used at airports for visibility measurements. However, at some airports, nephelometers are being used operationally. A nephelometer is an instrument in which a small volume of ambient air is illuminated by a narrow or wide beam of the light, depending on its construction. A photodetector measures the intensity of light scattered by the illuminated air sample at angles shifted relative to the direction of the incident light beam. As follows from Chapter 2, the amount of scattered light measured by a photodetector is related to the turbidity of the ambient air. Thus there is a correlation between the intensity of the angular scattering and the extinction coefficient inside the scattering volume. Different types of nephelometers have been developed and tested. Generally, four basic types of nephelometers are used: (1) a side-scattering nephelometer, in which a narrow light beam and a receiver with a small field of view are used (in such instruments, a light scattering angle is selected, typically either 45° or 60°); (2) an integrating nephelometer, in which a wide light beam and a receiver with a small field of view are used; in this instrument, the light scattering angle range extends from approximately 7° to 170° (Heintzenberg and Charlson, 1996; Anderson et al., 1996; Anderson and Ogden, 1998); (3) a forward-scattering instrument, in which the light scattering angle only slightly differs from 0° (VAISALA News, 2002); and (4) a backscattered light nephelometer, in which the scattering angle is close to 180° (generally, between 176° and 178°) (Doherty et al., 1999; Anderson et al., 2000; Masonis et al., 2002). At airports, only a forward scattering nephelometer is sometimes used. This instrument operates accurately under extremely poor visibilities only, for example, in heavy fogs. Therefore the use of a forward-scattering nephelometer is the only practical for visibility measurements in such weather conditions.

Unlike a transmissometer, the components of a nephelometer are not spatially separated. The instrument is generally constructed as a single unit. There are several basic assumptions that are made which may be sources of nephelometer measurement uncertainty. First, it is assumed that the total extinction coefficient of the atmosphere is uniquely related to the light scattering at a particular angle or over a selected angular range from a small scattering volume. Second, this relationship is assumed to be known or may be experimentally established during a calibration procedure. Third, this relationship is assumed to be the same for different types of atmospheric situations. This means that in any given visual range, no variation in the particulate size distribution or in the index of refraction will change the angular intensity of the scattered light. Obviously, these assumptions are not realistic for real atmospheres. This is the first principal disadvantage of these instruments. The small scattering volume is the second significant disadvantage of nephelometer measurements. The last feature may result in large fluctuations in the measured signal and large measurement uncertainties, especially in unstable atmospheres, for example, during fog or haze dissipation. Unlike a transmissometer, which can operate in both a scattering and an absorbing medium, the nephelometer measures only the scattering component of atmospheric extinction. Atmospheric heterogeneity significantly worsens the spread of nephelometer data. Therefore, the use of the nephelometer in the airport is quite restricted. In fact, a transmissometer remains the only instrument for the visibility measurements at most airports.

12.1.3. Methods of the Horizontal Visibility Measurement with Lidar

Lidars are the only instruments that can give information on atmospheric scattering properties in any direction along extended atmospheric paths. In a clear atmosphere, the length of the atmosphere examined by a lidar near the ground surface can extend up to tens kilometers. This provides significant advantages to elastic lidars compared with the instruments described in the previous section. The main advantages of the lidar are as follows:

(1) Unlike a transmissometer, a lidar is a monostatic instrument. Generally, it is a single-block unit, from which a beam can be pointed in any direction. This makes it possible to use a lidar for measurements in horizontal, slant, and vertical directions. These changes in the direction of lidar examination do not require special adjustment or readjustment of the instrument. Unlike a transmissometer, a change in the examined direction can be easily made without interrupting the measurement.

(2) A lidar allows determination of the profile of the atmospheric extinction over the examined path rather than only the mean value along the path.

(3) The operating measurement range of the lidar is not fixed as is that of a transmissometer. The length of the lidar operating range may be changed when the atmospheric transmittance changes. The range may be automatically increased when visibility improves, and vice versa. This makes it possible to optimize the distance over which the measurement is made for the particular conditions. This, in turn, makes it possible to determine atmospheric visibility over a wider range of atmospheric turbidity compared with a transmissometer.

(4) The signal from a nephelometer is related to the amount of scattering at a given angle, whereas the signal from a lidar is related both to the backscatter coefficient and to the atmospheric transmittance. Visibility is directly related to the transmittance, which is an integrated parameter that is less sensitive to local variations in particulate loading, size distribution, concentration, etc. The lidar can provide a stable mea-

surement even under conditions like snowfall and heavy rain, where conventional nephelometer operations are unsatisfactory because of large variations in the angular scattering. The lidar measurement of the extinction coefficient over an extended area is potentially much more accurate than a point measurement made with a nephelometer or a short-base transmissometer.

(5) Unlike the nephelometer data processing technique, which is based on an absolute instrument calibration, the lidar measurement technique makes it possible to avoid an absolute calibration of the lidar. The lidar measurement technique is generally based on a relative calibration.

The most significant impediment to the wide application of lidar for atmospheric measurements is the high cost of lidar systems and the complexity of lidar data processing. The latter problem is related to the uncertainty of the lidar equation. However, for horizontal measurements, this difficulty may be overcome by the application of reasonable assumptions, the validity of which can be easily checked by a posteriori analysis. An accurate determination of the visual range requires knowledge of the transmittance or the mean extinction coefficient over a spatially extended area. Because some degree of atmospheric heterogeneity is always present, the measurement accuracy is generally better if the visibility range and the measurement range of the instrument do not differ significantly. In other words, the lidar parameter z, defined similarly to the ratio z_{tr} in Eq. (12.14), should be not too large. This requirement stems from the fact that the measurement uncertainty increases with an increase in the ratio z_{tr} (Table 12.1). It should be stressed that the lidar is the only instrument that makes it possible to keep the ratio relatively constant when the atmospheric visibility changes significantly. This may be achieved by using a variable measurement range when processing lidar data obtained under different visibility.

In Section 5.1, a slope method was described to determine the extinction coefficient in a homogeneous atmosphere. It was pointed out that the method is sensitive to the presence of middle- or large-scale particulate heterogeneity. This method is most practical when the range-corrected signal profile is visualized directly by the instrument operator during lidar data processing. This allows the operator to exclude signals distorted by inhomogeneous particulate layering and thus avoid processing unreliable data. The slope method is more helpful when adjusting and testing a lidar rather than for atmospheric measurements. It can hardly be recommended for routine (especially automatic) lidar measurement of atmospheric visibility. Long-term field measurements of atmospheric visibility with a lidar, made in the U.S.S.R., in the vicinity of St. Petersburg, revealed that for routine measurements, the method based on the use of integrated values of the range-corrected signal is the most practical one (Baldenkov et al., 1988). Two variants of the method, used in these visibility measurements, are presented below.

In a homogeneous atmosphere, an approximate version of the lidar equation solution can be used, as given in Section 5.4. In this version, the lidar equation solution for a homogeneous two-component atmosphere can be obtained without determining the auxiliary function $Y(r)$. To apply the solution, the lidar signal needs only to be range corrected, the same as in the case of a single-component atmosphere. Two adjacent areas, $\Delta r_1 = r_1 - r_0$, and $\Delta r_2 = r_2 - r_1$, are selected within the maximum lidar operating range (r_0, r_2), where r_0 is the minimum distance of complete lidar overlap (see Section 3.2). The integrated Z_r functions for the ranges Δr_1 and Δr_2 are determined as (Fig. 12.2)

$$I_{r,1} = \int_{r_0}^{r_1} Z_r(r')dr' = \frac{1}{2} C_0 T_0^2 \Lambda [1 - \exp(-2\kappa_t \Delta r_1)] \tag{12.17}$$

and

$$I_{r,2} = \int_{r_1}^{r_2} Z_r(r')dr' = \frac{1}{2} C_0 T_0^2 \Lambda \exp(-2\kappa_t \Delta r_1)[1 - \exp(-2\kappa_t \Delta r_2)] \tag{12.18}$$

where [Eq. (5.87)]

$$\Lambda = \frac{\Pi_p \kappa_W}{\kappa_t}$$

Defining the two-way transmittance of the areas Δr_1 and Δr_2 as

$$T_1^2 = e^{-2\kappa_t \Delta r_1} \tag{12.19}$$

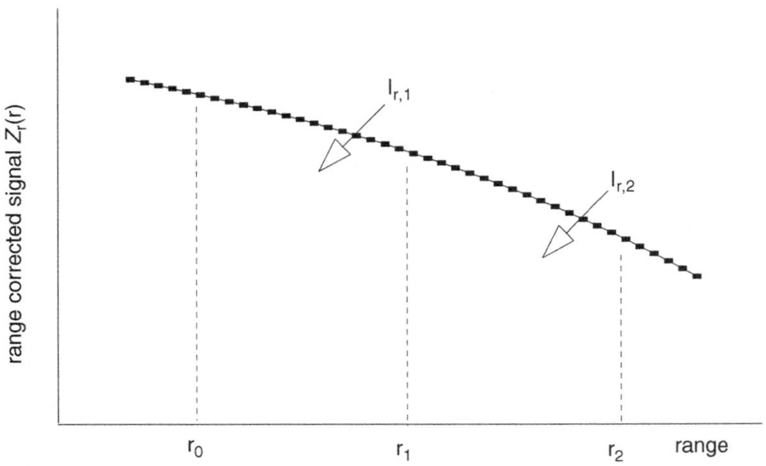

Fig. 12.2. Signal integration ranges in the horizontal visibility measurement.

and

$$T_2^2 = e^{-2\kappa_t \Delta r_2} \tag{12.20}$$

the relationship between the atmospheric transmission terms and the integral $I_{r,2}$ and $I_{r,1}$ ratio can be written as

$$\frac{T_1^2(1-T_2^2)}{1-T_1^2} = \frac{I_{r,2}}{I_{r,1}} \tag{12.21}$$

If the atmosphere is homogeneous, both transmittance terms in Eq. (12.21), T_1^2 and T_2^2, are functions of the same extinction coefficient κ_t. Therefore, the extinction coefficient can be determined through the calculation of the ratio of $I_{r,2}$ to $I_{r,1}$. Then, the transcendental equation above must be solved to find κ_t, from which the visual range is then calculated. However, simpler variants exist, which avoid this drawback and apply simple analytical solutions for κ_t. Two data processing variants are considered further, which are practical for visibility measurements in horizontal directions.

Method of Equal Ranges. A simple solution can be obtained for horizontal measurements if the areas Δr_1 and Δr_2 in Fig. 12.2 are selected to be of equal length, that is, $\Delta r_1 = \Delta r_2 = \Delta r$. Now $T_1^2 = T_2^2$, so that Eq. (12.21) is reduced to

$$T_1^2 = \frac{I_{r,2}}{I_{r,1}} \tag{12.22}$$

The mean extinction coefficient is then found as

$$\kappa_t = \frac{1}{2\Delta r} \ln\left(\frac{I_{r,1}}{I_{r,2}}\right) \tag{12.23}$$

and, accordingly, the meteorological optical range is

$$L = \frac{3}{\kappa_t} = \frac{6\Delta r}{\ln I_{r,1} - \ln I_{r,2}} \tag{12.24}$$

The method is practical for visibility measurements in clear and moderately turbid atmospheres. The principal requirement to obtain acceptable measurement accuracy is the absence of large-scale heterogeneous areas, whose length is comparable with the range increment Δr. Another requirement is that the backscatter-to-extinction ratio must not change, at least systematically, along the examined path. To avoid a systematic uncertainty in the visibility range measurement, the direction of lidar examination must be chosen with care. The lidar beam must not pass through a locally polluted area, such as a dusty road. As much as possible, the beam should be directed horizontally, especially in clear atmospheric conditions. This will avoid a significant difference in the examined volume heights above the ground surface over the areas Δr_1 and Δr_2.

The approximation of atmospheric homogeneity used in this variant may yield a significant measurement uncertainty. To avoid this, an automatic analysis of the recorded lidar signal profiles should be included in the lidar data processing procedure. This analysis should be made before the extraction of the extinction coefficient profile. An estimate of linearity of the square-corrected-signal logarithm, $[\ln Z_r(r)]$, might also be helpful.

Method of Asymptotic Approximation. According to a study by Baldenkov et al. (1988), this data processing method proved to be the most practical for horizontal visibility measurements in moderately polluted and turbid atmospheres. This conclusion was based on 2 years of lidar visibility measurements that included measurements in hazes, fogs, snowfalls, and rains. The method can even be used when some systematic differences occur in the scattering characteristics in the areas Δr_1 and Δr_2. Under such conditions, the asymptotic method is preferred for lidar data processing to the method of equal ranges, because it is less sensitive to spatial heterogeneities.

In the method of asymptotic approximation, the far-end range r_2 (Fig. 12.2) is selected to be the maximum operating distance, that is, $r_2 = r_{max}$, whereas the length of the range r_1 can be chosen arbitrarily. The general solution for this method can be obtained from Eq. (12.21) in the form

$$\frac{T_1^2 - T_{max}^2}{1 - T_{max}^2} = \frac{I_{r,2}}{I_{r,max}} \tag{12.25}$$

where

$$I_{r,max} = I_{r,1} + I_{r,2} = \int_{r_0}^{r_{max}} Z_r(r')dr' \tag{12.26}$$

and

$$T_{max}^2 = T_1^2 T_2^2 = \exp[-2\kappa_t(r_{max} - r_0)] \tag{12.27}$$

With Eq. (12.25), the mean extinction coefficient for the range $\Delta r_1 = r_1 - r_0$ can be determined as

$$\kappa_t = -\frac{1}{2\Delta r_1} \ln\left[\frac{I_{r,2}}{I_{r,max}}(1 - T_{max}^2) + T_{max}^2\right] \tag{12.28}$$

and the meteorological optical range is found as

$$L = \frac{3}{\kappa_t} = \frac{-6\Delta r_1}{\ln\left[\frac{I_{r,2}}{I_{r,\max}}(1 - T_{\max}^2) + T_{\max}^2\right]} \tag{12.29}$$

For mist and fog conditions, an approximate solution can be used. This solution is based on the existence of an asymptotic limit for integral $I_{r,\max}$ determined over the range (r_0, $r_{\max}$) as the upper range $r_{\max}$ tends to infinity. As shown in Chapter 5, the relationship between the maximum integral $I(r_0, r_{\max})$ and its theoretical limit, $I(r_0, \infty)$, can be written with Eqs. (5.53) and (5.57) as

$$I(r_0, r_{\max}) = I(r_0, \infty)(1 - T_{\max}^2) \tag{12.30}$$

Accordingly, the relationship between $T_{\max}^2$ and the integrals $I(r_0, r_{\max})$ and $I(r_0, \infty)$ is

$$T_{\max}^2 = \frac{I(r_0, \infty) - I(r_0, r_{\max})}{I(r_0, \infty)} \tag{12.31}$$

When $T_{\max}^2 \ll 1$, the integral $I(r_0, r_{\max})$ is close to its asymptotic limit $I(r_0, \infty)$. This takes place when the total optical depth of the atmospheric layer (r_0, $r_{\max}$) becomes larger than 1–1.5. Then the term $(1 - T_{\max}^2)$ in Eq. (12.30) is close to unity and the integral $I(r_0, r_{\max})$ can be used as an approximate estimate of its theoretical limit $I(r_0, \infty)$ (Kovalev, 1973 and 1973a; Platt, 1979). In Table 12.2, the systematic difference in % is given between $I(r_0, r_{\max})$ and $I(r_0, \infty)$ for different optical depths $\tau(r_0, r_{\max})$.

TABLE 12.2. Relative Difference Between $I(r_0, r_{\max})$ and $I(r_0, \infty)$ for Different Optical Depths $\tau(r_0, r_{\max})$

$\tau(r_0, r_{\max})$	0.5	1	1.5	2	2.5	3
Difference, %	36.8	13.5	5	1.8	0.67	0.25

One can see that the systematic difference between the calculated maximum integral $I(r_0, r_{\max})$ and its asymptotic limit is less than 5% if the total optical depth $\tau(r_0, r_{\max})$ exceeds 1.5. For the ranges where $T_1^2 \gg T_{\max}^2$, the latter can be ignored and one can obtain the approximate solution from Eq. (12.25) as

$$T_1^2 \approx \frac{I_{r,2}}{I_{r,\max}} \tag{12.32}$$

In this case, no a priori estimate of the boundary value $T_{\max}^2$ is required to calculate the extinction coefficient or the meteorological optical range. These characteristics can be determined by the simple formulas derived with Eqs. (12.19) and (12.32)

$$\kappa_t' \approx \frac{-1}{2\Delta r_1} \ln\left(\frac{I_{r,2}}{I_{r,\max}}\right) \tag{12.33}$$

and

$$L' \approx \frac{-6\Delta r_1}{\ln\left(\frac{I_{r,2}}{I_{r,\max}}\right)} \tag{12.34}$$

The relationship between the actual (L) and approximate (L') values can be found from Eqs. (12.29) and (12.34) as follows:

$$\frac{L'}{L} = \frac{2\tau(\Delta r_1)}{\ln[1 - T_{\max}^2] - \ln\{\exp[-2\tau(\Delta r_1)] - T_{\max}^2\}} \tag{12.35}$$

The behavior of the systematic difference between L' and L depends on the selection of the operating ranges $\Delta r_1 = r_1 - r_0$ and $\Delta r_{\max} = r_{\max} - r_0$. Both ranges can be either fixed or variable. In the latter case, the measurement range is selected to be proportional to the visibility range. In Fig. 12.3, the difference in percentage between the actual L and approximate L' obtained with Eq. (12.35) is shown. *Curves 1* and *2* are calculated for fixed Δr_1 and $\Delta r_{\max}$. *Curve 1* shows the systematic discrepancy $\delta L\%$ between L' and L for $\Delta r_1 = 0.15$ km and $\Delta r_{\max} = 1$ km, whereas *curve 2* shows the same for

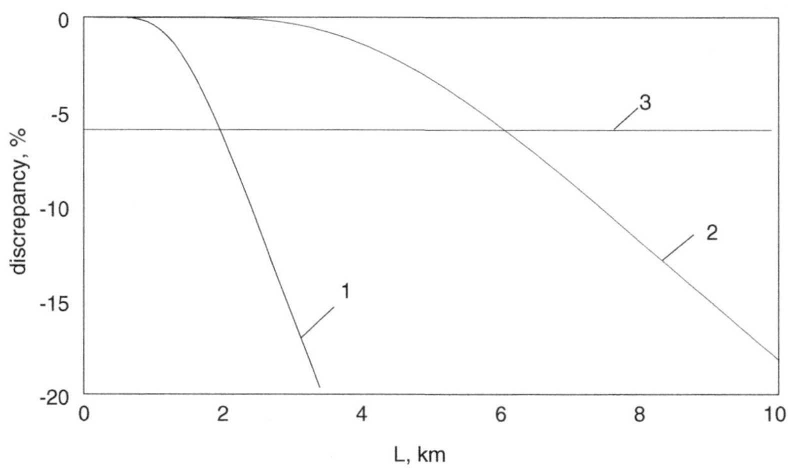

Fig. 12.3. Systematic shift in the measured meteorological optical range obtained with Eq. (12.34). *Curves 1* and *2* are calculated for the fixed ranges Δr_1 and $\Delta r_{\max}$. *Curve 1* shows the relative uncertainty for $\Delta r_1 = 0.15$ km and $\Delta r_{\max} = 1$ km, whereas *curve 2* shows the same uncertainty for $\Delta r_1 = 0.3$ km and $\Delta r_{\max} = 3$ km. *Curve 3* shows the systematic shift obtained with a fixed ratio of Δr_1 to $\Delta r_{\max}$.

$\Delta r_1 = 0.3$ km and $\Delta r_{max} = 3$ km. In both cases, the systematic discrepancies are small for small values of L and abruptly increase when L becomes large. (Note that for *curves 1* and *2*, $\delta L\%$ tends to zero when L decreases. This is because here only systematic contributions to uncertainty are analyzed. When all basic measurement uncertainty contributions are considered, the general U-shaped uncertainty dependence on the range takes place.)

Some additional comments are necessary to clarify the details of the asymptotic lidar measurement method. The first concerns the influence of multiple scattering when the lidar operates in fogs or hazes. As mentioned in Section 3.4.2, a multiple-scattering contribution becomes noticeable in the profile of the lidar return when the optical depth becomes larger than 1–1.5. However, when using the integral ratio to calculate atmospheric parameters [Eqs. (12.33) or (12.34)], its influence is significantly reduced (Zuev et al., 1976). Second, it is useful to point out the difference in the uncertainty behavior between a lidar and transmissometer. As shown in Section 12.1.1, the measurement uncertainty of a transmissometer is strictly related to the nondimensional parameter z_{tr}. This parameter is equal to the ratio of the optical depth over the range L ($\tau = 3$) to that over the instrument baseline [Eqs. (12.13) and (12.14)]. The baseline of the transmissometer is fixed; therefore, z_{tr} changes in proportion to the change in visibility. When the visibility increases, the optical depth over the instrument baseline decreases, so that z_{tr} becomes larger. This change in z_{tr} results in an increase of the measurement uncertainty (Table 12.1). As follows from Table 12.1, the increase in the uncertainty becomes significant when $z_{tr} > 6$. When a lidar is used for the visibility measurement, such an increase takes place only when the lidar measurement range Δr_1 is fixed. The case when Δr_1 is fixed is shown in Fig. 12.3 (*curves 1* and *2*). In this case, the ratio of L to Δr_1 increases proportionately to the increase in the visibility range. It causes the absolute value of the measurement uncertainty to increase similarly to that in transmissometer measurements. Thus the use of a fixed range, Δr_1, in lidar data processing reduces the lidar measurement capabilities to the level of those for a transmissometer.

Meanwhile, making visibility measurements with lidar, one can significantly decrease the measurement uncertainty by using variable rather than fixed ranges Δr_1 and Δr_{max}. The best results are achieved when these ranges are increased in proportion to the visibility range. (Obviously, such an increase is practical only within a restricted range of visibilities, until the requirements for the acceptable signal-to-noise ratio of lidar signals are met.) In a way similar to transmissometer measurements, the uncertainty in the visibility measurement with lidar depends on the atmospheric optical depth over the ranges Δr_1 and Δr_{max} rather than on their geometric distance. Analogously to the parameter z_{tr}, defined as the ratio of visibility range to the transmissometer baseline length, one can define such values for the lidar ranges Δr_1 and Δr_{max}

$$z_1 = \frac{L}{\Delta r_1} \tag{12.36}$$

and

$$z_{1,\max} = \frac{L}{\Delta r_{\max}} \tag{12.37}$$

Now the relationship between L and L' given in Eq. (12.35) can be written in the form

$$\frac{L'}{L} = \frac{6}{z_1}\left\{\ln\left[\frac{1-\exp\left(\frac{-6}{z_{1,\max}}\right)}{\exp\left(\frac{-6}{z_1}\right)-\exp\left(\frac{-6}{z_{1,\max}}\right)}\right]\right\}^{-1} \tag{12.38}$$

The question becomes: What values of z_1 and $z_{1,\max}$ can be considered to be optimum for visibility measurements? Ideally, the lidar measurement range should be as close as possible to the measured visibility range. This decreases the uncertainty caused by the extrapolation of the measurement result beyond the measurement range. Unfortunately, the maximum optical depth that can be measured by lidar is limited because of its finite dynamic range, the presence of the term r^{-2} in the lidar equation, the signal and background noise, multiple scattering, etc. Therefore, the lidar operating range will generally be less than the measured meteorological optical range L or visibility range L_M.

Numerical estimates made for the asymptotic method revealed that the optimum optical depth for the range Δr_1 must not exceed approximately unity (Zuev et al., 1978 and 1978a), so that the corresponding value of z_1 is $z_1 \geq 3$. On the other hand, as follows from Eqs. (12.35) and (12.38), the difference between the actual L and the approximate L' depends on the total transmittance $T^2_{\max}$, that is, on the total optical depth of the range $\Delta r_{\max}$. To keep the measurement uncertainty constant over the measurement range, it is necessary to keep $T^2_{\max} = const.$ This, in turn, requires that the range $\Delta r_{\max}$ should be variable and $z_{1,\max} = const.$ As follows from Eq. (12.38), when z_1 and $z_{1,\max}$ are constants, the ratio L'/L does not depend on the visibility range. This is important because a constant difference between L and L' can be considered to be a systematic measurement uncertainty and can be corrected. This case is shown in Fig. 12.3 with *curve 3*. The curve was obtained for variable ranges Δr_1 and $\Delta r_{\max}$, which correspond to $z_1 = 3$ and $z_{1,\max} = 1.5$. The constant discrepancy between L and L' is 6%.

A mobile lidar system in which the variable ranges Δr_1 and $\Delta r_{\max}$ were selected to be proportional to the visibility range is described in a study by Baldenkov et al. (1989). This instrument was developed to measure the horizontal meteorological optical range and slant visibility range along the airplane glide path under restricted visibility conditions. The analog lidar system operated for visibility ranges from 0.2 to approximately 10 km. The lidar signal was automatically range corrected. This was achieved by increasing the photomultiplier gain in proportion to the square of the time (t^2). This correction results in a range-corrected signal $Z_r(r) = P(r)r^2$ at the output of the photodetector rather than the raw signal $P(r)$. The lidar data were processed as follows. After the laser pulse emission, the signals were accumulated during two different times to obtain the integrals $I_{r,1}$ and $I_{r,2}$ [Eqs. (12.17) and (12.18)]. The first integral was accumulated from time t_0 to $t_1 = t_0 + \Delta t$, where the integration delay time $t_0 = 0.5\,\mu s$. This delay allowed the light pulse to travel through the zone of incomplete overlap before the signal was accumulated. The integration time Δt related to the range Δr_1 was variable; it automatically increased with the increase of visibility. The integration occurred during the time interval during which the range-corrected signal $Z_r(r)$ decreased by a factor of 10 compared with its initial value at t_0. For a homogeneous atmosphere, a monotonic decrease in the signal is caused only by the exponential term of the lidar equation [Eq. (5.85)]. In this case, a decrease in $Z_r(r)$ by a factor of 10 corresponds to an optical depth $\tau(\Delta r_1) \approx 1.15$ or $z_1 = 2.6$ [Eq. (12.36)]. The time integration of the integral $I_{r,2}$ was established to be from t_1 to $t_{max} = t_1 + \Delta t$, that is, the variable ranges $\Delta r_1 = r_0 + \Delta r_1$ and $\Delta r_2 = \Delta r_{max} - \Delta r_1$ were equal. The meteorological optical range was determined by Eq. (12.34), which was transformed into the form

$$L \approx \frac{3c\Delta t}{\ln\left(\dfrac{I_{r,2} + I_{r,1}}{I_{r,2}}\right)} \tag{12.39}$$

where c is the velocity of light. The lidar technique described above was developed and tested in 1986–1987. Long-term measurements of the horizontal and slant visibility were made and compared with the readings of a set of transmissometers placed along the lidar beam direction. The instruments showed good agreement with the transmissometers in all weather conditions, including snowfalls, rains, etc.

A variant of the asymptotic method in which the ranges Δr_1 and Δr_{max} are established in proportion to the visibility range meets with difficulty when applied to relatively clear atmospheres. In such atmospheres, it is difficult to increase Δr_1 and Δr_{max} to keep z_1 and $z_{1,max}$ invariant. The main reason that restricts increasing the lidar ranges in proportion to visibility is a poor signal-to-noise ratio in clear atmospheres. Here the intensity of the backscatter signal (and accordingly, the signal-to-noise ratio) dramatically decreases because of the small value of the backscatter coefficient and strong signal attenuation due to the factor r^{-2}. To maintain sensible constant values for z_1 and $z_{1,max}$ in clear atmospheres, it is necessary to measure the backscatter signal at large distances from the lidar. For example, for $z_1 \approx 1$, $z_{1,max} = 1.5$–2, and a visibility range $L_M = 30\,km$, the maximum operating range of the lidar must be approximately 20–25 km. For a typical ground-based elastic lidar, such ranges are not realistic. Generally, the maximum range of a ground-based tropospheric lidar system ranges from 3–5 to ~10 km. In clear atmospheres, the length of Δr_{max} cannot be increased indefinitely to maintain $z_{1,max} = const$. At best, Δr_{max} may be kept constant. In this case, the measurement uncertainty increases rapidly when visibility increases (Fig. 12.3, *curves 1* and *2*). This effect, known as an "edge effect" of the asymptotic method (Zuev et al., 1976 and 1978) can be decreased by using a correction procedure. For this, the maximum value of two-way transmittance, T^2_{max} in Eqs. (12.28) and (12.29) can in some way be estimated. The simplest way is the use of the information contained in the lidar signal itself. In homogeneous atmospheres, the two-way transmission term at the range Δr_{max} can be estimated by the simple formula

$$T^2_{max} = \frac{Z_r(r_{max})}{Z_r(r_0)} \tag{12.40}$$

Obviously, this type of estimate of T^2_{max} may contain considerable uncertainty; therefore, it can only reduce the edge effect. An inaccurate estimate of T^2_{max} will result in a systematic shift in the measured L. In Table 12.3, the shift in the calculated meteorological optical range caused by an inaccurate estimate of T^2_{max} is given as a function of the measured optical depth (Ignatenko and Kovalev, 1985). The actual two-way transmission term $T^2_{max} = 0.02$, its estimated value is 0.04, and the uncertainty of the range-corrected signal at r_0 is 0.01.

TABLE 12.3. The Systematic Shift δL% Due to Incorrect Estimate of T^2_{max} as a Function of the Optical Depth $\tau(\Delta r_1)$

$\tau(\Delta r_1)$	0.05	0.1	0.2	0.3	0.5	0.7	0.9	1.2
δL, %	−7.4	−7.7	−8.3	−9.1	−11.3	−14.0	−17.7	−26.0

The results given in Table 12.3 agree with the estimates made by Zuev et al. (1976 and 1978), which showed that the measurement errors increase rapidly when the optical depth of the measurement range becomes larger than unity.

12.2. VISUAL RANGE IN SLANT DIRECTIONS

12.2.1. Definition of Terms and the Concept of the Measurement

Interest in atmospheric path transmission in slant directions is primarily related to problems associated with airplane monitoring and photography of ground objects. Another problem is the determination of runway ground marker visibility at airports under poor visibility conditions. For slant or vertical visibility measurements, integrated atmospheric parameters over extended ranges, such as transmittance or optical depth, are the basic parameters of interest rather than range-resolved profiles of the extinction coefficient. Accordingly, the data processing techniques used for such measurements differ from those described in previous chapters.

The bulk of this section is devoted to the problem of slant visibility measurement at airports. As shown in Section 12.1.1, the main purpose behind determination of the horizontal visual range is to provide pilots and air traffic services with information on visibility along the runway. The information is obtained with a transmissometer or nephelometer. It determines the visual range close to the ground surface. These instruments provide the pilot with information that is useful either before aircraft take off or after plane landing, when the aircraft is rolling along the runway. However, to make an aircraft landing, the pilot needs to know the visibility that may be expected during the descent and approach to the landing strip. In particular, the pilot needs to know either the slant visual range or the height at which he/she will see the runway markers and lights during the aircraft descent.

Under conditions with high clouds and good visibility, the pilot can see the landing strip from great heights and far away from the airport. In bad weather conditions, the pilot may not be able to see the landing strip during the first stage of the landing, until the plane is close enough to the strip. In this case, the pilot should establish visual contact with the nearest ground marks when entering the landing approach zone. Particularly, the pilot must be able to see at least a short length of the runway markings or lights on the ground surface to have the proper spatial orientation with respect to the runway. In such conditions, the pilot should be provided with information on the *visual contact height* (Annex 3, 1995; Manual, 1995). The visual contact height is the maximum height at which the pilot on the descent glide can make reliable visual reference with the ground runway marks or lighting system. In Fig. 12.4, a schematic of the pilot's visibility conditions on the aircraft descent trajectory is shown. *Point A* is plane current location along the descent glide AB; h is the aircraft altitude relative to the ground surface $BCDE$, and *point B* is the

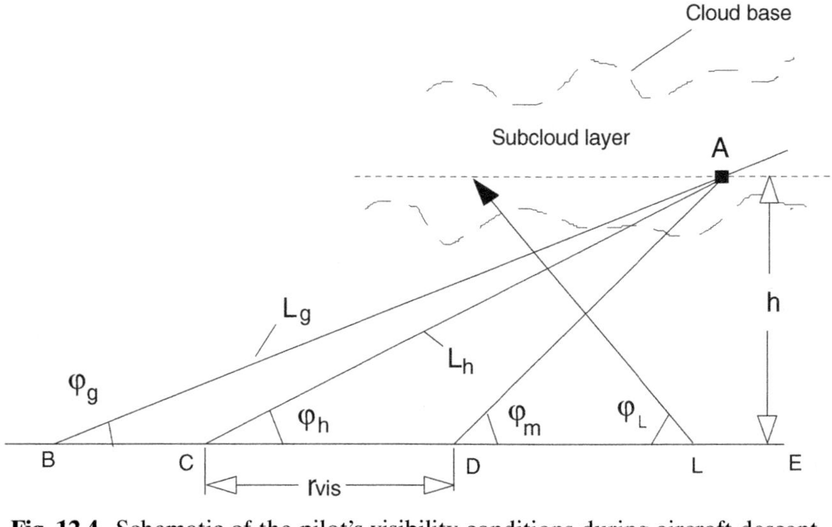

Fig. 12.4. Schematic of the pilot's visibility conditions during aircraft descent.

plane touchdown point near the landing strip threshold. Being at *point A*, the pilot may not see the threshold *B*, but only some restricted ground segment r_{vis}, with a chain of the approach lights on it. These lights allow the pilot to keep the right direction toward the strip. To make such orientation possible, some minimum number of lights must be simultaneously seen, so that the length of the visual segment r_{vis} must to be adequately extended.

According to existing regulations, a civil aircraft is permitted to land only if the visual contact height, assessed by the airport meteorological service, exceeds the pilot's personal *decision height* (DH). The decision height is established on the basis of the pilot's experience and is formally authorized. The decision height is the lowest altitude at which the aircraft pilot must either make the decision to land or interrupt the aircraft descent and go around for another attempt. In the former case, the pilot must see some minimum length, $r_{vis,min}$ to be able to continue the descent toward the landing strip. It is assumed that otherwise the pilot does not have a sufficiently reliable visual reference of the runway markings or lights and therefore must break off the plane descent.

The International Civil Aviation Organization (ICAO) has defined the lower limits for acceptable meteorological conditions in which aircraft landings may be permitted as Categories I, II and III.

These weather condition minima for the civil airports are (Manual, 1995):

- Category I: the decision height DH = 60 m and the runway visual range RVR = 800 m
- Category II: DH = 30 m and RVR = 400 m
- Category III: DH < 30 m and RVR < 400 m

As mentioned above, the visual contact height is the important piece of information on the visibility conditions that must be reported to the pilot before the plane landing. To obtain an accurate estimate of the visual contact height, information on the atmospheric turbidity is required about the layer from the ground to the height *h*. To determine the expected visual range r_{vis}, which will be seen by the pilot, one needs to know how the atmospheric transmittance (or optical depth) varies with height. Unfortunately, the meteorological services at airports have no commercial instrumentation that can determine the profile of the extinction coefficient in slant directions. The commercial ceilometer, used to determine the cloud base height, is the only instrument that is commonly used by air traffic services for the assessment of the visual contact height. This instrument operates in the same manner as conventional target-ranging radar, sometimes called a LADAR. The ceilometer emits a short light pulse in the vertical direction. Then the period is determined between the time at which the pulse is emitted and the time when the return pulse appears at the detector, reflected from the cloud base. Ceilometers can provide information on the visual contact height when the light pulse reflected by the cloud base is strong enough to be discriminated. In other words, a ceilometer can properly operate only if the cloud base is sufficiently well-defined to create a sharply reflected light pulse. This means that the operational use of a ceilometer requires a particular type of extinction coefficient vertical structure, specifically, a moderately clear atmosphere below the cloud and a sharp increase in the backscatter coefficient at the base of the cloud. Such a propitious situation usually occurs with high, dense clouds in which the cloud base usually is well-defined. The height of such clouds, generally, is not less than several hundred meters (Ratsimor, 1966; Lewis, 1976). In this case, the cloud base and visual contact heights coincide because the pilot is unlikely to make visual reference with ground lights when he is within a dense cloud. However, low-level clouds (especially, stratus) usually have no well-defined cloud base. Below the dense cloud body, these clouds generally have a subcloud layer, which is less dense and may extend from the cloud as far as the ground surface. In such situations, a slow degradation of the visibility with height occurs, so that there is no sharp boundary between the cloud and the underlying atmosphere.

In the 1970s, intensive airplane measurements of atmospheric optical parameters were carried out within the subcloud layer (Ratsimor, 1966). These measurements were made by an airborne backscattering nephelometer during plane horizontal flights within and below low clouds (stratus, nimbostratus, etc.). The analyses of the multitude arrayed data showed that the subcloud layer usually extends down to the ground surface if the cloud base height is less than 200 meters. The corresponding dependence of the horizontal visibility with altitude is shown as *curve 1* in Fig. 12.5. Note that the horizontal visibility decreases monotonically from the ground surface up to the cloud base.

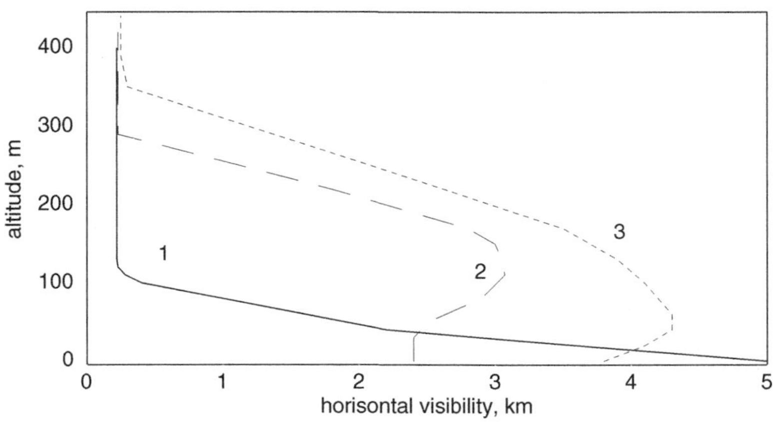

Fig. 12.5. Typical dependencies of horizontal visibility as a function of height for low-cloudiness conditions. *Curve 1* shows the visibility decrease with the height for stratus with a cloud base height from 100 to 150 m. *Curve 2* is the same but for stratus and cumulus with a cloud base height 150–300 m; *curve 3* is the same but for nimbus with a cloud base height more than 300 m. (Adapted from Ratsimor, 1966.)

(For simplicity, the cloud base is defined by the author of the study as the least height at which the horizontal visibility reaches some minimum value and above which it has no noticeable monotonic change). If the cloud base height is more than 200–300 m above the ground, the subcloud layer usually does not extend down to the ground surface. Here the horizontal visibility generally increases slightly near the ground, and then decreases monotonically toward the cloud base (*curves 2* and *3*). In Fig. 12.6, generalized dependencies of the vertical extinction coefficient profiles on the height are shown under low stratus and cumulonimbus, based on the study by Ratsimor (1966). This type of spatial structure of the extinction coefficient profile below low clouds creates great difficulties when attempting to provide pilots with accurate information on the visual contact height. With low-level clouds, as with heavy rains, snowfalls, and snowstorms, a conventional light pulse ceilometer has difficulty determining the cloud base boundary. Moreover, the definition of the cloud base boundary in such situations becomes an issue.

Unfortunately, even knowledge of the cloud base height (as defined above) cannot, by itself, solve the problem of the slant visibility determination. The presence of an extended subcloud region can seriously impede visibility through it in the flight direction (Fig. 12.4). The pilot may not be able to see the ground markings or lights through the subcloud layer, even after the plane has decended below the cloud base. On the other hand, it is impossible to determine the length of segment r_{vis} from the data obtained by a conventional light pulse ceilometer. This is a significant limitation of commercial ceilometers and requires consideration of alternative methods to determine the visual contact height. The determination of the vertical profile of the optical depth

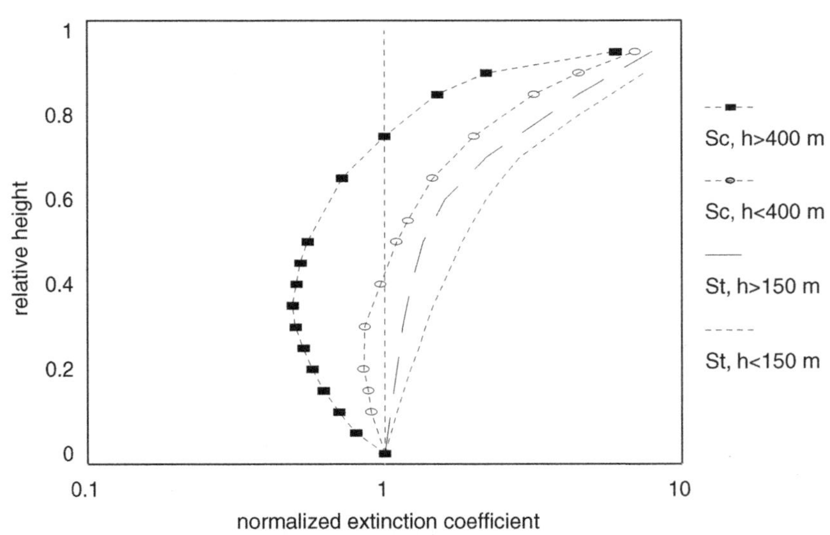

Fig. 12.6. Relative vertical extinction coefficient profiles under low clouds as a function of height. (Adapted from Ratsimor, 1966.)

is the only way to overcome this limitation. This makes lidar a potential instrument to determine the visual contact height. In fact, lidars are the only instruments that may be considered as practical for slant visibility measurements.

Before considering lidar data processing algorithms, let us consider what sort of visibility information can be extracted from the lidar signal. The theoretical basis for slant visibility measurements with lidar is much the same as for conventional horizontal visibility measurements (Kovalev, 1988). This means that the same formulas, such as Allard's law [Eq. (2.11)], can be used for the visual range assessment both in the horizontal and slant directions. For an inhomogeneous atmosphere, the transcendental Eq. (12.6) can be rewritten as

$$E_{\mathrm{T}} = \frac{I_{\mathrm{A}}}{L_{\mathrm{h}}^2} e^{-\int_0^{L_h} \kappa_t(x)\mathrm{d}x} \tag{12.41}$$

where I_{A} is the intensity of the runway approach lights, E_{T} is the visual threshold of illumination, and L_h is the slant visibility range from the altitude h. The integral in the exponent is the optical depth of the slant range $AC = L_h$ (Fig. 12.4). Note that this integral represents a limiting value for the optical depth along the distance AC when the light at point C can still be seen from the height h on a perception level. Any increase of the optical depth of the layer will make the light invisible to the pilot. Denoting the optical depth of the length L_h as

$$\tau_{\mathrm{B}}(L_h) = \int_0^{L_h} \kappa_{\mathrm{t}}(x)\mathrm{d}x \tag{12.42}$$

one can rewrite Eq. (12.41) in the form

$$\ln\frac{E_{\mathrm{T}}}{I_{\mathrm{A}}} + \tau_{\mathrm{B}}(L_h) + 2\ln L_h = 0 \tag{12.43}$$

To solve the transcendental Eq. (12.43) for the unknown L_h, it is necessary to know the optical depth $\tau_{\mathrm{B}}(L_h)$. There are two significant difficulties to be overcome, which are inherent in visibility measurements along slant paths. First, the integral in Eq. (12.42) has a variable upper limit L_h. To find the unknown optical depth $\tau_{\mathrm{B}}(L_h)$, the upper limit of the integration, that is, the unknown L_h, must be known. The unknowns $\tau_{\mathrm{B}}(L_h)$ and L_h are related to each other and must be determined simultaneously. The second difficulty deals with the restricted measurement range of the lidar as compared with the visibility range L_h. As stated in Section 12.1, the lidar measurement range is generally less than the measured visual range. Therefore, the application of a lidar for visibility measurements requires some extrapolation of the measured data beyond the lidar measurement range, in a way similar to that used for transmissometer measurements. Meanwhile, with slant directions, the assumption of atmospheric homogeneity cannot be applied, at least in the manner as was made for horizontal measurements. Slant measurement extrapolation should be based on more accurate presumptions.

This and other problems emerged when the first exploration of the operational utility of lidar for aircraft landing operations was made by Viezee et al. (1969). This study, made under conditions of low ceilings and poor visibility conditions, provided the researchers with information about difficulties associated with lidar measurements at airports. Because practical eye-safe lidars were developed more than 20 years later (Spinhirne, 1993 and 1995), the first requirement noted by the authors was the necessity to limit any likely hazard to the human eye. The lidar observations, made in the elevation angle range from approximately 6° to 65°, revealed that an unacceptably large attenuation of the lidar energy occurred along slant paths through fog at low elevation angles. High layering could be detected only at the highest elevation angles, where the path length through the lower-level clouds and the fog was not so significant. The authors were disappointed that the maximum range at which low clouds could be detected remained far below the distance at which the landing approach path intersected the cloud base (1–3 miles). It was established that under the prevailing weather conditions, the lidar was capable of describing the low-level cloud structure only over a range of 0.5–0.75 miles. Nevertheless, it was established that lidar can provide the vertical extinction profile through low-level clouds and describe the cloud ceiling spatial distribution when operating with variable elevation angles.

That the lidar maximum range is much less than the slant visibility range was the most discouraging revelation in the first attempts of slant visibility measurements. However, this drawback is inherent in both horizontal and slant visibility measurements. It was shown in the previous section that for horizontal visibility measurements, this problem is overcome by the extrapolation of the measured data beyond the instrument baseline, that is, beyond the measurement range. Comparisons of the lidar and transmissometer measurements revealed that horizontal low-level heterogeneity, which is typical for bad weather conditions, does not significantly worsen the accuracy of lidar measurements of the horizontal visibility (Baldenkov et al., 1989). This is because the visibility determination is based on the use of path-integrated optical parameters, such as the optical depth or transmission, over an extended area, which can be found more accurate than range-resolved parameters.

As with horizontal visibility measurements, the determination of the visual contact height with lidars can be based on the principle of horizontal extrapolation of integrated atmospheric characteristics as proposed by Spinhirne et al. (1980). The extrapolation of the integrated characteristics obtained by lidar must be made within the atmospheric layer defined by the ground surface and the height of the visual contact (Kovalev, 1988). This extrapolation is reasonable if the vertical optical depth of layer $(0, h)$ (Fig. 12.4) can be determined as the product of optical depth of this layer, measured in a slant direction, and the sine of the elevation angle [Eq. (9.15)]. In such atmospheres, the mean extinction coefficient of the atmospheric layer remains the same when it is measured with arbitrary elevation angles (Section 9.2). This allows the use of any arbitrary direction of lidar examination to determine the slant visibility. To explain the principle of the slant visibility measurement, we return to Fig. 12.4. The lidar system, located at point L, measures the mean extinction coefficient of the layer $(0, h)$. The measurement is made in a slant direction, at an elevation angle, ϕ_L. In the atmosphere, where the condition given in Eq. (9.15) is valid, the optical depth of the layer $(0, h)$ along the line of sight of the pilot, AC, can be determined as

$$\tau(h, \phi_h) = \tau(h, \phi_L)\left(\frac{\sin \phi_L}{\sin \phi_h}\right) \tag{12.44}$$

where $\tau(h, \phi_h)$ and $\tau(h, \phi_L)$ are the optical depths of the layer $(0, h)$ along the slope angles ϕ_h and ϕ_L, respectively. The relationship in Eq. (12.44) allows calculation of the slant visibility range using the lidar data measured over a range that can be much shorter than the visibility range L_h. This, in turn, makes it possible to select the direction of the lidar examination to be at an angle other than the pilot's line of sight.

The principal requirement that follows from Eq. (12.44) is the equality of the mean extinction coefficients of the layer h along the slant paths ϕ_h and ϕ_L, rather than equality of local values. When using this relationship, the local variations in the extinction coefficients are not influential. To illustrate this, let us split the layer h into m thin horizontal layers Δh_j, so that $h = m\Delta h_j$. Clearly, the presumption of horizontal homogeneity within every thin layer Δh_j is only some approximation of reality. In real atmospheres, the extinction coefficient κ_t within these layers Δh_j is not absolutely invariant, so that random fluctuations in the extinction coefficient κ_j always take place along the layer Δh_j. Let us denote the absolute deviation of the extinction coefficient along the pilot's line of sight as $\Delta\kappa_{j,h}$. If similar fluctuations occur at all altitudes, one can write the mean extinction coefficient of the layer $(0, h)$ along the pilot's line of sight as

$$\overline{\kappa_t}(h, \phi_h) = \frac{1}{m}\sum_{j=1}^{m}(\kappa_{t,j} \pm \Delta\kappa_{j,h}) \tag{12.45}$$

Similarly, the mean extinction coefficient along the lidar searching direction (ϕ_L) is

$$\overline{\kappa_t}(h, \phi_L) = \frac{1}{m}\sum_{j=1}^{m}(\kappa_{t,j} \pm \Delta\kappa_{j,L}) \tag{12.46}$$

where $\Delta\kappa_{j,L}$ denotes the random fluctuations of the extinction coefficient in the lidar examination direction. The difference in the mean extinction coefficient as measured by the lidar and that along the pilot's line of sight AC is

$$\delta\overline{\kappa_t}(h,\phi_h)=\frac{\sum_{j=1}^{m}\Delta\kappa_{j,L}-\sum_{j=1}^{m}\Delta\kappa_{j,h}}{m\overline{\kappa_t}(h,\phi_h)} \tag{12.47}$$

If the fluctuations $\Delta\kappa_{j,h}$ and $\Delta\kappa_{j,L}$ are randomly distributed relative to the mean extinction coefficient, the difference in Eq. (12.47) is small. It means that the value of $\bar{\kappa}_t(h, \phi_L)$, as determined by the lidar, is close to the mean extinction coefficient $\bar{\kappa}_t(h, \phi_h)$ over the line AC. Note that the relative fluctuations $\Delta\kappa_j/\kappa_{t,j}$ are generally higher when $\kappa_{t,j}$ is small. Therefore, the relative uncertainty, $\delta\kappa_t(h, \phi_h)$ decreases as the extinction coefficient, $\bar{\kappa}_t(h, \phi_h)$ becomes larger. This means that the extrapolation above yields more accurate results in bad visibility conditions. The long-term slant visibility measurements made in the USSR in 1987–1989 confirmed the potential of this method of extrapolation under poor visibility conditions (Rybakov et al., 1991).

As shown in Eq. (12.42), the optical depth $\tau_B(L_h)$ and the unknown slant visual range L_h are related to each other. This means that these values must be determined simultaneously rather than in sequence. Defining $\tau_B(L_h)$ through the mean extinction coefficient $\overline{\kappa_{t,max}}(0,h)$ of the layer $(0, h)$

$$\tau_B(L_h)=\overline{\kappa_{t,max}}(0,h)L_h \tag{12.48}$$

one can rewrite Eq. (12.43) as

$$\ln\frac{E_T}{I_A}+\overline{\kappa_{t,max}}(0,h)L_h+2\ln L_h=0 \tag{12.49}$$

The mean extinction coefficient $\overline{\kappa_{t,max}}(0, h)$, as defined in Eq. (12.48), has a simple physical interpretation, which is as follows. The extinction coefficient $\overline{\kappa_{t,max}}(0,h)$ defines the maximum level of atmospheric turbidity that still allows the ground lights at the distance $L_h = h \sin\phi_h$ to be seen. In other words, $\overline{\kappa_{t,max}}(0, h)$ is the maximum value of the extinction coefficient when a ground light with intensity I_R can be seen from the altitude h at the threshold of vision. When visibility worsens, L_h decreases. There is a least acceptable length for the distance L_h, below which safety requirements for aircraft landing are not met and, accordingly, the landing is not permitted. As follows from the definition of the visual contact height, some minimum length of the segment r_{vis} (Fig. 12.4) along the runway must be seen by the pilot to orient the plane during landing. As follows from the geometric scheme shown in Fig. 12.4, the relationship between L_h and h for the minimum visible area $r_{vis} = r_{vis,min}$ can be derived from the formula

$$L_h=\sqrt{h^2+\left(\frac{h}{\mathrm{tg}\,\phi_m}+r_{vis,min}\right)^2} \tag{12.50}$$

For civil aviation, the minimum visible area $r_{vis,min}$ must be at least 150–300 m. This length makes it possible for the pilot to see a line of lights that includes 6–11 lights separated by an interval of 30 m. With Eqs. (12.49) and (12.50), one can calculate the dependence of the extinction coefficient $\overline{\kappa_{t,max}}(0, h)$ on the height h, at which the established ground segment with the length $r_{vis,min}$ can be seen from the height h. The dependencies are shown in Fig. 12.7; here the length of $r_{vis,min}$ is chosen as 300 m and $I_A = 25000$ cd. *Curve 1* is calculated for a nighttime condition with the visual threshold of illumination $E_T = 10^{-6}$ lx, *curve 2* is calculated for a twilight condition with $E_T = 10^{-5}$ lx, and *curve 3* is calculated for a daytime condition with $E_T = 10^{-3}$ lx. These values of E_T are recommended by ICAO regulations (Manual, 1995). The dependencies in Fig. 12.7 can be treated as boundary curves to estimate the runway approach light visibility under different conditions of ambient illumination. The unknown visual contact height can be found by using two functions of height: (1) the vertical profile of the mean extinction coefficient $\overline{\kappa}_t(0, h)$ as measured by lidar and (2) the boundary profile of $\overline{\kappa_{t,max}}(0, h)$ calculated with Eq. (12.49) using the appropriate visual threshold of illumination E_T and light intensity I_A. The intersection of these curves indicates the height at which the mean extinction coefficient $\overline{\kappa}_t(0, h)$ of the layer $(0, h)$, as defined by lidar, is equal to the limiting value, $\overline{\kappa_{t,max}}(0, h)$. In other words, the intersection point determines the height at which the approach light chain can be seen at the minimum acceptable distance $r_{vis,min} = 300$ m. To clarify the application of the graph in Fig. 12.7, an imaginary profile of the extinction coefficient $\overline{\kappa}_t(h)$ is shown as *curve 4* over the altitude range from approximately, 65 to 110 m. When determining the intersection points of the curve with *curves 1, 2,* and *3*, one can establish the visual contact height h for different ambient illuminations. The height is equal to 100 m if landing is made during nighttime, ~90 m during twi-

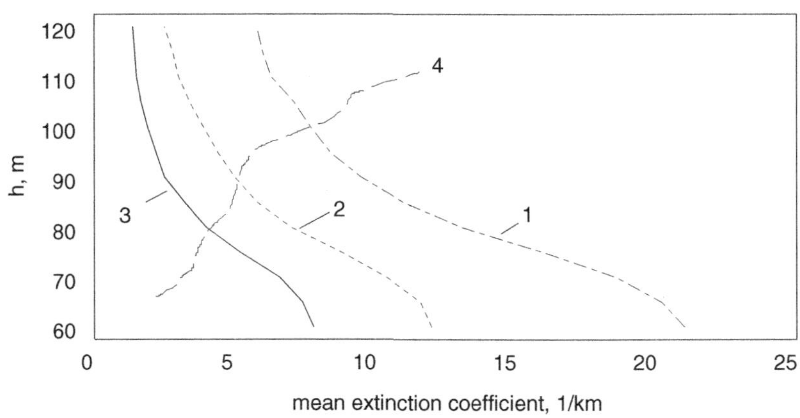

Fig. 12.7. Profiles of the extinction coefficient, $\overline{\kappa_{t,max}}(0, h)$, at which the lights are seen within the segment $r_{vis,min}$, given as the function of the altitude. *Curves 1, 2,* and *3* determine visibility conditions for night, twilight, and daytime, respectively. The length of the segment $r_{vis,min}$ is 300 m. The intensity of the approach lights is $I_R = 25{,}000$ cd.

light, and only ~80 m for daytime landing (the intersection points of *curve 4* with *curves 1, 2, 3*, respectively). It is assumed here that to facilitate plane landing, the runway lights are switched on even during the daytime. According to the ICAO regulations, this is generally done in poor visibility conditions, when runway lights can be seen better than other markings.

A similar approach may also be used to find the visibility range along the glide path, L_g (Fig. 12.4). To determine the height at which the runway threshold, B, may be seen by the pilot when descending, the visual range of the lights in that area should be known. Equation (12.49) is the transformed into the form

$$\ln\frac{E_T}{I'_R} + \overline{\kappa_{t,max}}(0, h)L_g + 2\ln L_g = 0 \tag{12.51}$$

where I'_R is intensity of the lights at the runway threshold and h is the maximum height at which the runway threshold lights at point B can be seen under existing atmospheric conditions; the slope visual range $L_g = h/\sin\phi_g$; where ϕ_g is the angle of the glide slope (generally, $\phi_g = 2°30'–2°40'$). Equation (12.48) is now reduced to

$$\tau_B(L_g) = \overline{\kappa_{t,max}}(0, h)L_g$$

Using the relationship between the length L_g and height h, one can transform Eq. (12.51) into a dependence between the maximum height h, from which the runway threshold lights are seen in existing weather conditions, and the corresponding mean extinction coefficient. The relationship is

$$\overline{\kappa_{t,max}}(0, h) = \frac{\sin\phi_g}{h}\left[\ln\frac{I'_R}{E_T} - 2\ln h + 2\ln\sin\phi_g\right] \tag{12.52}$$

The dependence between the maximum height h, at which the runway threshold lights are seen, and the corresponding mean extinction coefficients in the layer $(0, h)$ is shown in Fig. 12.8. All of the parameters involved are the same as for *curves 1–3* in Fig. 12.7.

Potentially, other characteristics describing the visibility conditions can be determined with lidar, for example, the visibility range from an established altitude. Unlike determining the visual contact height or visibility range along the glide path, the mean extinction coefficient profile is not required to determine the visibility from a fixed altitude. Here only the vertical optical depth or transmittance over the fixed layer of interest must be determined.

12.2.2. Asymptotic Method in Slant Visibility Measurement

In airports, instrumental visibility measurements are important only under poor visibility conditions, when the visibility is less than ~2 km. This limit can vary; however, the visibility range of interest is restricted to turbid atmos-

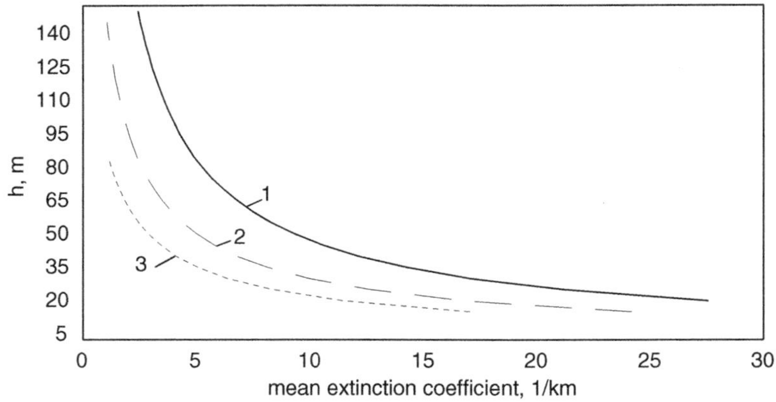

Fig. 12.8. Profiles of the extinction coefficient, $\overline{\kappa_{t,max}}(0, h)$, at which the pilot can establish visual contact with the runway edge lights, as functions of the altitude. *Curves 1*, *2*, and *3* are determined with the same values of E_T as that in Fig. 6.8.

pheres. In bad visibility conditions, molecular scattering is commonly negligible in comparison with particulate scattering, so that the approximation of a single-component atmosphere may be used. Another specific is that to determine visibility range, only integrated values are required to be measured rather than local parameters. The presence of significant multiple scattering, at least at the far end of the measurement range, which impedes the use of, for example, the Klett method, is the third specific of these measurements. Considering these, one can conclude that the above method of asymptotic approximation is the most appropriate for such visibility measurements. First, this method is less sensitive to multiple scattering in bad visibility conditions as compared with other methods (Zuev et al., 1978 and 1978a). Second, under poor visibility conditions, the boundary value for the lidar equation solution can be estimated from the lidar signal integrated over the maximum measurement range]. When the transmittance $T^2_{max} \ll 1$, the value of the range-corrected signal, integrated in the range from r_0 to r_{max}, is close to $I(r_0, \infty)$, which is the solution boundary value [Eq. (12.30)]. Therefore, the integral can be used as the boundary value for the lidar equation. The principal requirement to apply the asymptotic method is the need to measure the lidar signals over an extended range with a relatively large optical depth. Accordingly, the lidar must have an appropriate dynamic range and sensitivity to measure a lidar signal with an acceptable signal-to-noise ratio.

To find the mean extinction coefficient profile in the atmospheric layer of interest, the lidar must be directed into the atmosphere in a slope direction $\phi_L > \phi_h$ (Fig. 12.4). The lidar searching angle must be selected large enough to obtain the profile of the mean extinction coefficient over the vertical distance of interest. This altitude range must be larger or at least equal to the measured visual contact height h. With the lidar maximum range r_{max}, the upper height range is $h_{max} = r_{max} \sin\phi_L$, and the height h_{max} must be large enough to find the intersection points as shown in Fig. 12.7. Another issue is related to the minimum measurement range. Because of the incomplete overlap area of the lidar (r_0), useful returns can only be obtained for altitudes $h > r_0 \cos\phi_L$, rather than from the ground surface. Meanwhile, the determination of the visual range of the runway or approach lights requires the knowledge of the extinction coefficient over the atmospheric layer beginning from the ground surface. Therefore, the extinction coefficient in the lowest atmospheric layer should somehow be determined. As shown by Spinhirne et al. (1980), there are several ways to determine the extinction coefficient in the lower layer. This can be achieved by making additional lidar measurements with smaller elevation angles (Sasano, 1996). Another option is to extrapolate the measured extinction coefficient profiles down to the region (0, h_0). However, the particular details of the slant visibility measurements are beyond the scope of the general method outlined here.

Equation (12.25) can be rewritten as

$$T_1^2 = 1 - \frac{I_{r,1}}{I_{r,\max}}(1 - T_{\max}^2) \tag{12.53}$$

here $I_{r,max} = I_{r,1} + I_{r,2}$ is the total integral of the range-corrected signal Z_r over the range from r_0 to r_{max}. To apply Eq. (12.53) for slope visibility measurements, the terms T_1^2 and T_{max}^2 should be used here in their general form for heterogeneous atmosphere, so that Eq. (12.19) is written in the form

$$T_1^2 = e^{-2\int_{r_0}^{r_1} \kappa_t(r')dr'}$$

and T_{max}^2 is defined with Eq. (5.52). The mean extinction-coefficient $\bar{\kappa}(r_0, r_1)$ is found from Eq. (12.53) as

$$\overline{\kappa_t}(r_0, r_1) = \frac{\ln I_{r,\max} - \ln[I_{r,\max} - I_{r,1}(1 - T_{\max}^2)]}{2(r_1 - r_0)} \tag{12.54}$$

In Eqs. (12.53) and (12.54), the term $(1 - T_{max}^2)$ can be considered as the solution boundary value. The simplest way to determine T_{max}^2 is to use the information containied in the lidar signal itself. Basically, the same approach may be applied here as used in horizontal visibility measurement, that is, determining the ratio of $Z_r(r_{max})$ to $Z_r(r_0)$ [Eq. (12.40)]. A more accurate formula for a heterogeneous atmosphere is

$$1 - T_{\max}^2 = 1 - \left[\frac{Z_r(r_{\max})}{Z_r(r_0)}\right]\left[\frac{\beta_{\pi,p}(r_0)}{\beta_{\pi,p}(r_{\max})}\right] \tag{12.55}$$

With slant-direction measurements, the backscattering coefficients at r_0 and r_{max} cannot be taken to be the same value. Meanwhile, information on the backscattering coefficient $\beta_{\pi,p}(r_0)$ and $\beta_{\pi,p}(r_{max})$ ratio in Eq. (12.55) is not generally available. Two specifics facilitate the estimate of the term $(1 - T^2_{max})$. First, in turbid atmospheres, the relative uncertainty of the term $(1 - T^2_{max})$ is much less that that for T^2_{max}. In poor visibility conditions, the optical depth of the total range $\tau(r_0, r_{max})$ is large, and the value of T^2_{max} is small in comparison with the unit. Second, the extinction coefficient generally increases with height under reduced visibility and low cloudiness conditions (Figs. 12.5 and 12.6). In rainfalls or snowfalls, the vertical extinction coefficient usually remains relatively constant up to the cloud base. This is a good reason to assume that in poor visibility conditions the ratio of the backscattering terms in Eq. (12.55) obeys the condition

$$\frac{\beta_{\pi,p}(h_0)}{\beta_{\pi,p}(h_{max})} \leq 1 \tag{12.56}$$

Accordingly, one can reduce Eq. (12.55) to

$$T_m^2 \leq \frac{Z_r(r_{max})}{Z_r(r_0)} \tag{12.57}$$

Now the term $[1 - T^2_{max}]$ in Eqs. (12.53) and (12.54) may be estimated as

$$1 - T^2_{max} \approx \frac{Z_r(r_0) - Z_r(r_{max})}{Z_r(r_0)} \tag{12.58}$$

One can easily determine the behavior of the systematic uncertainty caused by an incorrect estimate of T^2_{max}. If instead of actual T^2_{max}, an inaccurate estimate of this quantity, $\langle T^2_{max}\rangle$ is used, the mean extinction coefficient is obtained with a systematic error $\Delta\bar{\kappa}(r_0, r)$, so that the calculated extinction coefficient is found with the formula

$$\overline{\kappa_t}(r_0, r_1) + \Delta\kappa_t(r_0, r_1) = \frac{\ln I_{r,max} - \ln[I_{r,max} - I_{r,1}(1 - \langle T^2_{max}\rangle)]}{2(r_1 - r_0)} \tag{12.59}$$

Note that a reasonable value of $\langle T^2_{max}\rangle$ is always be selected as a positive nonzero value. Subtracting Eq. (12.54) from Eq. (12.59), one can find the relationship between the absolute shift $\Delta T^2_{max} = \langle T^2_{max}\rangle - T^2_{max}$ and the systematic uncertainty in the derived extinction coefficient. To make such an estimate more general, it is reasonable to find the systematic shift in the measured optical depth, rather than in the extinction coefficient, which depends on the range r_1. After some algebraic manipulation, the systematic uncertainty $\Delta\tau_1$ in the obtained optical depth can be written in the form

$$\Delta\tau_1 = -\frac{1}{2}\ln\left[1+\left(\frac{\Delta T^2_{max}}{1-T^2_{max}}\right)\left(\frac{1-T_1^2}{T_1^2}\right)\right] \tag{12.60}$$

The relative uncertainty, $\Delta\tau_1/\tau_1$, incurred by the selection of an inaccurate value of T^2_{max} is shown in Fig. 12.9. The curves are calculated with different values of ΔT^2_{max}. *Curve 1* is calculated for the case when the actual $T^2_{max} = 0.02$ and the estimate used to determine the optical depth τ is $\langle T^2_{max}\rangle = 0.03$. *Curve 2* is found for $T^2_{max} = 0.03$ and $\langle T^2_{max}\rangle = 0.05$. *Curve 3* is calculated for $T^2_{max} = 0.05$ with an estimate $\langle T^2_{max}\rangle = 0.08$. One can conclude that in bad visibility conditions, an error in the determination of T^2_{max} results in an acceptable uncertainty in the retrieved optical depth τ_1 until the optical depth is less than 1.

When using the asymptotic method, one should discriminate between the lidar measurement range, where the atmospheric characteristics are determined, and the maximum range, over which the lidar signals should be measured. To obtain an acceptable estimate of $I(r_0, \infty)$, the lidar must measure signals over an extended range (r_0, r_{max}) with a total optical depth of not less than 1.5–2. However, this defines only the operating range (r_0, r_{max}) over which the lidar signal is integrated, rather than the lidar measurement range (r_0, r_1), where the profile of the mean extinction coefficient is determined. The optical depth of the range (r_0, r_1) generally does not exceed unity. Thus the asymptotic method allows one to determine the transmittance or the mean extinction coefficient profile over some range, which is much less than the total range (r_0, r_{max}), over which integral $I_{r,max}$ is defined.

The general solution given in Eq. (12.54) is derived with an assumed range-independent backscatter-to-extinction ratio $\Pi_p = const.$ In the analysis presented in this section, we maintain this approximation. However, in practical measurements, a range-dependent backscatter-to-extinction ratio may be used

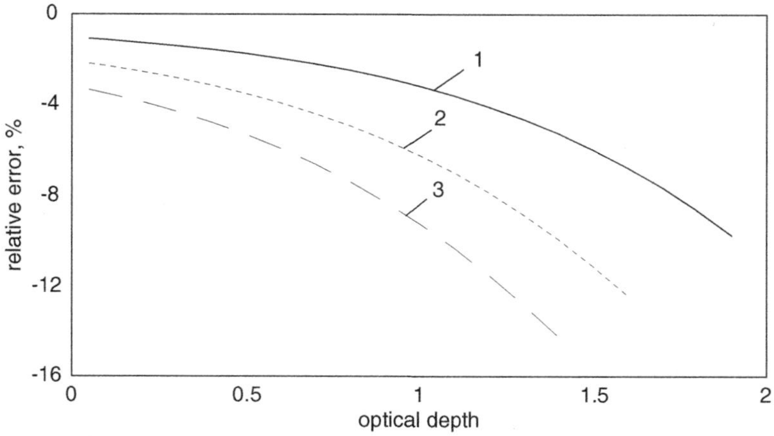

Fig. 12.9. Errors in the measured optical depth due to uncertainty in assumed T^2_{max}.

to improve the measurement accuracy. With slant visibility measurements, made in a cloudy atmosphere, the total operating range (r_0, r_{max}) often includes at least two zones with different type of backscattering. In the near zone, below the base of the cloud, backscattering occurs in moderately turbid or even clear air. In the far zone, the backscattering originates with large cloud particulates. The backscatter-to-extinction ratio is significantly different in these two zones. Ignoring this characteristic will result in increased measurement uncertainty. On the contrary, the use of a range-dependent Π_p makes it possible to obtain data with acceptable quality.

Field experiments made in the USSR in the late 1980s confirmed that the asymptotic method yields a reliable determination of slant visibility characteristics in bad weather conditions. In 1989, the mobile lidar instrument "Electronica-06R," developed to measure the visual contact height, underwent experimental tests at the airport in Uljuanovsk (Rybakov et al., 1991). The instrument automatically operated in a continuous manner. To process the lidar data, the methods described above to determine the visual contact height were used. The lidar was located at a distance of about a kilometer from the runway threshold, near the point where the airport operational ceilometer was set up. Data measured with the lidar were directly compared both with the ceilometer data and with visual observations made by the pilots during the aircraft descent. It was established that systematic discrepancies in the retrieved vertical extinction coefficient profiles may occur if the lidar data are processed with a range-independent backscatter-to-extinction ratio. The use of the constant Π_p caused systematic shifts in the vertical extinction coefficient profiles at the distant ranges. The shift disappeared when variable backscatter-to-extinction ratios were used (Kovalev et al., 1991).

12.3. TEMPERATURE MEASUREMENTS

Lidars have been used to measure atmospheric temperature by many investigators with a variety of methods. These may be broadly divided into four major classifications: rotational Raman, differential absorption, molecular (Rayleigh) scattering density measurements, and Doppler broadening of molecular scattering. The measurement of temperature in the atmosphere is among the earliest uses of lidars. Indeed, the methodology to convert density measurements to temperature with scattered light predates the invention of the laser (Elterman, 1951, 1953, 1954). The measurement of temperature using the molecular scattering of laser light was first demonstrated by Kent and Wright (1970). The development of additional methods quickly followed. The use of the rotational Raman spectrum of nitrogen was proposed by Strauch et al. (1971) and Cooney (1972) to obtain calibrated temperature measurements. With this technique, an accuracy of about 0.25°C can be achieved at low altitudes. Fiocco et al. (1971) used variations in the width of the Doppler broadening of molecular scattering to measure temperature. Kalshoven et al. (1981) demonstrated a differential absorption lidar method for temperature measurements. They used two laser wavelengths to measure the changes in oxygen absorption lines with temperature to infer the atmospheric temperature up to 1-km altitude with 1°C accuracy. Endemann and Byer (1981) reported simultaneous measurements of atmospheric temperature and humidity with a continuously tunable IR lidar. They used a three-wavelength differential absorption lidar technique with water vapor absorption lines. With this technique, a 2.3°C absolute accuracy was achieved. Today, lidars using a combination of these techniques continuously monitor the temperature of the atmosphere (see, for example, Hauchecorne et al., 1991, 1992; Keckhut et al., 1993, 1995, 1996; Chanin et al., 1990).

Four specific lidar techniques have been developed to measure temperature profiles in the middle and upper atmosphere. In addition to the use of molecular (Rayleigh) scattering to measure density, there are three differential absorption methods that make use of the existence of atomic metals (sodium, potassium, and iron) at high altitudes. Because these metals are found in a limited region of the atmosphere (roughly from 70 to 100 km), lidars using metallic fluorescence often extend the temperature measurements both higher and lower into the atmosphere with molecular scattering techniques. The measurement of temperature with molecular scattering is limited by relatively weak scattering cross sections and requires a power aperture product greater than about 100 W-m^2 to make useful temperature measurements at altitudes near 100 km (Meriwether et al., 1994). This requires a telescope with a diameter larger than a meter and 10–50 W of laser power. Although these systems are within current technology, they are large and have significant power requirements, making portable systems difficult. Narrow-band sodium lidars currently provide the highest resolution and the most accurate temperature measurements (Gardner and Papen, 1995; She et al., 2000; Chu et al., 2000). This is the result of a relatively high sodium density at high altitudes and a large fluorescent cross section that provides a strong signal from a moderately sized telescope (0.25–0.5 m). However, the requirement for precise wavelength control in sodium (and to a lesser extent in potassium) fluorescence lidars is difficult for mobile systems that may be subjected to rough handling. There may also be issues related to small signal amplitudes from molecular scattering below the sodium layers, making it difficult to determine temperatures over an extended area. Although narrow-band potassium lidar systems have been built (von Zahn and Hoffner, 1996), the density of potassium atoms is small so that the signals are always weak. Furthermore, the potassium resonance line is in the IR portion of the spectrum so that molecular scattering is also weak, making it difficult to extend the region in which temperature is measured with molecular scattering.

12.3.1. Rayleigh Scattering Temperature Technique

Temperature measurements using molecular scattering to determine molecular density have been performed for many years. They are a natural outgrowth of high-altitude density measurements first made in the early 1950s by

Elterman (1951) with a pulsed searchlight and a photomultiplier mounted at the focus of a large collecting mirror located several kilometers away from the searchlight. The methodology to convert density measurements to temperature was first developed and used by Elterman (1953, 1954). Kent and Wright (1970) were among the first to accomplish this using a laser as the light source. Many investigators have developed and improved the method (see, for example, Kent and Keenliside, 1975; Hauchecorne and Chanin, 1980; Shibata et al., 1986; Hauchecorne et al., 1991; Hauchecorne et al., 1992; Hauchecorne, 1995). Rayleigh scattering temperature measurements have been in continuous use since 1980 in studies of the upper atmosphere, particularly in the region from 30 to 90 km (Keckhut et al., 1990). Most middle-atmosphere Rayleigh lidars use a frequency-doubled Nd:YAG laser operating in the green region of the visible spectrum, at 532 nm. Typical systems employ telescopes with diameters near 1 m and lasers with average powers levels of 10–50 W. These systems typically have power aperture products of approximately 25 W-m^2.

In regions of the atmosphere where particulates are not present or are in low concentration, changes in the range-corrected signal of an elastic lidar are indicative of changes in the molecular density. If either the temperature or molecular density is known or can be assumed at some altitude, the temperature and density measurements can be extended over a larger region with the lidar data. The technique works best at stratospheric altitudes and for relatively short wavelengths, which maximize the return from molecular scattering and minimize the relative contribution from particulate scattering. For this situation, the range-corrected lidar signal from a vertically starting lidar system can be written as

$$P(h)h^2 = C\sigma_{\pi,\mathrm{m}}(\lambda)n_\mathrm{m}(h)\exp\left[-2\sigma_\mathrm{m}(\lambda)\int_{h_0}^{h} n_\mathrm{m}(h')\mathrm{d}h'\right] \tag{12.61}$$

where h is the altitude, $\sigma_{\pi,\mathrm{m}}(\lambda)$ is the angular molecular scattering cross section in the direction $\theta = 180°$ relative to the direction of the emitted laser light, $\sigma_\mathrm{m}(\lambda)$ is the total molecular extinction cross section, C is a system constant, and $n_\mathrm{m}(h)$ is the number density of molecules at the altitude h. A comparison of the signal from two altitudes, h_1 and h_2, results in

$$n_\mathrm{m}(h_2) = n_\mathrm{m}(h_1)\frac{P(h_2)h_2^2}{P(h_1)h_1^2}\exp\left[2\sigma_\mathrm{m}(\lambda)\int_{h_1}^{h_2} n_\mathrm{m}(h')\mathrm{d}h'\right] \tag{12.62}$$

The solution of this equation for a given set of lidar measurements, $P(h_1)$ and $P(h_2)$ requires iteration but converges rapidly. Combining Eq. (12.62) with the ideal gas law and the hydrostatic equations

$$p_\mathrm{atm}(h) = \kappa n_\mathrm{m}(h)T_\mathrm{atm}(h) \text{ and } -\frac{\mathrm{d}p_\mathrm{atm}(h)}{\mathrm{d}h} = Mn_\mathrm{m}(h)g(h) \tag{12.63}$$

one can obtain

$$T_{\text{atm}}(h_i) = \frac{Mg(h_i)\Delta h}{\kappa \ln\left[\dfrac{p_{\text{ref}}(h_0) + \sum_{j=0}^{i} Mn_{\text{m}}(h_j)g(h_j)\Delta h}{p_{\text{ref}}(h_0) + \sum_{j=0}^{i-1} Mn_{\text{m}}(h_j)g(h_j)\Delta h}\right]} \tag{12.64}$$

where $T_{\text{atm}}(h)$ is the absolute temperature, $p_{\text{atm}}(h)$ is the pressure, p_{ref} is the atmospheric pressure at some height, h_0, within the measurement range, and $g(h)$ is the acceleration due to gravity at altitude h; M is the weighted average mass of the air molecules, and k is the Boltzmann constant. A number of different versions of this equation are used, but all are variants of the result of the combination of the lidar equation, the ideal gas law, and the hydrostatic equation.

Sources of Uncertainty. The error analyses that have been done for this technique generally assume that photon statistics is the only or primary source of error. However, there are a number of assumptions that go into the derivation of Eq. (12.64), each of which is true to some degree. The first is the assumption that scattering from aerosols is negligible. At altitudes above 30 km, there are no large sources of particulates because the emissions from large surface sources (volcanoes, for example) seldom penetrate to these altitudes. Water vapor concentrations are also very low, so ice crystals are not common either. It is also assumed that molecular absorption is unimportant at the measured wavelength region. The molecules found above 30 km are primarily nitrogen, oxygen, and argon, none of which has strong absorption in the spectral region from 300 nm to 1000 nm. Similarly, the assumption of a constant molecular mixing ratio is also made. Hauchecorne and Chanin (1980) estimate that the molecular absorption coefficients are constant in this region of the atmosphere to an accuracy of 0.4% at visible wavelengths. The assumption of hydrostatic equilibrium in turn assumes that turbulence does not result in local density fluctuations. However, because these measurements have large spatial and temporal resolution because of their use of photon counting, any effects due to turbulence will tend to average out. These items are not included in an error analysis because it is difficult, if not impossible, to quantify their effects. Yet it is important to recognize that these assumptions are the limiting factors that ultimately determine how well the method works.

Assuming that photon statistics is the only source of error leads to (Hauchecorne and Chanin, 1980)

$$\frac{\Delta\rho}{\rho} = \frac{[P(h) + P_{\text{BGR}}(h)]^{1/2}}{P(h)} = \frac{\Delta X}{X} \tag{12.65}$$

where $P(h)$ in Eq. 12.65 is the lidar signal at height h, ρ is the density of the air, and X is defined as

$$X = \frac{\rho(h_i)g(h_i)\Delta\delta h}{P(h_i + \Delta h/2)} \tag{12.66}$$

where Δh is the height increment. This quantity is useful in the determination of the uncertainty of the temperature measurement as

$$\frac{\Delta T_{\text{atm}}}{T_{\text{atm}}} = \frac{\Delta X}{(1+X)\ln(1+X)} \tag{12.67}$$

Because this technique really measures the changes in temperature with altitude, it is clear that the lidar measured temperature can only be as accurate as the reference temperature or density. Model atmospheres can provide a starting point for these analyses but may be inaccurate by 10° or more for any given situation.

12.3.2. Metal Ion Differential Absorption

The existence of metal ions at high altitudes has has been examined with lidars for many years (Gault and Rundle, 1969; Felix et al., 1973; Megie et al., 1978). The metal ion differential absorption technique to determine temperature and vertical wind speeds is one of the few lidar methods that are used to consistently monitor the atmosphere (Frickle and Zahn, 1985; Gardner, 1989; Bills et al., 1991a,b; Kane and Gardner, 1993; von Zahn and Hoffner, 1996) These measurements have been done long enough to compile a climatology of the mesosphere (She et al., 2000). It is one of the successes of lidar technology. A great deal of science and understanding has been enabled with the metal ion temperature and wind measurement lidars (Gardner, 1989; Gardner et al., 1989, 1995, 1998; Gardner and Papen 1995; Chu et al., 2000a,b; States and Gardner, 2000a,b; Chu et al., 2001a,b; Gardner et al., 2001). The technique for temperature measurement with sodium and potassium ions relies on the temperature dependence of resonance fluorescence. Narrow-band resonance fluorescence temperature lidars exploit the fact that the absorption cross sections at wavelengths inside the absorption line of the atom change with temperature. The cross section of the sodium D_2 line is depicted in Fig. 12.10 for several temperatures. An increase in temperature results in broadening of the absorption line, while maintaining the total area under the line constant. To accurately measure the temperature of the ions, two laser frequencies are chosen, near the maximum (f_a) and minimum (f_c) of the absorption feature shown in Fig. 12.10 (Papen et al., 1995; Papen and Treyer, 1996). This choice of lines makes the ratio of the lidar returns at the lines, $R_T = P_{fc}/P_{fa}$, highly sensitive to temperature changes but insensitive to changes in the wind velocity (Bills et al., 1991). This choice also minimizes the sensitivity of the temperature measurement to frequency tuning errors.

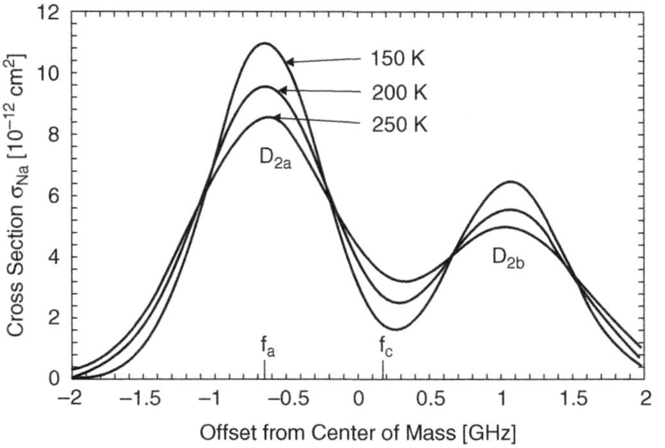

Fig. 12.10. The resonance fluoresence cross section of the sodium D_2 transition for three different temperatures. The wavelength of the centerline is 589.15826 nm (Papen and Treyer, 1996).

The amplitude of the lidar signal for a vertically pointed lidar is given by the lidar equation

$$P_{Na}(\lambda, h) = \frac{C_{Na} E n_{Na}(h) \sigma_{Na}(\lambda, T_{atm}, v_R, g, I) \exp\left\{-\int_0^h [\beta(h', \lambda) + \kappa_{A,Na}(h', \lambda)] dh'\right\}}{h^2} \tag{12.68}$$

where E is the laser energy per pulse, $n_{Na}(h)$ is the number density of sodium atoms at height h, $\sigma_{Na}(\lambda, T_{atm}, v_R, g, I)$ is the effective absorption cross section that depends on the laser wavelength, λ, the temperature T_{atm}, the radial wind velocity v_R, the line shape of the laser pulse, g, and its intensity I. This cross section is the integrated product of the laser line shape and the thermally Doppler-broadened atomic line; $\beta(h, \lambda)$ is the attenuation of the laser beam due to molecular and particulate scattering, and $\kappa_{A,Na}(h, \lambda)$ is the attenuation of the laser beam due to absorption by sodium.

For the two-frequency technique for temperature measurements, the ratio R_T of the lidar return at the two frequencies, f_a and f_c, is

$$R_T(h) = \frac{P_{fc}(h)}{P_{fa}(h)} = \frac{\sigma_{eff}(f_c, T_{atm}, v_R, g)}{\sigma_{eff}(f_a, T_{atm}, v_R, g)} \tag{12.69}$$

where f_a is a frequency near the peak of the sodium D_{2a} resonance and f_c is a frequency near the minimum between the D_{2a} and D_{2b} resonances. It has been assumed that (1) the lidar signals P_{fc} and P_{fa} are normalized by the emitted energy of each laser pulse, (2) there is no difference in the signal attenuation at the two frequencies, (3) the two lidar returns are measured simultaneously (so that the sodium density does not change), and (4) the response of the lidar is linear with light intensity for each wavelength. Because the spectroscopy of the sodium lines is known extremely accurately, the cross sections can be accurately calculated and the relationship between R_T and temperature can be established.

Papen et al. (1995) define the sensitivity as the normalized change in the ratio per degree of temperature change

$$S_T = \frac{1}{R_T(h,t)} \frac{\partial R_T}{\partial T} \tag{12.70}$$

Then a change in temperature, ΔT, can be determined from a change in the ratio, ΔR_T, found as

$$\Delta T = \Delta R_T \frac{\partial T}{\partial R_T} = \frac{1}{S_T} \frac{\Delta R_T}{R_T} \tag{12.71}$$

and assuming that the errors in the measured temperature are due only to photon statistical noise, the error in temperature can be calculated with

$$\Delta T = \frac{1}{S_T} \left(\frac{1 + 1/R_T}{P_{fa}} \right)^{1/2} = \frac{Q_T}{P_{fa}^{1/2}} \tag{12.72}$$

where the parameter Q_T is the number of counted photons required in the lidar signal P_{fa} to obtain a temperature with an accuracy of ΔT.

The analysis described above omits the complications that result from Doppler shifting of the lines due to motion of the molecules along the direction of the lidar beam. The effects of the Doppler shift can be seen in Fig. 12.11

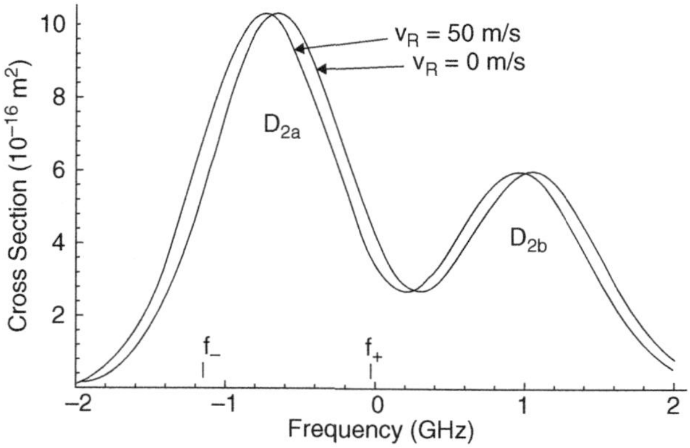

Fig. 12.11. The resonance fluoresence cross section of the sodium D_2 transition for two different velocities, showing the Doppler shift. The wavelength of the centerline is 589.15826 nm (Papen and Treyer, 1996).

for two velocities. This shift complicates the relationship between R_T and the local temperature. At least one more wavelength is required to solve simultaneously for the component of wind velocity along the lidar line of sight and the temperature. A pair of frequencies that could be used to determine the magnitude of the Doppler shift is shown in Fig. 12.11. To obtain the maximum sensitivity, the frequencies, f_+ and f_- are located symmetrically on either side of the D_{2a-} resonance. The considerations that go into the number and choice of optimal frequencies for use by sodium lidars are discussed in some detail by Papen et al. (1995). The availability of at least one more wavelength enables another ratio, R_W, to be constructed as

$$R_w(h) = \frac{P_{f_+}(h)}{P_{f_-}(h)} = \frac{\sigma_{\text{eff}}(f_+, T_{\text{atm}}, v_R/\lambda, g)}{\sigma_{\text{eff}}(f_-, T_{\text{atm}}, v_R/\lambda, g)} \tag{12.73}$$

where v_R/λ is the amount of the Doppler shift. For a given choice of wavelengths, the ratios R_T and R_W are functions only of the temperature T_{atm} and radial velocity v_R and can be calculated quite accurately. In practice, lookup tables are required and an iterative procedure is used to determine the temperature and radial velocity. It is possible, with judicious choices for the operating frequencies, to obtain a ratio R_T that is insensitive to the Doppler shift and a ratio R_W that is insensitive to changes in temperature to eliminate the requirement for an iteration (Papen et al., 1995).

In many ways, the potassium D_{2a} and D_{2b} resonances are very similar to those of sodium. They are different in that they are much closer together and are not resolved at the temperatures normally found in the upper atmosphere. As shown in Fig. 12.12, the lines form a single feature that is nearly Gaussian

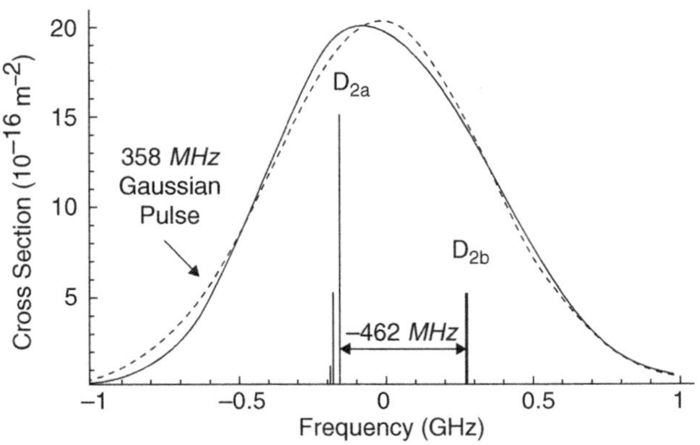

Fig. 12.12. A plot of the effective cross section for potassium at 200 K. A fitted single Gaussian curve with an rms width of 358 MHz is also shown. The six hyperfine lines that comprise the potassium D_2 line are shown. The wavelength of the centerline is 766 nm (Papen et al., 1995).

in shape. The amount of Doppler broadening is smaller than sodium because of the larger mass of the potassium atom and the longer wavelength of the absorption feature. This leads to a taller, narrower absorption feature in potassium in which the effective cross section of the peak is about twice that of sodium.

The method by which the radial velocity and temperature are found is similar to that for sodium. A frequency f_c can be found near the peak of the absorption feature at which the absorption cross section is insensitive to the amount of the Doppler shift due to motion of the atoms. Two frequencies, f_+ and f_- are located symmetrically on either side of the absorption resonance. Two ratios are constructed from the measured lidar returns, $P_{f+}(h, t)$, $P_{f-}(h, t)$, and $P_{fc}(h, t)$ as

$$R_w(h,t) = \frac{P_{f_+}(h,t)}{P_{f_-}(h,t)} \quad \text{and} \quad R_T(h,t) = \frac{P_{f_+}(h,t)+P_{f_-}(h,t)}{P_{f_c}(h,t)} \tag{12.74}$$

The construction of these two ratios in this way makes R_W insensitive to temperature changes and R_T insensitive to the magnitude of the Doppler shift. Thus only three laser frequencies are required for the potassium technique.

Because the shape of the potassium absorption feature is nearly Gaussian, an approximate analytical uncertainty analysis can be performed to determine the sensitivity of the derived temperature and radial velocity to the choice of laser wavelengths used and the measured parameters. The relative uncertainty in temperature, δT_{atm}, can be found from the approximate expression

$$\delta T_{atm} = \frac{\Delta T_{atm}}{T_{atm}} = \left[\frac{f_0^2}{2\sigma_D^2} - \frac{v_R f_0}{\lambda\sigma_D^2}\tanh\left(\frac{v_R f_0}{\lambda\sigma_D^2}\right)\right]^{-1} \frac{\Delta R_T}{R_T} \tag{12.75}$$

where f_0 is the difference in frequency between the centerline frequency and the frequencies f_+ and f_-, and σ_D^2 is a fitted parameter obtained from a comparison of the shape of the absorption feature to a Gaussian function; σ_D^2 is approximately equal to $\sigma_D^2 = 266.2 + 0.46T$ (Papen et al., 1995). Similarly, an estimate of the relative error of the radial wind velocity can be made from

$$\delta v_R = \left[\frac{\lambda\sigma_D^2}{2f_0}\right] \frac{\Delta R_w}{R_w} \tag{12.76}$$

Because of the complexity of the two expressions above, there is an optimal choice of f_0 (i.e., the separation between the frequencies used on either side of the peak) that simultaneously minimizes the uncertainties in both of the measured parameters. Papen et al. (1995) discuss the considerations required to optimally choose all three of the frequencies used in the technique.

Comparing the sodium and potassium methods, one can conclude that the sensitivity to temperature in both methods is nearly equal (S_T is 0.84 for sodium and 0.81 for potassium); however, a potassium system requires nearly 50% more photons to obtain the same temperature performance as a similar sodium system. On the other hand, the sensitivity to radial wind velocities in a potassium system is twice that of sodium (S_W is 0.85 for sodium and 1.9 for potassium), so that only 80% photons need be collected to obtain the same performance (Papen et al., 1995). Both methods require extremely fine control of the laser wavelength and line width. Because of this, it may be simpler to build potassium systems because they can use the fundamental lasing regions of Cr:LiSrAlF or Ti:sapphire laser systems. Sodium systems, in contrast, require some kind of frequency shifting technique (dyes or optical parametric oscillators). It should also be noted that both methods suffer from the potential saturation of the fluorescence with resulting nonlinear effects. Thus it is necessary to avoid the high laser powers and low-beam divergences that cause saturation. Unfortunately, it is these same characteristics that are necessary to make daytime observations (see, for example, Welsh and Gardener, 1989; Von der Gathen, 1991; She and Yu, 1995).

A third variant of the metal ion differential absorption method was proposed by Gelbwachs (1994). This type of system, known as an iron-Boltzman factor lidar, takes advantage of a layer of atomic iron from 80 to 100 km. The method uses iron as a fluorescence tracer and relies on the temperature dependence of the population difference of two closely spaced electronic transitions (Fig. 12.13). In thermal equilibrium, the ratio of the populations in the J = 3 and J = 4 sublevels in the ground-state manifold is given by the Maxwell–Boltzmann distribution law

$$\frac{n(\mathrm{J}=3)}{n(\mathrm{J}=4)} = \frac{g_2}{g_1}\exp\left(-\frac{\Delta E}{k_B T_{atm}}\right) \tag{12.77}$$

where $n(\mathrm{J}=3)$ and $n(\mathrm{J}=4)$ are the populations of the two states with degeneracy factors $g_1 = 9$ and $g_2 = 7$, ΔE is the energy difference between the two levels, $\Delta E = 416\,\mathrm{cm}^{-1}$, k_B is the Boltzmann constant, and T_{atm} is the atmospheric temperature. At 200 K, the ratio of the two populations is approximately 26. The temperature is then given by

$$T_{atm} = \frac{\Delta E/k_B}{\ln\left[\dfrac{g_2\, n(\mathrm{J}=4)}{g_1\, n(\mathrm{J}=3)}\right]} \tag{12.78}$$

The relative number density of the iron atoms in each of these two states can be measured with resonance fluorescence lidar techniques. Note that to determine the temperature, only the ratio of the densities need be found as opposed to the absolute number of atoms. The density of atoms in a given state is proportional to the number of backscattered photon counts from iron atoms

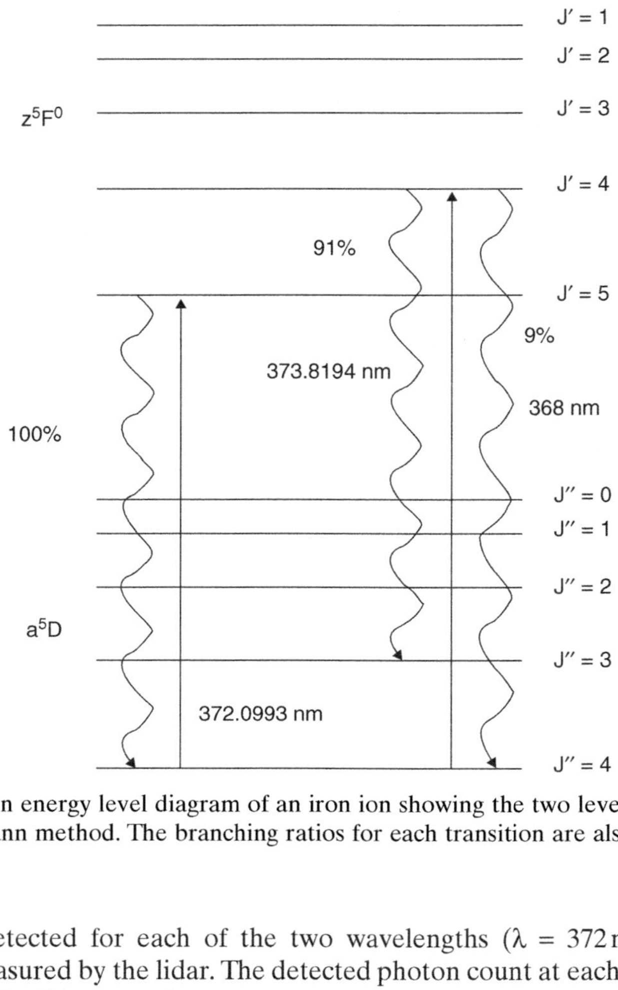

Fig. 12.13. An energy level diagram of an iron ion showing the two levels used in the iron-Boltzmann method. The branching ratios for each transition are also shown.

$P_{Fe}(\lambda, h)$ detected for each of the two wavelengths (λ = 372 nm and λ = 374 nm) measured by the lidar. The detected photon count at each wavelength is given by the lidar equation

$$P_{Fe}(\lambda, h) = \frac{C_{Fe} E n_{Fe}(h) \sigma_{Fe}(\lambda, T_{atm}, l_{laser}) R_{B\lambda} \exp\left\{-\int_0^h [\kappa_t(h', \lambda) + \kappa_t(h', \lambda_{Fe})] dh'\right\}}{h^2} \tag{12.79}$$

where E is power of the laser; $R_{B\lambda}$ is the branching ratio, ($R_{B374} = 0.9114$, $R_{B372} = 1$); and λ and λ_{Fe} are the laser and fluorescence wavelengths, respectively. Note that the fluorescence wavelengths in the above equation may have different values, (λ_{Fe} may be either 372 or 374 nm); $\sigma_{Fe}(\lambda, T_{atm}, l_{laser})$ is the effective absorption cross section of the Fe transition, which is a function of temperature T_{atm}, laser wavelength λ, and laser linewidth l_{laser}; $n_{Fe}(h)$ is the number density of iron atoms at height h, $\kappa_t(h, \lambda)$ and $\kappa_t(z, \lambda_{Fe})$ are the total extinction coefficients at the laser wavelength λ and at the fluorescence wavelength; C_{Fe} is the system coefficient that takes into account the effective area of the telescope, the transmission efficiency of the optical train, and the detector quantum efficiency at the desired wavelength. The effect of a possible atomic velocity on the absorption cross section has not been included but is negligible for vertical sounding lidars.

The method as implemented by Chu et al. (2002) uses two separate lasers and telescopes because the two iron lines are spectrally too far apart to use a single laser to generate them and are too close to be separated through the use of dichroic beam splitters. Because the amount of energy at each wavelength emitted by the laser may be different and the throughput at each wavelength may be different, it is necessary to normalize the photon counts at each wavelength. The normalized counts, R_{372} and R_{374}, are found by dividing the number of counts in the iron channels by the number of counts from molecular scattering at a common altitude. Using these values, the temperature at each altitude can be found from the formula

$$T_{\text{atm}}(h) = \frac{\Delta E / k_B}{\ln\left[\frac{g_2}{g_1}\frac{R_{B374}(h)}{R_{B372}(h)}\left(\frac{\lambda_{374}}{\lambda_{372}}\right)^{4.0117}\frac{R_E^2(h)R_\alpha}{R_T(h)}\right]} = \frac{598.44}{\ln\left[\frac{0.7221\, R_E^2(h)R_\alpha}{R_T(h)}\right]} \tag{12.80}$$

where the ratios R_T, R_E, and R_α are defined as

$$R_T(h) = \frac{P_{374}(h)}{P_{372}(h)} \quad R_E(h) = \frac{\kappa_{374}}{\kappa_{372}} \quad R_\alpha = \frac{\sigma_{\text{eff}}(374, T_{\text{atm}}, l_{\text{laser},374})}{\sigma_{\text{eff}}(372, T_{\text{atm}}, l_{\text{laser},372})} \tag{12.81}$$

$R_T(h)$ is the ratio of the normalized lidar signals at a given height, R_E is the ratio of the extinction coefficients at each of the two laser wavelengths, and R_α is the ratio of the effective iron absorption cross sections at a given temperature considering also the linewidth of the laser light at each wavelength ($l_{\text{laser},374}$ and $l_{\text{laser},372}$).

An alternate approach is presented by Paper and Treyer (1998). The approach is based on Eq. (12.77), so that a ratio of the lidar signals at each wavelength R_T is made such that

$$R_T = \frac{P_{374}}{P_{372}} = \frac{\sigma_{Fe}(374\text{nm}, T_{\text{atm}}, l_{\text{laser},374})g_2}{\sigma_{Fe}(372\text{nm}, T_{\text{atm}}, l_{\text{laser},372})g_1}\exp\left(-\frac{\Delta E}{k_B T_{\text{atm}}}\right) = C_1 \exp\left(-\frac{C_2}{T_{\text{atm}}}\right) \tag{12.82}$$

where C_1 and C_2 are constants that may be fit to a calibration data set or calculated from first principles if the laser lines and line widths are known to sufficient accuracy. An advantage of this approach is that it allows an analysis of the iron-Boltzmann method. From the equation above, the sensitivity follows directly as

$$S_T = \frac{1}{R_T(z,t)} \frac{\partial R_T}{\partial T} = \frac{C_2}{T_{atm}^2} \tag{12.83}$$

Although the exact values of the constants C_1 and C_2 are a function of the laser wavelengths and line shapes used, they are on the order of $C_1 \approx 0.725$ and $C_2 \approx 600$ (Paper and Treyer, 1998). It appears from Eq. (12.83) that the sensitivity would be higher for low temperatures. However, there are few atoms in the upper energy state at low temperatures, so that the number of returning photons is small and thus the uncertainty becomes large. The number of photons, Q_T, required to obtain an accuracy of 1 K can be found by substitution of Eq. (12.82) and (12.83) into Eq. (12.72) to obtain

$$Q_T = \frac{T_{atm}^2}{C_2}\left[1 + \frac{1}{C_1}\exp\left(\frac{C_2}{T_{atm}}\right)\right]^{1/2} \tag{12.84}$$

It can be seen that the number of photons required for some desired degree of accuracy is large both when T_{atm} is small (due to the exponential term) and when T_{atm} is large (due to the leading T_{atm}^2 term). For the iron-Boltzmann method, the minimum number of photons required is a minimum at about 150 K. A similar effect occurs in the sodium method of temperature measurement and is a minimum for that method near 80 K.

The biggest drawback to the iron-Boltzmann technique is the fact that the system is actually two complete lidar systems operating at 372 and 374 nm. The low signal level on the weak 374-nm channel limits the overall performance of the system. Typical iron densities in the ground state in the most dense portion of the iron layer vary from approximately 50 to 300 cm^{-3}. With densities this low, daytime observations are possible but difficult and require long integration times. The performance of an iron-Boltzmann system and a sodium system are similar if the total power of the iron system is about eight times that of the sodium system. Iron-Boltzmann lidar systems have a significant practical advantage in that the laser line widths that will give comparable performance can be an order of magnitude wider than those used in a sodium system. The larger line widths make the iron system less sensitive to frequency tuning errors. However, this insensitivity limits the ability of this kind of system to make wind measurements. (Papen and Treyer, 1998).

All three of the metal ion techniques are limited to the measurement of temperature in regions where the number density of the ion of interest is sufficiently dense to enable the technique. To make the system more useful, temperatures above and below the metal layers are found with Rayleigh scattering temperature techniques. The advantage of the metal ion methods is that they provide the absolute temperature reference information that is needed for the Rayleigh scattering method. The iron-Boltzmann technique uses light in the near ultraviolet in which molecular scattering is more than four times more intense than at 532 nm. Using the molecular scattering signal from both the

372- and 374-nm channels, temperatures could be measured down to 30 km, albeit with a longer integration time than most Rayleigh lidars. The use of molecular scattering becomes more difficult with sodium lidars (at 589 nm) and potassium lidars (766 nm) as the operating wavelength increases.

12.3.3. Differential Absorption Methods

The metal ion temperature methods described above are actually variants of a more general differential absorption method that exploits changes in the absorption cross sections of molecules with temperature. A change in temperature does two things to absorption cross sections; first, it widens the shape of individual absorption features in frequency space, reducing the intensity of the peak absorption, and second, it changes the relative population of the energy states available to the molecules. Temperature measuring systems can be based on either of these two effects. For example, the sodium and potassium methods above exploit the change in the shape of the absorption feature, and the iron-Boltzmann method uses the change in the relative population of two states. Both methods require strict control of the wavelength and line width of the emitted laser beams, but methods using changes in the shape of the absorption lines require extreme precision, a factor of 10–50 more precise than measurement of changes in population.

The differential method requires a molecule or atom that is plentiful in the atmosphere, is uniformly mixed in the atmosphere, and has absorption features at wavelengths for which there are laser transitions. In practice, the requirement for large number densities limits the useful molecules to nitrogen and oxygen. Water vapor and carbon dioxide, the next largest constituents of the atmosphere, are not well mixed and may vary considerably with altitude and time. Water vapor concentrations have been measured to vary by factors of several over relatively short distances in the atmosphere. The metal ion techniques work because the transitions used are resonance fluorescence lines with cross sections that are more than ten thousand times larger than normal absorption lines. It is possible to use molecules that are not uniformly mixed (for example, water vapor) to measure temperature if measurements are made using a sufficient number of frequencies to measure both the concentration and temperature in each range element. This makes an already complex system even more so, but it has been done. For lidars using atmospheric backscatter, the transitions must occur at wavelengths shorter than about 2 μm to have a reasonably sized backscatter cross section. Because many of these systems rely on photon counting, the usable wavelength range is much less, generally limiting to wavelengths less than 1 μm where detectors capable of photon counting are common (photomultipliers capable of measuring wavelengths as long as 1.7 μm have recently been introduced, albeit with low quantum efficiencies). As a result of these practical limitations, the number of options is severely limited, with oxygen bands at 680 and 760 nm receiving the largest amount of attention.

The first method exploits the change in the number density of molecules in various rotational quantum states. The method was first suggested for lidar use by Mason (1975). As the temperature increases, the population in the upper level states will increase while that of the lower level states will decrease. This causes the envelope of the absorption of each of the rotational lines to change as shown in Fig. 12.14. Measuring at least two of the lines allows one to determine the temperature. With an assumption of thermal equilibrium, the ratio of the populations in the two rotational states, J_1 and J_2, of the ground state is given by the Maxwell–Boltzmann distribution law

$$\frac{n(J_1)}{n(J_2)} = \frac{g_1}{g_2}\exp\left(-\frac{\Delta E_{1-2}}{k_B T_{atm}}\right) \tag{12.85}$$

where $n(J_1)$ and $n(J_2)$ are the populations of the two states, g_1 and g_2 are the degeneracy factors for each state, ΔE is the energy difference between the two levels, k_B is the Boltzmann constant, and T_{atm} is the atmospheric temperature. The ratio of the number density can be found from a ratio of the lidar signal at each of the two wavelengths. More detailed treatments can be found in the studies by Mason (1975) and Endemann and Byer (1981).

The method has been demonstrated experimentally by Murray et al. (1980), who used a CO_2 lidar to measure the average temperature along a 5-km path. This demonstration used only two laser lines and assumed that CO_2 is uniformly distributed in the air. Although the lidar-measured temperature correlated well with ambient temperature measurements, absolute errors on the order of 5°C were observed. This particular method requires the use of large range elements or a retroreflector because of the small size of the absorption

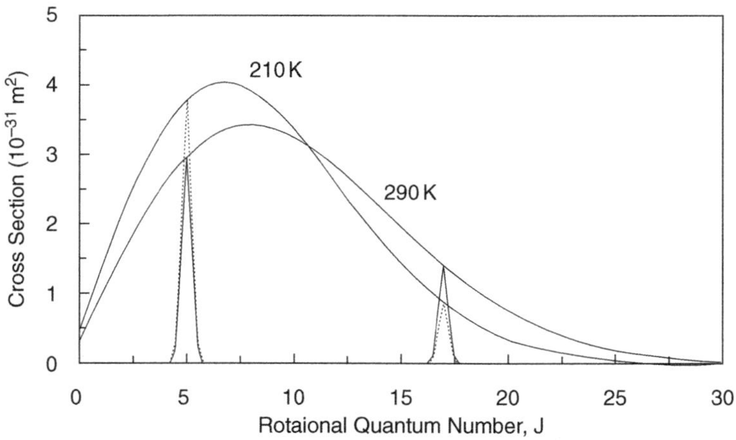

Fig. 12.14. The calculated absorption cross sections for two lines in the P branch of the oxygen molecule for two temperatures. The effect of a change in temperature for the two lines is clearly seen.

cross section of CO_2. An exceedingly weak atmospheric backscatter is also an issue limiting the range of such a system. Endemann and Byer (1981) used two water vapor lines near 1.9 μm over a 1-km path to achieve a 1.5°C uncertainty. It should be noted that each of these demonstrations used a retro-reflective target to increase the signal-to-noise ratio by several orders of magnitude above what it would be for a range-resolved system.

The shape of each of the individual absorption lines is also a function of temperature. As shown in Fig. 12.15, the absorption line becomes broader and shorter as temperature increases. Measurements at no less than three points are required to accurately determine the change in shape. This is because of the possibility of Doppler shifting of the line due to the relative motion between the molecule and lidar. Because the spectral width of an absorption line is small (a few hundredths of a wavenumber), the linewidths of the laser light used must also be small. The centerline wavelength of the laser must also be precisely controlled. This is often done with a cell filled with the appropriate gas to lock the laser line by a feedback technique. Because the temperature and concentration of the gas in the cell can be accurately known, the calibration constant can be determined simultaneously with data collection. Corrections must be made for the effects of collisional broadening, Doppler effects, pressure, and humidity. More details on the method can be found in Kalshoven et al. (1981) and Korb and Weng (1982).

The method has been demonstrated with two oxygen absorption lines at 770 nm over a 1-km path (Kalshoven et al., 1981). The relative error in this demonstration was 0.5°C. This method is particularly attractive because the required wavelengths can be easily generated by several tunable laser systems (for example, Ti:sapphire or alexandrite lasers). The demonstration by

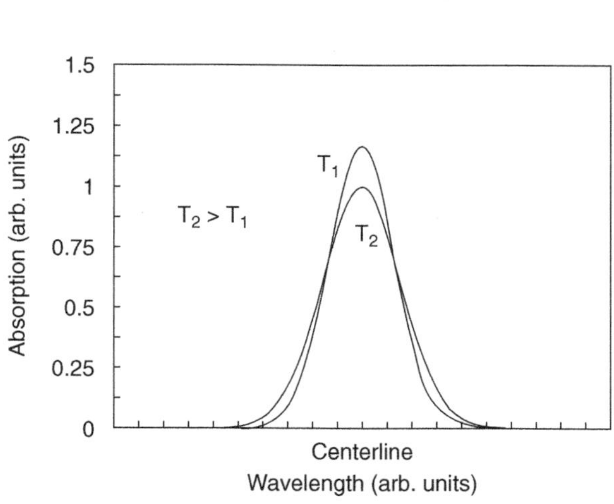

Fig. 12.15. The shape of an idealized absorption line calculated at two different temperatures. As temperature increases, absorption decreases near the center of the feature and increases at the wings.

Kalshoven et al. used a retroreflective target to increase the signal-to-noise ratio so that temperature measurements could be made.

12.3.4. Doppler Broadening of the Rayleigh Spectrum

The temperature dependence of Doppler broadening of the Rayleigh-scattered spectrum allows the measurement of atmospheric temperature by a high spectral resolution lidar (HSRL) (see Section 11.2). To invert data from an HSRL to obtain particulate extinction coefficients, the air density at each altitude is needed as an input. Thus the capability to measure temperature is desirable in an HSRL in that it provides a means of obtaining the needed densities without resort to radiosondes for reference measurements. Temperature measurements made with the variations in the spectral width of the molecular scattering spectrum were first reported by Fiocco et al. (1971). Temperature measurements made with a high-resolution lidar were first reported by She et al. (1992), followed soon after by Alvarez et al. (1993). In their temperature measurement, two barium absorption filters with different filter bandpass widths were used. The amount of light that passes through a molecular absorption cell is proportional to the width of the Doppler-broadened spectrum. Thus a comparison of the signal strength in two cells of different absorption width can be used to determine the range-resolved temperature of the atmosphere. The reported accuracy is 10°C for altitudes below 5 km.

Temperature measurements have also been made with the University of Wisconsin HSRL. Because of the leakage of a small amount of scattered light from particulates into the molecular channels, contamination of the molecular signal will occur in the presence of clouds or dense layers of particulates that will affect the temperature measurements. Thus temperature measurements are limited to areas in which the particulate content is small. The measurements of temperature by the University of Wisconsin HSRL used an iodine absorption filter. The Rayleigh-scattered signal from light passing through the iodine absorption cell is a convolution of the Doppler-broadened Rayleigh spectrum and shape of the iodine absorption spectrum. A Brillouin-modified approximation for the Doppler-broadened spectrum was used to calculate the molecular line shapes at temperatures ranging from −70 to +30°C with 1°C resolution. The calculated line shapes were adjusted to account for attenuation at each wavelength with a measured iodine absorption spectrum. A least-squares technique is used to fit the measured profile to each of the calculated profiles to determine the temperature. Light scattered from particulates that may contaminate the signal from molecular scattering in a particular range bin alters the measured spectrum in a way that underestimates the temperature.

If the scattering molecules are homonuclear, noninteracting, and randomly distributed, the shape of the backscattered spectrum is Gaussian. However, there are effects that may affect the shape of the backscattered spectrum that have nothing to do with changes in temperature. Because the changes in the spectral width of the molecular spectrum due to temperature changes are small, changes in the signal shape due to competing effects may lead to significant fitting errors. Fluctuations in the molecular density may lead to other signal components. For example, density fluctuations that are the result of propagating pressure fluctuations lead to Brillouin peaks in the scattered signal. Density fluctuations that are the result of isobaric fluctuations contribute to the Landau–Placzek band. A more complete discussion of the types of scattering that may occur can be found in Fiocco and DeWolf (1968). Schwiesow and Lading (1981) suggest that corrections to the Gaussian line shape must be made to achieve accuracies on the order of a few degrees. In addition to Rayleigh scattering by molecules, there is a Raman component to the signal straddling the laser line. The Stokes and anti-Stokes lines due to rotational transitions are located on both sides of the laser line. The relative intensity between the Stokes and anti-Stokes portions as well as the shapes of each are functions of temperature. To summarize, the determination of temperature to high accuracy (1°C or less) with an HSRL is limited by small competing effects that also change the shape of the measured spectrum. Achieving this kind of accuracy will require a more complex analysis method than has been used to date.

12.3.5. Rotational Raman Scattering

The use of Raman-scattered light to measure the temperature of the atmosphere was first suggested by Cooney (1972). The concept was first demonstrated and reported by Cohen et al. (1976), with improvements to the method made by Mason (1975), Gill et al. (1979), Arshinov et al. (1983), Mitev et al. (1985), and Vaughan et al. (1993). The method is ideal in that it can measure temperature in the lower part of the atmosphere (from the surface to 30 km) where the molecular scattering methods described in Section 12.3.1 cannot be used. Although the method is straightforward, it was not widely implemented until recent years because of technical difficulties, primarily associated with blocking light outside the desired band.

The origins of Raman scattering are discussed in Section 2.3. The ability to derive temperature information from Raman scattered light is due to the fact that the relative intensity of the various rotational scattering lines changes with temperature. As the temperature of the air, T_{atm}, increases, the populations of rotational states with higher rotational quantum number values, j, increases (Fig. 12.16). The shape of the envelope of the intensities of the Raman-scattered lines for linear molecules is described by

$$I(j,T) = I_0 \nu^4 g_j \frac{\beta hc}{kT} \omega_j N_0 (2j+1) S(j) \frac{\gamma^2}{(2I+1)} \exp\left[-\frac{\beta hc}{kT} j(j+1)\right] \quad (12.86)$$

where I_0 is the intensity of the incident light, I is the nuclear spin quantum number (1 for N_2, 0 for O_2), ν is the frequency of the incident light, N_0 is the

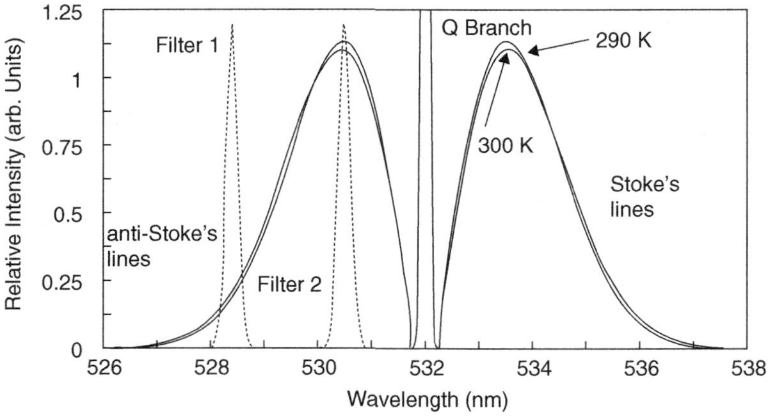

Fig. 12.16. The rotational Raman spectrum for an excitation wavelength of 532 nm. Shown are the spectra for two different temperatures. Also shown are two possible filter choices that could be used to measure the air tempeature. They are situated in regions of the spectrum that change rapidly with temperature.

number density of molecules in the atmosphere, g_j is a statistical weighting factor (for N_2, $g_j = 6$ if j is even and $g_j = 3$ if j is odd; for O_2, $g_j = 0$ if j is even and $g_j = 1$ if j is odd), $\beta = 1.83\,cm^{-1}$ (for N_2) is the molecular rotational constant, γ is the anisotropy of the molecular polarizability tensor, and k is the Boltzmann constant. The product of $S(j)$ and the degeneracy factor $(2j + 1)$ is

$$(2j+1)S(j) = \frac{(j+1)(j+2)}{(2j+3)} \quad \text{for the Stokes (S) branch}$$
$$(2j+1)S(j) = \frac{j(j-1)}{(2j-1)} \quad \text{for the anti-Stokes (O) branch} \tag{12.87}$$

ω_j is the magnitude of the Raman shift of line j and is given by

$$\omega_j = E_j = (4j+6)b_{n21} - D_0\left[6j+9+(2j+3)^3\right] \tag{12.88}$$

where b_{n21} (1.98958 for N_2, 1.43768 for O_2) is the rotational constant of the ground state vibrational level and D_0 ($5.48\ 10^{-6}\,cm^{-1}$ for N_2, $4.85\ 10^{-6}\,cm^{-1}$ for O_2) is the centrifugal distortion constant (Butcher et al., 1971). This value is also referred to as E_j, the energy shift from the central line (in inverse centimeters). It is not uncommon for researchers to deal with the envelope of lines rather than the individual lines. For purely rotational scattering, both oxygen and nitrogen lines contribute to the envelope along with a large number of trace gases. Each of these lines is pressure- and temperature broadened, so as to fill in the gaps between the individual lines (Nedeljkovic et al., 1993).

Along with the increase in signal intensity at lines far from the excitation wavelength, there is a decrease in the signal intensity at intermediate wavelengths. In the basic configuration, interference filters are used to measure the signal intensity in the regions where the signal either increases or decreases. A comparison of these signals can be used to determine the temperature. To obtain the greatest sensitivity, the lines transmitted by each of the filters must be chosen so that the populations change as much as possible in the range of temperatures likely to be measured. In fact, the population of the vibrational states is also a function of temperature, just as the rotational lines. Thus the amplitude of the Raman envelopes at each of the vibrational shifts is a function of temperature. A number of different schemes could be used to measure temperature.

There are several difficulties with the rotational Raman method. The most significant is the problem of rejection of the light from molecular and particulate (elastic) scattering that can contaminate the measured Raman signal. When using purely rotational scattering, some measures must be taken to filter or block the nearby elastically scattered particulate and molecular returns while transmitting lines that are less intense by a factor of about ten thousand. Blocking of at least 10^{-6} at the elastically scattered wavelengths is required to eliminate this component of the signal. The use of interference filters to accomplish this severely limits the maximum transmission in a system that already suffers from a limited signal intensity. Cohen et al. (1976) outlined a data collection and analysis method that could be used to eliminate or reduce the effects of elastic contamination; however, this has never been demonstrated with actual lidar measurements, to our knowledge. Two other recurring problems are maintenance of the long-term stability of the detector/amplifier/digitizer parameters and issues associated with the accurate inversion of the lidar data. Because the line with the maximum temperature sensitivity is only 0.2%/°K, this method requires measurement accuracies on the order of a few tenths of a percent to obtain temperature accuracies of less than a degree, so that exceptional stability is required of the electronics. In practice, this requires that the detectors be specially selected for compatibility and that the electronic components be temperature stabilized. Unfortunately, these actions address only short-term stability and not any long-term drifts. The paper by Vaughan et al. (1993) contains an excellent and thorough discussion of the many considerations that must be made to implement the method as well as estimates of the likely errors involved. Finally, as the discussion proceeds below, it is interesting to note that there are a variety of methods that have been used to analyze data taken in the manner suggested by Fig. 12.16. They are quite different, but each of the methods has some rationale behind its use. Each of the methods claims accuracies that are on the order of a few tenths of a degree.

Perhaps the most common method used to measure the changes in the envelope of the purely rotational shifts (as shown in Fig. 12.16) is to use interference filters (Arshinov et al., 1983; Nedeljkovic et al., 1993; Vaughan et al.,

1993; Behrendt and Reichardt, 2000). The advantage of this technique is that the intensity of the signal from the purely rotational lines is the largest of any of the possibilities. For example, the same technique could be used with the first vibrationally shifted, rotational lines. But for that case, the signal intensity is lower by a factor of 5–15. As previously mentioned, the difficulty with using purely rotational scattering is blocking the elastically scattered light. The width of the rotational envelope increases with increasing wavelength (the width in energy units is constant). It is also true that the longer the wavelength, the easier it is to obtain high-transmission, narrow-line width interference filters with strong out-of-band blocking. However, the cross section for Raman scattering is proportional to $1/\lambda^4$, so that the signal intensity decreases rapidly with longer wavelengths. It is most common to find this technique used with lasers such as XeF (351 nm) or doubled Nd:YAG (532 nm), although the technique has also been done with a ruby laser (694 nm), albeit with an energy of a joule per pulse.

The exact centerline wavelengths and spectral widths of the interference filters used by each researcher have been slightly different. The filters used by Nedeljkovic et al. (1993) are typical at 530.4 and 529.1 nm with a bandwidth of 0.7 nm for an excitation wavelength of 532.1 nm. As noted by Arshinov et al. (1983), the closest filter band should be at least 2 nm from the excitation wavelength to ensure sufficient blocking of the elastically scattered light. It should also be noted that the optimal filter wavelengths will vary with the temperature range that is measured. Nedeljkovic et al. (1993) obtain a response function, $R(T_{atm}, p)$, as the difference between the signal from the two filters normalized by the sum of the two signals. The temperature T_{atm} is obtained from a fitted function as

$$T_{atm} = \frac{a}{\ln(b) + \ln\left(\dfrac{1 - R(T_{atm}, p)}{1 + R(T_{atm}, p)}\right)} + c\left[\frac{a}{\ln(b) + \ln\left(\dfrac{1 - R(T_{atm}, p)}{1 + R(T_{atm}, p)}\right)}\right]^2 + d \qquad (12.89)$$

where a, b, c, and d are constants to be found by fitting the lidar data to a calibration data set. Note that the authors presume that the calibration is a function of the broadening of the lines that occurs as the temperature and pressure p changes. The authors present data showing an average temperature uncertainty of about 0.3 K.

Zeyn et al. (1996) have demonstrated a variant of the rotational Raman technique in which the output of a line-narrowed KrF laser (248 nm) was Raman shifted in hydrogen to 276.787 nm, a wavelength corresponding to a resonance absorption line of a thallium atomic vapor. The thallium filter is used to remove the returning light from particulate and molecular (elastic) scattering while passing the Stokes and anti-Stokes rotational lines. An echelle grating spectrometer is used to separate light at four separate wavelengths. Two wavelength bands are used in both the Stokes (277.65–278.03 nm and

276.94–277.33 nm) and anti-Stokes (275.41–275.84 nm and 276.21–276.60 nm) portions of the rotational Raman spectrum. It is thus quite similar to the basic technique except that it uses data from both sides of the laser line, offering increased sensitivity to temperature changes. The system has several advantages. First, the system operates in the ultraviolet portion of the spectrum so that considerably more Raman scattered light is available. Operation in the solar-blind portion of the spectrum means that there is negligible solar background and daytime operation is possible. The use of the grating is much more efficient in its use of the available photons than the beam splitters that are commonly used in the basic technique. A grating passes about 40% of the light at the relevant wavelengths as opposed to transmissions of about 5% from interference filters in the ultraviolet portion of the spectrum. A disadvantage of this technique is that the use of a Raman cell to shift the fundamental laser frequency results in a significant decrease in the intensity of the emitted light. The development of the tuned and line-narrowed, KrF laser and Raman cell required for the technique is described by Luckow et al. (1994). The demonstration system was capable of temperature measurements to distances of 2 km.

Yet another variation of the rotational Raman technique suggested by Heaps et al. (1997) uses the first vibrationally shifted, rotational spectrum from molecular nitrogen. The Q branch and the high rotational quantum number lines in the S branch are compared to determine the temperature. The signal level from a vibrationally shifted Q branch is more intense than that from the S or P branch and will have no contamination from elastic molecular or particulate scattering. Although the intensity of the vibrational-rotational Raman spectrum is smaller than the pure rotational Raman spectrum, measuring is simpler because the signals are spectrally farther from the molecular and particulate scattering lines, and thus the requirement for strong blocking at a nearby wavelength results in a higher transmission in the interference filters. Although it is not necessary to block the elastically scattered light, it is necessary to block the nitrogen Q branch signal when measuring the S branch lines. This line is only a factor of 20–50 times as intense as the measured line, so the blocking requirements are considerably relaxed. The change in the signal is estimated to be about 1.2% per degree Celsius. The data analysis method used by these researchers assumes that the ratio of the number of photons measured in the S branch to the number of photons measured in the Q branch and scaled for the relative intensity of the two signals is linear. The method used to scale the ratio is not specified by the authors. A least-squares fit to calibration data is used to measure the slope and intercept for a linear fit. A linear fit was also suggested but not demonstrated by Cooney (1972). He suggested using a differential amplifier to measure the difference in amplitude between the signal from the two filters inside the anti-Stokes rotational lines. The output of the differential amplifier is scaled by dividing it by the average amplitude of the two signals. The advantage of the differential amplifier is that it is extremely sensitive to the difference between the two signals, enabling maximum sensitivity. The use of the amplifier also removes most of the effects of background sunlight and possible contamination from elastically scattered light.

Arshinov et al. (1983) suggest an analysis method in which the ratio R of two individual lines with different rotational quantum numbers is

$$R(T_{atm}) = \frac{I(j_1, T_{atm})}{I(j_2, T_{atm})} = \exp\left(-\frac{\alpha}{T_{atm}} + \beta\right) \tag{12.90}$$

where $\alpha = [E_{rot}(j_2) - E_{rot}(j_1)]/k$, $\beta = \ln(S(j_1)) - \ln(S(j_2))$, E_{rot} is the energy of the state j_i, and $S(j_i)$ is the angular momentum quantum number of the state j_i. However, any real filter will encompass multiple rotational lines. To further compound the problem, the atmosphere is composed of multiple gases, so that lines from oxygen and nitrogen along with small contributions from atmospheric trace gases will all be measured simultaneously. Although an analytical solution of the form shown in Eq. (12.90) is impossible to derive, Arshinov et al. (1983) provide evidence that it is still approximately true. The values for α and β were experimentally determined by the authors and found to be $\alpha = 477.172$ (K) and $\beta = -0.9521$ for air. The method requires approximately a 3% change in the ratio of the lines to measure a 1 degree (Celsius) change in temperature. The method is also unusual in that it uses a double-grating monochrometer to separate the light. The double grating provides high rejection ($\sim 10^{-8}$) of the elastically scattered light while allowing a relatively high transmission at the desired wavelengths. The system was demonstrated to have an accuracy of 0.8°C for a 20-s integration time. Behrendt and Reichardt (2000) suggest an alternate formulation as

$$R(T_{atm}) = \frac{I(j_1, T_{atm})}{I(j_2, T_{atm})} = \exp\left(\frac{\alpha}{T_{atm}^2} + \frac{\beta}{T_{atm}} + \gamma\right) \tag{12.91}$$

where α, β, and γ are constants derived from a curve fit. The authors claim that the Eq. (12.91) fits synthetic data to an accuracy better than ±0.1 K whereas Eq. (12.90) has potential errors on the order of ±1 K.

An interferometric method has been suggested by several authors (Armstrong, 1975; Ivanova et al., 1993; Arshinov and Bobrovnikov, 1999) to determine the temperature (and in at least one variant, the pressure as well) (Ivanova et al., 1993). A Fabry–Perot interferometer is used to measure the intensity and width of the Raman-shifted lines. The Raman peaks are regularly spaced [Eq. (2.40)] on each side of the wavelength of the incident light. Each of these lines is temperature- and pressure broadened. A Fabry–Perot interferometer allows light to pass in a series of narrow bands that are regularly spaced. The interferometer can be matched to the Raman lines so that the free spectral range of the interferometer overlaps the spectral period of the Raman lines. In the matched condition, light from the Raman-shifted lines passes through the interferometer while the light scattered by molecules and particulates is rejected to high order. As the free spectral range is changed, some of the Raman lines pass through the filter while others are rejected. The spaces between lines can also be measured. In addition, the elastically scattered light is also passed when one of the interferometer lines coincides with those lines. The response function as the free spectral range is changed is complex but has been described by Armstrong (1974). The details of the shape of this function are a sensitive measure of the temperature and pressure of the atmosphere. Arshinov and Bobrovnikov (1999) detail a method to align the pass bands of the interferometer to the frequency-shifted Raman lines. They suggest that the étalon be set up and maintained so that the free spectral range matches the period of the Raman-shifted lines. Then the laser should be tuned so that these lines shift to the fixed pass bands of the interferometer. It seems clear that the line width and stability of the laser, the stability of the interferometer, and the ability to precisely tune the laser are all factors that are required to effect this method. The use of an interferometer has the benefit of a high transmission compared with interference filters, and passing all of the Raman-shifted lines simultaneously creates a much more intense signal than passing a narrow portion through an interference filter. Furthermore, the method is effective at blocking both the elastically scattered light and the ambient sunlight. To our knowledge, none of the variants of the method has ever been demonstrated.

12.4. BOUNDARY LAYER HEIGHT DETERMINATION

The planetary boundary layer (PBL) is the region of the atmosphere, near the surface, that is directly affected by processes or events that occur at the earth's surface. Thus the height and dynamics of the planetary boundary layer height are of great interest to meteorologists, environmentalists, and hydrologists. The parameters that describe the boundary layer vary with the amount of energy added to the atmosphere by the sun, the partitioning of that energy at the surface, the local wind, and changes in surface roughness. The dynamics at the top of the boundary layer has been shown to play a large role in the processes at the bottom of the boundary layer and thus is a major factor that governs pollutant concentrations and their long-range horizontal transport. Unfortunately, the height of the boundary layer is difficult to model accurately. Because of this, a great deal of effort has been invested in measuring the height and observing the dynamics of the boundary layer. Lidars have repeatedly proven themselves to be valuable tools in the study of entrainment and processes at the top of the boundary layer (see, for example, Kunkel et al., 1977; Boers et al., 1984; Boers and Eloranta, 1986; Crum et al., 1987; Boers, 1988; Hashmonay et al., 1991; Cooper and Eichinger, 1994).

A fair-weather convective boundary layer is characterized by warm, particulate-rich parcels of air rising from the surface and cooler, cleaner parcels of air moving toward the surface. These vertical motions cause irregularities at the top of the boundary layer that can be observed in lidar scans (for example, Fig. 12.17). Meteorologists use potential temperature and specific humidity profiles to estimate the height of the boundary layer (see Fig. 1.5). This height is taken to be the height at which the potential temperature is subject to an abrupt increase. A corresponding decrease in the specific humidity (and all other scalar quantities with their source at the surface) occurs at the same height. However, measurements with traditional point instruments are difficult in this type of situation. Measurements from a free balloon often lack sufficient resolution or may be made through the top of a plume or in a downwelling air parcel. In extreme cases, point instrument measurements of the boundary layer height may vary more than 100%. For many meteorological purposes, knowledge of the variations in height is also desirable in addition to the average height. As can be seen from the figures in this portion of the text, the variations in the height of the boundary layer in space and time may be considerable. Thus, to obtain meaningful height and entrainment zone depth estimates, some degree of either time or space averaging is required. Because these variables are not stationary in time, a spatial average is preferable to a temporal average.

The top of the convective boundary layer is marked by a large contrast between the backscatter signals from particulate-rich structures below and cleaner air above (Fig. 12.17). Because of this, boundary layer mean depths can be easily obtained from manual inspection of vertically staring, RHI, or vertical scans. Automated algorithms have proven more difficult. In part, this is the result of a lack of a specific definition of a phenomenon that extends

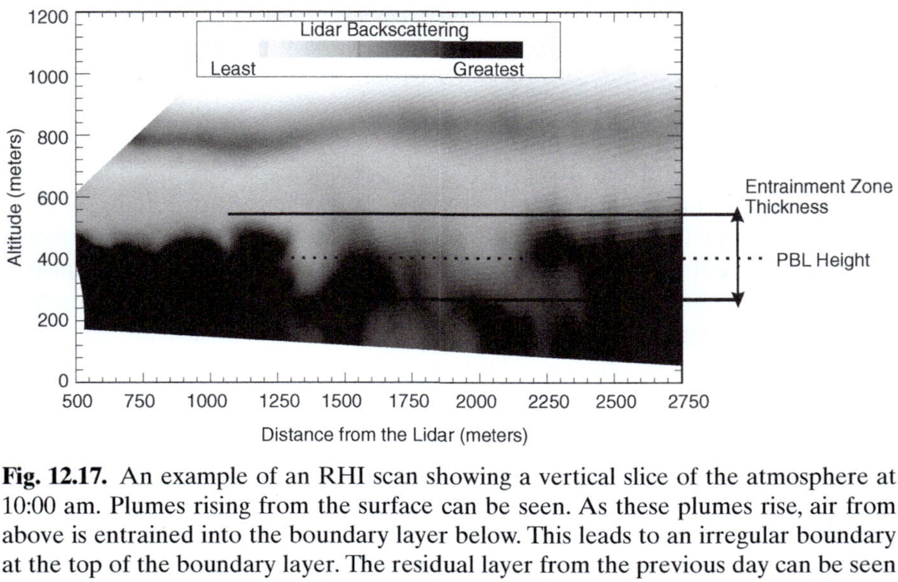

Fig. 12.17. An example of an RHI scan showing a vertical slice of the atmosphere at 10:00 am. Plumes rising from the surface can be seen. As these plumes rise, air from above is entrained into the boundary layer below. This leads to an irregular boundary at the top of the boundary layer. The residual layer from the previous day can be seen above the active convection. The current boundary layer is located at about 500 m.

over a finite altitude range, sometimes extending over 200m, even under ideal conditions. Table 12.4 is a collection of definitions of the height of the boundary layer in current use accumulated by Beyrich (1997).

The exact position of the boundary layer is not well specified, even for conventional meteorological soundings using one of the definitions in Table 12.4. The change in temperature at the top of the boundary layer and the drop in particulate concentration occur over a finite altitude range (Fig. 12.18), with the result that an uncomfortably large amount of interpretation of the data is often involved in the selection of a value for the boundary layer height. Considering the high range resolution of most lidars, a more definitive definition is desirable. Although it is not universal, the general definition of the bound-

TABLE 12.4. Definitions of the Planetary Boundary Layer (PBL) Height

PBL height definition based on profiles of mean variables (wind, temperature, humidity, chemical species concentrations)	PBL height definition based on profiles of turbulent variables [fluxes, variances, turbulent kinetic energy (TKE), structure parameters]
Convective Boundary Layer	
• Height of a zone with significant wind shear	• Height calculated from similarity methods using wind and temperature profile measurements within the mixing layer
• Base of an elevated inversion or stable layer	• Height at which the turbulent heat flux changes sign
• Height at which a rising parcel of air becomes neutrally buoyant during the day	• Height at which the turbulent heat flux has a negative maximum
• Height at which moisture or aerosol concentration sharply decreases	• Height at which the TKE dissipation rate or vertical velocity variance significantly decreases
• Height at which single plume vertical velocities vanishes.	• Height of an elevated maximum of acoustic/electromagnetic refractive index structure parameters
Stable Boundary Layer	
• Height of the first discontinuity in the temperature, humidity, aerosol, or trace gas concentration profiles	• Height at which some turbulence parameter has reduced to a few percent of its surface layer value or decreases below some threshold value
• Upper boundary of a layer of significant wind shear	• Height at which the Richardson number exceeds its critical value
• Top of the surface inversion or stable layer	• Height of maximum gradient or curvature in the vertical profiles of variances or structure parameters
• Height of the low-level jet	

Beyrich (1997).

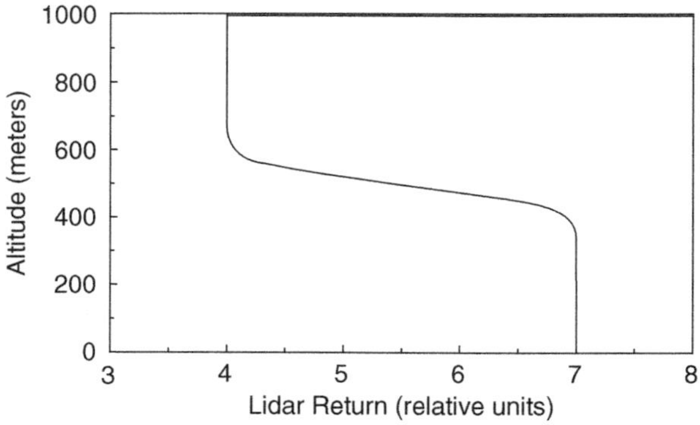

Fig. 12.18. An idealized plot of a range-corrected lidar return from a vertically staring system. A well-mixed boundary layer is shown below about 400 m along with a transition to the relatively clean air above. In this plot, the top of the boundary layer would be taken as 500 m with an entrainment zone depth of 200 m.

ary layer depth suggested by Deardorff et al. (1980) is most often used in lidar work. Deardorff et al. define the boundary layer height as the altitude where there are equal areas of clear air below and particulates above. A plot of an idealized range-corrected lidar signal with height is shown in Fig. 12.18. For such a lidar return, the location of the boundary layer top is taken to be the midpoint of the transition region between the areas of higher and lower backscattering. In the idealized model, this point corresponds to the location with the maximum slope in the lidar signal as well as the point of inflection in the signal. The question of how to determine this altitude in real signals is discussed in the next section.

Figure 12.19 is a plot of an actual range-corrected lidar signal with height above ground taken from the horizontal range interval between 2400 and 2450 m in Fig. 12.17. In this figure, the transition from high to low particulate concentrations occurs over a distance of about 150 m over the altitude range from 425 to 575 m. This represents the upper limit to the particulate matter lofted from the surface by convection at this time. A particulate-rich layer may exist above the boundary layer that remains from the previous day that is not directly affected by surface processes at that time. This layer is known as the residual layer (Stull, 1988). In Fig. 12.17, the residual layer encompasses the entire altitude range from about 500 m to 950 m. This layer above the convective layer may confuse lidar measurements made during the morning until it is fully entrained by the growing boundary layer. Note that there is a dense layer of particulates inside the residual layer that may also confuse automated estimates of the boundary layer height.

The vertical distance between the top of the highest plumes and lowest parts of downwelling air parcels is known as the entrainment zone (Fig. 12.17). The ratio of the depth of the entrainment zone to the boundary layer height is of great significance. It relates the amount of energy entrained from the

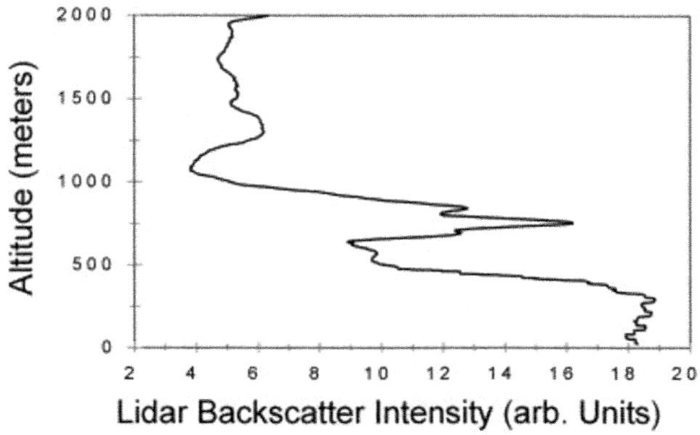

Fig. 12.19. A plot of the range-corrected backscatter return with height taken from Fig. 12.17 between the horizontal range interval 2400 and 2450 meters.

warm air above the boundary layer to the amount of energy injected into the boundary layer from solar heating at the bottom. The depth of the entrainment zone was defined by Deardorff et al. (1980) as "the depth being confined between the outermost height reached by only the most vigorous penetrating parcels and by the lesser height where the mixed layer fluid occupies usually some 90 to 95 percent of the total area." The depth of the entrainment zone may exceed the average depth of the boundary layer. Nelson et al. (1989) measured entrainment zone thicknesses ranging from 0.2 to 1.3 times the average depth of the boundary layer.

12.4.1. Profile Methods

Curve Fit Methods. The midpoint of the transition zone between areas of high and low backscattering is also the location of the inflection point. This point can be determined by a curve fit of some type, for example, by fitting the range-corrected backscatter return in the region of the entrainment zone with a fifth-order polynomial by a least-squares technique (Eichinger et al., 2002). The inflection point, where curvature changes from downward to upward, is used as the boundary layer height. The choice of a fifth-order polynomial is somewhat arbitrary. A polynomial fit using an odd order of at least three is required. A curve fit to a lower-order polynomial may not be able to accurately follow the shape of the backscatter distribution, whereas a higher-order polynomial will capture small variations in the signal that are of little consequence. Higher-order polynomials will also have a larger number of inflection points, complicating the selection of the point.

A better technique than a polynomial fit is to fit the backscatter profile to an assumed shape. The problem is to find a functional form in which the lowest altitudes will have a high backscatter with a sharp transition to lower levels of backscattering in the layers above (i.e., have a shape similar to that of Fig.

12.18). The functional form must be robust enough to accommodate the many variations in shape that may be found. Steyn et al. (1999) suggest the use of an error function of the form

$$Z(h) = \frac{(Z_m + Z_u)}{2} - \frac{(Z_m - Z_u)}{2}\mathrm{erf}\left[\frac{(h - h_m)}{s}\right] \tag{12.92}$$

where $Z(h)$ is the range corrected backscatter signal at height h, Z_m is the average level of the range corrected lidar signal in the mixed layer, Z_u is the average level of the range corrected lidar signal in the layer above the mixed layer (the subscript r is omitted for simplicity); h_m is the boundary layer height and the midpoint of the transition, and s is related to the width of the transition region. Taking the region between the 5% and 95% mixing ratio values as the total width of the transition region, the entrainment zone thickness (EZT), can be found as

$$\mathrm{EZT} = 2.77s \tag{12.93}$$

Fitting the function described by Eq. (12.92) involves a multidimensional minimization of the square of the difference between the function and the observed data. Steyn et al. (1999) suggested the use of "simulated annealing" routines found in Press et al. (1992).

Threshold Methods. A number of threshold methods to determine the boundary layer height have been proposed and used. Melfi et al. (1985), Boers and Melfi (1987), and Dupont et al. (1994) determined the height of the boundary layer as the highest data point where the backscatter intensity was some fraction higher than the average backscatter value in the free troposphere above. The use of a threshold suffers from the arbitrary nature of the choice of the threshold. Given the natural variability of the atmosphere, it is difficult to assign a value that clearly and consistently distinguishes between the boundary layer and the free air above in all cases. An inappropriate value will tend to bias the results. Batchvarova et al. (1997) attempted to overcome this weakness by defining the average values for the backscattering for the mixed layer and that for the free troposphere above by using all of the data for some period and then taking the critical value as the average of those two values. Determining the average values, however, presupposes that one has already identified the location of the mixed layer and the free air above so that these average values may be calculated. In practice, threshold methods will often misidentify particulate layers above or below the boundary layer as the top of the boundary layer and are thus not recommended.

Derivative Methods. Because of the abrupt drop in backscatter intensity at the top of the boundary layer, the use of a gradient to identify the height of the boundary layer would seem to be a good choice. A number of researchers have calculated the gradient of the signal with height and used the change in gradient as an indicator of the height. One may use a threshold value in the derivative to indicate the height of the boundary layer or use the point at which the derivative has a maximum value to indicate this height (Kaimal et al., 1982; Hoff et al., 1996; Hayden et al., 1997; Flamant et al., 1997). The location of the maximum derivative should also be the location of the inflection point and thus should identify boundary layer heights that are consistent with the curve-fitting methods above. Another mathematically similar method uses the minimum of the second-order derivative of the range-corrected signal with altitude (again, this is the location of the inflection point) as the height (Menut, 1999). Still another variant uses the location of the maximum value of the logarithmic derivative of the altitude-corrected lidar return

$$\text{logarithmic derivative} = -\frac{d}{dh}\ln[P(h)h^2] \tag{12.94}$$

as the height of the boundary layer (White et al., 1999). The use of the logarithmic derivative essentially measures the rate of the fractional change in the signal rather than the absolute change, and thus it could be argued that it is an improvement over methods based on the absolute size of the change in the signal. In general, inflection point or maximum derivative methods have the advantage of being independent of any arbitrary threshold values and show good accuracy when turbulent fluctuations are present (Menut et al., 1999). However, as a practical matter, running derivatives are difficult to calculate in the presence of noisy data, particularly at long ranges. Because of noise, point-to-point derivatives are not useful with derivative methods. Thus some type of spatial and/or temporal averaging is required. This averaging may significantly reduce the range resolution of the measurement and may also bias the result. Furthermore, particulate layers above or below the boundary layer often have sharp boundaries that are more well defined than those of the boundary layer. The change in backscatter with height is greater at the edges of these layers. The result is that derivative methods often falsely identify these particulate layers as the boundary level height.

Haar wavelets have also been used to identify the boundary layer height (Cohen et al., 1997; Davis et al., 1997). The height at which the maximum wavelet response occurs is used as the boundary layer height. The use of the Haar wavelet is equivalent to calculating a smoothed extended derivative and is thus not truly different from maximum derivative methods.

Entrainment Zone Measurement. Methods to determine the vertical extent of the entrainment zone are variations of either the threshold method or the cumulative probability method. Melfi et al. (1985) used a threshold method to determine the location of the top and bottom of the entrainment zone for instantaneous vertical measurements and compared them to the cumulative probability for the entire set. They determined that the bottom of the entrainment zone corresponds to a cumulative probability of 4% whereas the top corresponds to a cumulative probability of 98%. These values are similar to those found by Deardorff et al. (1980). Flamant et al. (1997) used a high pass filter on the set of boundary layer heights to filter out spatial wavelengths longer than 4 km (i.e., structures that have a size larger than 4 km) before analysis. The filter removes the effects of large-scale motions and gravity waves from local boundary layer motions. The result of the filtering was sets of instantaneous boundary layer height distributions that were narrow and symmetric. They determined that the cumulative probability corresponding to the bottom of the entrainment zone was 6.2%.

There is some confusion in the literature concerning the size of the entrainment zone and the meaning of the transition zone in an individual lidar scan. The entrainment zone is defined to be the area that stretches from the top of the upwelling plumes to the bottom of the downwelling (clean air) motions from the free troposphere above. In Fig. 12.19, this zone is from about 325 m to 500 m. Consider two extremes for vertical lidar data. If one has a vertical staring lidar that takes an average of laser pulses over a timescale on the order of seconds, the region of the signal over which the backscatter intensity decreases from the mixed layer average to the free tropospheric average is significantly smaller, on the order of 50–75 m. In this case, Eq. (12.94) indicates the depth of the local entrainment into an individual plume and not the depth of the entrainment into the boundary layer. On the other hand, if one averages over a timescale on the order of 15 min, this would incorporate the signal from several upwelling plumes and downdrafts. In this case, Eq. (12.94) would apply, because the width of the transition region is indicative of the distance over which larger-scale mixing occurs. Some interpretation of what an individual lidar scan represents is necessary before one can infer the meaning of the transition zone in that scan. The best solution to this problem is to take data with the highest spatial and temporal resolution possible and use the variations in the measured height of the boundary layer over some period of time or distance to determine the depth of the entrainment zone, for example, from the width of the probability distribution of the measured boundary layer heights. The use of Eq. (12.94) is discouraged unless the data must be taken with a long averaging time for some reason.

General Comments. When determining the height of the boundary layer, a simple vertically staring lidar is a substantial improvement over balloon-borne instruments. Because it can make continuous vertical observations, temporal averaging is easily accomplished. However, determination of the boundary layer height with the definitions or techniques described above is not always straightforward. This is particularly true early in the morning and late in the afternoon (Coulter, 1979). In both cases, residual layers of high particulate concentration may occur above the boundary layer. This type of situation is shown in Figs. 12.17 and 12.19. The residual layers confuse the determination of the boundary layer height for automated methods and often lead to heights that are systematically too high. In addition, the coverage of a vertically point-

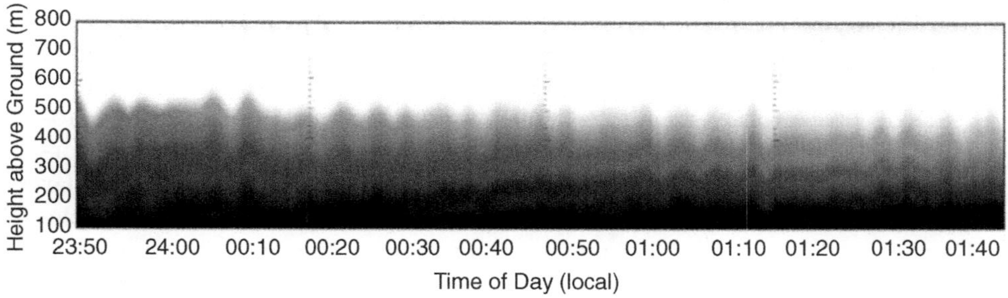

Fig. 12.20. An example of the signal from a vertically staring lidar system. Shown are a series of gravity waves over a period of about 2 h.

ing lidar depends on the time required for an upwelling parcel of air to drift over the lidar. Assuming a 4 m/s wind and a 1-km horizontal scale for a large plume, a parcel of air will take about 6 min to pass over the lidar. Averaging over enough plumes to obtain statistically meaningful boundary layer heights may take too long during times when the height is changing rapidly (during midmorning or late afternoon, for example). Visual inspection of multidimensional lidar data is always recommended as a check on automated techniques. On the other hand, the high sampling rates that may be achieved (a few seconds) make vertically staring systems ideal for the study of some types of phenomena, gravity waves, for example. Figure 12.20 is an example of several hours of gravity wave data. The ability to determine the height of the various layers is a powerful tool that can be used to determine many of the properties of the gravity waves.

12.4.2. Multidimensional Methods

In contrast to a vertically staring lidar system, a scanning lidar can cover a relatively large area quite quickly, allowing spatial averaging over many thermal structures. This is particularly true for three-dimensional scans that may cover many tens of square kilometers and average over 10–20 structures. The advantage of a scanning system is that a more instantaneous value of the properties of the boundary layer can be obtained. Measurements of a large number of structures can be made in minutes that would require hours of averaging by a vertically staring lidar. Scanning over the depth of the boundary layer allows far more information to be collected in a shorter period of time. Vertical or RHI scans are visibly rich in information on boundary layer structure. Two- or three-dimensional scans make it possible to visually distinguish between layers above the boundary layer and thermal structures that are connected to the ground. The issue with multidimensional scanning becomes how to best quantify the information gained.

Historically, visual estimates were made of the average boundary layer height from the RHI scans. Boers et al. (1984) suggested a procedure for estimating the height that is commonly used. Because visual estimates are subjective, the values for several successive scans are averaged and also repeated at a later time, after all of the data have been analyzed.

There are several variants to determine the boundary layer height automatically from RHI scans. Most of these methods use the range-squared corrected lidar signal in the analysis, but some have used an inverted lidar signal, the attenuation coefficient as the data to be analyzed (see, for example, Dupont et al., 1994). The first method is a variant of the curve-fitting method used in vertically staring systems. In this method, all of the data from a narrow horizontal region of an RHI scan are taken in the aggregate as if all of the data had been made at a single location. For example, all of the data taken at a horizontal distance between 2000 and 2025 m from the lidar for the scan in Fig. 12.21 have been plotted as a function of altitude to the right of the figure. Any of the single-shot types of analysis procedures may be used to determine the boundary layer height.

A second method for automated boundary height estimation uses the variance of the derivative of the range-squared corrected lidar signal. This method was described by Flamant et al. (1997), and Menut et al. (1999). They calculated the standard deviation of the slope of the lidar signal at each altitude. A threshold is defined to be a value that is three times the standard deviation of the slope in the free air above the boundary layer. The height of the boundary layer is taken to be the point where the standard deviation rises above the threshold.

Still another method for automated boundary height estimation uses the

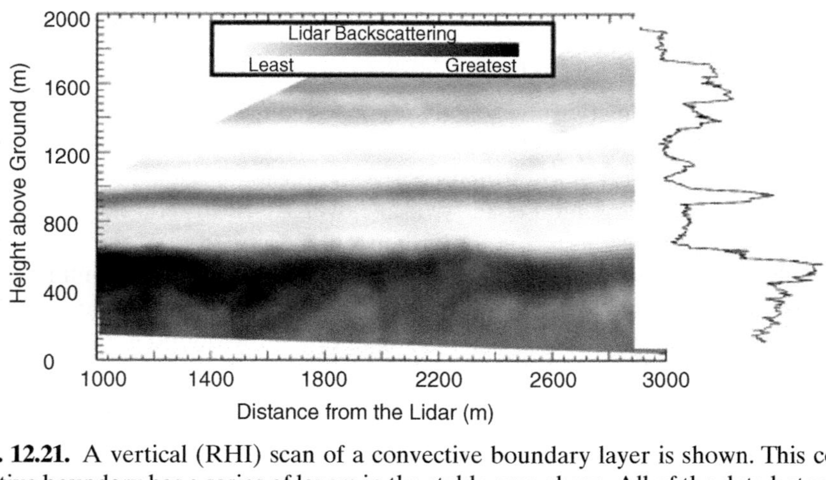

Fig. 12.21. A vertical (RHI) scan of a convective boundary layer is shown. This convective boundary has a series of layers in the stable area above. All of the data between 2000 and 2025 m distance from the lidar have been plotted as a function of height to the right. The dark area below 450 m at the left is the backscatter from an aerosol-rich residual from the previous day. The lighter area above 450 m is backscatter from the free atmosphere.

horizontal signal variance described by Hooper and Eloranta (1986). Horizontal variations of the particulate density and thus backscatter intensity are greatest at boundary layer height. This is due to the amount of contrast between the particulate-rich upwelling parcels of air and the relatively clean downwelling parcels of air. The result is a large horizontal variation in the backscatter signal at that region. When a two-dimensional lidar scan is used, all of the data inside a narrow interval about some height are used to calculate the variance at each height. The boundary layer height is taken to be the altitude at which the variance is greatest. The advantage of the variance technique is that it is insensitive to turbulent fluctuations throughout the depth of the boundary layer.

The method described by Piironen and Eloranta (1995) is applicable to three-dimensional data and is used with the University of Wisconsin volume imaging lidar (VIL). The method begins by high-pass filtering each shot in a volume scan with a 1-km-long median filter. This is done to reduce the effects of atmospheric extinction. The backscatter signals in the lidar coordinate system are then mapped to horizontal rectangular grids with 20-m vertical and 50-m horizontal resolution, known as constant altitude plan position indicators (CAPPI). Each of the CAPPI represents the backscatter return from a horizontal plane in the atmosphere. The variance of the backscatter returns in each of the CAPPI horizontal transects is calculated to generate a vertical profile of the variance.

The altitude of the lowest local maximum of the variance profile that is larger than the average variance of the profile is taken to be the height of the boundary layer. The search for the maximum value is accomplished working from the bottom upward to eliminate the possibility of a false identification caused by an aerosol layer above the boundary layer. Local maxima caused by particulate-rich air parcels are eliminated by the requirement that the variance be larger than the average variance of the entire profile. Random fluctuations due to signal noise may affect the detection of the maximum variance when the difference between the backscatter from boundary layer particulates and the free air is small. To reduce the effects of random local fluctuations, the variance of heights above and below the maximum point, h_{max}, are tested to ensure that the variance decreases smoothly on both sides. This is equivalent to

$$\sigma(h_{max} - 2\Delta h) < \sigma(h_{max} - \Delta h) < \sigma(h_{max}) > \sigma(h_{max} + \Delta h) > \sigma(h_{max} + 2\Delta h) \qquad (12.95)$$

where Δh is the difference in elevation between adjacent CAPPI. The method compares well to visually determined boundary layer heights and those determined from balloon measurements (Piironen and Eloranta, 1995). The principle advantage to the method is that the value obtained is a large area (~50–70 km^2) spatial average.

Comparisons between lidars, balloons, and sodars have shown that lidars tend to systematically overestimate the height of the boundary layer as measured by the height of the temperature inversion. This is believed to be due, at least in part, to the mixing of particles from parcels of air that overshoot the temperature inversion. These particles are mixed with the surrounding air when the air parcel reaches its maximum height and are then trapped in the stable layer above the temperature inversion (Russell et al., 1974; Coulter, 1979; Hanna et al., 1985; McElroy and Smith, 1991). The intensity of the lidar return may also increase because of the increase in relative humidity at the top of the boundary layer. At relative humidities above 90%, the particulates absorb large amounts of water, increasing the lidar signal significantly. Because of this, the scattering intensity from upwelling air may be greater than that from downwelling air parcels. The result is that the standard deviation peak is skewed upward (Menut et al., 1999). Differences between the various methods of measuring the height of the boundary layer are on the order of 10% in ideal conditions. The differences between the various definitions of the boundary layer contribute significantly to the differences in the measured heights. Under the condition of a weak capping inversion, or an imperfectly mixed boundary layer, differences in the various measurement techniques can exceed 25% (Beyrich, 1997). As has been noted, at certain times, particularly in the early morning or late afternoon, problems may occur with particles that remain from previous mixing to heights above the current height of the boundary layer (Coulter, 1979).

Efforts to automate the determination of boundary layer height suffer from several difficulties that are common to nearly all of the sensing methods (Beyrich, 1997). These difficulties include:

- The large number and types of patterns that may be observed makes it difficult to associate conditions with particular patterns so that one analysis technique cannot be used in all situations.
- The boundary layer is nonstationary, which complicates the interpretation of data averaged over time.
- Several different meteorological situations may lead to a given profile. Thus there is not a one-to-one correspondence between a measured profile and the events in the boundary layer that caused it. Residual layers may remain from the day before, or may be the result of shear in the boundary layer.
- The shapes of the measured profiles are seldom ideal (like that shown in Fig. 12.18), making it necessary to discriminate between features in the profile. During times when the contrast between scattering in the boundary layer and the air above is less than the established threshold of automatic discrimination, the methods may fail.

Even with multidimensional information, these problems may occur. A particular problem is that algorithms used in fully automated systems must be able to discriminate between the top of the boundary layer and layers above the boundary layer. Menut et al. (1999) note that there are advantages to the simultaneous use of both the variance and inflection point methods. Because they are sensitive to somewhat different conditions, they may complement the weaknesses of each other.

The presence of clouds at the top of the boundary layer may confuse an automatic boundary layer height calculation. The clouds will tend to dominate the variance and thus bias the estimate of the boundary layer height. Clouds will also cause the backscatter signal to increase at the top of the boundary layer rather than decrease, so inflection point methods will also fail. Figure 12.21 is a typical example of an RHI scan along with a signal profile. To compound the problem, when convective clouds dominate the boundary layer structure, the definition of the height of the boundary layer becomes unclear because convection may continue to several kilometers. Piironen and Eloranta (1995) suggested that heights from the variance technique are reliable if the fractional cloud coverage is not greater than 10%. As the cloud cover increases, the cloud base altitude should be taken as the boundary layer height. However, as they note, in these cases, the height of the boundary layer must be interpreted with caution. When low-altitude clouds are present, a manual inspection of the lidar scans provides a more reliable estimate of the boundary layer height.

12.5. CLOUD BOUNDARY DETERMINATION

Clouds are important for a wide variety of reasons in the study of meteorology, climate, weather prediction, and visibility, so that a number of methods to determine the location of bottom and top of cloud layers have been developed. In a fair-weather, high-pressure system, the wind divergence causes a lowering of the boundary layer height and generally only cumulus clouds are present. Conversely, in a low-pressure system, the wind convergence is associated with large-scale updrafts, which may transport air parcels from the boundary layer to high altitudes. Clouds that are associated with these updrafts may extend all the way to the top of the troposphere. Clouds scatter large amounts of light, so that there is a great deal of contrast in the lidar scans between them and the adjacent air in the free troposphere. Because of the high contrast in the lidar signal between the cloud and the surrounding air, and the fact that cloud boundaries occur over relatively short distances, the determination of the altitudes of cloud edges is relatively straightforward (except for cases of low-altitude clouds; see Section 12.2). However, with staring lidars (whether in vertical or slope directions) cloud top altitudes can be reliably determined only for optically thin clouds. Scanning the lidar in the vertical direction makes it possible to look through holes in the cloud cover to find cloud tops (see below). In addition, associations can be made with data at the same altitude, but at different angles (and thus different amounts of attenuation) to estimate the cloud top altitudes. As with boundary layer heights, there is a need for automated methods to determine these values. In contrast with boundary layer determination, there has been less work and fewer measurement methods have been developed (except for a great deal of effort done for airport measurements of the cloud baseline in poor weather conditions). Because of the sharp transition between the lidar return from clouds and the ambient air, the choice of method used to determine the location of that transition is less critical and differences between methods are small.

There are three basic measures of cloud geometry that have physical meaning, the cloud fractional coverage, and the altitudes of the cloud base and cloud top. The cloud base height is just the bottom of the cloud, the location where scattering rapidly increases with the height. Cloud base heights determined by lidar are compatible with measurements made by other methods. The cloud top is most often taken to be that altitude where the lidar signal decreases to that of the ambient air. This is, however, a poor definition. The reduction in signal may occur because the top of the cloud has been reached or because the lidar beam has been completely attenuated inside the cloud (this is arguably the most typical case). The cloud top altitude can only obtained with any degree of certainty when a signal from the air above the cloud can be seen. Carswell et al. (1995) suggest determining the signal-to-noise ratio at altitudes just above the suspected cloud top altitude to determine whether a signal is detected above the cloud. The location of the top of the cloud is often ambiguous, for example in Fig. 12.23, is the top of the cloud at 625 m, 750 m, or 950 m? Examination of Fig. 12.22 will allow one to

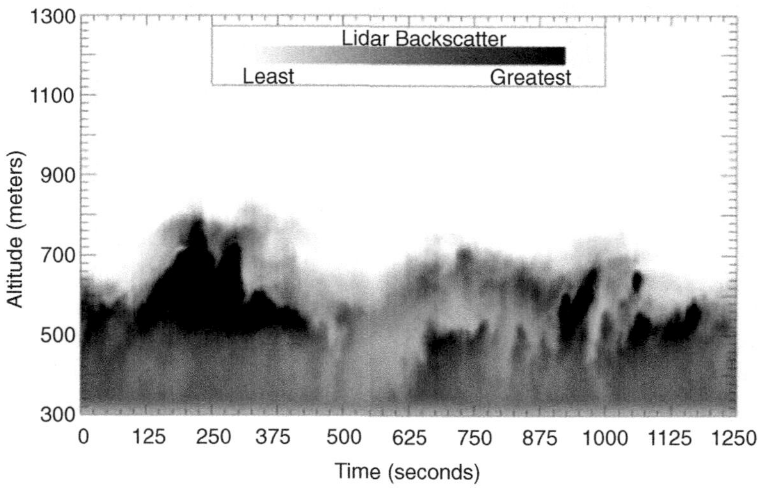

Fig. 12.22. A marine cloud-topped boundary layer as seen by a vertical staring lidar system. The dark areas above 430 m are the result of the large backscatter from clouds at the top of the boundary layer. These clouds are not optically thick so that aerosols can be seen above the clouds. Note that clouds often form at the top of upwelling air parcels.

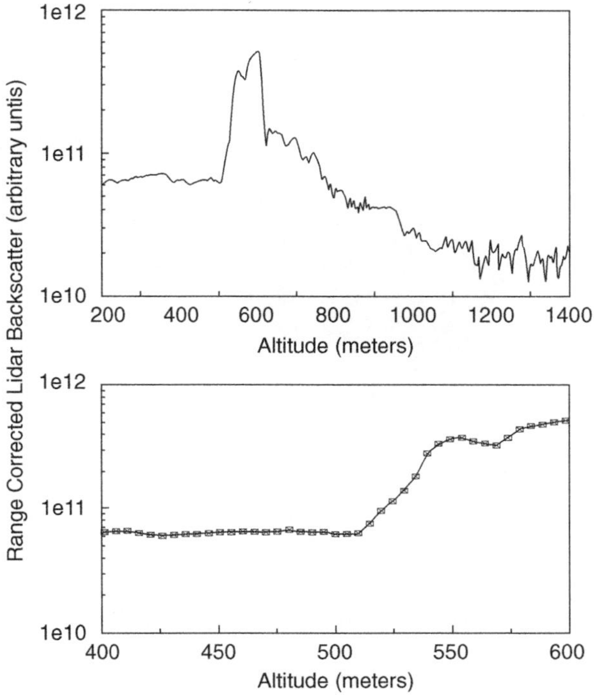

Fig. 12.23. The range-corrected lidar signal taken from the data shown in Fig. 12.22 at a time of 1150 s. The bottom panel shows that the size of the transition to the cloud is far larger than the variations found in the boundary layer. Most of the transition to the cloud occurs over a distance of less than 25 m.

conclude that it is the 625-m altitude, but this is not obvious from just a single trace.

Unfortunately, there is no general agreement how to use and compare measures made by lidars to other measures (Pal et al., 1992). To complicate the problem, the definitions of cloud boundaries may actually depend on the application of the data (Eberhard, 1987). Rotating beam ceilometers (RBC) used by the U.S. Weather Service determine the cloud base as the height at which the RBC signal reaches its maximum value. A detailed comparison of cloud base heights obtained from various types of measurements can be found in a paper by Eberhard (1987).

Most cloud boundary determination algorithms use some form of a threshold either of the signal magnitude or gradient to determine the location of the cloud bottom (Robinson and McKay, 1989). Threshold methods are more effective when used to determine the boundaries of a cloud than to determine the height of the boundary layer because in the former the change in the backscatter signal is larger and occurs over a shorter distance. However, as noted by Uttal et al. (1995), threshold methods may be limited by changes in

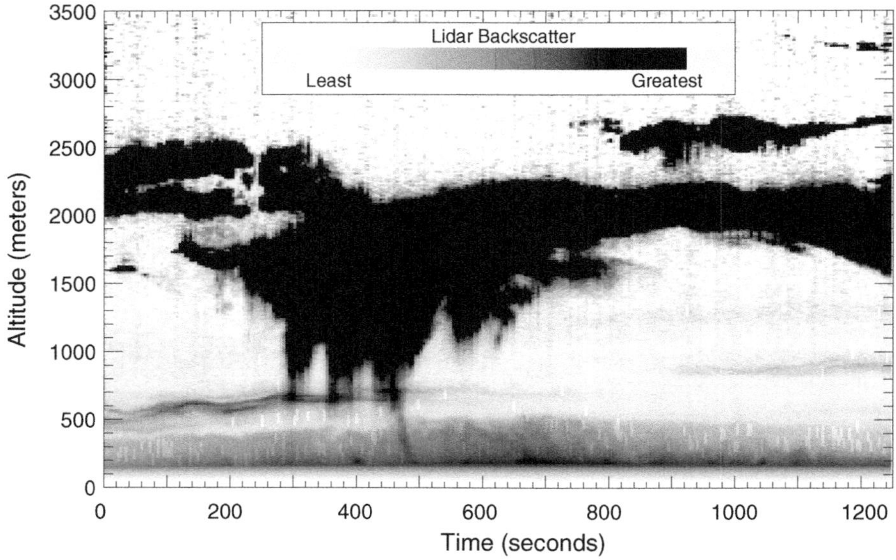

Fig. 12.24. High-level clouds above a marine boundary layer as seen by a vertical staring lidar system. The dark areas above 500 m are residual clouds from a large convective system. These clouds are not optically thick enough to preclude observation of aerosols above the clouds. However, the amount of noise in the data is much larger above the cloud layers.

the amount of ambient sunlight, background aerosols, laser power, detector amplification, the angle between the sun and lidar line of sight, and a host of other factors that change the relative signal level between areas of cloud and free air.

An automated algorithm was suggested by Pal et al. (1992) that used derivative methods. The cloud base is taken to be the location of the first zero crossing of the first derivative where the derivative changes from a negative value to a positive value. To reduce the effects of noise and spurious zero crossings, the derivative is determined as the slope of the least-squares fit to a set number of adjacent data points. The number of points used depends on the range resolution and must be small enough so that thin cloud layers can be detected. However, this method smooths the data, reducing the effective range resolution, and may bias the cloud base measurement. To avoid this bias, the algorithm searches for the changes in the lidar backscatter signal at locations near the identified zero crossings. Zero crossings are rejected in which the change in the signal is less than twice the noise level at that location.

For a vertically staring lidar, the fractional cloud coverage may be taken as the fraction of lidar shots in which a cloud is detected. For a scanning lidar, the situation is more complex because a single lidar line of sight at a low elevation angle (long range) represents a larger horizontal area than one at a high elevation angle. Thus, for two- or three-dimensional scanning lidars, each of the lidar lines of sight should be mapped to a uniform horizontal grid. Then a decision is made for each lidar line of sight whether a cloud is present and the appropriate point in the horizontal grid is annotated. The fractional coverage is then the fraction of the grid that is covered by clouds.

13

WIND MEASUREMENT METHODS FROM ELASTIC LIDAR DATA

The measurement of winds is one of the most developed techniques in use with elastic lidars. Because of the small physical size of the laser beam and the shortness of the laser pulses, the potential exists for lidars to make wind measurements with higher spatial and temporal resolution than current sodars or radars. There exist a number of methods to determine wind speed and direction as well as some turbulence parameters. None of the methods for wind retrieval requires an inversion of the lidar signal. Despite the number and capability of the methods outlined here, incoherent methods of wind measurement are not in widespread use. In part, this is due to the fact that they require lasers with relatively high pulse energies, with the result that they are generally not eye safe. Some of the methods require the capability to scan exceptionally fast. Thus they are ill-suited for routine measurements.

There is potentially a large market for high-resolution wind soundings and wind measurements over large areas. A wide variety of applications require reliable wind field information at increasingly small scales. The ability of lidars to provide wind information at distances and resolutions no other instrument can match suggests one of the most promising practical uses for lidars. Wind measurements are needed to deal effectively with urban air pollution and for a wide variety of long-range atmospheric transport problems. The ability to measure wind shear is badly needed for aviation purposes. Because conventional wind measurement methods measure at a point, measurements over large areas are costly and difficult. Another application that has received a

Elastic Lidar: Theory, Practice, and Analysis Methods, by Vladimir A. Kovalev and William E. Eichinger.
ISBN 0-471-20171-5

great deal of attention is the global measurement of wind from satellites with lidar. It has long been postulated that the measurement of tropospheric winds is the most important need for numerical weather forecasting (Atlas et al., 1985; Baker et al., 1995).

13.1. CORRELATION METHODS TO DETERMINE WIND SPEED AND DIRECTION

Because lidars can determine relative concentrations over large areas with high spatial resolution, they have the potential to map the spatial concentration of particulates as a function of time. The ability to track structures in time allows one to determine the wind speed. Although the use of correlation methods to determine wind velocities dates back to at least Mitra (1949), the first use of correlation methods with lidar was a feasibility demonstration by Derr and Little (1970), a more sophisticated demonstration with searchlights by Kreitzberg (1974), and a lidar experiment by Eloranta et al. (1975), followed closely by a horizontal lidar measurement by Armstrong et al. (1976). Correlation methods use elastic lidars to detect and track heterogeneities in the atmospheric particulate concentrations to measure wind velocities. This can be done over relatively large areas with reasonably fine spatial resolution. Several incoherent methods are discussed here, methods that do not use the mixing of light from a local oscillator to determine the size of the Doppler shift from the resulting beat frequency.

We begin with methods requiring the lidar to measure the particulate backscatter along several lines of sight. This may be done with multiple laser beams or by scanning the lidar in a regular pattern. These methods, collectively known as correlation methods, are common because they are relatively inexpensive and simple to implement. Correlation methods can measure the entire horizontal wind vector, something that Doppler systems cannot do directly. Doppler systems can measure only the radial component of the velocity. However, the accuracy of correlation methods is significantly less than Doppler methods and measurements are limited to parts of the atmosphere with significant numbers of discrete particulate structures. The need for contrast between atmospheric structures and their surroundings is a key limitation to these types of systems. The greater the contrast, the more spatial variability that exists, the better these systems will function. The requirement for contrast generally limits their use to the atmospheric boundary layer (1–2 km in altitude) and to unstable (convective) boundary layer conditions. Zuev et al. (1997) examined the quality of data provided by correlation methods, compared them with data gathered by more traditional methods, and concluded that wind data gathered with the correlation technique can be successfully used in modeling, in reconstruction of past events, and in short-term forecasting. This paper contains an excellent discussion of the probability and magnitude of errors as a function of wind velocity and altitude.

13.1.1. Point Correlation Methods

The correlation approach is quite simple in principle. It is an attempt to track the motion of discrete atmospheric structures by differences in their particulate backscatter. These structures have sizes on the order of 15–500 m in diameter and may be identified by large concentration gradients. By tracking the drift of these structures, one can determine the wind speed and direction. Consider two lines of sight oriented at two different elevation angles, θ_1 and θ_2, and horizontally such that the plane formed by the two lines of sight is parallel to the average wind direction as shown in Fig. 13.1. Although this is often done with a scanning lidar, it could be accomplished by the use of a wide field of view telescope and two lasers. At each point along each of the lines of sight, a time series of the particulate backscatter is developed. As structures containing aerosols advect horizontally with the wind and across the lines of sight, first one line of sight will detect it, and then the other at a later time, at the same height.

The time lag between detection the two lines of sight can be determined with the correlation function at a given height. The lidar takes data at a regular time interval, creating a plot of the backscatter with time and height for each angle. Using the range- and energy-corrected lidar signal

$$Z_{\theta_1}(r,t) = \frac{P(r,t)r^2}{E}$$

data are extracted at a given height, r, at all of the measured times. This gives an estimate of the backscatter variation at that height with time. This is done for each of the lines of sight. The time lag between detection of structures along two lines of sight can be determined with the correlation function. The correlation function is determined by

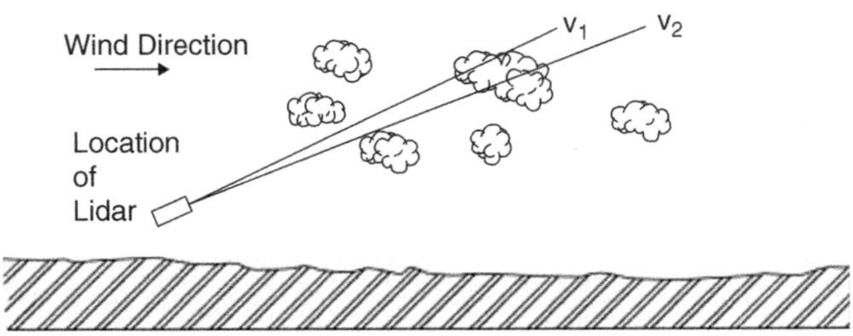

Fig. 13.1. The measurement geometry for multiangle wind measurements, looking horizontally, across the ground. Two or more lines of sight are oriented so that they are parallel to the average wind direction. At each point along each of the lines of sight, a time series of the particulate concentration is developed. As particulate structures advect with the wind and across the lines of sight, first one line of sight will detect it, then the other at a later time.

$$C(r,\Delta t)=\frac{\sum_{t=1}^{n}[Z_{\theta_1}(r,t)-\langle Z_{\theta_1}(r)\rangle][Z_{\theta_2}(r,t+\Delta t)-\langle Z_{\theta_2}(r)\rangle]}{\sqrt{\sum_{i=1}^{n}[Z_{\theta_1}(r,t)-\langle Z_{\theta_1}(r)\rangle]^2[Z_{\theta_2}(r,t)-\langle Z_{\theta_2}(r)\rangle]^2}} \tag{13.1}$$

where $Z_{\theta_1}(r, t)$ is the range and energy corrected lidar signal along the line of sight specified by θ_1 at range r and time t. The peak of the correlation function for a given range corresponds to the time delay between detection in the two lines of sight. Knowing the geometry and thus the distance between the measured points allows calculation of the velocity along each primary direction. Figure 13.2 shows an example of the signal from two lines of sight and the resulting correlation function. This calculation is repeated at each altitude so that a wind profile can be generated.

Perhaps the first successful demonstration of a multibeam correlation method was accomplished by Armstrong et al. (1976), although Derr and Little (1970) presented several methods by which wind velocity measurements could be made and data that suggested the method could be practical. A similar method was used by Eloranta et al. (1975) but correlated the movement of

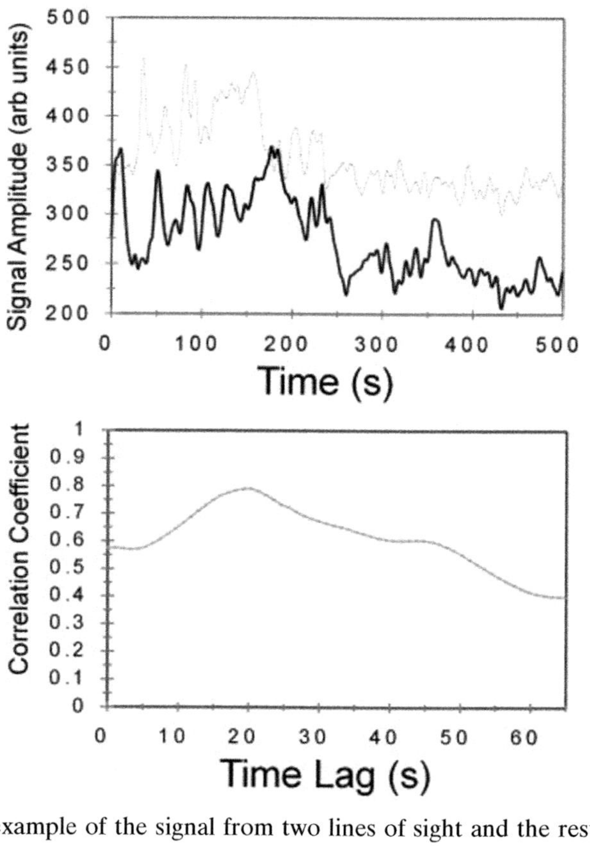

Fig. 13.2. An example of the signal from two lines of sight and the resulting correlation function.

structures for a single line of sight along a low elevation angle. The practical problem with this approach is that the lines of sight must be aligned with the direction of the mean wind. This is a problem because the direction of the mean wind may change rapidly and may be different for different altitudes.

A solution to this problem is to use the signals provided by three individual beams oriented near-vertically, in a "triple-beam sounding" technique, depicted in Fig. 13.3. Clemsha et al. (1981) demonstrated such a system for use in the upper troposphere as early as 1981. Each of the lidar signals provides the scattering intensity as a function of altitude and time. The problem is treated as if all of the structures at some height are planar and are transported horizontally. For any given altitude, the entire assembly will provide three separate intensities as a function of time, from three separate locations. If the beams are arranged in a right isosceles triangular arrangement, a cross section at some given altitude could be represented schematically by Fig. 13.4. The line connecting one beam pair has been designated the x-axis, with the axis connecting the other pair as the y-axis. The signal intensity as a function of time obtained from the vertex (at the specified altitude) has been denoted $Z_0(t)$, and the signal from the other two beams as $Z_{x1}(t)$ and $Z_{y1}(t)$. Because structures advect nearly horizontally (especially at altitudes greater than the surface layer), the correlation of two points at the same altitude makes sense for lines of sight at high elevation angles. If this is done, a minimum of three lines of sight are required to obtain the full horizontal wind vector.

The horizontal wind speed is designated as V and the wind orientation angle (measured counterclockwise with respect to the x-axis) as θ. Fluctuations in the lidar signals are generated by turbulence-induced fluctuations in the scattering intensity of the air (billows of dust). Turbulent structures at the scale of the beam spacing and smaller will cross one beam or another at random, and the correlations of these signals will produce primarily noise. However, larger-

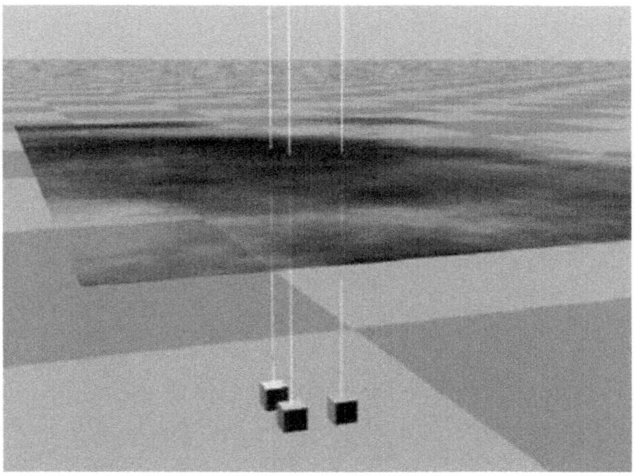

Fig. 13.3. The backscatter signal geometry for the triple-beam sounding approach. The reference signal is located at the origin of the coordinate system.

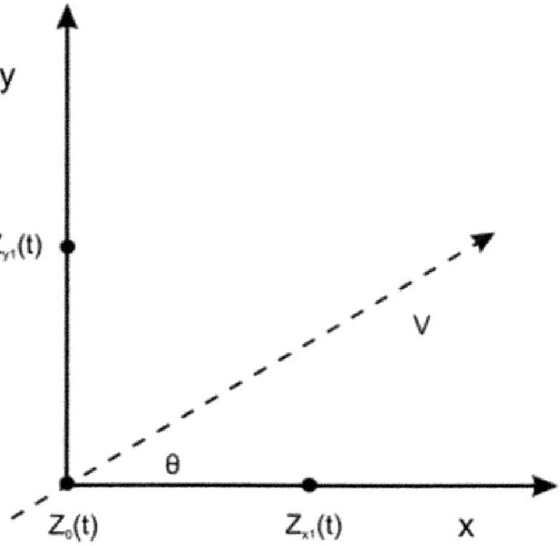

Fig. 13.4. In a triple-beam sounding approach, the beams are arranged in a right isosceles triangular arrangement. A cross section of the triple-beam sounding arrangement at some given altitude will be proportionately larger. The line connecting one of the beam pairs is designated as the x-axis, with the axis connecting the other pair as the y-axis.

scale structures will be observed by all three beams, and at different times depending on the wind speed and direction. In the ideal limit, turbulent fluctuations would be entirely one-dimensional along the line of motion and the three signals would be identical, except for the temporal offsets. (Deviations from this idealization are the source of much of the difficulty for all of the correlation methods. These techniques rely solely on the large-scale structures, whose fluctuations along the line of motion may be observed, but whose fluctuations transverse to it are not.) In this case

$$Z_x(t) = Z_o(t - \Delta t_x)$$
$$Z_y(t) = Z_o(t - \Delta t_y)$$

where Δt_x and Δt_y are the time lags of Z_x and Z_y with respect to Z_o. In the "triple-beam" approach to lidar-based wind profiling, these two time lags are calculated through the use of correlation functions for each pair of signals. The wind velocity components V_x and V_y are then calculated from the time lags and the beam separations x_1 and y_1

$$V_x = \frac{x_1}{\Delta t_x} \qquad V_y = \frac{y_1}{\Delta t_y}$$

The use of lidars to measure wind velocity has been around for some time, but vertically staring lidar-based profilers have received only scant attention to date. Among the few lidar profiling methods, described in the literature, the triple-beam near-vertical sounding technique was reported by Kolev et al.

(1988) and Parvanov et al. (1998). In these papers, the authors use three independent lidar devices, all pointed vertically along slightly divergent paths, to generate three separate lidar signals. These signals may then be correlated at each altitude to determine the beam-to-beam transit time of structures in the spatial particulate distribution and thus determine the transverse (horizontal) velocity vector.

13.1.2. Two-Dimensional Correlation Method

There is no particular requirement that the lidar be oriented near-vertically to perform a correlation. If the changes in wind speed are desired over a large area, scanning at horizontal or near-horizontal elevations can be used. When the lines of sight are near horizontal to the ground, there will be a lag in space as well as in time unless the lines of sight are perpendicular to the wind direction. In this case, the two-dimensional correlation technique is preferred. The most common application has been the measurement of two-dimensional velocity vectors (usually in a horizontal plane) through the use of scanning lidars and two-dimensional mathematical correlation (Kunkel et al., 1980; Sroga et al., 1980; Clemesha et al., 1981; Hooper and Eloranta, 1986).

It is possible to obtain two-dimensional wind vectors on timescales of minutes with a horizontal spatial resolution on the order of 250 m and vertical resolution of 50 m over distances of 6–8 km (depending on particulate loading) with a two-dimensional correlation technique (Barr et al., 1995). The two-dimensional methodology was originally developed at the University of Wisconsin (Sroga and Eloranta, 1980; Hooper and Eloranta, 1986; Barr et al., 1995). In this method, the lidar scans between several lines of sight that are parallel or near parallel to the ground, θ_1, then θ_2, then θ_3, then back to θ_1 to start the cycle over again. This produces relative concentration information along each of the lines of sight that is periodic in time. Figure 13.5 is an example of the relative particulate concentration in space and time along three different lines of sight. As a structure advects from one line of sight to the next it can be seen in the next plot, but at a different time and distance from the lidar. This method uses correlation to determine that time and distance difference. Instead of correlating individual points in space as was done in the previous method, portions of larger, two-dimensional plots of particulate concentration versus range and time are compared. A small portion of the data (on the order of 200–400 m in length) from line of sight 1 is compared with the data in the other two lines of sight. Equation (13.2) is used to calculate the correlation matrix using that portion of the signal from one line of sight, matrix $Z_{\theta 1}$, and the entire set from another line of sight, matrix $Z_{\theta 2}$.

$$C(\Delta r, \Delta t) = \frac{\sum_{i=0}^{n}\sum_{j=0}^{m}[Z_{\theta_1}(r_i+\Delta r, t_j+\Delta t) - \langle Z_{\theta_1}(r_i,t_j)\rangle][Z_{\theta_2}(r_i,t_j) - \langle Z_{\theta_2}(r_i,t_j)\rangle]}{\sqrt{\sum_{i=0}^{n}\sum_{j=0}^{m}[Z_{\theta_1}(r_i,t_j) - \langle Z_{\theta_1}(r_i,t_j)\rangle]^2[Z_{\theta_2}(r_i,t_j) - \langle Z_{\theta_2}(r_i,t_j)\rangle]^2}} \tag{13.2}$$

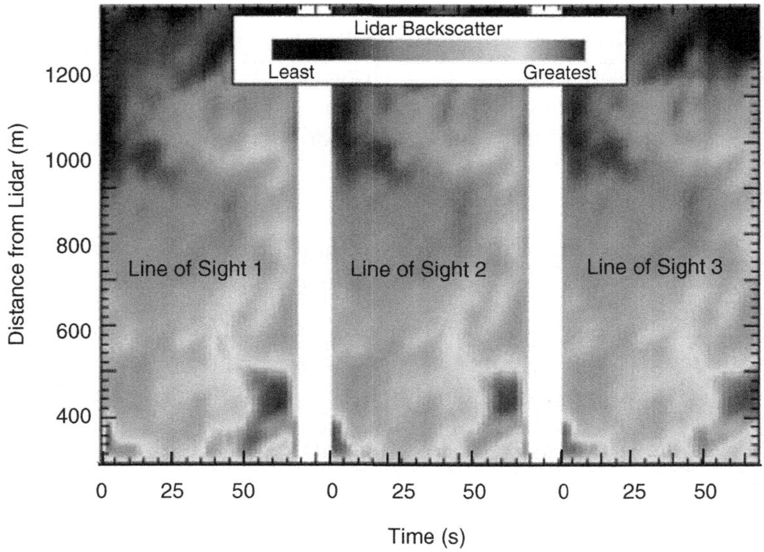

Fig. 13.5. An example of the relative particulate concentration in space and time along three different lines of sight, separated by 1.5°, that are horizontal to the ground.

The lag in space and time (Δr and Δt) at which the maximum value of the correlation matrix occurs is used to calculate the velocity at the location of the small segment used. One can see slight variations in the three lines of sight that indicate the transport of the structures across the lines of sight. Figure 13.6 is an example of a correlation done with some of the data from Fig. 13.5. Normally there will be just one correlation peak, as in Fig. 13.6. In determining the spatial distance traveled during the lagged time, one must account for the distance between the two lidar lines of sight at the correlated range in addition to the lag in range. The direction of motion is along the line between these two points (Fig. 13.7). In Fig. 13.7, the structure moves from point *a* to point *b*. The correlation will determine the lag in range, Δr, and the lag in time, Δt. The distance between the two lines of sight, Δy, must be determined from knowledge of the distance from the lidar to the structure, *r*, and the angle between the two lines of sight, $\Delta\theta$. The velocity is determined as

$$V = \frac{\sqrt{\Delta r^2 + \Delta y^2}}{\Delta t} = \frac{\sqrt{\Delta r^2 + (r\Delta\theta)^2}}{\Delta t}$$

The direction of the wind is found from the direction of *a* to *b*.

However, because of turbulence, spatial structures tend to deform and diffuse with time. This causes the maximum value of the correlation to decrease with time and distance. It can be shown that this can also cause the correlated lags to be smaller in magnitude than would be determined by the wind speed, that is, that the estimated wind speed is systematically underestimated by this technique (Kunkel et al., 1980). Kunkel et al. showed that the

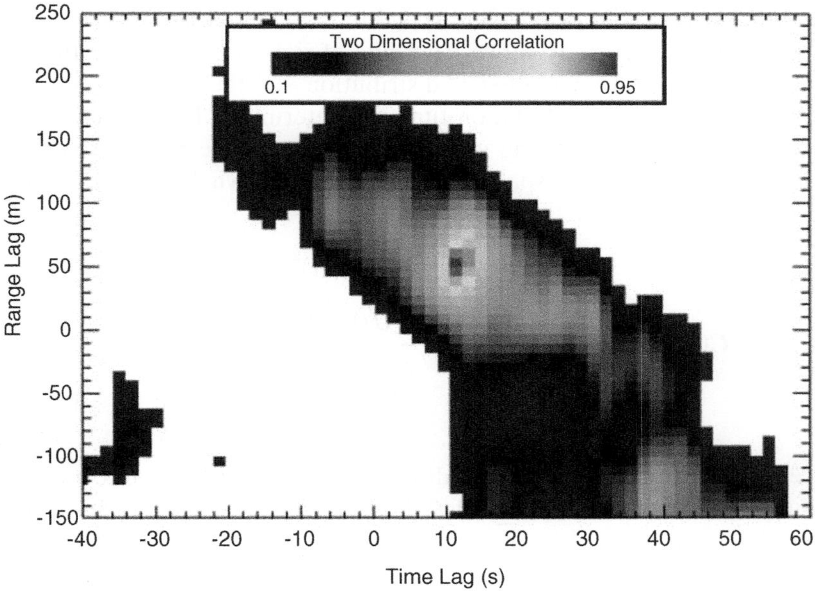

Fig. 13.6. An example of a two-dimensional correlation done using a portion of the data from Fig. 13.5.

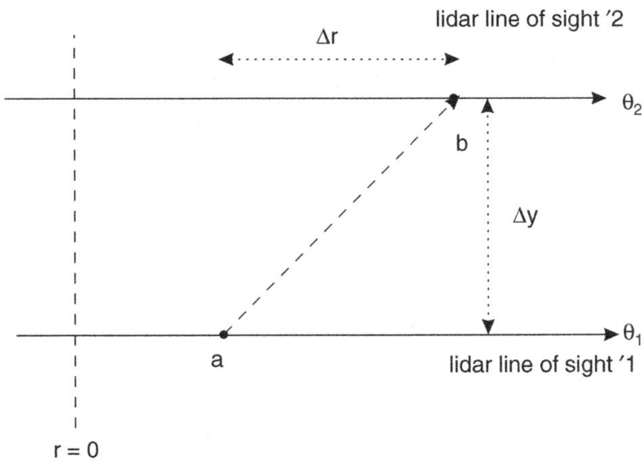

Fig. 13.7. The geometry of the wind analysis algorithm for the two-dimensional correlation.

width of the correlation function is determined by the size of the structure and the effects of turbulence. That portion of the width that is determined by the size of the structure can be estimated from the half-width of the correlation function at zero lag (the autocorrelation function), designated as σ_0. Because the width of a collection of particles will grow with time as $\sigma_y^2 = \sigma_v^2 t^2$, the relation between the size of the correlated structure between lines of sight θ_1 and θ_2, σ_1, can be written as $\sigma_v = (\sigma_1^2 - \sigma_0^2)^{1/2}/t_1$. σ_v is the root-mean-square deviation of that component of the wind speed in the v direction and t_1 is the time it takes for the structure to move from line of sight θ_1 to θ_2. From this information, and assuming a Gaussian distribution for the heterogeneities, the shape of the top of the correlation function determined by Eq. (13.3) can be predicted as a function of the velocity variances in the x and y directions, and the lags in time and space (Δy and Δx as shown by the geometry in Fig. 13.7) as

$$C(\Delta x, \Delta y, t) = \frac{\begin{array}{c}\pi^{3/2} B^2 \sigma_x^6 \exp\left[-(\Delta y - \overline{V}t)^2 / 4(\sigma_x^2 + 1/2t^2\sigma_v^2)\right] \\ \exp\left[-(\Delta x - \overline{U}t)^2 / 4(\sigma_x^2 + 1/2t^2\sigma_v^2)\right]\end{array}}{\left(\sigma_x^2 + \frac{1}{2}t^2\sigma_v^2\right)^{\frac{3}{2}}} \tag{13.3}$$

where B is a fitting constant, σ_x is half of the half-width of the autocorrelation function, σ_0, Δx and Δy are the lagged distances in space determined from the geometry shown in Fig. 13.7, $\overline{V}$ and $\overline{U}$ are the components of the velocity in directions perpendicular to and parallel to the lidar lines of sight, and t is the lagged time. Because at least three lines of sight are used, at least two estimates of the width of the correlation function can be determined. This allows σ_v to be estimated from

$$\sigma_v = \left[\frac{\sigma_2^2 - \sigma_1^2}{t_2^2 - t_1^2}\right]^{1/2} \tag{13.4}$$

To solve the problem, one calculates the correlation function from Eq. (13.2) and then equates it to Eq. (13.3) having estimated σ_x and σ_v and having determined Δx, Δy, and t from the highest value of the correlation function. From this, one can determine the wind velocity and make improved estimates of the correct spatial lags, iterating to a solution.

The improved method eliminates the errors associated with turbulent dissipation of the plumes and allows for subpixel resolution of the lags. This turns out to be an important factor in determining the minimum resolution with which the lidar can determine the wind velocity. The natural resolution of spatial lag is determined by the spatial resolution of the lidar, which is determined by the laser pulse length and digitizer sampling rate. Similarly, the resolution of the lag in time is determined by the time required for the lidar to complete a cycle through the three angles. The fractional error caused by this can be reduced to some extent by increasing the size of the angle between the lines of sight. This has the effect of increasing the time (and possibly the distance lag) required for a structure to pass through both lines of sight. This helps to some extent but increases the time required to make a cycle through the lines of sight and increases the amount of distortion caused by turbulence, reducing the significance of the correlation. In practice, the method is quite sensitive to the angle between the wind and the lidar lines of sight and the angular width between the lines of sight. Ideally, it would beneficial to be able to calculate wind vectors in real time and adjust the scan angles dynamically. To our knowledge, this has not yet been accomplished.

A wind vector can be determined from a scan over three angles that requires as little as 60–90 s to complete. By orienting these three angle sets in many directions, the wind field in a large area can be determined (see, for example, Barr et al., 1995). Despite the limitations of the method, this can be valuable in situations where the wind field is complex and cannot be effectively addressed with a limited number of fixed instruments or balloons. Figure 13.8 is an example of the wind pattern in the Rio Grande valley near El Paso, Texas, showing the complexity of the winds in the region of the pass through the mountain.

It should be noted that the analysis described here limits the method to three lines of sight differing in azimuth angle but at the same elevation angle. More lines of sight could be used to reduce the uncertainty in the measurements but would require an increase in the time required to complete a cycle in which data is collected at all of the angles. Some work has been done to explore the possibility of three-dimensional wind measurements using three lines of sight oriented horizontally with two additional lines of sight above and below the middle line of sight. To our knowledge, nothing has yet been published on results from more innovative scan configurations.

Because of the use of a two-dimensional correlation, the method is limited to near-horizontal elevation angles. For two horizontal lines of sight separated by some small angle (~1–3°), a structure traveling with the wind will intersect

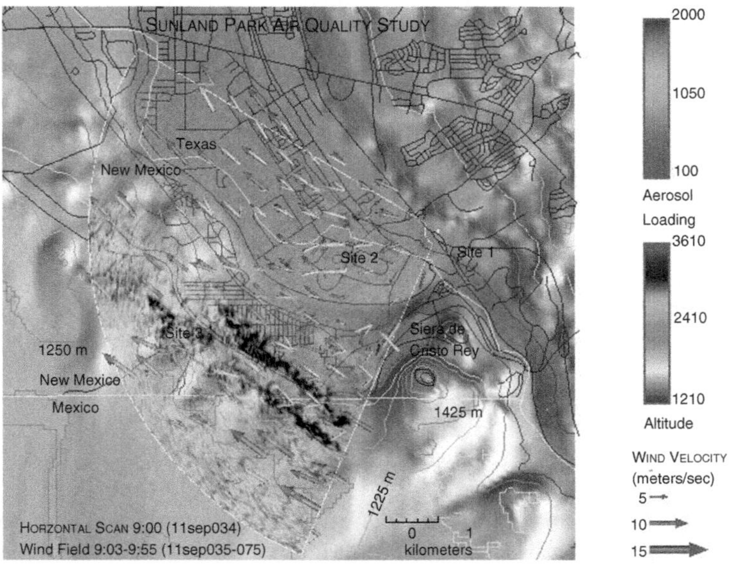

Fig. 13.8. An example of the wind pattern in the Rio Grande valley near El Paso, Texas, showing the complexity of the winds in the region of the pass through the mountain.

one, then the other, line of sight for nearly all wind directions. If the three lines of sight are at a high elevation angle, the only wind vectors that will intersect more than one line of sight at different ranges from the lidar are those oriented quite close to the plane of the lines of sight. Two-dimensional correlations require that a structure has a high probability of entering each of the lines of sight at a different distance from the lidar. This is certainly true for lines of sight oriented horizontally to the ground (any horizontal direction is, in principle, equally probable) but is not true for a vertical orientation (vertical wind speeds are nearly always much less than horizontal wind speeds so that structures travel nearly horizontally).

13.1.3. Fourier Correlation Analysis

Conventional correlation lidar devices, such as those developed by Eloranta et al. (1975), Kunkel et al. (1980), Hooper et al. (1986), and Kolev et al. (1988), compare signals at different places and times through the use of statistical correlations. Fourier transforms are sometimes involved, but only as a means of calculating the correlation function. This type of analysis retains the spatial particulate distribution information, which may be important if one is interested in calculating turbulence parameters but only serves to confuse velocity calculations. A mathematical technique using Fourier transforms may be applied to conventional correlation data, providing a simpler and more elegant method to determine the time lag directly, rather than applying some variety of peak-finding algorithm to a correlation function. Consider two identical signals offset by a given amount of time. According to the time-shifting theorem, the Fourier transform of one will equal the transform of the other multiplied by a phase factor

$$F[Z(t-\Delta t)] = e^{-i\omega\Delta t}F[Z(t)]$$

Thus

$$\begin{gathered} \frac{F[Z(t-\Delta t)]}{F[Z(t)]} = e^{-i\omega\Delta t} \\ \frac{i}{\omega}\ln\left\{\frac{F[Z(t-\Delta t)]}{F[Z(t)]}\right\} = \Delta t \end{gathered} \tag{13.5}$$

In this instance, the natural logarithm of the ratio of the Fourier transforms is directly proportional to frequency, with $i\omega$ multiplied by the time lag as the proportionality constant. The "curve fit" in this case simply amounts to multiplying all of the data points by i/ω and taking the average. Thus the logarithm of this ratio provides a simple way of comparing two signals to determine the time lag, without resorting to a lengthy correlation analysis. The same basic procedure can be applied in multiple dimensions as well, providing a method of calculating spatial lags as well as the temporal lags for various types of image correlation analysis. Note that this method calculates lag times that may be a fraction of the data sampling interval and an error estimate can be made using the standard deviation of the estimates. In a standard correlation calculation, the time between successive scans is one of the fundamental limitations on the accuracy of the method.

13.1.4. Three-Dimensional Correlation Method

The three-dimensional correlation method is also capable of determining the horizontal wind vectors with 250-m horizontal and 50-m vertical spatial resolution over about a 50-km^2 area. They are derived with two-dimensional cross correlations computed between a series of backscatter images derived from a volume image of relative particulate concentration. The algorithms to measure vertical profiles of the horizontal wind from a successive lidar images were first suggested and demonstrated by Sasano et al. (1982). Sasano et al. adapted a method used for some time to measure the motion of clouds from satellite photographs (Leese and Novak, 1971; Austin and Ballon; 1974; Asai et al., 1977) The method was further developed and extended to spatially resolved wind measurements in a three-dimensional volume by Schols and Eloranta (1992) and Piironen and Eloranta (1995). Early algorithms determined a single wind vector at every altitude representing the mean wind over the area of a scan. Recently, the algorithms have been improved to provide a vector wind field with a 250-m spatial resolution (Eloranta et al., 1999). This technique requires the combination of a high repetition rate, a high-power laser, a large telescope, and a fast scanning capability. The laser used in the University of Wisconsin volume imaging lidar (VIL) is capable of 1 J per pulse and 100 Hz. This and a large telescope are required so that the lidar signal from a single laser pulse has a sufficient signal-to-noise ratio that it can be used in the analysis. Fast scanning is required so that a large volume of space can be scanned on a timescale much shorter than the time required for a structure to move across the scanned volume. For a maximum range on the order of 10 km, a horizontal angular range of 45°, and a vertical angular range of 30°, the scan must be completed in about 30 s. Data collection at these rates results in severe requirements for data storage, generating on the order of a gigabyte of data per hour. The wind profiling method is based on following the movements of structures inside the scanned volume from subsequent horizontal scans. The method used by the University of Wisconsin group is quite complex and is covered here only in general terms. A more complete explanation can be found in Piironen (1994), which is available on the Internet (hppt://lidar.ssec.wisc.edu/papers).

The wind speed and direction are derived from a spatial, two-dimensional cross-correlation computed between portions of a larger three-dimensional volume. Each of these smaller portions are 250 m on a side. Correlations are

University of Wisconsin Volume Imaging Lidar (VIL)

Transmitter		Receiver	
Wavelength	1064 nm	Type	Schmidt–Cassegrain
Pulse length	~10 ns	Diameter	0.5 m
Pulse repetition Rate	100 Hz	Maximum range	15 km
		Range resolution	15 m

computed between portions of every other scan so that left-moving and right-moving scans are compared with the same scan direction and thus the time interval between laser profiles in each part of successive images is similar. In high-wind conditions, particulate structures may be advected out of the 250-m portion during the time between scans. To minimize this problem, the second image used in the correlation is chosen to be from a position displaced downwind from the first image by the distance the structure may be expected to move during the time between scans. This allows the correlation to take place with approximately the same air mass that was present in the first image. The displacement of the image position is added to the displacement of the correlation peak when computing the wind vector.

The method relies on the comparison of constant altitude plan position indicator (CAPPI) scans, which are two-dimensional horizontal maps of the relative particulate concentration. The mean motion of particulate structures is determined by calculating the location of the maximum of the correlation function between successive CAPPI to determine the average wind speed and direction in the area covered by the CAPPI. CAPPI scans at each height are extracted from the three-dimensional volume scans. The creation of a CAPPI begins with filtering the data from each of the lidar lines of sight to eliminate the effects of variable atmospheric attenuation, scan angle-dependent background level, and shot-to-shot variations in laser energy with a 2-km-long high-pass filter.

Because the wind moves the structures during the time it takes to make a lidar scan, the measured patterns in the lidar signal are distorted from what was actually present at any instant in time. Thus the location of the backscatter signal must be corrected by moving it a distance, ut, upwind where u is the mean wind vector and t is the time elapsed from the beginning of the volume scan. This correction is repeated when each new wind vector is determined, creating a new set of CAPPI from which a new estimate of the wind vector is determined. Piironen (1994) reports that if no correction is made for the wind on the first iteration, only one more iteration of the wind analysis loop is required to achieve convergence.

A CAPPI represents the lidar backscatter in a rectangular grid with some vertical resolution. Because the lidar takes data in a spherical coordinate system, all of the data inside each cell are used to determine the one value for the cell. When the grid spacing is small, some grid cells at long ranges remain unsampled. The values for the backscatter in the cells in which no actual data are taken are determined by linearly interpolating the closest sampled cells. It is important to preserve the coherence between adjacent and subsequent CAPPI planes. If a sparse grid spacing is used to avoid empty pixels, the spatial resolution is reduced in the region close to the lidar, where the quality of the signal is best.

When the CAPPI is extracted, an average of five consecutive scans is subtracted from each scan to minimize the influence of stationary structures. Near the surface, structures are often found to be "attached" to the surface and do not advect with the wind. These structures will result in an erronous zero lag. The scan is then histogram equalized. Each of the pixels in the CAPPI is sorted into N number of amplitudes, and the amplitudes in the scan are changed so that the probability density of the amplitudes is uniform. The modifications to the amplitudes are done in a way that maintains the relative magnitudes of the amplitudes in the scan. Histogramming reduces the influence that any one structure might have on the final correlation. Reducing the number of amplitudes that are used in the histogram will reduce the contrast in the CAPPI and broaden the correlation function. The average intensity is subtracted from each pixel before calculating the correlation function to reduce the effects of correlations with zero spatial lag.

To determine the lags in space and time, the maximum value of the correlation function is found. Regions in the correlation that have an amplitude within a factor of $1/e$ of the maximum are then identified. Each region is weighted by the sum of all the pixels contained inside the region. The region with the largest weighting factor is assumed to contain the correlation maximum that corresponds to the desired lags. The exact location of the peak is determined by a least-squares fit of a two-dimensional quadratic polynomial in a five by five pixel region about the highest point in the selected region. The fitted function is

$$F(x, y) = a_0 + a_1 x + a_2 y + a_3 x^2 + a_4 y^2 + a_5 xy$$

where x and y denote coordinates in correlation plane and the coefficients a_n are fitting parameters. The maximum value of the fitted function is used as the peak position. This is done to achieve a resolution in space finer than the resolution of the pixels that were used in the calculation. The fitting also interpolates the points near the maximum to minimize the effects of noise. The constants a_i are found from a least-squares analysis. The desired lags are then found from

$$x_{\max} = \frac{(2a_4 a_1 - a_5 a_2)}{(a_5^2 - 4a_3 a_4)} \qquad y_{\max} = \frac{(2a_3 a_2 - a_5 a_1)}{(a_5^2 - 4a_3 a_4)} \tag{13.6}$$

The mean wind speed and direction can be found from:

$$u = \frac{1}{\Delta t}\sqrt{x_{max}^2 + y_{max}^2}$$

$$\cos\theta = \frac{-x_{max}}{\sqrt{x_{max}^2 + y_{max}^2}} \qquad \sin\theta = \frac{-y_{max}}{\sqrt{x_{max}^2 + y_{max}^2}} \tag{13.7}$$

where Δt is the time separation between subsequent volume scans.

Time-averaged wind estimates are found by averaging the cross-correlation functions over some length of time and determining the velocity and direction from the average of the cross-correlation functions. Averaging the correlation functions minimizes the contributions from noise because random correlations average to zero. With averaging of the correlations in time, even weak correlation peaks dominate after sufficient averaging. Estimating the average wind speed by averaging each of the velocities determined from the individual correlations can result in large fluctuations in wind speed and direction between nearby points because of spurious results being averaged with more accurate results.

Because the CAPPI scans are constructed from three-dimensional lidar scans, the average wind speed is determined at a series of heights. These kinds of measurements near the surface are especially valuable for atmospheric scientists and for studies of surface transport. Wind measurements are difficult to make at altitudes above a few meters. Although balloons can make these measurements, the altitude at which measurements are made is not well resolved, and measurements over time require many balloons (Mayor and Eloranta, 2001).

13.1.5. Multiple-Beam Technique

The multiple-beam method is related to the correlation methods presented above, yet is a different approach to the measurement of the transverse wind vector with a vertically staring lidar-based profiler. It is similar to the triple-beam approach in that it relies on turbulent structures in the air to generate fluctuations in the lidar signals, which may then be used to "track" the structures or "correlate" at various locations and thus determine a wind velocity. For this reason, it is also similar in its dependence on atmospheric conditions and the time and length scales necessary to make a measurement. This method involves the emission of several lidar beams simultaneously, imaging the scattering light from all of the beams on a single detector, and seeks corresponding patterns in the lidar signals. The horizontal wind vector may be determined with only two lasers and two lidar signals, rather than three. A unique mathematical analysis technique is used to extract the wind information from the multiple-beam lidar.

In this technique, a number of beams aligned in a plane are propagated simultaneously and imaged on a single detector. Heterogeneities in the particulate concentration in the atmosphere modulate the amplitude of the lidar signal as they pass through the series of lidar beams. The Fourier transform of these signals will produce frequencies corresponding to the component of the wind velocity in the plane of the lidar beams and the beam spacing. Two arrays of beams in a plane are projected vertically and orthogonally to each other. The horizontal wind speed and direction can be determined as the vector sum of the wind speeds in the two orthogonal directions represented by the arrays. This technique offers the possibility of sampling the wind velocities fast enough to obtain measurements of turbulent kinetic energy and shear stress with spatial resolutions on the order of a meter or less.

The multiple-beam wind lidar uses two Nd:YAG lasers operating at 1.064 μm with an energy of 100 mJ at 50 Hz. The lasers are attached to a plate that also supports a 25-cm, f/10, Cassegrain telescope inside the housing. The light from each laser follows one of two paths, each of which has a series of five beam splitters. The beam splitters are a sequence of 20%, 25%, 33%, 50%, and 100% reflectivity mirrors, so that the outgoing beams will have the same intensity. The series of beam splitters are mounted below the exit windows mounted on the top of the lidar. The lidar is operated in a vertical staring mode to determine the horizontal wind components.

Behind the telescope, the light passes through an interference filter and a lens system that focuses the light on a 3-mm diameter, IR-enhanced silicon avalanche photodiode. The signal from all of the beams in an array are imaged on the one detector. The laser in the second array is triggered to fire about 150 ms after the first laser fires. This makes the two signals nearly simultaneous, yet the signal from the first laser pulse will have decayed away and has no influence on the second.

The technique can provide near-instantaneous velocities as well as average velocities. Thus some turbulence quantities (e.g., turbulent intensities, Reynolds stresses, and higher moments or statistics) could be derived. In addition, particulate-related quantities can also be measured to obtain such quantities as cloud height and optical depth/reflectivity or boundary layer height and relative particulate loading with altitude. The current system can provide wind measurements every 5 m in altitude throughout the depth of the bound-

University of Iowa Multiple-Beam Lidar

Transmitter (2 each)		Receiver	
Wavelength	1064 nm	Type	Cassegrain
Pulse length	~10 ns	Diameter	0.27 m
Pulse repetition rate	50 Hz each direction	Focal length	2.5 m
Pulse energy	120 mJ maximum	Filter bandwidth	3.0 nm
Beam divergence	~1 mrad	Field of view	>80 mrad
		Range resolution	1.5, 2.5, 5.0, 7.5 m

ary layer (generally 1–2 km in altitude). Wind velocities can be determined on time scales as short as 20 s. Longer-term averaging is also possible, resulting in more precise wind measurements.

Consider Fig. 13.9, which represents two orthogonal arrays of simultaneous lidar beams at a given altitude, each array providing a signal $Z_x(t)$ and $Z_y(t)$. The intersection of the two arrays has been taken to be the geometric origin, the axes of the two arrays have been denoted as the x- and y-axes, and the wind angle is measured counterclockwise from the x-axis, as before. Because the beams in each array are emitted and observed simultaneously, the observed signal from each array at any altitude and instant in time will be the sum of contributions from the individual beams. The x-array produces a signal $Z_x(t)$ which can be written as

$$Z_x(t) = Z_{x1}(t) + Z_{x2}(t) + \ldots = \sum_{\text{beams}} Z_{xi}(t)$$

In the ideal limit of negligible turbulent fluctuations transverse to the line of motion, the contributions from all of the beams will be identical except for a temporal offset, Δt_{xi}, in each. Conceptually, the resulting signal is the sum of the same signal, $Z_o(t)$, as would be obtained from a hypothetical single beam, placed at the origin and possessing unit intensity, summed five times with different offsets in time:

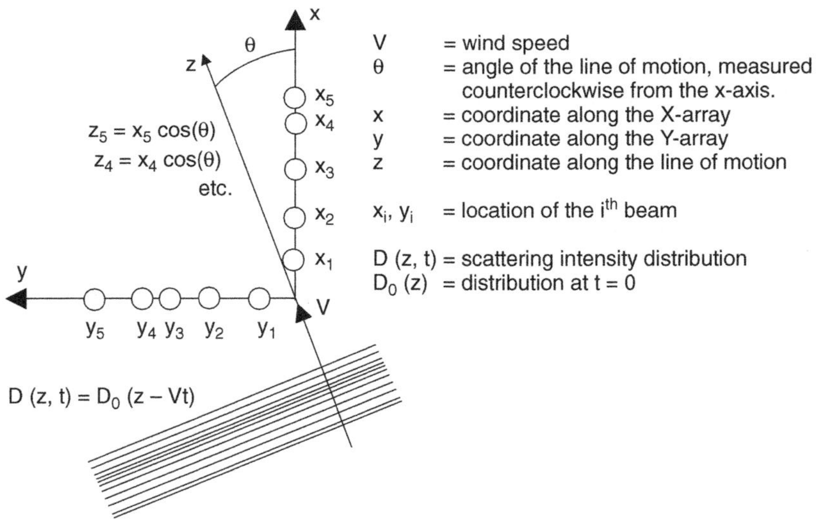

Fig. 13.9. The geometry of the two orthogonal arrays of five simultaneous lidar beams in the multibeam method, at some altitude. Each array creates signals $Z_x(t)$ and $Z_y(t)$. The intersection of the two arrays is taken to be the origin, the axes define the x- and y-directions. The wind angle is measured counterclockwise from the x-axis.

$$
\begin{aligned}
Z_x(t) &= A_{x1} Z_o(t - \Delta t_{x1}) + A_{x2} Z_o(t - \Delta t_{x2}) + \ldots \\
&= \sum_{\text{beams}} A_{xi} Z_o(t - \Delta t_{xi})
\end{aligned}
$$

The beam strength factors A_{xi} have been included to allow differing beam strengths within the array. The motive for observing multiple beams simultaneously is to produce temporal patterns in the lidar signals. As turbulent structures pass through an array of beams, their features will be reproduced in succession in the array's total signal, and the pattern of the repetition will correspond to the spatial placement of the beams. Furthermore, the speed of the pattern will be related to the wind speed; the faster structures appear to propagate along the array, the faster the signal patterns will be. (The "apparent" speed is important here, because of the relative orientation between the array and the wind speed. For example, if one-dimensional structures, plane waves, cross an array at an angle of nearly 90°, the structures will "appear" to propagate along the array very rapidly.)

If the beams are regularly spaced, the "pattern" in the signal will have a regular periodicity and the frequency of the periodicity as determined with a power spectrum would reveal the apparent propagation speed along the beams. If the beams are placed in an asymmetric distribution, however, the orientation of the pattern will also reveal the direction of the wind. If the pattern of beam locations is observed "forward" in time, the wind will be passing one direction along the array; if it is observed "backward" in time, the wind will be passing in the opposite direction. Finally, the most useful mathematical tool for dealing with patterns, the Fourier transform, has very efficient algorithms available for computation. Thus, to seek out the patterns in the observed multibeam lidar signal, the Fourier transform of the signal is taken. Using the definition of the transform as

$$
F[Z(t)] = \int_{-\infty}^{\infty} Z(t) e^{-i\omega t} \mathrm{d}t
$$

and because Fourier transforms are linear functions

$$
\begin{aligned}
F[Z_x(t)] &= F\left[\sum_{\text{beams}} A_{xi} Z_o(t - \Delta t_{xi})\right] \\
&= \sum_{\text{beams}} A_{xi} F[Z_o(t - \Delta t_{xi})]
\end{aligned}
$$

The time-shifting theorem states that the Fourier transform of a time-shifted signal equals the transform of the unshifted signal multiplied by a phase factor. Thus

$$
F[Z_x(t)] = F[Z_o(t)] \sum_{\text{beams}} A_{xi} e^{-i\omega t_{xi}} \tag{13.8}
$$

According to Eq. (13.8), the Fourier transform of a lidar signal from an array will be the same as the transform of a single beam's contribution, multiplied by a sum of phase factors. The signal from a single beam corresponds to unmodulated particulate fluctuations passing through the array, the amplitude of which is, in general, unknown and irrelevant for velocity calculations. The phase factor sum contains the time offsets, and thus the wind information. The unknown and irrelevant information can be eliminated by combining the information from two arrays in a ratio. If the time lags in both arrays are referred to the same hypothetical signal at the origin $Z_o(t)$, then

$$\frac{F[Z_x(t)]}{F[Z_y(t)]}=\frac{F[Z_o(t)]\sum\limits_{\text{beams}} A_{xi}e^{-i\omega t_{xi}}}{F[Z_o(t)]\sum\limits_{\text{beams}} A_{yi}e^{-i\omega t_{yi}}}=\frac{\sum\limits_{\text{beams}} A_{xi}e^{-i\omega t_{xi}}}{\sum\limits_{\text{beams}} A_{yi}e^{-i\omega t_{yi}}} \tag{13.9}$$

In other words, the ratio of transforms of the signals from two arrays will equal a ratio of sums of the phase factors, each phase factor depending on the wind vector and the relative position of each beam.

The time lags (relative to the hypothetical signal at the origin) Δt_i will depend on the apparent positions of each beam along the line of motion and can be expressed as functions of the beam positions x_i and y_i and the windspeed V and the angle θ:

$$\Delta t_{xi}=\frac{x_i\cos\theta}{V}$$

$$\Delta t_{yi}=\frac{y_i\sin\theta}{V}$$

For convenience, the factors are defined

$$c_x=\frac{\cos\theta}{V} \quad \text{and} \quad c_y=\frac{\sin\theta}{V} \tag{13.10}$$

These parameters contain the wind information and represent the reciprocals of the apparent velocity of the structures that are moving with the wind across the arrays. When multiplied by the beam positions, they serve as "scaling factors" reducing the size of the arrays from their full dimension to their apparent size when projected onto the line of motion. They scale the patterns in the signals from each of the arrays in time with the wind speed and the angle between the wind and the array. The wind velocity may be calculated from these parameters using θ = arctan(c_x/c_y) and $V=1/\sqrt{(c_x^2+c_y^2)}$. With these substitutions

$$\frac{F[Z_x(t)]}{F[Z_y(t)]}=\frac{\sum\limits_{\text{beams}} A_{xi}e^{-i\omega x_i c_x}}{\sum\limits_{\text{beams}} A_{yi}e^{-i\omega y_i c_y}}$$

The function on the left of this equation represents a ratio of Fourier transforms of the two-multibeam lidar signals, which may easily be calculated from the data. The quantity on the right is a function of frequency, with the known beam strengths and positions and the desired wind constants as parameters. With the exception of certain special beam arrangements, such as a symmetric array, the quantity on the right will be a unique function of wind speed and angle. The collected lidar data may be fitted over all meaningful frequencies to determine the best-fit values for c_x and c_y, to determine the wind speed and angle.

As a practical matter, it should be noted that the ability to adjust the intensity of the lidar beams is highly desirable, meaning that the constants A_{xi} will vary. However, a convenient arrangement for the production of a multiple beam array is to pass a single beam through a series of beam splitters, the reflectivities of which will in general be known, so the relative beam strengths within an array will be fixed and known. On the other hand, if the two arrays are powered by two separate lasers, the relative array strengths will still be arbitrary. Let A_x is defined as the sum of the strengths of all beams in the x-array (i.e., the total array strength), a_{xi} as the normalized beam strength, and R as the relative array strengths

$$a_{xi} \equiv \frac{A_{xi}}{\sum A_{xi}} = \frac{A_{xi}}{A_x}$$

$$R \equiv \frac{A_x}{A_y}$$

With these definitions, Eq. (13.10) can be written

$$\frac{F[Z_x(t)]}{F[Z_y(t)]} R^{-1} = \frac{\sum\limits_{\text{beams}} a_{xi} e^{-i\omega x_i c_x}}{\sum\limits_{\text{beams}} a_{yi} e^{-i\omega y_i c_y}} \tag{13.11}$$

All of the quantities in the function on the right are fixed and known, except for the independent variable ω and the desired wind parameters. The normalization constant R could be calculated from the laser settings and laser calibration curves, but a more accurate and convenient way of normalizing the Fourier transform ratio is simply to divide by the first transformed data point. Because the first data point in a discrete Fourier transform corresponds to the zero-frequency component of the signal, it is simply a sum of all untransformed data points and for sufficiently long signals will be proportional to the laser intensity. Thus the first data point in the series, $F[Z_x]/F[Z_y]$ will simply equal R.

Performing the calculations in this way has the benefit of allowing the power of the two lasers to be varied arbitrarily and independently, without additional manual input into the data analysis. Equation (13.11) provides a way of calculating wind velocity from two multiple-beam lidar signals, by fitting the function on the right to the data on the left. It was derived, however, based on the idealization of one-dimensional turbulent structure along the line of motion. In reality, fluctuations transverse to the line of motion will pollute the signals and generate large amounts of noise in the data. This noise may be significantly reduced by averaging together data from multiple time intervals and by eliminating high-frequency data from the curve fit. The latter is effective because much of the noise will be contributed by turbulent fluctuations having scales on the order of the array size or smaller, corresponding to the higher frequencies in the transform.

The ratio of sums of phase factors in Eq. (13.11) is highly nonlinear and computationally inconvenient. A more tractable expression may be obtained by expanding each phase factor in an infinite series

$$e^{-ix_ic_x} = \sum_{n=0}^{\infty} \frac{(-i\omega x_i c_x)^n}{n!}$$

Expressing the phase factors as sums allows the sum of phase factors to be rewritten as a single infinite sum

$$\sum_{\text{beams}} a_{xi} e^{-ix_ic_x} = \sum_{\text{beams}} a_{xi} \sum_{n=0}^{\infty} \frac{(-i\omega x_i c_x)^n}{n!} = \sum_{n=0}^{\infty} \left[\frac{(-i\omega c_x)^n}{n!} \sum_{\text{beams}} a_{xi} x_i^n \right] \tag{13.12}$$

Each term in the infinite sum now contains a sum over beams of the normalized beam strength multiplied by the beam position raised to the nth power. This inner sum is nothing more than the nth moment of the beam distribution. [The phase factor sum amounts to the Fourier transform of the beam distribution, and Eq. (13.11) is an example of expanding the Fourier transform of a distribution function in moments of the function.] Defining the nth moment of the x-array beam distribution as

$$\mu_{n,x} = \sum_{\text{beams}} a_{xi} x_i^n$$

the phase factor sum can be written

$$\sum_{\text{beams}} a_{xi} e^{-i\omega x_i c_x} = \sum_{n=0}^{\infty} \frac{(-i\omega c_x)^n}{n!} \mu_{n,x} \tag{13.13}$$

Using this series expansion for the phase factor sum, Eq. (13.11) can be written

$$\frac{F[Z_x(t)]}{F[Z_y(t)]} R^{-1} = \frac{\sum_{n=0}^{\infty} \frac{(-i\omega c_x)^n}{n!} \mu_{n,x}}{\sum_{n=0}^{\infty} \frac{(-i\omega c_y)^n}{n!} \mu_{n,y}} \tag{13.14}$$

This ratio of series expansions can be greatly simplified using the definition of the cumulants k_n from statistical mathematics, defined by the expression

$$\sum_{n=1}^{\infty}\frac{(-ik)^n}{n!}\kappa_n \equiv \ln\left[\sum_{n=0}^{\infty}\frac{(-ik)^n}{n!}\mu_n\right] \tag{13.15}$$

The nth cumulant may be calculated from the nth and lower order moments, as shown by Kenney and Keeping (1951), for example. Taking the logarithm of Eq. (13.14) and using the definition of the cumulants

$$\ln\left[\frac{F[Z_x(t)]}{F[Z_y(t)]}R^{-1}\right] = \sum_{n=0}^{\infty}\frac{(-i\omega)^n}{n!}(c_x^n\kappa_{n,x} - c_y^n\kappa_{n,y}) \tag{13.16}$$

Equation (13.16) may be regarded as the central equation to multibeam lidar signal analysis. It is a complex function of frequency relating the multiple beam lidar signals to the horizontal wind vector. The function on the left may be calculated in a straightforward manner given two signals, $Z_x(t)$ and $Z_y(t)$. This function, the natural logarithm of a ratio of Fourier transforms, has some interesting properties when used to compare two functions, and for reasons discussed elsewhere (Krieger, 2000) may be designated as the "relative modulation function" or RMF. The quantity on the right of Eq. (13.16) may be fitted over all relevant frequencies to determine the best-fit values for c_x and c_y, and thus the wind velocity. The series must be truncated at some order in the computations, but the higher frequencies will correspond to the smaller scale turbulent structures and will contain much of the noise anyway. Calculation of the cumulants is quite difficult for higher orders, but this may be done beforehand, and if only low orders are kept in the expansion, this is not a significant problem. The first three cumulants, in fact, are identical to the first three central moments (Kenney and Keeping, 1951).

Solving Eq. (13.16) for the c_x and c_y at each height, the wind speed and direction can be found from Eq. (13.10). This will provide the horizontal wind velocity at each range bin of the lidar. At the time of this writing, only preliminary measurements have been made with this technique, but these tests have shown the method to provide consistent results.

13.1.6. Uncertainty in Correlation Methods

A complete uncertainty analysis for correlation methods has not been done. A limited analysis and comparisons with more conventional methods have been done in studies by Piironen (1994), Piironen and Eloranta (1995), and Zuev (1997). Despite the seeming simplicity of correlation methods, there are a large number of effects that complicate a correlation analysis. Atmospheric structures are not simply transported horizontally with the wind, they distort, rotate, and evolve as well. They may also have a velocity component in the vertical direction so that the correlation is done between different parts of the structure. During daylight hours, most of the signal noise is due to photon noise from background sunlight. This leads to spatially uncorrelated noise in the CAPPI scans. Deformation or rotation of the structures due to turbulence or traveling waves will distort the correlation functions, leading to erroneous wind velocities. False correlations may occur between two different structures, leading to erroneous correlation peaks.

The deformation of the particulate spatial shape is a significant error source. For two-dimensional scanning, an equation can be written to correct for this effect [Eq. (13.3)], at least to some extent. However, as the data are processed to remove stationary structures, and other effects that lead to erroneous correlations, the data are also distorted. Thus the data that are actually correlated are not the structures that were actually there on an instantaneous basis, so that the application of Eq. (13.3) is questionable. Moreover, this analysis does not correct for rotation of the structure or transport in the direction orthogonal to the plane of the correlation. The presence of gravity waves or strong vertical wind shear will tend to either move the structures in ways not anticipated by the concept or may systematically deform the structures.

Correlations may also occur from random correlations between two different structures. This is a particular problem with the two-dimensional method, because structures will often follow, one after another, and are often periodic. An intense signal in one of the images that is not present in the other (the passage of a bird, for example) may also lead to a strong random correlation. Normally, a cross-correlation function is dominated by a single peak, but fluctuations due to random noise or different structures may lead to additional peaks that may be stronger than the true correlation peak. Because the wind speed and direction are calculated from the strongest peak, a random error occurs.

Piironen and Eloranta (1995) have developed an error analysis for the effects of random fluctuations in the lidar signal. Although this is valuable, it certainly underestimates the uncertainty in the measurement. An additional source of uncertainty is the range and time resolutions of the measurement. Although the lags can be interpolated between the correlation values, this cannot be done with high resolution.

Piironen and Eloranta (1995) examined the wind profiles determined from the 1989 FIFE data and determined that 76% of hourly averaged wind estimates in the convective boundary layer were reliable. The wind profiles determined with the three-dimensional correlation compare well with traditional wind measurements made with radiosondes or surface weather stations. The differences between lidar wind profiles and traditional measurements are dominated by natural wind fluctuations and the fact that lidar measurements represent an average over an area. This makes it difficult to determine the error in the lidar measurements with a simple comparison to measurements made by other instruments. Inside the boundary layer, error estimates made by Piironen and Eloranta (1995) are relatively constant with altitude and are about 0.2 m/s in speed and 3° in direction. Above the boundary layer, the errors grow rapidly because the calculated correlations become poorer due to the large reduction in particulate intensity (and thus contrast) with altitude. As the averaging time increases, the influence of random correlations decreases, and thus the measurement errors also decrease.

A detailed experimental examination of the effects of all the sources of uncertainty in the correlation method is not likely. Such a study would require in situ measurements with an instrument that can directly measure the motion of an air mass over some area. At this time, the lidar is the only instrument that even approaches this capability.

13.2. EDGE TECHNIQUE

The edge technique is an incoherent method that uses the Doppler shift in the scattered light to measure the wind speed. Conventional Doppler lidars mix the scattered light with light from the master oscillator to produce a beat frequency that is the difference between the frequency of the emitted light and the frequency of the scattered light. The velocity of the scatterer can be found from this frequency difference $\Delta\nu$. For a monostatic lidar system,

$$v = \frac{c}{2}\frac{\Delta\nu}{\nu}$$

where c is the speed of light and ν is the frequency of the scattered light. Incoherent methods, by way of contrast, attempt to measure the change in frequency with some other method. The edge technique uses high-resolution optical filters in such a way that a small change in the frequency results in a large change in the measured signal amplitude. There are several advantages to the edge technique. It is relatively insensitive to the spectral width of the laser if the width of the edge filter is larger than the spectral width of the laser. It is claimed that it is possible to measure the Doppler shift to an accuracy better than 100 times the spectral width of the laser (Korb et al., 1992). Because direct detection of the scattered light is used, the divergence of the laser beam does not have to be narrow and the field of view of the telescope can be large. The magnitude of the lidar return is also larger in comparison to most coherent Doppler lidars, because short wavelengths can be used. This means that the system requires considerably less laser power, an important consideration for satellite applications. There are several variants of the edge technique that may be generally grouped by whether they use the particulate return (Korb et al., 1992; Gentry and Korb, 1994) or the molecular return to determine the shift.

Single-Edge Technique. The amplitude of the "elastic" return from the atmosphere is shown in Fig. 13.10 as a function of wavelength. The basic edge technique takes advantage of the large change in transmission of a filter with frequency. The filter is characterized by the centerline frequency and the half-width at half-maximum (FWHM), α. Figure 13.11 shows a filter and the locations of the outgoing laser frequency and the frequency of the scattered light. Any instrument or technique that can produce a large change in transmission for a small change in frequency could be used as the edge filter. A molecular

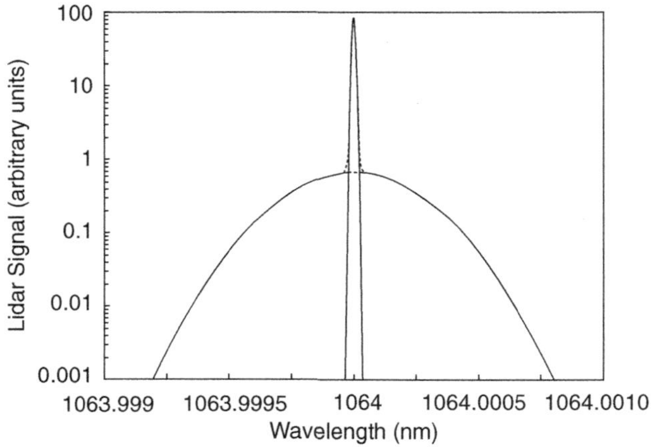

Fig. 13.10. The relative amplitude of the lidar return for a 1.064-μm (Nd:YAG) lidar as a function of wavelength. The narrow central peak is the Doppler-broadened return from particulates and the wider peak is the Doppler-broadened peak from molecular scattering. The relative amplitudes of the two peaks is a function of the wavelength and particulate loading.

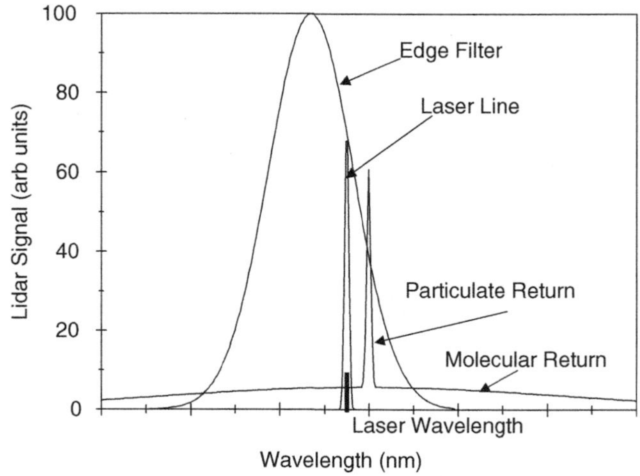

Fig. 13.11. The spectral location of the edge filter and the locations of the outgoing laser frequency and the frequency distribution of the "elastically" scattered light.

or atomic absorption line, a prism or grating could be used, although a Fabry–Perot étalon is most common. A variation of the edge method using an iodine molecular filter has been described (Liu et al., 1997) and demonstrated to an altitude of 45 km (Friedman et al., 1997). There are no particular restrictions on the wavelengths that can be used, although some will work better than others depending on the details of the method that is used. If the narrow shift in the particulate return is used, an infrared wavelength that maximizes the particulate return as opposed to the molecular return is preferred. For this case, the molecular return, which is at least a factor of 10 wider than the particulate return, is essentially a constant background in the signal and requires compensation. If the wider molecular return is used, the magnitude of the signal return can be increased by moving toward the ultraviolet (molecular scattering is proportional to $1/\lambda^4$). Thus 355 nm is often suggested because it can be generated with high efficiency and represents a balance between maximizing the molecular return and avoiding the high attenuation that occurs deeper in the ultraviolet.

The transmission of the filter at the frequency of the outgoing laser light, ν_{laser}, is measured as the laser pulse is emitted from the lidar. The transmission of the filter at the frequency of the scattered light at ν_{ret} is measured as the scattered light is collected by the telescope. Knowing the properties of the filter, the change in frequency between the outgoing and the scattered light (i.e., the Doppler shift) can be determined. Because the amplitude of the scattered light changes as a function of time because of changes in particulate loading and range attenuation, the relative amplitude of the lidar signal at each time must also be measured. The ratio of the Doppler-shifted lidar signal through the edge filter, I_{Edge}, to the signal measured by an energy monitor, I_{EM}, is the normalized shifted signal

$$I_{\text{N}}(\nu+\Delta\nu) = \frac{I_{\text{Edge}}}{I_{\text{EM}}} = CF(\nu) \tag{13.17}$$

where $F(\nu)$ is the spectral response of the edge filter and C is a calibration constant. The calibration constant can in principle be measured by comparing the signals from a fixed target both with and without the edge filter. Because the frequency of the laser may drift, the outgoing laser wavelength must also be monitored to obtain an $I_N(\nu)$.

The difference between the normalized, shifted signal at a given range r and the normalized laser value can be used to determine the radial velocity v at range r as

$$v = \frac{c}{2\nu}\left[\frac{I_{\text{N}}(\nu+\Delta\nu) - I_{\text{N}}(\nu)}{C\beta(\nu, \nu+\Delta\nu)}\right] = \frac{c}{2\nu}\left[\frac{\Delta I_{\text{N}}}{C\beta(\nu, \nu+\Delta\nu)}\right] \tag{13.18}$$

where ν is the laser frequency and $\beta(\nu + \Delta\nu)$ is the average slope of the transmission of the edge filter in the frequency range, from ν to $\nu + \Delta\nu$, and ΔI_N is the change in the normalized signal between the two frequencies. This equation is limited to small Doppler shifts in which $\Delta\nu <$ FWHM/4. Beyond this, the edge technique could be used, but the changes in the slope of the filter would have to be accounted for. An additional advantage of using the difference between the normalized signals in this way is that the system is insensitive to small variations in the frequency of the laser. This will be true as long as the changes in frequency are not large enough to change $\beta(\nu + \Delta\nu)$.

The sensitivity of the measurement is an important parameter in this lidar. The sensitivity is defined as the fractional change in the normalized measurement quantity, ΔI_N, for a unit change in velocity. The sensitivity is thus

$$\theta = \frac{1}{V_0}\frac{\Delta I_\mathrm{N}}{I_\mathrm{N}} \tag{13.19}$$

where V_0 is the velocity. The sensitivity governs the precision with which the velocity can be measured. A comparison of Eqs. (13.18) and (13.19) shows that θ must be proportional to the change in transmission of the edge filter that frequency. This implies that the sensitivity of the system is proportional to the spectral width of the edge filter. It is often claimed that the edge method is insensitive to the spectral width of the laser. However, an unnarrowed laser requires a wider edge filter that will have a decreased sensitivity and thus will result in decreased precision in the system. The fractional uncertainty in the velocity, δV, is related to the fractional uncertainty in the normalized velocity signal, $\delta\Delta I_\mathrm{N}$, as

$$\delta V = \frac{\delta\Delta I_\mathrm{N}}{\theta} \approx \frac{1}{\frac{\mathrm{S}}{\mathrm{N}}\theta} \tag{13.20}$$

where S/N is the signal-to-noise ratio of the lidar measurements. The primary source of error is the accuracy with which the normalized signals can be measured. The precision of the measurement is also related to the sensitivity, which is in turn proportional to the rate of change in transmission with frequency of the edge filter.

The most precise measurements are made when the signal to noise is large and then the sensitivity is largest, that is, the edge filter is narrowest. Thus infrared lidars using the particulate return would be the preferred operating system in the boundary layer, where particulate concentrations are high. Measurements higher into the troposphere, where the particulate loading is considerably less, would likely use a near-ultraviolet wavelength to maximize the molecular return at the cost of a decrease in the precision of the measurements. With either system, the uncertainty in the measured wind is a strong function of distance from the lidar, increasing at least as fast as r^2.

Wind measurements using the edge technique were demonstrated by Korb et al. (1997), using an infrared Nd:YAG lidar. The laser was injection seeded to obtain a spectral width of 35–40 MHz and operated with an energy of 120 mJ per pulse at 10 Hz. A small portion of the outgoing laser pulse is used to make a reference measurement of the laser frequency on the edge filter for each outgoing laser pulse. A fiber-optic cable is used to transfer the light from the focal plane of the telescope to the focal plane of a collimating lens. This lens collimates the light for a planar Fabry–Perot étalon, which is used as the edge filter. A beam splitter is used to divert 30% of the light into a conventional detector that is used to measure the amplitude of the signal with the same time resolution as the edge-filtered signal to determine the relative amplitude of these two signals. Solid-state silicon avalanche photodiodes with 3.3-MHz bandwidth amplifiers are used as detectors. The Fabry–Perot étalon has a plate separation of 5 cm and a clear aperature of 5 cm, yielding a free spectral range of 0.1 cm. The étalon plates have a reflectivity of 93.5%, resulting in a finesse of 47 and a spectral resolution (FWHM) of 65 MHz. The sensitivity of this system is about 3.8% /(m/s) when the system is operated at the half-transmission portion of the étalon. A feedback system is used to lock the edge of the etalon to the frequency of the laser.

Hard-target measurements were made to provide a zero-velocity calibration for the lidar. These measurements of a stationary target had a mean value of 0.19 m/s and a standard deviation of 0.17 m/s. To measure winds, the lidar makes measurements at four lines of sight, separated by 90 degrees in azimuth at a fixed elevation angle. The profiles are measured for at intervals of 10 s. The line-of-sight winds from each of the four quadrants are combined to form two orthogonal line-of-sight wind measurements that are used to determine the horizontal components of the wind vector. The lidar wind measurements were compared to pilot balloons and rawinsondes. The standard deviation for

NASA Edge Lidar (Korb et al., 1997)

Transmitter		Receiver	
Wavelength	1064 nm	Type	Newtonian
Pulse length	~15 ns	Diameter	0.406 m
Pulse repetition rate	10 Hz		
Pulse energy	120 mJ	Filter bandwidth	5 nm
Laser bandwidth	40 MHz	Maximum range	boundary layer
		Range resolution	22–26 m
Detector		**Étalon**	
Type	1.5-mm Si avalanche	Aperture	50 mm
Responsivity	35 A/W	Étalon spacing	50 mm
Bandwidth	3.3 MHz	Free spect. range	3 GHz
Digitizer	60 MHz, 12 bit	Spectral width	100 MHz (FWHM)
		Plate reflectivity	93.5%

the four lidar profiles is less than 1.9 m/s, with an average value for all altitudes of 1.16 m/s. The effects of atmospheric temporal variability dominate the standard deviation for the data. The standard deviation of the lidar data calculated with the difference between adjacent points in the vertical direction for a given profile is less than 0.3 m/s, indicating that the internal consistency of the lidar is far greater than the variability of the wind. As with most lidars, the uncertainty is a function of the averaging time and distance. The instrumental uncertainty is estimated by the authors to be 0.40 m/s for a 10-shot average, and 0.11 m/s for a 500-shot average, which compares favorably to conventional point wind sensors. The maximum range of the instrument is limited by the particulate concentrations. Although this limits the useful region of the atmosphere to the boundary layer and areas immediately above, studies in this portion of the atmosphere can take advantage of the high spatial resolution offered by this instrument.

A more detailed discussion of the design requirements for an edge filter lidar may be found in McKay (1998). This paper includes a discussion of design trade-offs and issues related to the design. For example, the finesse of the étalon places requirements on the field of view of the telescope so that the characteristics of the étalon cannot be determined totally on the basis of the desired spectral resolution.

Double-Edge Technique. The double-edge technique is a variation of the general edge technique. It uses two edge filters with opposite slopes located on both sides of the laser frequency. The laser frequency is located at approximately the half-width of each filter (Fig. 13.12). A Doppler shift in the return-

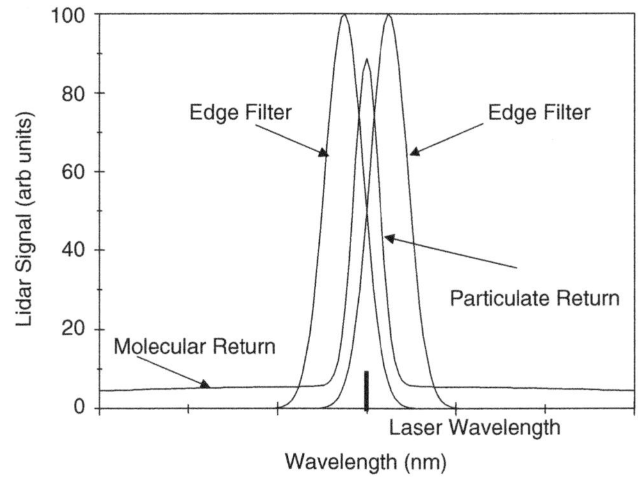

Fig. 13.12. A representation of the particulate and molecular backscattered portions of the lidar signal and the location of the filters used in a double-edge method with the particulate backscatter peak. The particulate/molecular return is shown for the case when the wind velocity is zero.

ing light will produce an increase in the signal from one edge filter and a decrease in the signal from the other filter of approximately the same magnitude. The result is that the change in the signal is twice the amount that is would be for a single-filter system for the same Doppler shift. This results in an improvement in the measurement accuracy by a factor of about 1.6 as compared with the single-edge technique. The use of two high-resolution edge filters also reduces the effects of Rayleigh scattering on the measurement by more than an order of magnitude. The use of two filters also eliminates the requirement to measure the energy of the returning light to normalize the signal.

The theory behind the double-edge technique was described by Korb et al. (1997). The technique may be applied to either the particulate return or the molecular return. The particulate method uses two high-resolution filters with a width that is less than one-tenth of the width of the thermally broadened Rayleigh spectrum. This greatly reduces the effects of Rayleigh background on the measurement, which increases the signal-to-noise ratio because of the reduction in the background, particularly in cases where the particulate signal is small.

The frequency of the laser is located at the midpoint of the region between the peaks of two overlapping edge functions (Fig. 13.12). A portion of the outgoing laser pulse is directed to the edge filter and compared to the atmospheric backscatter measured by each edge filter. The frequency of the outgoing light from the laser is locked so that the signal in each filter is the same. The particulate spectrum is spectrally narrow relative to the width of the laser. The amount of broadening due to thermal motion of atmospheric particulates is less than 1 MHz. Because even line-narrowed lasers have spectral widths much larger than this, the backscatter spectrum from particulates is essentially the same as the spectral width and shape of the laser. The edge filters should be approximately twice as wide (FWHM) as the laser spectral width and should overlap near the half-transmission points. This maximizes the change in signal for small changes in frequency, increasing the sensitivity and precision of the instrument. The use of a spectrally narrow laser line and narrow filters will decrease the effect of the molecular scattering signal on the measurement. The molecular background is not negligible compared with the particulate signal, so that corrections for this background must be made. However, with a measurement made of the entire lidar return, both particulate and molecular, it is possible to calculate the amplitude of the particulate return.

Obtaining a wind velocity requires an iterative procedure in which a small Doppler shift is assumed so that a molecular correction can be calculated. This molecular correction is used to calculate a new Doppler shift and so on until convergence is obtained. Details of this iterative procedure may be found in Korb et al. (1997). The authors claim that the error after just two iterations is less than 0.05%. As with a single edge system, the sensitivity, θ, is important for precision and uncertainty analysis. However, the sensitivity of the double edge method is due to the use of the two filters, that is, the sensitivity of this kind of system is doubled.

The usefulness of the system is limited by the spectral region over which the edge filters have a dramatic change in transmission, Thus there is a limited dynamic range for the system. However, this range is greater than is likely to occur in most applications near the surface and is on the order of 60 m/s (Korb et al., 1997). A knowledge of the convolution of the edge filter characteristic and the molecular return is required to perform the iteration. This in turn requires knowledge of the temperature of the air at each point. The width of the molecular return is a function of the square root of the atmospheric temperature. An error will occur if the value used for the molecular correction due to temperature is not the actual atmospheric temperature. The size of the temperature error in the Doppler shift is also a function of the size of the Doppler shift but is generally less than 0.5 m/s for a 5 K temperature error.

The molecular scattering signal can also be used to measure the wind with a double-edge filter. The general theory is outlined in a paper by Flesia and Korb (1999). In this case, wider edge filters must be used. They would be located at each side of the molecular signal in a manner similar to that for the particulate signal (Fig. 13.13). The laser is line-narrowed for this type of lidar. The amount of narrowing is not as important for a molecular scattering wind lidar but is a natural byproduct of the need to stabilize the frequency of the laser. For wind measurements using a molecular signal, the particulate return is a contaminant. Thus the filters must be spectrally located so that the sensitivity of a wind measurement from the molecular signal is the same as the

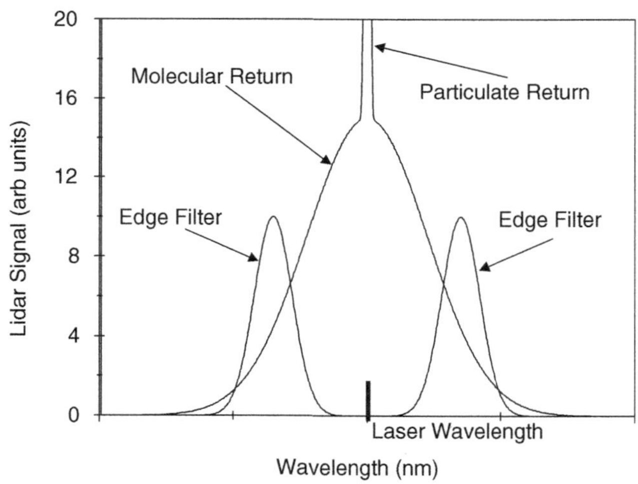

Fig. 13.13. A representation of the particulate and molecular backscattered portions of the lidar signal and the location of the filters used in a double-edge method with the molecular backscatter peak. The particulate/molecular return is shown for the case when the wind velocity is zero.

sensitivity of a wind measurement from the particulate signal (Garnier and Chanin, 1992; Chanin et al., 1994; Flesia and Korb, 1999). This desensitizes the measurement to effects from the particulate signal. The use of molecular backscatter to measure winds is desirable because particulate loading is small in the boundary layer in many parts of the world and is always small in the troposphere. Thus it is a logical method to explore for satellite application. However, the sensitivity of a wind-measuring lidar using molecular backscatter is approximately a factor of 10 less than a similar system using particulate backscatter.

The analysis of a molecular backscatter data is much simpler than that for a particulate backscatter wind measurement. Defining the function $f(\Delta\nu)$ as

$$f(\Delta\nu) = \frac{I_1(\nu_1, \nu_1 + \Delta\nu)}{I_2(\nu_2, \nu_2 + \Delta\nu)} \tag{13.22}$$

where $I_1(\nu_1, \nu_1 + \Delta\nu)$ is the signal from edge filter 1, located at a frequency of ν_1, measuring a Doppler shifted frequency of $\nu_1 + \Delta\nu$. I_2 is the signal from the second edge filter. The wind velocity can be found from

$$V = \frac{c}{2\nu} \frac{[f(\Delta\nu) - f(0)]}{f(0)(\theta_1 + \theta_2)} \tag{13.23}$$

where $f(0)$ is the ratio of signals that would be received from a stationary source. Flesia and Korb (1999) describe a method by which this factor could be determined for each laser pulse by taking a portion of the outgoing laser light and directing it though the edge filters. This light can also be used in a feedback mechanism to stabilize the laser wavelength. An alternate method which uses measurements at three vertical angles is described by Friedman et al. (1997). The determination of $f(0)$ on a shot-to-shot basis is desirable to correct for shot-to-shot jitter in the frequency. The frequency of the laser must be locked to the frequencies of the étalon filters.

Wind measurements using molecular backscatter have been demonstrated by Gentry et al. (2000) and by Flesia et al. (2000). The systems are essentially the same. Each uses a single étalon that is layered to provide three different transmission bands. Two form the two edge filters, and one is used to lock the laser to the desired frequency. Each also uses a beam splitter to measure the energy of the returning light through a standard interference filter. This requires splitting the collected backscatter light into at least four channels, considerably reducing the amount of available light in each. The biggest difference between the two systems is the laser energy. The demonstration by Flesia et al. (2000) used an effective energy of 5 mJ per pulse. This enabled measurements up to an altitude of about 10 km with a standard deviation of 1–2 m/s. The demonstration by Gentry et al. (2000) used an effective energy of 70 mJ per pulse. This enabled measurements up to an altitude of about

35 km with an uncertainty that varies from 0.4 m/s at 7 km to 4 m/s at 20 km. The errors at each altitude are a function of the number of laser pulses that are averaged.

13.3. FRINGE IMAGING TECHNIQUE

An alternative to the edge technique for direct, optical measurement of the Doppler shift of lidar backscatter is the fringe imaging technique. This technique, which actually predates the edge technique, images the backscatter signal with a Fabry–Perot interferometer so as to create a classic circular fringe pattern. The results of a demonstration were first published in 1992 by a team from the University of Michigan (Abreu et al., 1992). The Doppler shift is found by measurement of the physical displacement of a fringe with an imaging detector. As with all of the incoherent methods, knowledge of the spectral position of the zero-wind signal is required to find the amount of frequency change. Thus a reference measurement must be made of the outgoing signal with each laser pulse. As with the edge technique, Doppler wind lidars may be based on the measurement of either the molecular or the particulate backscatter peaks. Particulates are preferred from an ideal point of view because the particulate scattered signal is not significantly broadened by the thermal motion of the particulates, so that a high degree of precision is possible. However, the real value in this method lies in its ability to determine velocities from the molecular backscatter signal. Although the wind speeds determined with the molecular signal are significantly broadened with respect to the amount of the Doppler shift so that the sensitivity is considerably reduced (as opposed to using the particulate signal), the ability to determine winds from just the molecular signal make it attractive for use by space platforms. The bulk of the troposphere contains limited amounts of particulates, so that measurements that rely on the presence of particulates are, at best difficult, if not impossible.

In the following analysis, the University of Michigan lidar is used as an example of how such a device might be constructed (Abreu et al., 1992; Fischer et al., 1995; McGill et al., 1997; McGill et al., 1997). Backscattered light is collected by the telescope and transferred via a fiber-optic cable to a collimating lens. The light is then directed to a high-resolution Fabry–Perot étalon (HRE) and then imaged by a second lens (Fig. 13.14). Not shown in the figure is a low-resolution étalon (LRE) that is used to reduce the amount of solar background. It is important that the two étalons be matched and stabilized together to accurately measure the Doppler shift (McGill and Skinner, 1997). The high-resolution étalon will produce the classic pattern of fringes. This fringe pattern is imaged onto a 32-channel image plane detector (IPD). The image plane detector is a photomultiplier-type device that has concentric ring anodes of equal areas that are designed to match the étalon fringe pattern. Each of the rings responds in a way that is similar to a separate photomultiplier. The

University of Michigan Fringe Imaging Lidar

Transmitter		Reciever	
Wavelength	532 nm	Type	Newtonian
Pulse length	~6 ns	Diameter	0.445 m
Pulse repetition Rate	50 Hz	Field of view	0.8 mrad
Pulse energy	60 mJ	Filter bandwidth	Low-resolution étalon 0.05 nm
Laser bandwidth	0.0045 cm^{-1}	Maximum range	boundary layer
		Range resolution	150 m
Detector		**Étalon**	
Type	Image plane detector	Aperture	100 mm
Channels	32 channels	Étalon spacing	10-cm air gap
Size	1.225-cm radius	Free spect. range	1.5 GHz
Velocity shift/channel	36.66 m/s	Spectral width	100 MHz (FWHM)

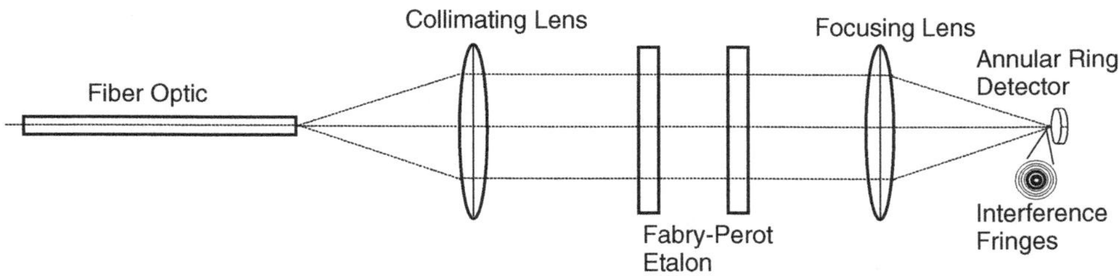

Fig. 13.14. A schematic diagram of the optical hardware used to determine the change in frequency for a fringe imaging lidar system. Light from the telescope is collimated and passed through an étalon, generating a fringe pattern that is measured.

transmission $A(\Delta\lambda)$ through a Fabry–Perot interferometer into a ring of width $\Delta\theta$ is given by

$$A(\Delta\lambda) = \left(\frac{1-R}{1+R}\right) \left[1 + 2\sum_{n=1}^{\infty} R^n \cos 2\pi n\left(\frac{\Delta\lambda}{\mathrm{FSR}} + \frac{\lambda_0\theta_0^2}{2\mathrm{FSR}} + \frac{\lambda_0\Delta\theta^2}{8\mathrm{FSR}}\right) + \sin c\left(\frac{n\lambda_0\theta_0\Delta\theta}{\mathrm{FSR}}\right)\right] \tag{13.24}$$

where R is the plate reflectivity, FSR is the free spectral range, λ_0 is the central wavelength and θ_0 is the average angle corresponding to the average wavelength being transmitted through the ring. A consequence of Eq. (13.24) is that a change in frequency is related to two rings with angles θ_1 and θ_2 as

$$\Delta\nu = \frac{c}{2\lambda}(\theta_1^2 - \theta_2^2)$$

with the result that the component of the wind speed along the lidar line of sight is

$$V = \frac{c}{4}(\theta_1^2 - \theta_2^2)$$

so that an angular measurement can be directly transformed to a velocity measurement. The widths of the rings in the detector are chosen so that the spatial scan will be linear with wavelength. Equal wavelength intervals in the fringe pattern result in equal areas in the detector. The width of the detector rings (i.e., the frequency intervals) is small enough that the étalon transmittance can be considered as constant.

The output of the étalon is a complex convolution of a Gaussian laser spectrum, scattered by molecules and particulates, temperature broadened and Doppler shifted, and the response of an étalon. This convolution has been examined in detail by McGill et al. (1997). The system response is modeled as

$$\begin{aligned} P(r,i) = {} & \left(\frac{E_T \lambda \varepsilon \Delta t}{hc}\right) O_A(r) \frac{A_T}{4\pi r^2} \Delta h Q_E T_0 T_F(\nu) T_{LRE}(i,\nu) \\ & \times \frac{\eta(i)}{n_c} \sum_{n=0}^{\infty} A_{n,i} \,\mathrm{sinc}\left[\frac{n}{N_{FSR}}\right] \\ & \times \exp\left(\frac{\pi^2 n^2 \Delta\nu_L^2}{\Delta\nu_{FSR}^2}\right) \cos 2\pi n \left(\frac{i - i_0(r)}{N_{FSR}}\right) \\ & \times \left[\alpha(r) + \omega(r) \exp\left(-\frac{\pi^2 n^2 \Delta\nu_M^2}{\Delta\nu_{FSR}^2}\right)\right] \end{aligned} \tag{13.25}$$

where i is the detector channel number, r is the range from the lidar (m), $N(r, i)$ is the number of detected photons on channel i at range r, E_T is the pulse energy of the laser (J), ε is the pulse repetition frequency, Δt is the total integration time (s), $O_A(r)$ is the fractional overlap between the telescope and laser beam, A_T is the area of the telescope (m^2), Δh is the range resolution (m), Q_E is the quantum efficiency of the detector, T_0 is the transmission of the optical train (excluding the filters), $T_F(\nu)$ is the transmission of the filters, T_{LRE} is the transmission of the low-resolution étalon, n_c is the number of detector channels, $\eta(j)$ is a detector normalization coefficient, $\Delta\nu_L$ and $\Delta\nu_M$ are the 1/e widths of the laser and molecular linewidths (cm^{-1}), N_{FSR} is the number of detector channels per HRE FSR (free spectral range) and $\Delta\nu_{FSR}$ is the wave number change per HRE FSR (cm^{-1}). The data analysis procedure is essentially a spectral curve fit with Eq. (13.25). There are three parameters that this fit will determine, the Doppler shift, the particulate signal, and the molecular signal. The inversion is simplified somewhat because the spectral signatures of each of these are distinctly different and mathematically orthogonal (McGill et al., 1997).

Because the signals are small, photon-counting techniques are required for each of the IPD channels. Corrections are also required for dead time in the IPD. Because the number of counted photons is necessarily limited, the uncertainty in the measured velocity is a function of photon-counting statistics in each of the channels as well as the system parameters. There is a close connection between measurement precision and the characteristic spectral bandwidth of the instrument. The performance of the system improves as the free spectral range decreases as long as the finesse is held constant. The maximum sensitivity is achieved with the minimum feasible étalon bandwidth. Thus the uncertainty of the measurement decreases as the étalon passband width is decreased. However, with the edge technique, the Doppler-shifted backscatter signal must remain within the passband of the étalon, which sets a limit on the minimum spectral width of the edge filter. No similar limitation applies to a fringe imager, in which the Doppler-shifted backscatter need only remain within the free spectral range of the étalon. As long as the order number transitions can be counted, there is no limitation on the étalon spectral width. Thus the étalon width can be decreased to the limit of the available technology, without conflicting with wind speed dynamic range requirements. This consideration is particularly important for potential satellite applications, where the dynamic range of wind speed variations is on the order of 7000 m/s (the orbital velocity of the spacecraft).

A related issue is the behavior of the precision of the measurement of the Doppler shift as the wind speed increases. The response of the fringe imaging technique is linear in Doppler shift if a detector with multiple rings of equal area is used. Thus the precision of the measured winds will be similar across the entire range of measurements. This contrasts strongly with the edge technique because the slope of the étalon transmission changes rapidly with frequency and for large Doppler shifts, the response is highly nonlinear.

A Fabry–Perot étalon is characterized by a small acceptance angle. The angle to one free spectral range is

$$\theta = \left(\frac{\lambda}{h}\right)^{1/4}$$

where h is the distance from the etalon to the image plane.

Ideally, one free spectral range would be illuminated by the fiber optic. The relationships between the finesse of the étalon and the etendue can impose minimum aperture requirements on the étalon. This requirement may be significant for large-aperture lidar systems, such as those used in satellite applications. The result is that the étalon diameter may be forced to large values and the field of view of the telescope may be required to be as small as 20 mrad (McKay and Rees, 2000).

13.4. KINETIC ENERGY, DISSIPATION RATE, AND DIVERGENCE

In addition to calculating the mean winds in an area or in profile, there are a number of wind characteristics or parameters that are of interest to modelers and researchers or for practical application. Many of the atmospheric models used in intermediate scales are known as K-ε models. These models close the Navier–Stokes equations through assumed relationships between the turbulent kinetic energy K and the dissipation rate of turbulent kinetic energy ε. Measurements of these values would be valuable to modelers, particularly in urban areas and in complex terrain where conditions are not ideal and spatially homogeneous. σ_v^2 is the average value of the square of the turbulent velocity fluctuations in the wind direction. Under conditions of isotropic turbulence, this is one-third of the specific turbulent kinetic energy. Very near the surface, the assumption of isotropy does not hold, but for reasons related to eye safety and clear lines of sight, lidars are not likely to be used to measure winds that close to the surface.

The wind velocity determination method developed by Kunkel et al. (1980) described in Section 13.1.2 allows the calculation of several turbulence parameters of interest. The kinetic energy dissipation rate can be calculated by using the normalized power spectrum of wind velocity. Kaimal et al. (1976) showed that the power spectra of wind velocity, $\phi_V(\nu)$, reduces to a single function when normalized by $(\varepsilon h_i)^{2/3}$ and plotted as a function of a nondimensional frequency, $\nu_n = \nu h/V$; h_i is the height of the atmospheric boundary layer and V is the average wind velocity. Thus the function, $\dfrac{\nu\phi_V(\nu)}{(\varepsilon h_i)^{2/3}}$ can be assumed to be known. The power spectrum of wind velocity and the square of the fluctuations in wind speed can also be related, resulting in

$$\sigma_V^2 = \int_{\nu_a}^{\infty} \phi_V(\nu)d\nu = (\varepsilon z_i)^{2/3} \int_{\nu_a}^{\infty} \left| \frac{\nu\phi_V(\nu)}{(\varepsilon h_i)^{2/3}} \right| \frac{d\nu}{\nu} \tag{13.28}$$

Rearranging, one can obtain an expression for the dissipation rate of turbulent kinetic energy

$$\varepsilon = \frac{\sigma_V^3}{h_i} \left[\int_{\nu_a}^{\infty} \left| \frac{\nu\phi_V(\nu)}{(\varepsilon h_i)^{2/3}} \right| \frac{d\nu_n}{\nu_n} \right]^{-3/2} \tag{13.29}$$

To calculate ε, one can obtain h_i from a lidar measurement of the boundary layer altitude, σ_V is found from the wind measurement technique, and ν_a is equal to t_a^{-1}, t_a is the averaging time over which the measurement of σ_V is determined.

Young and Eloranta (1995) have demonstrated the ability of a scanning lidar to determine the divergence of the wind velocity over an area. Using successive CAPPI scans, the amount that the wind stretches or compresses heterogeneities in a horizontal slice of the atmosphere can be determined. Consider two horizontal slices of the atmosphere showing the particulate concentrations at the same place but at two times separated by Δt. The particulates in the second scan will be advected along the mean wind as well as distorted by the divergence of the mean wind. If the scan covers some rectangular area A with sides L_x and L_y, then in the second scan it will occupy an area A' such that

$$A' = L_x\left(1+\frac{\partial V_x}{\partial x}\Delta t\right) \times L_y\left(1+\frac{\partial V_y}{\partial y}\Delta t\right)$$

where V_x and V_y are the components of the wind velocity in the x and y directions, respectively. Young and Eloranta calculate the cross-correlation function between the original scan and a second scan that has been distorted by some $\partial V_x/\partial x$ and $\partial V_y/\partial y$. The maximum of the correlation function is calculated for a range of $\partial V_x/\partial x$ and $\partial V_y/\partial y$. The largest value of the set of correlation maxima is found by fitting a two-dimensional quadratic fit to the data. This value corresponds to the lags in space that determine the wind velocity, but also the values of $\partial V_x/\partial x$ and $\partial V_y/\partial y$ that best approximate the distortion. The wind divergence is then found from

$$\vec{\nabla}_h \cdot \vec{v} = \frac{\partial V_x}{\partial x} + \frac{\partial V_y}{\partial y}$$

The precision with which the divergence can be determined is a function of the spatial resolution of the horizontal images. Typical values are on the order of $3 \times 10^{-5}\,\mathrm{s}^{-1}$. The lidar used by Young and Eloranta scans a three-dimensional volume from which the individual CAPPI scans are constructed. This enables them to determine the divergence as a function of altitude as well as an individual height.

BIBLIOGRAPHY

Abreu, V. J. (1979) "Wind Measurements from an Orbital Platform using a Lidar with Incoherent Detection: An Analysis," *Appl. Opt.*, **18**, 2992–2997.

Abreu, V. J., J. E. Barnes, and P. B. Hays (1992) "Observations of Winds with an Incoherent Lidar Detector," *Appl. Opt.*, **31**, 4509–4514.

Ackerman, J. (1998) "The Extinction to Backscatter Ratio of Tropospheric Aerosol: A Numerical Study," *J. Atm. Ocean. Tech.*, **15**, 1043–1050.

Ackerman, S., E. Eloranta, C. Grund, R. Knuteson, H. Revercomb, W. Smith, and D. Wylie (1993) "University of Wisconsin Cirrus Remote Sensing Pilot Experiment," *Bull. of Amer. Meteor. Soc.*, **74**, 1041–1049.

Ackermann, J. (1990) "Analytical Solution of the Two-Frequency Lidar Inversion Technique," *Appl. Opt.*, **38**, 7414–7418.

Ackermann, J. (1997) "Two-Wavelength Lidar Inversion Algorithm for a Two-component Atmosphere," *Appl. Opt.*, **36**, 5134–5143.

Adriani, A., G. Gobbi, F. Congeduti, and G. Didonfrancesco (1991) "Lidar Observations of Stratospheric and Mesospheric Temperature—November 1988–November 1989," *Ann. Geoph. Atmos. Hydrospheres and Space Sci.*, **9**, 252–258.

Agishev, R. R., and A. Comeron (2002) "Spatial Filtering Efficiency of Monostatic Biaxial Lidar: Analysis and Applications," *Appl. Opt.*, **36**, 7516–7521.

Alduchov, O. A., and R. E. Eskridge (1996) "Improved Magnus Form Approximation of Saturation Vapor Pressure," *J. Appl. Met.*, **35**, 601–609.

Allen R. J., and W. E. Evans (1972) "Lase Radar (LIDAR) for Mapping Aerosol Structure, *Rev Sci. Instrum.*, **43**, 1422–1426.

Elastic Lidar: Theory, Practice, and Analysis Methods, by Vladimir A. Kovalev and William E. Eichinger.
ISBN 0-471-20171-5

Allen, R. J., and C. M. R. Platt (1977) "Lidar for Multiple Backscattering and Depolarization Observations," *Appl. Opt*, **16**, 3193–3199.

Alvarez, J. M., W. H. Fuller, and R. M. Lawrence (1998) "A Radar Safety Device for Lidars Probing the Atmosphere," in Conference Abstracts, Nineteenth International Laser Radar Conference, U. Singh, S. Ismail, and G. Schwemmer, eds. (NASA, Langley Research Center, Hampton, VA, 1998), pp. 983–986.

Alvarez R., L. Caldwell, P. Wolyn, D. Krueger, T. B. McKee, and C. Y. She (1993) "Profiling Temperature, Pressure, and Aerosol Properties Using a High Spectral Resolution Lidar Employing Atomic Blocking Filters," *J. Atmos. and Ocean. Tech.*, **10**, 546–556.

Alvarez II, R. J., W. L. Eberhard, J. M. Intrieri, S. P. Sandberg, and K. W. Koenig (1998) "Cloud Backscatter and Phase Measurements in the Arctic Using NOAA/ETL'S DABUL lidar," in *Fourth International Symposium on Tropospheric Profiling: Needs and Technologies*, Vol. 1 (Snowmass, Colorado, USA, 1998), 7–9.

Ancellet, G., A. Papayannis, J. Pelon, and G. Megie (1989) "DIAL Tropospheric Measurement Using a ND:YAG Laser and the Raman Shifting Technique," *J. Atmos. and Ocean. Tech.*, **6**, 832–839.

Ancellet, G. M., M. J. Kavaya, R. T. Menzies, and A. M. Brothers (1986) "Lidar Telescope Overlap Function and Effects of Misalignment for Unstable Resonator Transmitter and Coherent Receiver," *Appl. Opt.*, **25**, 2886–2890.

Anderson, G. P., S. A. Clough, F. X. Kneizys, J. H. Chetwynd, E. P. Shettle (1986) "AFGL Atmospheric Constituent Profiles (0–120 km)," AFGL-TR-86-0110, (NTIS AD A175173).

Anderson, S. G. (1995) "Diode-Pumped Parametric Oscillator/Laser Simplifies Lidar System," *Laser Focus World*, **31**(4), 46–47.

Anderson, T. L., S. J. Masonis, D. S. Covert, R. J. Charlson , and M. J. Rood (2000) "In situ Measurement of the Aerosol Extinction-to-Backscatter Ratio at a Polluted Continental Site," *J. Geoph. Res.*, **105**, 26907–26915.

Anderson, T. L., and J. A. Ogden (1998) "Determining Aerosol Radiative Properties Using the TSI 3563 Integrating Nephelometer," *Aerosol Sci. and Technol.*, **29**, 57–69.

Anderson, T. L., D. S. Covert, S. F. Marshall, M. L. Laucks, R. J. Charlson, A. P. Waggoner, J. A. Ogren, R. Caldow, R. L. Holm, F. R. Quant, G. J. Sem, A. Wiedensohler, N. A. Ahlquist, and T. S. Bates (1996) "Performance Characteristics of a High-Sensitivity Three-Wavelength, Total Scatter/backscatter Nephelometer," *J. Atmos. Oceanic Technol.*, **13**, 967–986.

Angstrom, A. (1951) "Actinometric Measurements," in *Compendium of Meteorology*, T. F. Malone ed., AMS, Boston.

Annex 3 *Meteorological Service for International Air Navigation*. International Civil Aviation Organization, 1995, 104 pp.

ANSI Z136.1-1993, American National Standards Institute (1993) "American National Standard for the Safe Use of Lasers," American National Standards Institute Inc., Orlando, FL.

Ansmann, A. (1985) "Errors in Ground-Based Water-Vapor DIAL Measurements Due to Doppler-broadened Rayleigh Backscattering," *Appl. Opt.*, **21**, 3476–3480.

Ansmann A., M. Riebesell, and C. Weitkamp (1990) "Measurement of Atmospheric Aerosol Extinction Profiles with a Raman Lidar," *Optics Lett.*, **15**, 746–748.

Ansmann A., M. Riebesell, U. Wandinger, C. Weitkamp, and W. Michaelis (1991) "Combined Raman Elastic Backscatter Lidar for the Independent Measurement of Aerosol Backscatter and Extinction Profiles," GKSS-Forschungszentrum Geesthacht GmbH, Geesthacht, GKSS 91/E/42.

Ansmann A., U. Wandinger, M. Riebesell, C. Weitkamp, and W. Michaelis (1992) "Independent Measurement of Extinction and Backscatter Profiles in Cirrus Clouds by Using a Combined Raman Elastic-Backscatter Lidar," *Appl. Opt.*, **31**, 7113–7131.

Ansmann A., M. Riebesell, U. Wandinger, C. Weitkamp, E. Voss, W. Lahman, and W. Michaelis (1992a) "Combined Raman Elastic Backscatter Lidar for Vertical Profiling of Moisture, Aerosol Extinction, Backscatter and Lidar Ratio," *Appl. Phys.*, B **55**, 18–28.

Argall, P. S., and F. Jacka (1996) "High-Pulse-Repetition-Frequency Lidar System Using a Single Telescope for Transmission and Reception," *Appl. Opt.*, **35**, 2619–2629.

Armstrong, R. (1974) "Analysis of Fabry-Perot Interferograms of Periodic Structures," *J. Opt. Soc. Amer.*, **64**, 871–876.

Armstrong, R. (1975) "Rotational Raman Interferometric Technique to Measure Gas Temperature," *Appl. Opt.*, **14**, 383–387.

Armstrong, R., J. Mason, and T. Barber (1976) "Detection of Atmospheric Aerosol Flow using a Transit Time Lidar Velocimeter.," *Appl. Opt.*, **15**, 2891–2895.

Arshinov, Y. F., and S. M. Bobrovnikov (1999) "Use of a Fabry-Perot Interferometer to Isolate Pure Rotational Raman Spectra of Diatomic Molecules", *Appl. Opt.*, **38**, 4635–4638.

Arshinov, Y. F., S. M. Bobrovnikov, V. E. Zuev, and V. M. Mitev (1983) "Atmospheric Temperature Measurements Using a Pure Rotational Raman Lidar", *Appl. Opt.*, **22**, 2984.

Asai, T., M. Yoshizaki, and K. Ishikawa (1977) "Some Results of an Objective Analysis for Tracking Radar Echoes of Convective Clouds," *J. Met. Soc. Japan.*, **55**, 553–558.

Atlas, R., E. Kalnay, and M. Halem (1985) "Impact of Satellite Temperature Sounding and Wind Data on Numerical Weather Prediction," *Opt. Eng.*, **24**, 341–346.

Austin, G., and A. Bellon (1974) "The Use of Digital Weather Radar Records for Short Term Precipitation Forecasting," *Quart. J. Roy. Met. Soc.*, **100**, 658–664.

Baker, B. C. (1997) "Keeping the Signal Clean in Photosensing Instrumentation," *SENSORS*, p. 12.

Baker P. W. (1983) "Atmospheric Water Vapor Differential Absorption Measurement on Vertical Paths with a CO_2 Lidar," *Appl. Opt.*, **22**, 2257–2264.

Baker, W. E., G. D. Emmitt, F. Robertson, R. M. Atlas, J. E. Molinari, D. A. Bowdle, J. Paegle, R. M. Hardesty, R. T. Menzies, T. N. Krishnamurti, R. A. Brown, M. J. Post, J. R. Anderson, A. C. Lorenc, and J. McElroy (1995) "Lidar-Measured Winds from Space: A Key Component for Weather and Climate Prediction," *Bull. Am. Meteorol. Soc.*, **76**, 869–888.

Baldenkov, G. N., Yu. S. Balin, V. A. Kovalev, V. I. Kozintsev, I. A. Movchan, E. A. Mozharov, and E. E. Rybakov (1988) "Lidar Visibility Meter 'Electronics-03' for Slant Atmospheric Paths," *Atmos. Opt.*, **1**, 68–74 (in Russian).

Balin, Y. S., S. I. Kavkyanov, G. M. Krekov, and I. A. Rasenkov (1987) "Noise-proof Inversion of Lidar Equation," *Opt. Lett.*, **12**, 13–15.

Balin, Y. S., and I. A. Rasenkov (1993) "Laser Monitoring of Aerosol Pollution at an Air Basin of Industrial Centers," *Atmos. and Oceanic Opt.*, **6**, 104–114.

Banakh, V. A., I. N. Smalikho, F. Kopp, and C. Werner (1995) "Representativity of the Wind Measurements by a CW Doppler Lidar in the Atmospheric Boundary Layer," *Appl. Opt.*, **34**, 2055–2067.

Baray, J. L., J. Leveau, J. Porteneuve, J. Ancellet, P. Keckhut, F. Posny, and S. Baldy (1999) "Description and Evaluation of a Tropospheric Ozone Lidar Implemented on an Existing Lidar in the Southern Subtropics," *Appl. Opt.*, **38**, 6808–6817.

Barber, P. W., and S. C. Hill (1990) *Light Scattering by Particles: Computational Methods*, World Scientific Publishing, Singapore, 261 pp.

Barber, T., and D. R. Larson (1985) "Visibility Related to Backscatter at 1.06 μm," *Appl. Opt.*, **24**, 3523–3525.

Barr, S., W. Buttler, D. Clark, W. Cottingame, and W. Eichinger (1995) "Lidar-Observed Wind Patterns in the New Mexico-Texas Border Region, 9–11 September, 1994." *Alliance for Transportation Research* Report.

Barret, E. W., and O. Ben-Dov (1967) "Application of Lidar to Air Pollution Measurements," *J. Appl. Meteorol.*, **6**, 500–515.

Barrick, J. D. (1986) "Gating Characteristics of Photomultiplier Tubes for Lidar Applications", NASA Tech. Memo, 8769S.

Barteneva, O. D. (1960) "Scattering Functions of Light in the Boundary Layer of the Atmosphere," *Bull. Acad. Sci. U.S.S.R.*, Geoph. Ser., No. 12, 1852–1865 (in Russian).

Barteneva, O. D., E. N. Dovgyallo, and E. A. Polyakova (1967) "Experimental Investigation of the Optical Properties of the Aerosol in the Bottom Layer," *Proceedings of the Main Geophysical Observatory*, E. A. Polyakova, ed. (Main Geophysical Observatory, Leningrad, 1967), No. 220, 153 (in Russian).

Bas, C. F., T. J. Kane, and C. S. Ruf (1991) "Enhanced-Resolution Lidar," *J. Opt. Soc. Am.*, A **14**, 1044–1050.

Bass, A. M., A. E. Ledford, and A. H. Lauffer, "Extinction Coefficients of NO_2 and N_2O_4," *J. Res. Nat. Bur. Stand.*, A **80**, 143–166.

Batchvarova, E., D. Steyn, X. Cat, S. Gryning, and M. Baldi (1997) "Modeling the Internal Boundary Layer in the Lower Frasier Vallye, B.C.," *Proc. EURSAP Workship on the Determination of Mixing Height-Current Progress and Problems*, Risoe, Denmark, EURSAP, 141–144.

Behrendt, A., and J. Reichardt (2000) "Atmospheric Temperature Profiling in the Presence of Clouds with a Pure Rotational Raman Lidar by the Use of an Interference Filter Based Polychromator, *Appl. Opt.*, **39**, 1372–1378.

Behrendt, A., T. Nakamura, M. Onishi, R. Baumgart, and T. Tsuda (2002) "Combined Raman Lidar for the Measurement of Atmospheric Temperature, Water Vapor, Particle Extinction Coefficient, and Particle Backscatter Coefficient," *Appl. Opt.*, **41**, 7657–7666.

Belove, C., ed. *Handbook of Modern Electronics and Electrical Engineering*, 2nd ed., John Wiley & Sons, New York, 1986, pp. 434.

Berk, A., L. S. Bernstein, D. C. Robertson (1989) "MODTRAN: A Moderate Resolution Model for LOWTRAN 7," AFGL-TR-89-0122.

Bertone, N., P. Webb, and B. Dion (1996) "Photodiode Design Strategies Squelch the Noise," *Photonics Spectra*, July, 1996, 122–124.

Bevington, P. R., and D. K. Robinson (1992) "*Data Reduction and Error Analysis for the Physical Sciences*," 2nd ed., McGraw-Hill, Inc., 328.

Beyerle, G., and S. McDermid (1999) "Altitude Range Resolution of Differential Absorption Lidar Ozone Profiles," *Appl. Opt.*, **38**, 924–927.

Beyrich, F. (1997) "Mixing Height Estimation from Sodar Data—A Critical Discussion," *Atmos. Environ.*, **31**, 3941–3953.

Bills, R. E., C. S. Gardner, and C. Y. She (1991) "Narrowband Lidar Technique for Sodium Temperature and Doppler Wind Observations of the Upper Atmosphere," *Opt. Eng.*, **30**, 13–21.

Bills, R. E., C. S. Gardner, and S. J. Franke (1991) "Na Doppler/Temperature Lidar: Initial Mesopause Region Observations and Comparison with the Urbana Medium Frequency Radar," *J. Geoph. Res.*, **96**, 22701–22707.

Bissonnette, L. (1996) "Multiple Scattering Lidar Equation," *Appl. Opt.*, **35**, 6449–6465.

Bissonnette, L. R. (1986) "Sensitivity Analysis of Lidar Inversion Algorithms," *Appl. Opt.*, **25**, 2122–2125.

Bissonnette, L. R., and D. L. Hutt (1990) "Multiple Scattering Lidar," *Appl. Opt.*, **29**, 5045–5046.

Bissonnette, L. R., and D. L. Hutt (1995) "Multiply Scattered Aerosol Lidar Returns: Inversion Method and Comparison with in Situ Measurements," *Appl. Opt.*, **34**, 6959–6975.

Bissonnette, L. R., and G. Roy (2000) "Retrieval of Cloud Water Content and Efficient Droplet Diameter from Multiply Scattered Lidar Returns," Proceedings of the Twentieth International Laser Radar Conference (Vishi, France, 2000), Eds. J. Pelon and A. Dabas, 251–254.

Bissonnette, L. R., G. Roy, L. Poutier, S. G. Cober, and G. A. Isaac (2002) "Multiple-Scattering Lidar Retrieval Method: Tests on Monte Carlo Simulations and Comparisons with in Situ Measurements," *Appl. Opt.* **41**, 6307–6324.

Bissonnette, L. R., P. Bruscaglioni, A. Ismaelli, G. Zaccanti, A. Cohen, Y. Benayahu, M. Kleiman, S. Egert, C. Flesia, P. Schwendimann, A. V. Starkov, M. Noormohammadian, U. G. Oppel, D. M. Winkler, E. P. Zege, I. L. Katsev, and I. N. Polonsky (1995) "Lidar Multiple Scattering from the Clouds," *Appl. Phys.*, B **60**, 355–362.

Bissonnette, L. R. (1995) "Multiple Scattering of Narrow Light Beams in Aerosols," *Appl. Phys.*, B **60**, 315–323.

Bockmann, C., J. Biele, and R. Neuber (1998) "Analysis of Multi-Wavelength Lidar Data by Inversion with Mollifier Method," *Pure Appl. Opt.*, **7**, 827–836.

Boers, R. (1988) "A Parameterization of the Depth of the Entrainment Zone," *J. Appl. Meteor.*, **23**, 107–111.

Boers, R., and E. W. Eloranta (1986) "Lidar Measurements of the Atmospheric Entrainment Zone and the Potential Temperature Jump across the Top of the Mixed Layer," *Boundary-Layer Meteorology*, **34**, 357–375.

Boers, R., E. W. Eloranta, and R. L. Coulter (1984) "Lidar Observations of Mixed Layer Dynamics: Tests of Parameterized Entrainment Models of Mixed Layer Growth Rate," *J. Cli. Appl. Meteor.*, **23**, 247–266.

Boers, R., and H. Melfi (1987) "Cold Air Outbreak during MASEX: Lidar Observations and Boundary Layer Model Test," *Bound. Layer Met.*, **39**, 41–51.

Bohren, C. (1987) *Clouds in a Glass of Beer*, John Wiley & Sons, New York, 195 pp.

Bohren F. C., and D. R. Huffman (1983) *Absorption and Scattering of Light by Small Particles*, John Wiley & Sons, Inc., New York, 530 pp.

Bohren F. C. (1989) ed., *Scattering in the Atmosphere*, SPIE Optical Engineering Press, Bellingham, WA, 641 pp.

Born, M., and E. Wolf (1999) *Principles of Optics*, Cambridge University Press, Cambridge, UK.

Bosenberg, J. (1998) "Ground-Based Differential Absorption Lidar for Water Vapor and Temperature Profiling," *Appl. Opt.*, **37**, 3845–3861.

Bouble, R., D. Fox, D. Turner, and A. Stern (1994) *Fundamentals of Air Pollution*. Academic Press, New York, 574 pp.

Brassington, D. J. (1981) "Sulfur Dioxide Absorption Cross-Section Measurements from 290 nm to 317 nm," *Appl. Opt.*, **20**, 3774–3779.

Brock, C. A., L. F. Radke, and P. V. Hobbs (1990) "Sulfur in Particles in Arctic Hazes Derived from Airborne in Situ and Lidar Measurements," *J. Geoph. Res.*, **95**, 22369–22387.

Browell, E. V, S. Ismail, and S. T. Shipley (1985) "Ultraviolet DIAL Measurements of O_3 Profiles in Regions of Spatially Inhomogeneous Aerosols," *Appl. Opt.*, **24**, 2827–2836.

Brown, R. G., and R. Y. C. Hwang (1992) *Introduction to Random Signals and Aplied Kalman Filtering*, John Wiley & Sons, Inc., New York.

Brown, R. T. (1973) "A New Lidar for Meteorological Application," *J. Appl. Meteorol.*, **12**, 698–708.

Brown, R. W., R. Jones, J. Rarity, and K. Ridley (1986) "Characterization of Silicon Avalanche Photodiodes for Photon Correlation Measurements. 1. Passive Quenching." *Appl. Opt.*, **25**, 4122–4129.

Brown, R. W., R. Jones, J. Rarity, and K. Ridley (1987) "Characterization of Silicon Avalanche Photodiodes for Photon Correlation Measurements. 2. Active Quenching.," *Appl. Opt.*, **26**, 2383-2389.

Bruscaglioni, P., A. Ismaelli, G. Zaccanti, M. Gai, and M. Gurioli (1995) "Polarization of Lidar Returns from Water Clouds: Calculations and Laboratory Scaled Measurements," *Opt. Rev.*, **2**, 312–318.

Bruscaglioni, R. P., A. Ismaelli, and G. Zaccanti, "Monte-Carlo Calculations of LIDAR Returns: Procedure and Results," *Appl. Phys.*, B **60**, 325–330.

Brutsaert, W. (1982) *Evaporation Into the Atmosphere*, Reidel Pub. Comp., Dordrecht, Holland, 299 pp.

Brutsaert, W. (1998) "Land-Surface Water Vapor and Sensible Heat Flux: Spatial Variability, Homogeneity, and Measurement," *Water Resources Res.*, **34:10**, 2433–2442.

Bucci, O. M., L. Crocco, T. Isernia, and V. Pascazio (2000) "Inverse Scattering Problems with Multifrequency Data: Reconstruction Capabilities and Solution Strategies," *IEEE Trans. Geosci. Remote Sens.*, **38**, 1749–1756.

Budde, W. (1983) *Optical Radiation Measurements*. Vol. 4, "Physical Detectors of Optical Radiation," Academic Press, New York.

Bundschuh, B., D. Schneider, M. Grindel, and E. Becker (1996) "Simulation of a Low Cost Correlation Lidar," Advances in Atmospheric Remote Sensing, *Proceedings of 18th International Laser Radar Conference*, Springer, Berlin, 22–26 July 1996.

Butcher, R., D. Willetts, and W. Jones (1971) "On the Use of a Fabry-Perot Etalon for the Determination of Rotational Constants of Simple Molecules—Pure Rotational Raman Spectra of Oxygen and Nitrogen," *Proc. Roy Soc. London*, A **324**, 231–245.

Cambell, L. (1992) "Afterpulse Measurement and Correction," *Rev. Sci. Instrum.*, **63**, 5794–5798.

Campbell, J. R., D. L. Hlavka, D. D. Turner, J. D. Spinhirne (1998) "Operational Multiple Cloud Boundary Detection and Analysis from Micro Pulse Lidar Data," ARM Science Team Meeting, Tucson, AZ.

Candy, B. H. (1985) "Photomultiplier Characteristics and Practice Relevant to Photon Counting," *Rev. Sci. Instrum.*, **56**, 183–193.

Carlton, H. R. (1980) "Mass Extinction Coefficients Estimated for Nonabsorbing Spherical Aerosol Particles in the Geometric Scattering Regime," *Appl. Opt.*, **19**, 1891–1892.

Carnuth, W., U. Kempfer, and T. Trickl (2002) "Highlights of the Tropospheric Lidar Studies at IFU within the TOR Project," *Tellus*, **54B**, 163–185.

Carnuth, W., and R. Reiter (1986) "Cloud Extinction Profile Measurements by Lidar Using Klett's Inversion Method," *Appl. Opt.*, **25**, 2899–2907.

Carnuth, W., and T. Trickl (1994) A Powerful Eyesafe Infrared Aerosol Lidar: Application of Stimulated Raman Backscattering at 1.06 Mm Radiation, *Rev. Sci. Instrum.*, **65**, 3324–3331.

Carswell, A. I., A. Fong, S. R. Pal, and I. Pribluda (1995) "Lidar-Derived Distribution of Cloud Vertical Location and Extent," *J. Appl. Meteorol.*, **34**, 107–120.

Caughey, S. J., and C. J. Readings (1975) "An Observation of Waves and Turbulence in the Earth's Boundary Layer," *Bound.-Layer Meteor.*, **9**, 279–296.

Chaikovsky, A. P., and V. N. Shcherbakov (1989), "Calculation of the Aerosol Extinction in Data Analysis of the Stratospheric Searching," *Atmos. Opt.* (Institute of Atmospheric Optics, Tomsk, U.S.S.R., 1989), **2**, 552–554 (in Russian).

Chanin, M. L., A. Garnier, A. Hauchecorne, and J. Porteneuve (1989) "A Doppler Lidar for Measuring Winds in the Middle Atmosphere," *Geoph. Res. Lett.*, **16**, 1273–1276.

Chanin, M. L., A. Hauchecorne, and N. Smires (1990) "Contributions to the New Reference Atmosphere from Ground Based Lidar," *Advances in Space Res.*, **12**, 211–216.

Chazette, P. (2003) "The Monsoon Aerosol Extinction Properties at Goa During INDOEX as Measured with Lidar," *J. Geoph. Res.*, **108**, 4187, doi:10.1029/2002JD002074.

Chen, H., M. A. White, D. A. Krueger, and C. Y. She (1996) "Daytime Mesopause Temperature Measurements with a Sodium-Vapor Dispersive Faraday Filter in a Lidar Receiver," *Opt. Lett.*, **21**, 1093–1099.

Chu, X., C. S. Gardner, and G. Papen (2001a) "Lidar Observations of Polar Mesospheric Clouds at South Pole: Seasonal Variations," *Geoph. Res. Lett.*, **28**, 1203–1206.

Chu, X., C. S. Gardner, and G. Papen (2001b) "Lidar Observations of Polar Mesospheric Clouds at South Pole: Diurnal Variations," *Geoph. Res. Lett.*, **28**, 1937–1940.

Chu, X., A. Z. Liu, G. Papen, C. S. Gardner, M. Kelley, J. Drummond, and R. Fugate (2000a) "Lidar Observations of Elevated Temperatures in Bright Chemiluminescent Meteor Trails During the 1998 Leonid Shower," *Geoph. Res. Lett.*, **27**, 1815–1818.

Chu, X., W. Pan, G. Papen, C. S. Gardner, G. Swenson, and P. Jenniskens (2000b) "Characteristics of Fe Ablation Trails Observed During the 1998 Leonid Meteor Shower," *Geoph. Res. Lett.*, **27**, 1807–1810.

Chu, X., W. Pan, G. C. Papen, C. S. Gardner, and J. A. Gelbwachs (2002) " Fe Boltzmann Temperature Lidar: Design, Error Analysis, and Initial Results at the North and South Poles," *Appl. Opt.*, **41**, 4400–4410.

Chu, Z., U. N. Singh, and T. D. Wilkerson (1991) "Multiple Stokes Wavlength Generation in H_2, D_2 and CH_4 for Lidar Aerosol Measurements," *Appl. Opt.*, **30**, 4350–4357.

Chu, Z., U. N. Singh, and T. D. Wilkerson (1990) "A Self-Seeded SRS System for the Generation of 1.54 μm Eyesafe Radiation," *Opt. Comm.*, **75**, 173–178.

Chudamani, S., J. Spinhirne, and A. Clarke (1996) "Lidar Aerosol Backscatter Cross-Sections in the 2 micron near Infrared Wavelength Region," *Appl. Opt.*, **35**, 4812–4819.

Clemesha, B., V. Kirchoff, and D. Simonich (1981) "Remote Measurement of Tropospheric and Stratospheric Wind by Ground Based Lidar," *Appl. Opt.*, **20**, 2907–2910.

Clemesha, B. R., P. P. Batista, and D. M. Simonich (1996) "Formation of Sporadic Sodium Layers," *J. Geoph. Res.*, **101(A9)** 19701–19706.

Clemesha, B. R., P. P. Batista, and D. M. Simonich (1997) "Long-Term and Solar Cycle Changes in the Atmospheric Sodium Layer," *J. Atmos. Solar-Terrest. Phys.*, **59**, 1673–1678.

Clothiaux, E. E., G. G. Mace, T. P. Ackerman, T. J. Kane, J. D. Spinhirne, and V. S. Scott (1998) "An Automated Algorithm for Detection of Hydrometeor Returns in Micropulse Lidar Data," *J. Atmos. Oceanic Technol.*, **15**, 1035–1042.

Coates, P. B. (1973a) "The Origin of Afterpulses in Photomultipliers," *J. Phys. D: Appl. Phys.*, **6**, 1159–1166.

Coates, P. B. (1973b) "A Theory of Afterpulse Formation in Photomultipliers and the Pulse Height Distribution," *J. Phys. D: Appl. Phys.*, **6**, 1862–1869.

Cohen, W. C., S. Grund, and W. Angevine (1997) "Boundary Layer Height and Vertical Velocity Measurements at LIFT," *12th Symph. on Boundary Layers and Turbulence*, Vancouver, B.C., Canada, Amer. Meteor. Soc., 5–6.

Cohen, A., J. Cooney, and K. Gellar (1976) "Atmospheric Temperature Profiles from Lidar Measurements from Rotational Raman and Elastic Scattering," *Appl. Opt.*, **15**, 2896–2901.

Collis, R. T. H. (1969) "Lidar," in *Atmospheric Exploration by Remote Probes*, in Proceedings of the Scientific Meetings of the Panel on Remote Atmospheric Probing (Panel on Remote Atmospheric Probing, April 18–20 and 16–17, 1968), **2**, 147–171.

Collis, R. T. H. (1966) "Lidar: a New Atmospheric Probe," *Q. J. R. Meteorol. Soc.*, **92**, 220–230.

Collis, R. T. H. (1970) "Lidar," *Appl. Opt.*, **9**, 1782–1788.

Connell, B., and D. Miller (1995) "An Interpretation of Radiosonde Errors in the Atmospheric Boundary Layer," *J. Appl. Met.*, **34**, 1070–1081.

Cook, C. S., G. W. Bethke, and W. D. Conner (1972) "Remote Measurement of Smoke Plume Transmittance Using Lidar," *Appl. Opt.*, **11**, 1742–1748.

Cooney, J. (1968) "Measurements on the Raman Component of Laser Atmospheric Backscatter," *Appl. Phys. Lett.*, **12**, 40–42.

Cooney, J. (1970) "Remote Measurements of Atmospheric Water-Vapor Profiles Using the Raman Component of Laser Backscatter, *J. Appl. Meteorol.*, **9**, 182–184.

Cooney, J. (1972) "Measurement of Atmospheric Temperature Profiles by Raman Backscatter," *Appl. Meteorol.*, **11**, 108–112.

Cooney, J. (1976) "Laser Radar Measurements of Atmospheric Temperature Profiles by Use of Raman Rotational Backscatter," *Appl. Opt.*, **15**, 602–603

Cooney, J. (1983) "Uses of Raman Scattering for Remote Sensing of Atmospheric Properties of Meteorological Significance," *Opt. Eng.*, **22**, 292–301.

Cooney, J. (1986) "Lidar Method of Measurement of Atmospheric Extinction and Ozone Profiles," *Appl. Opt.*, **25**, 2035–2036.

Cooney, J. (1987) "Acquisition of Atmospheric Extinction Profiles by Lidar: Further Notes," *Appl. Opt.*, **26**, 3487–3489.

Cooney, J., J. Orr, and C. Tomasetti (1969) "Measurements Separating the Gaseous and Aerosol Components of Laser Atmospheric Backscatter," *Nature.*, **224**, 1098–1099.

Cooney, J., K. Petri, and A. Salik (1985) "Measurements of High Resolution Atmospheric Water-Vapor Profiles by the Use of a Solar-Blind Raman Lidar," *Appl. Opt.*, 104–108.

Cooper, D. I., and W. E. Eichinger (1994) "Structure of the Atmosphere in an Urban Planetary Boundary Layer from Lidar and Radiosonde Observations," *J. Geoph. Res.*, **99**, 22937–22948.

Cooper, D. I., W. E. Eichinger, M. V. Hynes, D. F. Keller, C. F. Lebeda, and D. A. Poling (1996) "High Resolution Properties of the Equatorial Pacific Marine Atmospheric Boundary Layer from Lidar and Radiosonde Observations," *J. Atmos. Sci.*, **53**, 2054–2075.

Coulter, R. L. (1979) "A Comparison of Three Methods for Measuring Mixing-Layer Height," *J. Appl. Meteor.*, **18**, 1495–1499.

Cova S., A. Longoni, and G. Ripamonti (1982) "Active Quenching and Gating Circuits for Single-Photon Avalanche Photodiodes (SPAD)," *IEEE Trans. Nucl. Sci.*, **NS-29**, 599–612.

Crum, T. D., R. B. Stull, and E. W. Eloranta (1987) "Coincident Lidar and Aircraft Observations of Entrainment into Thermals and Mixed Layers", *J. Climate and Appl. Meteorol.*, **26**, 774–788.

Csanady, G. T. (1973) *Turbulent Diffusion in the Environment*, Reidel Publishing Company, Dordrecht, Holland.

Curcio, J. A., and K. A. Durbin (1959) "Atmospheric Transmittance in the Visible," NRL Report, 5368, Washington D.C.

Curcio, J. A., and C. L. Knestric (1958) "Correlation of the Atmospheric Transmission and Backscattering," *J. Opt. Soc. Amer.*, **48**, 686–689.

Danichkin, S. A., and I. V. Samokhvalov (1976) "Laser Radar Equation for Extended Scattering Medium with the Consideration of Lidar Parameters," in *Laser Sensing of the Atmosphere*, V. E. Zuev, ed. (Nauka, Novosibirsk, 1976), 104–110 (in Russian).

Dao, P., W. Klemetti, D. Sipler, W. Moskowitz, and G. Davidson (1989) "Density Measurement with Combined Raman-Rayleigh Lidar," *Proc. SPIE Int. Soc. Opt. Eng.*, **1062**, 138–143.

Darland J., and S. Crouch (1972) "Pulse (Photon) Counting: Determination of Optimum Measurement System Parameters," *Anal. Chem.*, **51**, 240–245.

Davis, P. A. (1969) "The Analysis of Lidar Signatures of Cirrus Clouds," *Appl. Opt.*, **8**, 2044–2102.

Davis, K., D. Lenshow, S. Oncley, C. Kiemle, G. Ehret, A. Giez, and J. Mann (1997) "Role of Entrainment in Surface-Atmosphere Interactions over the Boreal Forest," *J. Geoph. Res.*, **102**, 29219–29230.

Deardorff, J. W., G. E. Willis, and B. H. Stockton (1980) "Laboratory Studies of the Entrainment of a Convective Mixed Layer," *J. Fluid Mech.*, **100**, 41–64.

DeBruin, H. A. R., W. Kohsiek, and B. Van den Hurk (1993) "A Verification of Some Methods to Determine the Fluxes of Momentum, Sensible Heat, and Water Vapor Using Standard Deviation and Structure Parameter of Scalar Meteorological Quantities," *Bound. Lay. Met.*, **63**, 231–257.

Deirmendjian, D. (1963) "Tables of Mie Scattering Cross Sections and Amplitudes," Rep. R-407-PR, Rand Corporation, Santa Monica, CA.

Deirmendjian, D. (1964) "Scattering And Polarization Properties of Water Clouds and Hazes in the Visible and Near Infrared," *Appl. Opt.*, **3**, 187–196.

Deirmendjian, D. (1969) "*Electromagnetic Scattering on Spherical Polydispersions*, American Elsevier, NY.

Del Guasta, M. (1998) "Errors in the Retrieval of Thin-cloud Optical Parameters Obtained with a Two-Boundary Algorithm," *Appl. Opt.*, **37**, 5522–5540.

Del Guasta, M., M. Morandi, L. Stefanutti, J. Brechet, and J. Piquad (1993) "One Year of Cloud Lidar Data from Dumont D'urville (Antarctica). 1. General Overview of Geometrical and Optical Properties," *J. Geoph. Res.*, **98**, 18575–18587.

Dellago C., and H. Horvath (1993) "On the Accuracy of the Size Distribution Information Obtained from Light Extinction and Scattering Measurements," *J. Aerosol Sci.*, **24**, 129–141.

Derbyshire, S. H. (1995) "Stable Boundary Layers: Observations, Models and Variability. Part I: Modeling and Measurements," *Bound.-Layer Meteor.*, **74**, 19–54.

Dereniak, E. L., and D. G. Crowe. (1984) *Optical Radiation Detectors*, John Wiley & Sons, New York.

Derr, V. E. (1980) Estimation of the Extinction Coefficient of Clouds from Multiwavelength Lidar Backscatter Measurements," *Opt.*, **19**, 2310–2314.

Derr, V., and C. Little (1970) "A Comparison of Remote Sensing of the Clear Atmosphere by Optical, Radio, and Acoustic Radar Techniques," *Appl. Opt.*, **9**, 1972–1976.

Devara, P. C. S., P. E. Raj, and G. Pandithurai (1994) Lidar-Observed Long-Term Variations in Aerosol Characteristics and their Connection with Meteorological Parameters, *Int. J. Climatology*, **14**, 581–591.

Devara P. C., P. E. Raj, B. S. Murthy, G. Pandithurai, S. Sharma, and K.G. Vernekar (1995b) "Intercomparison of Nocturnal Lower Atmospheric Structure Observed with Lidar and Sodar Techniques at Pune, India," *J. Applied Meteorol.*, **34**, 1375–1383.

Dho, S. W., Y. J. Park, and H. J. Kong (1997) "Experimental Determination of the Geometrical Form Factor in the Lidar Equation for Inhomogeneous Atmosphere," *Appl. Opt.*, **36**, 6009–6010.

Doherty, S. J., T. L. Anderson, and R. J. Charlson (1999) "Measurement of the Lidar Ratio for Atmospheric Aerosols with a 180^0 Backscatter Nephelometer," *Appl. Opt.*, **38**, 1823–1832.

Donovan, D. P., and A. I. Carswell (1997) "Principal Component Analysis Applied to Multiwavelength Lidar Aerosol Backscatter and Extinction Measurements," *Appl. Opt.*, **36**, 9406–9424.

Donovan, D. P., and A. C. A. P. van Lammeren (2001) "Cloud Effective Particle Size and Water Content Profile Retrievals Using Combined Lidar and Radar Observations. 1. Theory and Examples," *J. Geoph. Res.*, **106**, 27425–27448.

Donovan, D. P., A. C. A. P. van Lammeren, R. J. Hogan, H. W. J. Russchenberg, A. Apituley, P. Francis, J. Testud, J. Pelon, M. Quante, and J. Goddard (2001a) "Cloud Effective Particle Size and Water Content Profile Retrievals Using Combined Lidar and Radar Observations. 2. Comparison with IR Radiometer and in Situ Measurements of Ice Clouds," *J. Geoph. Res.*, **106**, 27449–27464.

Donovan, D.P., J. A. Whiteway, and A. I. Carswell (1993) "Correction for Nonlinear Photon-Counting Effects in Lidar Systems," *Appl. Opt.*, **322**, 6742–6753.

Dreischuh, T. N., L. L. Gurdev, and D. V. Stoyanov (1995) "Effect of Pulse Shape Uncertainty on the Accuracy of Deconvoluted Lidar Profiles," *J. Opt. Soc. Am., A* **12**, 301–306.

Dubinsky, R. H., A. I. Carswell, and S. R. Pal (1985) "Determination of Cloud Microphysical Properties by Laser Backscattering and Extinction Measurements," *Appl. Opt.*, **24**, 1614–1622.

Dupont, E., J. Pelon, and C. Flamant (1994) "Study of the Moist Convective Boundary-Layer by Backscattering Lidar," *Boundary Layer Meteorol.*, **69**, 1–25.

Durrani, T.S., and C. A. Greated (1975) "Spectral Analysis and Cross-Correlation Techniques for Photon Counting Measurements on Fluid Flows," *Appl. Opt.*, **14**, 778–786.

Durieux, E. and L. Fiorani (1998) "Measurement of the Lidar Signal Fluctuations with a Shot-per-shot Instrument,," *Appl. Opt.*, **37**, 7128–7131.

Dyer, P., W. Eichinger, and D. Holtkamp (1992) "Linear Depolarization Lidar Measurements of Dugway Clouds," Los Alamos National Laboratory, LA-UR-92-1217.

Eberhard, W. (1987) "Cloud Signals from Lidar and Rotating Beam Ceilometer Compared with Pilot Ceilings," *J. Atmos. Oceanic Technol.*, **3**, 499–512.

Eberhard, W. L., R. J. Alvarez II, J. M. Intrieri, S. P. Sandberg, K. W. Koenig, and K. R. Healy (1998) "Automated All-Weather Lidar with Scanning Options," *Proceedings of the Nineteenth International Laser Radar Conference*, U. N. Singh, S. Ismail, and G. K. Schwemmer, eds. (Annapolis, Maryland, USA, 1998), 639–642.

Eberhard, W. L., G. T. McNice, and S. W. Troxel (1987) "Lidar Sensing of Plume Dispersion: Analysis Method and Product Quality for Light-Scattering Tracer Particles," *J. Atmos. Oceanic Technol.*, **4**, 674–689.

Edlen, B. (1953) "Dispersion of Standard Air," *J. Opt. Soc. Am.*, **43**, 339–344.

Edlen, B. (1966) "The Refractive Index of Air," *Meteorol.*, **2**, 71–80.

Einaudi, F., and J. J. Finnigan (1981) "The Interaction Between an Internal Gravity Wave and the Planetary Boundary Layer. Part I: The Linear Analysis". *Quart. J. Roy. Meteor. Soc.*, **107**, 793–806.

Elbaum, M., and P. Diament (1976) "SNR in Photocounting Images of Rough Objects in Partially Coherent Light," *Appl. Opt.*, **15**, 2268–2275.

Electro-Optics Handbook (1974) RCA Solid State Division, Electro-Optics and Devices, Lancater, PA, 250pp.

Eloranta, E. W. (1972) "Calculation of Doubly Scattered Lidar Returns," Ph. D. Dissertation (University of Wisconsin, Madison, Wisconsin).

Eloranta, E. W. (1998) "Practical Model for the Calculation of Multiply Scattered Lidar Returns," *Appl. Opt.*, **37**, 2464–2472.

Eloranta, E. W., and R. Boers (1986) "Lidar Observations of the Atmospheric Entrainment Zone and the Potential Temperature Jump Across the Top of the Mixed Layer," *Bound. Lay. Met.*, **34**, 357–375.

Eloranta, E. W., and D. K. Forrest (1992) "Volume Imaging Lidar Observations of the Convective Structure Surrounding the Flight Path of a Flux Measuring Air Craft," *J. Geoph. Res.*, **97**, 18383–18393.

Eloranta, E. W, J. M. King, and J. A. Weinman (1975) "The Determination of Wind Speeds in the Boundary Layer by Monostatic Lidar," *J. Appl. Meteor.*, **14**, 1485–1489.

Eloranta, E. W., S. D. Mayor, and J. P. Garcia (1999) "Lidar Measurements of Vector Wind Fields", The Pacific Rim Conference on Lasers and Electro-Optics, Aug. 30–Sept. 3, 1999, Seoul, Korea.

Eloranta, E. W., and S. T. Shipley (1982) "A Solution for Multiple Scattering," *Atmospheric Aerosols: Their Formation, Optical Prop[erties and Effects*, A. Deepak, ed. (Spectrum, Hampton, Va.).

Elterman, L. (1951) "The Measurement of Stratospheric Density Distribution with the Searchlight Technique," *J. Geoph. Res.*, **56**, 509–520.

Elterman, L. (1953) "A Series of Stratospheric Temperature Profiles Obtained with the Searchlight Technique," *J. Geoph. Res.*, **58**, 519–530.

Elterman, L. (1954) "Seasonal Trends of Temperature, Density and Pressure to 67.6 km Obtained with the Searchlight Probing Technique," *J. Geoph. Res.*, **59**, 351–358.

Elterman, L. (1962) "Searchlight Probing Technique for Upper Atmospheric Measurements," in *Encyclopedic Dictionary of Physics*, J. Thewliss, ed. (Perg. Press, N.Y., 1962), 81–101.

Elterman, L. (1966) "Aerosol Measurements in the Troposphere and Stratosphere," *Appl. Opt.*, **5**, 1760–1772.

Elterman, L. (1968) "Atmospheric Attenuation Model in the Ultraviolet, Visible, and Infrared Regions for Altitudes to 50 km." A. F. Cambridge Research Laboratories, Hanscom Field, MA., AFCRL-68-0153.

Emery, Y., and C. Flesia (1996) "Effect of Signal Dynamics: A Comparative Study between PRN and Pulse Illuminated Lidar Systems," Advances in Atmospheric

Remote Sensing, *Proceedings of 18th International Laser Radar Conference*, Springer, Berlin, 22–26 July 1996.

Emery, Y., and C. Flesia (1998) "Use of the A1- and the A2-Sequences to Modulate Continuous-Wave Pseudorandom Noise Lidar, *Appl. Opt.*, **37**, 2238–2241.

Endemann, M., and R. L. Byer (1981) "Simultaneous Measurements of Atmospheric Temperature and Humidity Using a Continuously Tunable IR Lidar," *Appl. Opt.*, **20**, 3211–3217.

Erbrink, H. J. (1994) "Plume Rise in Different Atmospheres: a Practical Scheme and Some Comparisons with Lidar Measurements," *Atmos. Environ.*, **28**, 3625–3636.

Evans, B. T. N. (1988) "Sensitivity of the Backscatter/Extinction Ratio to Changes in Aerosol Properties: Implications for Lidar," *Appl. Opt.*, **27**, 3299–3306.

Felix, F., W. Keenliside, G. Kent, and M. C. W. Sandford (1973) "Laser Radar Observations of Atmospheric Potassium," *Nature*, **246**, 345–346.

Fenn, R. W. (1966) "Correlation Between Atmospheric Backscattering and Meteorological Visual Range," *Appl. Opt.*, **4**, 293–295.

Fensterer A., J. Leblanc, W. Taylor, and H. Johns (1973) "Linearity and Fatigue in Photomultipliers," *Rev. Sci. Instrum.*, **44**, 689–694.

Ferguson, J. A., and D. H. Stephens (1983) "Algorithm for Inverting Lidar Returns," *Appl. Opt.*, **22**, 3673–3675.

Fernald, F. G. (1984) "Analysis of Atmospheric Lidar Observations: Some Comments," *Appl. Opt.*, **23**, 652–653.

Fernald, F. G., B. M. Herman, and J. A. Reagan (1972) "Determination of Aerosol Height Distribution by Lidar," *J. Appl. Meteorol.*, **11**, 482–489.

Ferrare, R. A., S. H. Melfi, D. N. Whiteman, and K. D. Evans (1992) "Raman Lidar Measurements of Pinatubo Aerosols over Southeastern Kansas During November–December 1991, *Geoph. Res. Let.*, **19**, 1599–1602.

Ferrare R., T. McGee, D. Whiteman, J. Burris, M. Owens, J. Butler, R. Barnes, F. Schmidlin, W. Komhyr, P. Wang, M. McCormick, and A. Miller (1995) "Lidar Measurements of Stratospheric Temperature During STOIC," *J. Geoph. Res.-Atmospheres,* **100**, 9303–9312.

Ferrare, R. A., S. H. Melfi, D. N. Whiteman, and K. D. Evans (1995) "A Comparison of Water-Vapor Measurements Made by Raman Lidar and Radiosondes," *J. Atm. Ocean. Tech.*, **12**, 1177–1195.

Ferrare, R. A., S. H. Melfi, D. N. Whiteman, K. D. Evans, and R. Leifer (1998) "Raman Lidar Measurements of Aerosol Extinction and Backscattering. I. Methods and Comparisons," *J. Geoph. Res.*, **103**, 19663–19672.

Ferrare, R. A., S. H. Melfi, D. N. Whiteman, K. D. Evans, M. Poellot, and Y. Kaufman (1998) "Raman Lidar Measurements of Aerosol Extinction and Backscattering, II. Derivation of Aerosol Real Refractive Index, Single Scattering Albedo, and Humidification Factor Using Raman Lidar and Aircraft Size Distribution Measurements," *J. Geoph. Res.*, **103**, 19673–19689.

Finger F. (1993) "Evaluation of NMC Upper-stratospheric Temperature Analyses Using Rocketsonde and Lidar Data," *Bulletin of The American Meteorological Society*, **74**, 1116.

Finger F., M. Gelman, J. Wild, M. Chanin, A. Hauchecorne, and A. Miller (1993) "Evaluation of NMC Upper-stratospheric Temperature Analyses Using Rocketsonde and Lidar Data," *Bull. of the Amer. Meteorol. Soc.*, **74**, 789–799.

Finnigan, J. J., and F. Einaudi (1981) "The Interaction Between an Internal Gravity Wave and the Planetary Boundary Layer. Part II: Effect of the Wave on the Turbulence Structure," *Quart. J. Roy. Meteor. Soc.*, **107**, 807–832.

Finnigan, J. J., F. Einaudi, and D. Fua (1984) "The Interaction between an Internal Gravity Wave and Turbulence in the Stably-Stratified Nocturnal Boundary Layer," *J. Atmos. Sci.*, **41**, 2409–2436.

Fiocco G., G. Benedetti-Michelangeli, K. Maischberger, and E. Madonna (1971), *Nature, London Phys. Sci.*, **229**, 78.

Fiocco, G., and J. B. DeWolf (1968) "Frequency Spectrum of Laser Echoes from Atmospheric Constituents and Determination of Aerosol Content of Air," *J. Atm. Sci.*, **25**, 488–496.

Fiocco G., and L. D. Smullins (1963) "Detection of Scattering Layers in the Upper Atmosphere (60–140 km) by Optical Radar," *Nature*, **199**, 1275–1276.

Fiorani, L., B. Calpini, L. Jaquet, H. Van den Bergh, and E. Durieux (1997) "Correction Scheme for Experimental Biases in Differential Absorption Lidar Tropospheric Ozone Measurements Based on the Analysis of Shot per Shot Data Samples," *Appl. Opt.* **36**, 6857–6863.

Fiorani, L., B. Calpini, L. Jaquet, H. Van Den Bergh, and E. Durieux (1998) "A Combined Determination of Wind Velocities and Ozone Concentrations for a First Measurement of Ozone Fluxes with a DIAL Instrument During the Medcaphot-trace Campaign," *Atmos. Environ*, **32**, 2151–2159.

Fischer, K. W., V. Abreu, W. R. Skinner, J. E. Barnes, M. J. McGill, and T. D. Irgang (1995) "Visible Wavelength Doppler Lidar for Measurement of Wind and Aerosol Profiles During Day and Night," *Opt. Eng.* **34**, 499–511.

Fitzgerald, J. W. (1984) "Effect of Relative Humidity on the Aerosol Backscattering Coefficient at 0.694- and 10.6-μm Wavelengths," *Appl. Opt.*, **23**, 411–422.

Flamant, C., J. Pelon, P. H. Flamant, and P. Durant (1997) "Lidar Determination of the Entrainment Zone Thickness and the Top of the Unstable Marine Atmospheric Boundary Layer," *Boundary-Layer Meteorol.*, **83**, 247–284.

Flamant, C., J. Pelon, P. Chazette, V. Trouillet, P. K. Quinn, R. Frouin, D. Bruneau, J. F. Leon, T. S. Bates, J. Johnson, and J. Livinston (2000) "Airborne Lidar Measurements of Aerosol Spatial Distribution and Optical Properties over the Atlantic Ocean During a European Pollution Outbreak of ACE-2," *Tellus*, **52B**, 662–677.

Flesia, C., and C. L. Korb (1999) "Theory of the Double Edge Molecular Technique for Doppler Lidar Wind Measurement," *Appl. Opt.*, **38**, 432–440.

Flesia, C., C. L. Korb, and C. Hirt (2000) "Double-Edge Molecular Measurement of Lidar Wind Profiles at 355 nm," *Opt. Lett*, **25**, 1466–1468.

Flesia, C., and P. Schwendimann (1995) "Analytical Multiple-Scattering Extension of the Mie Theory: The LIDAR Equation," *Appl. Phys.*, B **60**, 331–334.

Foitzik, L., and H. Zschaeck (1953) "Measurement of Spectral Scattering Coefficient of Air near Ground Level for Good Visibility, Haze, and Fog," *J. Meteorol.*, **7**, No. 1, 1–19.

Frejafon, E., J. Kasparian, P. Rambaldi, J. Yu, B. Vezin, and J. P. Wolf (1998) "Three Dimensional Analysis of Urban Aerosols by Use of a Combined Lidar, Scanning Electron Microscopy, and X-ray Microanalysis," *Appl. Opt.*, **37**, 2231–2237.

Freudenthaler, V., F. Homberg, and H. Jaeger (1996) "Optical Parameters of Contrails from Lidar Measurements: Linear Depolarization," *Geoph. Res. Lett.*, **23**, 3715–3718.

Frickle, K. H., and U. von Zahn (1985) "Mesopause Temperatures Derived from Probing the Hyperfine Structure of the D_2 Resonance Line of Sodium by Lidar," *J. Atmos. Terr. Phys.*, **47**, 499–512.

Friedman, J. S., C. A. Tepley, P. A. Castleberg, and H. Roe (1997) "Middle-Atmospheric Doppler Lidar Using an Iodine-Vapor Edge Filter," *Opt. Lett.*, **22**, 1648–1650.

Fritts, D. C., J. F. Garten, and O. Andreassen (1996b) "Wave Breaking and Transition to Turbulence in Stratified Shear Flows," *J. Atmos. Sci.*, **53**, 1057–1085.

Fritts, D. C., J. R. Isler, and O. Andreassen (1994) "Gravity Wave Breaking in Two and Three Dimensions. 2. Three-Dimensional Evolution and Instability Structure," *J. Geoph. Res.*, **99**, 8109–8123.

Fritts, D. C., J. R. Isler, J. H. Hecht, R. L. Walterscheid, and O. Andreasson (1997) "Wave Breaking Signatures in OH Airglow and Sodium Densities and Temperatures. 2. Simulation of Wave and Instability Structures," *J. Geoph. Res.*, **102**, 6669–6684.

Fritts, D. C., T. L. Palmer, O. Andreassen, and I. Lie (1996a) "Evolution and Breakdown of Kelvin-Helmholtz Billows in Stratified Compressible Flows. I: Comparison of Two- and Three-Dimensional Flows," *J. Atmos. Sci.*, **53**, 3173–3191.

Fritts, D. C., and P. K. Rastogi (1985) "Convective and Dynamical Instabilities due to Gravity Wave Motions in the Lower and Middle Atmosphere: Theory and Observations," *Radio Sci.*, **20**, 1247–1277.

Fujimoto, T., O. Uchino, and T. Nagai (1994) "Estimation of DIAL Algorithms for Stratospheric Ozone," *Proceedings of the Seventeenth International Laser Radar Conference*, M. P. McCormick, ed. (Chiba University, Sendai, Japan, 1994), pp. 392–395.

Funck, E. (1986) "Dead Time Effects from Linear Amplifiers and Discriminators in Single Detector Systems," *Nuclear Instr. and Meth.*, **A245**, 519–524.

Fymat, A. L., and K. D. Mease (1978) "Complex Refractive Index of Atmospheric Aerosols: a Size Distribution Independent Retrieval Approach Using Multispectral Transmission Ratios," in *Remote Sensing of the Atmosphere: Inversion Methods and Applications*, A. L. Fymat and V. E. Zuev ed., Elsevier Sci. Publish. Co., 233–256.

Gambling, D. J., and K. Bartusek (1972) "Lidar Observations of Tropospheric Aerosols," *Atmos. Environ.*, **6**, 181–189.

Gao, Z., and J. W. Meriwether (1998) "Mesoscale Airborne in situ and Lidar Observations of Variance and Spatial Structure in the Troposphere and Stratosphere Regions," *J. Geoph. Res.*, **103**, 6391–6396.

Gao, Z., J. W. Meriwether, V. B. Wickwar, T. D. Wilkerson, and S. Collins (1998) "Rayleigh Lidar Measurements of the Temporal Frequency and Vertical Wavenumber Spectra in the Mesosphere over the Rocky Mountain Region," *J. Geoph. Res.*, **103**, 6405–6416.

Gardner, C. S. (1989) "Sodium Resonance Fluorescence Lidar Applications in Atmospheric Science and Astronomy," *IEEE Proc.*, **77**, 408–418.

Gardner, C. S., S. J. Franke, W. Yang, X. Tao, and J. R. Yu (1998) "Interpretation of Gravity Waves Observed in the Mesopause Region at Starfire Optical Range, New Mexico: Strong Evidence for Nonseparable Intrinsic (M, Omega) Spectra," *J. Geoph. Res.*, **103**, 8699–8713.

Gardner, C. S., G. C. Papen, X. Chu, and W. Pan (2001) "First Lidar Observations of Middle Atmosphere Temperatures, Fe Densities, and Polar Mesospheric Clouds over the North and South Poles," *Geoph. Res. Lett.*, **28**, 1199–1202.

Gardner, C. S., and G. C. Papen (1995) "Mesospheric Na Wind Temperature Lidar," *Laser Rev.*, **23**, 131–134.

Gardner, C. S., D. C. Senft, and K. H. Kwon (1989) "Lidar Observation of Substantial Sodium Depletion in the Summertime Arctic Mesosphere," *Nature*, **332**, 142–144.

Gardner, C. S., and M. J. Taylor (1998) "Observational Limits for Lidar, Radar, and Airglow Imager Measurements of Gravity Wave Parameters," *J. Geoph. Res.*, **103**, 6247–6438.

Gardner, C. S., X. Tao, and G. C. Papen (1995) "Simultaneous Lidar Observations of Vertical Wind, Temperature, and Density Profiles in the Upper Mesosphere: Evidence for Nonseparability of Atmospheric Perturbation Spectra," *Geoph. Res. Lett.*, **22**, 2877–2880.

Gault, W. A., and H. N. Rundle (1969) "Twilight Observations of Upper Atmospheric Sodium, Potassium and Lithium," *Can. J. Phys.*, **47**, 85–98.

Gavrilov, V. A. (1966) "*Atmospheric Visibility*," 1st ed. (Gidrometeoizdat, Leningrad) 324, (in Russian).

Gelbwachs J. A. (1994) "Iron Boltzmann Factor Lidar—Proposed New Remote-sensing Technique For Mesospheric Temperature," *Appl. Opt.*, **33**, 7151–7156.

Gelbwachs, J. A. (1996) "Lidar for Remote Measurement of Ozone in the Exhaust Plumes of Launch Vehicles," *Appl. Opt.*, **35**, 2630–2636.

Gentry, B. M., and C. L. Korb (1994) "Edge Technique for High Accuracy Doppler Velocimetry," *Appl. Opt.*, **33**, 5770–5777.

Gentry, B. M., H. Chen, and S. X. Li (2000) "Wind Measurements with 355 nm Molecular Doppler Lidar," *Opt. Lett*, **25**, 1231–1233.

Gill, R., K. Geller, J. Farina, and J. Cooney (1979) "Measurement of Atmospheric Temperature Profiles Using Raman Lidar," *J. Appl. Meteorol.*, **18**, 225–227.

Gillespie, J., and J. Lindberg (1992) "Seasonal and Geographic Variations in Imaginary Refractive Index of Atmospheric Particulate Matter," *Appl. Opt.*, **31**, 2107–2111.

Gillespie, J., and J. Lindberg (1992a) "Ultraviolet and Visible Imaginary Refractive Index of Strongly Absorbing Atmospheric Particulate Matter," *Appl. Opt.*, **31**, 2112–2115.

Gimmestad, G. G. (2001) "Comment on Two Wavelength Lidar Inversion Techniques," *Appl. Opt.*, **40**, 2004–2009.

Girard, T. (1995) "Understanding Effective Bits, Application Note AN95091, Signatec Corp.

Girolamo, P., R. Gagliardi, G. Pappalardo, N. Spinelli, R. Velotta, and V. Bernardi (1995) "Two Wavelength Lidar Analysis of Stratospheric Aerosol Size Distribution," *J. Aerosol Sci.*, **26**, 989–1001.

Gobbi G., C. Souprayen, F. Congeduti, G. Didonfrancesco (1995) "Lidar Observations of Middle Atmosphere Temperature Variability," *Annales Geophysicae-Atmos.*, **13**, 648–655.

Gobbi, G. P. (1995) "Lidar Estimation of Stratospheric Aerosol Properties: Surface, Volume, and Extinction to Backscatter Ratio," *J. Geoph. Res.*, **100**, 11219–11235.

Gobbi, G. P. (1998) "Parameterization of Stratospheric Aerosol Physical Properties on the Basis of Nd:YAG Lidar Observations," *Appl. Opt.*, **37**, 4712–4721.

Godin, S., A. I. Carswell, D. P. Donovan, H. Claude, W. Steinbrecht, I. S. McDermid, T. I. McGee, M. R. Gross, N. Nakane, D. P. J. Swart, H. B. Bergwerff, O. Uchino, P. Gathen, and R. Neuber (1999) "Ozone Differential Absorption Lidar Algorithm Intercomparison," *Appl. Opt.*, **38**, 6225–6236.

Golberg, M. A. (1968) "Relationship Between Backscatter and Total Scattering Coefficients," *Proceedings of the Institute of Hydrology and Meteorology Instrumentation*, U. S. S. R., Moscow, No. 18, 61–69, (in Russian).

Golberg, M. A. (1971) "Theory of the Meteorological Visibility Range Measurement by Means of the Backscattered-Light Intensity Determination," *Proceedings of the Institute of Hydrology and Meteorology Instrumentation*, U. S. S. R., Moscow, No. 25, 26–29, (in Russian).

Goody, R. M., and Y. L. Yung, *Atmospheric Radiation: Theoretical Basis*, Oxford University Press, New York, 1989.

Gorchakov, G. I., and A. A. Isakov (1976) "Some Results of the Light Scattering Investigation in the Region of Radar Viewing Angles," *Proceedings of the All Union Workshop on Atmos. Opt.*, (Institute of Atmospheric Optics, Tomsk, U. S. S. R., 1976) Part 1, 218–221, (in Russian).

Goryshin, V. I., and V. A. Kovalev (1978) "A Methodology of Instrumental Measurements of the Runway-Light Visibility," *Proceedings of the Main Geophysical Observatory*, G. P. Gushchin, ed. (Main Geophysical Observatory, Leningrad, 1978) No. 406, 38–45, (in Russian).

Gossard, E., R. Chadwick, W. Neff, and K. Moran (1982) "The Use of Ground-Based Doppler Radars to Measure Gradients, Fluxes, and Structure Parameters in Elevated Layers," *J. Appl. Met.*, **21**, 211–226.

Grainger, R. G., A. Lambert, C. D. Rodgers, F. W. Taylor, and T. Deshler (1995) "Stratospheric Aerosol Effective Radius, Surface Area and Volume Estimated from Infrared Measurements," *J. Geoph. Res.*, **100**, 16507–16518.

Grant, W. B., A. M. Brothers, and J. R. Bogan (1988) "Differential Absorption Lidar Signal Averaging," *Appl. Opt.*, **27**, 1934–1938.

Grant, W. B., E. V. Browell, N. S. Higdon, and S. Ismail (1991) "Raman Shifting of KrF Laser Radiation for Tropospheric Ozone Measurements," *Appl. Opt.*, **30**, 2628–2633.

Grewal, M. S., and A. P. Andrews (1993) *Kalman Filtering: Theory and Practice* (Prentice-Hall, Englewood Cliffs, N. J.).

Gross, M. R., T. J. McGee, R. Ferrare, U. N. Singh, and P. Kimvilakani (1997) "Temperature Measurements Made with a Combined Rayleigh-Mie and Raman Lidar," *Appl. Opt.*, **36**, 5987–5995.

Gross, M. R., T. J. McGee, U. N. Singh, and P. Kimvilakani (1995) "Measurements of Stratospheric Aerosols with a Combined Elastic-Raman Backscatter Lidar," *Appl. Opt.*, **34**, 6915–6924.

Grund, C. J., F. L. Roesler, and E. W. Eloranta (1988) "Thermally Induced Spectral Drift Cancellation in Pressure-tuned Fabry-Perot Etalons," *Appl. Opt.*, **27**, 662–663.

Grund, C. J., and E. W. Eloranta (1990) "The 27–28 October 1986 IFO Cirrus Case Study: Cloud Optical Properties by High Spectral Resolution Lidar," *Mon. Weather Rev.*, **118**, 2344–2355.

Grund, C. J., and E. W. Eloranta (1991) "Fiber Optic Scrambler Reduces the Bandpass Range Dependence of Fabry-Perot Etalons Used for Spectral Analysis of Lidar Backscatter," *Appl. Opt.*, **30**, 312–326.

Grund, C. J., and E. W. Eloranta (1991) "University of Wisconsin High Spectral Resolution Lidar," *Opt. Engineering.*, **30**, 6–12.

Grund C. J., and S. P. Sandberg (1996) "Depolarization and Backscatter Lidar for Unattended Operation," in *Advances in Atmospheric Remote Sensing with Lidar*, A. Ansman, R. Neuber, P. Rairoux, and U. Wandinger eds. (Selected Papers of the 18th International Laser Radar Conference, Berlin, Germany, 1996), 3–6.

Gu, Y. Y., C. S. Gardner, P. A. Castleberg, G. C. Papen, and M. C. Kelley (1997) "Validation of the Lidar In-space Technology Experiment: Stratospheric Temperature and Aerosol Measurements," *Appl. Opt.*, **36**, 5148–5157.

Guelle, W., Y. Balkanski, M. Schulz, F. Dulac, J. M. Prospero, D. Winker, R. Hoff, and P. Chazette (1997) "Modeling the Vertical and Horizontal Distribution of the Mineral Aerosol: Comparison with Observations," *J. Aerosol Sci.*, **28**, S 449–S 450.

Gurdev, L. L., Dreischuh, T. N., and D. V. Stoyanov (1993) "Deconvolution Techniques for Improving the Resolution of Long Pulse Lidars," *J. Opt. Soc. Am., A* **10**, 2296–2306.

Gutkowicz-Krusin, D. (1993) "Multiangle Lidar Performance in the Presence of Local Horizontal Inhomogeneities in Atmospheric Extinction and Scattering," *Appl. Opt.*, **18**, 3266–3272.

Hall, F. F., Cupp R. E., and S. W. Troxel (1988) "Cirrus Cloud Transmittance in the Infrared Measured with a CO_2 Lidar," *Appl. Opt.*, **12**, 2510–2516.

Hall, F. F., and H. Y. Ageno (1970) "Absolute Calibration of a Laser System for Atmospheric Probing," *Appl. Opt.*, **9**, 1820–1824.

Halldorsson T., and J. Langerholc (1978) "Geometrical Form Factors for the Lidar Function," *Appl. Opt.*, **17**, 240–244.

Hamilton, P. M. (1969) "Lidar Measurement of Backscatter and Attenuation of Atmospheric Aerosol," *Atmos. Environ.*, **3**, 221–223.

Hamming, R. W. (1989) *Digital Filters*, 3rd ed. (Prentice-Hall, Englewood Cliffs, N. J.).

Hanser, D. A., and I. S. McDermid (1990) "Stimulated Raman Shifting in the Nd:YAG Fourth Harmonic (266 nm) in H2, HD and D2," *IEEE J. Quantum Electron*, **26**, 1292–1298.

Harms, J. (1979) "Lidar Return Signal for Coaxial and Noncoaxial Systems with Central Obstruction, *Appl. Opt.*, **18**, 1559–1566.

Harms, J., W. Lahmann, and C. Weitkamp (1978) "Geometrical Compression of Lidar Return Signals," *Appl. Opt.*, **17**, 1131–1135.

Hashmonay, R., A. Cohen, and U. Dayan (1991) "Lidar Observation of the Atmospheric Boundary Layer in Jerusalem," *J. Appl. Meteorol.*, **30**, 1228–1236.

Hauchecorne, A. (1995) "Lidar Temperature Measurements in the Middle Atmosphere," *Rev. Laser Engineering*, **23**, 119–123, 1995.

Hauchecorne, A., and M. Chanin (1980) "Density and Temperature Profiles Obtained by Lidar between 35 and 70 km," *Geoph. Res. Letters*, **7**, 565–568.

Hauchecorne, A., M. Chanin, and P. Keckhut (1991) "Climatology and Trends of the Middle Atmospheric Temperature (33–87 km) as Seen by Rayleigh Lidar over the South of France," *J. Geoph. Res. Atmos.*, **96**, 15297–15309.

Hauchecorne, A., M. Chanin, P. Keckhut, and D. Nedeljkovic (1992) " Lidar Monitoring of the Temperature in the Middle and Lower Atmosphere," *Appl. Phys. B—Photophysics and Laser Chemistry*, **55**, 29–34.

Hayashida, S., and Y. Sasano (1993) "Stratospheric Aerosol Change in the Early Stage of Volcanic Disturbance by the Pinatubo Eruption Observed over Tsukuba, Japan," *Geoph. Res. Lett.*, **20**, 575–578.

Hayden, K., K. Anlauf, R. Hoff, J. Strapp, J. Bottenheim, H. Wiebe, F. Froude, J. Martin, D. Steyn, and I. McKendry (1997) "The Vertical Chemical and Meteorological Structure of the Boundary Layer in the Lower Frasier Valley During PACIFIC'93," *Atmos. Environ.*, **31**, 2089–2106.

Heaps, W. S., and J. Burris (1996) "Airborne Raman Lidar," *Appl. Opt.*, **35**, 7128–7135.

Heaps, W. S., J. Burris, and J. A. French (1997) "Lidar Technique for Remote Measurement of Temperature by Use of Vibrational-Rotational Raman Spectroscopy," *Appl. Opt.*, **36**, 9402–9404.

Hecht, J. H., R. L. Walterscheid, D. C. Fritts, J. R. Isler, D. C. Senft, C. S. Gardner, and S. J. Franke (1997) "Wave Breaking Signatures in OH Airglow and Sodium Densities and Temperatures 1. Airglow Imaging, Na Lidar, and MF Radar Observations," *J. Geoph. Res.*, **102**, 6655–6668.

Heintzenberg, J., and R. J. Charlson (1996) "Design and Application of the Integrating Nephelometer: a Review," *J. Atmos. Oceanic Technol.*, **13**, 987–1000.

Heintzenberg, J., H. Müller, H. Quenzel, and E. Thomalla (1981) "Information Content of Optical Data with Respect to Aerosol Properties: Numerical Studies with a Randomized Minimization-Search—Technique Algorithm," *Appl. Opt.*, **20**, 1308–1315.

Helstrom, C. W. (1984) "Output Distributions of Electrons in a Photomultiplier," *J. Appl. Phys.*, **55**, 2786–2792.

Hinkley, E. D. ed. (1976) *Laser Monitoring of the Atmosphere*, Springer-Verlag, New York, 378 pp.

Hobbs, P. V. (1993) *Aerosol-Cloud-Climate Interactions*, Academic Press, NY, 233 pp.

Hoff, R. M., L. Guise-Bagley, R. M. Staebler, H. A. Wiebe, J. Brook, B. Georgi, and T. Düsterdiek (1996) "Lidar, Nephelometer, and in Situ Aerosol Experiments in Southern Ontario," *J. Geoph. Res.*, **101**, 19199–19209.

Hoff, R., M. Harwood, A. Sheppard, F. Froude, B. Martin, and W. Strapp (1997) "Use of Airborne Lidar to Determine Aerosol Sources and Movement in the Lower Frasier Valley (LFV), British Columbia," *Atmos. Environ.*, **31**, 2123–2134.

Holmen, B., W. Eichinger, and R. Flocchini (1998) "Application of Elastic Lidar to PM10 Emissions from Agricultural NonPoint Sources," *Env. Sci. and Techn.*, **32**, 3068–3076.

Hooper, W. P., and E. Eloranta (1986) "Lidar Measurements of Wind in the Planetary Boundary Layer: The Method, Accuracy and Results from Joint Measurements with Radiosonde and Kytoon," *J. Climate Appl. Meteorol.*, **25**, 990–1001.

Hopkins, F. K. (1998) "Military Laser Application," *Opt. Photonics News*, **9(2)**, 32–38.

Horvath, H. (1981) "Atmospheric Visibility," *Atm. Environ.*, **15**, No. 10–11.

Houghton, J. T. (1977) *The Physics of Atmospheres*, 2nd ed., Cambridge University Press, New York, 271 pp.

Hoyt, D. V. (1977) "A Redetermination of Rayleigh Optical Depth and its Application to Selected Solar Radiation Problems," *J. Appl. Meteorol.*, **16**, 432–436.

Huges, H. G., J. A. Ferguson, and D. H. Stephens (1985) "Sensitivity of a Lidar Inversion Algorithm to Parameters Relating Atmospheric Backscatter and Extinction," *Appl. Opt.* **24**, 1609–1613.

Hunt, W. H., and S. K. Poultney (1975) "Testing the Linearity of Response of Gated Photomultipliers in Wide Dynamic Range Laser Radar Systems," *IEEE Trans. Nucl. Sci.*, **NS-22**, 116–120.

Hutt, D. L., L. R. Bisonnette, and L. Durand (1994) "Multiple Field of View Lidar Returns from Atmospheric Aerosols," *Appl. Opt.*, **33**, 2338–2348.

Igerson, T. E., R. J. Kearney, and R. L. Coulter (1983) "Photon Counting with Photodiodes." *Appl. Opt.*, **22**, 2013–2021.

Ignatenko, V. M. (1985a) "Experimental Determination of the Lidar Geometrical Function," *Proceedings of the Main Geophysical Observatory*, G. P. Gushchin, ed. (Main Geophysical Observatory, Leningrad, 1991), No. 499, 91–95, (in Russian).

Ignatenko, V. M. (1985c) "Instrumental Errors of the Extinction Coefficient Measurement with Lidar," *Proceedings of the Main Geophysical Observatory*, G. P. Gushchin, ed. (Main Geophysical Observatory, Leningrad, 1985), No. 487, 72–76, (in Russian).

Ignatenko, V. M. (1991) "Method of Data Processing for the Lidar Searching along Slope Ways," *Proceedings of the Main Geophysical Observatory*, G. P. Gushchin, ed. (Main Geophysical Observatory, Leningrad, 1991), No. 533, 76–78, (in Russian).

Ignatenko, V. M., and V. A. Kovalev (1985b) "Analysis of Measurement Errors in the Extinction Coefficient Retrieval by the Asymptotic Method," *Proceedings of the Main Geophysical Observatory*, G. P. Gushchin, ed. (Main Geophysical Observatory, Leningrad, 1985), No. 499, 85–90, (in Russian).

Ignatenko, V. M., V. A. Kovalev, and E. E. Rybakov (1980) "Influence of Instrumental Errors on the Accuracy of Atmospheric Transparency Measurements with Lidar," *Proceedings of the Main Geophysical Observatory*, G. P. Gushchin, ed. (Main Geophysical Observatory, Leningrad, 1980), No. 419, 89–99, (in Russian).

Ignatenko, V., and V. Kovalev (1985) "Method for Determining Atmospheric Transparency," U. S. S. R. Pat., SU 1163217 A (U.S.S.R Bulletin No. 23, 1985).

Inaba, H., and T. Kobayasi (1972) "Laser Raman Radar—Laser Raman Scattering Methods for Remote Detection and Analysis of Atmospheric Pollution," *Opto-Electronics*, 101–123.

Ivanova, I., L. Gurdev, and M. Mitev (1993) "Lidar Technique for Simultaneous Temperature and Pressure Measurement Based on Rotational Raman Scattering," *J. Modern Opt.*, **40**, 367–371.

Jager, H., and D. Hofmann (1991) "Midlatitude Lidar Backscatter to Mass, Area, and Extinction Conversion Model Based on in Situ Aerosol Measurements from 1980 to 1987," *Appl. Opt.*, **30**, 127–138.

Jaenicke, R. (1993) "Tropospheric Aerosols," in *Aerosol-Cloud-Climate Interactions*, P. Hobbs ed., Academic Press, New York, 233 pp.

Jennes, J., D. Lysak, and C. R. Philbrick (1997) "Design of a Lidar Receiver with a Fiber-Optic Output," *Appl. Opt.*, **36**, 4278–4289.

Jeys, T. H. (1992) "Development of Mesospheric Laser Beacon for Atmospheric Adaptive Optics," Abstr. of Sixteenth International Laser Radar Conf., MIT, Cambridge, Ma., July 20–24, 1992, *NASA Conf. Pub.*, 3158, pp. 297–299.

Jinhuan, Q. (1988) "Sensitivity of Lidar Equation Solution to Boundary Values and Determination of the Values," *Adv. Atmos. Sci.* **2(5)**, 229–241 (1988).

Jinhuan, Q. (1994) "An Ozone-Aerosol Lidar and a Three-wavelength Inversion Technique," *Proceedings of the Seventeenth. International Laser Radar Conference, (*Sendai International Center, Sendai, Japan, 1994), 119–121.

Jones, D. P., and G. S. Kent (1974) "Measurement of Overload Effects in a Photomultiplier," *J. Phys.* E., **7**, 744–746.

Junge, C. E. (1960) "Aerosols," *Handbook on Geophysics*, C. F. Chapman et al., eds., Macmillan, NY.

Junge, C. E. (1963) *Air Chemistry and Radioactivity*, Academic Press, NY.

Kaestner, M. (1986) "Lidar Inversion with Variable Backscatter Extinction Ratios: Comment," *Appl. Opt.* **25**, 833–835.

Kaimal, J., N. Abshire, R. Chadwick, M. Decker, W. Hooke, R. Kropfli, W. Neff, F. Pasqualucci, and P. Hildebrand (1982) "Estimating the Depth of the Daytime Convective Boundary Layer," *J. Appl. Meteor.*, **21**, 1123–1129.

Kaimal, J., and J. Finnigan (1994) *Atmospheric Boundary Layer Flows*, Oxford Univ. Press, New York, 289 pp.

Kaiser, J., and W. Reed (1977) "Data Smoothing Using Low Pass Digital Filters," *Rev. Sci. Instrum.*, **48**, 1447–1457.

Kalshoven, J. E. Jr., C. L. Korb, G. K. Schemmer, and M. Dombrowski (1981) "Laser Remote Sensing of Atmospheric Temperature by Observing Resonant Absorption of Oxygen," *Appl. Opt.*, **20**, 1967–1971.

Kane, T. J., and C. S. Gardner (1993) "Structure and Seasonal Variability of the Nighttime Mesospheric Fe Layer at Mid-Latitudes," *J. Geoph. Res.*, **98**, 16875–16886.

Kano, M. (1968) "On the Determination of Backscattered and Extinction Coefficient of the Atmosphere by Using a Laser Radar," *Papers Meteorol. And Geoph.*, **19**, 121–129.

Kaul, B. V. (1977) *Laser Sensing the Aerosol Pollution in the Atmosphere*, Dissertation (Institute of Atmospheric Optics, Tomsk, 1977, 176), (in Russian).

Kaul, B. V., and I. V. Samokhvalov (1975) "Lidar Equation That Includes the Estimate of the Doubly Scattered Light," *Izvestija Vuzov*, No. 8, 109–113, (in Russian).

Kavaya, M., and R. T. Menzies (1985) "Lidar Aerosol Backscatter Measurement: Systematic Modeling, and Calibration Error Considerations," *Appl. Opt.*, **24**, 3444–3453.

Keckhut, P., M. E. Gelman, J. D. Wild, F. Tissot, A. J. Miller, A. Hauchecorne, M. L. Chanin, E. F. Fishbein, J. Gille, J. M. Russell III, and F. W. Taylor (1996) "Semidiurnal and Diurnal Temperature Tides (30–55 km): Climatology and Effect on UARS-lidar Data Comparisons," *J. Geoph. Res.*, **101**, 10299–10310.

Keckhut, P., A. Hauchecorne, and M. L. Chanin (1995) "Midlatitude Long-term Variability of the Middle Atmosphere: Trends and Cyclic and Episodic Changes," *J. Geoph. Res.*, **100**, 18887–18897.

Keckhut, P., A. Hauchecorne, and M. L. Chanin (1993) "A Critical Review of the Database Acquired for the Long-Term Surveillance of the Middle Atmosphere of the French Rayleigh Lidars," *J. Atm. Ocean. Techn.*, **10**, 850–867.

Keckhut, P., M. Chanin, and A. Hauchecorne (1990) "Stratosphere Temperature Measurement Using Raman Lidar," *Appl. Opt.*, **29**, 5182–5186.

Keen, J. P. (1963) "Fatigue and Saturation in Photomultipliers," *Rev. Sci. Instrum.*, **34**, 1220–1222.

Kempfer, U., W. Carnuth, R. Lotz, and T. Trickl (1994) "A Wide-Range Ultraviolet Lidar System for Tropospheric Ozone Measurements: Development and Application," *Rev. Sci. Instrum.*, **65**, 3145–3164.

Kenney, J. F., and E. S. Keeping (1951) *Mathematics of Statistics: Part Two,* 2nd ed., D. Van Nostrand Company, Inc., NY.

Kent, G., and W. Keenliside (1975) "A Review of Laser Radar Measurements of Atmospheric Properties," *J. Atmos. Terr. Phys.*, **32**, 917–943.

Kent, G. S., and G. M. Hansen (1999) "Scanning Lidar with a Coupled Radar Safety System," *Appl. Opt.*, **38**, 6383–6387.

Kent, G. S., and G. M. Hansen (1998) "Multiwavelength Lidar Observations of the Decay Phase of the Stratospheric Aerosol Layer Produced by the Eruption of Mount Pinatubo in June 1991," *Appl. Opt.*, **37**, 3861–3872.

Kerscher, M., W. Drichbaumer, M. Noormohammadian, and U. G. Oppel (1995) "Polarized Multiply Scattered Lidar Signals," *Opt. Rev.*, **12**, 304–307.

Khrgian, A. (1963) *Cloud Physics*, Israel Program for Scientific Translation, U. S. Dept. of Commerce, Washington, D. C. .

Killinger, D. K., and N. Menyuk (1981) "Effect of Turbulence Induced Correlation on Laser Remote Sensing Errors," *Appl. Phys. Lett.*, **38**, 968–970.

Kjelaas, A., and G. Ochs (1974) "Study of Divergence in the Boundary Layer Using Optical Propagation Techniques," *J. Appl. Meteor.*, **13**, 242–248.

Klett, J. D. (1981) "Stable Analytical Inversion Solution for Processing Lidar Returns," *Appl. Opt.*, **20**, 211–220.

Klett, J. D. (1985) "Lidar Inversion with Variable Backscatter/extinction Ratios," *Appl. Opt.* **24**, 1638–1643.

Kmetec, J., T. Kubo, T. Kane, and C. Grund (1994) "Laser Performance of Diode-Pumped Thulium-Doped $Y_3Al_5O_{12}$, $(YLu)_3Al_5O_{12}$ and $Lu_3Al_5O_{12}$ Crystals," *Opt. Lett.*, **19**, 186–188.

Kneizys, F. X., E. P. Shettle, L. W. Abreu, J. H. Chetwynd, G. P. Anderson, W. O. Gallery, J. E. A. Selby, and S. A. Clough (1988) "Users Guide to LOWTRAN 7," AFGL-TR-88-0177, (NTIS AD A206773).

Knoll, G. (1979) *Radiation Detection and Measurement*, John Wiley & Sons, New York.

Kobayashi, T., T. Taira, T. Yamamoto, A. Hori, and T. Kitada (1992) "Rotational Raman Lidar for Lower Tropospheric Temperature Profiling," *NASA Conference Publication 3158, Sixteenth International Laser Radar Conference*, Cambridge, Massachusetts, 205–208.

Kohl, R. H. (1978) "Discussion of the Interpretation Problem Encountered in Single-Wavelength Lidar Transmissometer," *J. Appl. Meteorol.*, **17**, 1034.

Kolev, I., O. Parpanov, and B. Kaprielov (1988) "Lidar Determination of Winds by Aerosol Inhomogeneities: Motion Velocity in the Planetary Boundary Layer," *Appl. Opt.*, **27**, 2524–2531.

Kolev, I., O. Parvanov, B. Kaprielov, V. Mitev, V. Simeonov, and I. Grigorov (1992) "Lidar Observation of Aerosol and Temperature Stratification over Urban Area During the Formation of a Stable Atmospheric PBL," *NASA Conference Publication 3158, Sixteenth International Laser Radar Conference*, Cambridge, Massachusetts, 213–216.

Kondo, J., O. Kanechika, and N. Yasuda (1978) "Heat and Momentum Transfers under Strong Stability in the Atmospheric Surface Layer," *J. Atmos. Sci.*, **35**, 1012–1031.

Korb, C. L., B. M. Gentry, and C. Y. Weng (1992) "Edge Technique: Theory and Application to the Lidar Measurement of Wind," *Appl. Opt.*, **31**, 4202–4213.

Korb, C. L., B. M. Gentry, and S. X. Li (1997) "Edge Technique Doppler Lidar Wind Measurements with High Vertical Resolution," *Appl. Opt.* **36**, 5976–5983.

Korb, C. L., B. M. Gentry, S. X. Li, and C. Flesia (1998) "Theory of the Double-Edge Technique for Doppler Lidar Wind Measurement," *Appl. Opt.*, **38**, 3097–3104.

Korb, C. L., and C. Weng (1982) "A Theoretical Study of a Two-Wavelength Lidar Technique for the Measurement of Atmospheric Temperature Profiles," *Appl. Opt.*, **21**, 1346–1355.

Koschmider, H. (1924) "Theorie der Horisontalen Sichtweite," *Beitr. Phys. Freien Atmos.*, **12**, 33–53, 171–181, (in German).

Kostko, O., S. Khmelevtsov, and G. Kalyagina (1992) "Temperature Measurements in the Upper Stratosphere with Lidar," *Bull. Acad. Sci. U. S. S. R*, FAO, **28**, 506–511.

Kovalev, V. A., R. A. Susott, and Wei Min Hao (2003) "Inversion of Lidar Signals from Dense Smokes Contaminated with Multiple Scattering," in *Laser Radar Technology for Remote Sensing*, Proceedings of SPIE 5240, C. Werner ed., 214–222.

Kovalev, V. A. (1973) "New Method of Lidar Return Processing," *Proceedings of the Main Geophysical Observatory*, E . A. Poljakova, ed. (Main Geophysical Observatory, Leningrad, 1973), No. 312, 128–133, (in Russian).

Kovalev, V. A. (1973a) "A Method for Determining Atmospheric Transparency," U. S. S. R. Pat., SU 390401, U.S.S.R Bulletin No. 30, (in Russian).

Kovalev, V. A. (1988) "*Visibility in the Atmosphere and its Determination*," 1st ed. (Gidrometeoizdat, Leningrad), 216, (in Russian).

Kovalev, V. A. (1991) "On the Extrapolation of Lidar Data in Determining the Atmospheric Visibility along Slant Paths," *Atmos Opt.*, No. 4, 089–094, (New York, Amer. Instit. of Phys.).

Kovalev, V. A. (1993) "Lidar Measurement of the Vertical Aerosol Extinction Profiles with Range Dependent Backscatter-to-Extinction Ratios," *Appl. Opt.*, **32**, 6053–6065.

Kovalev, V. A. (1995) "Sensitivity of the Lidar Equation Solution to Errors in the Aerosol Backscatter-to-Extinction Ratio: Influence of a Monotonic Change in the Aerosol Extinction Coefficient," *Appl. Opt.* **34**, 3457–3462.

Kovalev, V. A. (2003) "Stable Near-End Solution of the Lidar Equation for Clear Atmospheres," *Appl. Opt.*, **42**, 585–591.

Kovalev, V. A. (2003a) "Near-end Solution for Lidar Signals Containing a Multiple Scattering Component," *Appl. Opt.*, **42**, 7215–7224.

Kovalev, V. A. (2002) "Analytical Differentiation of the Differential-Absorption-Lidar Data Distorted by Noise," *Appl. Opt.* **41**, 1156–1162.

Kovalev, V. A., and V. M. Ignatenko (1980) "The Influence of Phase Function Fluctuations on the Accuracy of the Atmospheric Transparency Determination with Lidar," *Proceedings of the Main Geophysical Observatory*, G. P. Gushchin ed. (Main Geophysical Observatory, Leningrad, 1980), No. 445, 63–68, (in Russian).

Kovalev, V. A., and V. M. Ignatenko (1985) "Method for Determining Atmospheric Transparency," U. S. S. R. Pat., SU 1163217, Bull. No. 23 (in Russian).

Kovalev, V. A., G. N. Baldenkov, and V. I. Kozintsev (1987) "Relationship Between Total Scattering and Backscattering in the Atmosphere," *Bull. Acad. Sci. U. S. S. R.*, Geoph. Ser., No. 23, 611–615, (in Russian).

Kovalev, V. A., E. E. Rybakov, and V. M. Ignatenko (1991) "Retrieving the Profile of Attenuation Factor from Data of One-Angle Lidar Sensing," *Atmos. Opt.*, **4**, 588–592, (in Russian).

Kovalev, V. A., E. E. Rybakov, and V. M. Ignatenko (1990) "Extrapolation of the Lidar Measurement Data for Slant Visibility Determination," *Proceedings of the Fifteenth International Laser Radar Conference* (Institute of Atmospheric Optics, Tomsk, U.S.S.R, 1990), Part 1, 282–283.

Kovalev, V. A., and H. Moosmüller (1992) "A Variable Phase Function Approach for the Inversion of Lidar Return Signals," *Proceedings of the Sixteenth International Laser Radar Conference*, M. P. McCormick, ed. (MIT, Cambridge., Mass., 1992), Part 2, 607–610.

Kovalev, V. A., and H. Moosmüller (1994) "Distortion of Particulate Extinction Profiles Measured with Lidar in a Two-Component Atmosphere," *Appl. Opt.*, **33**, 6499–6507.

Kovalev, V. A., and J. L. McElroy (1994) "Differential Absorption Lidar Measurement of Vertical Ozone Profiles in the Troposphere that Contains Aerosol Layers with Strong Backscattering Gradients: a Simplified Version," *Appl. Opt.*, **33**, 8393–8401.

Kovalev, V. A., J. L. McElroy, and W. E. Eichinger (1996a) "Lidar-inversion Technique for Monitoring and Mapping Localized Aerosol Plumes and Thin Clouds," *Proceedings of International Soc. Opt. Eng (SPIE).*, A. J. Sedlacek, ed. (Denver, Co., 1996), **2833**, 251–258.

Kovalev, V. A., and M. P. Bristow (1996) "Compensational Three-Wavelength Differential-absorption Lidar for Reducing the Influence of Differential Scattering on Ozone-Concentration Measurements," *Appl. Opt.* **35**, 4790–4797.

Kovalev, V. A., M. P. Bristow, and J. M. McElroy (1996) "Nonlinear-Approximation Technique for Determining Vertical Ozone-Concentration Profiles with a Differential-Absorption Lidar," *Appl. Opt.* **35**, 4803–4811.

Kovalev, V. A., R. A. Susott, W. E. Eichinger, and W. M. Hao (2002) "Application of a Stable Near-End Solution to Determine the Extinction Coefficient of Smoke Aerosols from Biomass Fires," in *Proceedings of the Twenty First International Laser Radar Conference*, L. R. Bissonnette, G. Roy, and G. Vallee eds. (Quebec, Canada, 2002), Part 1, 247–250.

Kovalev, V. A., M. Pahlow, and M. B. Parlange (2002a) "Elimination of Asymmetry in the Two-angle Lidar-equation Solution for Aerosol Extinction Profiles," *Proceedings of the Twenty-First International Laser Radar Conference*, L. R. Bissonnette, G. Roy, and G. Vallee eds. (Quebec, Canada, 2002), Part 2, pp. 621–624.

Krieger, J. K. (2000) "A Multiple Beam Lidar For Wind Measurement," Ph. D. Thesis, (The University of Iowa, Department of Mechanical Engineering, December 2000).

Kreitzberg, C. (1974) AFCRL Report, AFCRL-TR-74-0253, Air Force Cambridge Research Laboratories, Bedford, MA.

Krekov, G. M., and R. F. Rakhimov (1986) "Discrimination of the Molecular and Aerosol Scattering Components with Two-Wavelength Searching of the Stratosphere," V. E. Zuev ed., in *Remote Sensing in the Atmosphere*, (Nauka, Novosibirsk, 1978), 96–104, (in Russian).

Krekov, G. M., and M. M. Krekova (1998) "Structure of Spaceborne Lidar Signal Reflected by Upper-Level Clouds. II. Optically Inhomogeneous Clouds," *Atm. Oceanic. Opt.* **11**, 51–54.

Krueger, D. A., L. M. Caldwell, R. J. Alvarez II, and C. Y. She (1993) "Self-Consistent Method of Determining Vertical Profiles of Aerosol and Atmospheric Properties Using a High Spectral Resolution Rayleigh-Mie Lidar," *J. Atm. Ocean. Techn.* **10**, 533–545.

Kruse, P. W., L. D. McGlauchlin, and R. B. McQuistan (1963) *Elements of Infrared Technology*, John Wiley & Sons, New York.

Kuettner, J. P., P. A. Hildebrand, and T. L. Clark (1987) "Convection Waves: Observations of Gravity Wave Systems over Convectively Active Boundary Layers," *Q. J. R. Meteorol. Soc.*, **113**, 445–467.

Kunkel, K. E., E. W. Eloranta, and S. T. Shipley (1977) "Lidar Observations of the Convective Boundary Layer," *J. Appl. Meteor.*, **16**, 1306–1311.

Kunkel, K. E., E. W. Eloranta, and J. A. Weinman (1980) "Remote Determination of Winds, Turbulence Spectra, and Energy Dissipation Rates in the Boundary Layer from Lidar Measurements," *J. Atm. Sci.*, **37**, 978–985.

Kunkel, K. E., and D. L. Walters (1982) "Intermittent Turbulence in Measurements of the Temperature Structure Parameter under Very Stable Conditions," *Bound. - Layer Meteor.*, **22**, 49–60.

Kunz, G. J. (1999) "Two-Wavelength Lidar Inversion Algorithm," *Appl. Opt.*, **38**, 1015–1020.

Kunz, G. J. (1996) "Transmission as an Input Boundary Value for an Analytical Solution of a Single-Scatter Lidar Equation" *Appl. Opt.*, **35**, 3255–3260.

Kunz, G. J., and G. Leeuw (1993) "Inversion of Lidar Signals with the Slope Method," *Appl. Opt.*, **32**, 3249–3256.

Kuze, H., H. Kinjo, Y. Sakurada, and N. Takeuchi (1998) "Field of View Dependence of Lidar Signals by Use of Newtonian and Cassegrainian Telescopes," *Appl. Opt.*, **37**, 3128–3132.

Kwon, S. A., Y. Iwasaka, and T. Shibata (1994) "Depolarization Ratio of Air Molecules in the Free Troposphere—Lidar Measurement in Spring 1994," *J. Geomag. Geoelect.*, **48**, 415–419.

Kwon, S., Y. Iwasaka, T. Shibata, and T. Sakai (1997) "Vertical Distribution of Atmospheric Particulates and Water Vapor Densities in the Free Troposphere: Lidar Measurements in Spring and Summer in Nagoya, Japan," *Atmos. Envir.*, **31**, 1459–1465.

Lacaita, A., F. Zappa, S. Cova, and P. Lovati (1996) "Single Photon Detection Beyond 1 μm: Performance of Commercially Available InGaAs/InP Detectors," *Appl. Opt*, **35**, 2986–2996.

Land, P. L. (1971) "A Discussion of the Region of Linear Operation of Photomultipliers," *Rev. Sci. Instrum.*, **42**, 420–425.

Larcheveque, G., I. Balin, R. Nessler, P. Quaglia, V. Simeonov, H. van den Bergh, and B. Calpini (2002) "Development of a Multiwavelength Aerosol and Water-Vapor Lidar at the Jungfraujoch Alpine Station (3580 m Above Sea Level) in Switzerland," *Appl. Opt.*, **41**, 2781–2790.

Leblanc, T., I. S. McDermid, A. Hauchecorne, and P. Keckhut (1998) "Evaluation of Optimization of Lidar Temperature Analysis Algorithms Using Simulated Data," *J. Geoph. Res.*, **103**, 6177–6187.

Lee, H. S., G. K. Schwemmer, C. L. Korb, M. Dombrowski, and C. Prasad (1990) "Gated Photomultiplier Response Characterization for DIAL Measurements," *Appl. Opt.*, **29**, 3303–3315.

Leese, J., and C. Novak (1971) "An Automated Technique for Obtaining Cloud Motion from Geosynchronous Satellite Data using Cross Correlation," *J. Appl. Met.*, **10**, 118–132.

Leeuw, G., G. J. Kunz, and C. W. Lamberts (1986) "Humidity Effects on the Backscatter/Extinction Ratio," *Appl. Opt.*, **25**, 3971–3974.

Lehmann, S., V. Wulfmeyer, and J. Boesenberg (1997) "A Time Dependent Attenuator for Dynamic Range Reduction of Lidar Signals," *Appl. Opt.*, **36**, 3469–3474.

Lenschow, D. H., X. S. Li, C. J. Zhu, and B. B. Stankov (1988) "The Stably Stratified Boundary Layer over the Great Plains. Part I: Mean and Turbulence Structure," *Bound.-Lay. Meteor.*, **42**, 95–121.

Lenschow, D. H., S. F. Zhang, and B. B. Stankov (1988) "The Stably Stratified Boundary Layer over the Great Plains. Part II: Horizontal Variations and Spectra," *Bound.-Lay. Meteor.*, **42**, 123–135.

Leonard D. A. (1967) "Observation of the Raman Scattering from the Atmosphere Using a Pulsed Nitrogen Ultraviolet Laser," *Nature*, **216**, 142–143.

Levine, B., and C. Berthea (1984) "Detection of 1.3 micron Photons at 45 Mbits/s," *Electron. Lett*, **20**, 289–294.

Lewis, L. (1976) "Vertical Fog Classifications and Slant Runway Visual Range Ratios," *Proceedings of the Seventh Conference on Aerospace and Aeronautical Meteorology and Remote Sensing from Satellite*, (Melbourne, 1976), 237–241.

Liao, K. H., and R. Gupta (1978) "I_2 Vapor Cell as Narrow Band Optical Filter," *Rev. Sci. Instr.*, **49**, 867–869.

Ligon, D., J. B. Gillespie, and P. Pellegrino (2000) "Aerosol Properties from Spectral Extinction and Backscatter Estimated by an Inverse Monte Carlo Method," *Appl. Opt.*, **39**, 4402–4410.

Lindberg, J. (1975) "The Composition and Optical Absorption Coefficient of Atmospheric Particulate Matter," *Opt. Quantum Electron.*, **7**, 131–139.

Lindberg, J., and J. Gillespie (1977) "Relationship Between Particle Size and Imaginary Reflective Index in Atmospheric Dust," *Appl. Opt.*, **16**, 2628–2630.

Lindberg, J., and L. Laude (1974) "Measurement of the Absorption Coefficient of Atmospheric Dust, " *Appl. Opt.*, **13**, 1923–1927.

Lindberg, J. D., W. J. Lenz, E. M. Measure, and R. Rubio (1984) "Lidar Determinations of Extinctions in Clouds," *Appl. Opt.*, **23**, 2172–2177.

Liou, K. N. (1992) *Radiation and Cloud Processes in the Atmosphere*, Oxford University Press, New York, 487 pp.

Liou, K. N., and H. Lahore (1974) "Laser Sensing of Cloud Composition: a Backscatter Depolarization Technique," *J. Appl. Meteor.*, **13**, 257–263.

Liou, K. N., and R. M. Schotland (1971) "Multiple Scattering and Depolarization from Water Clouds for a Pulsed Lidar System," *J. Atm. Sci.*, 1971.

Liu, Y., W. P. Arnott, and J. Hallett (1999) "Particle Size Distribution Retrieval from Multispectral Optical Depth: Influences of Particle Nonsphericity and Refractive Index," *J. Geoph. Res.*, **104**, 31753–31762.

Liu, Z., N. Sugimoto, and T. Murayama (2002) "Extinction-to-Backscatter Ratio of Asian Dust Observed with High-Spectral-Resolution Lidar and Raman Lidar," *Appl. Opt.*, **41**, 2760–2767.

Liu, Z. S., W. B. Chen, T. L. Zhang, J. W. Hair, and C. Y. She (1997) "An Incoherent Doppler Lidar for Ground-Based Atmospheric Wind Profiling," *Appl. Phys.,* B **64**, 561–566.

Liu, Z. Y., and T. Kobayashi (1996) "Differential Discrimination Technique for Doppler Lidar to Measure Atmospheric Wind and Backscatter Ratio," *Opt. Rev.*, **3**, 47–52.

Lubken, F., W. Hiller, G. Lehmacher, and U. Vonzahn (1994) "Intercomparison of Density and Temperature Profiles Obtained by Lidar, Ionization Gauges, Falling Spheres, Datasondes and Radiosondes During the Dyana Campaign," *J. Atmos. Terr. Phys.*, **56**, 1969–1984.

Luckow F., E. Voss, J. Zeyn, W. Lahmann, C. Weitkamp, and W. Michaelis (1994) "Raman-Shifted KrF Laser Radiation with Low Amplified Spontaneous Emission for a Rotational Raman Daytime-Temperature Lidar," *Opt. Lett.*, **19**, 1049–1051.

Luers, J. (1990) "Estimating the Temperature Error of the Radiosonde Rod Thermistor under Different Conditions," *J. Atmos. Oceanic Tech.*, **7**, 882–895.

Lush, H. J. (1965) "Photomultiplier Linearity," *J. Sci. Instrum.*, **42**, 597–602.

Lyscev, V. E. (1978) "Some Results of Nephelometer Measurements in Clouds and Hazes," *Informative Abstracts on Hydrology and Meteorology Instruments and Observation Methods*, U. S. S. R., Moscow, No. 76, 5–9, (in Russian).

Machol, J. L. (1997) "Comparison of the Pseudorandom Noise Code and Pulsed Direct-Detection Lidars for Atmospheric Probing," *Appl. Opt.*, **36**, 6021–6023.

Macke, A. (1993) "Scattering of Light by Polyhedral Ice Crystals," *Appl. Opt.*, **32**, 2780–2788.

Mahrt, L. (1998) "Stratified Atmospheric Boundary Layers and Breakdown of Models," *J. Theor. Comp. Fluid Dynam.*, **11**, 263–280.

Mahrt, L. (1999) "Stratified Atmospheric Boundary Layer," *Boundary-Layer Meteorol.*, **90**, 375–396.

Mahrt, L., J. Sun, W. Blumen, T. Delany, G. McClean, and S. Oncley (1998) "Nocturnal Boundary-Layer Regimes," *Boundary-Layer Meteorol.*, **88**, 255–278.

Manual of Runway Visual Range Observing and Reporting Practices (1981). Doc 9328, International Civil Aviation Organization, 1995, 84.

Marenco F., V. Santacesaria, A. F. Baus, D. Balis, A. di Sarra, A. Papayannis, and C. Zerefos (1997) "Optical Properties of Tropospheric Aerosols Determined by Lidar and Spectrophotometric Measurements (Photochemical Activity and Solar Ultraviolet Radiation Campaign)," *Appl. Opt.*, **36**, 6875–6886.

Mason, J. (1975) "Lidar Measurement of Temperature: A New Approach," *Appl. Opt.*, **14**, 76–78.

Masonis, J., K. Franke, A. Ansmann, D. Müller, D. Althausen, J. A. Ogden, A. Jefferson, and P. J. Sheridan (2002) "An Intercomparison of Aerosol Light Extinction and 180° Backscatter as Derived Using In Situ Instruments and Raman Lidar During the INDOEX Field Campaign," *J. Geophys Res.* 107(D19), 8014, doi:10.1029/2001JD000035.

Matiss, I., A. Ansmann, D. Althausen, V. Jaenisch, U. Wandinger, D. Müller, Y. Arshinov, S. Bobrovnikov, and I. Serikov (2002) "Relative Humidity Profiling in the Troposphere with a Raman Lidar," *Appl. Opt.*, **41**, 6451–6462.

Matsui, I., H. Kubomura, H. Imoto, and N. Sugimoto (1994) "Eye-Safe Compact Mie Scattering Lidar Using a Diode-Laser-Pumped Nd:YAG Laser for Measuring the Atmospheric Boundary Layer," *Jap. J. Appl. Phys., Part I*, **33**, 6569–6571.

Matsui, I., N. Sugimoto, and Y. Sasano (1990) "Wind Profiling by a Conical-Scanning Time-Correlation Lidar," *Jap. J. Appl. Phys., Part 1*, **29**, 441–444.

Matsumoto, M., and N. Takeuchi (1994) "Effects of Misestimated Far-End Boundary Values on Two Common Lidar Inversion Solutions," *Appl. Opt.*, **33**, 6451–6455.

Max, C. E., and S. Olivier (1997) "Image Improvement from a Sodium-Layer Laser Guide Star Adaptive Optics System," *Science*, **277**, 1649–1652.

Mayor, S., and E. Eloranta (2001) "Two Dimensional Vector Wind Fields from Volume Imagine Lidar Data," *J. Appl. Met.*, **40**, 1331–1346.

McCartney, E. J. (1977) *Optics of Atmosphere. Scattering by Molecules and Particles*, John Wiley & Sons, Inc., New York.

McClatchey, R. A., R. W. Fenn, J. E. A. Selby, F. E. Volz, and J. S. Garing (1972) "Optical Properties of the Atmosphere," 3rd edition, AFCRL-72-0497, (AD A753075).

McDermid, I. S., S. M. Godin, and L. O. Lindqvist (1990) "Ground-Based Laser DIAL System for Long-Term Measurements of Stratospheric Ozone," *Appl. Opt.*, **29**, 3603–3612.

McDermid, I. S., T. D. Walsh, A. Deslis, and M. L. White (1995) "Optical Systems Design for a Stratospheric Lidar System" *Appl. Opt.*, **34**, 6201–6210.

McDermid, I. S., S. M. Godin, and D. T. Walsh (1989) "Lidar Measurements of Stratospheric Ozone, Intercomparisons and Validation," *Appl. Opt.*, **28**, 4914–4923.

McDonald, A. J., L. Thomas, and D. P. Wareing (1998) "Night-to-Night Changes in the Characteristics of Gravity Waves at Stratospheric and Lower-Mesospheric Heights," *Ann. Geoph.—Atmospheres Hydrospheres and Space Sciences*, **16**, 229–237.

McElroy, J. L., and T. B. Smith (1986) "Vertical Pollutant Distribution and Boundary Layer Structure Observed by Airborne Lidar near the Complex Southern California Coastline," *Atmos. Environ.*, **20**, 1555–1566.

McElroy, J., and T. Smith (1991) "Lidar Descriptions of Mixed Layer Thickness Characteristics in a Complex Terrain/Coastal Environment," *J. Appl. Meteor.*, **30**, 585–597.

McGill, M. J., and W. Skinner (1997) "Multiple Fabry-Perot Interferometers in an Incoherent Doppler Lidar," *Opt. Eng.*, **36**, 139–145.

McGill, M. J., W. Skinner, and T. Irgang (1997) "Analysis Techniques for the Recovery of Winds and Backscatter Coefficients from a Multiple Channel Incoherent Doppler Lidar," *Appl. Opt.*, **36**, 1253–1268.

McGill, M. J., W. R. Skinner, and T. D. Irgang (1997) "Validation of Wind Profiles Measured Using Incoherent Doppler Lidar," *Appl. Opt.*, **36**, 1928–1939.

McIntyre, R. J. (1966) "Multiplication Noise in Uniform Avalanche Diodes," *IEEE Trans. On Elect. Dev.*, **13**, 123–148.

McKay, J. A. (1998) "Modeling of Direct Detection Doppler Wind Lidar. I. The Edge Technique," *Appl. Opt.*, **37**, 6480–6486.

McKay, J. A. (1998) "Modeling of Direct Detection Doppler Wind Lidar. II. The Fringe Imaging Technique," *Appl. Opt.*, **37**, 6487–6493.

McKay, J. A., and D. Rees (2000) "Space-Based Doppler Wind Lidar: Modeling of Edge Detection and Fringe Imaging Doppler Analyzers," *Adv. Space Res.*, **26**, 883–891.

McNider, R. T., D. E. England, M. J. Friedman, and X. Shi (1995) "Predictability of the Stable Atmospheric Boundary Layer," *J. Atmos. Sci.*, **52**, 1602–1623.

Mead, M. L. (1981) "Instrumentation Aspects of Photon Counting Applied to Photometry," *J. Phys. E: Sci Instrum.*, **14**, 909–918.

Measures, R. M. (1984) *Laser Remote Sensing*, John Wiley & Sons, Inc., New York, 510.

Megie, G. J., G. Ancellet, and J. Pelon (1985) "Lidar Measurements of Ozone Vertical Profiles," *Appl. Opt.*, **24**, 3454–3463.

Megie, G., F. Bos, J. E. Blamont, and M. L. Chanin (1978) "Simultaneous Nighttime Lidar Measurements of Atmospheric Sodium and Potassium," *Planet. Space Sci.*, **26**, 27–35.

Megie, G. J., and R. T. Menzies (1980) "Complementarity of UV and IR Differential Absorption Lidar for Global Measurements of Atmospheric Species," *Appl. Opt.*, **19**, 1173–1183.

Melas, D. (1993) "Similarity Methods to Derive Turbulence Quantities and Mixed Layer Depth from SODAR Measurements in the Convective Boundary Layer: A Review," *Appl. Phys.*, B **57**, 11–17.

Melfi S. H., J. D. Lawrence, and M. P. McCormick (1969) "Observation of Raman Scattering by Water-Vapor in the Atmosphere," *Appl. Phys. Lett.*, **15**, 295–297.

Melfi S. H. (1972) "Remote Measurements of the Atmosphere Using Raman Scattering," *Appl. Opt.*, **11**, 1605–1611.

Melfi, S. H., J. D. Spinhire, S. H. Chou, and S. P. Palm (1985) "Lidar Observations of Vertically Organized Convection in the Planetary Boundary Layer over the Ocean," *J. Clim. Appl. Meteor.*, **24**, 806–821.

Melfi S. H., K. Evans, J. Li, D. Whiteman, R. Ferrare, and G. Schwemmer (1997) "Observation of Raman Scattering by Cloud Droplets in the Atmosphere," *Appl. Opt.*, **36**, 3551–3559.

Melfi S. H., J. Li, K. Evans, D. Whiteman, R. Ferrare, and G. Schwemmer (1996) "Observation of Anomalous Raman Scattering Associated with Clouds," *Advances in Atmospheric Remote Sensing with Lidar*, Ansmann, Neuber, Rairoux and Wandinger, eds., 95–98.

Menut, L., C. Flamant, J. Pelon, and P. H. Flamant (1999) "Urban Boundary-Layer Height Determination from Lidar Measurements over the Paris Area," *Appl. Opt.*, **38**, 945–954.

Menyuk, N., D. K. Killinger, and C. R. Menyuk (1982) "Limitations of Signal Averaging due to Temporal Correlation in Laser Remote-Sensing Technique," *Appl. Opt.*, **21**, 3377–3383.

Menyuk, N., and D. K. Killinger (1983) "Assessment of Relative Error Sources in IR DIAL Measurement Accuracy," *Appl. Opt.*, **22**, 2690–2698.

Menyuk, N., D. K. Killinger, and W. E. Defeo (1983) "Experimental Comparison of Heterodyne and Direct Detection for Pulsed Differential Absorption CO_2 Lidar," *Appl. Opt.*, **22**, 682–690.

Meriwether J., P. Dao, R. Mcnutt, and W. Klemetti (1994) "Rayleigh Lidar Observations of Mesosphere Temperature Structure," *J. Geoph. Res.-Atmos.*, **99**, 16973–16987.

Mie, G. (1908) "A Contribution to the Optics of Turbid Media, Especially Colloidal Metallic Suspensions," *Ann. Phys.*, **25**, 377–445.

Miles, N., J. Verlinde, and E. Clothiaux (2000) "Cloud Droplet Size Distributions in Low Level Stratiform Clouds," *J. Atmos. Sci.*, **57**, 295–311.

Milton, M. J. T., and P. T. Woods (1987) "Pulse Averaging Methods for a Laser Remote Monitoring System Using Atmospheric Backscatter," *Appl. Opt.*, **13**, 2598–2603.

Mishchenko, M. I., D. J. Wieland, and B. E. Carlson (1997) "T-Matrix Computations of Zenith Enhanced Lidar Backscatter from Horizontally Oriented Ice Plates," *Geoph. Res. Lett.*, **24**, 771–774.

Mischenko, M. I., and K. Sassen (1998) "Depolarization of Lidar Returns by Small Ice Crystals: an Application to Contrails," *Geoph. Res. Lett.*, **25**, 309–312.

Mitchenkov V. M., and A. V. Solodukhin (1990) "Three Wavelength Lidar with a Single Laser Source for Probing the Optical Tropospheric Parameters," *Proceedings of the Fifteenth International Laser Radar Conference* (Institute of Atmospheric Optics, Tomsk, 1990) Pt. **2**, 345–346.

Mitev, V. M., I. V. Grigorov, and V. B. Simeonov (1992) "Lidar Measurements of Atmospheric Aerosol Extinction Profiles: A Comparison between two Techniques—Klett Inversion and Pure Rotational Raman Scattering Methods," *Appl. Opt.*, **31**, 6469–6474.

Mitev, V., V. Simeonov, L. Ivanov, Y. Arshinov, and S. Bobrovnikov (1985) "Raman Lidar Measurements of Atmospheric Temperature Profiles," *Bulgar. J. Phys.*, **12**, 429–433.

Mitra, S. N. (1949) *Proc. Inst. Electr. Eng., Part 3*, **96**, 441.

Molina, L. T., and M. J. Molina (1986) "Absolute Absorption Cross Sections of Ozone in the 185- to 350-nm Wavelength Range," *J. Geoph. Res.*, **91**, 14501–14508.

Moosmüller, H., D. Diebel, D. H. Bundy, M. P. Bristow, C. M. Edmonds, R. M. Turner, V. A. Kovalev, R. P. Haas, and J. L. McElroy (1991) "The U. S. EPA Airborne UV-DIAL System," *Optical Remote Sensing of the Atmosphere Technical Digest, 1991* (Optical Society of America, Washington, D. C., 1991), **18**, 253–255.

Moosmüller, H., R. J. Alvarez II, R. M. Jorgensen, C. M. Edmonds, D. H. Bundy, and J. L. McElroy (1993) "An Airborne UV-DIAL System for Ozone Measurements: Field Use and Verification," *Proceedings of the Optical Sensing for Environmental Monitoring*, J. L. Hudson, ed. (Atlanta, Georgia, 1993), **SP-89**, 413–422.

Moosmüller, H., and T. D. Wilkerson (1997) "Combined Raman Elastic-Backscatter Lidar Method for the Measurement of Backscatter Ratios," *Appl. Opt.*, **36**, 5144–5147.

Mulders, J. M. (1984) "Algorithm for Inverting Lidar Signals: Comment," *Appl. Opt.*, **23**, 2855–2857.

Müller, H., and H. Quenzel (1985) "Information Content of Multispectral Lidar Measurements with Respect to the Aerosol Size Distribution," *Appl. Opt.*, **24**, 648–654.

Müller, D., K. Franke, F. Wagner, D. Althausen, A. Ansmann, and J. Heintzenberg (2001) "Vertical Profiling of Optical and Physical Particle Properties over the Tropical Indian Ocean with Six-Wavelength Lidar. 2. Case Studies," *J. Geoph. Res.*, **106**, 28577–28595.

Müller, D., K. Franke, A. Ansmann, D. Althausen, and F. Wagner (2003) "Indo-Asian Pollution During INDOEX: Microphysical Particle Properties and Single-Scattering Albedo Inferred from Multiwavelength Lidar Observations," *J. Geophys Res.* 108(D19), 4600, doi:10.1029/2003JD003538.

Müller, D., U. Wandinger, D. Althausen, I. Mattis, and A. Ansmann (1998) "Retrieval of Physical Particle Properties from Lidar Observations of Extinction and Backscatter at Multiple Wavelengths," *Appl. Opt.*, **37**, 2260–2263.

Müller, D., U. Wandinger, D. Althausen, and M. Fiebig (2001a) "Comprehensive Particle Characterization from Three-Wavelength Raman-Lidar Observations: Case Study," *Appl. Opt.*, **40**, 4863–4869.

Müller, D., U. Wandinger, and A. Ansmann (1999a) "Microphysical Particle Parameters from Extinction and Backscatter Lidar Data by Inversion with Regularization: Simulation," *Appl. Opt.*, **38**, 2358–2368.

Müller, D., U. Wandinger, and A. Ansmann (1999) "Microphysical Particle Parameters from Extinction and Backscatter Lidar Data by Inversion with Regularization: Theory," *Appl. Opt.*, **38**, 2346–2357.

Müller, D., F. Wagner, U. Wandinger, A. Ansmann, M. Wendisch, D. Althausen, and W. von Hoyningen-Huene (2000) "Microphysical Particle Parameters from Extinction and Backscatter Lidar Data by Inversion with Regularization: Experiment," *Appl. Opt.*, **39**, 1879–1892.

Muller, J. W. (1973) "Dead-Time Problems," *Nucl. Instrum. Methods*, **112**, 47–57.

Mulnonen, K., K. Lumme, J. Peltoniemi, and W. Irvine (1989) "Light Scattering by Randomly Oriented Crystals," *Appl. Opt.*, **28**, 3051–3060.

Murayama, T., M. Furushima, A. Oda, and N. Twasaka (1996) "Depolarization Ratio Measurements in the Atmospheric Boundary Layer by Lidar in Tokyo," *J. Meteorol. Soc. Japan*, **74**, 571–578.

Murray, E., D. Powell, and J. Van der Laan (1980) "Measurement of Average Atmospheric Temperature Using a CO2 Lidar" *Appl. Opt.*, **19**, 1794–1801.

Nagasawa, C., M. Abo, H. Yamamoto, and O. Uchino (1990) "Random Modulation CW Lidar Using New Random Sequence," *Appl. Opt.*, **29**, 1466–1470.

Namboothiri, S. P., T. Tsuda, M. Tsutsumi, T. Nakamura, C. Nagawawa, and M. Abo (1996) "Simultaneous Observations of Mesospheric Gravity Waves with the MU Radar and a Sodium Lidar," *J. Geoph. Res.*, **101**, 4057–4063.

Nappo, C. (1991) "Sporadic Breakdowns of Stability in the PBL over Simple and Complex Terrain," *Bound.-Layer Meteor.*, **54**, 69–87.

National Oceanic and Atmospheric Administration, *US Standard Atmosphere,* Washington, D. C., Natl. Aeron. Space Admin., 1976.

Nedeljkovic, D., A. Hauchecorne, and M. Chanin (1993) "Rotational Raman Lidar to Measure the Atmospheric Temperature from the Ground to 30 km," *IEEE Transactions on Geoscience and Remote Sensing*, **31**, 90–101.

Neff, W. D., and C. W. King (1987) "Observations of Complex Terrain Flows Using Acoustic Sounders: Experiments, Topography and Winds," *Bound.-Layer Meteor.*, **40**, 363–392.

Nelson, E., R. B. Stull, and E. W. Eloranta (1989) "A Prognostic Relationship for Entrainment Zone Thickness," *J. Appl. Meteor.*, **28**, 885–903.

Nicolas, F., L. R. Bisonnette, and P. H. Flamant (1997) "Lidar Effective Multiple-Scattering Coefficients in Cirrus Clouds," *Appl. Opt.*, **36**, 3458–3468.

Nieuwstadt, F. T. M. (1984a) "The Turbulent Structure of the Stable, Nocturnal Boundary Layer," *J. Atmos. Sci.*, **41**, 2202–2216.

Nieuwstadt, F. T. M. (1984b) "Some Aspects of the Turbulent Stable Boundary Layer," *Bound.-Layer Meteor.*, **30**, 31–55.

Noonkester, V., D. Jensen, J. Richter, W. Viezee, and R. Collis (1974) "Concurrent FM-CW Radar and Lidar Observations of the Boundary Layer, *J. Appl. Meteor.*, **13**, 249–256.

Omote K. (1990) "Dead-Time Effects in Photon Counting Distributions," *Nucl. Instrum. Methods*, **293**, 582–588.

Oppenheim A., and R. Schafer (1989) *Discrete-Time Signal Processing*, Prentice-Hall, New Jersey, 879 pp.

Orgill, M. M., and R. I. Schreck (1985) "An Overview of the ASCOT Multi-Laboratory Field Experiments in Relation to Drainage Winds and Ambient Flow," *Bull. Amer. Meteor. Soc.*, **66**, 1263–1277.

Overbeck, J. A., M. S. Salisbury, and R. Richmond (1996) "Airborne Lidar System Profiles Wind Fields," *Laser Focus World*, **32**(**4**), 89–92.

Owens, P., J. Rarity, P. Tapster, and P. Townsend (1994) "Photon Counting with Passively Quenched Germanium Avalanche Diodes," *Appl. Opt*, **33**, 6895–6901.

Pace, G. M. Cacciani, A. di Sarra, G. Fiocco, and D. Fua (2003) "Lidar Observations of Equatorial Cirrus Clouds at Mahe Seychelles," *J. Geoph. Res.*, **108**, 4236, doi:10.1029/2002JD002710.

Pahlow, M. (2002) "Atmospheric Boundary Layer Dynamics and Inversion Technologies to Obtain Extinction Coefficient Profiles in the Atmosphere from Elastic Lidar," Dissertation (Johns Hopkins University, Baltimore, USA, 210, 2002).

Pal, S. R., and A. I. Carswell (1973) "Polarization Properties of Lidar Backscattering from Clouds,", *Appl. Opt.*, **12**, 1530–1535.

Pal, S. R., and A. I. Carswell (1978) "Polarization Properties of Lidar Scattering from Clouds at 347 nm and 694 nm," *Appl. Opt.*, **17**, 2321–2328.

Pal, S. R., W. Steinbrecht, and A. I. Carswell (1992) "Automated Method for Lidar Determination of Cloud-Base Height and Vertical Extent," *Appl. Opt.*, **31**, 1488–1494.

Palm, S. P., D. Hagan, G. Schwemmer, and S. H. Melfi (1998) "Inference of Marine Atmospheric Boundary Layer Moisture and Temperature Structure Using Airborne Lidar and Infrared Radiometer Data," *J. Appl. Meteorol.*, **37**, 308–324.

Palm, S. P., and S. H. Melfi (1992) "On the Existence of Convectively Produced Gravity Waves," *NASA Conference Publication 3158, Sixteenth International Laser Radar Conference*, Cambridge, Massachusetts, 389–392.

Pandithurai, G., P. C. S. Devara, P. E. Raj, and S. Sharma (1996) "Aerosol Size Distribution and Refractive Index from Bistatic Lidar Angular Scattering Measurements in the Surface Layer," *Remote Sens. Environ.*, **56**, 87–96.

Papayannis, A., G. Ancellet, J. Pelon, and G. Meggie (1990) "Multiwavelength Lidar for Ozone Measurements in the Troposphere and the Lower Stratosphere," *Appl. Opt.*, **29**, 467–476.

Papen, G. C., and D. Treyer (1998) "Comparison of an Fe Boltzmann Temperature Lidar with a Na Narrow-Band Lidar," *Appl. Opt.*, **37**, 8477–8481.

Papen, G. C., C. S. Gardner, and W. M. Pfenninger (1995) "Analysis of a Potassium Lidar System for Upper-Atmospheric Wind-Temperature Measurements," *Appl. Opt.*, **34**, 6950–6958.

Papen, G. C., W. M. Pfenninger, and D. M. Simonich (1995) "Sensitivity Analysis of Na Narrowband Wind-Temperature Lidar Systems," *Appl. Opt.*, **34**, 480–498.

Parameswaran, K., K. O. Rose, and B. V. K. Murthy (1991) "Relationship Between Backscattering and Extinction Coefficients of Aerosols with Application to Turbid Atmosphere," *Appl. Opt.*, **30**, 3059–3071.

Park, Y. J., S. W. Dho, and H. J. Kong (1997) "Deconvolution of Long-Pulse Lidar Signals with Matrix Formulation," *Appl. Opt.*, **36**, 5158–5161.

Park, Y. J., and H. J. Kong (1997) "Iterative Deconvolution Algorithm for Improving Range Resolution of a CO_2-Laser Based Lidar System," *Opt. Rev.*, **4**, 605–609.

Parker, A. (1997) "NASA Study of Wind Patterns May Save Airports Time, Money," Virginian Pilot (Norfolk, Va.), pp. D1–D2, March 21.

Parvanov, O., I. Kolev, and B. Kaprielov (1998) "Lidar Measurements of Wind Velocity Profiles in the Low Troposphere," *19th International Laser Radar Conference*, Annapolis, Maryland.

Patterson, E. M., D. W. Roberts, and G. G. Gimmestad (1989) "Initial Measurements Using a 1.54 μm Eyesafe Raman Shifted Lidar," *Appl. Opt.*, **28**, 4978–4980.

Pelon, J., and G. Megie (1982) "Ozone Monitoring in the Troposphere and Lower Stratosphere: Evaluation and Operation of a Ground-Based Lidar Station," *J. Geoph. Res.*, **87**, 4947–4955.

Penndorf, R. (1957) "Tables of the Refractive Index for Standard Air and the Rayleigh Scattering Coefficient for the Spectral Region between 0.2 and 20 Microns and Their Application to Atmospheric Optics, *J. Opt. Soc. Amer.*, **47**, 176–182.

Penney, C., R. S. Peters, and M. Lapp (1974) "Absolute Rotational Raman Cross Sections for N_2, O_2, and CO_2," *J. Opt. Soc. Am.*, **64**, 712–716.

Petri, K., A. Salik, and J. Cooney (1982) "Variable Wavelength Solar Blind Raman Lidar for Remote Measurement of Atmospheric Water Vapor Concentration and Temperature," *Appl. Opt.*, **21**, 1212–1218.

Pettifer, R. E. W. (1975) "Signal Induced Noise in Lidar Measurements," *J. Atm. Terr. Phys.*, **37**, 669–673.

Photomultiplier Handbook, Burle Industries, Lancaster, PA, 1980, 177 pp.

Piironen, A. K., "Analysis of Volume Imaging Lidar Signals," Ph. D. Thesis, University of Joensuu, Department of Physics, Vaisala Laboratory, Finland, 1994, available at Univ. of Wis. Website.

Piironen, A., and E. W. Eloranta (1995) "Convective Boundary Layer Mean Depths, Cloud Base Altitudes, Cloud Top Altitudes, Cloud Coverages, and Cloud Shadows Obtained From Volume Imaging Lidar Data," *J. Geoph. Res.*, **100**, 25569–25576.

Piironen, A., and E. W. Eloranta (1995) "An Accuracy Analysis of the Wind Profiles Calculated from Volume Imaging Lidar Data," *J. Geoph. Res.*, **100**, 25559–25567.

Piironen, P., and E. W. Eloranta (1994) "Demonstration of a High-Spectral-Resolution Lidar Based on a Iodine Absorption Filter". *Optics Letters*, **19**, 234–236.

Ping, L. N., W. D. Neff, and J. C. Kaimal (1983) "Wave and Turbulence Structure in a Disturbed Nocturnal Inversion," *Bound.-Layer Meteor.*, **26**, 141–155.

Pinnick, R. G., S. G. Jennings, P. Chylek, C. Ham, and W. T. Grandy (1983) "Backscatter and Extinction in Water Clouds," *J. Geoph. Res.*, **88**, 6787–6796.

Pitz, E. (1979) "Nonlinearity Measurements in Photometric Devices," *Appl. Opt.*, **18**, 1360–1362.

Plane, J., R. Cox, and R. Rollason (1999) "Metallic Layers in the Mesopause and Lower Thermosphere Region," *Adv. Space Res.*, **24**, 1559–1570.

Platt, C. M. R. (1973) "Lidar and Radiometric Observations of Cirrus Clouds," *J. Atm. Sci.*, **30**, 1191–1204.

Platt, C. M. R. (1979) "Remote Sounding of High Clouds: I. Calculation of Visible and Infrared Optical Properties from Lidar and Radiometer Measurements," *J. Appl. Meteorol.*, **18**, 1130–1143.

Platt, C. M. R., and T. Takashima (1987) "Retrieval of Water Cloud Properties from Carbon Dioxide Lidar Soundings," *Appl. Opt.*, **26**, 7, 1257–1263.

Pollard, B., S. Khanna, S. Frasier, J. Wyngaard, D. Thomson, and R. McIntosh (2000) "Local Structure of the Convective Boundary Layer from a Volume Imagine Radar," *J. Atmos, Sci.*, **57**, 2281–2296.

Post, M. J. (1996) "A Graphical Technique for Retrieving Size Distribution Parameters from Multiple Measurements: Visualization and Error Analysis," *J. Atmos. Ocean. Technol.*, **13**, 863–873.

Post, M. J., C. J. Grund, A. M. Weickmann, K. R. Healy, and R. J. Willis (1996) "A Comparison of the Mt. Pinatubo and El Chichon Volcanic Events: Lidar Observations at 1.06 and 0.69 μm, *J. Geoph. Res.*, **101**, 3929–3940.

Post, M. J., C. J. Grund, D. Wang, and T. Deshler (1997) "Evolution of Mount Pinatubo's Aerosol Size Distributions over the Continental United States: Two Wavelength Lidar Retrievals and In Situ Measurements," *J. Geophy. Res.*, **102**, 13535–13542.

Potter, J. F. (1987) "Two-Frequency Lidar Inversion Technique," *Appl. Opt.*, **26**, 1250–1256.

Press, W., S. Teukolsky, W. Vettering, and B. Flannery (1992) *Numerical Recipes in Fortran. The Art of Scientific Computing.*, Cambridge University Press, 963 pp.

Proffitt, M. H., and A. O. Langford (1997) "Ground-Based Differential Absorption Lidar System for Day or Night Measurements of Ozone Throughout the Free Atmosphere," *Appl. Opt.*, **36**, 2568–2585.

Pruppacher, H., and J. Klett (1997) *Microphysics of Clouds and Precipitation*, Kluwer Acad. Norwell, Mass.

Quing, P., H. Nakane, Y. Sasano, and S. Kitamura (1989) "Numerical Simulation of the Retrieval of Aerosol Size Distribution from Multiwavelength Laser Radar Measurements," *Appl. Opt.*, **28**, 5259–5265.

Rabiner, L. R., and B. Gold (1975) "*Theory and Application of Digital Signal Processing*, Prentice-Hall, Englewood Cliffs, N. J.

Raj, P. E., and P. C. S. Devara (1995) "Scattering Angle Distribution of Laser Return Signal Strength in the Lower Atmosphere," *J. Aerosol Science*, **26**, 51–59.

Raj, P. E., and P. C. S. Devara (1992) "On the Stable Stratification of the Nocturnal Lower Troposphere Inferred from Lidar Observations over Pune, India," *Bound. Layer Met.*, **65**, 197–205.

Raj, P. E., P. C. S. Devara, R. S. Maheskumar, G. Pandithurai, and K. K. Dani (1997) "Lidar Measurements of Aerosol Column Content in an Urban Nocturnal Boundary Layer," *Atmos. Res.*, **45**, 201–216.

Rajeev, K., and K. Parameswaran (1998) "Iterative Method for the Inversion of Multiwavelength Lidar Signals to Determine Aerosol Size Distribution," *Appl. Opt.*, **37**, 4690–4701.

Rarity, J., T. Wall, K. Ridley, P. Owens, and P. Tapster (2000) "Single Photon Counting for the 1300–1600 nm Range by Use of Peltier-Cooled and Passively Quenched InGaAs Avalanche Photodiodes," *Appl. Opt.*, **39**, 6746–6753.

Ratsimor, M. Ya. (1966) "Vertical Distribution of Horizontal Visibility in and Below Clouds," *Proceedings of the Central Institute of Forecast* (Central Institute of Forecast, Moscow, 1966), No. **157**, 53–57, (in Russian).

Reagan, J., J. Spinhirne, T. Bruhns, and R. Peck (1976) "Digital Detector for Monitoring Lidar Output Energy," *Rev Sci. Instrum.*, **47**, 1527–1530.

Reagan, J., D. Byrne, M. King, J. Spinhirne, and B. Herman (1980) "Determination of the Complex Refractive Index from Bistatic-Monostatic Lidar and Solar Radiometer Measurements," J. Geoph. Res., **85**, 1591–1599.

Reagan, J., D. Byrne, and B. Herman (1982a) "Bistatic Lidar: A Tool for Characterizing Atmospheric Particulates—Part I—The Remote Sensing Problem," *IEEE Trans. Geosci. Remote Sens.*, **20**, 229–235.

Reagan, J., D. Byrne, and B. Herman (1982b) "Bistatic Lidar: A Tool for Characterizing Atmospheric Particulates—Part II—The Inverse Problem," *IEEE Trans. Geosci. Remote Sens.*, **26**, 236–243.

Reagan, J. A., M. V. Apte, A. Ben-David, and B. M. Herman (1988) "Assessment of Aerosol Extinction to Backscatter Ratio Measurements Made at 694.3 Nm in Tucson, Arizona," *Aerosol Sci. Technol.*, **8**, 215–226.

Rees, D., and I. S. McDermid (1990) "Doppler Lidar Atmospheric Wind Sensor: Reevaluation of a 355 nm Incoherent Doppler Lidar," *Appl. Opt.*, **29**, 4133–4144.

Renault, D., C. Pourney, and R. Capitini (1980) "Daytime Raman Lidar Measurements of Water Vapor," *Opt. Lett.*, **5**, pp. 232–235.

Renault, D., and R. Capitini (1988) "Boundary-Layer Water-Vapor Probing with a Solar-Blind Raman Lidar: Validations, Meteorological Observations and Prospects," *J. Atmos. Oceanic Tech.*, **5**, 5–15.

Riebesell A., and C. Weitkamp (1990) Measurement of Atmospheric Aerosol Extinction Profiles with a Raman Lidar," *Opt. Let.*, **15**, 746–748.

Reichardt, J., U. Wandinger, M. Serwazi, and C. Weitkamp (1996) "Combined Raman Lidar for Aerosol, Ozone, and Moisture Measurements," *Opt. Eng.*, **35**, 1457–1465.

Riley, R. J., and A. G. Wright (1977) "The Effect of Photomultiplier Afterpulse in Coincidence Systems," *J. Phys.*, E **10**, 873–874.

Robinson, C., and D. McKay (1989) "A Practical Cloud Algorithm for Automatic Weather Stations," *Instruments and Observing Report #35*, Fourth World Meteorological Organization Technical Conference on Instruments and Methods of Observation, WMO/TD 303, World Meteorological Organization, Geneva.

Rocadenbosch, F., A. Comeron, and L. Albiol (2000) "Statistics of the Slope-Method Estimator," *Appl. Opt.*, **39**, 6049–6057.

Rocadenbosch, F., C. Soriano, A. Comeron, and J. M. Baldasano (1999) "Lidar Inversion of Atmospheric Backscatter and Extinction-to-Backscatter Ratios by Use of a Kalman Filter," *Appl. Opt.*, **38**, 3175–3189.

Rocadenbosch, F., A. Comeron, and D. Pineda (1998) "Assessment of Lidar Inversion Errors for Homogeneous Atmospheres," *Appl. Opt.*, **37**, 2199–2206.

Rosen, J. M., R. G. Pinnick, and D. M. Garvey (1997) "Measurement of Extinction-to-Backscatter Ratio for Near-Surface Aerosols," *J. Geoph. Res.*, **102**, 6017–6024.

Rosen, J. M., and N. T. Kjome (1997) "Balloon-Borne-Measurements of the Aerosol Extinction-to-Backscatter Ratio," *J. Geoph. Res.*, **102**, 11165–11169.

Rothermel, J., D. Bowdle, J. Vaughan, and M. J. Post (1989) "Evidence of a Tropospheric Aerosol Backscatter Background Mode," *Appl. Opt.*, **28**, 1040–1042.

Rowell, R. L., and G. M. Aval (1971) "Rayleigh-Raman Depolarization of Laser Light Scattered by Gases," *J. Chem. Phys.*, **54**, 1960–1964.

Roy, G., Vallee G., and M. Jean (1993) "Lidar-Inversion Technique Based on Total Integrated Backscatter Calibrated Curves," *Appl. Opt.*, **32**, 6754–6763.

Rue, B. J., and R. M. Hardesty (1989) "Nonlinear Kalman Filtering Techniques for Incoherent Backscatter Lidar: Return Power and Logarithm Power Estimation," *Appl. Opt.*, **28**, 3908–3917.

Russell, P., E. Uthe, F. Ludwig, and N. Shaw (1974) "A Comparison of Atmospheric Structure as Observed with Monostatic Acoustic Sounder and Lidar Techniques," *J. Geoph. Res.*, **79**, 5555–5566.

Russel, P. B., T. J. Swissier, and M. P. McCormick (1979) "Methodology for Error Analysis and Simulation of Lidar Aerosol Measurements," *Appl. Opt.*, **18**, 3783–3797.

Rybakov, E. E., V. A. Kovalev, A. S. Pak, E. E. Mozharov, and G. N. Baldenkov (1991) "Results of the Lidar Determination of Slant Visibility Characteristics at the Airport," *Meteorol. and Hydro.*, **9**, Gidrometeoizdat, Leningrad, U. S. S. R., 18–25, (in Russian).

Saiki, E. M., C. H. Moeng, and P. P. Sullivan (1999) "Large-Eddy Simulation of the Stably Stratified Planetary Boundary Layer," *Proceedings of 13th Symposium on Boundary Layers and Turbulence*, 211–214.

Saiki, E. M., C. H. Moeng, and P. P. Sullivan (2000) "Large Eddy Simulation of the Stably Stratified Planetary Boundary Layer," *Boundary-Layer Meteorol.*, **9**, 1–30.

Sakai, T., T. Shibata, and Y. Iwasaka (1998) "Relative Humidity, Backscattering Ratio and Depolarization Ratio as Derived from Raman Lidar Observations," *J. Meteor. Soc. Japan*, **75**, 1179–1185.

Samokhvalov, I. V. (1979) "Double Scattering Approximation of Lidar Equation for Inhomogeneous Atmosphere," *Opt. Lett.*, **4**, 12–14.

Sanford, M. C. W. (1967a) "Laser Scatter Measurements in Mesosphere and Above," *J. Atmos. And Terr. Phys.*, **29**, 1657–1664.

Sanford, M. C. W. (1967) "Optical Radar Performance in Atmospheric Scattering," *J. Atmos. and Terr. Phys.*, **29**, 1651–1655.

Sasano, Y., H. Shimizu, N. Takeuchi, and M. Okuda (1979) "Geometrical Form Factor in the Laser Radar Equation: an Experimental Determination," *Appl. Opt.*, **18**, 3908–3910.

Sasano, Y., H. Hirohara, T. Yamasaki, H. Shimizu, N. Takeuchi, and T. Kawamura (1980) "Horizontal Wind Vector Determination from the Displacement of Aerosol Distribution Patterns Observed by a Scanning Lidar", *J. Appl. Meteor.*, **21**, 1516–1523.

Sasano, Y., H. Shimizu, N. Sugimoto, I. Matsui, N. Takeuchi, and M. Okuda (1980) "Diurnal Variation of the Atmospheric Boundary Layer Observed by a Computer-Controlled Laser Radar", *J. Meteor. Soc. of Japan*, **58**, 143–148.

Sasano, Y., and H. Nakane (1984) "Significance of the Extinction/Backscatter Ratio and the Boundary Value Term in the Solution for the Two-Component Lidar Equation," *Appl. Opt.* **23**, 11–13.

Sasano, Y., E. V. Browell, and S. Ismail (1985) "Error Caused by Using a Constant Extinction/Backscattering Ratio in the Lidar Solution," *Appl. Opt.*, **24**, 3929–3932.

Sasano, Y. (1985) "Simultaneous Determination of Aerosol and Gas Distribution by DIAL Measurements," *Appl. Opt.*, **27**, 2640–2641.

Sasano, Y., and H. Nakano (1987) "Quantitative Analysis of RHI Data by an Iterative Adjustment of the Boundary Condition Term in the Lidar Solution," *Appl. Opt.*, **26**, 615–616.

Sasano, Y., and E. V. Browell (1989) "Light Scattering Characteristics of Various Aerosol Types Derived from Multiple Wavelength Lidar Observation," *Appl. Opt.* **28**, 1670–1679.

Sasano, Y. (1996) "Tropospheric Aerosol Extinction Coefficient Properties Derived from Scanning Lidar Measurements over Tsukiba, Japan, from 1990 to 1993," *Appl. Opt.*, **35**, 4941–4952.

Sassen, K., and K. N. Liou (1979) "Scattering of Polarized Light by Water Droplet, Mixed Phase and Ice Cristal Clouds: I. Angular Depolarizing and Multiple Scattering Behavior," *J. Atm. Sci.*, **36**, 852–861.

Sassen, K., and G. C. Dodd (1982) "Lidar Crossover Function and Misalignment Effects," *Appl. Opt.*, **21**, 3162–3165.

Sassen, K., and R. L. Petrilla (1986) "Lidar Depolarization from Multiple Scattering in Marine Stratus Clouds," *Appl. Opt.*, **25**, 1450–1459.

Sassen, K., M. K. Griffin, and G. C. Dodd (1989) "Optical Scattering and Microphysical Properties of Subvisual Cirrus Clouds, and Climatic Implications," *J. Appl. Meteorol.*, **28**, 91–98.

Sassen, K. (1991) "The Polarization Lidar Technique for Cloud Research: A Review and Current Assessment," *Bull. Am. Meteor. Soc.*, **72**, 1848–1866.

Sassen, K., and B. S. Cho (1992) "Subvisual-Thin Cirrus LIDAR Dataset for Satellite Verification and Climatological Research," *J. Appl. Meteorol.*, **31**, 1275–1285.

Sassen, K. (1994) "Advances in Polarization Lidar for Cloud Remote Sensing," *Proc. IEEE*, **82**, 1907–1914.

Sassen, K., and C. Y. Hsueh (1998) "Contrail Properties Derived from High-Resolution Polarization Lidar Studies During SUCCESS," *Geoph. Res. Lett.*, **25**, 1165–1168.

Schols, J. L., and E. W. Eloranta (1992) "The Calculation of Area-Averaged Vertical Profiles of the Horizontal Wind Velocity from Volume Imaging Lidar Data," *J. Geoph. Res.*, **97**, 18395–18407.

Schotland, R. M. (1974) "Errors in the Lidar Measurements of Atmospheric Gases by Differential Absorption," *J. Appl. Meteorol.*, **13**, 71–77.

Schotland, R. M., K. Sassen, and R. Stone (1971) "Observations by Lidar of Linear Depolarizations of Hydrometeors," *J. Appl. Meteor.*, **10**, 1011–1017.

Schuster, B. G., and T. G. Kyle (1980) "Pollution Plume Transport and Diffusion Studies using Fluorescence Lidar," *Appl. Opt.*, **19**, 2524–2528.

Schwiesow, R. L., and L. Lading (1981) "Temperature Profiling by Rayleigh-Scattering Lidar," *Appl. Opt.*, **20**, 1972–1979.

Seinfeld, J. H., R. J. Charlson, P. A. Durkee, D. Hegg, D. J. Huebert, J. Kiehl, M. P. McCormick, J. A. Ogren, J. E. Penner, V. Ramaswamy, and W. G. Slinn, eds. (1996) *Aerosol Radiative Forcing and Climate Change*, National Research Council (National Academy Press, Washington D. C., 1996).

Seippel, R. G. (1981) *Optoelectronics*, Reston Publishing Company, Inc.

Sentrayan, K., A. Michael, U. N. Singh, and V. S. Kushawaha (1997) "Effect of Buffer Gas on the Generation of Eye-Safe 1.54 um Radiation Using the Stimulated Raman Scattering Technique in CH4," *Opt. Laser Technol.*, **29**, 275–279.

Sharma, A., and J. Walker (1992) "Paralyzable and Nonparalyzable Deadtime Analysis in Spatial Photon Counting," *Rev. Sci. Instrum.*, **63**, 5784–5793.

She, C. Y., J. R. Yu, H. Latifi, and R. E. Bills (1992) "High-Spectral Resolution Lidar for Mesospheric Sodium Temperature Measurements," *Appl. Opt.*, **31**, 2095–2106.

She, C. Y., R. J. Alvarez II, L. M. Caldwell, and D. A. Krueger (1992) "High-Spectral-Resolution Rayleigh-Mie Lidar Measurements of Aerosol and Atmospheric Profiles," *Opt. Lett.*, **17**, 541–543.

She, C. Y., and J. R. Yu (1995) "Doppler-Free Saturation Fluorescence Spectroscopy of Na Atoms for Atmospheric Applications," *Appl. Opt.*, **34**, 1063–1075.

She, C. Y., S. Chen, Z. Hu, J. Sherman, J. D. Vance, V. Vasoli, M. A. White, J. Yu, and D. A. Krueger (2000) "Eight-Year Climatology of Nocturnal Temperature and Sodium Density in the Mesopause Region (80 to 105 km) over Fort Collins, CO (41 N, 105 W)," *Geoph. Res. Lett.*, **27**, 3289–3292.

She, C. Y. (2001) "Spectral Structure of Laser Light Scattering Revisited: Bandwidths of Nonresonant Scattering Lidars," *Appl. Opt.*, **40**, 4875–4884.

Shettle, E. P., and R. W. Fenn (1979) "Models of Aerosols of the Lower Atmosphere and the Effect of Humidity Variations on Their Optical Properties," (U. S. Air Force Geophysics Laboratory, Hansom Air Force, Mass., 1979).

Shibata, T., M. Kobuchi, and M. Maeda (1986) "Measurement of Density and Temperature Profiles with a XeF Lidar," *Appl. Opt.*, **25**, 685–688.

Shimizu, H., S. A. Lee, and C. Y. She (1983) "High Spectral Resolution Lidar System with Atomic Blocking Filters for Measuring Atmospheric Parameters," *Appl. Opt.*, **22**, 1373–1381.

Shipley, S. T., D. H. Tracy, E. W. Eloranta, J. T. Trauger, J. T. Sroga, F. L. Roesler, and J. A. Weinman (1983) "High Spectral Resolution Lidar to Measure Optical Scattering Properties of Atmospheric Aerosols. 1: Theory and Instrumentation," *Appl. Opt.*, **22**, 3716–3724.

Sicard, M., P. Chazette, J. Pelon, J. G. Won, and S. C. Yoon (2002) "Variational Method for the Retrieval of the Optical Thickness and the Backscatter Coefficient from Multiangle Lidar Profiles," *Appl. Opt.*, **41**, 493–502.

Simeonov, V., B. Calpini, and H. van den Bergh (2002) "New Raman-Shifted Sources for Ozone DIAL Applications," *Proceedings of the Twenty First International Laser Radar Conference*, L. R. Bissonnette, G. Roy and G. Vallee eds. (Quebec, Canada, 2002), Pt. 1, 19–22.

Singh, U. N., J. A. Williams-Byrd, N. P. Barnes, J. Yu. M. Petros, G. E. Lockard, and E. A. Modlin (1997) "Diode-Pumped 2-um Solid State Lidar Transmitter for Wind Measurements," *Proc. SPIE*, **3104**, 173–178.

Singh, U. N., P. Keckhut, T. J. McGee, M. R. Gross, A. Hauchecorne, E. F. Fishbein, J. W. Waters, J. C. Gille, A. E. Roche, and J. M. Russell III (1996) "Stratospheric Temperature Measurements by Two Collocated NDSC Lidar During UARS Validation Campaign," *J. Geoph. Res.*, **101**, 10287–10297.

Skinner, W. R., and P. B. Hays (1994) "Incoherent Doppler Lidar for Measurement of Atmospheric Winds," *Proc. SPIE*, **2266**, 383–394.

Smith, H. J. P., D. J. Dube, M. E. Gardner, S. A. Clough, F. X. Kneizys, and L. S. Rothman (1978) "FASCODE-Fast Atmospheric Signature Code (Spectral Transmittance and Radiance)," *Air Force Geophysics Laboratory Technical Report* AFGL-TR-78-0081, Hanscom AFB, MA.

Sorbjan, Z., *Structure of the Atmospheric Boundary Layer*, Prentice Hall, New Jersey, 1989, 317 pp.

Spinhirne, J., and J. Reagan (1976) "Gain Switching Amplifier for Compression of Optical Radar Return Signals," *Rev Sci. Instrum.*, **47**, 437–439.

Spinhirne, J. D., J. A. Reagan, and B. M. Herman (1980) "Vertical Distribution of Aerosol Extinction Cross Section and Inference of Aerosol Imaginary Index in the Troposphere by Lidar Technique," *J. Appl. Meteorol.*, **19**, 426–438.

Spinhirne, J. (1993) "Micro Pulse Lidar," *IEEE Transactions on Geoscience and Remote Sensing*, **31**, 48–55.

Spinhirne, J., J. A. R. Rall, and V. S. Scott (1995) "Compact Eye Safe Lidar Systems," *Rev. Laser Eng.*, **23**, 112–118.

Spinhirne, J. (1996) "Eye-Safe Lidar Provides Full-Time Atmospheric Monitoring," *Photonics Spectra*, **30**(3), 98–99.

Spinhirne, J. D., S. Chudamani, J. F. Cavanaugh, and J. L. Bufton (1997) "Aerosol and Cloud Backscatter at 1.06, 1.54, and 0.53 Mm by Airborne Hard-Target Calibrated Nd:YAG/methane Raman Lidar," *Appl. Opt.*, **36**, 3475–3490.

Sroga, J. T., E. W. Eloranta, and T. Barber (1980) "Lidar Measurement of Wind Velocity Profiles in the Boundary Layer," *Journ. Appl. Meteorol.*, **19**, 598–605.

Sroga, J. T., E. W. Eloranta, S. T. Shipley, F. L. Roesler, and P. J. Tryon (1983) "High Spectral Resolution Lidar to Measure Optical Scattering Properties of Atmospheric Aerosols. 2: Calibration and Data Analysis," *Appl. Opt.*, **22**, 3725–3732.

Stann, B. L., W. C. Ruff, and Z. G. Sztankay (1996) "Intensity-Modulated Diode Laser Radar Using Frequency-Modulation/Continuous-Wave Ranging Techniques," *Opt. Eng.*, **35**, 3270–3278.

Starkov, A. V., M. Noormohammadian, and U. G. Oppel (1995) "A Stochastic Model and a Variance-Reduction Monte-Carlo Method for the Calculation of the Light Transport," *Appl. Phys.*, B **60**, 335–340.

States, R. J., and C. S. Gardner (2000a) "Thermal Structure of the Mesopause Region 80–115 Km at 40 N Latitude. Part I: Seasonal Variations," *J. Geoph. Res.* **57**, 66–77.

States, R. J., and C. S. Gardner (2000b) "Thermal Structure of the Mesopause Region 80–115 Km at 40 N Latitude. Part II: Diurnal Variations," *J. Geoph. Res.* **57**, 78–92.

Stefanutti, L., F. Castagnoli, M. Del Guasta, M. Morandi, V. M. Sacco, L. Zuccagnoli, S. Godin, G. Megie, and J. Porteneuve (1992) "The Antarctic Ozone LIDAR System," *Appl. Phys.*, B **55**, 3–12.

Steinbrecht, W., K. W. Rothe, and H. Walther (1989) "Lidar Setup for Daytime and Nighttime Probing of Stratospheric Ozone and Measurements in Polar and Equatorial Regions," *Appl. Opt.*, **28**, 3616–3624.

Stelmaszczyk, K., A. Czyzewski, A. Szymanski, A. Pietruczuk, S. Chudzynski, K. Ernst, and T. Stacewicz (2000) "New Method of Elaboration of the Lidar Signal," *Appl. Phys.*, B **70**, 295–299.

Stepanenko, V. D. (1973a) "Determination of Atmospheric Transparency with Optical Radar," *Proceedings of the Main Geophysical Observatory*, V. D. Stepanenko, ed. (Main Geophysical Observatory, Leningrad, 1973), No. 281, 101–107, (in Russian).

Stepanenko, V. D. (1973) *Radar in Meteorology*, 1st ed. (Gidrometeoizdat, Leningrad), p. 209, (in Russian).

Stephens, G. (1994) *Remote Sensing of the Lower Atmosphere*, Oxford University Press, New York.

Stevens, N. B., M. H. Horman, and E. E. Dodd (1957) "The Determination of Atmospheric Transmissivity by Backscatter from a Pulsed-Light System," Report AFCRL, TR-57-201, (Riverside, CA, 1957), 82p.

Steyn, D., M. Baldi, and R. Hoff (1999) "The Detection of Mixed Layer Depth and Entrainment Zone Thickness from Lidar Backscatter Profiles," *J. Ocean. Atm. Tech.*, **16**, 953–959.

Strauch, R. G., V. E. Derr, and R. E. Cupp (1971) "Atmospheric Temperature Measurement Using Raman Backscatter," *Appl. Opt.*, **10**, 2665–2669.

Stull, R. B. (1988) *An Introduction to Boundary Layer Meteorology*, Kluwer Academic Publishers, Dordrecht.

Sugawara, A., and S. Yip (1967) "Kinetic Model Analysis of Light Scattering by Molecular Gases," *Physics of Fluids*, **10**, 1911–1921.

Sugimoto, N., N. Sims, and K. Chan (1990) "Eye-Safe 2.1-μm Ho Lidar for Measuring Atmospheric Density Profiles," *Opt. Lett.*, **15**, 302–313.

Sunesson, J. A., A. Apituley, and D. P. J. Swart (1994) "Differential Absorption Lidar System for Routine Monitoring of Tropospheric Ozone," *Appl. Opt.*, **33**, 7045–7058.

Sutton, S., and M. Bennett (1994) "Measurement of Wind Speed Using a Rapid-Scanning Lidar," *Int. J. Remote Sens.*, **15**, 375–380.

Sze, S. M. (1981) *Physics of Semiconductor Devices*. 2nd ed., John Wiley & Sons, New York.

Takamura, T., Y. Sasano, and T. Hayasaka (1994) "Tropospheric Aerosol Optical Properties Derived from Lidar, Sun Photometer, and Optical Particle Counter Measurements," *Appl. Opt.*, **33**, 7132–7140.

Takamura T., and Y. Sasano (1987) "Ratio of Aerosol Backscatter to Extinction Coefficients as Determined from Angular Scattering Measurements for Use in Atmospheric Lidar Applications," *Optical and Quantum Electronics*, **19**, 293–302.

Takeuchi, N., H. Baba, K. Sakurai, and T. Ueno (1986) "Diode Laser Random Modulation CW Lidar," *Appl. Opt.*, **25**, 63–67.

Takeuchi, N., N. Sugimoto, H. Baba, and K. Sakurai (1983) "Random Modulation CW Lidar," *Appl. Opt.*, **22**, 1382–1386.

Teillet, P. M. (1990) "Rayleigh Optical Depth Comparisons from Various Sources, *Appl. Opt.*, **29**, 1897–1900.

Taylor, J. R. (1982) "*An Introduction to Error Analysis*," University Science Books.

Thayer, J. P., N. B. Nielsen, R. E. Warren, C. J. Heinselman, and J. Sohn (1997) "Rayleigh Lidar System for Middle Atmosphere Research in the Arctic," *Opt. Eng.*, **36**, 2045–2061.

Theopold, F., and J. Bosenberg (1988) "Evaluation of DIAL Measurements in the Presence of Signal Noise," *Proceedings of the 14th International Laser Radar Conference*, Florence, Italy, 209–211.

Theopold, F. A., and J. Bosenberg (1993) "Differential Absorption Lidar Measurements of Atmospheric Temperature Profiles—Theory and Experiment," *J. Atmos. Oceanic Tech.*, **10**, 165–179.

Thomey, S. (1977) *Introduction to the Mathematics of Inversion in Remote Sensing and Indirect Measurementrs*, Elsevier, NY.

Tikhomirov, A. A. (1976) "Estimation of the Geometrical Factor of Lidar," *Proceedings of the Fourth All Union Symposium on Laser Searching of the Atmosphere* (Institute of Atmospheric Optics, Tomsk, U. S. S. R., 1976, 301–304), (in Russian).

Tikhonov, A. N., and V. Y. Arsenin (1977) *Solution of Ill-Posed Problems*, John Wiley & Sons, Inc., New York.

Tomine, K., C. Hiramaya, K. Michimoto, and N. Takeuchi (1989) "Experimental Determination of the Crossover Function in the Laser Radar Equation for Days with a Light Mist," *Appl. Opt.*, **28**, 2194–2195.

Tomine, K., K. Michimoto, I. Hikiji, S. Abe, and N. Takeuchi (1991) "The Vertical Structure of Fog Observed with a Lidar System at Misawa Airbase, Japan," *J. Appl. Meteorol.*, **30**, 1088–1096.

Tonna, G. (1991) "Backscattering, Extinction, and Liquid Water Content in Fog: a Detailed Study of Their Relations for Use in Lidar Systems," *Appl. Opt.* **30**, 1132–1140.

Trakhovsky, E., A. Ben-Shalom, U. P. Openheim, A. D. Devir, L. S. Balfour, and M. Engel (1989) "Contribution of Oxigen to Attenuation in the Solar Blind UV Spectral Region," *Appl. Opt.*, **28**, 1588–1591.

Uchino, O., and T. Fujimoto (1992) "Lidar Measurements of Ozone and Temperature in the Stratosphere at Tsukuba (36.1-degrees N, 140.1-degrees-E) During Dyana Campaign," *J. Geomag. Geoelect.*, **44**, 1061–1070.

Uchino, O., and I. Tabata (1991) "Mobile Lidar for Simultaneous Measurements of Ozone, Aerosols, and Temperature in the Stratosphere," *Appl. Opt.*, **30**, 2005–2012.

US Standard Atmosphere, 1976, (1976) US GPO, Washington, D. C.

Uthe, E. E. (1978) "Remote Sensing of Aerosol Properties Using Laser Radar Techniques, *Proceedings of the Society of Photo-Optical Instrumentation Engineers, SPIE*, **142**, Optical Properties of the Atmosphere, 67–77.

Uthe, E. E., and J. M. Livingston (1986) "Lidar Extinction Methods Applied to Observations of Obscurant Events," *Appl. Opt.*, **25**, 678–684.

Uttal, T., E. E. Clothiaux, T. P. Ackerman, J. M. Intrieri, and W. L. Eberhard (1995) "Cloud Boundary Statistics During FIRE II," *J. Atmos. Sci.*, **52**, 4276–4284.

VAISALA News (2002) "New Technology for Runway Visual Range Applications: FS11 Visibility Sensor Launched," (VN159, 25, VAISALA Group, Finland, 2002).

Valero, F., and P. Pilewski (1992) "Latitudinal Survey of Spectral Optical Depths of the Pinatubo Aerosol Cloud-Derived Particle Sizes, Columnar Mass Loadings, and Effects on Planetary Albedo," *Geoph. Res. Let.*, **19**, 163–166.

Van de Hulst, H. C. (1957) *Light Scattering by Small Particles*, John Wiley & Sons, Inc., New York, 470.

Vaughan, G., D. P. Wareing, S. J. Pepler, L. Thomas, and V. Mitev (1993) "Atmospheric Temperature Measurements by Rotational Raman Scattering," *Appl. Opt.*, **32**, 2758–2764.

Velotta, R., B. Bartoli, R. Capobianco, L. Fiorani, and N. Spinelli (1998) "Analysis of the Receiver Response in Lidar Measurements," *Appl. Opt.*, **37**, 6999–7007.

Veselovskii, I., A. Kolgotin, V. Griaznov, D. Müller, U. Wandinger, and D. N. Whiteman (2002) "Inversion with Regularization for the Retrieval of Tropospheric Aerosol Parameters from Multiwavelength Lidar Sounding," *Appl. Opt.*, **41**, 3685–3699.

Viezee, W., E. E. Uthe, and R. T. H. Collis (1969) "Lidar Observations of Airfield Approach Conditions: an Exploratory Study," *J. Appl. Meteorol.*, **8**, 274–283.

Volkov, S. N., and B. V. Kaul (1994) "A Method for Determining the Light Backscatter and Attenuation Coefficients in Aerosol Layers of the Troposphere by Lidar Operating at the Elastic and Raman Scattering Frequencies," Optika Atmosfery i Okeana, 7, 1592–1602, (in Russian).

von der Gathen, P. (1991) "Saturation Effects in Na Lidar Temperature Measurements," *J. Geoph. Res.*, **96**, 3679–3690.

von der Gathen, P. (1995) "Aerosol Extinction and Backscatter Profiles by Means of a Multiwavelength Raman Lidar: A New Method Without a priori Assumptions," *Appl. Opt.*, **34**, 463–466.

von Zahn, U., and J. Hoeffner (1996) "Mesopause Temperature Profiling by Potassium Lidar," *Geoph. Res. Lett.*, **23**, 141–144.

Voss, E., C. Weitkamp, and W. Michaelis (1994) "Lead-Vapor Filters for High-Spectral-Resolution Temperature Lidar," *Appl. Opt.*, **33**, 3250–3260.

Voss, K. J., E. I. Welton, P. K. Quinn, J. Johnson, A. M. Thompson, and H. R. Gordon (2001) "Lidar Measurements During Aerosol99," *J. Geoph. Res.*, **106**, 20821–20831.

Wade, C. (1994) "An Evaluation of Problems Affecting the Measurements of Low Relative Humidity on the United States Radiosonde," *J. Atmos. Oceanic Tech.*, **11**, 687–700.

Wandinger, U. (1998) "Multiple Scattering Influence on Extinction and Backscatter Coefficient Measurements with Raman and High Spectral Resolution Lidars," *Appl. Opt.*, **37**, 417–427.

Wandinger, U., A. Ansmann, J. Reichardt, and T. Deshler (1995) "Determination of Stratospheric-Aerosol Microphysical Properties from Independent Extinction and Backscattering Measurements with a Raman Lidar," *Appl. Opt.*, **34**, 8315–8329.

Wang, R. T., and H. van de Hulst (1995) "Application of the Exact Solution for Scattering by an Infinite Cylinder to the Estimation of Scattering by a Finite Cylinder," *Appl. Opt.*, **34**, 2811–2821.

Wang, Z., H. Hu, and J. Zhou (1994) "Dual-DIAL Measurements of Ozone Profiles," *Proceedings of the Seventeenth International Laser Radar Conference*, (Sendai International Center, Sendai, Japan, 1994), 54–57.

Wang, Z., H. Nakane, H. Hu, and J. Zhou (1997) "Three-Wavelength Dual-Differential Absorption Lidar Method for Stratospheric Ozone Measurements in the Presence of Volcanic Aerosols," *Appl. Opt.*, **36**, 1245–1252.

Ward, G., K. Kushing, R. McPeters, and A. Green (1973) "Altitude Aerosol Index of Refraction and Size-Altitude Distribution from Bistatic Laser Scattering and Solar Aureole Measurements," *Appl. Opt.*, **12**, 2585–2592.

Warren, R. E. (1987) "Adaptive Kalman-Bucy Filter for Differential Absorption Lidar Time Series Data," *Appl. Opt.*, **26**, 4755–4760.

Weaver, H. L. (1990) "Temperature and Humidity Flux Variance Relations Determined by One Dimensional Eddy Correlation," *Bound.-Layer Meteor.*, **53**, 77–91.

Weinman, J. A. (1988) "Derivation of Atmospheric Extinction Profiles and Wind Speed over the Ocean from a Satellite-Borne Lidar," *Appl. Opt.*, **27**, 3994–4001.

Welsh, B. M., and C. S. Gardner (1989) "Nonlinear Resonant Absorption Effects on the Design of Resonance Fluorescence Lidars and Laser Guide Stars," *Appl. Opt.*, **28**, 4141–4153.

Werner, C. (1981) "Slant Range Visibility Determination from Lidar Signatures by the Two-Point Method," *Opt. Laser. Technol.*, **13**, 27–36.

Werner, C., J. Streicer, H. Hermann, and H. G. Dahn (1992) "Multiple-Scattering Lidar Experiments," *Opt. Eng.*, **31**, 1731–1745.

Werne, J., and D. C. Fritts (1999) "Stratified Shear Turbulence: Evolution and Statistics," *Geoph. Res. Lett.*, **24**, 439–442.

Werne, J., and D. C. Fritts (2001) "Anisotropy in a Stratified Shear Layer," *Phys. Chem. Earth*, B **26**, 263–268.

Wesely, M. L. (1988) "Use of Variance Techniques to Measure Dry Air-Surface Exchange Rates," *Bound.-Layer Meteor.*, **44**, 13–31.

Westwater, E. R. (1997) "Remote Sensing of Tropospheric Temperature and Water Vapor by Integrated Observing Systems," *Bull. Am. Meteorol. Soc.*, **78**, 1991–2006.

White, A., C. Senff, and R. Banta (1999) "A Comparison of Mixing Depths Observed by Ground-Based Wind Profilers and an Airborne Lidar," *J. Atmos. Ocean. Tech.*, **16**, 584–590.

Whiteman D. N., S. H. Melfi, and R. A. Ferrare (1992) "Raman Lidar System for the Measurement of Water Vapor and Aerosols in the Earth's Atmosphere," *Appl. Opt.*, **31**, 3068–3082.

Whiteman, D. N., W. F. Murphy, N. W. Walsh, and K. Evans (1993) "Temperature Sensitivity of an Atmospheric Raman Lidar System Based on a XeF Excimer Laser," *Opt. Lett.*, **18**, 247–249.

Whiteman, D. N., G. E. Walrafen, W. H. Yang, and S. H. Melfi (1999) "Measurement of an Isosbestic Point in the Raman Spectrum of Liquid Water by Use of a Backscattering Geometry," *Appl. Opt.*, **38**, 2614–2615.

Whiteman, D. N., and S. H. Melfi (1999) "Cloud Liquid Water, Mean Droplet Radius, and Number Density Measurements Using a Raman Lidar," *J. Geoph. Res.*, **104**, 31411–31419.

Whiteman, D. N. (1999) "Application of Statistical Methods to the Determination of Slope in Lidar Data," *Appl. Opt.*, **15**, 3360–3369.

Wilkerson, T. D., and H. Moosmüller (1994) "New Lidar Method Utilizing Elastic and Raman Scattering for the Measurement of Backscattering and Extinction Profiles," in *Tunable Diode Laser Spectroscopy, Lidar, and DIAL Techniques*, H. I. Schiff, A. Fried, and T. D. Wilkerson, eds., Proc. SPIE, **2112**, 177–186.

Wilson, R., A. Hauchecorne, and M. Chanin (1991) "Gravity Waves in Middle Atmosphere Observed by Rayleigh Lidar, Part I: Case Studies," *J. Geoph. Res.*, **96**, 5153–5167.

Wilson, R., A. Hauchecorne, and M. Chanin (1991) "Gravity Waves in Middle Atmosphere Observed by Rayleigh Lidar, Part II: Climatology," *J. Geoph. Res.*, **96**, 5169–5183.

Winkler, D. M., and L. R. Poole (1995) "Monte-Carlo Calculations of Cloud Returns for Ground-Based and Space-Based LIDARS," *Appl. Phys.*, B **60**, 341–344.

WMO, Tech. Rep. 37 (1994) "Scientific Assessment on Ozone Depletion," *Global Ozone Research Monitoring Project* (World Health Organization, Geneva, 1994).

Wu, Y., Y. Zhao, F. Yang, A. Ziong, J. Tang, and L. Zheng (1995) "New Method for Acquiring a High-Resolution Atmospheric Rayleigh-Mie Spectrum, *Opt. Eng.*, **34**, 1195–1199.

Wuerker, R. F. (1997) "Bistatic Liquid Mirror Telescope Lidar Alignment, *Opt. Eng.*, **36**, 1421–1424.

Wulfmeyer, V. (1998) "Ground-Based Differential Absorption Lidar for Water Vapor and Temperature Profiling: Development and Specifications of a High-Performance Laser Transmitter," *Appl. Opt.* **37**, 3804–3824.

Wylie, C. R., and L. B. Barret (1982) *Advanced Engineering Mathematics*, 5th ed., McGraw-Hill, Inc., p 36.

Wyngaard, J. C. (1973) "On Surface Layer Turbulence. Workshop on Micrometeorology," D. A. Haugen, ed., *Amer. Meteor. Soc.*, 109–149.

Wyngaard, J. C., and B. Kosovic (1994) "Similarity of Structure-Function Parameters in the Stably Stratified Boundary Layer," *Bound.-Layer Meteor.*, **71**, 277–287.

Wyngaard, J. C., and L. J. Peltier (1996) "Experimental Micrometeorology in an Era of Turbulence Simulation," *Bound.-Layer Meteor.*, **78**, 71–86.

Yang, G., R. Billmers, P. Herczfeld, and V. Contarino (1997) "Temporal Characteristics of Narrow-Band Optical Filters and Their Application in Lidar Systems," *Opt. Lett.*, **22**, 414–416.

Yariv, A. (1976) *Introduction to Optical Electronics*, Holt, Rinehart and Winston, New York.

Yip, S., and M. Nelking (1964) "Application of Kinetic Model to Time-Dependent Density Correlations in Fluids," *Physical Review*, **135**, A1241–A1247.

Yee, E., P. R. Kosteniu, and G. Roy (1992) "Remote Biodetection Performance of a Pulsed Monostatic Lidar System," *Appl. Opt.*, **31**, 2900–2904.

Yoshiyama, H., A. Ohi, and K. Ohta (1996) "Derivation of the Aerosol Size Distribution from a Bistatic System of a Multiwavelength Laser with the Singular Value Decomposition Method," *Appl. Opt.*, **35**, 2642–2648.

Young, A. T. (1980) "Revised Depolarization Corrections for Atmospheric Extinction," *Appl. Opt.*, **19**, 3427–3428.

Young, A. T. (1981) "Rayleigh Scattering," *Appl. Opt.*, **20**, 533–535.

Young, A. T. (1981a) "On the Rayleigh Scattering Optical Depth of the Atmosphere," *J. Appl. Meteorol.*, **20**, 328–330.

Young, S. A. (1995) "Analysis of Lidar Backscatter Profiles in Optically Thin Clouds," *Appl. Opt.*, **34**, 7019–7031.

Young, P., and E. Eloranta (1995) "Calculation of Divergence and Vertical Motion from Volume Imaging Lidar Data," *J. Geoph. Res*, **100**, 25577–25583.

Yu, J. R., and C. Y. She (1993) "Lidar-Observed Temperature Structures and Gravity-Wave Perturbations of the Mesopause Region in the Springs of 1990–1992 over Fort Collins, Co.," *Appl. Phys.*, B **57**, 231–238.

Yu, J. R., H. Latifi, C. Y. She, and J. H. Yee (1991) "Simultaneous Lidar and Airglow Temperature Measurements in the Mesopause Region," *Geoph. Res. Lett.*, **18**, 1361–1363.

Yu, J., S. Arshinov, M. Bobronvnikov, V. Zuev, and V. Mitev (1983) "Atmospheric Temperature Measurements Using a Pure Rotational Raman Lidar," *Appl. Opt.*, **22**, 2984–2990.

Yue, G. K. (2000) "Retrieval of Stratospheric Aerosol Size Distribution and Integral Properties from Simulated Lidar Backscatter Measurements," *Appl. Opt.*, **39**, 5446–5455.

Zege, E. P., A. P. Ivanov, and I. L. Katsev (1971) "Determination of the Extinction and Scattering Parameters of the Water Medium and the Atmosphere by the Analysis of Erosion of the Reflected Pulse Signal," *Bull. Acad. Sci. U. S. S. R.*, Geoph. Ser., No. 7, 750–754, (in Russian).

Zege, E. P., I. L. Katsev, and I. N. Polonsky (1995) "Analytical Solution to LIDAR Return Signals from Clouds with Regard to Multiple Scattering," *Appl. Phys.*, B **60**, 345–354.

Zeyn, J., W. Lahmann, and C. Weitkamp (1996) "Remote Daytime Measurements of Tropospheric Temperature Profiles with a Rotational Raman Lidar," *Opt Lett.*, **21**, 1301–1303.

Zhao, Y., and R. Hardesty (1988) "Technique for Correcting Effects of Long CO_2 Laser Pulses in Aerosol Backscattered Coherent Lidar Returns," *Appl. Opt.*, **27**, 2719–2729.

Zhao, Y., T. Lea, and R. Schotland (1988) "Correction Function for the Lidar Equation and Some Techniques for Incoherent CO_2 Lidar Data Reduction," *Appl. Opt.*, **27**, 2730–2740.

Zhao, Y., R. M. Hardesty, and M. J. Post (1992) "Multibeam Transmitter for Signal Dynamic Range Reduction in Incoherent Lidar Systems," *Appl. Opt.*, **31**, 7623–7632.

Zhao, Y., J. N. Howell, and R. M. Hardesty (1993) "Transportable Lidar for the Measurement of Ozone Concentration and Aerosol Profiles in the Lower Troposphere," *A&WMA/SPIE International Symposium on Optical Sensing for Environmental Monitoring*, October 11–14, 1993, Atlanta, Georgia, 310–320.

Zhao, Y. (1999) "Signal-Induced Fluorescence in Photomultipliers in Differential Absorption Lidar Systems," *Appl. Opt.*, **38**, 4639–4648.

Zhong, C., H. Zhengde, and C. Qun (1989) "Gain Stability of Photomultipliers at High Variable Counting Rates," *Nucl. Instr. and Meth. Phys. Res.*, **A281**, 384–387.

Zuev, V. E., G. M. Krekov, M. M. Krekova, and I. E. Naats (1976) "Theoretical Aspects of Cloud Searching with Lidar," in *Problems of Laser Sensing in the Atmosphere*, V. E. Zuev ed. (Nauka, Novosibirsk, 1976), 3–33, (in Russian).

Zuev, V. E., Yu. M. Vorevodin, G. G. Matvienko, and I. V. Samokhvalov (1977) "Investigation of Structure and Dynamics of Aerosol Inhomogeneities in the Ground Layer of the Atmosphere," *Appl. Opt.*, **16**, 2231–2235.

Zuev, V. E., G. M. Krekov, and M. M. Krekova (1978) "Laser Searching of Atmospheric Aerosol: Theoretical Aspects," in *Remote Sensing in the Atmosphere*, V. E. Zuev ed. (Nauka, Novosibirsk, 1978), 3–46, (in Russian).

Zuev, V. E., G. O. Zadde, S. I. Kavkjanov, and B. V. Kaul (1978a) "Interpretation of Lidar Return Signals from the Regions of Large Optical Depths," in *Remote Sensing in the Atmosphere*, V. E. Zuev ed. (Nauka, Novosibirsk, 1978), 60–68, (in Russian).

Zuev, V. E., and I. E. Naats (1983) *Inverse Problems of Lidar Sensing of the Atmosphere*, Springer-Verlag, Berlin.

Zuev, V. E., Yu. S. Makushkin, V. N. Marichev, A. A. Mitsel, and V. V. Zuev (1983) "Lidar Differential Absorbing and Scattering Technique: Theory," *Appl. Opt.*, **22**, 3733–3741.

Zuev, V. E., and G. M. Krekov (1986) *Optical Models of the Atmosphere*, Gidrometeoizdat, Leningrad, 145–157, (in Russian).

Zuev, V. E., V. S. Komarov, and A. V. Kreminskii (1997) "Application of Correlation Lidar Data to Modeling and Prediction of Wind Components," *Appl. Opt.*, **36**, 1906–1914.

INDEX

Elastic Lidar: Theory, Practice, and Analysis Methods, by Vladimir A. Kovalev and William E. Eichinger.
ISBN 0-471-20171-5